Regression

Ludwig Fahrmeir • Thomas Kneib • Stefan Lang •
Brian Marx

Regression

Models, Methods and Applications

Springer

Ludwig Fahrmeir
Department of Statistics
University of Munich
Munich
Germany

Stefan Lang
Department of Statistics
University of Innsbruck
Innsbruck
Austria

Thomas Kneib
Chair of Statistics
University of Göttingen
Göttingen
Germany

Brian Marx
Experimental Statistics
Louisiana State University
Baton Rouge, LA
USA

ISBN 978-3-642-34332-2 ISBN 978-3-642-34333-9 (eBook)
DOI 10.1007/978-3-642-34333-9
Springer Heidelberg New York Dordrecht London

Library of Congress Control Number: 2013934096

Printed on acid-free paper

Springer is part of Springer Science+Business Media (www.springer.com)

Preface

Regression is the most popular and commonly used statistical methodology for analyzing empirical problems in social sciences, economics, and life sciences. Correspondingly, there exist a large variety of models and inferential tools, ranging from conventional linear models to modern non- and semiparametric regression. Currently available textbooks mostly focus on particular classes of regression models, however, strongly varying in style, mathematical level, and orientation towards theory or application. Why then another book on regression? Several introductory textbooks are available for students and practitioners in diverse fields of applications, but they deal almost exclusively with linear regression. On the other hand, most texts concentrating on modern non- and semiparametric methods primarily address readers with strong theoretical interest and methodological background, presupposing a correspondingly high-level mathematical basis. They are therefore less accessible to readers from applied fields who need to employ these methods.

The aim of this book is an applied and unified introduction into parametric, non-, and semiparametric regression that closes the gap between theory and application. The most important models and methods in regression are presented on a solid formal basis, and their appropriate application is shown through many real data examples and case studies. Availability of (user-friendly) software has been a major criterion for the methods selected and presented. In our view, the interplay and balance between theory and application are essential for progress in substantive disciplines, as well as for the development of statistical methodology, motivated and stimulated through new challenges arising from multidisciplinary collaboration. A similar goal, but with somewhat different focus, has been pursued in the book Semiparametric Regression by Ruppert, Wand, and Carroll (2003).

Thus, our book primarily targets an audience that includes students, teachers, and practitioners in social, economic, and life sciences, as well as students and teachers in statistics programs and mathematicians and computer scientists with interests in statistical modeling and data analysis. It is written at an intermediate mathematical level and assumes only knowledge of basic probability, calculus, and statistics. Short parts in the text dealing with more complex details or providing additional information start with the symbol  and end with . These parts may be omitted in a first reading without loss of continuity. The most important definitions and statements are concisely summarized in boxes. Two appendices describe required matrix algebra, as well as elements of probability calculus and statistical inference.

Depending on the particular interests, parts of the book can be read independently of remaining parts or also in modified order:

- Chapter 2 provides an introductory overview on parametric, non-, and semiparametric regression models, deliberately omitting technical details and inferential tools.
- Chapters 1–4 can be read as an introduction to linear models.
- Linear mixed models (Sects. 7.1–7.4) can be studied immediately after Chaps. 1–4 and before reading Chaps. 5 and 6.
- Sections 10.1 and 10.2 on linear quantile regression can be read immediately after Chaps. 1–4.
- Chapters 1–4, Sects. 7.1–7.4, and Chaps. 8–10 can be read as an introduction to parametric and semiparametric regression for continuous responses (including semiparametric quantile regression).
- Chapters 1–6 comprise parametric regression models for continuous and discrete responses.

An overview of possible reading alternatives is given in the following table (chapters in brackets [. . .] could be omitted):

Description	Chapters
Linear models	1, 2, 3, 4, [Sects. 10.1, 10.2]
Linear mixed models	1, 2, 3, 4, Sects. 7.1–7.4, [Sects. 10.1–10.2]
Variable selection in linear models	3, Sects. 4.2–4.4
Generalized linear models	1, 2, 3, 4, 5, [6]
Generalized linear mixed models	1, 2, 3, 4, 5, [6], 7
Semiparametric regression for continuous responses (excluding mixed models)	1, 2, 3, 4 , 8 (excl. pages 481 ff.), 9 (excl. Sects. 9.4, 9.6.2), [10]
Semiparametric regression for continuous responses (including mixed models)	1, 2, 3, 4, Sects. 7.1–7.4, 8, 9, [10]

Many examples and applications from diverse fields illustrate models and methods. Most of the data sets are available via the url http://www.regressionbook.org/ and the symbol ⇩ added to an example indicates the availability of corresponding software code from the web site. This facilitates independent work and studies through real data applications and small case studies. In addition, the web site provides information about statistical software for regression.

Highlights of the book include:

- An introduction of regression models from first principles, i.e., a complete and comprehensive introduction to the linear model in Chaps. 3, 4, and 10
- A coverage of the entire range of regression models starting with linear models, covering generalized linear and mixed models and also including (generalized) additive models and quantile regression

- A presentation of both frequentist and Bayesian approaches to regression
- The inclusion of a large number of worked out examples and case studies
- Although the book is written in textbook style suitable for students, the material is close to current research on advanced regression analysis

This book is partly based on a preceding German version that has been translated and considerably extended. We are indebted to Alexandra Reuber for translating large parts of the German version. We also thank Herwig Friedl, Christian Heumann, Torsten Hothorn, and Helga Wagner for acting as referees of the book. They all did a great job and were very helpful in improving the manuscript. Many thanks to Jesus Crespo Cuaresma, Kathrin Dallmeier, Martin Feldkircher, Oliver Joost, Franziska Kohl, Jana Lehmann, Lorenz Oberhammer, Cornelia Oberhauser, Alexander Razen, Helene Roth, Judith Santer, Sylvia Schmidt, Nora Seiwald, Iris Burger, Sven Steinert, Nikolaus Umlauf, Janette Walde, Elisabeth Waldmann, and Peter Wechselberger for support and assistance in various ways. Last but not least we thank Alice Blanck, Alphonseraja Sagayaraj, Ulrike Stricker-Komba and Niels Peter Thomas from Springer Verlag for their continued support and patience during the preparation of the manuscript.

München, Germany — Ludwig Fahrmeir
Göttingen, Germany — Thomas Kneib
Innsbruck, Austria — Stefan Lang
Baton Rouge, LA — Brian Marx
January 2013

Table of Contents

1 **Introduction** ... 1
1.1 Examples of Applications ... 4
1.2 First Steps ... 11
1.2.1 Univariate Distributions of the Variables ... 11
1.2.2 Graphical Association Analysis ... 13
1.3 Notational Remarks ... 19

2 **Regression Models** ... 21
2.1 Introduction ... 21
2.2 Linear Regression Models ... 22
2.2.1 Simple Linear Regression Model ... 22
2.2.2 Multiple Linear Regression ... 26
2.3 Regression with Binary Response Variables: The Logit Model ... 33
2.4 Mixed Models ... 38
2.5 Simple Nonparametric Regression ... 44
2.6 Additive Models ... 49
2.7 Generalized Additive Models ... 52
2.8 Geoadditive Regression ... 55
2.9 Beyond Mean Regression ... 61
2.9.1 Regression Models for Location, Scale, and Shape ... 62
2.9.2 Quantile Regression ... 66
2.10 Models in a Nutshell ... 68
2.10.1 Linear Models (LMs, Chaps. 3 and 4) ... 68
2.10.2 Logit Model (Chap. 5) ... 68
2.10.3 Poisson Regression (Chap. 5) ... 68
2.10.4 Generalized Linear Models (GLMs, Chaps. 5 and 6) ... 69
2.10.5 Linear Mixed Models (LMMs, Chap. 7) ... 69
2.10.6 Additive Models and Extensions (AMs, Chaps. 8 and 9) ... 70
2.10.7 Generalized Additive (Mixed) Models (GA(M)Ms, Chap. 9) ... 70
2.10.8 Structured Additive Regression (STAR, Chap. 9) ... 71
2.10.9 Quantile Regression (Chap. 10) ... 71

3 The Classical Linear Model 73
3.1 Model Definition 73
3.1.1 Model Parameters, Estimation, and Residuals 77
3.1.2 Discussion of Model Assumptions 78
3.1.3 Modeling the Effects of Covariates 86
3.2 Parameter Estimation 104
3.2.1 Estimation of Regression Coefficients 104
3.2.2 Estimation of the Error Variance 108
3.2.3 Properties of the Estimators 110
3.3 Hypothesis Testing and Confidence Intervals 125
3.3.1 Exact F-Test 128
3.3.2 Confidence Regions and Prediction Intervals 136
3.4 Model Choice and Variable Selection 139
3.4.1 Bias, Variance and Prediction Quality 142
3.4.2 Model Choice Criteria 146
3.4.3 Practical Use of Model Choice Criteria 150
3.4.4 Model Diagnosis 155
3.5 Bibliographic Notes and Proofs 168
3.5.1 Bibliographic Notes 168
3.5.2 Proofs 168

4 Extensions of the Classical Linear Model 177
4.1 The General Linear Model 177
4.1.1 Model Definition 177
4.1.2 Weighted Least Squares 178
4.1.3 Heteroscedastic Errors 182
4.1.4 Autocorrelated Errors 191
4.2 Regularization Techniques 201
4.2.1 Statistical Regularization 202
4.2.2 Ridge Regression 203
4.2.3 Least Absolute Shrinkage and Selection Operator 208
4.2.4 Geometric Properties of Regularized Estimates 211
4.2.5 Partial Regularization 216
4.3 Boosting Linear Regression Models 217
4.3.1 Basic Principles 217
4.3.2 Componentwise Boosting 218
4.3.3 Generic Componentwise Boosting 222
4.4 Bayesian Linear Models 225
4.4.1 Standard Conjugate Analysis 227
4.4.2 Regularization Priors 237
4.4.3 Classical Bayesian Model Choice (and Beyond) 243
4.4.4 Spike and Slab Priors 253
4.5 Bibliographic Notes and Proofs 257
4.5.1 Bibliographic Notes 257
4.5.2 Proofs 258

5 Generalized Linear Models ... 269
5.1 Binary Regression ... 270
5.1.1 Binary Regression Models ... 270
5.1.2 Maximum Likelihood Estimation ... 279
5.1.3 Testing Linear Hypotheses ... 285
5.1.4 Criteria for Model Fit and Model Choice ... 287
5.1.5 Estimation of the Overdispersion Parameter ... 292
5.2 Count Data Regression ... 293
5.2.1 Models for Count Data ... 293
5.2.2 Estimation and Testing: Likelihood Inference ... 295
5.2.3 Criteria for Model Fit and Model Choice ... 297
5.2.4 Estimation of the Overdispersion Parameter ... 297
5.3 Models for Nonnegative Continuous Response Variables ... 298
5.4 Generalized Linear Models ... 301
5.4.1 General Model Definition ... 301
5.4.2 Likelihood Inference ... 306
5.5 Quasi-likelihood Models ... 309
5.6 Bayesian Generalized Linear Models ... 311
5.6.1 Posterior Mode Estimation ... 313
5.6.2 Fully Bayesian Inference via MCMC Simulation Techniques ... 314
5.6.3 MCMC-Based Inference Using Data Augmentation ... 316
5.7 Boosting Generalized Linear Models ... 319
5.8 Bibliographic Notes and Proofs ... 320
5.8.1 Bibliographic Notes ... 320
5.8.2 Proofs ... 321

6 Categorical Regression Models ... 325
6.1 Introduction ... 325
6.2 Models for Unordered Categories ... 329
6.3 Ordinal Models ... 334
6.3.1 The Cumulative Model ... 334
6.3.2 The Sequential Model ... 337
6.4 Estimation and Testing: Likelihood Inference ... 343
6.5 Bibliographic Notes ... 347

7 Mixed Models ... 349
7.1 Linear Mixed Models for Longitudinal and Clustered Data ... 350
7.1.1 Random Intercept Models ... 350
7.1.2 Random Coefficient or Slope Models ... 357
7.1.3 General Model Definition and Matrix Notation ... 361
7.1.4 Conditional and Marginal Formulation ... 365
7.1.5 Stochastic Covariates ... 366
7.2 General Linear Mixed Models ... 368

7.3 Likelihood Inference in LMMs ... 371
7.3.1 Known Variance–Covariance Parameters ... 371
7.3.2 Unknown Variance–Covariance Parameters ... 372
7.3.3 Variability of Fixed and Random Effects Estimators ... 378
7.3.4 Testing Hypotheses ... 380
7.4 Bayesian Linear Mixed Models ... 383
7.4.1 Estimation for Known Covariance Structure ... 384
7.4.2 Estimation for Unknown Covariance Structure ... 385
7.5 Generalized Linear Mixed Models ... 389
7.5.1 GLMMs for Longitudinal and Clustered Data ... 389
7.5.2 Conditional and Marginal Models ... 392
7.5.3 GLMMs in General Form ... 394
7.6 Likelihood and Bayesian Inference in GLMMs ... 394
7.6.1 Penalized Likelihood and Empirical Bayes Estimation ... 395
7.6.2 Fully Bayesian Inference Using MCMC ... 397
7.7 Practical Application of Mixed Models ... 401
7.7.1 General Guidelines and Recommendations ... 401
7.7.2 Case Study on Sales of Orange Juice ... 403
7.8 Bibliographic Notes and Proofs ... 409
7.8.1 Bibliographic Notes ... 409
7.8.2 Proofs ... 410

8 Nonparametric Regression ... 413
8.1 Univariate Smoothing ... 415
8.1.1 Polynomial Splines ... 415
8.1.2 Penalized Splines (P-Splines) ... 431
8.1.3 General Penalization Approaches ... 446
8.1.4 Smoothing Splines ... 448
8.1.5 Random Walks ... 452
8.1.6 Kriging ... 453
8.1.7 Local Smoothing Procedures ... 460
8.1.8 General Scatter Plot Smoothing ... 468
8.1.9 Choosing the Smoothing Parameter ... 478
8.1.10 Adaptive Smoothing Approaches ... 490
8.2 Bivariate and Spatial Smoothing ... 500
8.2.1 Tensor Product P-Splines ... 503
8.2.2 Radial Basis Functions and Thin Plate Splines ... 512
8.2.3 Kriging: Spatial Smoothing with Continuous Location Variables ... 515
8.2.4 Markov Random Fields ... 521
8.2.5 Summary of Roughness Penalty Approaches ... 527
8.2.6 Local and Adaptive Smoothing ... 529
8.3 Higher-Dimensional Smoothing ... 530
8.4 Bibliographic Notes ... 531

9 Structured Additive Regression 535
9.1 Additive Models 536
9.2 Geoadditive Regression 540
9.3 Models with Interactions 543
9.3.1 Models with Varying Coefficient Terms 544
9.3.2 Interactions Between Two Continuous Covariates 547
9.4 Models with Random Effects 549
9.5 Structured Additive Regression 553
9.6 Inference 561
9.6.1 Penalized Least Squares or Likelihood Estimation 561
9.6.2 Inference Based on Mixed Model Representation 566
9.6.3 Bayesian Inference Based on MCMC 568
9.7 Boosting STAR Models 573
9.8 Case Study: Malnutrition in Zambia 576
9.8.1 General Guidelines 576
9.8.2 Descriptive Analysis 580
9.8.3 Modeling Variants 583
9.8.4 Estimation Results and Model Evaluation 584
9.8.5 Automatic Function Selection 589
9.9 Bibliographic Notes 594

10 Quantile Regression 597
10.1 Quantiles 599
10.2 Linear Quantile Regression 601
10.2.1 Classical Quantile Regression 601
10.2.2 Bayesian Quantile Regression 609
10.3 Additive Quantile Regression 612
10.4 Bibliographic Notes and Proofs 616
10.4.1 Bibliographic Notes 616
10.4.2 Proofs 618

A Matrix Algebra 621
A.1 Definition and Elementary Matrix Operations 621
A.2 Rank of a Matrix 626
A.3 Block Matrices and the Matrix Inversion Lemma 628
A.4 Determinant and Trace of a Matrix 629
A.5 Generalized Inverse 631
A.6 Eigenvalues and Eigenvectors 631
A.7 Quadratic Forms 633
A.8 Differentiation of Matrix Functions 635

B Probability Calculus and Statistical Inference 639
B.1 Some Univariate Distributions 639
B.2 Random Vectors 645

B.3 Multivariate Normal Distribution 648
B.3.1 Definition and Properties 648
B.3.2 The Singular Multivariate Normal Distribution 650
B.3.3 Distributions of Quadratic Forms 651
B.3.4 Multivariate t-Distribution 651
B.3.5 Normal-Inverse Gamma Distribution 652
B.4 Likelihood Inference 653
B.4.1 Maximum Likelihood Estimation 653
B.4.2 Numerical Computation of the MLE 660
B.4.3 Asymptotic Properties of the MLE 662
B.4.4 Likelihood-Based Tests of Linear Hypotheses 662
B.4.5 Model Choice 664
B.5 Bayesian Inference 665
B.5.1 Basic Concepts of Bayesian Inference 665
B.5.2 Point and Interval Estimation 669
B.5.3 MCMC Methods 670
B.5.4 Model Selection 676
B.5.5 Model Averaging 679

Bibliography 681

Index 691

Introduction 1

Sir Francis Galton (1822–1911) was a diverse researcher, who did pioneering work in many disciplines. Among statisticians, he is especially known for the Galton board which demonstrates the binomial distribution. At the end of the nineteenth century, Galton was mainly interested in questions regarding heredity. Galton collected extensive data illustrating body height of parents and their grown children. He examined the *relationship* between body heights of the children and the average body height of both parents. To adjust for the natural height differences across gender, the body height of women was multiplied by a factor of 1.08. In order to better examine this relationship, he listed all his data in a contingency table (Table 1.1). With the help of this table, he was able to make the following discoveries:

- Column-wise, i.e., for given average heights of the parents, the heights of the adolescents approximately follow a normal distribution.
- The normal distributions in each column have a common variance.
- When examining the relationship between the height of the children and the average height of the parents, an approximate linear trend was found with a slope of 2/3. A slope with value less than one led Galton to the conclusion that children of extremely tall (short) parents are usually shorter (taller) than their parents. In either case there is a tendency towards the population average, and Galton referred to this as *regression* towards the mean.

Later, Galton illustrated the data in the form of a scatter plot showing the heights of the children and the average height of the parents (Fig. 1.1). He visually added the trend or the *regression line*, which provides the average height of children as (average) parent height is varied.

Galton is viewed as a pioneer of regression analysis, because of his regression analytic study of heredity. However, Galton's mathematical capabilities were limited. His successors, especially Karl Pearson (1857–1936), Francis Ysidro Edgeworth (1845–1926), and George Udny Yule (1871–1951) formalized his work. Today, linear regression models are part of every introductory statistics book. In modern terms, Galton studied the systematic influence of the *explanatory variable*

L. Fahrmeir et al., *Regression*, DOI 10.1007/978-3-642-34333-9_1,

Table 1.1 Galton heredity data: contingency table between the height of 928 adult children and the average height of their 205 set of parents

Height of	Average height of parents											
children	64.0	64.5	65.5	66.5	67.5	68.5	69.5	70.5	71.5	72.5	73.0	Total
73.7	0	0	0	0	0	0	5	3	2	4	0	14
73.2	0	0	0	0	0	3	4	3	2	2	3	17
72.2	0	0	1	0	4	4	11	4	9	7	1	41
71.2	0	0	2	0	11	18	20	7	4	2	0	64
70.2	0	0	5	4	19	21	25	14	10	1	0	99
69.2	1	2	7	13	38	48	33	18	5	2	0	167
68.2	1	0	7	14	28	34	20	12	3	1	0	120
67.2	2	5	11	17	38	31	27	3	4	0	0	138
66.2	2	5	11	17	36	25	17	1	3	0	0	117
65.2	1	1	7	2	15	16	4	1	1	0	0	48
64.2	4	4	5	5	14	11	16	0	0	0	0	59
63.2	2	4	9	3	5	7	1	1	0	0	0	32
62.2	–	1	0	3	3	0	0	0	0	0	0	7
61.7	1	1	1	0	0	1	0	1	0	0	0	5
Total	14	23	66	78	211	219	183	68	43	19	4	928

The unit of measurement is inch which has already been used by Galton (1 inch corresponds to 2.54 cm)

Source: Galton (1889)

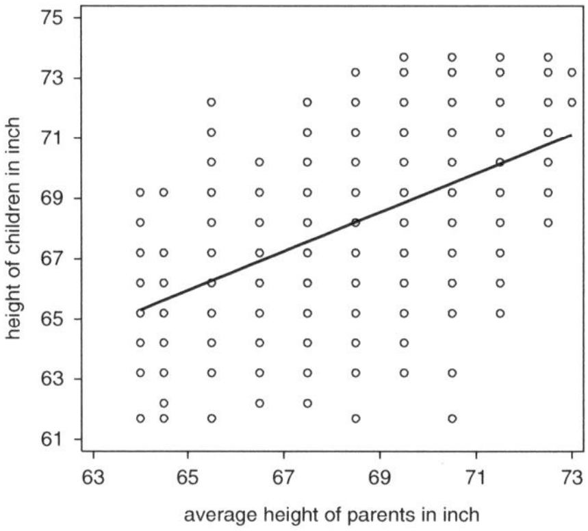

Fig. 1.1 Galton heredity data: scatter plot including a regression line between the height of children and the average height of their parents

x = "average size of the parents" on the *response variable* y = "height of grown-up children." Explanatory variables are also known as *independent variables*, *regressors*, or *covariates*. Response variables are also known as *dependent variables* or *target variables*. The fact that the linear relationship is not exact, but rather depends on random errors, is a main characteristic for regression problems. Galton assumed the most simple regression model,

$$y = \beta_0 + \beta_1 x + \varepsilon,$$

where the systematic component $\beta_0 + \beta_1 x$ is linear and ε constitutes the random error. While Galton determined the parameters β_0 and β_1 of the regression line in an ad hoc manner, nowadays these regression parameters are estimated via the *method of least squares*. The parameters β_0 and β_1 are estimated using the data pairs (y_i, x_i), $i = 1, \ldots, n$, so that the sum of the squared deviations

$$\sum_{i=1}^{n} (y_i - \beta_0 - \beta_1 x_i)^2$$

of the observations y_i from the regression line $\beta_0 + \beta_1 x_i$ is minimized. If we apply this principle to Galton's data, the estimated slope of the regression line is 0.64, a value that is fairly close to Galton's visually determined slope of 2/3.

Interestingly, the method of least squares was already discovered prior to Galton's study of heredity. The first publication by the mathematician Adrien Marie Legendre (1752–1833) appeared in 1805 making the method of least squares one of the oldest general estimation concepts in statistics. In the eighteenth and nineteenth century, the method was primarily used to predict the orbits of asteroids. Carl Friedrich Gauß (1777–1855) became famous for the prediction of the orbit of the asteroid Ceres, which was discovered in the year 1801 by the astronomer Giuseppe Piazzi. After forty days of observation, the asteroid disappeared behind the sun. Since an exact calculation of the asteroid's orbit was very difficult at that time, it was impossible to relocate the asteroid. By using the method of least squares, the twenty-four-year-old Gauß was able to give a feasible prediction of the asteroid's orbit. In his book "Theoria Motus Corporum Coelestium in Sectionibus Conicis Solem Ambientium" (1809), Gauß claimed the discovery of the method of least squares. Sometime later, Gauß even stated to have used this method since 1795 (as an eighteen year old), which provoked a priority dispute between Legendre and Gauß. Fact is that Gauß's work is the basis of the modern linear regression model with Gaussian errors.

Since the discovery of the method of least squares by Legendre and Gauß and the first regression analysis by Francis Galton, the methodology of regression has been improved and developed in many ways, and is nowadays applied in almost all scientific disciplines. The aim of this book is to give a modern introduction of the most important techniques and models of regression analysis and their application. We will address the following models in detail:

- *Regression models:* In Chap. 2, we present the different model classes without technical details; the subsequent chapters provide a thorough presentation of each of these models.

- *Linear models:* In Chaps. 3 and 4, we present a comprehensive introduction into linear regression models, including recent developments.
- *Generalized linear models:* In Chaps. 5 and 6, we discuss generalized linear models. These are especially suitable for problems where the response variables do not follow a normal distribution, including categorical response variables or count data.
- *Mixed models:* In Chap. 7, we present mixed models (models with random effects) for clustered data. A main focus in this chapter will be the analysis of panel and longitudinal data.
- *Univariate, bivariate, and spatial smoothing:* In Chap. 8, we introduce univariate and bivariate smoothing (nonparametric regression). These semiparametric and nonparametric methods are suitable to estimate complex nonlinear relationships including an automatic determination of the required amount of nonlinearity. Methods of spatial smoothing will also be discussed in detail.
- *Structured additive regression:* In Chap. 9, we present a unifying framework that combines the methods presented in Chap. 8 into one all-encompassing model. Structured additive regression models include a variety of special cases, for example, nonparametric and semiparametric regression models, additive models, geoadditive models, and varying-coefficient models. This chapter also illustrates how these models can be put into practice using a detailed case study.
- *Quantile regression:* Chapter 10 presents an introduction to quantile regression. While the methods of the previous chapters are more or less restricted to estimating the (conditional) mean depending on covariates, quantile regression allows to estimate the (conditional) quantiles of a response variable depending on covariates.

For the first time, this book presents a comprehensive and practical presentation of the most important models and methods of regression analysis. Chapter 2 is especially innovative, since it illustrates all model classes in a unified setting without focusing on the (often complicated) estimation techniques. The chapter gives the reader an overview of modern methods of regression and, at the same time, serves as a guide for choosing the appropriate model for each individual problem.

The following section illustrates the versatility of modern regression models to examine scientific questions in a variety of disciplines.

1.1 Examples of Applications

This book illustrates models and techniques of regression analysis via several applications taken from a variety of disciplines. The following list gives an overview:

- *Development economics:* Analysis of socioeconomic determinants of childhood malnutrition in developing countries
- *Hedonic prices:* Analysis of retail prices of the VW-Golf model
- *Innovation research:* Examination of the probability of opposition against patents granted by the European patent office

- *Credit scoring:* Analysis of the creditability of private bank customers
- *Marketing research:* Analysis of the relationship between the weekly unit sales of a product and sales promotions, particularly price variations
- *Rent index:* Analysis of the dependence between the monthly rent and the type, location, and condition of the rented apartment
- *Calculation of risk premium:* Analysis of claim frequency and claim size of motor vehicle insurance in order to calculate the risk premium
- *Ecology:* Analysis of the health status of trees in forests
- *Neuroscience:* Determination of the active brain area when solving certain cognitive tasks
- *Epidemiologic studies and clinical trials:*
  - Impact of testosterone on the growth of rats
  - Analysis of the probability of infection after Caesarean delivery
  - Study of the impairment to pulmonary function
  - Analysis of the life span of leukemia patients
- *Social science:* Analysis of speed dating data

Some of the listed examples will play a central role in this book and will now be discussed in more detail.

Example 1.1 Munich Rent Index

Many cities and communities in Germany establish rent indices in order to provide the renter and landlord with a market review for the "typical rent for the area." The basis for this index is a law in Germany that defines the "typical rent for the area" as the common remuneration that has been stipulated or changed over the last few years for price-maintained living area of comparable condition, size, and location within a specific community. This means that the average rent results from the apartment's characteristics, size, condition, etc. and therefore constitutes a typical regression problem. We use the net rent—the monthly rental price, which remains after having subtracted all running costs and incidentals—as the response variable. Alternatively, we can use the net rent per square meter as the response.

Within the scope of this book and due to data confidentiality, we confine ourselves to a fraction of the data and variables, which were used in the rent index for Munich in the year 1999. We use the 1999 data since more recent data is either not publicly available or less adequate for illustration purposes. The current rent index of Munich including documentation can be found at www.mietspiegel.muenchen.de (in German only).

Table 1.2 includes names and descriptions of the variables used in the subsequent analyses. The data of more than 3,000 apartments were collected by representative random sampling.

The goal of a regression analysis is to model the impact of explanatory variables (living area, year of construction, location, etc.) on the response variable of net rent or net rent per square meter. In a final step, we aim at representing the estimated effect of each explanatory variable in a simpler form by appropriate tables in a brochure or on the internet.

In this book, we use the Munich rent index data mainly to illustrate regression models with continuous responses (see Chaps. 2–4, 9, and 10). In doing so, we use simplified models for illustration purposes. This implies that the results do not always correspond to the official rent index.

△

Example 1.2 Malnutrition in Zambia

The World Health Organization (WHO) has decided to conduct representative household surveys (demographic and health surveys) in developing countries on a regular basis. Among others, these surveys consist of information regarding malnutrition, mortality, and

Table 1.2 Munich rent index: description of variables including summary statistics

Variable	Description	Mean/ frequency in %	Std.-dev.	Min/max
rent	Net rent per month (in Euro)	459.43	195.66	40.51/1,843.38
rentsqm	Net rent per month per square meter (in Euro)	7.11	2.44	0.41/17.72
area	Living area in square meters	67.37	23.72	20/160
yearc	Year of construction	1,956.31	22.31	1918/1997
location	Quality of location according to an expert assessment			
	1 = average location	58.21		
	2 = good location	39.26		
	3 = top location	2.53		
bath	Quality of bathroom			
	0 = standard	93.80		
	1 = premium	6.20		
kitchen	Quality of kitchen			
	0 = standard	95.75		
	1 = premium	4.25		
cheating	Central heating			
	0 = without central heating	10.42		
	1 = with central heating	89.58		
district	District in Munich			

health risks for children. The American institute Macro International collects data from over 50 countries. This data is freely available at www.measuredhs.com for research purposes. In this book, we look at an exemplary profile of a data set for Zambia taken in the year 1992 (4,421 observations in total). The Republic of Zambia is located in the south of Africa and is one of the poorest and most underdeveloped countries of the world.

One of the most serious problems of developing countries is the poor and often catastrophic nutritional condition of a high proportion of the population. Immediate consequences of malnutrition are reduced productivity and high mortality. Within the scope of this book, we will analyze the nutritional condition of children who are between 0 and 5 years old. The nutritional condition of children is usually determined by an anthropometric measure called Z-score. A Z-score compares the anthropometric status of a child, for example, a standardized age-specific body height, with comparable measures taken from a reference population. Until the age of 24 months, this reference population is based on white US-American children from wealthy families with a high socioeconomic status. After 24 months, the reference population changes and then consists of a representative sample taken from all US-American children. Among several possible anthropometric indicators, we use a measure for chronic malnutrition, which is based on body height as indication for the long-term development of the nutritional condition. This measure is defined as

$$zscore_i = \frac{h_i - mh}{\sigma},$$

for a child i, where h_i represents the height of the child, mh represents the median height of children belonging to the reference population of the same age group, and σ refers to the corresponding standard deviation for the reference population.

Table 1.3 Malnutrition in Zambia: description of variables including summary statistics

Variable	Description	Mean/ frequency in %	Std-dev.	Min/max
zscore	Child's Z-score	−171.19	139.34	−600/503
c_gender	Gender			
	1 = male	49.02		
	0 = female	50.98		
c_breastf	Duration of breast-feeding in months	11.11	9.42	0/46
c_age	Child's age in months	27.61	17.08	0/59
m_agebirth	Mother's age at birth in years	26.40	6.87	13.16/48.66
m_height	Mother's height in centimeter	158.06	5.99	134/185
m_bmi	Mother's body mass index	21.99	3.32	13.15/39.29
m_education	Mother's level of education			
	1 = no education	18.59		
	2 = primary school	62.34		
	3 = secondary school	17.35		
	4 = higher education	1.72		
m_work	Mother's work status			
	1 = mother working	55.25		
	0 = mother not working	44.75		
region	Region of residence in Zambia			
	1 = Central	8.89		
	2 = Copperbelt	21.87		
	3 = Eastern	9.27		
	4 = Luapula	8.91		
	5 = Lusaka	13.78		
	6 = Northern	9.73		
	7 = North western	5.88		
	8 = Southern	14.91		
	9 = Western	6.76		
district	District of residence in Zambia (55 districts)			

The primary goal of the statistical analysis is to determine the effect of certain socioeconomic variables of the child, the mother, and the household on the child's nutritional condition. Examples for socioeconomic variables are the duration of breastfeeding (*c_breastf*), the age of the child (*c_age*), the mother's nutritional condition as measured by the body mass index (*m_bmi*), and the mother's level of education as well as her work status (*m_education* and *m_work*). The data record also includes geographic information such as region or district where the mother's place of residence is located. A description of all available variables can be found in Table 1.3.

With the help of the regression models presented in this book, we will be able to pursue the aforementioned goals. Geoadditive models (see Sect. 9.2) are employed in particular. These also allow an adequate consideration of spatial information in the data. The data are analyzed within a comprehensive case study (see Sect. 9.8), which illustrates in detail the practical application of many techniques and methods presented in this book. △

Table 1.4 Patent opposition: description of variables including summary statistics

Variable	Description	Mean/ frequency in %	Std-dev.	Min/max
opp	Patent opposition			
	1 = yes	41.49		
	0 = no	58.51		
biopharm	Patent from biotech/pharma sector			
	1 = yes	44.31		
	0 = no	55.69		
ustwin	US twin patent exists			
	1 = yes	60.85		
	0 = no	39.15		
patus	Patent holder from the USA			
	1 = yes	33.74		
	0 = no	66.26		
patgsgr	Patent holder from Germany, Switzerland, or Great Britain			
	1 = yes	23.49		
	0 = no	76.51		
year	Grant year			
	1980	0.18		
	⋮	⋮		
	1997	1.62		
ncit	Number of citations for the patent	1.64	2.74	0/40
ncountry	Number of designated states for the patent	7.8	4.12	1/17
nclaims	Number of claims	13.13	12.09	1/355

Example 1.3 Patent Opposition

The European Patent Office is able to protect a patent from competition for a certain period of time. The Patent Office has the task to examine inventions and to declare patent if certain prerequisites are fulfilled. The most important requirement is that the invention is something truly new. Even though the office examines each patent carefully, in about 80 % of cases competitors raise an objection against already assigned patents. In the economic literature the analysis of patent opposition plays an important role as it allows to (indirectly) investigate a number of economic questions. For instance, the frequency of patent opposition can be used as an indicator for the intensity of the competition in different market segments.

In order to analyze objections against patents, a data set with 4,866 patents from the sectors biotechnology/pharmaceutics and semiconductor/computer was collected. Table 1.4 lists the variables contained in this data set. The goal of the analysis is to model the probability of patent opposition, while using a variety of explanatory variables for the binary response variable "patent opposition" (yes/no). This corresponds to a regression problem with a binary response.

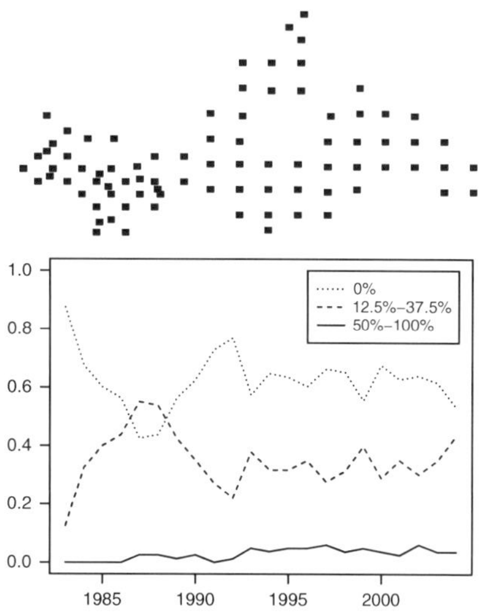

Fig. 1.2 Forest health status: The *top panel* shows the observed tree locations where the center constitutes the town of Rothenbuch. The *bottom panel* displays the temporal trend of defoliation degree

A possible explanatory variable is how often a patent has been cited in succeeding patents (variable *ncit*). Citations of patents are somewhat comparable to citations of scientific papers. Empirical experience and economic arguments indicate that the probability of an objection against a patent increases the more often it is cited. Regression models for binary response variables can formulate and examine this particular and other hypotheses. In this book the data set on patent opposition is primarily used to illustrate regression models with binary responses; see Chaps. 2 and 5.

△

Example 1.4 Forest Health Status

Knowledge about the health status of trees in a forest and its influencing factors is important from an ecological and economical point of view. This is the reason why Germany (and many other countries) conducts annual surveys regarding the condition of the forest. The data in our example come from a specific project in the forest of Rothenbuch (Spessart), which has been carried out by Axel Göttlein (Technical University, Munich) since 1982. In comparison to the extensive official land surveys, the observations, i.e., the locations of the examined trees, are much closer to each other. Figure 1.2 visualizes the 83 examined locations in Rothenbuch forest. Five tree species are part of this survey: beech, oak, spruce, larch, and pine. Here we will restrict ourselves to beech trees. Every year, the condition of beech trees is categorized by the response variable "defoliation" (*defol*) into nine ordinal categories 0 %, 12.5 %, 25 %, 37.5 %, 50 %, 62.5 %, 75 %, 87.5 %, and 100 %. Whereas the category 0 % signifies that the beech tree is healthy, the category 100 % implies that the tree is dead.

In addition to the (ordinal) response variable, explanatory variables are collected every year as well. Table 1.5 includes a selection of these variables including summary statistics. The mean values and frequencies (in percent) have been averaged over the years (1982–2004) and the observation points.

The goal of the analysis is to determine the effect of explanatory variables on the degree of defoliation. Moreover, we aim at quantifying the temporal trend and the spatial effect of geographic location, while adjusting for the effects of the other regressors. Additionally to the observed locations Fig. 1.2 presents the temporal trend of relative frequencies for the degree of defoliation of three (aggregated) categories.

Table 1.5 Forest health status: description of variables including summary statistics

Variable	Description	Mean/ frequency in %	Std-dev.	Min/max
id	Location identification number			
year	Year of data collection	1,993.58	6.33	1983/2004
defol	Degree of defoliation, in nine ordinal categories			
	0 %	62.07		
	12.5 %	24.26		
	25 %	7.03		
	37.5 %	3.79		
	50 %	1.62		
	62.5 %	0.89		
	75 %	0.33		
	87.5 %	0.00		
	100 %	0.00		
x	x-coordinate of location			
y	y-coordinate of location			
age	Average age of trees at the observation plot in years	106.17	51.38	7/234
canopyd	Canopy density in percent	77.31	23.70	0/100
gradient	Gradient of slope in percent	15.45	11.27	0/46
alt	Altitude above see level in meter	387.04	58.86	250/480
depth	Soil depth in cm	24.63	9.93	9/51
ph	pH-value in 0–2 cm depth	4.29	0.34	3.28/6.05
watermoisture	Level of soil moisture in three categories			
	1 = moderately dry	11.04		
	2 = moderately moist	55.16		
	3 = moist or temporarily wet	33.80		
alkali	Fraction of alkali ions in soil in four categories			
	1 = very low	19.63		
	2 = low	55.10		
	3 = moderate	17.18		
	4 = high	8.09		
humus	Thickness of humus layer in five categories			
	0 = 0 cm	25.71		
	1 = 1 cm	28.56		
	2 = 2 cm	21.58		
	3 = 3 cm	14.84		
	4 = more than 3 cm	9.31		

(continued)

Table 1.5 (continued)

Variable	Description	Mean/ frequency in %	Std-dev.	Min/max
type	Type of forest			
	0 = deciduous forest	50.31		
	1 = mixed forest	49.69		
fert	Fertilization			
	0 = not fertilized	80.87		
	1 = fertilized	19.13		

To analyze the data we apply regression models for multi-categorical response variables that can simultaneously accommodate nonlinear effects of the continuous covariates, as well as temporal and spatial trends. Such complex categorical regression models are illustrated in Chaps. 6 and 9.

△

The next section shows the first exploratory steps of regression analysis, which are illustrated using the data on the Munich rent index and the Zambia malnutrition data.

1.2 First Steps

1.2.1 Univariate Distributions of the Variables

The first step when conducting a regression analysis (and any other statistical evaluation) is to get an overview of the variables in the data set. We pursue the following goals for the initial descriptive and graphical univariate analysis:

- Summary and exploration of the distribution of the variables
- Identification of extreme values and outliers
- Identification of incorrect variable coding

To achieve these goals, we can use descriptive statistics, as well as graphical visualization techniques. The choice of appropriate methods depends on the individual type of variable. In general, we can differentiate between continuous and categorical variables.

We can get a first overview of continuous variables by determining some descriptive summary statistics, in particular the arithmetic mean and the median as typical measures of location, the standard deviation as a measure of variation, and the minimum and maximum of variables. Furthermore, it is useful to visualize the data. Histograms and box plots are most frequently used, but smooth nonparametric density estimators such as kernel densities are useful alternatives to histograms. Many introductory books, e.g., Veaux, Velleman, and Bock (2011) or Agresti and Finlay (2008), give easily accessible introductions to descriptive and exploratory statistics.

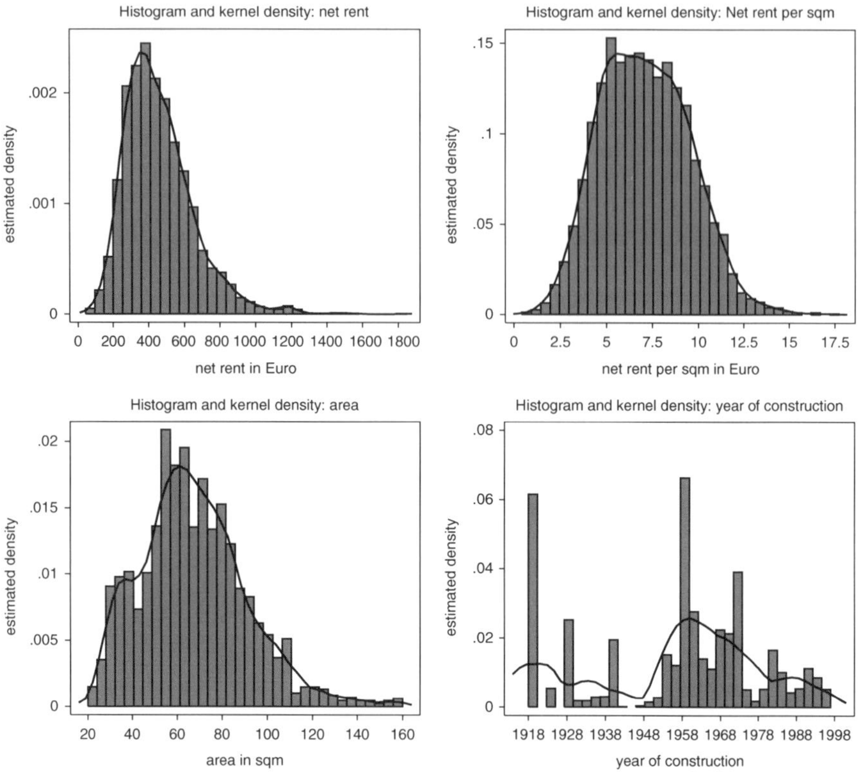

Fig. 1.3 Munich rent index: histograms and kernel density estimators for the continuous variables *rent*, *rentsqm*, *area* and *yearc*

Compared to continuous variables, it is easier to get an overview of the distribution of categorical variables. Here, we can use simple frequency tables or their graphical counterparts, particularly bar graphs.

Example 1.5 Munich Rent Index—Univariate Distributions

Summary statistics for the continuous variables *rent*, *rentsqm*, *area*, and *yearc* are already listed in Table 1.2 (p. 6). Figure 1.3 displays histograms and kernel density estimators for these variables. To give an example, we interpret summary statistics and graphical representations for the two variables "net rent" and "year of construction":

The monthly net rent roughly varies between 40 and 1,843 Euro with an average rent of approximately 459 Euro. For the majority of apartments, the rent varies between 50 and 1,200 Euro. For only a few apartments the monthly rent is higher than 1,200 Euro. This implies that any inference from a regression analysis regarding expensive apartments is comparably uncertain, when compared to the smaller and more modest sized apartments. Generally, the distribution of the monthly net rent is asymmetric and skewed towards the right.

The distribution of the year of construction is highly irregular and multimodal, which is in part due to historical reasons. Whereas the data basis for apartments for the years of the economic crises during the Weimar Constitution and the Second World War is rather limited, there are much more observations for the later years of reconstruction (mode near 1960). Starting in the mid-1970s the construction boom stopped again. Altogether the data range from 1918 until 1997. Obviously, the 1999 rent index does not allow us to draw conclusions about new buildings after 1997 since there is a temporal gap of more than one year between data collection and the publication of the rent index. Particularly striking is the relative accumulation of apartments constructed in 1918. However, this is a data artifact since all apartments that were built prior to 1918 are antedated to the year 1918.

We leave the interpretation of the distribution of the other continuous variables in the data set to the reader.

Table 1.2 also shows frequency tables for the categorical variables. We observe, for example, that most of the apartments (58 %) are located in an average location. Only about 3 % of the apartments are to be found in top locations.

△

Example 1.6 Malnutrition in Zambia—Univariate Distributions

In addition to Table 1.3 (p. 7), Fig. 1.4 provides a visual overview of the distribution of the response variable and selected continuous explanatory variables using histograms and kernel density estimators. We provide detailed interpretations in our case study in Sect. 9.8. Note that for some variables (duration of breast-feeding and child's age in months) the kernel density estimate shows artifacts in the sense that the density is positive for values lower than zero. However, for the purpose of getting an overview of the variables, this somewhat unsatisfactory behavior is not problematic.

△

1.2.2 Graphical Association Analysis

In a second step, we can graphically investigate the relationship between the response variable and the explanatory variables, at least for continuous responses. By doing so, we get a first overview regarding the type (e.g., linear versus nonlinear) and strength of the relationship between the response variable and the explanatory variables. In most cases, we focus on bivariate analyses (between the response and one explanatory variable). In the following we assume a continuous response variable.

The appropriateness of graphical tools depends on whether the explanatory variable is continuous or categorical.

Continuous Explanatory Variables

As already used by Galton at the end of the nineteenth century, simple scatter plots can provide useful information about the relationship between the response variable and the explanatory variables.

Example 1.7 Munich Rent Index—Scatter Plots

Figure 1.5 shows for the rent index data scatter plots between net rent or net rent per square meter and the continuous explanatory variables living area and year of construction. A first impression is that the scatter plots are not very informative which is a general problem with large sample sizes (in our case more than 3,000 observations). We do find some evidence of an approximately linear relationship between net rent and living area. We also notice that the variability of the net rent increases with an increased living area. The relationship

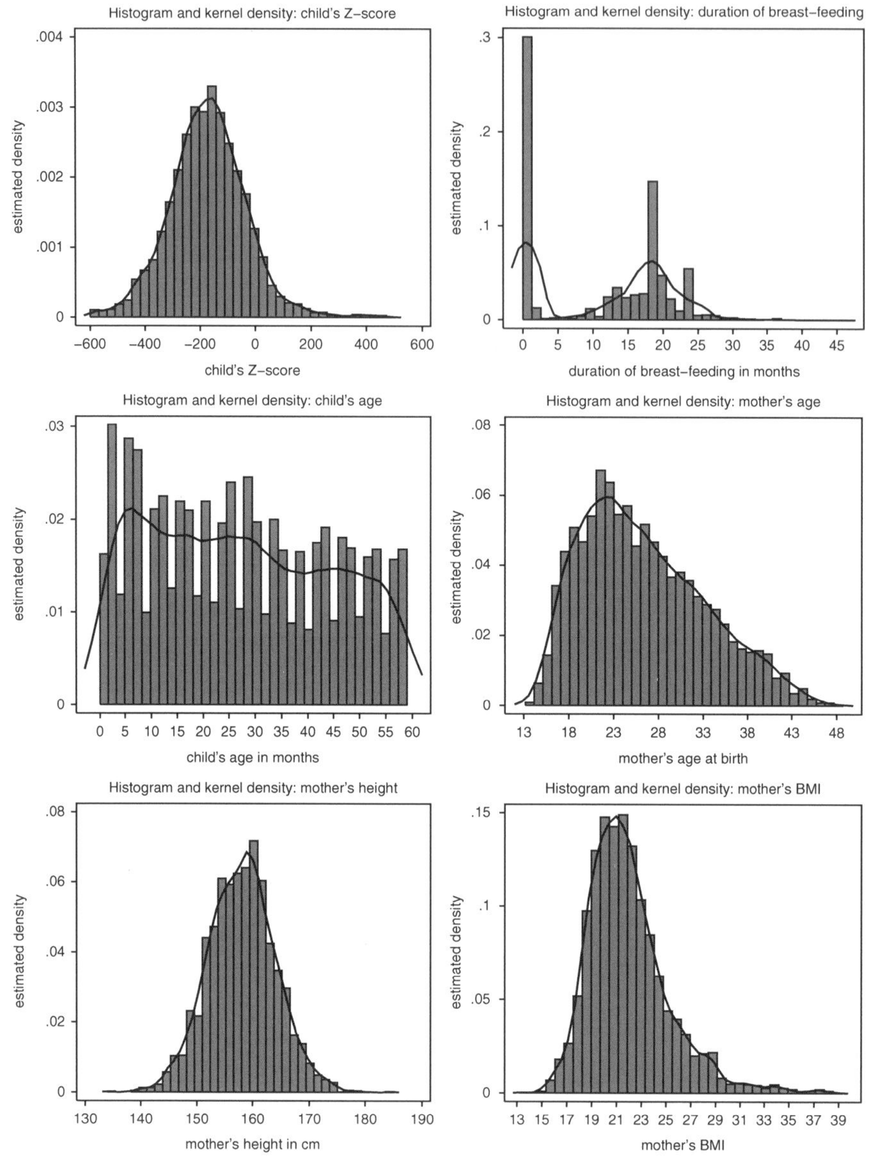

Fig. 1.4 Malnutrition in Zambia: histograms and kernel density estimators for the continuous variables

between net rent per square meter and living area is more difficult to determine. Generally the net rent per square meter for larger apartments seems to decrease. It is however difficult to assess the type of relationship (linear or nonlinear). The relationship of either of the two response variables and the year of construction is again hardly visible (if it exists at all), but

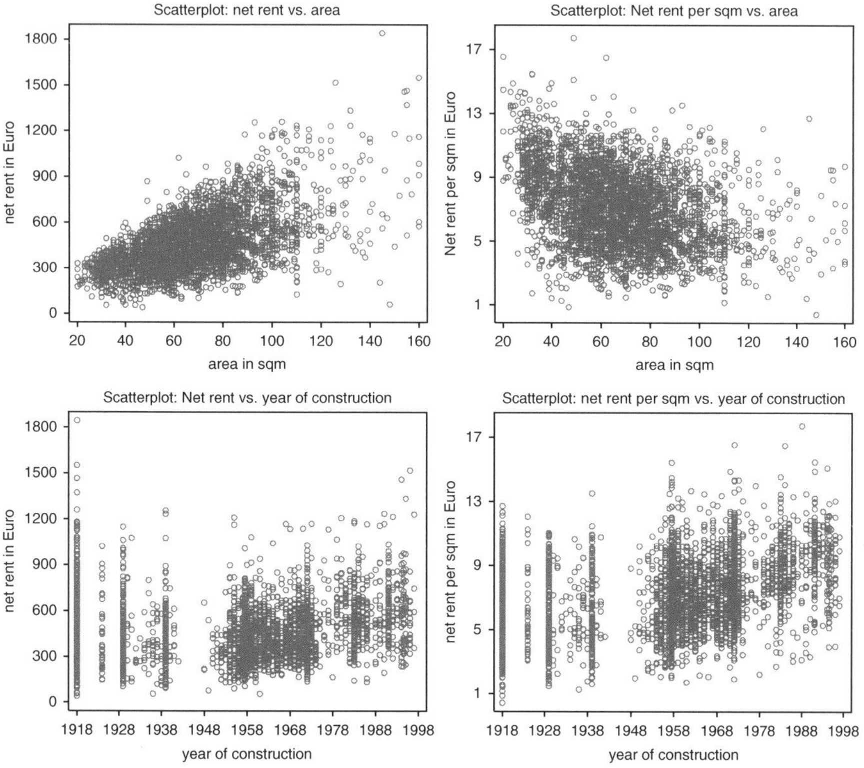

Fig. 1.5 Munich rent index: scatter plots between net rent (*left*) / net rent per sqm (*right*) and the covariates area and year of construction

there is at least evidence for a monotonic increase of rents (and rents per square meter) for flats built after 1948.

△

The preceding example shows that for large sample sizes simple scatter plots do not necessarily contain much information. In this situation, it can be useful to *cluster* the data. If the number of *different* values of the explanatory variable is relatively small in comparison to the sample size, we can summarize the response with the mean value and the corresponding standard deviation for each observed level of the explanatory variable and then visualize these in a scatter plot. Alternatively we could visualize the cluster medians together with the 25 % and 75 % quantiles (or any other combination of quantiles). The resulting data reduction often makes it easier to detect relationships. If the number of different levels of the explanatory variables is large relative to the sample size, it can be useful to cluster or categorize the data. More specifically, we divide the range of values of the explanatory variable into small intervals and calculate mean and standard deviation of the aggregated

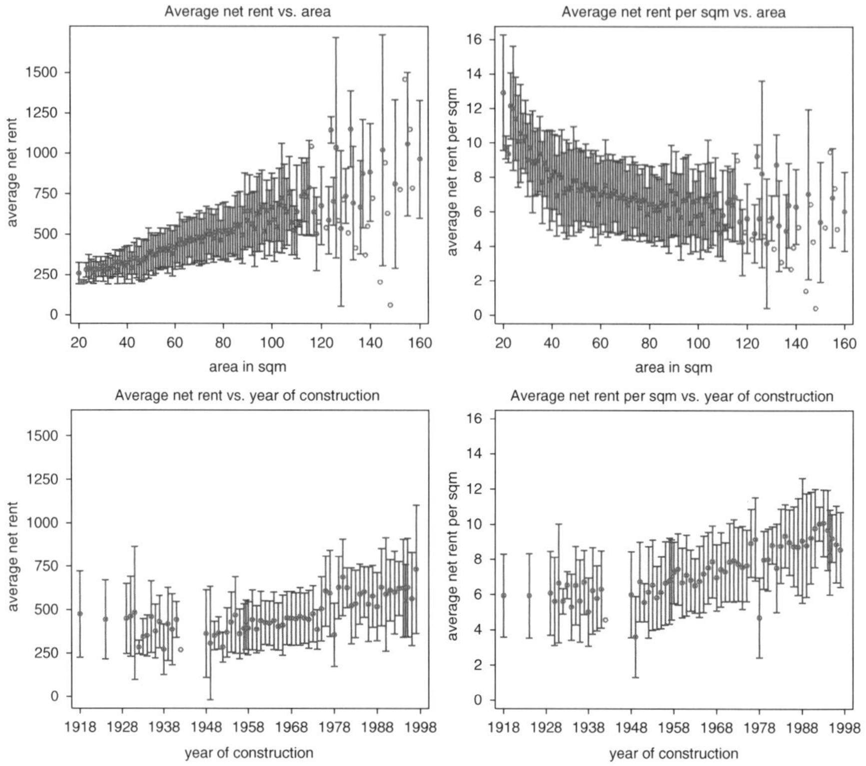

Fig. 1.6 Munich rent index: average net rent (*left*) and net rent per sqm (*right*) plus/minus one standard deviation versus area and year of construction

response for each interval separately. The cluster mean plus/minus one standard deviation is next combined into a scatter plot.

Example 1.8 Munich Rent Index—Clustered Scatter Plots

Living area and year of construction are measured in square meters and years, respectively. In both cases the units of measurement provide a natural basis for clustering. It is thus possible to calculate and visualize the mean values and standard deviations for either of the net rent responses clustered either by living area or year of construction (see Fig. 1.6). Compared to Fig. 1.5 it is now easier to make statements regarding possible relationships that may exist. If we take, e.g., the net rent per square meter as the response variable, a clear nonlinear and monotonically decreasing relationship with the living area becomes apparent. For large apartments (120 square meters or larger), we can also see a clear increase in the variability of average rents.

It also appears that there exists a relationship between the year of construction and the net rent per square meter, even though the relationship seems to be much weaker. Again the relationship is nonlinear: for apartments that were constructed prior to 1940, the rent per square meter is relatively constant (about 6 Euro). On average the rent appears somewhat lower for the few apartments from the sample taken from the years of the war. After

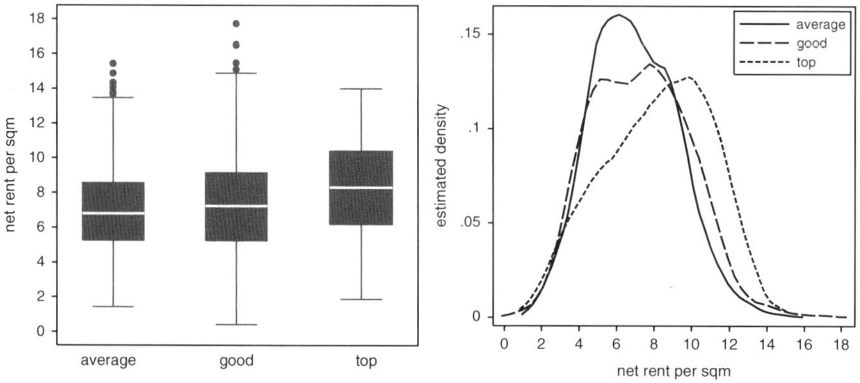

Fig. 1.7 Munich rent index: distribution of net rent per sqm clustered according to location

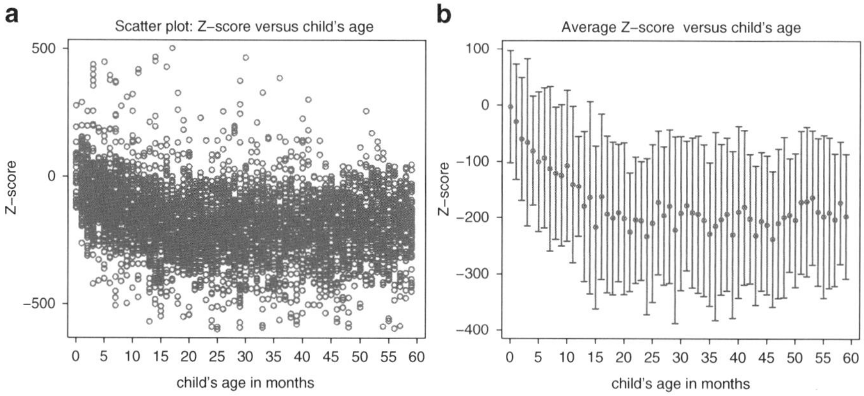

Fig. 1.8 Malnutrition in Zambia: different visualizations of the relationship between Z-score and child's age

1945, the average rent per square meter shows a linearly increasing trend with year of construction.

△

Categorical Explanatory Variables

Visualizing the relationship between a continuous response variable and categorical explanatory variables can be obtained by summarizing the response variable at each level of the categorical variable. Histograms, box plots, and (kernel) density estimators are all adequate means of illustration. In many cases, box plots are best suited as differences in mean values (measured through the median) can be well detected.

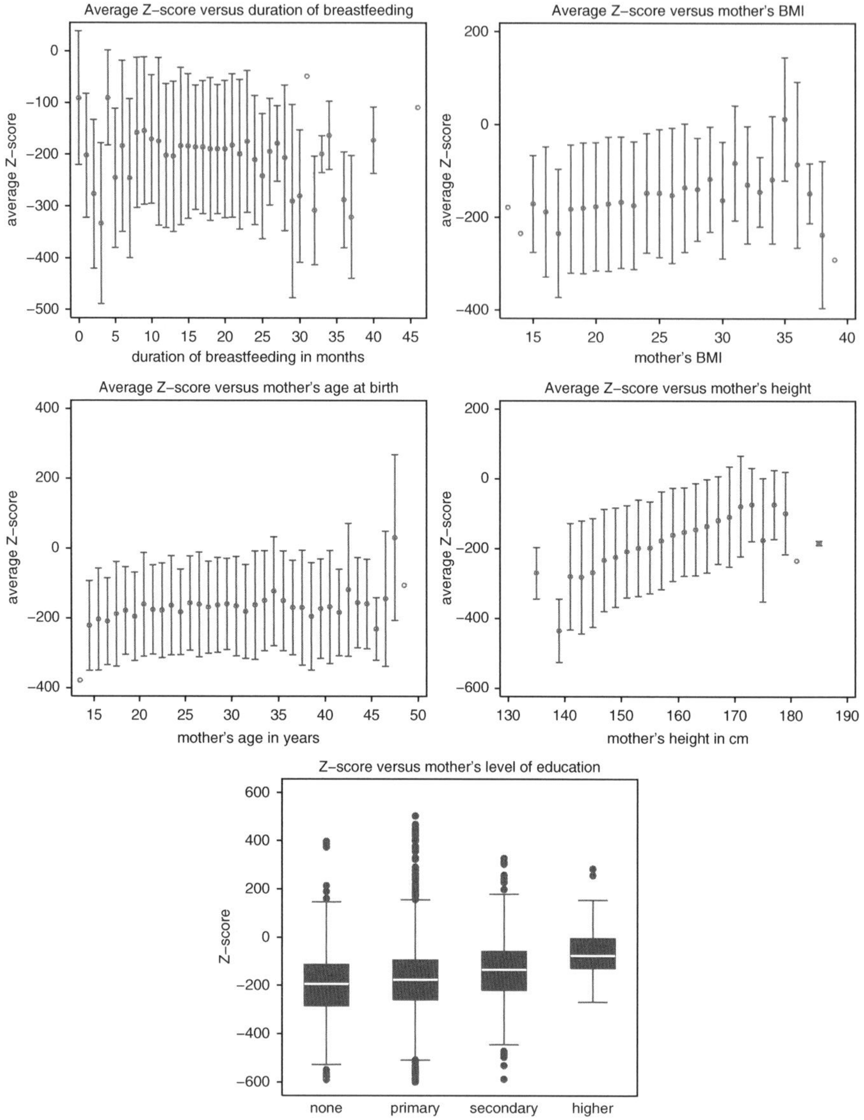

Fig. 1.9 Malnutrition in Zambia: visualization of the relationship between Z-score and selected explanatory variables

Example 1.9 Munich Rent Index—Categorical Explanatory Variables

Figure 1.7 illustrates the distribution of the net rent per square meter as the location (average, good, top) of the apartment is varied. The left panel uses box plots for illustration, and the right panel uses kernel density estimators. The box plots clearly show how the median rent (as well as the variation) increases as the location of the apartment improves.

Even though the smooth density estimators offer similar information, the visualization of these findings is not as obvious as for box plots.

△

Example 1.10 Malnutrition in Zambia—Graphical Association Analysis

Figures 1.8 and 1.9 offer a graphical illustration of the relationship between Z-scores and various explanatory variables. Similar to the rent data, the relationship between the Z-score and the age of the child is difficult to visualize (Fig. 1.8, left panel). A better choice of illustration is obtained when clustering the Z-scores by monthly age of the children (0 to 59 months). For each month, the mean plus/minus standard deviation of Z-scores is computed and plotted (right panel), which provides a much clearer picture of the relationship between Z-score and age. This type of illustration is also used for the other continuous explanatory variables, see Fig. 1.9. We will provide detailed interpretations of Figs. 1.8 and 1.9 in our case study on malnutrition in Zambia in Sect. 9.8.

△

1.3 Notational Remarks

Before we give an overview of regression models in the next chapter some remarks on notation are in order.

In introductory textbooks on statistics authors usually distinguish notationally between random variables and their realizations (the observations). Random variables are denoted by upper case letters, e.g., X, Y, while realizations are denoted by lower case letters, e.g., x, y. However, in more advanced textbook, in particular books on regression analysis, random variables and their realizations are usually *not* distinguished and both denoted by lower case letters, i.e., x, y. It then depends on the context whether y denotes the random variable or the realization. In this book we will keep this convention with the exception of Appendix B which introduces some concepts of probability and statistics. Here we will distinguish between random variables and realizations notationally in the way described above, i.e., by denoting random variables as capital letters and realizations as lower case letters.

Regression Models

2

2.1 Introduction

All case studies that have been discussed in Chap. 1 have one main feature in common: We aim at modeling the effect of a given set of explanatory variables $x_1, \ldots, x_k$ on a variable y of primary interest. The variable of primary interest y is called *response* or *dependent variable* and the explanatory variables are also called *covariates*, *independent variables*, or *regressors*. The various models differ mainly through the type of response variables (continuous, binary, categorical, or counts) and the different kinds of covariates, which can also be continuous, binary, or categorical. In more complex situations, it is also possible to include time scales, variables to describe the spatial distribution or geographical location, or group indicators.

A main characteristic of regression models is that the relationship between the response variable y and the covariates is not a deterministic function $f(x_1, \ldots, x_k)$ of $x_1, \ldots, x_k$ (as often is the case in classical physics), but rather shows random errors. This implies that the response y is a random variable, whose distribution depends on the explanatory variables. Galton's data set on heredity exemplified that, even though we know the exact height of the parents, we are unable to predict the exact height of their children. We can rather only estimate the *average size of the children* and the degree of dispersion from the mean value. Similar statements are valid for all problems discussed in Chap. 1. One main goal of regression is to analyze the influence of the covariates on the mean value of the response variable. In other words, we model the (conditional) expected value $\mathrm{E}(y \mid x_1, \ldots, x_k)$ of y depending on the covariates. Hence, the expected value is a function of the covariates:

$$\mathrm{E}(y \mid x_1, \ldots, x_k) = f(x_1, \ldots, x_k).$$

It is then possible to decompose the response into

$$y = \mathrm{E}(y \mid x_1, \ldots, x_k) + \varepsilon = f(x_1, \ldots, x_k) + \varepsilon,$$

L. Fahrmeir et al., *Regression*, DOI 10.1007/978-3-642-34333-9_2,

where ε is the random deviation from the expected value. The expected value $E(y \mid x_1, \ldots, x_k) = f(x_1, \ldots, x_k)$ is often denoted as the *systematic component* of the model. The random deviation ε is also called *random or stochastic component*, *disturbance*, or *error term*. In many regression models, in particular in the classical linear model (see Sect. 2.2 and Chap. 2), it is assumed that the error term does not depend on covariates. This may not be true, however, in general. The primary goal of regression analysis is to use the data $y_i, x_{i1}, \ldots, x_{ik}$, $i = 1, \ldots, n$, to estimate the systematic component f, and to separate it from the stochastic component ε.

The most common class is the linear regression model given by

$$y = \beta_0 + \beta_1 x_1 + \ldots + \beta_k x_k + \varepsilon.$$

Here, the function f is linear so that

$$E(y \mid x_1, \ldots, x_k) = f(x_1, \ldots, x_k) = \beta_0 + \beta_1 x_1 + \ldots + \beta_k x_k$$

holds, i.e., the (conditional) mean of y is a linear combination of the covariates. Inserting the data yields the n equations

$$y_i = \beta_0 + \beta_1 x_{i1} + \ldots + \beta_k x_{ik} + \varepsilon_i, \qquad i = 1, \ldots, n,$$

with unknown parameters or regression coefficients $\beta_0, \ldots, \beta_k$. The linear regression model is especially applicable when the response variable y is continuous and shows an approximately normal distribution (conditional on the covariates). More general regression models are, e.g., required when either the response variable is binary, the effect of covariates is nonlinear, or if spatial or cluster-specific heterogeneity has to be considered. Starting from the classical linear regression model, the following sections of this chapter describe regression models of increasing flexibility and complexity. Examples taken from various fields of application provide an overview of their usefulness. A more detailed presentation of the different regression models, and especially the corresponding statistical inference techniques, will be given in the chapters to follow.

2.2 Linear Regression Models

2.2.1 Simple Linear Regression Model

Example 2.1 Munich Rent Index—Simple Linear Regression

We start by analyzing only the subset of apartments built after 1966. This sample is divided into three location strata: average, good, and top location. The left panel of Fig. 2.1 shows the scatter plot between the response variable net rent and the explanatory variable living area for apartments in average location. The scatter plot displays an approximate linear relationship between *rent* and *area*, i.e.,

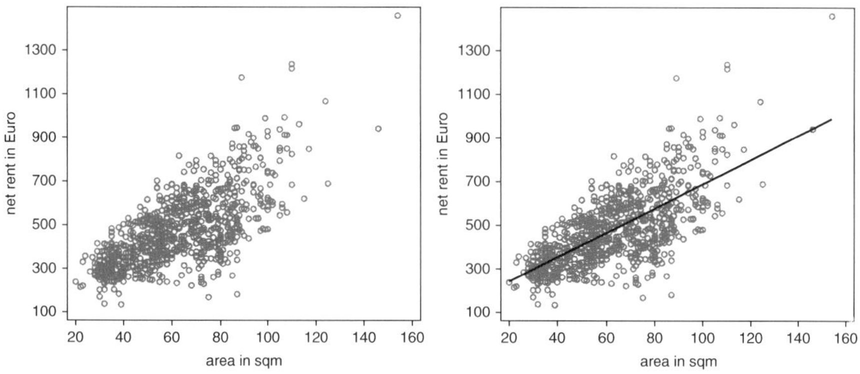

Fig. 2.1 Munich rent index: scatter plot between net rent and area for apartments in average location built after 1966 (*left panel*). In the *right panel*, a regression line is additionally included

$$rent_i = \beta_0 + \beta_1 \cdot area_i + \varepsilon_i. \quad (2.1)$$

The errors ε_i are random deviations from the regression line $\beta_0 + \beta_1 \cdot area$. Since systematic deviations from zero are already included in the parameter β_0, $\mathrm{E}(\varepsilon_i) = 0$ can be assumed. An alternative formulation of Eq. (2.1) is

$$\mathrm{E}(rent \mid area) = \beta_0 + \beta_1 \cdot area.$$

This means that the expected net rent is a linear function of the living area. △

The example is a special case of the simple linear regression model

$$y = \beta_0 + \beta_1 x + \varepsilon,$$

where the expected value $\mathrm{E}(y \mid x) = f(x)$ is assumed to be linear in the general relationship

$$y = f(x) + \varepsilon = \mathrm{E}(y \mid x) + \varepsilon.$$

This implies that $\mathrm{E}(y \mid x) = f(x) = \beta_0 + \beta_1 x$. More specifically, for the standard model of a simple linear regression, we assume

$$y_i = \beta_0 + \beta_1 x_i + \varepsilon_i, \quad i = 1, \ldots, n, \quad (2.2)$$

with independent and identically distributed errors ε_i, such that

$$\mathrm{E}(\varepsilon_i) = 0 \quad \text{and} \quad \mathrm{Var}(\varepsilon_i) = \sigma^2.$$

The property of constant variance σ^2 across errors ε_i is also called *homoscedasticity*. In particular, this implies that the errors are independent of the covariates. When constructing confidence intervals and statistical tests, it is convenient if the

2.1 Standard Model of Simple Linear Regression

Data

(y_i, x_i), $i = 1, \dots, n$, with continuous variables y and x.

Model

$$y_i = \beta_0 + \beta_1 x_i + \varepsilon_i\,, \quad i = 1, \dots, n\,.$$

The errors $\varepsilon_1, \dots, \varepsilon_n$ are independent and identically distributed (i.i.d.) with

$$\mathrm{E}(\varepsilon_i) = 0, \quad \mathrm{Var}(\varepsilon_i) = \sigma^2.$$

We can interpret the estimated regression line $\hat{f}(x) = \hat{\beta}_0 + \hat{\beta}_1 x$ as an estimate $\widehat{\mathrm{E}(y|x)}$ for the conditional expected value of y given the covariate value x. We can, thus, predict y through $\hat{y} = \hat{\beta}_0 + \hat{\beta}_1 x$.

additional assumption of Gaussian errors is reasonable:

$$\varepsilon_i \sim \mathrm{N}(0, \sigma^2).$$

In this case, the observations of the response variable follow a (conditional) normal distribution with

$$\mathrm{E}(y_i) = \beta_0 + \beta_1 x_i, \quad \mathrm{Var}(y_i) = \sigma^2,$$

and the y_i are (conditionally) independent given covariate values x_i. The unknown parameters β_0 and β_1 are estimated according to the method of least squares (LS): the estimated values $\hat{\beta}_0$ and $\hat{\beta}_1$ are determined as the minimizers of the sum of the squared deviations

$$\mathrm{LS}(\beta_0, \beta_1) = \sum_{i=1}^{n} (y_i - \beta_0 - \beta_1 x_i)^2$$

for given data (y_i, x_i), $i = 1, \dots, n$. Section 3.2 will present the method of least squares in detail. Inserting $\hat{\beta}_0, \hat{\beta}_1$ into the conditional mean, the estimated regression line

$$\hat{f}(x) = \hat{\beta}_0 + \hat{\beta}_1 x$$

results. The regression line is to be understood as an estimate $\widehat{\mathrm{E}(y|x)}$ for the conditional mean of y given the covariate value x. Thus, the regression line can also be used to predict y for a given x. The predicted value of y is usually denoted by $\hat{y}$, i.e., $\hat{y} = \hat{\beta}_0 + \hat{\beta}_1 x$.

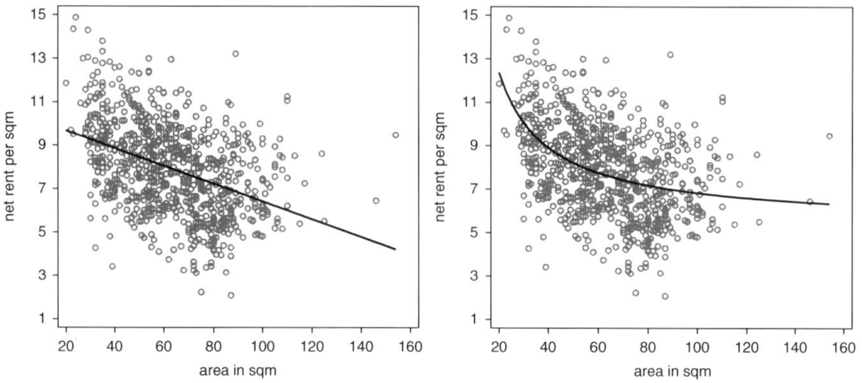

Fig. 2.2 Munich rent index: scatter plots between net rent per square meter and area. Included is the estimated effect $\hat{f}$ of living area for a linear (*left*) and reciprocal (*right*) area effect

Example 2.2 Munich Rent Index—Simple Linear Regression

We illustrate the simple linear regression model using the data shown in Fig. 2.1 and the corresponding model (2.1). The data gives rise to doubts about the assumption of equal variances $\text{Var}(\varepsilon_i) = \text{Var}(y_i) = \sigma^2$ across observations since variability in rent seems to increase as living area increases. For the moment, we will ignore this problem, but Sect. 4.1.3 will illustrate how to deal with problems associated with unequal variances. See also Sect. 2.9.2 and Chap. 10 on quantile regression. According to the method of least squares, the parameter estimates for model (2.1) are $\hat{\beta}_0 = 130.23$ and $\hat{\beta}_1 = 5.57$ implying the estimated regression line

$$\hat{f}(area) = 130.23 + 5.57 \cdot area$$

illustrated in the right panel of Fig. 2.1. The slope parameter $\hat{\beta}_1 = 5.57$ can be interpreted as follows: If the living area increases by 1 m^2, the rent increases about 5.57 Euro on average.

If we choose the rent per square meter instead of the rent as response variable, the scatter plot illustrated in Fig. 2.2 (left) results. It is quite obvious that the relationship between rent per square meter and living area is nonlinear. This is also supported by the estimated regression line

$$\hat{f}(area) = 10.5 - 0.041 \cdot area.$$

The fit to the data is poor for small and large living area. A better fit can be achieved by defining the new explanatory variable

$$x = \frac{1}{area}$$

that yields a simple linear regression of the form

$$rentsqm_i = \beta_0 + \beta_1 x_i + \varepsilon_i = \beta_0 + \beta_1 \frac{1}{area_i} + \varepsilon_i. \tag{2.3}$$

With the help of the transformed covariate, Eq. (2.3) is again a simple linear regression, and we can still use the method of least squares to estimate the parameters β_0 and β_1 of the function

$$f(area) = \beta_0 + \beta_1 \cdot \frac{1}{area}.$$

We obtain

$$\hat{f}(area) = 5.44 + 138.32 \cdot \frac{1}{area}.$$

The corresponding curve in Fig. 2.2 (right) shows a better fit to the data. It reveals that on average the net rent per square meter declines nonlinearly as living area increases. A given living area, e.g., 30 m^2, corresponds to an estimated average rent per square meter of

$$\widehat{rentsqm} = 5.44 + 138.32 \cdot \frac{1}{area}.$$

If the living area increases by 1 m^2, the average rent decreases and is now given by

$$\widehat{rentsqm} = 5.44 + 138.32 \frac{1}{area + 1}.$$

Figure 2.2 (right) shows that the decline is nonlinear. It can be computed by inserting the specific values (e.g., 30 and 31 m^2):

$$\widehat{rentsqm}(30) - \widehat{rentsqm}(31) = 138.32/30 - 138.32/31 \approx 0.15 \text{ Euro}.$$

An apartment of 60 m^2 shows a decline of the average rent per square meter by

$$\widehat{rentsqm}(60) - \widehat{rentsqm}(61) \approx 0.038 \text{ Euro}.$$

△

In general, the application of a linear regression model requires a relationship between the response and the covariate that is *linear in the coefficients* β_0 and β_1. The regressor x- and also the response y-can be transformed to achieve linearity in the parameters, as has been illustrated in the above example. However, the question remains how to find an appropriate transformation for the covariate. Nonparametric regression models offer flexible and automatic approaches; see Sect. 2.5 for a first impression and Chap. 8 for full details.

2.2.2 Multiple Linear Regression

Example 2.3 Munich Rent Index—Rent in Average and Good Locations

We now add apartments in good location to the analysis. Figure 2.3 shows the data for rents in average and good locations. In addition to the estimated regression line for apartments in average location, there is another estimated regression line for apartments in a good location. Alternatively, both strata can be analyzed within a single model that shows parallel regression lines. This can be achieved through the model

$$rent_i = \beta_0 + \beta_1 \, area_i + \beta_2 \, glocation_i + \varepsilon_i. \tag{2.4}$$

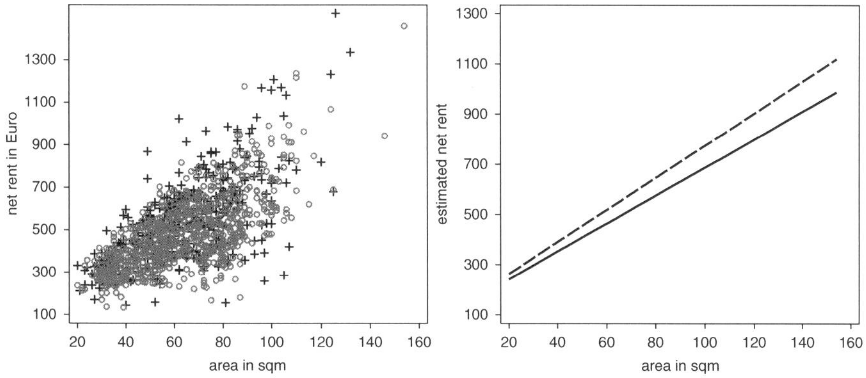

Fig. 2.3 Munich rent index: The *left panel* shows a scatter plot between net rent and area for apartments in average (*circles*) and good location (*plus signs*). The *right panel* displays separate regression lines for apartments in average (*solid line*) and good location (*dashed line*)

The variable *glocation* is a binary *indicator variable*

$$glocation_i = \begin{cases} 1 & \text{if the } i\text{th apartment is in good location,} \\ 0 & \text{if the } i\text{th apartment is in average location.} \end{cases}$$

The least squares method produces the estimated regression equation

$$\widehat{rent} = 112.69 + 5.85 \cdot area + 57.26 \cdot glocation.$$

Because of the 1/0 coding of *glocation*, we obtain the equivalent formulation

$$\widehat{rent} = \begin{cases} 112.69 + 5.85 \cdot area & \text{for average location,} \\ 169.95 + 5.85 \cdot area & \text{for good location.} \end{cases}$$

Figure 2.4 shows both parallel lines. The coefficients can be interpreted as follows:

- For apartments in a good and average location, the increase of living area by 1 m^2 leads to an average increase of rent of about 5.85 Euro.
- The average rent for an apartment in a good location is about 57.26 Euro higher than for an apartment of the same living area in an average location.

△

Model (2.4) is a special case of a multiple linear regression model for k regressors or covariates $x_1, \ldots, x_k$:

$$y_i = \beta_0 + \beta_1 x_{i1} + \ldots + \beta_k x_{ik} + \varepsilon_i,$$

where x_{ij} is the value of the jth covariate, $j = 1, \ldots, k$, for the ith observation, $i = 1, \ldots, n$. The covariates can be continuous, binary, or multi-categorical (after an appropriate coding, see below). Similar to the simple linear regression, x-variables can also be attained via transformation of original covariates. The same assumptions are made for the error variables ε_i in a multiple linear regression model as those in a simple linear regression model. In the case of Gaussian errors, the response variable

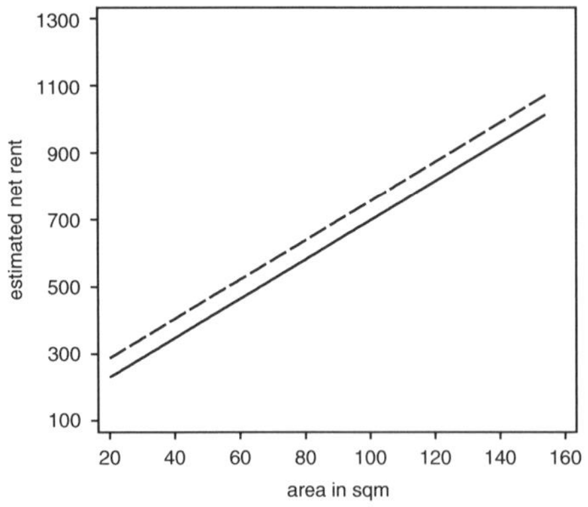

Fig. 2.4 Munich rent index: estimated regression lines for apartments in average (*solid line*) and good location (*dashed line*) according to model (2.4)

is (conditionally) independent and normally distributed given the covariates

$$y_i \sim \mathrm{N}(\mu_i, \sigma^2),$$

with

$$\mu_i = \mathrm{E}(y_i) = \beta_0 + \beta_1 x_{i1} + \ldots + \beta_k x_{ik}.$$

The model is also called the *classical linear regression model*. A summary is given in Box 2.2. For notational convenience we omit here (and elsewhere) the dependence of expressions on the covariates, i.e., $\mathrm{E}(y_i)$ is to be understood as an abbreviation for $\mathrm{E}(y_i \mid x_{i1}, \ldots, x_{ik})$.

The following examples illustrate the flexible usage of a multiple linear regression model through appropriate transformation and coding of covariates.

Example 2.4 Munich Rent Index—Nonlinear Influence of Living Area

As in Example 2.2, we transform the living area to $x = \frac{1}{area}$ and formulate the linear model

$$rentsqm_i = \beta_0 + \beta_1 \cdot \frac{1}{area_i} + \beta_2 \cdot glocation_i + \varepsilon_i. \tag{2.5}$$

The estimated model for the average rent per square meter is

$$\widehat{rentsqm} = 5.51 + 134.72 \cdot \frac{1}{area} + 0.9 \cdot glocation.$$

Figure 2.5 shows both graphs for the average rent per square meter:

$$\widehat{rentsqm} = \begin{cases} 5.51 + 134.72 \cdot \frac{1}{area} & \text{for average location,} \\ 6.41 + 134.72 \cdot \frac{1}{area} & \text{for good location.} \end{cases}$$

The nonlinear effect of the living area can be interpreted as in Example 2.2. △

2.2 Classical Linear Regression Model

Data

$(y_i, x_{i1}, \ldots, x_{ik})$, $i = 1, \ldots, n$, for a continuous variable y and continuous or appropriately coded categorical regressors $x_1, \ldots, x_k$.

Model

$$y_i = \beta_0 + \beta_1 x_{i1} + \ldots + \beta_k x_{ik} + \varepsilon_i\,, \quad i = 1, \ldots, n\,.$$

The errors $\varepsilon_1, \ldots, \varepsilon_n$ are independent and identically distributed (i.i.d.) with

$$\mathrm{E}(\varepsilon_i) = 0\,, \quad \mathrm{Var}(\varepsilon_i) = \sigma^2.$$

The estimated linear function

$$\hat{f}(x_1, \ldots, x_k) = \hat{\beta}_0 + \hat{\beta}_1 x_1 + \ldots + \hat{\beta}_k x_k$$

can be used as an estimator $\widehat{\mathrm{E}}(y|x_1, \ldots, x_k)$ for the conditional expected value of y given the covariates $x_1, \ldots, x_k$. As such it can be used to predict y, denoted as $\hat{y}$.

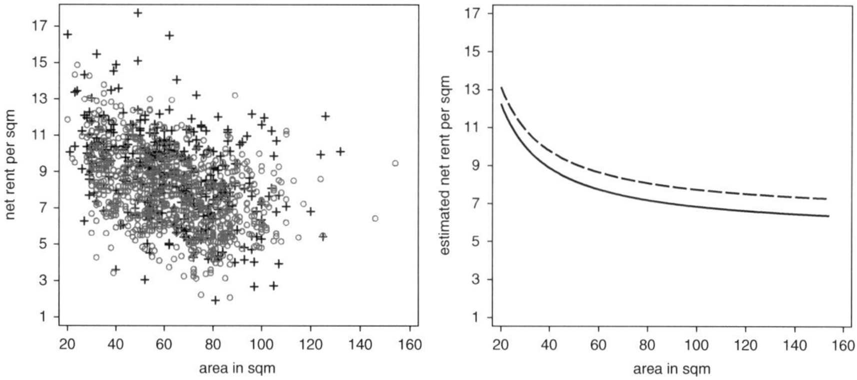

Fig. 2.5 Munich rent index: The *left panel* shows a scatter plot between net rent per square meter and area for apartments in average (*circles*) and good location (*plus signs*). The *right panel* shows estimated regression curves for apartments in normal (*solid line*) and good location (*dashed line*)

Examples 2.3 and 2.4 assume an additive effect of the location. Both models show that an apartment in a good location leads to an increase in rent (or rent per square meter) when compared to an apartment in average location with equal living area. In Example 2.3, the increase in rent is 57.26 Euro and in the previous example 0.9 Euro per square meter. The assumption of a solely additive effect in model

(2.4) implies the parallel lines in Fig. 2.4. However, comparing Figs. 2.3 and 2.4, the validity of this assumption is questionable. Including an interaction between the two covariates living area and location relaxes the assumption of parallel regression lines.

Example 2.5 Munich Rent Index—Interaction Between Living Area and Location

In order to include an interaction between living area and location in model (2.4), it is necessary to define an interaction variable by multiplying the covariates *area* and *glocation*

$$inter_i = area_i \cdot glocation_i .$$

It follows

$$inter_i = \begin{cases} 0 & \text{for average location,} \\ area_i & \text{for good location.} \end{cases}$$

We now extend the model (2.4) by adding the interaction effect $inter = area \cdot glocation$ to the two main effects and obtain

$$rent_i = \beta_0 + \beta_1\, area_i + \beta_2\, glocation_i + \beta_3\, inter_i + \varepsilon_i\,. \tag{2.6}$$

Because of the definition of *glocation* and *inter*, an equivalent formulation of the model is given by

$$rent_i = \begin{cases} \beta_0 + \beta_1\, area_i + \varepsilon_i & \text{for average location,} \\ (\beta_0 + \beta_2) + (\beta_1 + \beta_3)\, area_i + \varepsilon_i & \text{for good location.} \end{cases}$$

There is no interaction effect if $\beta_3 = 0$, and we retain the assumption of parallel lines with common slope β_1 as in model (2.4). If $\beta_3 \neq 0$, the effect of the living area, i.e., the slope of the line for apartments in a good location, changes by an amount of β_3 when compared to apartments in average location. In contrast to Fig. 2.3 (right), the least squares estimates for the regression coefficients are not obtained separately for the two locations, but rather simultaneously for model (2.6). We obtain

$$\hat\beta_0 = 130.23, \quad \hat\beta_1 = 5.57, \quad \hat\beta_2 = 5.20, \quad \hat\beta_3 = 0.82.$$

Figure 2.6 shows the estimated regression lines for apartments in average and good location. Whether or not an inclusion of an interaction effect is necessary can be statistically tested using the hypothesis

$$H_0 : \beta_3 = 0 \quad \text{against} \quad H_1 : \beta_3 \neq 0;$$

see Sect. 3.3.

△

As described in Example 1.1 (p. 5), in the entire data set, the location of apartments is given in three categories:

$$\begin{aligned} 1 &= \text{average location,} \\ 2 &= \text{good location,} \\ 3 &= \text{top location.} \end{aligned}$$

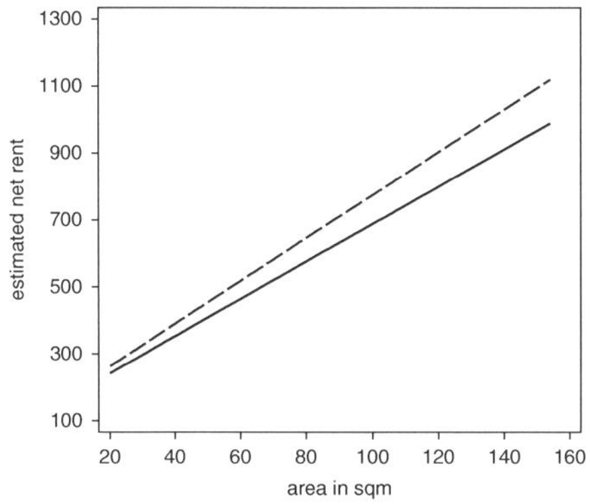

Fig. 2.6 Munich rent index: estimated regression lines for average (*solid line*) and good location (*dashed line*) based on the interaction model (2.6)

Since *location* is categorical and not continuous, it is not possible to include the effect of the location in the form of $\beta \cdot location$ in a linear regression model with the integer values 1, 2, or 3 for the location. The arbitrary coding of *location* would have considerable impact on the estimation results. The chosen coding automatically implies that the effect of apartments in a good location would be twice as high as in average location and the effect of apartments in top location would be three times as high. These relations change automatically with a different coding, especially if the distance between the arbitrarily coded covariate values is altered. For example, with a coding of 1, 4, and 9 for average, good, and top location, the effect would be four times or nine times as high for apartments in a good or top location as for apartments in average location, and the incremental impact varies when comparing average and good or good and top location. Further, not all categorical covariates have ordinal structure. Similar to the previous coding of *location* via *one* binary indicator variable expressed in Example 2.3, a coding using *two* binary variables is now necessary. In order to do so, one of the three location categories must be defined as the *reference category*. In the case of average location as the reference category, the two 1/0-indicator variables for good location and top location are defined as

$$glocation_i = \begin{cases} 1 & \text{if apartment } i \text{ is in good location,} \\ 0 & \text{otherwise,} \end{cases}$$

$$tlocation_i = \begin{cases} 1 & \text{if apartment } i \text{ is in top location,} \\ 0 & \text{otherwise.} \end{cases}$$

An apartment in the reference category (average location) is, thus, defined as $glocation = tlocation = 0$. The effect of each of these two binary variables is always directly interpreted in relation to the reference category in the regression model, as demonstrated in the next example. This type of 1/0 coding of a *multi-categorical variable* is also called dummy or indicator coding. In general, dummy coding is defined as follows for a variable x with c categories, $x \in \{1, \ldots, c\}$: A

Table 2.1 Munich rent index: estimated coefficients in the multiple regression model

Variable	Estimated coefficient
$1/area$	137.504
$yearc$	−3.801
$yearc^2$	0.001
$glocation$	0.679
$tlocation$	1.519
$bath$	0.503
$kitchen$	0.866
$cheating$	1.870

reference category must be defined, e.g., c. The variable x can be then coded with $c - 1$ dummy variables $x_1, \dots, x_{c-1}$:

$$x_j = \begin{cases} 1 & \text{if } x = j, \\ 0 & \text{otherwise,} \end{cases} \qquad j = 1, \dots, c - 1\,.$$

For the reference category c we obtain

$$x_1 = \dots = x_{c-1} = 0.$$

Section 3.1.3 describes the coding of categorical covariates in more detail.

Example 2.6 Munich Rent Index—Multiple Regression Model

For illustration, we now use the entire data set, including all explanatory variables mentioned in Example 1.1, in a multiple regression model for the response variable rent per square meter (*rentsqm*). The nonlinear effect of the living area is modeled via the transformed variable $1/area$ and the effect of location via dummy coding as described above. Since the effect of the year of construction may also be nonlinear, an additional quadratic polynomial is specified. We obtain the following model without interaction:

$$\begin{aligned} rentsqm_i &= \beta_0 + \beta_1 \cdot (1/area_i) + \beta_2\, yearc_i + \beta_3\, yearc_i^2 + \beta_4\, glocation_i + \beta_5\, tlocation_i \\ &\quad + \beta_6\, bath_i + \beta_7\, kitchen_i + \beta_8\, cheating_i + \varepsilon_i\,. \end{aligned}$$

The binary regressors *bath*, *kitchen*, and *cheating* (central heating system) are dummy coded, as shown in Table 1.2 (p. 6). Table 2.1 contains the estimated coefficients $\hat{\beta}_1$ to $\hat{\beta}_8$ for the regressors. Figure 2.7 shows the estimated nonlinear effects of living area and year of construction. The average effect plots (solid lines) are obtained by inserting different values for living area into the predicted rent per square meter

$$\begin{aligned} \widehat{rentsqm} &= 3684.991 + 137.5044 \cdot (1/area) - 3.8007\, yearc + 0.0098\, yearc_i^2 \\ &\quad + 0.6795\, glocation_i + 1.5187\, tlocation_i + 0.5027\, bath_i + 0.8664\, kitchen \\ &\quad + 1.8704\, cheating, \end{aligned}$$

while the other covariates are held constant at their mean values (apart from year of construction). As expected, the effect on the net rent per square meter decreases nonlinearly with an increase of living area. For a detailed comparison of two apartments, e.g., with a

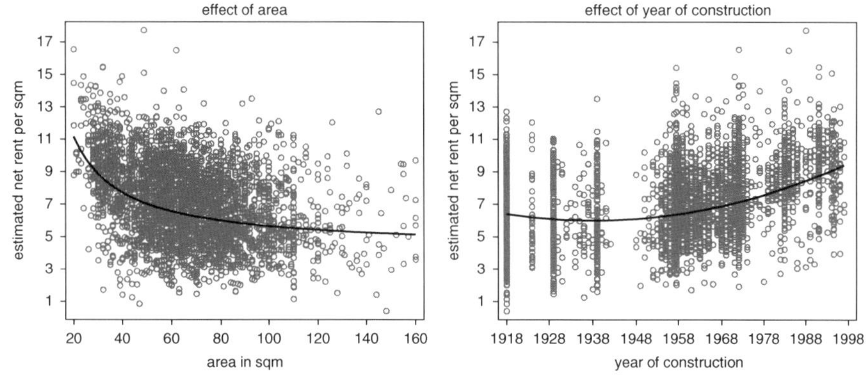

Fig. 2.7 Munich rent index: effects of area (*left*) and year of construction (*right*)

living area of 60 and 100 m^2, but with otherwise identical values for the year of construction, location, bath room, and central heating system indicators, we obtain a difference of $\hat{\beta}_1(1/60) - \hat{\beta}_1(1/100) = 137.504\,(1/60 - 1/100) = 0.92$ Euro for the average rent per square meter. The effect of year of construction is almost constant until 1945 and increases linearly thereafter. The effects of the indicator variables shown in Table 2.1 are interpreted as the difference in net rent per square meter compared to the reference category. The average rent per square meter increases, for example, by 0.68 Euro, if the apartment is in a good location (relative to one in average location).

△

2.3 Regression with Binary Response Variables: The Logit Model

The linear regression model is well suited for continuous response variables, which show—possibly after an appropriate transformation—an approximate normal distribution (conditional on the covariates). However, many applications have binary or more general categorical response variables.

Example 2.7 Patent Opposition

During the validation of a patent application, it is possible that objections are raised, e.g. see Example 1.3 (p. 8). The response variable patent opposition (*opp*) is binary and coded by

$$opp_i = \begin{cases} 1 & \text{opposition against patent } i, \\ 0 & \text{otherwise.} \end{cases}$$

The decision for an opposition against a patent may depend on various covariates. Some of these variables are continuous, for example, the year of the application (variable *year*), the number of citations (*ncit*), and the number of designated states (*ncountry*). Other covariates are binary, as given in Table 1.4 (p. 8).

△

The expected value of a binary variable y is given by

$$\mathrm{E}(y) = \mathrm{P}(y = 0) \cdot 0 + \mathrm{P}(y = 1) \cdot 1 = \mathrm{P}(y = 1).$$

The aim of a regression analysis with binary responses $y \in \{0, 1\}$ is to model the expected value $\mathrm{E}(y)$ or in other words the probability

$$\mathrm{P}(y = 1) = \mathrm{P}(y = 1 \,|\, x_1, \ldots, x_k) = \pi$$

in the presence of covariates. The classical linear regression model

$$y_i = \mathrm{P}(y_i = 1) + \varepsilon_i = \beta_0 + \beta_1 x_{i1} + \ldots + \beta_k x_{ik} + \varepsilon_i,$$

with $\varepsilon_i \sim \mathrm{N}(0, \sigma^2)$ is not applicable for several reasons:

- In contrast to the left-hand side, the right-hand side is not binary.
- Even if the assumption of normality is relaxed for ε_i, the error variance $\mathrm{Var}(\varepsilon_i)$ cannot be homoscedastic, i.e., $\mathrm{Var}(\varepsilon_i) = \sigma^2$. Since y_i would have a Bernoulli distribution with $\pi_i = \beta_0 + \beta_1 x_{i1} + \ldots + \beta_k x_{ik}$, it follows that

$$\mathrm{Var}(y_i) = \pi_i (1 - \pi_i)$$

  depends on the values of the covariates and the parameters $\beta_0,\ldots,\beta_k$, and thus cannot have the same value σ^2 for all observations i.
- The linear model allows values $\pi_i < 0$ or $\pi_i > 1$ for $\pi_i = \mathrm{P}(y_i = 1)$ which are impossible for probabilities.

These problems can be avoided by assuming the model

$$\pi_i = \mathrm{P}(y_i = 1) = F(\beta_0 + \beta_1 x_{i1} + \ldots + \beta_k, x_{ik}),$$

where the domain of the function F is restricted to the interval [0, 1]. For reasons of interpretability it is sensible if we restrict ourselves to monotonically increasing functions F. Hence cumulative distribution functions (cdfs) are a natural choice for F. Choosing the logistic distribution function

$$F(\eta) = \frac{\exp(\eta)}{1 + \exp(\eta)}$$

yields the logit model

$$\pi_i = \mathrm{P}(y_i = 1) = \frac{\exp(\eta_i)}{1 + \exp(\eta_i)}, \qquad (2.7)$$

with the *linear predictor*

$$\eta_i = \beta_0 + \beta_1 x_{i1} + \ldots + \beta_k x_{ik}.$$

Analogous to the linear regression model, the binary response variables y_i are assumed to be (conditionally) independent given the covariates $x_i = (x_{i1}, \ldots, x_{ik})'$. Even though the predictor is linear, the interpretation changes compared to the linear model: If the value of the predictor η increases to $\eta + 1$, the probability for $y = 1$ increases in a *nonlinear* way from $F(\eta)$ to $F(\eta + 1)$. An alternative interpretation is obtained by solving the model equation (2.7) for η using the inverse function $\eta = \log\{\pi/(1-\pi)\}$ of the logistic cdf $\pi = \exp(\eta)/\{1 + \exp(\eta)\}$. We obtain

$$\log\left(\frac{\pi_i}{1-\pi_i}\right) = \log\left(\frac{\mathrm{P}(y_i = 1)}{1 - \mathrm{P}(y_i = 1)}\right) = \beta_0 + \beta_1 x_{i1} + \ldots + \beta_k x_{ik} \qquad (2.8)$$

or alternatively (because of $\exp(a + b) = \exp(a) \cdot \exp(b)$)

$$\frac{\pi_i}{1-\pi_i} = \frac{\mathrm{P}(y_i = 1)}{\mathrm{P}(y_i = 0)} = \exp(\beta_0)\exp(\beta_1 x_{i1}) \cdot \ldots \cdot \exp(\beta_k x_{ik}). \qquad (2.9)$$

The left-hand side of Eq. (2.9), i.e., the ratio of the probabilities for $y = 1$ and $y = 0$, is referred to as *odds*. The left-hand side of Eq. (2.8), thus, corresponds to logarithmic odds (*log-odds*) for the outcome of $y = 1$ relative to $y = 0$. Here, we obtain a *multiplicative model* for the odds: A unit increase of the value x_{i1} of the covariate x_1 leads to a multiplication of the ratio (2.9) by the factor $\exp(\beta_1)$. Specifically,

$$\begin{aligned} \frac{\mathrm{P}(y_i = 1 \mid x_{i1} + 1, \ldots)}{\mathrm{P}(y_i = 0 \mid x_{i1} + 1, \ldots)} &= \exp(\beta_0)\exp(\beta_1(x_{i1} + 1)) \cdot \ldots \cdot \exp(\beta_k x_{ik}) \\ &= \frac{\mathrm{P}(y_i = 1 \mid x_{i1}, \ldots)}{\mathrm{P}(y_i = 0 \mid x_{i1}, \ldots)} \exp(\beta_1). \end{aligned} \qquad (2.10)$$

In the special case of a binary covariate x_1 the result is

$$\frac{\mathrm{P}(y_i = 1 \mid x_{i1} = 1, \ldots)}{\mathrm{P}(y_i = 0 \mid x_{i1} = 1, \ldots)} = \frac{\mathrm{P}(y_i = 1 \mid x_{i1} = 0, \ldots)}{\mathrm{P}(y_i = 0 \mid x_{i1} = 0, \ldots)} \exp(\beta_1). \qquad (2.11)$$

This implies an increase of the odds $\mathrm{P}(y_i = 1)/\mathrm{P}(y_i = 0)$ for $\beta_1 > 0$, a decrease for $\beta_1 < 0$, and no change for $\beta_1 = 0$. For the log-odds (2.8) the usual interpretations of the parameters as in the classical linear regression model apply: if x_1, say, increases by 1 unit, the log-odds change by β_1. Since the assumptions for the linear regression model are not met, the parameters will not be estimated via the least squares method, but rather using the method of maximum likelihood (ML); see Sect. 5.1. A general introduction to likelihood-based inference is given in Appendix B.4.

Example 2.8 Patent Opposition

Prior to analyzing the probability of patent opposition, we take a look at Fig. 2.8, which presents histograms and kernel density estimators for the continuous covariates number of patent claims (*nclaims*) and number of citations (*ncit*). The distributions of both variables show an extreme skewness to the right. The majority of the observations of *nclaims* are between 0 and 60, with only very few observations between 61 and the maximum value of

2.3 The Logit Model for Binary Response Variables

Data

$(y_i, x_{i1}, \dots, x_{ik})$, $i = 1, \dots, n$, for a binary response variable $y \in \{0, 1\}$ and for continuous or appropriately coded covariates $x_1, \dots, x_k$.

Model

For the binary response variables $y_i \in \{0, 1\}$ the probabilities $\pi_i = P(y_i = 1)$ are modeled by

$$\pi_i = \frac{\exp(\eta_i)}{1 + \exp(\eta_i)}$$

with the linear predictor

$$\eta_i = \beta_0 + \beta_1 x_{i1} + \ldots + \beta_k x_{ik}.$$

An equivalent formulation is given by assuming the multiplicative model

$$\frac{P(y_i = 1)}{P(y_i = 0)} = \frac{\pi_i}{1 - \pi_i} = \exp(\beta_0) \cdot \exp(\beta_1 x_{i1}) \cdot \ldots \cdot \exp(\beta_k x_{ik})$$

for the odds $\pi_i/(1 - \pi_i)$.

355. The variable *ncit* varies mainly between 0 and 15 with only a handful of observations between 15 and the maximum value of 40. Hence, it is impossible to make any reliable statements regarding the probability of patent opposition for observations with *nclaims*>60 or *ncit* > 15. Consequently, these extreme cases are excluded from all analyses to follow. This example demonstrates the importance of the descriptive analysis of data prior to the application of more complex statistical tools.

We next divide the data into two groups: *biopharm* = 0 and *biopharm* = 1. For the subset *biopharm* = 0, i.e., the patents derived from the semiconductor/computer industry, a logit model

$$P(opp_i = 1) = \frac{\exp(\eta_i)}{1 + \exp(\eta_i)}$$

is estimated with the main effects linear predictor

$$\eta_i = \beta_0 + \beta_1\, year_i + \beta_2\, ncit_i + \beta_3\, nclaims_i + \beta_4\, ustwin_i + \beta_5\, patus_i$$
$$+ \beta_6\, patgsgr_i + \beta_7\, ncountry_i.$$

Table 2.2 contains the estimated coefficients $\hat{\beta}_j$, $j = 0, \dots, 7$, together with the corresponding odds ratios $\exp(\hat{\beta}_j)$. In multiplicative form (2.9) we obtain

$$\frac{P(\text{opposition})}{P(\text{no opposition})} = \exp(201.74) \cdot \exp(-0.102 \cdot year_i) \cdot \ldots \cdot \exp(0.097 \cdot ncountry_i).$$

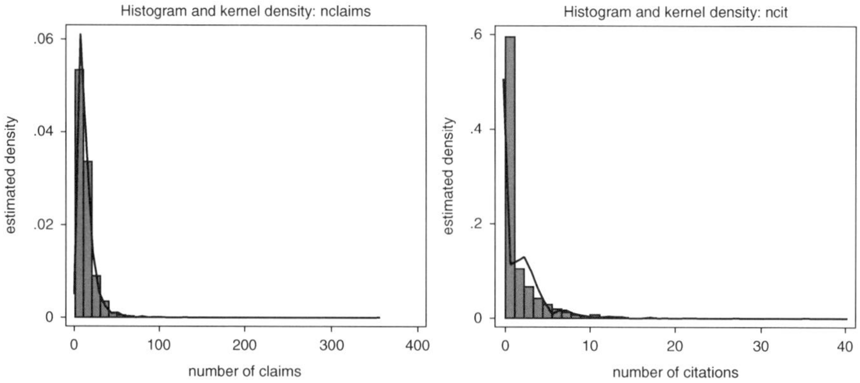

Fig. 2.8 Patent opposition: histogram and kernel density estimator for the continuous covariates *nclaims* (*left*) and *ncit* (*right*)

Table 2.2 Patent opposition: estimated coefficients and odds ratios for the logit model

Variable	Estimated coefficient	Estimated odds ratio
intercept	$\hat{\beta}_0 = 201.74$	
year	$\hat{\beta}_1 = -0.102$	$\exp(\hat{\beta}_1) = 0.902$
ncit	$\hat{\beta}_2 = 0.113$	$\exp(\hat{\beta}_2) = 1.120$
nclaims	$\hat{\beta}_3 = 0.026$	$\exp(\hat{\beta}_3) = 1.026$
ustwin	$\hat{\beta}_4 = -0.402$	$\exp(\hat{\beta}_4) = 0.668$
patus	$\hat{\beta}_5 = -0.526$	$\exp(\hat{\beta}_5) = 0.591$
patgsgr	$\hat{\beta}_6 = 0.196$	$\exp(\hat{\beta}_6) = 1.217$
ncountry	$\hat{\beta}_7 = 0.097$	$\exp(\hat{\beta}_7) = 1.102$

We observe, for instance, an increase in the odds of opposition against a patent from Germany, Switzerland, or Great Britain ($patgsgr = 1$) by the factor $\exp(0.196) = 1.217$ relative to a patent from the United States with the same values of the other covariates. A prediction of the odds P(opposition) / P(no opposition) for a new patent is obtained by inserting the observed covariate values into the estimated model. Similar to linear regression models, we have to decide whether the effect of a continuous covariate is linear or nonlinear. As an example, we model the effect of the number of countries (*ncountry*) using a cubic polynomial

$$\beta_7\, ncountry + \beta_8\, ncountry^2 + \beta_9\, ncountry^3.$$

The parameter estimates are given by

$$\hat{\beta}_7 = 0.3938 \quad \hat{\beta}_8 = -0.0378 \quad \hat{\beta}_9 = 0.0014.$$

Figure 2.9 shows the estimated polynomial and, for comparison, the linear effect (left panel). As before, the values of the remaining covariates are held fixed at their respective average values. Both the estimated regression coefficients and the visualized functions suggest that a linear effect of *ncountry* is sufficient in this case. This hypothesis can be formally tested using the statistical tests described in Sect. 5.1. The right panel of Fig. 2.9 shows

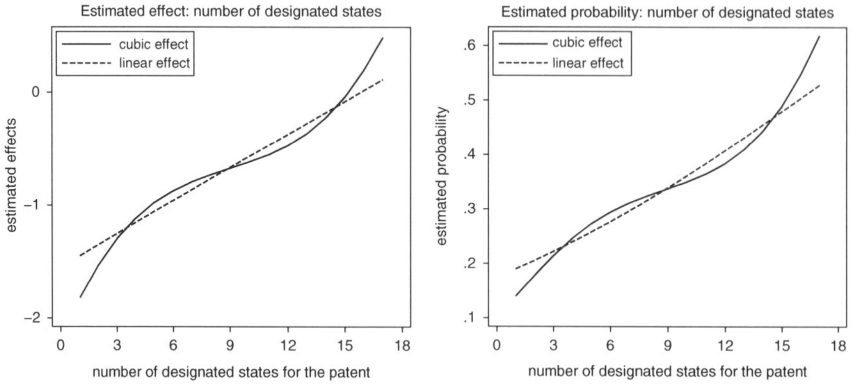

Fig. 2.9 Patent opposition: estimated linear and cubic effect of covariate *ncountry* (*left panel*) as well as estimated probabilities (*right panel*). For the probability plot, the values of the remaining covariates are held fixed at their respective mean values

the *estimated probabilities* π corresponding to the estimated *ncountry* effects. This is an alternative to the effect plots as the probability plots provide an intuition about the variability of the probability of patent opposition as the number of designated states increases and the remaining covariates are kept fixed at their mean values. Specifically, the graph is obtained by plotting $\hat{\pi}(\eta(ncountry))$ against *ncountry*. Thereby

$$\eta(ncountry) = \hat{\beta}_0 + \hat{\beta}_1\, \overline{year} + \cdots + \hat{\beta}_6\, \overline{patgsgr} + \hat{\beta}_7\, ncountry + \hat{\beta}_8\, ncountry^2 + \hat{\beta}_9\, ncountry^3,$$

with $\overline{year}, \ldots, \overline{patgsgr}$ being the mean values of the remaining covariates. In our case the probability of patent opposition varies approximately between 0.15 and 0.6 as *ncountry* increases.

△

In addition to the logit model, other regression models for binary responses exist. Different models result when the logistic distribution function is replaced by an alternative distribution function. For instance, assuming $F = \Phi$, where Φ is the cdf of the standard normal distribution, yields the probit model (see Sect. 5.1 for more details).

In addition to binary response variables, other types of discrete response variables are possible in applications. For these applications, linear regression models are not appropriate. An example is a response y that represents counts $\{0, 1, 2, \ldots\}$, e.g., the amount of damage events reported to an insurance company (see Example 2.12), or a multi-categorical response variable, e.g., with the categories poor, average, and good. Chapters 5 and 6 describe regression models for such discrete response variables in full detail.

2.4 Mixed Models

The regression models presented so far are particularly useful for the analysis of regression data resulting from cross-sectional studies, where the regression coefficients $\beta_0, \ldots, \beta_k$ are unknown population ("fixed") parameters. Regression

Table 2.3 Hormone therapy with rats: number of observations per time point and dose group

Age (in days)	Control	Low	High	Total
50	15	18	17	50
60	13	17	16	46
70	13	15	15	43
80	10	15	13	38
90	7	12	10	29
100	4	10	10	24
110	4	8	10	22

problems also occur when analyzing longitudinal data, where a number of subjects or objects are repeatedly observed over time. In such a case, regression models for longitudinal data allow to model and estimate both the fixed population parameters and subject- or object-specific effects. The latter are called "random effects," since they often belong to individuals who have been selected randomly from the population. Closely related to the random effects models with temporal structure are models for clustered data. Here, the response and covariates are collected repeatedly on several subjects, selected from primary units (clusters). An example for clusters are selected schools, in which certain tests for a subsample of students are conducted.

Mixed models include both the usual fixed population effects $\beta_0, \ldots, \beta_k$ and subject- or cluster-specific random effects in the linear predictor. Mixed modeling allows estimation and analysis on a subject-specific level, which is illustrated in the following example in the case of longitudinal data.

Example 2.9 Hormone Therapy with Rats

Researchers at the Katholieke Universiteit Leuven (KUL, Belgium) performed an experiment to examine the effect of testosterone on the growth of rats. A detailed description of the data and the scientific questions of the study can be found in Verbeke and Molenberghs (2000). A total of 50 rats were randomly assigned to either a control group or to one of two therapy groups. The therapy consisted of either a low or high dose of Decapeptyl, an agent to inhibit the production of testosterone in rats. The therapy started when the rats were 45 days old. Starting with the 50th day, the growth of the rat's head was measured every tenth day via an X-ray examination. The distance (measured in pixels) between two well-defined points of the head served as a measure for the head height and was used as the response variable. The number n_i of repeated measures y_{ij}, $j = 1, \ldots, n_i$, $i = 1, \ldots, 50$, of the response was different for each rat. Only 22 rats in total had the complete seven measurements until the age of 110 days. Four rats were actually only measured once when they were 50 days old. Table 2.3 summarizes the resulting design of the study. Figure 2.10 shows the individual time series $\{y_{ij}, j = 1, \ldots, n_i\}$ of rats $i = 1, \ldots, 50$ separated for the three treatment groups.

To formulate regression models, we define the transformed age

$$t = \log(1 + (age - 45)/10)$$

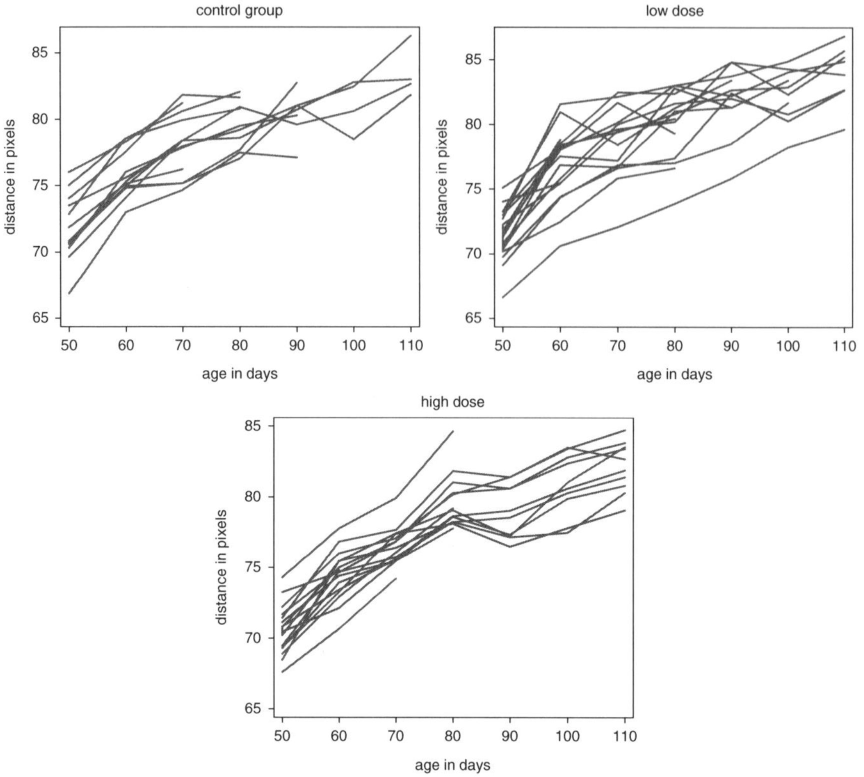

Fig. 2.10 Hormone therapy with rats: time series stratified for dose groups

as a covariate, analogous to Verbeke and Molenberghs (2000). The value $t = 0$ corresponds to the initiation of the treatment (age $=$ 45 days). For the three therapy groups we define the indicator variables L, H, and C by

$$L_i = \begin{cases} 1 & \text{rat } i \text{ in low-dose group,} \\ 0 & \text{otherwise,} \end{cases}$$

$$H_i = \begin{cases} 1 & \text{rat } i \text{ in high-dose group,} \\ 0 & \text{otherwise,} \end{cases}$$

$$C_i = \begin{cases} 1 & \text{rat } i \text{ in control group,} \\ 0 & \text{otherwise.} \end{cases}$$

Using the transformed age t as time scale and $t = 0$ as the initiation of the treatment, we can formulate simple linear regression models according to the three groups:

$$y_{ij} = \begin{cases} \beta_0 + \beta_1 t_{ij} + \varepsilon_{ij} & i \text{ in low-dose group,} \\ \beta_0 + \beta_2 t_{ij} + \varepsilon_{ij} & i \text{ in high-dose group,} \\ \beta_0 + \beta_3 t_{ij} + \varepsilon_{ij} & i \text{ in control.} \end{cases}$$

For $t = 0$, all three treatment groups have $\mathrm{E}(y_{ij}) = \beta_0$, i.e., β_0 is the *population mean* at treatment initiation. The coefficients β_1, β_2, and β_3 correspond to the different slopes associated with the variable t, i.e., the effects of the (transformed) age in the three linear models. The three models can be combined into the single model

$$y_{ij} = \beta_0 + \beta_1 L_i \cdot t_{ij} + \beta_2 H_i \cdot t_{ij} + \beta_3 C_i \cdot t_{ij} + \varepsilon_{ij}, \tag{2.12}$$

with 1/0-indicator variables L, H, and C for the three groups. Similar to β_0, the parameters β_1, β_2, and β_3 are *population effects*, which do not capture any individual differences between the rats. However, Fig. 2.10 reveals some obvious differences of the individual curves in the intercept, as well as possible differences in the slope. Moreover, the variability within the individual curves is notably less than the total variation of the data in any of the three group-specific scatter plots. In particular, these findings show that the observations are partly *correlated*, whereas so far we have always assumed independence among observations. While the observations between subjects can still be assumed independent, observations within rats are clearly correlated. The inclusion of subject-specific information will consider the correlation and therefore improve the quality of the estimates. In order to incorporate subject-specific effects, we extend the regression models mentioned above and obtain

$$y_{ij} = \begin{cases} \beta_0 + \gamma_{0i} + (\beta_1 + \gamma_{1i})t_{ij} + \varepsilon_{ij} & i \text{ in low-dose group,} \\ \beta_0 + \gamma_{0i} + (\beta_2 + \gamma_{1i})t_{ij} + \varepsilon_{ij} & i \text{ in high-dose group,} \\ \beta_0 + \gamma_{0i} + (\beta_3 + \gamma_{1i})t_{ij} + \varepsilon_{ij} & i \text{ in control group,} \end{cases}$$

or equivalently

$$y_{ij} = \beta_0 + \gamma_{0i} + \beta_1 L_i \cdot t_{ij} + \beta_2 H_i \cdot t_{ij} + \beta_3 C_i \cdot t_{ij} + \gamma_{1i} \cdot t_{ij} + \varepsilon_{ij}. \tag{2.13}$$

The model contains subject-specific deviations γ_{0i} from the population mean β_0 as well as subject-specific deviations γ_{1i} from the population slopes β_1, β_2, and β_3. In contrast to the "fixed" effects $\boldsymbol{\beta} = (\beta_0, \beta_1, \beta_2, \beta_3)'$, the subject-specific effects $\boldsymbol{\gamma}_i = (\gamma_{0i}, \gamma_{1i})'$ are considered as random effects, because the rats have been randomly selected from a population. We assume that the random effects are independent and identically distributed and follow a normal distribution, i.e.,

$$\gamma_{0i} \sim \mathrm{N}(0, \tau_0^2), \quad \gamma_{1i} \sim \mathrm{N}(0, \tau_1^2). \tag{2.14}$$

Without loss of generality the expected values can be set to zero, because the population mean values are already included in the fixed effects $\boldsymbol{\beta}$. At first sight, a more natural approach to consider subject-specific effects is the inclusion of *subject-specific dummy variables* without a random effects distribution. In principal, such an approach is possible for a small or moderate number of subject-specific effects. However, since usually a large or even huge number of dummy variables are necessary for the subject-specific effects, the resulting estimates would be highly unstable. As we will see in full detail in Chap. 7, the random effects distribution Eq. (2.14), in particular, the common variances τ_0^2 and τ_1^2 across subjects, *stabilizes* estimation.

For the errors ε_{ij}, we make the same assumptions as in the classical linear model, i.e., the errors are independent and identically distributed as

$$\varepsilon_{ij} \sim \mathrm{N}(0, \sigma^2). \tag{2.15}$$

Note, however, that correlation within individuals is considered through the subject-specific effects. Since the model (2.13) includes fixed effects as in the classical linear regression

2.4 Linear Mixed Models for Longitudinal and Clustered Data

Data

For each of the $i = 1, \ldots, m$ subjects or clusters, n_i repeated observations

$$\left(y_{ij}, x_{ij1}, \ldots, x_{ijk}\right), \qquad j = 1, \ldots, n_i,$$

for a continuous response variable y and continuous or appropriately coded covariates $x_1, \ldots, x_k$ are given.

Model

For the linear mixed model, we assume

$$y_{ij} = \beta_0 + \beta_1 x_{ij1} + \ldots + \beta_k x_{ijk} + \gamma_{0i} + \gamma_{1i} u_{ij1} + \ldots + \gamma_{qi} u_{ijq} + \varepsilon_{ij},$$

$i = 1, \ldots, m$, $j = 1, \ldots, n_i$. Thereby, the $\beta_0, \ldots, \beta_k$ are fixed *population effects* and $\boldsymbol{\gamma}_i = (\gamma_{0i}, \gamma_{1i}, \ldots, \gamma_{qi})'$ are *subject- or cluster-specific effects.* We assume that the random effects $\boldsymbol{\gamma}_i$ are independent and identically distributed according to a (possibly multivariate) normal distribution.

model (2.12), as well as random effects $\gamma_{0i}, \gamma_{1i}, i = 1, \ldots, 50$, it is called a *linear mixed model* or a regression model with both fixed and random effects.

△

In the case of longitudinal data, some of the covariates $x_{ij1}, \ldots, x_{ijk}$ can be time-varying, as, e.g., the transformed age in the rats example. They can also be time-constant; examples are the indicator variables L_i, H_i, and C_i. For cluster data, this means that in cluster i the covariate value depends on object j or alternatively that the covariate contains only cluster-specific information.

A general notation for linear mixed models for longitudinal and cluster data is given by

$$y_{ij} = \beta_0 + \beta_1 x_{ij1} + \ldots + \beta_k x_{ijk} + \gamma_{0i} + \gamma_{1i} u_{ij1} + \ldots + \gamma_{qi} u_{ijq} + \varepsilon_{ij}, \quad (2.16)$$

where $i = 1, \ldots, m$ is the individual or cluster index and $j = 1, \ldots, n_i$ indicates the jth measurement for individual or cluster i. In the case of repeated measurements over time, the observed (not necessarily equally spaced) time points for individual i are denoted by $t_{i1} < \ldots < t_{ij} < \ldots < t_{in_i}$. The fixed parameters $\beta_0, \ldots, \beta_k$ in Eq. (2.16) measure population effects, while the random parameters $\boldsymbol{\gamma}_i = (\gamma_{0i}, \gamma_{1i}, \ldots, \gamma_{qi})'$ describe subject- or cluster-specific effects. The additional design variables $u_{ij1}, \ldots, u_{ijq}$ often consist of some of the covariates $x_{ij1}, \ldots, x_{ijk}$, as the transformed age t_{ij} in Example 2.9.

In most situations, similar assumptions are made for the random errors as in the classical linear regression model, i.e., the ε_{ij} are independently and identically (normally) distributed with $\mathrm{E}\left(\varepsilon_{ij}\right) = 0$ and $\mathrm{Var}\left(\varepsilon_{ij}\right) = \sigma^2$. It is also possible to model correlations between the errors ε_{ij}, $j = 1, \ldots, n_i$, of repeated observations within individuals or clusters; see Chap. 7. Such correlated errors are necessary if there is extra correlation not taken into account by the subject-specific effects. For the random effects, it is also often assumed that the γ_{li}, $l = 0, \ldots, q$, are independent and identically distributed according to separate normal distribution, as in Example 2.9. Again, more general formulations with correlated random effects are possible. Then the vector of individual random effects $\boldsymbol{\gamma}_i$ is assumed to be i.i.d. according to a multivariate normal distribution with possibly non-diagonal covariance matrix; see Chap. 7 for details.

Analyses with mixed models for longitudinal or cluster data have the following advantages:

- Correlations between observations of the same individual or cluster are taken into account (at least to a certain extent).
- Subject-specific effects can serve as surrogates for unobserved covariate effects, which either have been measured insufficiently or not measured at all. Since the observations differ due to such unobserved covariates, there is an implied unobserved heterogeneity.
- The inclusion of subject-specific information often leads to more precise estimates for the fixed effects, i.e., less variable estimators, when compared to standard regression models. In any case, mixed models ensure that inference regarding the regression coefficients is correct in the sense that we obtain correct standard errors, confidence intervals, and tests.
- Mixed models stabilize the estimators of random effects by assuming a common random effects distribution.
- The estimated individual curves further allow for individual-specific predictions, which are not available in standard regression models.

Statistical inference for fixed and random effects, as well as for the error and random effects variances, is accomplished using likelihood approaches or Bayesian inference as outlined in Chap. 7.

Example 2.10 Hormone Therapy with Rats

First consider model (2.13), which comprises subject-specific deviations γ_{0i} from the overall population intercept β_0 and the subject-specific slope parameters γ_{1i}. We estimate the fixed effects, the variance parameters σ^2, τ_0^2, and τ_1^2, and the random effects using inference techniques described in Chap. 7 and implemented in function `lmer` of the R package `lme4`. Table 2.4 contains the estimated fixed effects and the variance parameters. Since the estimated value $\hat{\tau}_1^2$ for $\mathrm{Var}(\gamma_{1i})$ is very small, we also consider a simpler model that does not contain the subject-specific terms $\gamma_{1i} t_{ij}$. The results can be found again in Table 2.4. As we can see, the estimates are all very similar.

For the simpler model without random slope, Fig. 2.11 shows a kernel density estimator for the estimated values $\hat{\gamma}_{0i}$, $i = 1, \ldots, 50$ together with a superimposed normal density and a normal quantile plot in a separate graph. We do not find any serious deviations from the assumed normal distribution.

△

Table 2.4 Hormone therapy with rats: estimation results for the mixed model (2.13) and the simplified model without individual-specific slope parameter

	Parameter	Model (2.13) estimated value	Simplified model estimated value
Intercept	β_0	68.607	68.607
Low-dose	β_1	7.505	7.507
High-dose	β_2	6.874	6.871
Control	β_3	7.313	7.314
$\text{Var}(\gamma_{0i})$	τ_0^2	3.739	3.565
$\text{Var}(\gamma_{1i})$	τ_1^2	<0.001	
$\text{Var}(\varepsilon_{ij})$	σ^2	1.481	1.445

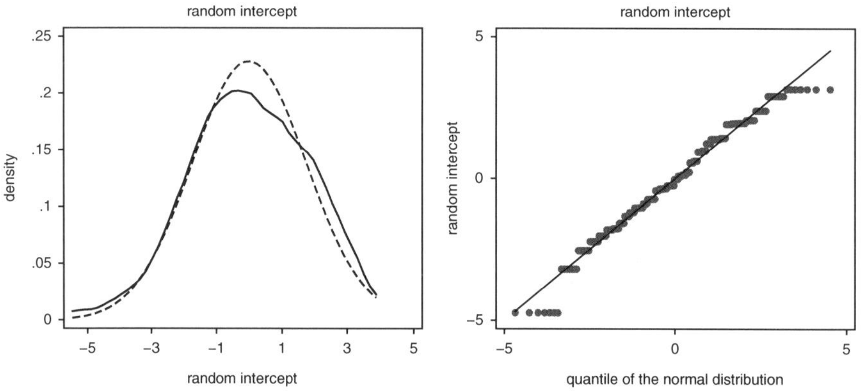

Fig. 2.11 Hormone therapy with rats: The *left plot* shows for the random intercept a kernel density estimator (*solid line*) and the density of an adapted normal distribution (*dashed line*). The *right panel* displays a normal quantile plot

In Chap. 7, we present generalizations of the basic linear mixed models, including mixed models for binary and discrete response variables. This more general group of models can also be used as a basis for inference in nonparametric and semiparametric regression models outlined in the next section and presented in detail in Chaps. 8 and 9.

2.5 Simple Nonparametric Regression

Figure 2.2 (p. 25) shows the scatter plot of the two variables *rentsqm* and *area* for the rent index data discussed in Sect. 2.2. The scatter plot reveals that the living area has a nonlinear effect on the net rent per square meter. Consequently, in Example 2.2, the effect of living area was modeled nonlinearly by

$$f(area) = \beta_0 + \beta_1 / area. \tag{2.17}$$

Figure 1.8 (p. 17) presents the scatter plot between the Z-score (as a measure of chronic undernutrition) of a child in Zambia and the age of the child (see Example 1.2 on p. 5). Again, we observe that the Z-score depends on the age of the child in a nonlinear way.

In fact, in most of the various applications presented in Chap. 1, nonlinear effects are present. It is often very difficult to model these with ad hoc parametric approaches as for instance in Eq. (2.17). Moreover, in most cases other transformations are reasonable, e.g., $f(area) = \beta_0 + \beta_1 \log(area)$ or $f(area) = \beta_0 + \beta_1 (area)^{\frac{1}{2}}$. In more complex applications with more continuous regressors, searching for suitable transformations becomes very difficult or intractable even for very experienced researchers.

Non- and semiparametric regression models allow for flexible estimation of nonlinear effects. They do not require any restrictive assumptions regarding a certain parametric functional form. In the case of just one continuous covariate x, the *standard model for nonparametric regression* is defined as

$$y_i = f(x_i) + \varepsilon_i. \tag{2.18}$$

For the error variable ε_i, the same assumptions as in the simple linear regression model (2.2) are made.

The function f is assumed to be sufficiently smooth, but no specific parametric form is specified. It is estimated in a data-driven way through nonparametric approaches. Chapter 8 describes several techniques of how to estimate the unknown function f. To give the reader a first impression of nonparametric regression models, Figs. 2.12 and 2.13 demonstrate an easily comprehensible estimation concept using the Zambia malnutrition data. For illustration purposes, we restrict our analysis to the observations of a specific district in Zambia; see Fig. 2.12a. The goal is to find an estimator $\hat{f}(c\_age)$ for the (nonlinear) relationship between the Z-score and the age of a child. Figure 2.12a shows that a simple regression line cannot produce a satisfactory fit. However, we find that a linear model is *locally* justified, i.e., if the analysis is restricted to appropriately defined intervals; see Fig. 2.12b, c.

Based on these observations, we obtain the following approach that is known as the *nearest neighbor estimator*:

1. Determine a number of values

$$c\_age_1 < c\_age_2 < \ldots < c\_age_m$$

   within the support of c_age for which estimates $\hat{f}(c\_age_j)$, $j = 1, \ldots, m$, of f are to be computed.
2. When estimating f at c_age_j, use a predetermined number of observations in the *neighborhood* of c_age_j. In Fig. 2.12b, c the nearest 70 observations, either to the right or left, of $c\_age_j = 11$ (panel b) and $c\_age_j = 28$ (panel c) were used.

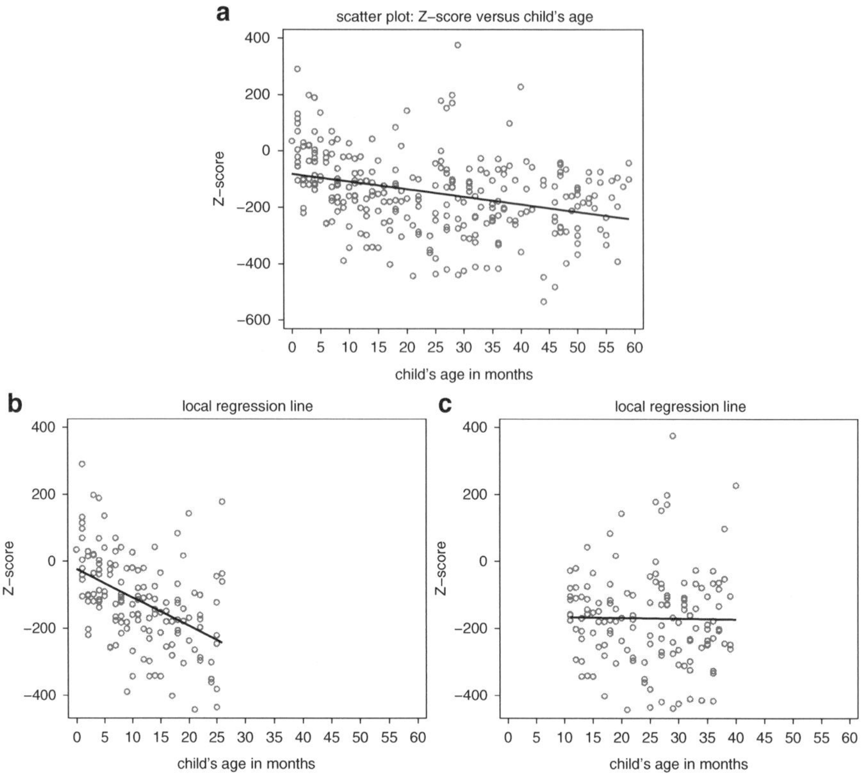

Fig. 2.12 Malnutrition in Zambia: illustration of a global and local regression. (**a**) shows a global regression line for the relationship between Z-score and child's age. (**b**) and (**c**) show local regression lines based on a subset of the data

3. Estimate a local regression line based on the observations taken in step 2. to obtain the estimate $\hat{f}(c\_age_j) = \hat{\beta}_0 + \hat{\beta}_1 c\_age_j$. Note that for every value of c_age_j, a separate regression line is estimated. This implies that the regression coefficients $\hat{\beta}_0$ and $\hat{\beta}_1$ vary according to *c_age*.
4. Combine the obtained estimates $\hat{f}(c\_age_1), \ldots, \hat{f}(c\_age_m)$ and visualize the estimated curve.

An illustration of the nearest neighbor estimator is given in Fig. 2.13.

Figure 2.12 also suggests another approach. Instead of estimating a *global* regression line, the domain of the covariate, in our example the child's age, could first be subdivided into several non-overlapping intervals. In a second step, a separate regression line could then be estimated using the data in each interval. This procedure is illustrated in Fig. 2.14a. Here the range of values was divided into the three intervals $[0, 19)$, $[19, 39)$ and $[39, 59]$. In contrast to the global regression line, we obtain a satisfactory fit to the data. There is, however, a flaw: the separate regression lines induce discontinuities at the interval boundaries. An obvious solution is to impose the additional constraint that the function is globally

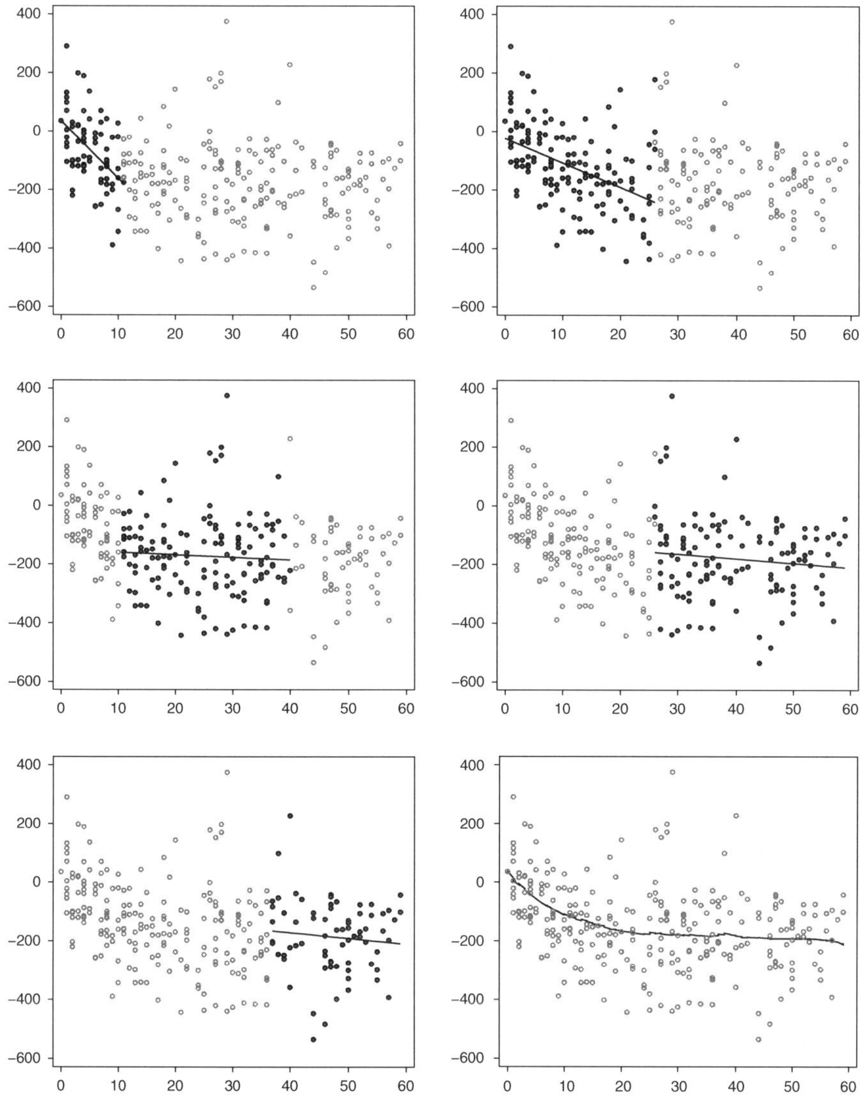

Fig. 2.13 Malnutrition in Zambia: illustration for a nearest neighbor estimator based on the nearest 70 observations (either to the *right* or *left*) of the relationship between Z-score and child's age

continuous. This implies that the regression lines merge continuously at the interval boundaries. Taking this requirement into consideration, we obtain the estimates shown in Fig. 2.14b, which is a special case of *polynomial splines*. Splines are piecewise polynomials, which fulfill certain smoothness conditions at the interval boundaries (also called the knots of the spline). Flexible regression based on splines

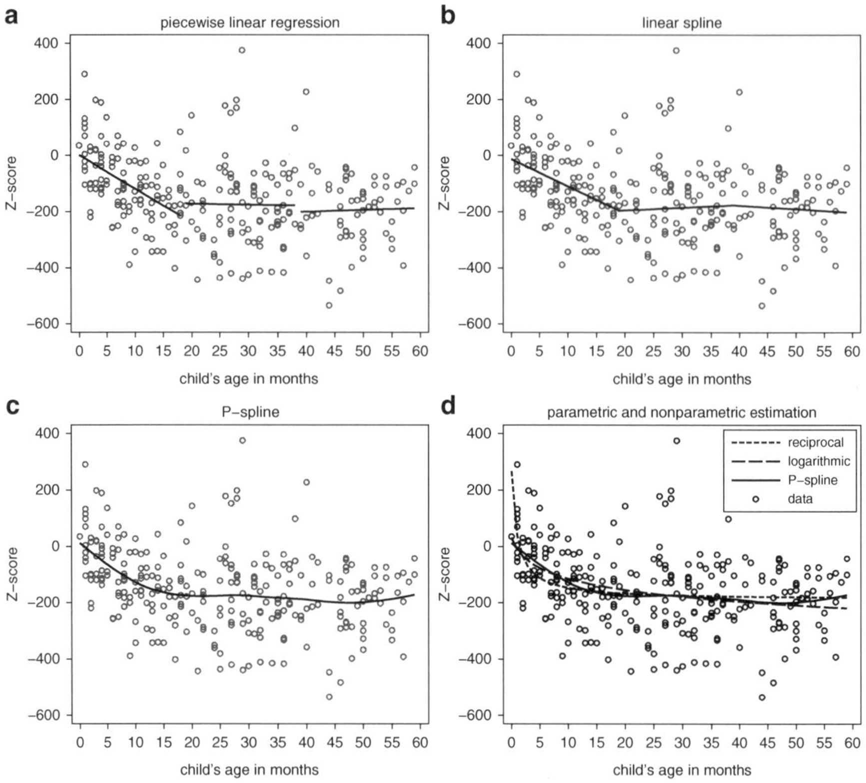

Fig. 2.14 Malnutrition in Zambia: piecewise linear regression [panel (**a**)], linear spline [panel (**b**)], and cubic P-spline [panel (**c**)] for estimating the relationship between Z-score and child's age. Panel (**d**) shows a comparison of parametric and nonparametric regressions

will play a major role in Chaps. 8 and 9. There we will discuss the most important questions and problems regarding spline regression, for example, how can splines be represented mathematically? How many knots are required to get a satisfactory fit? Where should we place the knots? Figure 2.14c gives a first flavor of the capabilities of spline estimators. It shows a flexible fit to the data based on so-called *P(enalized)-splines*.

On the basis of a nonparametric fit to the data, it is reasonable to search for a simpler parametric functional form that conserves the key features of the function. In this sense we can also understand nonparametric and semiparametric regression as a means of exploratory data analysis that helps to find satisfactory parametric forms. Figure 2.14d shows a comparison of the nonparametric fit with two parametric specifications given by

$$zscore = \beta_0 + \beta_1 \log(age + 1) + \varepsilon$$

2.5 Standard Nonparametric Regression Model

Data

(y_i, x_i), $i = 1, \ldots, n$, with continuous response variable y and continuous covariate x.

Model

$$y_i = f(x_i) + \varepsilon_i, \qquad i = 1, \ldots, n.$$

We do not assume a simple parametric form for function f. We rather assume certain smoothness characteristics, for example, continuity or differentiability. The same assumptions as in the classical linear regression model apply for errors ε_i.

and

$$zscore = \beta_0 + \beta_1/(age + 1) + \varepsilon.$$

Despite similar functional forms, the nonparametric estimated curve still fits better, particularly in the range of 0–20 months.

2.6 Additive Models

In most applications, as in the examples on the Munich rent index and on malnutrition in Zambia, a moderate (or even large) number of continuous or categorical covariates are available.

Example 2.11 Malnutrition in Zambia

The continuous covariates are *c_age* (age of the child), *c_breastf* (duration of breastfeeding), *m_bmi* (mother's body mass index), *m_height* (mother's height), and *m_agebirth* (mother's age at birth). As in the linear regression model, the categorical covariates *m_education* (mother's education), *m_work* (mother's professional status), and *region* (place of residence) must be dummy coded. Category 2 = "primary school" is chosen as the reference category for the education level. The dummy variables *m_education1*, *m_education3*, and *m_education4* correspond to the education levels "no education," "secondary education," and "higher education," respectively. For the place of residence, we choose the region Copperbelt (*region* = 2) as reference category. The variables *region1*, *region3*, ..., *region9* serve as dummy variables for the remaining regions.

For some of the continuous covariates nonlinear effects on the Z-score can be expected (see the scatter plots in Figs. 1.8 and 1.9). For this reason, we specify the *additive model*

$$\begin{aligned} zscore = {} & f_1(c\_age) + f_2(m\_bmi) + f_3(m\_agebirth) + f_4(m\_height) \\ & + \beta_0 + \beta_1\, m\_education1 + \ldots + \beta_{11}\, region9 + \varepsilon \end{aligned} \tag{2.19}$$

rather than a linear model. For the moment, the duration of breast-feeding (*c_breastf*) is excluded from the model, since it is highly correlated with the child's age (*c_age*). See the case study in Sect. 9.8 for more details on modeling the effects of the correlated covariates *c_breastf* and *c_age*.

The interpretation of the intercept β_0 and the regression coefficients $\beta_1, \beta_2, \ldots$ of the categorical covariates *m_education*, *m_work*, and *region* is identical to the linear regression model.

Similar to the possibly nonlinear function f in the basic nonparametric regression model (2.18), the functions f_1, f_2, f_3, and f_4 remain unspecified and are also estimated in a nonparametric way, together with the regression coefficients $\beta_0, \beta_1, \ldots$. Even though the model is no longer linear due to the nonlinear effects $f_1, \ldots, f_4$, it remains additive. As there are no interaction terms between the covariates, it is called an additive main effects model.

Additive models exhibit the following *identification problem*: If we change, e.g., $f_1(c\_age)$ to $\tilde{f}_1(c\_age) = f_1(c\_age) + a$ by adding an arbitrary constant a, and at the same time change β_0 to $\tilde{\beta}_0 = \beta_0 - a$ by subtracting a, the right-hand side of Eq. (2.19) remains unchanged. Hence, the level of the nonlinear function is not identified, and we are forced to impose additional identifiability conditions. This is done, for instance, by imposing the constraints

$$\sum_{i=1}^{n} f_1\left(c\_age_i\right) = \ldots = \sum_{i=1}^{n} f_4\left(m\_height_i\right) = 0,$$

i.e., each nonlinear function is centered around zero. In Fig. 2.15, the estimated functions are constrained in this way. Visualization of the estimated curves as is done in Fig. 2.15 is also the best way to interpret the estimated effects. The effect of the child's age can be interpreted as follows: The average Z-score decreases linearly as the child gets older until 18 months, then stabilizes. There is slight evidence that children older than three years may even show a slight improvement in nutrition condition. Figure 2.15 shows also 80 % and 95 % confidence intervals for the estimated effects. They can be understood as measures for the uncertainty of the effects. In case of the age effect the confidence intervals become wider as age increases. To a large extent the width of confidence intervals reflects the distribution of covariates. Densely populated areas of the covariate domain typically show narrower confidence intervals than sparsely populated areas. For the covariate age the number of observations (slightly) decreases as age increases.

In Sect. 9.8, we will interpret the remaining effects within the scope of a detailed case study.

△

The general form of an additive model (without interaction) is

$$y_i = f_1(z_{i1}) + \ldots + f_q(z_{iq}) + \beta_0 + \beta_1 x_{i1} + \ldots + \beta_k x_{ik} + \varepsilon_i, \tag{2.20}$$

with the same assumptions for the error term as in the linear regression model. The smooth functions $f_1, \ldots, f_q$ represent the (main) effects of the continuous covariates $z_1, \ldots, z_q$ and are estimated using nonparametric techniques; see Chap. 9. The covariates $x_1, \ldots, x_k$ are categorical or continuous having linear effects. Additive main effect models of the form Eq. (2.20) can be expanded with the inclusion of interaction terms. For two continuous covariates z_1 and z_2, this can be achieved by adding a smooth two-dimensional function $f_{1,2}(z_1, z_2)$. Compared to (2.20) this leads to the extended predictor

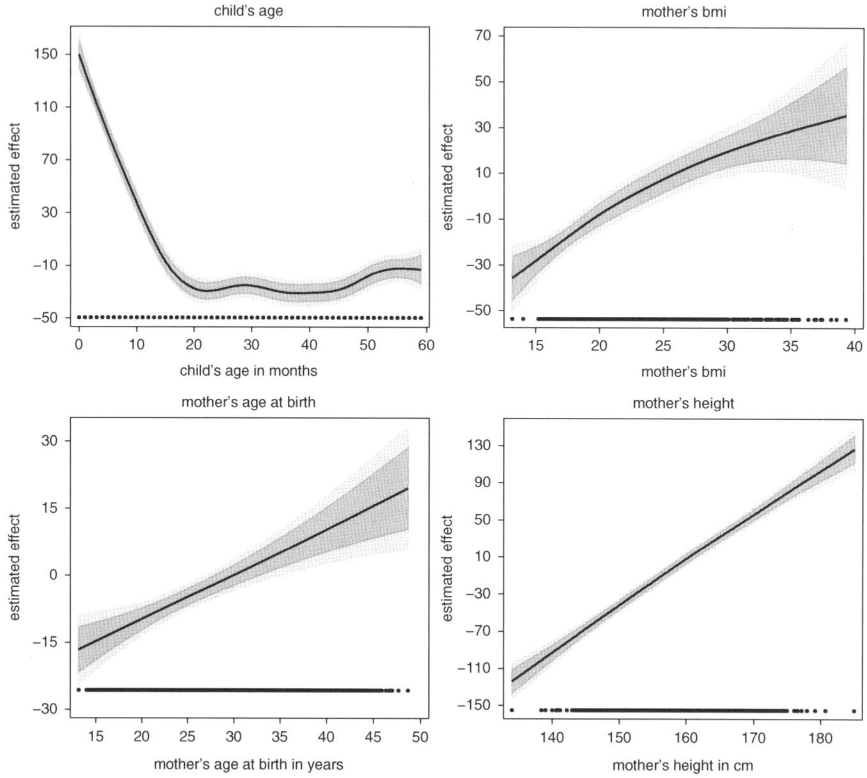

Fig. 2.15 Malnutrition in Zambia: estimated nonlinear functions including 80 % and 95 % pointwise confidence intervals. The *dots* in the lower part of the figures show the distribution of covariate values. Estimation has been carried out using `remlreg objects` of the software package `BayesX`

$$\eta_i = f_1(z_{i1}) + f_2(z_{i2}) + f_{1,2}(z_{i1}, z_{i2}) + \ldots.$$

Hence, the interaction effect $f_{1,2}$ modifies the main effects f_1 and f_2 of both covariates. Estimation of the smooth surface $f_{1,2}$ results from the extension of nonparametric techniques for one-dimensional functions to the bivariate case; see Sect. 8.2 for details. An interaction between a continuous covariate z_1 and a binary covariate x_1 is modeled by extending the predictor to

$$\eta_i = f_1(z_{i1}) + \ldots + f_q(z_{iq}) + \beta_0 + f_{x_1}(z_{i1})\, x_{i1} + \ldots.$$

The interaction term $f_{x_1}(z_1)\, x_1$, with a smooth function f_x, can be interpreted as a varying effect of x_1 over the domain of z_1. Models with parametric interactions are covered in detail in section on "Interactions Between Covariates" of Sect. 3.1.3, while Sect. 9.3 focuses on models with nonparametric interactions.

2.6 Standard Additive Regression Models

Data

$(y_i, z_{i1}, \ldots, z_{iq}, x_{i1}, \ldots, x_{ik})$, $i = 1, \ldots, n$, with y and $x_1, \ldots, x_k$ similar to those in linear regression models and additional continuous covariates $z_1, \ldots, z_q$.

Model

$$y_i = f_1(z_{i1}) + \ldots + f_q(z_{iq}) + \beta_0 + \beta_1 x_{i1} + \ldots + \beta_k x_{ik} + \varepsilon_i.$$

For the errors ε_i the same assumptions as in the classical linear regression model are made. The functions $f_1(z_1), \ldots, f_q(z_q)$ are assumed to be "smooth" and represent nonlinear effects of the continuous covariates $z_1, \ldots, z_q$.

A possible approach to estimate additive models is via an iterative procedure, called backfitting, with the simple smoothers (nearest neighbor, splines, etc.) as building blocks. Details will be given in Chap. 9.

2.7 Generalized Additive Models

Nonlinear effects of continuous covariates can also occur in regression models for binary and other non-normal response variables. Similar to the additive models presented in the previous section, it is often preferable to allow for flexible nonparametric effects of the continuous covariates rather than assuming restrictive parametric functional forms. Approaches for flexible and data-driven estimation of nonlinear effects become even more important for non-normal responses, as graphical tools (e.g., scatter plots) are often not applicable to get an intuition about the relationship between responses and covariates.

Example 2.12 Vehicle Insurance

We first illustrate the usage of generalized additive models with the analysis of vehicle insurance data for Belgium in 1997; see Denuit and Lang (2005) for a complete description of the data.

The calculation of vehicle insurance premiums is based on a detailed statistical analysis of the risk structure of the policyholders. An important part of the analysis is the modeling of the claim frequency, which generally depends on the characteristics of the policyholder and the vehicle type. Typical influencing factors of the claim frequency are the policyholder's age (*age*), the age of the vehicle (*age_v*), the engine capacity measured in horsepower (*hp*), and the claim history of the policyholder. In Belgium, the claim history is measured

2.7 Poisson Additive Model

A Poisson additive model $y_i \sim \text{Po}(\lambda_i)$ is defined via the rate

$$\lambda_i = \text{E}(y_i) = \exp(\eta_i)$$

and the additive predictor

$$\eta_i = f_1(z_{i1}) + \ldots + f_q(z_{iq}) + \beta_0 + \beta_1 x_{i1} + \ldots + x_{ik}.$$

Poisson additive models are a special case of generalized additive models for non-normal responses (Chap. 9).

with the help of a 23-step bonus malus score (*bm*). The higher the score, the worse is the insurant's claims history. The statistical analysis is based on regression models with the claim frequency (within 1 year) as the response variable. Since the claim frequency is restricted to the discrete values $0, 1, 2, \ldots$, regression models for continuous response variables are not appropriate.

△

For count data, the Poisson distribution is often assumed for the response, i.e., $y \sim \text{Po}(\lambda)$, with $\lambda = \text{E}(y)$ as the expected number of claims; see Definition B.4 in Appendix B.1 for the Poisson distribution. Our goal is to model the expected number of claims λ as a function of the covariates. Similar to binary responses, the obvious choice of $\lambda = \eta$ with a linear or additive predictor η is problematic, since we cannot guarantee that the estimated expected claim frequency $\hat{\lambda}$ is positive. We therefore assume $\lambda = \exp(\eta)$ in order to ensure a positive expected claim frequency. When using a linear predictor, we obtain a multiplicative model for the expected claim frequency that leads us to a similar interpretation as we already obtained in the logit model:

$$\lambda = \exp(\beta_0 + \beta_1 x_1 + \ldots + \beta_k x_k) = \exp(\beta_0) \cdot \exp(\beta_1 x_1) \cdot \ldots \cdot \exp(\beta_k x_k).$$

A unit increase in one of the covariates, e.g., x_1, leads to a change of the expected claim frequency by a factor of $\exp(\beta_1)$. For an additive predictor, we obtain

$$\lambda = \exp(f_1(z_1) + \ldots + f_q\left(z_q\right) + \beta_0 + \beta_1 x_1 + \ldots + \beta_k x_k).$$

Depending on the form of the nonlinear function, the expected count increases or decreases with a unit increase in a covariate. Moreover, in contrast to the purely linear predictor, the change is also dependent on the value of the covariate (because of the nonlinearity). Typically, an increase in x_1, e.g., from 20 to 21, causes a

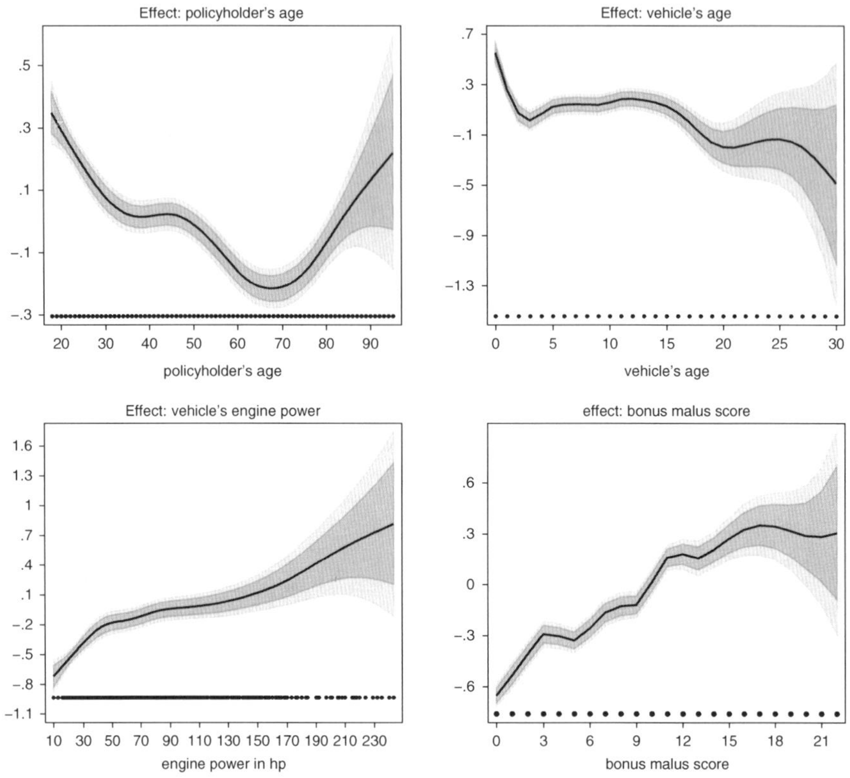

Fig. 2.16 Vehicle insurance: estimated nonlinear functions including 80 % and 95 % pointwise confidence intervals. The *dots* in the lower part of the figures show the distribution of covariate values. Estimation has been carried out using `remlreg objects` of the software package `BayesX`

different change in the expected count when compared to that of an increase, e.g., from 30 to 31.

Example 2.13 Vehicle Insurance—Additive Model

We model the claim frequencies of the Belgian insurance data using an additive predictor with possibly nonlinear functions of the variables *age*, *age_v*, *hp*, and *bm*:

$$\eta_i = f_1(age_i) + f_2(age\_v_i) + f_3(hp_i) + f_4(bm_i) + \beta_0 + \beta_1 gender_i + \ldots.$$

The dots indicate that the predictor may contain other categorical covariates in addition to the continuous variables, e.g., *gender*. We estimated the nonlinear functions and the regression coefficients using the methods presented in detail in Chap. 9. Figure 2.16 shows the estimates $\hat{f}_1, \ldots, \hat{f}_4$ for the insurance data. The function related to policyholder's age is notably nonlinear. Initially, the effect on the expected frequency is almost linear until the policyholder reaches the age of 40, then the effect remains nearly constant for several years until the age of 50, then decreases until approximately 70, followed by a rapid increase for

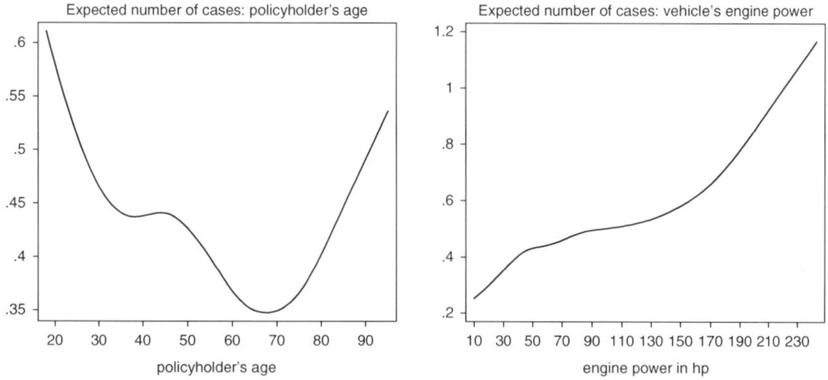

Fig. 2.17 Vehicle insurance: expected number of cases per year among 1,000 policyholders depending on the policyholder's age and vehicle's engine power. The effects of the remaining covariates are held fixed at their mean values

older policyholders. Since we do not have much data on the elderly policyholders, we must be careful with interpretation. This is also reflected by the wide confidence intervals.

Since the expected claim frequency $\lambda = \exp(\eta)$ is a nonlinear function of the covariates, it is not easy to decipher the effect of a particular covariate on λ. It is therefore advisable to plot the estimated expected claim frequency $\hat{\lambda}$ against the covariates. To do so, we plot the estimated rate $\hat{\lambda}$ separately against every continuous covariate while keeping the effects of the remaining covariates fixed at their mean value. See Fig. 2.17 which demonstrates such plots with the covariates *age* and *hp*. Since the expected frequencies are quite low, we plotted $1000\hat{\lambda}$, i.e., the expected claim frequency per year among 1,000 policyholders, rather than the estimated rate $\hat{\lambda}$ itself. For instance, we have plotted

$$1000 \exp\left(\hat{f}_1(age) + \hat{f}_2(\overline{age\_v}) + \hat{f}_3(\overline{hp}) + \hat{f}_4(\overline{bm}) + \beta_0 + \beta_1 \overline{gender} + \ldots\right)$$

against *age* with $\overline{age\_v}$, $\overline{hp}$, $\overline{bm}$, and $\overline{gender}$ being the respective covariate means. We observe that the expected number of insurance cases varies between 0.35 and 0.6 per 1,000 policyholders for the *age* variable and between 0.2 and 1.2 for *hp*.

△

2.8 Geoadditive Regression

In addition to the values $(y_i, x_{i1}, \ldots, x_{ik}, z_{i1}, \ldots, z_{iq})$, $i = 1, \ldots, n$, of the response and covariates, many applications contain small-scale geographical information, for example, the residence (address), zip code, location, or county for the individual or unit. For the examples discussed so far, this applies to the data regarding the Munich rent index, malnutrition in Zambia, vehicle insurance, and the health status of trees. In these applications, it is often important to appropriately include geographic information into the regression models in order to capture spatial heterogeneity not covered by the other covariates.

Example 2.14 Malnutrition in Zambia—Geoadditive Model

Example 2.11 (p. 49) already included regional effects using dummy variables for the regions. The corresponding region effects were estimated as categorical "fixed" effects. This conventional approach has two disadvantages. First, information regarding regional closeness or the neighborhood of regions is not considered. Second, if we wish to use small scale information about the district the mother resides, the analogous district-specific approach is difficult or even impossible. Including a separate fixed effect dummy variable for every district results in a model with a large number of parameters, causing a high degree of estimation inaccuracy. Hence, it is better to understand the geographic effect of the variable *district* as an unspecified function $f_{geo}(district)$ and to consider the geographical distance of the districts appropriately when modeling and estimating f_{geo}. A typical assumption is that the regression parameters of two neighboring districts sharing a common boundary should be "similar in size" (a more precise definition of this concept is given in Sect. 8.2). Conceptually, this is very similar to the nonparametric estimation of a smooth function f of a continuous covariate, as, for example, $f(c\_age)$. Hence, we reanalyze the data with the following *geoadditive model*:

$$\begin{aligned} zscore = {} & f_1(c\_age) + f_2(m\_bmi) + f_3(m\_agebirth) + f_4(m\_height) \\ & + f_{geo}(district) + \beta_0 + \beta_1\, m\_education + \ldots + \beta_4\, m\_work + \varepsilon. \end{aligned}$$

We used `remlreg objects` of the software `BayesX` for estimation. In comparison to Example 2.11, the linear part of the predictor no longer contains any region-specific dummy variables. Figure 2.18 shows the map of Zambia, which is divided into color-coded district-specific effects. The geographic effect can now be interpreted similar to a nonlinear effect of a continuous covariate. An effect of, e.g., 40 implies an average Z-score increase of 40 points relative to a district with a zero effect.

The district-specific pattern shows that geographic or spatial effects do not have much in common with the administrative boarders. Other causes must be responsible for the spatial effects. In the current situation, the visible north–south divide is due to climatic differences, with a much better nutrition situation in the north. The climatic conditions in the south are worse than in the north, since the southern regions have a much lower altitude compared to the northern ones. In this sense geoadditive models can be understood as an exploratory tool for data analysis: The estimated spatial effects may help to identify geographic covariates that explain the geographic variations.

△

In general, *geoadditive regression* is useful, if in addition to the response variable and continuous or categorical covariates, a *location variable* s_i, $i = 1, \ldots, n$, is observed for every unit i. This location variable s can be a location index, as in Example 2.14, with a finite domain $s \in \{1, \ldots, S\}$, comprising, e.g., counties, districts etc. In addition, neighborhood information, on the basis of a geographical map or a graph, is available. In other applications, for example, the data on the health status of trees (see Example 1.4), s is a continuous variable containing precise information about the position or the location through geographic coordinates. For flexibly modeling the function f_{geo}, several alternative approaches are available. The choice of a particular model in part depends on whether the location variable is discrete or continuous; see Sect. 8.2 for details.

Geoadditive regression analyses can also be conducted for non-normally distributed responses, especially binary, categorical, or discrete response variables, as in the analysis of the health status of trees or the claim frequency of vehicle insurance policies. In these cases, we expand the predictor η_i in additive logit

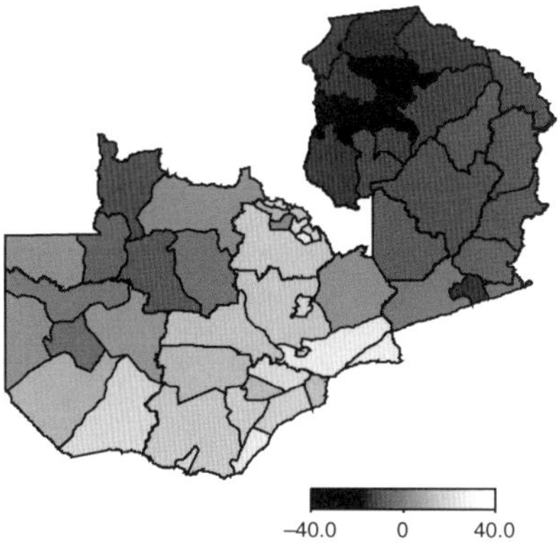

Fig. 2.18 Malnutrition in Zambia: estimated spatial effect

or Poisson models or in generalized additive models to a so-called *geoadditive predictor*

$$\eta_i = f_1(z_{i1}) + \ldots + f_q(z_{iq}) + f_{geo}(s_i) + \beta_0 + \beta_1 x_{i1} + \ldots + \beta_k x_{ik}.$$

Example 2.15 Vehicle Insurance

It is known that claims associated with vehicle insurance can widely vary across geographic areas. For this reason, many insurance companies report geographically heterogeneous insurance premiums, i.e., the insurance rates differ depending on the policyholder's residence. A realistic modeling of the claim frequency, thus, requires an adequate consideration of the spatial heterogeneity of claim frequencies. In order to do so, we extend the additive predictor of Example 2.13 to

$$\eta_i = f_1(age_i) + f_2(age_v) + f_3(hp_i) + f_4(bm_i) + f_{geo}(district_i) + \ldots,$$

where $f_{geo}(district)$ represents a spatial district-specific effect. Figure 2.19 shows the estimated geographic effect obtained using `remlreg objects` of the software `BayesX`: the darker the color, the higher is the estimated effect. The hatched areas mark districts for which we do not have any observations. In comparison to a region with an effect of zero, an effect of approximately 0.3 implies an increase of the expected frequency by a factor of $\exp(0.3) = 1.35$. We find three areas where the expected claims frequencies are clearly higher: the metropolitan areas around Brussels in the center, Antwerp in the North, and Liége in the East of Belgium. The sparsely populated regions in the South show on average lower frequencies.

△

In the following example, we will look at an application taken from survival analysis.

Example 2.16 Survival Analysis of Patients Suffering from Leukemia

The goal of this application is the analysis of covariate effects on the survival time of patients who are diagnosed with a specific type of leukemia. The geographic variation of the survival time is of particular interest, as it might give us information about other risk

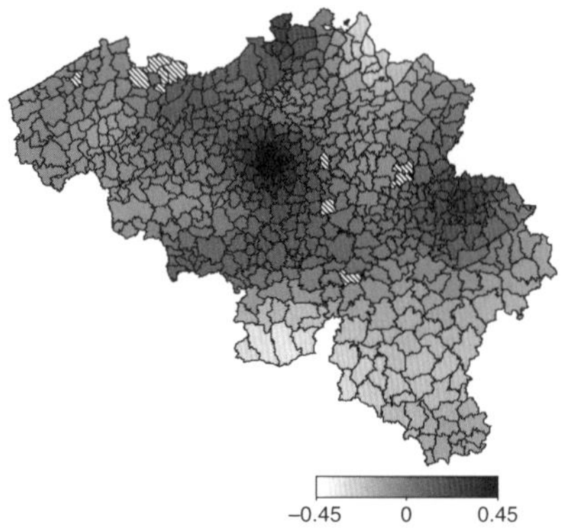

Fig. 2.19 Vehicle insurance: estimated spatial effect

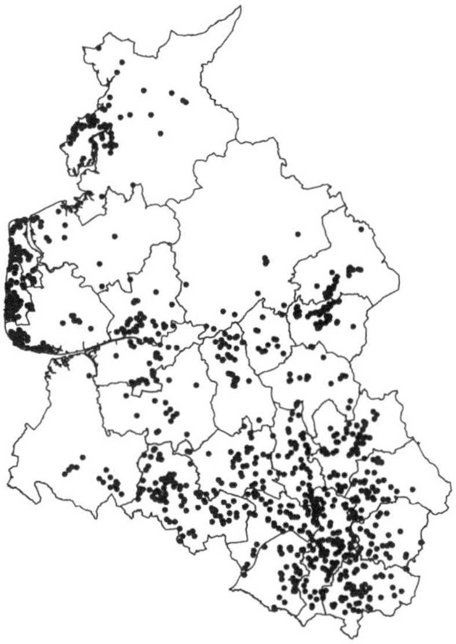

Fig. 2.20 Leukemia data: spatial distribution of observations in Northwest England. Every point corresponds to one observation

factors that so far are unknown. The geographical effect may also be closely related to the quality of the health care system in a certain area.

The application studies the survival time of 1,043 patients from the Northwest of England who were diagnosed with acute myeloid leukemia during 1982 through 1998. The data are taken from the British Northwest leukemia register. In addition to the survival time of the patients, there is also information about the following covariates: gender (1 = female, 0 = male), patient's age at the time of diagnosis (*age*), the amount of leucocytes (*lc*), and the Townsend Index (*ti*) that specifies a poverty index of a patient's residential district. A higher value of the Townsend Index reflects a poorer residential district. Geographic information is also included for each patient, as we know the exact coordinates (longitude and latitude) of the patient's residence in Northwest England. Moreover, the patient's residence can be assigned to the particular district of Northwest England. Figure 2.20 shows the geographic distribution of the observations. Approximately 16 % of the patients were censored, i.e., they survived the end of the study.

2.8 Geoadditive Models

Data

In addition to the continuous response variable y, the continuous covariates $z_1, \ldots, z_q$, and the remaining covariates $x_1, \ldots, x_k$, there is information about the geographic location s available.

Model

$$y_i = f_1(z_{i1}) + \ldots + f_q(z_{iq}) + f_{geo}(s_i) + \beta_0 + \beta_1 x_{i1} + \ldots + \beta_k x_{ik} + \varepsilon_i.$$

We make the same assumptions for the error variable ε_i as in the classical linear regression model. The unknown smooth functions $f_1, \ldots, f_q, f_{geo}$ and the parametric effects are to be estimated on the basis of the given data.

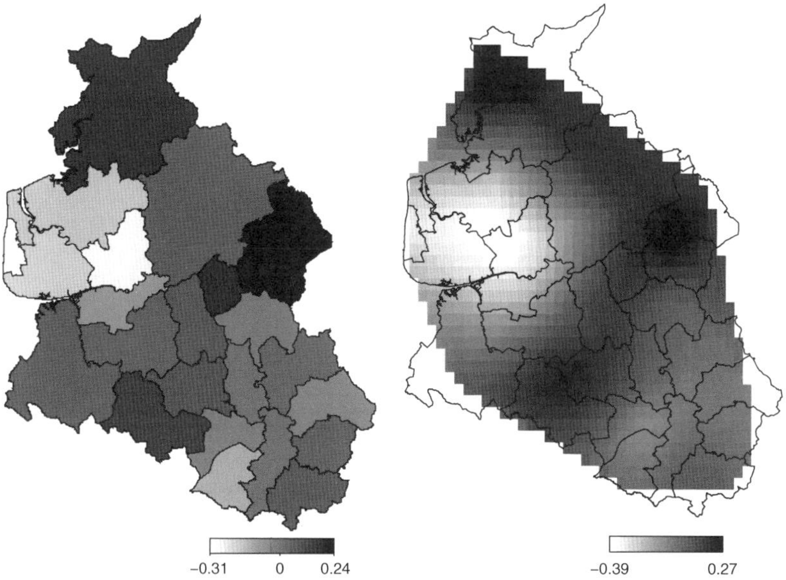

Fig. 2.21 Leukemia data: estimated spatial effect based on districts (*left*) and exact coordinates of the observations (*right*)

In order to estimate the effect of the covariates on the survival time T_i of an individual i, we use *hazard rate models*. The hazard rate $\lambda_i(t)$ of the survival time for individual i is defined as the limit

$$\lambda_i(t) = \lim_{\Delta t \to 0} \frac{P(t \le T_i \le t + \Delta t \mid T_i \ge t)}{\Delta t}.$$

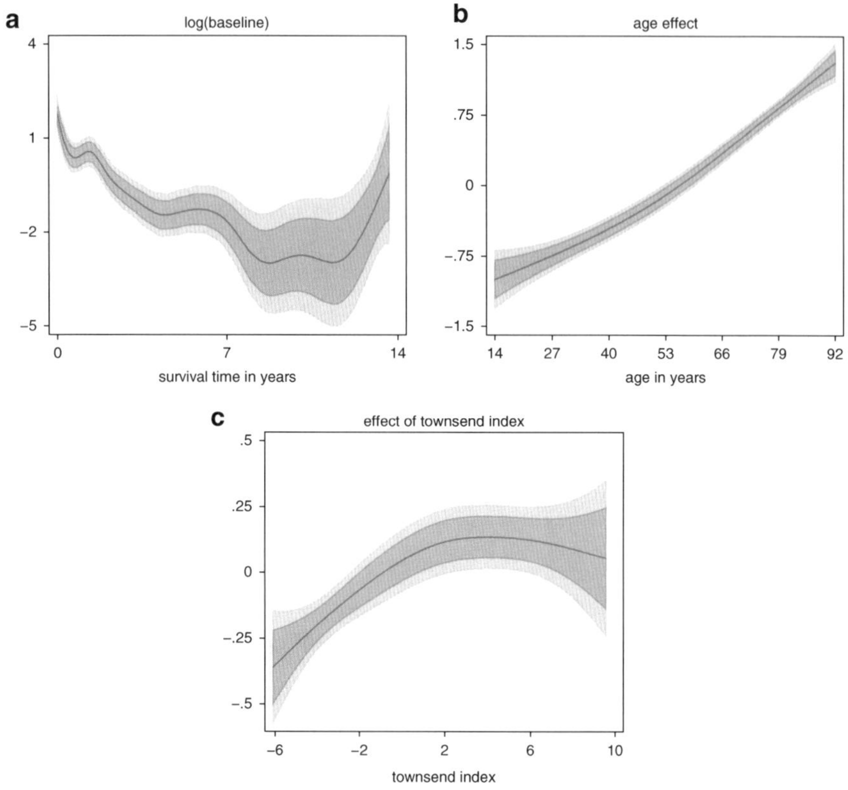

Fig. 2.22 Leukemia data: estimated nonlinear covariate effects with 80 % and 95 % pointwise confidence intervals

The hazard rate $\lambda_i(t)$ therefore characterizes the conditional probability of survival in the interval $[t, t + \Delta t]$, given the individual survived until time t, relative to the interval length Δt. In our application, we use a geoadditive hazard rate model, which links the hazard rate $\lambda_i(t)$ with a geoadditive predictor over the exponential function (similar to the Poisson model of Examples 2.13 and 2.15):

$$\lambda_i(t) = \exp\left[g(t) + f_1(age_i) + f_2(ti_i) + f_{geo}(s_i) + \beta_0 + \beta_1 lc_i + \beta_2 gender_i\right].$$

This model can be viewed as a generalization of the popular Cox model with simple linear predictors. The model contains nonparametric effects of the continuous covariates *age* and *ti*, linear effects of *lc* and *gender*, as well as a spatial effect, which can either be defined by being based on the exact coordinates or on the districts. We will outline different possibilities of how to model various types of spatial effects in Sect. 8.2. Furthermore, the model also has a time-dependent component $g(t)$, which models the temporal variation of the mortality risk from the time of diagnosis. We refer to the function $g(t)$ as the *log baseline hazard rate* and $\lambda_0(t) = \exp[g(t)]$ as the *baseline hazard rate*.

The following results are based on `remlreg objects` of the software `BayesX`. Figure 2.21 shows the estimated spatial effects and displays obvious geographic variation of the mortality risk. Presumably the geographic effects are surrogates for unobserved covariates, which could to some extent explain the geographic variation. Figure 2.22 shows the estimated functions $g(t)$, $f_1(age)$, and $f_2(ti)$. The log-baseline hazard rate reflects a decreasing nonlinear trend in mortality risk, up to approximately eight years after the first diagnosis, then followed by an increasing trend. The effect of age has a monotone, almost linear trend, whereas the effect of the Townsend Index indicates that the mortality risk increases in poorer areas (corresponding to higher values in the index) and then remains relatively constant. The estimated effect of the number of leucocytes lc is positive with $\hat{\beta}_1 = 0.003$, but apparently very low. Only when lc is large, the effect of $\hat{\beta}_1 lc_i$ becomes important in size. The estimated effect of gender is very small with $\hat{\beta}_2 = 0.073$; upon further testing, we conclude that gender has little effect on the hazard rate.

△

Even though regression models for the analysis of survival times play an important role in many fields, this book will not give a detailed presentation, but Sect. 5.8 lists references for further reading. The methodology for the presented example is described in Kneib and Fahrmeir (2007) and Fahrmeir and Kneib (2011).

2.9 Beyond Mean Regression

In the models considered so far, we have restricted ourselves to modeling the (conditional) mean of the response y in dependence of covariates. For example, in the multiple linear regression model of Sect. 2.2.2, we assume independent and normally distributed responses $y_i \sim \mathrm{N}(\mu_i, \sigma^2)$, $i = 1, \ldots, n$, where the expected value μ_i depends linearly on the covariates in the form

$$\mu_i = \beta_0 + \beta_1 x_{i1} + \ldots + \beta_k x_{ik}.$$

Other parameters of the response distribution (in case of the normal distribution the variance σ^2) are explicitly assumed to be independent of covariates. In a number of applications, this assumption might not be justified as we will illustrate through the data on the Munich rent index.

Example 2.17 Munich Rent Index—Heterogeneous Variances

Consider Fig. 2.23 which shows scatter plots between the net rent in Euro and the living area (left panel) and year of construction (right panel). Additionally included are estimated regression lines between the response and the covariates. At least the scatter plot between net rent and living area suggests that, additional to the expected value μ, also the variance σ^2 of net rents depends on the covariates. We observe increasing variability as living area increases.

△

In the next two sections, we present regression models that allow the modeling of other parameters of the response distribution in dependence of covariates, in addition to the expected value. In Sect. 2.9.1 we introduce models with normally distributed responses where the mean *and* the variance depend on covariates. In Sect. 2.9.2 we even go one step further by dropping the normality assumption and modeling the *quantiles* of the response distribution in dependence of covariates.

Another important class of regression models beyond the mean are *hazard regression models* for durations or lifetimes, which are briefly considered in Example 2.16. In fact, it can be shown that a complete specification of the hazard

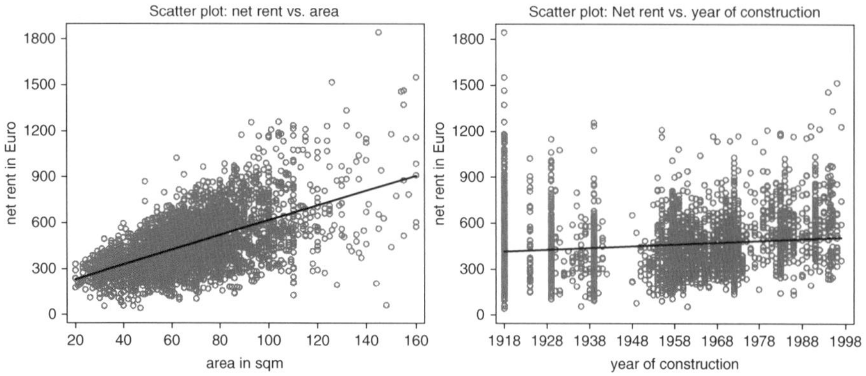

Fig. 2.23 Munich rent index: scatter plots of rents in Euro versus living area (*left panel*) and year of construction (*right panel*) together with estimated regression lines

rate implies a complete specification of the distribution of a lifetime in dependence of covariates. We do not further discuss hazard rate regression in this book, but refer to the literature cited in Sect. 5.8.

2.9.1 Regression Models for Location, Scale, and Shape

A straightforward approach that extends the multiple linear regression model to cope with variances depending on covariates is to assume $y_i \sim \mathrm{N}(\mu_i, \sigma_i^2)$, where in addition to the means

$$\mu_i = \beta_0 + \beta_1 x_{i1} + \ldots + \beta_k x_{ik},$$

the standard deviations (alternatively the variances)

$$\sigma_i = \alpha_0 + \alpha_1 x_{i1} + \ldots + \alpha_k x_{ik} \tag{2.21}$$

depend linearly on the covariates. Similar to logit or probit models and Poisson regression, assumption (2.21) is problematic as it does not guarantee positive standard deviations. Therefore, we replace Eq. (2.21) by

$$\sigma_i = \exp\left(\alpha_0 + \alpha_1 x_{i1} + \ldots + \alpha_k x_{ik}\right) = \exp(\alpha_0)\exp(\alpha_1 x_{i1}) \cdots \exp(\alpha_k x_{ik}), \tag{2.22}$$

to ensure that the standard deviations are positive. For notational simplicity, we assume exactly the same set of covariates for the expected values μ_i as for the standard deviations σ_i. Of course, this limitation can easily be dropped in practice to allow for different covariates in the mean and the variance equation.

Example 2.18 Munich Rent Index—Linear Model for Location and Scale

We take the data on the Munich rent index and assume the model $rent_i \sim N(\mu_i, \sigma_i^2)$ with

$$\mu_i = \beta_0 + \beta_1 area_i + \beta_2 yearc_i, \qquad \sigma_i = \exp\left(\alpha_0 + \alpha_1 area_i + \alpha_2 yearc_i\right),$$

for the expected value and standard deviation of the net rents. For simplicity, we restrict ourselves to the two covariates living area and year of construction. Using the R package `gamlss` we obtain the estimates

$$\hat{\mu}_i = -4617.6889 + 5.1847 \cdot area_i + 2.4162 \cdot yearc_i$$

and

$$\hat{\sigma}_i = \exp\left(8.5235 + 0.0141 area_i - 0.0023 yearc_i\right).$$

The results for the mean can be interpreted in the usual way as outlined in Sect. 2.2.2:

- Increasing the living area by 1 m^2 leads to an average increase of the net rent of about 5.18 Euro.
- Modern flats are on average more expensive than older flats. Every year increases the average net rent by 2.42 Euro.

Of course, this interpretation is only meaningful if the chosen linear model is justified (apart from the question whether the net rent per square meter is more appropriate than the plain net rent as a response variable). Figure 2.23 suggests that the linearity assumption is at least problematic for the effect of the year of construction.

Interpretation of the results for the standard deviation is slightly more complicated due to the nonlinearity induced by the exponential link but is similar to Poisson regression (see Sect. 2.7):

- A unit increase of the living area increases the standard deviation by a (small) factor of $\exp(0.014094) = 1.0141938$. This is in line with our observation from the scatter plot in Fig. 2.23 which shows increased variability of net rents as the living area increases.
- A unit increase of the year of construction decreases the standard deviation of net rents by the factor $\exp(-0.002347) = 0.99765575$ which is again close to unity. This estimate is not easily verified through the scatter plot in Fig. 2.23.

△

It is straightforward to generalize mean and variance estimation in Gaussian regression models to nonlinear covariate effects as in additive or geoadditive models. An additive model for location and scale is obtained by generalizing the equations for the mean and standard deviation to

$$\mu_i = f_1(z_{i1}) + \ldots + f_q(z_{iq}) + \beta_0 + \beta_1 x_{i1} + \ldots + \beta_k x_{ik}$$

and

$$\sigma_i = \exp(g_1(z_{i1}) + \ldots + g_q(z_{iq}) + \alpha_0 + \alpha_1 x_{i1} + \ldots + \alpha_k x_{ik}).$$

Here, $g_1, \ldots, g_k$ are additional smooth functions of the covariates $z_1, \ldots, z_q$. Additive modeling is even more important for the standard deviation as the type of effect (linear or nonlinear) of a certain covariate is much harder (if not impossible) to detect through graphical aids as with mean regression. For instance, the scatter plot between the net rent and the year of construction in the right panel of Fig. 2.23 does not provide clear guidance how to model the effect of year of construction on the standard deviation of the net rent. The following example shows additive models for location and scale in action.

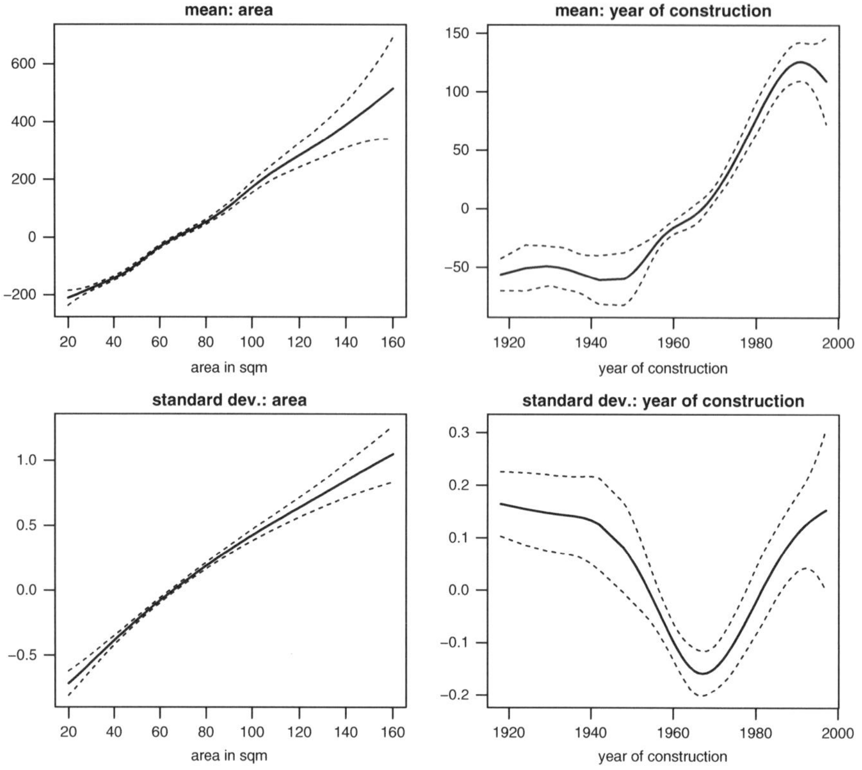

Fig. 2.24 Munich rent index: estimated effects of living area and year of construction for the mean and standard deviation

Example 2.19 Munich Rent Index—Additive Model for Location and Scale

We continue the previous example and assume now possibly nonlinear effects of living area and year of construction in the equations for the mean and the standard deviation:

$$\mu_i = f_1(area_i) + f_2(yearc_i) + \beta_0 \qquad \sigma_i = \exp\left(g_1(area_i) + g_2(yearc_i) + \alpha_0\right).$$

The resulting estimates including pointwise confidence intervals have been obtained using the R package `gamlss` and are provided in Fig. 2.24. We see that the linear effects of Example 2.18 are (ex post) justified for the area effects but not for the effects of year of construction. The upper left panel of Fig. 2.24 confirms our previous finding that larger flats are on average more expensive than smaller flats with more or less linearly increasing rents. The effect of the year of construction is almost constant until the post-World War II era indicating that flats built before 1945 with otherwise identical living area are on average equally expensive. After World War II, the rents increase almost linearly as the year of construction increases. The effect of living area on the standard deviation is again almost linear implying increasing variability of net rents as the living area increases. The effect of the year of construction is approximately U-shaped with lower variability of net rents in the

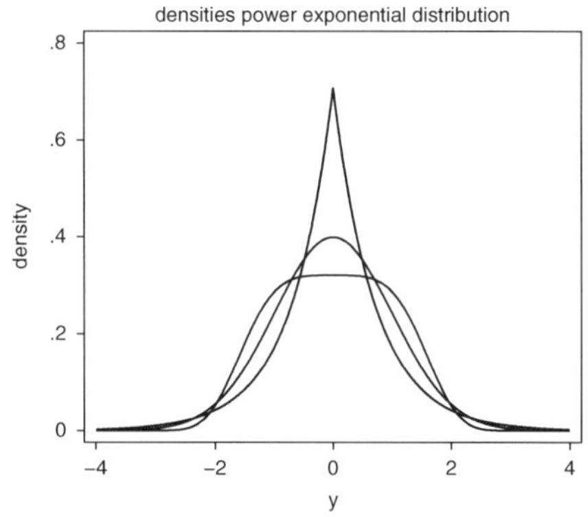

Fig. 2.25 Some densities of the power exponential distribution for $\nu = 1, 2, 4$

1960s and 1970s. This can be explained by a boom in construction building in these years, with flats having comparably homogeneous (typically poor) quality.

△

So far we have modeled the mean and the standard deviation of responses as a function of covariates. This type of modeling is a special case of an even more general approach for linear or additive modeling of location, scale, and shape. Generalized additive models for location, scale, and shape (GAMLSS) have been proposed by Rigby and Stasinopoulos (2005). Meanwhile the approach has been fully developed including professional software and inference; see Rigby and Stasinopoulos (2009) and the GAMLSS homepage http://gamlss.org/ for a full introduction. The approach and corresponding software are able to deal with a huge variety of continuous and discrete distributions for regression modeling. An example is the so-called power exponential distribution whose probability density contains, additional to the mean $\mu \in \mathbb{R}$ and variance $\sigma^2 > 0$, a shape parameter $\nu > 0$ controlling the shape of the density. The probability density is given by

$$f(y) = \frac{\nu \exp\left(-|\frac{z}{c}|^{\nu}\right)}{2c\Gamma(1/\nu)},$$

where $c^2 = \Gamma(1/\nu)\Gamma(3/\nu)^{-1}$ is a constant depending on the shape parameter ν, $z = (y - \mu)/\sigma$, and Γ is the gamma function. Compared to the normal distribution, the parameter ν gives the density some extra flexibility to control the shape (to a certain extent). Figure 2.25 shows the density of the power exponential distribution for the three choices $\nu = 1, 2, 4$ of the shape parameter and fixed mean $\mu = 0$ and variance $\sigma^2 = 1$. For $\nu = 2$ we obtain the normal distribution as a special case. Using GAMLSS, we are able to assign additive predictors for each of the three parameters μ, σ^2, and ν of the power exponential distribution. It is beyond the scope of this book to cover GAMLSS modeling in full detail. However, the GAMLSS literature is readily accessible once the material on additive models and extensions described in Chaps. 8 and 9 has been studied.

2.9.2 Quantile Regression

The GAMLSS framework allows to model the most important characteristics of the response distribution as a function of covariates. However, we still rely on a specific parametric probability distribution like the normal or power exponential distribution. In contrast, quantile regression aims at directly modeling the *quantiles* of the response distribution in dependence of covariates without resorting to a specific parametric distribution family. For $0 < \tau < 1$ let q_τ be the τ-quantile of the response distribution, e.g., $q_{0.75}$ is the 75 % quantile. Then in linear quantile regression we assume

$$q_{\tau,i} = \beta_{\tau,0} + \beta_{\tau,1} x_{i1} + \ldots + \beta_{\tau,k} x_{ik},$$

i.e., the quantile q_τ of the response distribution is a linear combination of the covariates as in the multiple linear regression model. Generalizations to additive or geoadditive predictors are conceptually straightforward (although estimation is truly a challenge). The response distribution is implicitly determined by the estimated quantiles q_τ provided that quantiles for a reasonable dense grid of τ-values are estimated. In contrast to the GAMLSS framework, a specific *parametric* distribution is not specified a priori which makes quantile regression a distribution-free approach. The following example gives a flavor of the capabilities of quantile regression. Full details are given in the last chapter of the book (but note that large portions of Chap. 10 on quantile regression are accessible immediately after reading the parts on the classical linear model in Chap. 3 and Sect. 4.1).

Example 2.20 Munich Rent Index—Quantile Regression

We take the rent index data and estimate a linear effect of living area on 11 quantiles q_τ, $\tau = 0.05, 0.1, \ldots, 0.9, 0.95$, of the net rent in Euro, i.e.,

$$q_{\tau,i} = \beta_{\tau,0} + \beta_{\tau,1} \cdot area_i .$$

The top left panel of Fig. 2.26 shows a scatter plot of the net rents versus living area together with estimated quantile regression lines. From top to bottom the lines correspond to the 95 %, 90 %, 80 %, ..., 10 %, and 5 % quantiles. The results are based on the R package `quantreg`. We observe a clear change of the slope (and to a lesser extent also the intercept) of the regression lines with the quantile τ. For higher quantiles, the regression lines are comparably steep indicating a strong effect of the living area on the respective quantile. Note that higher quantiles correspond to the high-price segment of the rent market. As τ decreases, the slopes of the regression lines decrease more and more. For the lowest quantiles, corresponding to the low-price segment of the rent market, the regression lines are almost parallel to a constant line. That is, in the low-price segment, the rents increase very slowly.

We finally point out that estimates for the quantiles of the response distribution can be obtained also on the basis of the linear models based on normally distributed responses. Assuming the simple linear model $rent_i = \beta_0 + \beta_1 area_i + \varepsilon_i$ with $\varepsilon_i \sim \mathrm{N}(0, \sigma^2)$ implies $rent_i \sim \mathrm{N}(\beta_0 + \beta_1 area_i, \sigma^2)$ and the quantiles $q_{\tau,i} = \beta_0 + \beta_1 area_i + \sigma \cdot z_\tau$ where z_τ is the τ-quantile of the N(0, 1) distribution. Thus, assuming a simple linear model with normal errors implies quantile curves that are *parallel* to each other. Assuming a model with linear predictors for location and scale, i.e.,

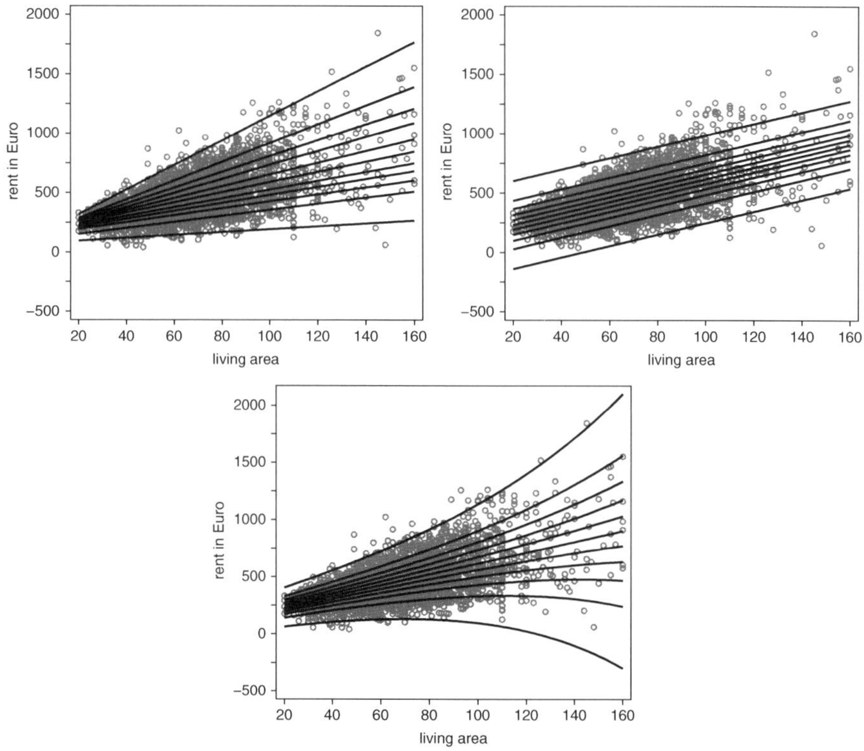

Fig. 2.26 Munich rent index: scatter plots of the rents in Euro versus living area together with linear quantile regression fits for 11 quantiles (*top left panel*), quantiles determined from a classical linear model (*top right panel*), and quantiles determined from a linear model for location and scale (*bottom panel*)

$$rent_i \sim N(\beta_0 + \beta_1 area_i, \sigma_i^2), \qquad \sigma_i = \exp(\alpha_0 + \alpha_1 area_i),$$

results in the quantiles

$$q_{\tau,i} = \mu_i + \sigma_i z_\tau = \beta_0 + \beta_1 area_i + \exp(\alpha_0 + \alpha_1 area_i) z_\tau,$$

which are no longer parallel to each other because the standard deviations of the rents depend on the living area. For comparison with completely distribution-free quantile regression, the estimated quantiles based on linear models with normal errors are also included in Fig. 2.26. The top right panel shows estimated quantiles in the simple linear model; the bottom panel displays results if the standard deviation is additionally modeled in dependence of the living area. While the parallel quantile lines of the simple linear model are clearly not adequate, the linear model for location and scale shows reasonable estimated quantiles that are not too far away from the distribution-free estimated quantile curves in the top left row. The largest differences can be observed for very large and low quantiles (95 %, 90 %, 5 %, 10 %). Our comparison shows that parametric regression models, in our case normal regression models for location *and* scale, may well be an alternative to completely distribution-free quantile regression. Particularly promising are the models from the GAMLSS family of regression models.

△

2.10 Models in a Nutshell

We summarize the regression models of this chapter in concise form and indicate in which chapters they are described in more detail. In this way, the common general structure of all models will also become more transparent.

2.10.1 Linear Models (LMs, Chaps. 3 and 4)

- *Response:* Observations y_i are continuous with

$$y_i = \eta_i + \varepsilon_i, \quad i = 1, \ldots, n.$$

  Errors $\varepsilon_1, \ldots, \varepsilon_n$ are i.i.d. with

$$\mathrm{E}(\varepsilon_i) = 0, \quad \mathrm{Var}(\varepsilon_i) = \sigma^2.$$

- *Mean:*

$$\mathrm{E}(y_i) = \beta_0 + \beta_1 x_{i1} + \ldots + \beta_k x_{ik} = \eta_i^{lin}.$$

- *Predictor:*

$$\eta_i = \beta_0 + \beta_1 x_{i1} + \ldots + \beta_k x_{ik} = \eta_i^{lin}.$$

2.10.2 Logit Model (Chap. 5)

- *Response:* Observations $y_i \in \{0, 1\}$ are binary and independently $\mathrm{B}(1, \pi_i)$ distributed.
- *Mean:*

$$\mathrm{E}(y_i) = \mathrm{P}(y_i = 1) = \pi_i = \frac{\exp\left(\eta_i^{lin}\right)}{1 + \exp\left(\eta_i^{lin}\right)}.$$

- *Predictor:*

$$\eta_i = \beta_0 + \beta_1 x_{i1} + \ldots + \beta_k x_{ik} = \eta_i^{lin}.$$

2.10.3 Poisson Regression (Chap. 5)

- *Response:* Observations $y_i \in \{0, 1, 2, \ldots\}$ are count data, indicating how often some event of interest has been observed in a certain period of time. In a Poisson model it is assumed that the y_i are independently $\mathrm{Po}(\lambda_i)$ distributed.

- *Mean:*

$$\mathrm{E}(y_i) = \lambda_i = \exp\left(\eta_i^{lin}\right).$$

- *Predictor:*

$$\eta_i = \beta_0 + \beta_1 x_{i1} + \ldots + \beta_k x_{ik} = \eta_i^{lin}.$$

2.10.4 Generalized Linear Models (GLMs, Chaps. 5 and 6)

- *Response:* Observations y_i are continuous, categorical, or count data. Depending on the measurement scale and distributional assumptions, they are (realizations of) independent Gaussian, binomial, Poisson, or gamma random variables.
- *Mean:*

$$\mathrm{E}(y_i) = \mu_i = h\left(\beta_0 + \beta_1 x_{i1} + \ldots + \beta_k x_{ik}\right) = h\left(\eta_i^{lin}\right),$$

where h is a (known) response function, such as $h(\eta) = \exp(\eta)/(1 + \exp(\eta))$ in a logit model.
- *Predictor:*

$$\eta_i = \beta_0 + \beta_1 x_{i1} + \ldots + \beta_k x_{ik} = \eta_i^{lin}.$$

- *Remark:* Generalized linear models are a broad class of models, with linear models, logit models, and Poisson models as special cases. Extensions to categorical responses are presented in Chap. 6.

2.10.5 Linear Mixed Models (LMMs, Chap. 7)

- *Response:* Observations y_{ij}, $i = 1, \ldots, m$, $j = 1, \ldots, n_i$ are continuous with

$$y_{ij} = \eta_{ij} + \varepsilon_{ij}.$$

They are structured in form of longitudinal or clustered data for m individuals or clusters, with n_i observations per individual or cluster. For errors ε_{ij}, we usually make the same assumptions as for linear models. More general error assumptions, taking correlations within individual- or cluster-specific observations into account, are possible.
- *Mean:*

$$\begin{aligned}\mathrm{E}(y_{ij}) &= \beta_0 + \beta_1 x_{ij1} + \ldots + \beta_k x_{ijk} + \gamma_{0i} + \gamma_{1i} u_{ij1} + \ldots + \gamma_{qi} u_{ijq} \\ &= \eta_{ij}^{lin} + \gamma_{0i} + \gamma_{1i} u_{ij1} + \ldots + \gamma_{qi} u_{ijq}.\end{aligned}$$

The individual- or cluster-specific random effects γ_{li}, $l = 0, \ldots, q$, are assumed to be i.i.d. Gaussian random variables. Alternatively the vector $\boldsymbol{\gamma}_i = (\gamma_{0i}, \ldots, \gamma_{qi})'$ is i.i.d. multivariate Gaussian with possibly non-diagonal covariance matrix.

- *Predictor:*

$$\eta_{ij} = \eta_{ij}^{lin} + \gamma_{0i} + \gamma_{1i} u_{ij1} + \ldots + \gamma_{qi} u_{ijq}.$$

- *Remark:* LMM with correlated random effects, as well as generalized linear mixed models (GLMMs), will also be considered in Chap. 7.

2.10.6 Additive Models and Extensions (AMs, Chaps. 8 and 9)

- *Response:* Observations y_i are continuous with

$$y_i = \eta_i + \varepsilon_i.$$

For errors ε_i, the same assumptions are made as for linear models.

- *Mean:*

$$\mathrm{E}(y_i) = f_1(z_{i1}) + \ldots + f_q(z_{iq}) + \eta_i^{lin} = \eta_i^{add}.$$

- *Predictor:*

$$\eta_i = f_1(z_{i1}) + \ldots + f_q(z_{iq}) + \eta_i^{lin} = \eta_i^{add}.$$

- *Remark:* Additive models can be extended to include interactions, spatial effects, and random effects. For interactions the predictor is extended to

$$\eta_i = \eta_i^{add} + f_1(z_1, z_2) + \ldots$$

or

$$\eta_i = \eta_i^{add} + f(z_1)\, x_1 + \ldots.$$

In geoadditive models the predictor is extended to

$$\eta_i = \eta_i^{add} + f_{geo}(s_i)$$

with the spatial effect $f_{geo}(s)$ of the location variable s. Incorporation of random effect results in the predictor

$$\eta_{ij} = \eta_{ij}^{add} + \gamma_{0i} + \gamma_{1i} u_{ij1} + \ldots.$$

The additive model then becomes an additive mixed model (AMM), generalizing linear mixed models.

2.10.7 Generalized Additive (Mixed) Models (GA(M)Ms, Chap. 9)

- *Response:* Observations y_i are continuous, categorical, or count data. Depending on the measurement scale and distributional assumptions, they are (realizations of) independent Gaussian, binomial, multinomial, Poisson, or gamma random variables.

- *Mean:*
$$E(y_i) = \mu_i = h\left(\eta_i^{add}\right)$$
with (known) response function h.
- *Predictor:*
$$\eta_i = f_1(z_{i1}) + \ldots + f_q(z_{iq}) + \eta_i^{lin} = \eta_i^{add}.$$
- *Remark:* Interactions, spatial effects, and random effects can be included as for additive (mixed) models.

2.10.8 Structured Additive Regression (STAR, Chap. 9)

- *Response:* Observations y_i are continuous, categorical, or count data. Depending on the measurement scale and distributional assumptions, they are (realizations of) independent Gaussian, binomial, multinomial, Poisson, or gamma random variables.
- *Mean:*
$$E(y_i) = \mu_i = h(\eta_i)$$
with response function h.
- *Predictor:*
$$\eta_i = f_1(v_{i1}) + \ldots + f_q(v_{iq}) + \eta_i^{lin}.$$
The arguments $v_1, \ldots, v_q$ are scalar or multivariate variables of different type, constructed from the covariates. Correspondingly, the functions $f_1, \ldots, f_q$ are of different type. Some examples are:

$f_1(v_1) = f(z_1),$	$v_1 = z_1,$	nonlinear effect of z_1
$f_2(v_2) = f_{geo}(s),$	$v_2 = s,$	spatial effect of the location variable s
$f_3(v_3) = f(z)x,$	$v_3 = (z, x),$	effect of x varying over the domain of z
$f_4(v_4) = f_{1,2}(z_1, z_2),$	$v_4 = (z_1, z_2),$	nonlinear interaction between z_1 and z_2
$f_5(v_5) = \gamma_i u,$	$v_5 = u,$	random effect of u.

- *Remark:* Structured additive regression reflects the fact that the predictor includes effects of different type in structured additive form. All model classes discussed so far are special cases of STAR models.

2.10.9 Quantile Regression (Chap. 10)

- *Response:* Observations y_i are continuous and independent with generally unspecified distribution.
- *Quantiles:* Quantile regression models the quantiles q_τ, $0 < \tau < 1$, of the response distribution using a linear or additive predictor. The most general

predictor is a STAR predictor, i.e.,

$$q_{\tau,i} = f_{\tau,1}(v_{i1}) + \ldots + f_{\tau,q}(v_{iq}) + \eta^{lin}_{\tau,i}$$

with variables v_j and functions f_j as in Sect. 2.10.8.

3 The Classical Linear Model

The following two chapters will focus on the theory and application of *linear regression models*, which play a major role in statistics. We already studied some examples in Sect. 2.2. In addition to the direct application of linear regression models, they are also the basis of a variety of more complex regression methods. Examples are generalized linear models (Chap. 5), mixed models (Chap. 7), or semiparametric models (Chaps. 8 and 9).

In this chapter, we will focus exclusively on the classical linear model $y = \beta_0 + \beta_1 x_1 + \ldots + \beta_k x_k + \varepsilon$, with independent and identically distributed errors, which were already introduced in Sect. 2.2.2. Section 3.1 defines the model, discusses model assumptions, and illustrates how covariate effects are modeled. The following Sects. 3.2 and 3.3 describe the theory of classical estimation and testing within the linear model framework. The method of least squares plays a major role for estimating the unknown regression coefficients $\beta_0, \ldots, \beta_k$.

This chapter's last part (Sect. 3.4) discusses methods and strategies for model choice and diagnostics, as well as the application of the models and inference techniques presented so far. A careful and detailed case study on prices of used cars will show how linear models can be used in practice.

The concluding Sect. 3.5 gives suggestions for further reading and provides proofs of the theorems that have been omitted in the main text to improve readability.

In order to circumvent limitations of classical linear models, several extensions have been proposed in the literature. These extensions are devoted to the next chapter.

3.1 Model Definition

Suppose we are given a variable of primary interest y and we aim to model the relationship between this response variable y and a set of regressors or explanatory variables $x_1, \ldots, x_k$. In general, we model the relationship between y and $x_1, \ldots, x_k$ with a function $f(x_1, \ldots, x_k)$. This relationship is not exact, as it is affected by random noise ε. In practice, we usually assume additive errors and thus obtain

L. Fahrmeir et al., *Regression*, DOI 10.1007/978-3-642-34333-9_3,

$$y = f(x_1, \ldots, x_k) + \varepsilon.$$

Our goal is to estimate the unknown function f, i.e., to separate the systematic component f from the random noise ε. Within the framework of linear models, the following specific assumptions regarding the unknown function f and the noise ε are made:

1. *The systematic component f is a linear combination of covariates*

   The unknown function $f(x_1, \ldots, x_k)$ is modeled as a linear combination of covariates, i.e.,

$$f(x_1, \ldots, x_k) = \beta_0 + \beta_1 x_1 + \ldots + \beta_k x_k.$$

   The parameters $\beta_0, \beta_1, \ldots, \beta_k$ are unknown and need to be estimated. The parameter β_0 represents the intercept. If we combine the covariates and the unknown parameters each into $p = (k+1)$ dimensional vectors, $\boldsymbol{x} = (1, x_1, \ldots, x_k)'$ and $\boldsymbol{\beta} = (\beta_0, \ldots, \beta_k)'$, then

$$f(\boldsymbol{x}) = \boldsymbol{x}'\boldsymbol{\beta}.$$

   Note that the intercept β_0 in the model implies that the first element of the vector $\boldsymbol{x}$ equals one.

   At first glance, the assumption of a linear function f, i.e., a linear relation between y and $\boldsymbol{x}$, appears to be very restrictive. As already demonstrated in Sect. 2.2 (p. 22), nonlinear relationships can also be modeled within the framework of linear models. Section 3.1.3 will elaborate on this aspect in even more detail.

2. *Additive Errors*

   Another basic assumption of the linear model is additivity of errors, which implies

$$y = \boldsymbol{x}'\boldsymbol{\beta} + \varepsilon.$$

   Even though this appears to be very restrictive, this assumption is reasonable for many practical applications. Moreover, problems, which at first do not show additive error structure, can be specified by models with additive errors after a transformation of the response variable y (refer to Sect. 3.1.2, p. 83).

   In order to estimate the unknown parameters $\boldsymbol{\beta}$, we collect data y_i and $\boldsymbol{x}_i = (1, x_{i1}, \ldots, x_{ik})'$, $i = 1, \ldots, n$, and for every observation construct the equation

$$y_i = \beta_0 + \beta_1 x_{i1} + \ldots + \beta_k x_{ik} + \varepsilon_i = \boldsymbol{x}_i'\boldsymbol{\beta} + \varepsilon_i. \tag{3.1}$$

If we define the vectors

$$\boldsymbol{y} = \begin{pmatrix} y_1 \\ \vdots \\ y_n \end{pmatrix} \quad \text{and} \quad \boldsymbol{\varepsilon} = \begin{pmatrix} \varepsilon_1 \\ \vdots \\ \varepsilon_n \end{pmatrix}$$

and the *design matrix* $\boldsymbol{X}$,

$$\boldsymbol{X} = \begin{pmatrix} 1 & x_{11} & \cdots & x_{1k} \\ \vdots & \vdots & & \vdots \\ 1 & x_{n1} & \cdots & x_{nk} \end{pmatrix} = \begin{pmatrix} \boldsymbol{x}'_1 \\ \vdots \\ \boldsymbol{x}'_n \end{pmatrix},$$

then the n equations in Eq. (3.1) can be compactly summarized as

$$\boldsymbol{y} = \boldsymbol{X}\boldsymbol{\beta} + \boldsymbol{\varepsilon}.$$

For the remaining chapter we assume that $\boldsymbol{X}$ is of full column rank, i.e., $\text{rk}(\boldsymbol{X}) = k + 1 = p$, implying the columns of $\boldsymbol{X}$ are linearly independent. A necessary requirement is that the number of observations n must at least be equal to (or larger than) the number p of regression coefficients. The assumption is violated if one of the covariates is a linear transformation of another covariate, for example, $x_1 = a + b\,x_2$. For example, the variable x_1 could represent the height of a person in meters and x_2 the height in centimeters, implying $a = 0$ and $b = 1/100$. In general, the assumption of linear independence is always violated when at least one of the explanatory variables can be represented as a linear combination of the other covariates, implying redundancy of information. We will see that linear independence of the columns in the design matrix $\boldsymbol{X}$ is necessary in order to obtain *unique* estimators of the regression coefficients in $\boldsymbol{\beta}$.

Our model is finalized by appropriate assumptions on the error terms ε_i. Within the *classical linear model*, the following assumptions for the vector $\boldsymbol{\varepsilon}$ of errors are made:

1. *Expectation of the errors*

   The errors have mean or expectation zero, i.e., $\text{E}(\varepsilon_i) = 0$ or in matrix notation $\text{E}(\boldsymbol{\varepsilon}) = \boldsymbol{0}$.

2. *Variances and correlation structure of the errors*

   We assume a constant error variance σ^2 across observations, that is homoscedastic errors with $\text{Var}(\varepsilon_i) = \sigma^2$. The errors are called heteroscedastic when the variances vary among observations, i.e., $\text{Var}(\varepsilon_i) = \sigma_i^2$. In addition to homoscedastic variances, we assume that errors are uncorrelated, which means $\text{Cov}(\varepsilon_i, \varepsilon_j) = 0$ for $i \neq j$. The assumption of homoscedastic and uncorrelated errors leads to the covariance matrix $\text{Cov}(\boldsymbol{\varepsilon}) = \text{E}(\boldsymbol{\varepsilon}\boldsymbol{\varepsilon}') = \sigma^2\boldsymbol{I}$.

3. *Assumptions about the covariates and the design matrix*

   We distinguish between two situations. On one hand, explanatory variables $x_1, \ldots, x_k$ may be deterministic or non-stochastic, e.g., as is the case in designed experiments. In most cases, and also in almost every application presented in this book, both the response as well as the covariates are stochastic (observational data). For example, the rent data for apartments in Munich are a random sample of all apartments in the city of Munich. Hence, all characteristics of an apartment, for example, its area or year of construction, are realizations of random variables. In the case of stochastic regressors, the observations $(y_i, \boldsymbol{x}'_i)$, $i = 1, \ldots, n$,

3.1 The Classical Linear Model

The model

$$y = X\beta + \varepsilon$$

is called the classical linear regression model, if the following assumptions are true:

1. $\mathrm{E}(\varepsilon) = \mathbf{0}$.
2. $\mathrm{Cov}(\varepsilon) = \mathrm{E}(\varepsilon\varepsilon') = \sigma^2 I$.
3. The design matrix X has full column rank, i.e., $\mathrm{rk}(X) = k + 1 = p$.

The classical normal regression model is obtained if additionally

4. $\varepsilon \sim \mathrm{N}(\mathbf{0}, \sigma^2 I)$

holds. For stochastic covariates these assumptions are to be understood conditionally on X.

can be understood as realizations of a random vector (y, x'), and all model assumptions are conditional on the design matrix, as, for example, $\mathrm{E}(\varepsilon \mid X) = \mathbf{0}$ (instead of $\mathrm{E}(\varepsilon) = \mathbf{0}$) or $\mathrm{Cov}(\varepsilon \mid X) = \sigma^2 I$ (instead of $\mathrm{Cov}(\varepsilon) = \sigma^2 I$). The latter implies that ε_i and x_i are (stochastically) independent. The assumption that errors and stochastic covariates are independent can be relaxed, for example, by allowing $\mathrm{Var}(\varepsilon_i \mid x_i)$ to depend on covariates, i.e., $\mathrm{Var}(\varepsilon_i \mid x_i) = \sigma^2(x_i)$ (see in Chap. 4, Sect. 4.1.3).

In any case, we assume that the design matrix has full column rank, i.e. $\mathrm{rk}(X) = k + 1 = p$.

4. *Gaussian errors*

To construct confidence intervals and hypothesis tests for the regression coefficients, we often assume a normal distribution for the errors (at least approximately). Together with assumptions 1 and 2, we obtain $\varepsilon_i \sim \mathrm{N}(0, \sigma^2)$ or in matrix notation $\varepsilon \sim \mathrm{N}(\mathbf{0}, \sigma^2 I)$. With stochastic covariates we have $\varepsilon_i \mid x_i \sim \mathrm{N}(0, \sigma^2)$ and $\varepsilon \mid X \sim \mathrm{N}(\mathbf{0}, \sigma^2 I)$, implying that ε_i and x_i are independent.

For notational simplicity we usually suppress the dependence of terms on the design matrix X in case of stochastic covariates.

From our model assumptions it immediately follows that

$$\begin{aligned}
\mathrm{E}(y_i) &= \mathrm{E}(x_i'\beta + \varepsilon_i) = x_i'\beta = \beta_0 + \beta_1 x_{i1} + \ldots + \beta_k x_{ik}, \\
\mathrm{Var}(y_i) &= \mathrm{Var}(x_i'\beta + \varepsilon_i) = \mathrm{Var}(\varepsilon_i) = \sigma^2, \\
\mathrm{Cov}(y_i, y_j) &= \mathrm{Cov}(\varepsilon_i, \varepsilon_j) = 0,
\end{aligned}$$

for the mean and the variance of y_i, as well as the covariance between y_i and y_j, respectively. In matrix notation we obtain

$$\mathrm{E}(y) = X\beta \quad \text{and} \quad \mathrm{Cov}(y) = \sigma^2 I.$$

If we additionally assume normally distributed errors, we have

$$\mathbf{y} \sim \mathrm{N}(\mathbf{X}\boldsymbol{\beta}, \sigma^2 \mathbf{I}).$$

Note that covariates do only affect the mean of $\mathbf{y}$. The variance σ^2 of y_i or the covariance matrix $\sigma^2 \mathbf{I}$ of $\mathbf{y}$ is independent of the covariates; however we relax this assumption in Chap. 4, Sect. 4.1.3.

3.1.1 Model Parameters, Estimation, and Residuals

Prior to discussing the model assumptions for the classical linear model, we introduce some notation and terminology. We distinguish the model parameters from their estimates by a "hat," which means estimates of $\boldsymbol{\beta}$ and σ^2 are denoted by $\hat{\boldsymbol{\beta}}$ and $\hat{\sigma}^2$, respectively. This distinction is necessary, since it is more or less impossible to estimate the "true" parameter vector $\boldsymbol{\beta}$ without error; thus $\hat{\boldsymbol{\beta}} \neq \boldsymbol{\beta}$ in general. Regression parameters are usually estimated using the method of least squares, which we have begun to introduce in Sect. 2.2 for the simple linear model with just one regressor (see p. 24). Section 3.2 will outline estimation of $\boldsymbol{\beta}$ and σ^2 in full detail. Based on an estimator $\hat{\boldsymbol{\beta}}$ for $\boldsymbol{\beta}$, a straightforward estimator of the mean $\mathrm{E}(y_i)$ of y_i is given by

$$\widehat{\mathrm{E}(y_i)} = \hat{\beta}_0 + \hat{\beta}_1 x_{i1} + \cdots + \hat{\beta}_k x_{ik} = \mathbf{x}_i' \hat{\boldsymbol{\beta}}.$$

We usually refer to this estimator as $\hat{y}_i$, i.e., $\hat{y}_i = \widehat{\mathrm{E}(y_i)}$. The estimated error, i.e., the deviation $y_i - \mathbf{x}_i'\hat{\boldsymbol{\beta}}$ between the true value y_i and the estimated value $\hat{y}_i$, is called residual and denoted by $\hat{\varepsilon}_i$. We have

$$\hat{\varepsilon}_i = y_i - \hat{y}_i = y_i - \mathbf{x}_i'\hat{\boldsymbol{\beta}}.$$

Defining the vector of residuals $\hat{\boldsymbol{\varepsilon}} = (\hat{\varepsilon}_1, \ldots, \hat{\varepsilon}_n)'$, we obtain

$$\hat{\boldsymbol{\varepsilon}} = \mathbf{y} - \mathbf{X}\hat{\boldsymbol{\beta}}.$$

It is important to understand that the residuals $\hat{\varepsilon}_i$ are not identical to the errors ε_i. These are, like the parameter vector $\boldsymbol{\beta}$, unknown. In fact the residuals $\hat{\varepsilon}_i$ can be seen as estimates (or more precisely predictions) of ε_i.

The residuals contain the variation in the data that could not be explained by covariates. In some situations, especially for model diagnostics, we compute *partial residuals*, which quantify the removal of the effect of most, but not of all, covariates. Partial residuals with respect to the jth covariate x_j are defined as follows:

$$\begin{aligned}
\hat{\varepsilon}_{x_j,i} &= y_i - \hat{\beta}_0 - \ldots - \hat{\beta}_{j-1} x_{i,j-1} - \hat{\beta}_{j+1} x_{i,j+1} - \ldots - \hat{\beta}_k x_{ik} \\
&= y_i - \mathbf{x}_i'\hat{\boldsymbol{\beta}} + \hat{\beta}_j x_{ij} \\
&= \hat{\varepsilon}_i + \hat{\beta}_j x_{ij}.
\end{aligned}$$

Here, the effect of all covariates with the exception of x_j is removed. We will see that the partial residuals are helpful for model choice, model diagnostics, and visualization.

3.1.2 Discussion of Model Assumptions

This section has two goals: First, the assumptions of the classical linear model are critically discussed. Secondly, we direct the reader to particular sections within this book, in which various restrictive assumptions of classical linear models will be relaxed to yield more flexible modeling approaches. For illustration, we usually use the simple regression model $y = \beta_0 + \beta_1 x + \varepsilon$ with one explanatory variable.

Linearity of Covariate Effects

At first, the assumption of linear covariate effects may appear quite restrictive. However, within the scope of linear models, nonlinear relations are possible as well. Consider, for instance, the following model, in which the effect of the explanatory variable z_i is logarithmic:

$$y_i = \beta_0 + \beta_1 \log(z_i) + \varepsilon_i .$$

Defining a new variable $x_i = \log(z_i)$ yields the linear model $y_i = \beta_0 + \beta_1 x_i + \varepsilon_i$. In general, nonlinear relationships can be connected to linear models provided that they are *linear in the parameters*. An example for a model, which is nonlinear in the parameter β_2, is given by

$$y_i = \beta_0 + \beta_1 \sin(\beta_2 z_i) + \varepsilon_i .$$

This book discusses several techniques on how to model nonlinear relationships between a response variable y and an explanatory variable x. Section 3.1.3 will illustrate two simple techniques in detail: the modeling of nonlinear relationships through variable transformation and also through polynomials. Example 2.2 (p. 25) already presented an example for variable transformation. We also already used polynomial regression in Example 2.6 (p. 32). Both of these simple techniques do have the disadvantage that the functional form of the relationship must be known in advance. In Chaps. 8 and 9, we will discuss methods for estimating nonlinear covariate effects without special assumptions on the functional form.

Homoscedastic Error Variances

Homoscedastic error variances imply that the variance of ε_i does not systematically vary across individuals, e.g., by an increase or decrease with one or more covariates x_j. Figure 3.1 illustrates the difference between homo- and heteroscedastic variances with simulated data. Panel (a) displays the typical ideal setting of homoscedastic errors: All observations fluctuate with a stable variability around the

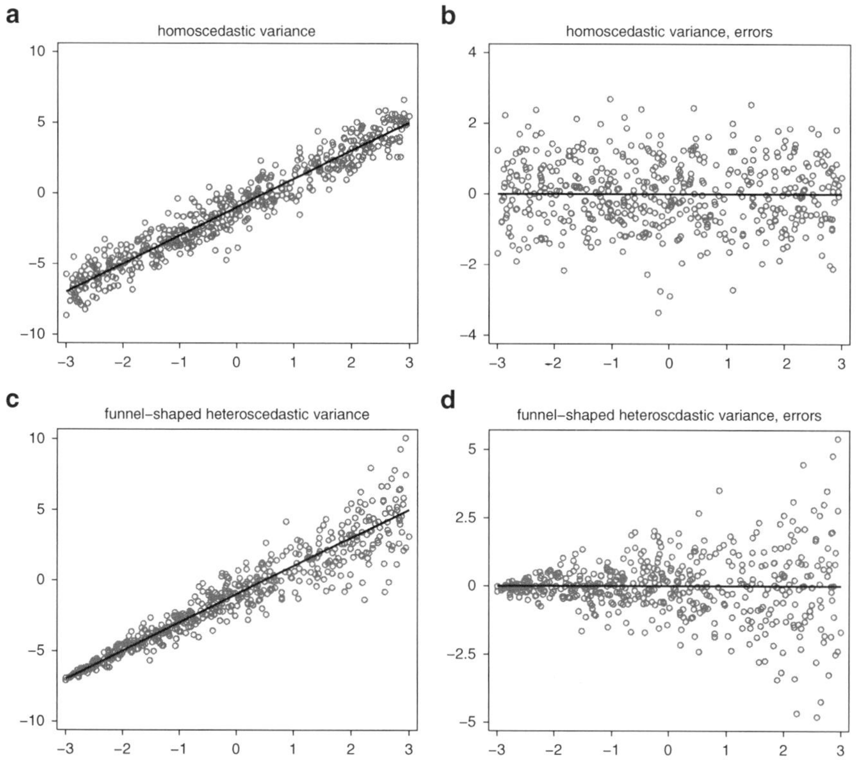

Fig. 3.1 Illustration for homo- and heteroscedastic variances: The graphs on the *left* show simulated data together with the true regression line. In the graphs on the *right* the corresponding errors are displayed. The data are based on the model $y_i \sim \mathrm{N}(-1 + 2x_i, 1)$ [panels (**a**)—homoscedastic variance and (**b**)—homoscedastic variance, errors] and $y_i \sim \mathrm{N}(-1 + 2x_i, (0.1 + 0.3(x_i + 3))^2)$ [panels (**c**)—funnel-shaped heteroscedastic variance and (**d**)—funnel-shaped heteroscedastic variance, errors]

regression line. Panel (b) shows the associated errors centered at zero. Since these data are simulated, we know the true regression coefficients and also the regression line and error values. In real data situations, we have to rely on estimators of the regression parameters. Since the errors $\varepsilon_i = y_i - \boldsymbol{x}_i'\boldsymbol{\beta}$ depend on the unknown regression parameters, we estimate the errors with the residuals $\hat{\varepsilon}_i = y_i - \boldsymbol{x}_i'\hat{\boldsymbol{\beta}}$. When verifying the assumption of homoscedastic errors, we rely on the examination of the residuals. Section 3.4.4 provides further details. Panels (c) and (d) in Fig. 3.1 provide a typical example for heteroscedastic variances. With an increasing x, the variation around the regression line also increases. The error values in panel (d) still fluctuate around zero, but the variances increase with increasing x.

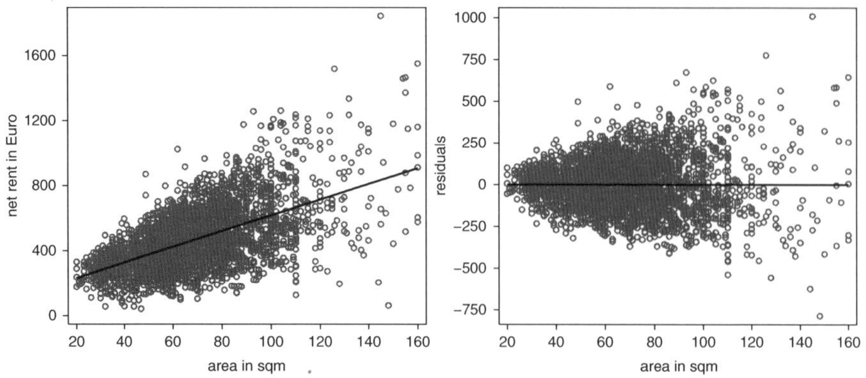

Fig. 3.2 Munich rent index: illustration of heteroscedastic variances. The *left panel* shows a scatter plot between net rent and area together with the estimated regression line. The *right panel* displays the corresponding residuals versus area

Example 3.1 Munich Rent Index—Heteroscedastic Variances

The funnel-shaped trend of the errors in Fig. 3.1c d is typical for many real data situations. As an example we take the Munich rent data; see Fig. 3.2 which shows the scatter plot between the net rent and living area together with the estimated regression line (left panel). The right panel of Fig. 3.2 shows a scatter plot of the corresponding residuals as a function of living area. The observed net rent scatters with an increasing variance around the plotted regression line. Clearly, a wider range of rent is found for larger living areas than for smaller ones. The scatter plot of the residuals shows the funnel-shaped trend mentioned above. We already encountered this phenomenon in Example 2.1 (p. 22), where we examined the relationship between net rent and area of apartments in average location that were built after 1966 (Fig. 2.1 on p. 23). We obtain a very similar pattern as presented in Fig. 3.2. The variability is, however, less pronounced, which is due to the fact that data for the apartments in average location are more homogeneous.

△

An obvious question arises: What are the consequences of ignoring heteroscedastic variances? In particular, we will see that the variances of estimated regression coefficients $\hat{\boldsymbol{\beta}}$ are not properly estimated. This in turn may cause problems for hypothesis tests and confidence intervals for the regression coefficients.

Section 4.1 (p. 177) of the next chapter will discuss heteroscedastic variances in more detail. There we will generalize the classical linear model to allow for heteroscedastic variances of the errors.

Uncorrelated Errors

In some cases, especially in time series and longitudinal data, the assumption of uncorrelated errors is not realistic. Many applications show *autocorrelated errors*. For example, first-order autocorrelation reveals a linear relationship $\varepsilon_i = \rho\varepsilon_{i-1} + u_i$ between the errors ε_i at time i and the errors of the previous time period, ε_{i-1}. The u_i are independent and identically distributed random variables. Second-order or more generally lth-order autocorrelation is given if $\varepsilon_i = \rho_1\varepsilon_{i-1} + \rho_2\varepsilon_{i-2} + u_i$ or $\varepsilon_i = \rho_1\varepsilon_{i-1} + \ldots + \rho_l\varepsilon_{i-l} + u_i$, respectively.

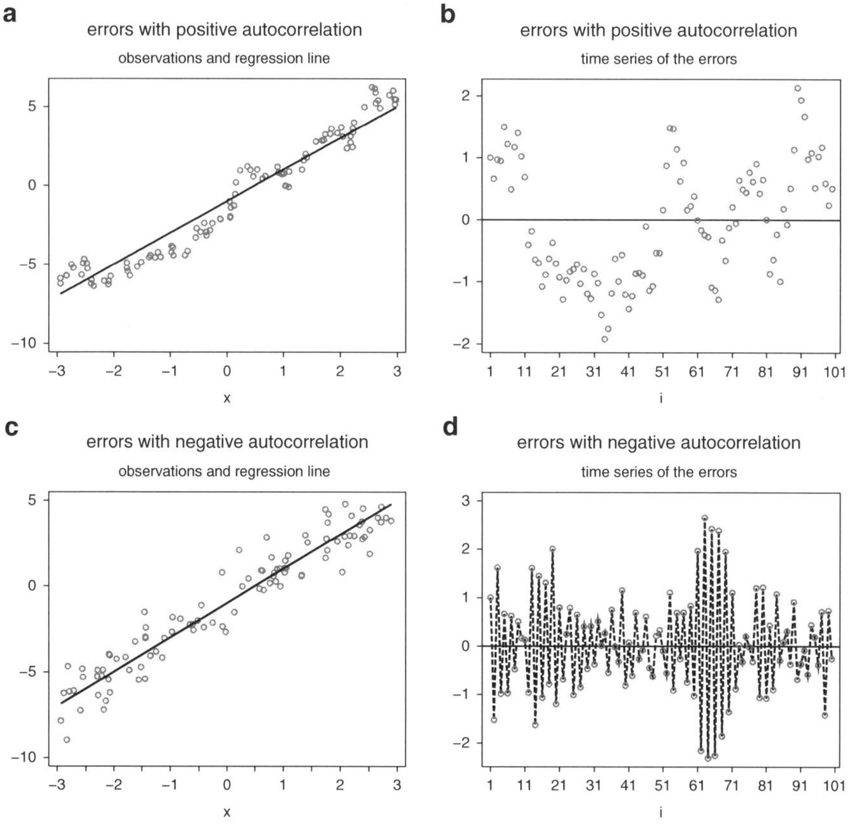

Fig. 3.3 Illustration for autocorrelated errors: Panels (**a**) and (**b**) show errors with positive autocorrelation and panels (**c**) and (**d**) correspond to negative autocorrelation. The respective graphs on the *left* show the (simulated) data including the (true) regression line. The graphs on the *right-hand side* display the corresponding errors. In case of negative autocorrelation, observations are connected in order to emphasize the changing algebraic sign. The data with positive correlation are simulated according to the model $y_i = -1 + 2x_i + \varepsilon_i$ where $\varepsilon_i = 0.9\varepsilon_{i-1} + u_i$ and $u_i \sim \mathrm{N}(0, 0.5^2)$. The data with negative correlation in the errors are simulated according to $y_i = -1 + 2x_i + \varepsilon_i$ where $\varepsilon_i = -0.9\varepsilon_{i-1} + u_i$ and $u_i \sim \mathrm{N}(0, 0.5^2)$

An example for autocorrelated errors is presented in Fig. 3.3. Panels (a) and (b) display simulated data with errors that have a first-order *positive autocorrelation*, which means that a positive (negative) error is likely to be followed by yet another positive (negative) error. Panels (c) and (d) illustrate the contrary, showing errors with negative autocorrelation. Positive (negative) errors are typically followed by negative (positive) errors, i.e., we frequently observe alternating signs of the errors.

Autocorrelated errors usually appear when the regression model is misspecified, e.g., a covariate is missing or the effect of a continuous covariate is nonlinear rather than linear. Figure 3.4 exemplifies such a situation. Panel (a) shows data

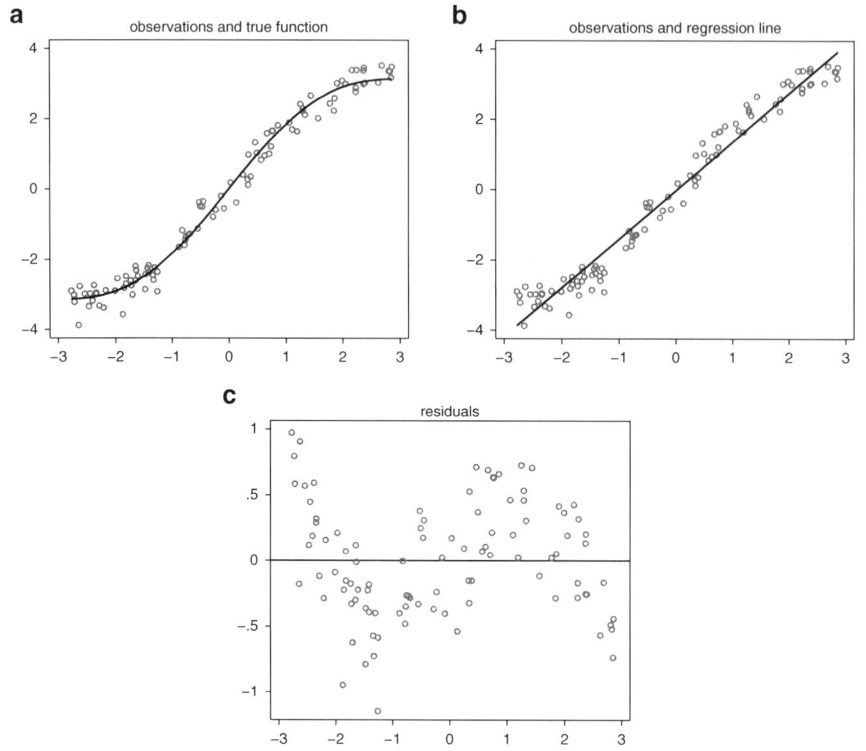

Fig. 3.4 Illustration for correlated residuals when the model is misspecified: Panel (**a**) displays (simulated) data based on the function $\text{E}(y_i \mid x_i) = \sin(x_i) + x_i$ and $\varepsilon_i \sim \text{N}(0, 0.3^2)$. Panel (**b**) shows the estimated regression line, i.e., the nonlinear relationship is ignored. The corresponding residuals can be found in panel (**c**)

simulated from the model $y_i = \sin(x_i) + x_i + \varepsilon_i$. The conditional mean of y_i is $\text{E}(y_i \mid x_i) = \sin(x_i) + x_i$, which is a nonlinear function of x; see Fig. 3.4a. In panel (b), a simple linear regression line has been fitted, which means that the estimated model is misspecified. The corresponding residuals in panel (c) show positive autocorrelation.

Autocorrelated errors are most often encountered in time series or longitudinal data. In many cases, relevant covariates cannot be included in the model because they cannot be observed. If the unobserved but relevant covariates show a temporal or seasonal trend, correlated errors are induced. We again use simulated data for illustration. Panels (a) and (b) of Fig. 3.5 show a time series plot of variables x_1 and x_2. Whereas x_1 is apparently subject to a clear temporal trend, x_2 fluctuates randomly around zero. Now consider the regression model $y_i = -1 + x_1 - 0.6\,x_2 + \varepsilon_i$, $i = 1, \ldots, 100$, with independent and identically distributed normal errors $\varepsilon_i \sim \text{N}(0, 0.5^2)$. This is a classical linear regression model, which complies with the assumptions stated in Box 3.1 on p. 76. We obtain the estimates $\hat{y}_i =$

$-1.03 + 0.98x_{i1} - 0.61x_{i2}$. Figure 3.5c shows the corresponding residuals, which does not reveal any conspicuous behavior. Moreover, we do not see any evidence of autocorrelation. If we estimate a regression model that does not include covariate x_2, we obtain $\hat{y}_i = -1.01 + 0.95 \cdot x_{i1}$. The corresponding time series of residuals is presented in Fig. 3.5d. Even though covariate x_2 is missing, there does not appear to be any anomalies in the fit. The estimated regression coefficients are close to their true values, and the residual variation over time appears to be random. In case that we neglect to model the variable x_1 fluctuating about some trend, we obtain the estimates $\hat{y}_i = -2.22 - 0.60 \cdot x_{i2}$. Figure 3.5e shows the residuals, which now suggest presence of autocorrelation. Why are the residuals rather inconspicuous when x_2 is not considered and why do the residuals show autocorrelation when neglecting x_1? The reason is that x_1 shows a distinct temporal trend, whereas x_2 does not. We can explain these relationships as follows: The omitted effects $\beta_1 x_1$ or $\beta_2 x_2$ can be absorbed into the error term. If x_1 (x_2) is omitted in the model, we can denote the corresponding errors as $\tilde{\varepsilon} = \beta_1 x_1 + \varepsilon$ ($\tilde{\varepsilon} = \beta_2 x_2 + \varepsilon$). Since x_2 essentially shows no time trend, the residuals $\tilde{\varepsilon}$ are still uncorrelated. However, if the effect of x_1 is absorbed into $\tilde{\varepsilon}$, the trend in x_1 is reflected in the errors and autocorrelated residuals are obtained.

Analogous to heteroscedastic errors, we also have to examine the consequences of ignoring correlated errors. Section 4.1 (p. 177) of the next chapter will discuss this in detail including estimation methods in the presence of correlated errors. Intuitively, it is clear that ignorance regarding correlation implies loss of information. Suppose we are interested in predicting the response based on a new observation $\boldsymbol{x}_{n+1}$ of the covariates at time $n+1$. An obvious estimator in this situation is given by $\hat{y}_{n+1} = \boldsymbol{x}'_{n+1}\hat{\boldsymbol{\beta}}$. For correlated errors using this estimator results in a loss of information. If, for instance, errors are positively correlated, then positive residuals are more likely to be followed by other positive residuals. That is, if the residual ε_n at time n is positive, the residual ε_{n+1} in the next time period is likely positive again. The stronger the correlation, the less $\hat{\varepsilon}_n$ differs from $\hat{\varepsilon}_{n+1}$. Thus, it makes sense to predict a *higher* value for y_{n+1} than $\boldsymbol{x}'_{n+1}\hat{\boldsymbol{\beta}}$. On the other hand we should predict a *smaller* value than $\boldsymbol{x}'_{n+1}\hat{\boldsymbol{\beta}}$ if the residual $\hat{\varepsilon}_n$ is negative. Section 4.1 (p. 177) will show that $\hat{y}_{n+1} = \boldsymbol{x}'_{n+1}\hat{\boldsymbol{\beta}} + \hat{\rho}\hat{\varepsilon}_n$ represents an optimal prediction of y_{n+1}, where $\hat{\rho}$ is the (empirical) correlation coefficient between $\hat{\varepsilon}_i$ and $\hat{\varepsilon}_{i-1}$.

Additivity of Errors

In principle, many different models for the structure of the errors are conceivable. In the vast majority of cases, two alternative error structures are assumed: additive errors and multiplicative errors. An example for multiplicative errors is the exponential model

$$\begin{aligned} y_i &= \exp(\beta_0 + \beta_1 x_{i1} + \ldots + \beta_k x_{ik} + \varepsilon_i) \\ &= \exp(\beta_0)\exp(\beta_1 x_{i1}) \cdot \ldots \cdot \exp(\beta_k x_{ik})\exp(\varepsilon_i), \end{aligned} \tag{3.2}$$

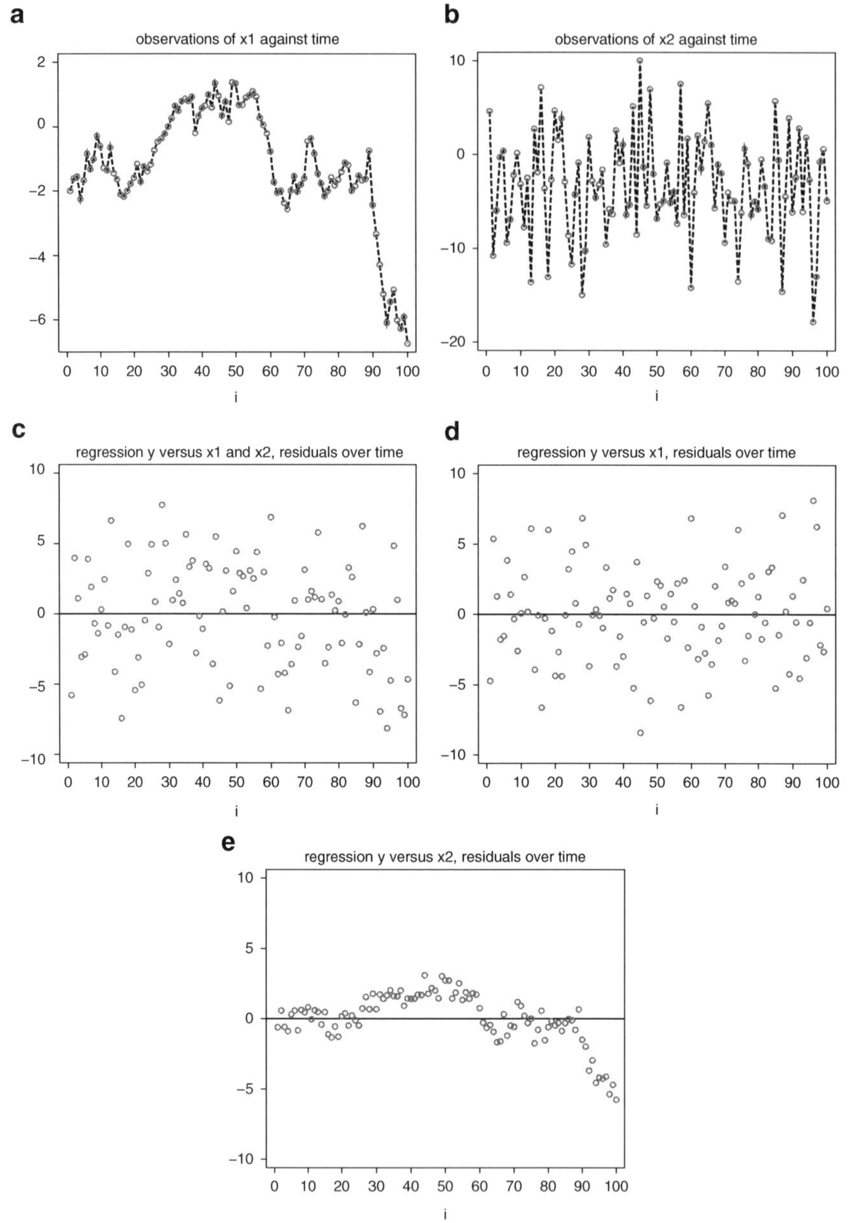

Fig. 3.5 Illustration for autocorrelated errors if relevant covariates showing a temporal trend are ignored. Panels (**a**) and (**b**) show the covariates over time. Panels (**c–e**) display the residuals for the regression models $y_i = \beta_0 + \beta_1 x_{i1} + \beta_2 x_{i2} + \varepsilon_i$ (correct model), $y_i = \beta_0 + \beta_1 x_{i1} + \varepsilon_i$ (x_2 ignored), and $y_i = \beta_0 + \beta_1 x_{i2} + \varepsilon_i$ (x_1 ignored)

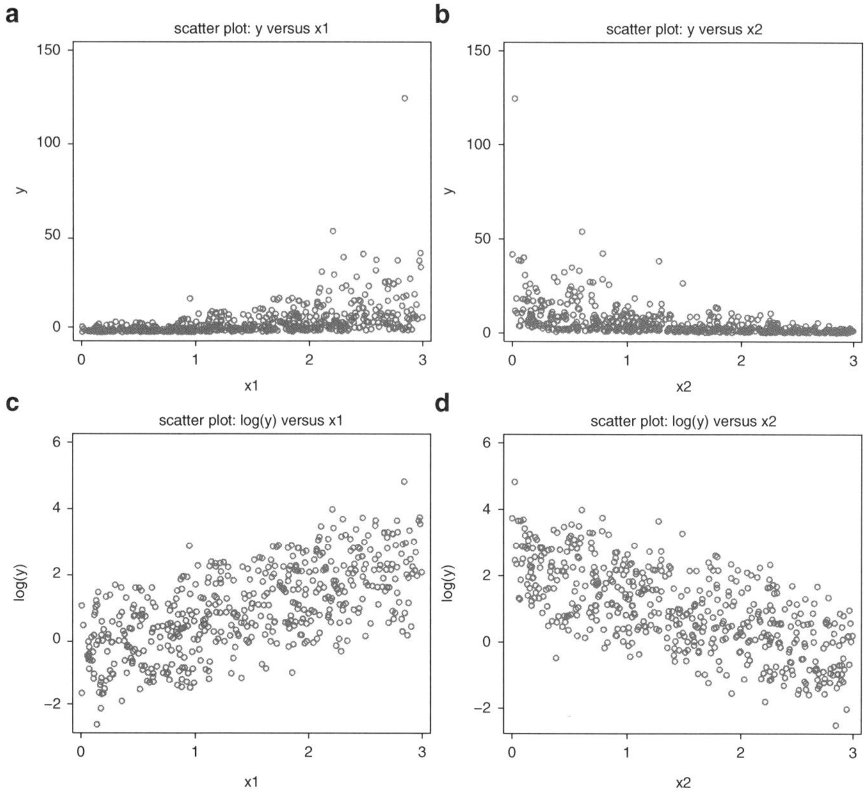

Fig. 3.6 Example for a multiplicative model: Panels (**a**) and (**b**) show scatter plots between simulated data y and x_1, respectively, x_2 based on the model $y_i = \exp(1 + x_{i1} - x_{i2} + \varepsilon_i)$ with $\varepsilon_i \sim \mathrm{N}(0, 0.4^2)$. Panels (**c**) and (**d**) display scatter plots of $\log(y)$ versus x_1 and x_2, respectively

with multiplicative errors $\tilde{\varepsilon}_i = \exp(\varepsilon_i)$. Models with multiplicative error structure are more plausible for exponential relationships since the errors are proportional to the mean value of y. Figure 3.6 shows simulated data with multiplicative errors. Panels (a) and (b) display scatter plots of a response y with two explanatory variables x_1 and x_2. The data are generated according to the model

$$y_i = \exp(1 + x_{i1} - x_{i2} + \varepsilon_i), \tag{3.3}$$

with $\varepsilon_i \sim \mathrm{N}(0, 0.4^2)$. It is difficult to make a statement about the strength and type of the relationship between y and x_1 or x_2 on the basis of the scatter plots. Many popular models with multiplicative errors can be expressed as linear models with additive errors through a simple transformation of variables. A logarithmic transformation of (3.2) results in the linear model

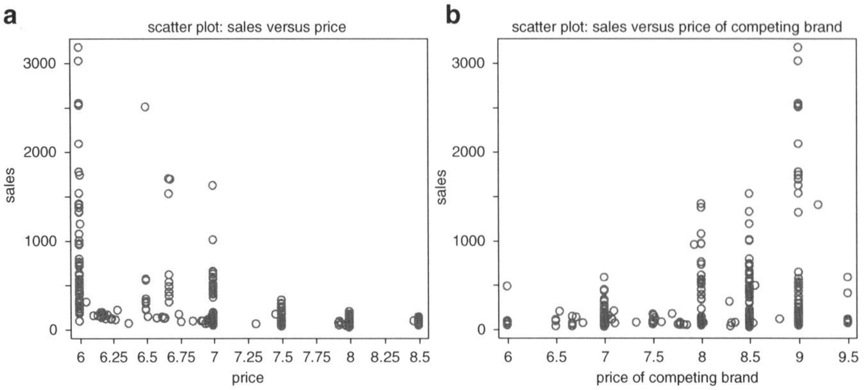

Fig. 3.7 Supermarket scanner data: scatter plot between the sales of a particular brand and its price [panel (**a**)] and the price of a competing brand [panel (**b**)], respectively

$$\log(y_i) = \beta_0 + \beta_1 x_{i1} + \ldots + \beta_k x_{ik} + \varepsilon_i.$$

Hence, we can treat an exponential model within the scope of linear models by taking the logarithm of the response variable. Panels (c) and (d) in Fig. 3.6 show scatter plots between the logarithmic response value $\log(y)$ and the covariates x_1 and x_2 for the simulated model (3.3), which provides clear evidence of linear relationships under transformation. Provided that the errors are normally distributed, the response y is *log-normally distributed* (see Appendix B, p. 641) resulting in

$$\mathrm{E}(y_i) = \exp(\boldsymbol{x}_i'\boldsymbol{\beta} + \sigma^2/2)$$

for the (conditional) mean of y_i.

Example 3.2 Supermarket Scanner Data

Figure 3.7 shows scatter plots for real data that look similar to the simulated data in Fig. 3.6. The graphs show scanner data, which are collected routinely during checkout at supermarkets. Here, we want to examine the dependence of weekly sales of a product (in our case a particular brand of coffee) on its own price and the price of a competing brand, respectively. Panel (a) shows the scatter plot between sales and the brand price; panel (b) shows a similar scatter plot between sales and the price for a competing brand of coffee. Based on the scatter plots alone, it is difficult to verify the hypothesis that a higher price leads to a decrease in sales and a higher price for the competing brand leads to an increase in sales.

△

3.1.3 Modeling the Effects of Covariates

Modeling the effects of covariates has already been explored in Sect. 2.2 (p. 22) and in connection with the discussion of model assumptions (last paragraph of Sect. 3.1.2). This section examines this aspect in more detail.

Continuous Covariates

As mentioned in our discussion of the model assumptions, we can fit nonlinear relationships within the scope of linear models. When dealing with continuous explanatory variables, a nonlinear specification is often necessary, and in this section we examine two simple methods to do so: simple *variable transformations* and *polynomial regression*. Chapter 8 will examine more flexible and automated methods.

Modeling with variable transformation or use of polynomials can be best demonstrated through examples. We use the Munich rent index data for illustration and start by modeling the relationship between net rent and area.

Example 3.3 Munich Rent Index—Variable Transformation

In connection with homo- and heteroscedastic variances in Sect. 3.1.2, a regression line between net rent and area has already been fit (see Fig. 3.2). Here, the assumption of a linear relationship seems to be justified. If we rather use the net rent per square meter as the response variable, we obtain the fit $\widehat{rentsqm}_i = 9.47 - 0.035\,area_i$. Figure 3.8, panel (a) shows the scatter plot between net rent per square meter and living area, along with the fitted regression line. Panels (c) and (e) show the corresponding residuals and average residuals for every square meter.

We find that the residuals are mostly positive for smaller apartments ($area < 30$) indicating a nonlinear relationship between net rent per square meter and area. Variable transformations offer one possibility to specify such a nonlinear relationship. To do so, we consider the regression model

$$rentsqm_i = \beta_0 + \beta_1 \cdot f(area_i) + \varepsilon_i, \tag{3.4}$$

where f is an arbitrary function. The function f must be appropriately specified, i.e., f itself is fixed in advance and not estimated. Obviously, this model is a generalization of the linear model $rentsqm_i = \beta_0 + \beta_1 area_i + \varepsilon_i$. By defining the variable $x_i = f(area_i)$, we again obtain a linear model $rentsqm_i = \beta_0 + \beta_1 x_i + \varepsilon_i$ as a special case. A transformation that seems to be appropriate for modeling the nonlinear relationship between *rentsqm* and *area* is given by $f(area_i) = 1/area_i$. We also chose this transformation in Example 2.2, but using only a subset of the data (apartments built after 1966 in average location). The least squares estimates for the entire data set result in

$$\widehat{rentsqm}_i = 4.73 + 140.18 \cdot f(area_i) = 4.73 + 140.18 \cdot 1/area_i.$$

The design matrix for this model is

$$X = \begin{pmatrix} 1 & 1/30 \\ 1 & 1/37 \\ \vdots & \vdots \\ 1 & 1/73 \\ 1 & 1/73 \end{pmatrix}.$$

The second column of the design matrix contains observations $x_1 = f(area_1) = 1/area_1 = 1/30$, $x_2 = 1/37, \ldots$ of the variable x, which represents the transformed *area* variable.

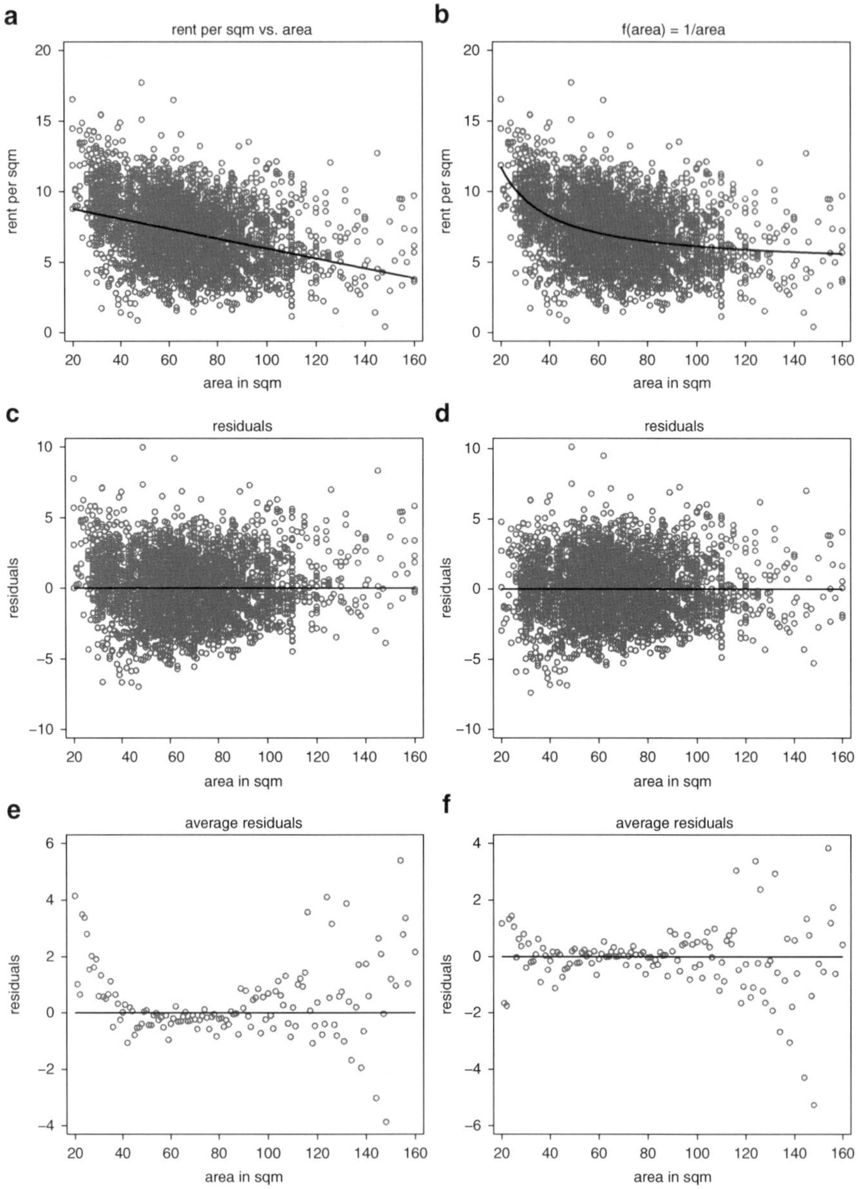

Fig. 3.8 Munich rent index: illustration for modeling nonlinear relationships via variable transformation. The *left* column shows the estimated regression line including observations [panel (**a**)], corresponding residuals [panel (**b**)], and average residuals for every distinct covariate value [panel (**c**)]. The *right* column displays the estimated nonlinear relationship $\widehat{rentsqm}_i = 4.73 + 140.18 \cdot 1/area_i$ [panel (**d**)] and the corresponding residual plots [panels (**e**) and (**f**)]

The interpretation of nonlinear relationships is best done by visualizing the estimated curves. Figure 3.8b plots the estimated relationship $\widehat{rentsqm} = 4.73 + 140.18 \cdot 1/area$ as a function of living area together with the observations $(rentsqm_i, area_i)$. Note that the curve represents the (conditional) mean of the response variable *rentsqm* conditional on the living area. Overall the average net rent per square meter decreases with an increase of living area. The decrease of net rent is highest for small apartments up to about 40 to 45 square meters, then the curve levels off. Apartments of about 100 square meters or more have almost the same average net rent per square meter. When looking at the residuals $\hat{\varepsilon}_i = rentsqm_i - 4.73 - 140.18 \cdot 1/area_i$ in Fig. 3.8d or the average residuals in Fig. 3.8f, we can see that the chosen nonlinear curve fits the data well. Of course, transformations other than $f(area_i) = 1/area_i$ are possible, e.g., $f(area_i) = \log(area_i)$. In many cases different transformations lead to very similar fits. Hence, we do not have a panacea when choosing an appropriate transformation. We depend on the visual inspection of scatter plots or residuals in order to find reasonable and suitable variable transformations.

△

Polynomial fitting offers another simple way to specify nonlinear relationships. In this case, we approximate the effect of a continuous covariate with a polynomial. We illustrate polynomial modeling again using the data of the Munich rent index.

Example 3.4 Munich Rent Index—Polynomial Regression

Assuming a second-order polynomial for the effect of area, we obtain the following regression model:

$$rentsqm_i = \beta_0 + \beta_1 \cdot area_i + \beta_2 \cdot area_i^2 + \varepsilon_i.$$

A polynomial of third degree leads to

$$rentsqm_i = \beta_0 + \beta_1 \cdot area_i + \beta_2 \cdot area_i^2 + \beta_3 \cdot area_i^3 + \varepsilon_i.$$

Typically, polynomials of a low degree are preferred. We rarely use polynomials of degree higher than three, as the resulting estimates become very unstable and exhibit high variability particularly at the borders of the covariate domain. Similar to using variable transformation, the nonlinear models can be connected to linear models. By defining new variables $x_{i1} = area_i$, $x_{i2} = area_i^2$ and $x_{i3} = area_i^3$, we obtain

$$rentsqm_i = \beta_0 + \beta_1 \cdot x_{i1} + \beta_2 \cdot x_{i2} + \varepsilon_i,$$

when using a polynomial of degree 2 and

$$rentsqm_i = \beta_0 + \beta_1 \cdot x_{i1} + \beta_2 \cdot x_{i2} + \beta_3 \cdot x_{i3} + \varepsilon_i$$

when using a polynomial of degree 3. The corresponding design matrices are defined, respectively, by

$$X = \begin{pmatrix} 1 & 30 & 30^2 \\ 1 & 37 & 37^2 \\ \vdots & \vdots & \vdots \\ 1 & 73 & 73^2 \\ 1 & 73 & 73^2 \end{pmatrix} = \begin{pmatrix} 1 & 30 & 900 \\ 1 & 37 & 1369 \\ \vdots & \vdots & \vdots \\ 1 & 73 & 5329 \\ 1 & 73 & 5329 \end{pmatrix} \qquad (3.5)$$

and

$$X = \begin{pmatrix} 1 & 30 & 30^2 & 30^3 \\ 1 & 37 & 37^2 & 37^3 \\ \vdots & \vdots & \vdots & \vdots \\ 1 & 73 & 73^2 & 73^3 \\ 1 & 73 & 73^2 & 73^3 \end{pmatrix} = \begin{pmatrix} 1 & 30 & 900 & 27000 \\ 1 & 37 & 1369 & 50653 \\ \vdots & \vdots & \vdots & \vdots \\ 1 & 73 & 5329 & 389017 \\ 1 & 73 & 5329 & 389017 \end{pmatrix}.$$

The estimated curves are given by

$$\begin{aligned} \widehat{rentsqm}_i &= 11.83 - 0.106 \cdot x_{i1} + 0.00047 \cdot x_{i2} \\ &= 11.83 - 0.106 \cdot area_i + 0.00047 \cdot area_i^2, \end{aligned}$$

when using a polynomial of degree 2 and

$$\begin{aligned} \widehat{rentsqm}_i &= 14.28 - 0.217 \cdot x_{i1} + 0.002 \cdot x_{i2} - 0.000006 \cdot x_{i3} \\ &= 14.28 - 0.217 \cdot area_i + 0.002 \cdot area_i^2 - 0.000006 \cdot area_i^3 \end{aligned}$$

for the polynomial of degree 3. To interpret results, we again visualize the estimated effects. Figure 3.9, panel (a) shows the estimated curve between net rent per square meter and living area for a second-degree polynomial. Panel (c) illustrates the corresponding residuals associated with this model. Panels (b) and (d) show the fit when using a third-degree polynomial. We obtain quite similar results compared to variable transformation $f(area_i) = 1/area_i$. Using visualization, we cannot decide as to which of the models is more appropriate. Here, we need more formal techniques for variable selection and model choice; see Sect. 3.4.

△

Thus far we have only considered one variable. The extension to more covariates is straightforward, as the next example demonstrates.

Example 3.5 Munich Rent Index—Additive Models

It is straightforward to include additional covariates with a linear effect into the model. For instance, we can extend the model (3.4) by adding the apartment's year of construction. We obtain

$$rentsqm_i = \beta_0 + \beta_1 \cdot f(area_i) + \beta_2 \cdot yearc_i + \varepsilon_i. \tag{3.6}$$

When using the transformation $f(area) = 1/area$, we obtain the fit

$$\widehat{rentsqm}_i = -65.41 + 119.36 \cdot 1/area_i + 0.036 \cdot yearc_i.$$

Alternatively, we can also estimate the effect of the year of construction nonlinearly. A model using a polynomial of degree 3 is given by

$$rentsqm_i = \beta_0 + \beta_1 \cdot 1/area_i + \beta_2 \cdot yearc_i + \beta_3 \cdot yearc_i^2 + \beta_4 \cdot yearc_i^3 + \varepsilon_i. \tag{3.7}$$

The corresponding fit is

$$\begin{aligned} \widehat{rentsqm}_i = {} & 29113.6 + 129.57 \cdot 1/area_i - 42.835 \cdot yearc_i + 0.020949 \cdot yearc_i^2 \\ & -0.00000340 \cdot yearc_i^3. \end{aligned}$$

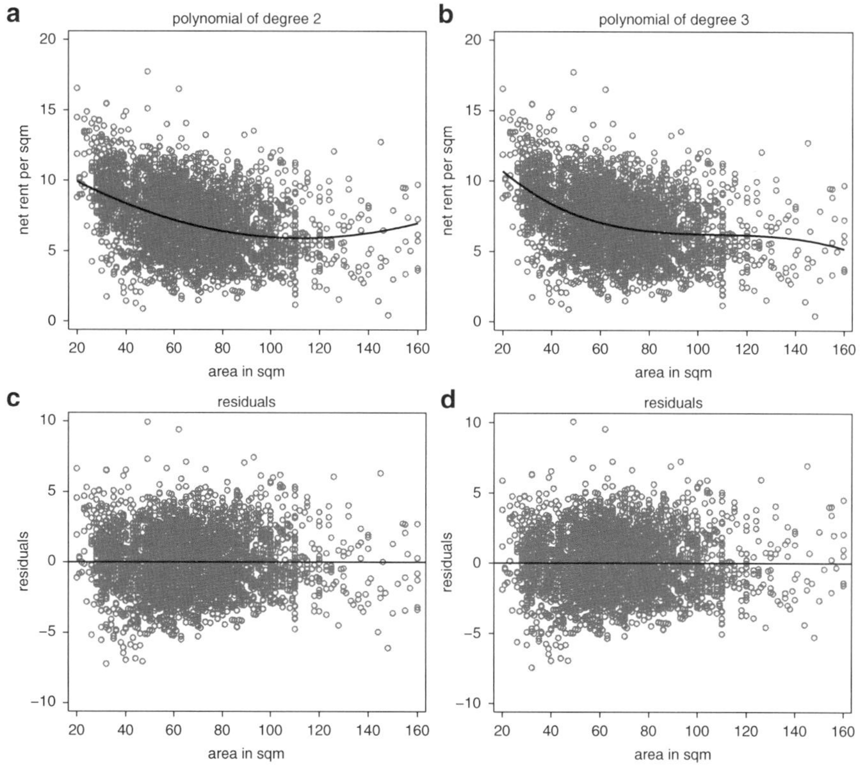

Fig. 3.9 Munich rent index: illustration for modeling nonlinear relationships using polynomials. The *upper panels* show fitted quadratic and cubic polynomials including observations. The *lower panels* display the corresponding residuals

The results are obtained with STATA. Note that with other statistics packages we sometimes obtained slightly different results due to rounding errors.

Combining the effects of the living area and the year of construction in the functions

$$f_1(area) = \beta_1 \cdot 1/area$$

and

$$f_2(yearc) = \beta_2 \cdot yearc + \beta_3 \cdot yearc^2 + \beta_4 \cdot yearc^3$$

we obtain the model

$$rentsqm_i = \beta_0 + f_1(area_i) + f_2(yearc_i) + \varepsilon_i.$$

The estimated functions are

$$\hat{f}_1(area) = 129.57 \cdot 1/area$$

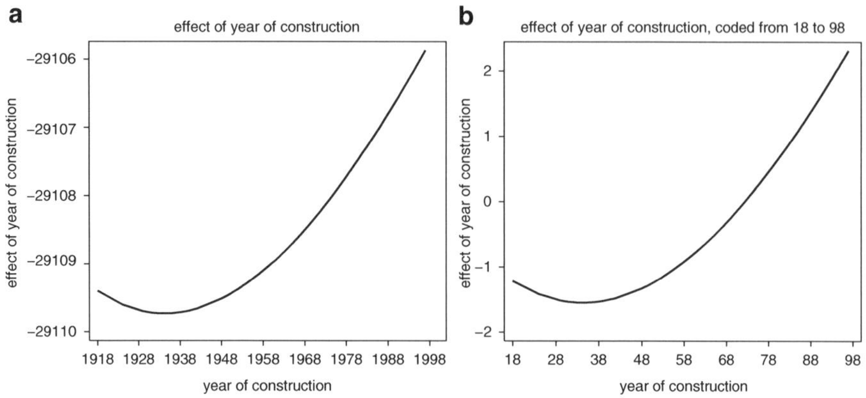

Fig. 3.10 Munich rent index: plots of the estimated nonlinear effect $\hat{f}_2(yearc)$ of the year of construction

and

$$\hat{f}_2(yearc) = -42.835 \cdot yearc + 0.020949 \cdot yearc^2 - 0.00000340 \cdot yearc^3,$$

respectively. This is a special case of an additive model as in Chap. 2, Sect. 2.6.

The interpretation of the effects can again be achieved most successfully through visualization. We plot the estimated effects $\hat{f}_1(area)$ and $\hat{f}_2(yearc)$ for living area or year of construction. Figure 3.10a provides a visualization of the effect of year of construction. We first notice that the range of values on the vertical axis is approximately between $-29,110$ and $-29,106$. The reason for the unusual range is the scale of the year of construction, having values between 1918 and 1998. If we rather specify the year of construction with the values $18,\ldots,98$ (subtracting 1,900 of the original values), we obtain

$$\widehat{rentsqm}_i = 5.42 + 129.57 \cdot 1/area_i - 0.094 \cdot yearc_i + 0.0016 \cdot yearc_i^2 - 0.0000034 \cdot yearc_i^3.$$

Looking at the parameter estimates, it appears that the redefined effect of the year of construction differs clearly from the original effect. However, Fig. 3.10b shows that the effect is only shifted vertically, otherwise remaining the same. In fact, the level of the nonlinear function is not identifiable (as already pointed out in Sect. 2.6). If we add any constant a to $\hat{f}_2$ and subtract the same constant from $\hat{\beta}_0$, the estimated rent per square meter $\widehat{rentsqm}$ remains unchanged. This implies that the level of the nonlinear function can be arbitrarily changed, e.g., by transforming the explanatory variable.

For the sake of better interpretability, it is often useful to require that all functions "are centered around zero." This condition is automatically fulfilled if all covariates are centered around zero. In case of the living area, we would replace the variable $areainv = 1/area$ by the centered version $areainvc = areainv - \overline{areainv}$, where $\overline{areainv}$ is the arithmetic mean of the variable $areainv$. We could do the same with the year of construction by replacing the variables $yearc$, $yearc2 = yearc^2$ and $yearc3 = yearc^3$ by $yearcc = yearc - \overline{yearc}$, $yearc2c = yearc2 - \overline{yearc2}$ and $yearc3c = yearc3 - \overline{yearc3}$.

In case of polynomial modeling, we can go a step further and use the so-called *orthogonal polynomials*, where we replace the usual basis $yearc, yearc^2, yearc^3$ for representing polynomials with an orthogonal polynomial basis. This implies that the columns of the design matrix corresponding to the variables $yearc$, $yearc^2$, and $yearc^3$ are centered and

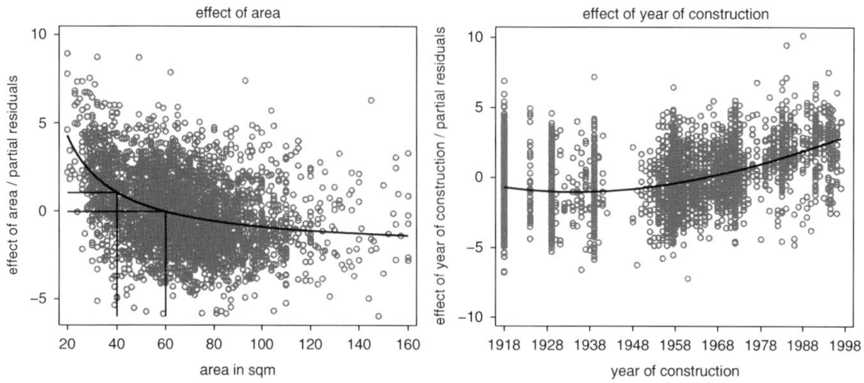

Fig. 3.11 Munich rent index: plots of the nonlinear and centered effects of living area and year of construction including partial residuals

orthogonal. Apart from the fact that the corresponding covariate effects are centered, orthogonal polynomials have other desirable properties, such as more stable computation of the least squares estimators. Typically, orthogonal polynomials are computed automatically using built-in functions provided by most statistical packages (e.g., in STATA the function `orthpol`). In Example 3.9 (p. 111), however, we show how to compute orthogonal polynomials using properties of the least squares method.

Using the orthogonal variables *yearco*, *yearco2*, *yearco3*, and the centered variable *areainvc*, we obtain the model fit

$$\widehat{rentsqm}_i = 7.11 + 129.57 \cdot areainvc_i + 0.79 \cdot yearco_i + 0.50 \cdot yearco2_i - 0.032 \cdot yearco3_i.$$

Figure 3.11 displays the corresponding effects $\hat{f}_1(area)$ and $\hat{f}_2(yearc)$ together with the partial residuals $\hat{\varepsilon}_{area,i}$ and $\hat{\varepsilon}_{yearc,i}$, which are defined by

$$\hat{\varepsilon}_{area,i} = rentsqm_i - \hat{\beta}_0 - \hat{f}_2(yearc_i) = \hat{\varepsilon}_i + \hat{f}_1(area_i)$$

and

$$\hat{\varepsilon}_{yearc,i} = rentsqm_i - \hat{\beta}_0 - \hat{f}_1(area_i) = \hat{\varepsilon}_i + \hat{f}_2(yearc_i),$$

respectively. The partial residuals for area, $\hat{\varepsilon}_{area,i}$, account for the remaining variation due to area, where all of the other effects are eliminated (in this case only the year of construction). Analogously, in the partial residuals for year of construction, $\hat{\varepsilon}_{yearc,i}$, the effect of the area, but not that of year of construction, is eliminated. By plotting the partial residuals in Fig. 3.11, we are able to visually assess whether or not the nonlinear fit is adequate.

We can interpret the effects of the living area and the year of construction accordingly: The plotted area effect specifies the influence of the living area on the average net rent per square meter, if the other covariates (in our case the year of construction) are held constant. For apartments with living area of 40 and 60 square meters, we obtain, for example, the effects $\hat{f}_1(40) \approx 1$ and $\hat{f}_1(60) \approx 0$. This implies that on average 60-square-meter-sized apartments are 1 Euro less expensive than apartments of size 40 square meters, assuming that both apartments were constructed in the same year. If we compare 60- and 80-square-meter-sized apartments, the difference in price is only 0.5 Euro. Even though in both examples we have a difference of 20 square meters of living space, the difference is cut in half in relation to the average net rent. The reason for this can be seen through the clear

3.2 Modeling Nonlinear Covariate Effects Through Variable Transformation

If the continuous covariate z has an approximately nonlinear effect $\beta_1 f(z)$ with known transformation f, then the model

$$y_i = \beta_0 + \beta_1 f(z_i) + \ldots + \varepsilon_i$$

can be transformed into the linear regression model

$$y_i = \beta_0 + \beta_1 x_i + \ldots + \varepsilon_i,$$

where $x_i = f(z_i) - \bar{f}$. By subtracting

$$\bar{f} = \frac{1}{n} \sum_{i=1}^{n} f(z_i),$$

the estimated effect $\hat{\beta}_1 x$ is automatically centered around zero. The estimated curve is best interpreted by plotting $\hat{\beta}_1 x$ against z (instead of x).

nonlinear effect, which becomes nearly constant for large values of the living area. We can interpret the effect of the year of construction in a similar fashion. Apartments that were constructed prior to World War II show roughly the same price (for the same living area). Apartments that were constructed after 1945 show an approximately linear price increase. △

Categorical Covariates

To this point, we have discussed modeling the effect of continuous covariates. In this section, we will discuss categorical covariates and their characteristics. We will illustrate the methodology with the help of the Munich rent index data. More specifically, we will discuss appropriate modeling of the variable *location* with the three categories, 1 = average location, 2 = good location, and 3 = top location. In a first (naive) attempt we treat *location* as if it were continuous and obtain the model

$$rentsqm_i = \beta_0 + \beta_1 \cdot location_i + \varepsilon_i.$$

For simplicity, we have omitted all other covariates in this illustrative model. Using the least squares method, we obtain the fit

$$\widehat{rentsqm}_i = 6.54 + 0.39 \cdot location.$$

3.3 Modeling Nonlinear Covariate Effects Through Polynomials

If the continuous covariate z has an approximately polynomial effect $\beta_1 z + \beta_2 z^2 + \ldots + \beta_l z^l$ of degree l, then the model

$$y_i = \beta_0 + \beta_1 z_i + \beta_2 z_i^2 + \ldots + \beta_l z_i^l + \ldots + \varepsilon_i$$

can be transformed into the linear regression model

$$y_i = \beta_0 + \beta_1 x_{i1} + \beta_1 x_{i2} + \ldots + \beta_l x_{il} + \ldots + \varepsilon_i,$$

where $x_{i1} = z_i$, $x_{i2} = z_i^2, \ldots, x_{il} = z_i^l$.
The centering (and possibly orthogonalization) of the vectors $\boldsymbol{x}^j = (x_{1j}, \ldots, x_{nj})'$, $j = 1, \ldots, l$, to $\boldsymbol{x}^1 - \bar{\boldsymbol{x}}_1, \ldots, \boldsymbol{x}^l - \bar{\boldsymbol{x}}_l$ with the mean vector $\bar{\boldsymbol{x}}_j = (\bar{x}_j, \ldots, \bar{x}_j)'$ facilitates interpretation of the estimated effects. A graphical illustration of the estimated polynomial is a useful way to interpret the estimated effect of z.

Due to the chosen coding, the effect of a good location would be twice as high as it would be for an average location (0.39 Euro versus $2 \cdot 0.39 = 0.78$ Euro). The effect for top location would be three times as high (0.39 Euro versus $3 \cdot 0.39 = 1.17$ Euro). If we coded the location with 2 = average location, 6 = good location, and 8 = top location, apartments in a good location would be three times as expensive as apartments in an average location; apartments in a top location would be four times as expensive as apartments in an average location. This shows that the results are dependent on the arbitrarily chosen coding of the categorical covariate. The problem is that we cannot interpret the distances between the categories in a reasonable way. A good location is not twice as good (or three times as good in the second coding) as an average location. A remedy is to define new covariates, so-called *dummy variables*, and estimate a separate effect for each category of the original covariate. In the case of location, we define the following three dummy variables:

$$alocation = \begin{cases} 1 & location = 1 \text{ (average location)}, \\ 0 & \text{otherwise}, \end{cases}$$

$$glocation = \begin{cases} 1 & location = 2 \text{ (good location)}, \\ 0 & \text{otherwise}, \end{cases}$$

$$tlocation = \begin{cases} 1 & location = 3 \text{ (top location)}. \\ 0 & \text{otherwise}, \end{cases}$$

We obtain the model

$$rentsqm_i = \beta_0 + \beta_1 \cdot alocation_i + \beta_2 \cdot glocation_i + \beta_3 \cdot tlocation_i + \varepsilon_i .$$

If we look more closely, we find another difficulty when taking this approach: The regression parameters are not identifiable. To fully understand the identification problem, we inspect the definition of the location effects. For an average location ($alocation_i = 1$, $glocation_i = 0$, $tlocation_i = 0$), we obtain the effect $\beta_0 + \beta_1$. For good or top locations the effects result in $\beta_0 + \beta_2$ and $\beta_0 + \beta_3$, respectively. If we now add an arbitrary value a to the intercept β_0 and, at the same time, subtract a from the coefficients β_1, β_2, and β_3, we obtain the same total effects. Hence we cannot uniquely determine the regression parameters. Even with the help of the least squares method, we would not obtain unique estimators. This identification problem can be resolved in one of two ways: Either we use a model without intercept, or we remove one of the dummy variables *alocation*, *glocation*, or *tlocation* from the model. Since an intercept is usually included in the model, we restrict ourselves to the second option. If we do not consider the variable *glocation*, we obtain the model

$$rentsqm_i = \beta_0 + \beta_1 \cdot alocation_i + \beta_2 \cdot tlocation_i + \varepsilon_i .$$

The effects are now given by $\beta_0 + \beta_1$ for an average location, β_0 for good location, and $\beta_0 + \beta_2$ for top location. The regression parameters are uniquely determined and are thus identifiable. Using the method of least squares, we obtain the fit

$$\widehat{rentsqm}_i = 7.27 - 0.31 \cdot alocation_i + 0.90 \cdot tlocation_i .$$

Care must be taken when interpreting the estimated regression coefficients. They have to be interpreted with respect to the omitted category, in our example apartments in good location. We obtain an estimated effect of $\hat{\beta}_0 + \hat{\beta}_1 = 7.27 - 0.31 = 6.96$ for average location, $\hat{\beta}_0 = 7.27$ for good location, and $\hat{\beta}_0 + \hat{\beta}_2 = 7.27 + 0.90 = 8.17$ for the top location. *Compared to* apartments in good location, apartments in average location are 0.31 Euro less expensive and apartments in top location are 0.90 Euro more expensive. The category of good location omitted in the model and used for comparison is also called the *reference category*. Note that the interpretation of the estimated regression coefficients depends on the chosen reference category. However, all models are equivalent, which means that the estimated regression coefficients for a particular reference category can be computed from the regression coefficients of the original model. For example, if we use the average location as reference category, we obtain the fitted model

$$\widehat{rentsqm}_i = 6.96 + 0.31 \cdot glocation_i + 1.21 \cdot tlocation_i .$$

In comparison to apartments in average location, apartments in good location are 0.31 Euro more expensive and apartments in top location are 1.21 Euro more expensive. The results are fully consistent with the previous coding.

3.4 Dummy Coding for Categorical Covariates

For modeling the effect of a covariate $x \in \{1, \ldots, c\}$ with c categories using dummy coding, we define the $c - 1$ dummy variables

$$x_{i1} = \begin{cases} 1 & x_i = 1, \\ 0 & \text{otherwise}, \end{cases} \quad \ldots \quad x_{i,c-1} = \begin{cases} 1 & x_i = c - 1, \\ 0 & \text{otherwise}, \end{cases}$$

for $i = 1, \ldots, n$, and include them as explanatory variables in the regression model

$$y_i = \beta_0 + \beta_1 x_{i1} + \ldots + \beta_{i,c-1} x_{i,c-1} + \ldots + \varepsilon_i.$$

For reasons of identifiability, we omit one of the dummy variables, in this case the dummy variable for category c. This category is called reference category. The estimated effects can be interpreted by direct comparison with the (omitted) reference category.

In principle, any of the categories of a categorical variable could be chosen as the reference. In practice, we usually choose the category which makes most sense for interpretation, for example, the most common category found in the data set. For an arbitrary categorical covariate x with c categories, dummy coding is summarized in Box 3.4.

Note that there is more than one coding scheme for categorical covariates. Another popular scheme is *effect coding*, which is defined by

$$x_{i1} = \begin{cases} 1 & x_i = 1, \\ -1 & x_i = c, \\ 0 & \text{otherwise}, \end{cases} \quad \ldots \quad x_{i,c-1} = \begin{cases} 1 & x_i = c - 1, \\ -1 & x_i = c, \\ 0 & \text{otherwise}. \end{cases}$$

In contrast to dummy coding, effect coding produces new variables that are coded with -1 for the reference category yielding a sum to zero constraint as explained in the following example that illustrates the difference between the two coding schemes, using the Munich rent index data.

Example 3.6 Munich Rent Index—Effect Coding

If we choose an average location as the reference category, we obtain the two variables

$$glocation = \begin{cases} 1 & location = 2 \text{ (good location)}, \\ -1 & location = 1 \text{ (average location)}, \\ 0 & \text{otherwise}, \end{cases}$$

$$tlocation = \begin{cases} 1 & location = 3 \text{ (top location)}, \\ -1 & location = 1 \text{ (average location)}, \\ 0 & \text{otherwise}, \end{cases}$$

in effect coding. For the regression model

$$rentsqm_i = \beta_0 + \beta_1 \cdot glocation_i + \beta_2 \cdot tlocation_i + \varepsilon_i,$$

we obtain the fitted model

$$\widehat{rentsqm}_i = 7.47 - 0.19 \cdot glocation_i + 0.71 \cdot tlocation_i.$$

Due to the specific coding, we are able to compute an additional regression coefficient associated with the reference category. It is obtained as the negative sum of the other parameters, i.e., $\hat{\beta}_3 = -\hat{\beta}_1 - \hat{\beta}_2 = 0.19 - 0.71 = -0.52$. This results in estimated effects $\hat{\beta}_0 + \hat{\beta}_3 = 7.47 - 0.52 = 6.95$, $\hat{\beta}_0 + \hat{\beta}_1 = 7.47 - 0.19 = 7.28$, and $\hat{\beta}_0 + \hat{\beta}_2 = 7.47 + 0.71 = 8.18$ for average, good, and top locations, respectively. △

Interactions Between Covariates

An interaction between covariates exists if the effect of a covariate depends on the value of at least one other covariate. To start with, consider the following simple model:

$$y = \beta_0 + \beta_1 x + \beta_2 z + \beta_3 x z + \varepsilon \tag{3.8}$$

between a response y and two other explanatory variables x and z. The term $\beta_3 x z$ is called an *interaction* between x and z. The terms $\beta_1 x$ and $\beta_2 z$ depend on only one variable and are called *main effects*. We can understand the impact of an interaction term when considering how $\mathrm{E}(y)$ changes when one variable, e.g., x, changes by an amount d. We have

$$\begin{aligned} \mathrm{E}(y \mid x + d, z) - \mathrm{E}(y \mid x, z) &= \beta_0 + \beta_1 (x + d) + \beta_2 z + \beta_3 (x + d) z \\ &\quad - \beta_0 - \beta_1 x - \beta_2 z - \beta_3 x z \\ &= \beta_1 d + \beta_3 d z. \end{aligned}$$

Now we can distinguish between the two cases $\beta_3 = 0$ and $\beta_3 \neq 0$:

- In the case $\beta_3 = 0$, the interaction is dropped from the model and the expected change $\beta_1 d$ is *independent* from the specific value of the second covariate z.
- In the case $\beta_3 \neq 0$, the expected change $\beta_1 d + \beta_3 d z$ does not only depend on the amount d but also on the value of the covariate z.

Therefore, an interaction is always needed if the effect of the change of a variable is also dependent on the value of another variable. The specification of interactions is dependent on the type of the variables involved. In the sections to follow, we discuss interactions between two categorical variables, i.e. between a continuous and a categorical variable, as well as between two continuous variables.

Interactions Between Categorical Variables

We first consider the simplest case, the interaction between two binary variables x and z. In context of the rent index example, the value $x = 1$ or $z = 1$ could imply the existence of, e.g., a premium kitchen or bathroom. In this case, the coefficient β_1

in Eq. (3.8) measures the effect of a premium kitchen on the net rent, β_2 measures the effect of a premium bathroom, and β_3 measures the additional effect of having both a premium kitchen and a premium bathroom. The interpretation depends on the values of the coefficients. In our example, all three coefficients are most likely positive, leading to the following interpretation: The existence of either a premium kitchen ($x = 1$) or a premium bathroom ($z = 1$) will increase the average net rent. If the apartment has both a premium kitchen and a bathroom, there is an additional rent increase.

Our example shows that the interaction between two binary variables quantifies the effect of both characteristics associated with x and z occurring simultaneously. In the case of two arbitrary categorical variables x and z with c and m categories, respectively, modeling of interactions is more complicated. We next illustrate the situation of two categorical covariates, each with three categories: Define the dummy variables x_1, x_2 that correspond to x and the dummy variables z_1, z_2 corresponding to z. We choose the last category as the reference for both x and z. For modeling the interaction effect we have to consider all possible combinations of the values of x and z (with the exception of the reference categories), specifically x_1z_1, x_1z_2, x_2z_1, and x_2z_2. We now obtain

$$y = \beta_0 + \beta_1 x_1 + \beta_2 x_2 + \beta_3 z_1 + \beta_4 z_2 + \beta_5 x_1 z_1 + \beta_6 x_1 z_2 + \beta_7 x_2 z_1 + \beta_8 x_2 z_2 + \varepsilon.$$

The coefficients can be interpreted as follows:

β_0: effect if $x = 3$ (reference) and $z = 3$ (reference),
$\beta_0 + \beta_1$: effect if $x = 1$ and $z = 3$ (reference),
$\beta_0 + \beta_2$: effect if $x = 2$ and $z = 3$ (reference),
$\beta_0 + \beta_3$: effect if $x = 3$ (reference) and $z = 1$,
$\beta_0 + \beta_4$: effect if $x = 3$ (reference) and $z = 2$,
$\beta_0 + \beta_1 + \beta_3 + \beta_5$:effect if $x = 1$ and $z = 1$,
$\beta_0 + \beta_1 + \beta_4 + \beta_6$:effect if $x = 1$ and $z = 2$,
$\beta_0 + \beta_2 + \beta_3 + \beta_7$:effect if $x = 2$ and $z = 1$,
$\beta_0 + \beta_2 + \beta_4 + \beta_8$:effect if $x = 2$ and $z = 2$.

As is common with categorical covariates, the effects have to be interpreted as difference relative to the reference categories $x = 3$ and $z = 3$. The quantity $\beta_2 + \beta_4 + \beta_8$, e.g., measures the effect of the combination $x = 2$, $z = 2$ when compared to the combination $x = 3$, $z = 3$. Note that the interaction between x and z can also be modeled by defining the new categorical variable w, whose categories consist of all possible combinations of the values of x and z:

$$w = \begin{cases} 1 & x = 1,\ z = 1, \\ 2 & x = 1,\ z = 2, \\ 3 & x = 1,\ z = 3, \\ \vdots & \vdots \qquad \vdots \\ 9 & x = 3,\ z = 3. \end{cases}$$

Of course, one of the categories of w must be specified as the reference category when defining the eight dummy variables to include in the regression model. In many cases this approach is preferable because interpretation of the effects is more straightforward.

Example 3.7 Munich Rent Index—Interaction with Quality of Kitchen

The rent index is updated every two years, with the collection of new data. For financial reasons, the update during 2001 only consisted of data for 1,500 apartments. Due to the smaller sample size, a complete redesign of the rent index was not possible. Instead the following procedure was chosen: The same characteristics of the 1999 rent index were used, implying that the structure of the rent index did not change. For all characteristics possible changes of their effects compared to 1999 have been examined. To do so, both data sets of 1999 and 2001 were analyzed simultaneously, and changes in covariate effects have been investigated with the help of interactions. We will illustrate the approach using the quality of the kitchen (*kitchen*), with categories "kitchen below average" (reference category), "normal kitchen" (dummy variable *nkitchen*), and premium kitchen (dummy variable *pkitchen*). Note that we measure in this example the quality of the kitchen in three categories rather than two categories as in Table 1.2 (p. 6) and all examples so far. Our starting point is the model (3.7) which was developed in Example 3.5 (p. 90). Here, we modeled the effect of the living area with the help of the transformation $1/area$ and the effect of the year of construction through a third-degree polynomial. We now extend this model with an interaction term between the quality of the kitchen and the survey year. We obtain the estimate

$$\begin{aligned}\widehat{rentsqm} &= \cdots - 0.26\, year01 + 0.91\, nkitchen + 1.09\, pkitchen \\ &\quad + 0.41\, nkitchen \cdot year01 + 0.74\, pkitchen \cdot year01.\end{aligned}$$

The dummy variable *year01* specifies whether an observation has been taken from the year 2001 ($year01 = 1$) or from the year 1999 ($year01 = 0$). For interpretation purposes, we compare apartments with the same living area and year of construction. The results can be summarized as follows, on a per square meter basis:

- In 2001, apartments with a below average kitchen are approximately 0.26 Euro per square meter cheaper than apartments in 1999.
- Apartments with a normal kitchen are approximately $-0.26 + 0.41 = 0.15$ Euro more expensive in 2001 than in 1999.
- Apartments with a premium kitchen are approximately $-0.26 + 0.74 = 0.48$ Euro more expensive in 2001 than in 1999.

In comparison to the original rent index, the estimated surcharge is very high for average and above average kitchens. This results partly from the fact that the model in this example does not contain all relevant covariates. At this point it is impossible to determine if the interaction is necessary. The required inference techniques will be developed in the following sections; in particular, see Sects. 3.3 and 3.4.

△

Interactions with Continuous and Categorical Variables

In the illustrative Example 2.5 (p. 30) of Chap. 2, we were already concerned with the modeling of an interaction between continuous and categorical variables. There, we used data from apartments that were built after 1966 and that were found in an average or good location. Our goal was to model the relationship of the net rent with the explanatory variables living area and location. Figure 2.3 (p. 27) gives the impression that the main effects model

$$rent = \beta_0 + \beta_1\, area + \beta_2\, glocation + \varepsilon$$

is not adequate, as the regression line between net rent and living area for apartments in a good location ($glocation = 1$) is very steep relative to the reference of average location. As a result, we replaced the main effect model by

$$rent = \beta_0 + \beta_1\, area + \beta_2\, glocation + \beta_3\, inter + \varepsilon,$$

with the interaction variable $inter = area \cdot glocation$ and obtained the estimate

$$\widehat{rent} = 130.23 + 5.57\, area + 5.20\, glocation + 0.82\, inter.$$

We can interpret the terms as follows:

- $\beta_1\, area$: Effect of the living area for apartments in average location ($glocation = 0$).
- $\beta_1\, area + \beta_2 + \beta_3\, inter = \beta_2 + (\beta_1 + \beta_3) area$: Effect of the area for apartments in a good location ($glocation = 1$). The two linear area effects for normal and good location can be best interpreted through visualization; see Fig. 2.6 (p. 31).
- $\beta_2 + \beta_3\, inter = \beta_2 + \beta_3\, area$: This is the difference effect of apartments in good location in comparison to apartments in average location. This effect varies depending on the area. Due to the positive coefficients $\hat{\beta}_2 = 5.20$ and $\hat{\beta}_3 = 0.82$, apartments in good location are always more expensive than apartments in average location (when having the same living area). The difference is even more evident when the living area is larger. Every additional square meter of living area increases the difference to apartments in average location by 0.82 Euro.

Interactions for any continuous and categorical covariates x and z can be handled analogously. To keep things simple we first assume $z \in \{0, 1\}$ is binary. If the main effects and interaction effects are linear, we have the model

$$y = \beta_0 + \beta_1 z + \beta_2 x + \beta_3 xz + \ldots + \varepsilon.$$

The dots indicate that the model may contain additional terms (of other covariates). The terms in the model can be interpreted both with respect to the continuous covariate x and the binary variable z. Referring to x, we obtain the following interpretation:

- $\beta_2 x$: Linear effect of the continuous covariate x, if $z = 0$.
- $\beta_1 + (\beta_2 + \beta_3)\, x$: Linear effect of the continuous covariate x, if $z = 1$.

On the contrary, we obtain the following interpretation with respect to z:

- $\beta_1 + \beta_3 x$: Difference effect for observations with $z = 1$ relative to $z = 0$. The difference effect varies depending on the value of x (unlike a constant difference effect in models without interaction).

To this point, we assumed that the main effect of the continuous variable x and the interaction effect are linear. Modeling nonlinear interaction effects within the scope of our current modeling framework is also possible. To do so, we express the model in terms of the functions f_1 and f_2:

$$y = \beta_0 + \beta_1 z + f_1(x) + f_2(x) z + \ldots + \varepsilon. \qquad (3.9)$$

In the case of linear modeling, we obtain the special cases $f_1(x) = \beta_2 x$ and $f_2(x) = \beta_3 x$. Nonlinear relationships can be achieved using transformation of variables or through polynomials. For example, using polynomials of degree 2 yields the functions $f_1(x) = \beta_2 x + \beta_3 x^2$ and $f_2(x) = \beta_4 x + \beta_5 x^2$ and the model

$$\begin{aligned} y &= \beta_0 + \beta_1 z + \beta_2 x + \beta_3 x^2 + (\beta_4 x + \beta_5 x^2) z + \varepsilon \\ &= \beta_0 + \beta_1 z + \beta_2 x + \beta_3 x^2 + \beta_4 xz + \beta_5 x^2 z + \varepsilon. \end{aligned}$$

We can interpret the terms analogously to linear interaction effects:

- $f_1(x)$: Nonlinear effect of the continuous covariate x when $z = 0$.
- $\beta_1 + f_1(x) + f_2(x)$: Nonlinear effect of the continuous covariate x when $z = 1$. The two curves $f_1(x)$ and $\beta_1 + f_1(x) + f_2(x)$ show the strength of interaction. In an extreme case of no interaction both curves run parallel to each other, i.e., they only differ by the constant β_1.
- $\beta_1 + f_2(x)$: Nonlinear difference effect for observations with $z = 1$ relative to $z = 0$. The effect varies nonlinearly, depending on x. We call this a varying coefficient term, since the effect of z varies in a way that depends on the value of the other variable x. The variable x "modifies" the effect of z and thus is also called an effect modifier. In Sect. 9.3.1, we will deal with methods for flexibly and automatically estimating nonlinear effects f_2.

To facilitate interpretation, again it is convenient to center the main effect $f_1(x)$ around zero. Automatic centering can be achieved when using centered design vectors. We do not have to center the function f_2, since it serves as the *difference effect* to the main effect. The estimated effects $\hat{f}_1$, $\hat{\beta}_1 + \hat{f}_1 + \hat{f}_2$ and $\hat{\beta}_1 + \hat{f}_2$ can be plotted together against x for interpretation.

Example 3.8 Munich Rent Index—Interaction Between Living Area and Location

For illustration, we use the data for the Munich rent index. For simplicity, we only consider apartments in average or top locations and model the relationship between net rent per square meter and the explanatory variables living area and location. In previous examples (e.g., Example 2.4, p. 28), we modeled the effect of the living area using the transformation $1/area$. If we assume a linear interaction effect between area and location, we have

$$\begin{aligned} rentsqm &= \beta_0 + \beta_1\, tlocation + \beta_2\, areainvc + \beta_3\, area \cdot tlocation + \varepsilon \\ &= \beta_0 + \beta_1\, tlocation + f_1(area) + f_2(area) \cdot tlocation + \varepsilon, \end{aligned}$$

where *areainvc* denotes the transformation $1/area$ centered around zero. The dummy variable *tlocation* is one for apartments in top location, zero otherwise. The function $f_1(area) = \beta_2\, areainvc$ represents the nonlinear effect of living area and is automatically centered around zero. The function $f_2(area) = \beta_3\, area \cdot tlocation$ represents the linear interaction effect. We obtain

$$\widehat{rentsqm} = 6.94 + 0.77\, tlocation + 143.12\, areainvc + 0.01\, area \cdot tlocation.$$

To interpret the estimated results, we visualize:

- The estimated effect $\hat{f}_1(area)$ of the living area for apartments in an average location (Fig. 3.12a, solid line)

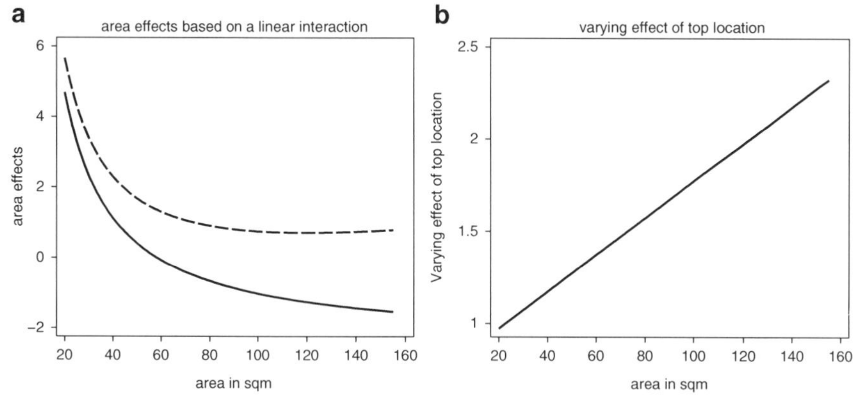

Fig. 3.12 Munich rent index: Panel (**a**) visualizes the area effect for average (*solid line*) and top location (*dashed line*). Panel (**b**) shows the effect of top location varying with respect to area

- The effect $\hat{\beta}_1 + \hat{f}_1(area) + \hat{f}_2(area)$ of the living area for apartments in a top location (Fig. 3.12a, dotted line)
- The varying effect $\hat{\beta}_1 + \hat{f}_2(area)$ of apartments in top location (relative to average location), depending on the living area (Fig. 3.12b)

These figures can be interpreted both with respect to the living area or the location:

- *Living area:* The effects of area in average or top location in Fig. 3.12a are similar. However, the fitted curves are not parallel to each other implying an interaction effect. If the living area increases, the average rent per square meter decreases. The decrease in rent is at first very steep, but the slope moderates for apartments of about 80 square meters or larger. Apartments in top location are always more expensive than apartments in average location, regardless of size. For smaller apartments this difference is much smaller than for larger apartments. This implies that with an increase in living area, we find an increase in the difference of the net rent per square meter for apartments in average relative to top location.
- *Location:* Regardless of living area, apartments in top location are more expensive than apartments in average location. The difference increases linearly with an increase in living area.

Thus far, we are unable to answer whether the inclusion of an interaction is significant and necessary. This question can be answered with inference techniques described in the sections to follow, specifically with statistical hypothesis testing (Sect. 3.3) and methods of model choice (Sect. 3.4).

△

Next, we briefly illustrate the case of an interaction involving multi-categorical rather than binary variables. To simplify notation, we limit ourselves to a variable with three categories 1, 2, and 3. Extensions to more than three categories are straightforward. Let z_1 and z_2 denote the dummy variables derived from z, with $z_1 = 1$ if $z = 1$, and $z_2 = 1$ if $z = 2$, and zero otherwise. The third level $z = 3$ serves as the reference category. Analogous to Eq. (3.9), an interaction model can be written as

$$y = \beta_0 + \beta_1 z_1 + \beta_2 z_2 + f_1(x) + f_2(x) z_1 + f_3(x) z_2 + \ldots + \varepsilon,$$

where f_1, f_2, and f_3 are again linear or nonlinear functions. The model consists of the following effects:

- $f_1(x)$: Effect of the continuous variable x, if $z = 3$ (reference category).
- $\beta_1 + f_1(x) + f_2(x)$: Effect of the continuous variable x, if $z = 1$.
- $\beta_2 + f_1(x) + f_3(x)$: Effect of the continuous variable x, if $z = 2$.
- $\beta_1 + f_2(x)$: Varying effect of the level $z = 1$ depending on x. As always with categorical variables, the effect has to be interpreted as a difference relative to the reference category $z = 3$.
- $\beta_2 + f_3(x)$: Varying effect of the level $z = 2$, depending on x.

We can interpret the estimated results most easily if all the estimated effects $\hat{f}_1(x)$, $\hat{\beta}_1+\hat{f}_1(x)+\hat{f}_2(x)$, and $\hat{\beta}_2+\hat{f}_1(x)+\hat{f}_3(x)$ are visualized in one figure. The varying effects $\hat{\beta}_1 + \hat{f}_2(x)$ and $\hat{\beta}_2 + \hat{f}_3(x)$ can be plotted and analogously interpreted.

If we model all functions linearly, then we obtain the linear model

$$y = \beta_0 + \beta_1 z_1 + \beta_2 z_2 + \beta_3 x + \beta_4 x z_1 + \beta_5 x z_2 + \ldots + \varepsilon.$$

Using quadratic polynomials results in

$$y = \beta_0+\beta_1 z_1+\beta_2 z_2+\beta_3 x+\beta_4 x^2+\beta_5 x z_1+\beta_6 x^2 z_1+\beta_7 x z_2+\beta_8 x^2 z_2+\ldots+\varepsilon.$$

Interactions Between Continuous Variables

When estimating interactions between two continuous covariates, we need to model two-dimensional functions, e.g., by using two- or higher-dimensional polynomials. Here we reach the limit within the scope of linear models, since it is very difficult if not impossible to find adequate models through the inspection of (three-dimensional) scatter plots. In this case, automated methods are clearly superior, and we refer to Sects. 8.2 and 9.3.2.

3.2 Parameter Estimation

In this section we develop estimators for the unknown parameters $\boldsymbol{\beta}$ and σ^2 of the linear model and derive their statistical properties. Section 3.2.1 that follows addresses estimation of the regression coefficients $\boldsymbol{\beta}$, and Sect. 3.2.2 discusses estimation of the variance σ^2. In Sect. 3.2.3, we examine the statistical properties of these estimators.

3.2.1 Estimation of Regression Coefficients

The theory of estimation for the regression coefficients of linear models is closely connected to the method of least squares, first discovered by Legendre in 1806. In addition to the method of least squares, other principles for parameter estimation are possible, in particular the maximum likelihood estimator; see Appendix B.4 for a general introduction to likelihood-based inference. We will see that the maximum likelihood estimator of the regression coefficients coincides with the least squares estimator with Gaussian errors. Alternatives include *robust methods*, which reduce the influence of outlying observations, compared to the least squares method.

The Method of Least Squares

According to the principle of least squares, the unknown regression coefficients $\boldsymbol{\beta}$ are estimated by minimizing the *sum of the squared deviations*

$$\mathrm{LS}(\boldsymbol{\beta}) = \sum_{i=1}^{n} (y_i - \boldsymbol{x}_i'\boldsymbol{\beta})^2 = \sum_{i=1}^{n} \varepsilon_i^2 = \boldsymbol{\varepsilon}'\boldsymbol{\varepsilon}, \tag{3.10}$$

with respect to $\boldsymbol{\beta} \in \mathbb{R}^p$. Perhaps, at first thought, a more reasonable principle is minimizing the *sum of absolute deviations*

$$\mathrm{SM}(\boldsymbol{\beta}) = \sum_{i=1}^{n} |y_i - \boldsymbol{x}_i'\boldsymbol{\beta}| = \sum_{i=1}^{n} |\varepsilon_i|.$$

In fact, this method is historically much older than the method of least squares. It was first proposed around 1760, by Rudjer Joseph Boscovich (1711–1787). In more modern terminology, we refer to this approach as *median regression*, since the minimizing solution yields an estimate for the median of the response variable conditional on the covariates. Median regression is a special case of *quantile regression*, which aims at estimating an arbitrary quantile q_τ, $\tau \in (0,1)$ of the response variable conditional on the covariates. More on quantile regression can be found in Chap. 10.

Yet, the method of least squares remains the most common method for estimating the regression coefficients $\boldsymbol{\beta}$. This is mainly due to the following reasons: On the one hand, the use of the least squares principle is relatively simple from a mathematical point of view. It is, for example, possible to differentiate $\mathrm{LS}(\boldsymbol{\beta})$, in contrast to $\mathrm{SM}(\boldsymbol{\beta})$, with respect to $\boldsymbol{\beta}$. Additionally, estimators that rely on the least squares method have a number of desirable statistical properties; see Sect. 3.2.3. Figure 3.13 illustrates the difference between the least squares principle and the minimization of absolute differences by means of the simple model $y = \beta_0 + \beta_1 x + \varepsilon$. The illustration shows that observations with large deviations have relatively large impact as a matter of taking squares in the minimization criterion. Unlike median regression, least squares estimators have the disadvantage of being highly sensitive to outliers.

In order to determine the minimum of $\mathrm{LS}(\boldsymbol{\beta})$, we rearrange Eq. (3.10) and obtain

$$\begin{aligned} \mathrm{LS}(\boldsymbol{\beta}) &= \boldsymbol{\varepsilon}'\boldsymbol{\varepsilon} \\ &= (\boldsymbol{y} - \boldsymbol{X}\boldsymbol{\beta})'(\boldsymbol{y} - \boldsymbol{X}\boldsymbol{\beta}) \\ &= \boldsymbol{y}'\boldsymbol{y} - \boldsymbol{\beta}'\boldsymbol{X}'\boldsymbol{y} - \boldsymbol{y}'\boldsymbol{X}\boldsymbol{\beta} + \boldsymbol{\beta}'\boldsymbol{X}'\boldsymbol{X}\boldsymbol{\beta} \\ &= \boldsymbol{y}'\boldsymbol{y} - 2\,\boldsymbol{y}'\boldsymbol{X}\boldsymbol{\beta} + \boldsymbol{\beta}'\boldsymbol{X}'\boldsymbol{X}\boldsymbol{\beta}. \end{aligned} \tag{3.11}$$

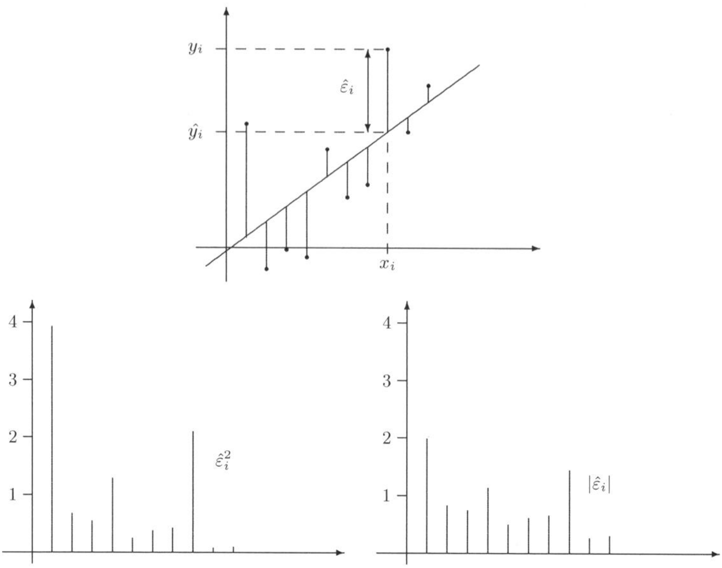

Fig. 3.13 Illustration of the least squares method: The *upper panel* shows some observations (y_i, x_i), scattered around a straight line. According to the principle of least squares, the regression line is chosen such that the sum of squared differences in the *lower left panel* is minimized. As discussed before, squaring the residuals yields comparably large values for observations with larger deviations from the regression line. For comparison, the *lower right panel* displays absolute deviations where this effect is much smaller

We exploit the fact that the terms $\boldsymbol{\beta}'\boldsymbol{X}'\boldsymbol{y}$ and $\boldsymbol{y}'\boldsymbol{X}\boldsymbol{\beta}$ (as well as $\boldsymbol{y}'\boldsymbol{y}$ and $\boldsymbol{\beta}'\boldsymbol{X}'\boldsymbol{X}\boldsymbol{\beta}$) are scalars. Thus $\boldsymbol{\beta}'\boldsymbol{X}'\boldsymbol{y}$ is equal to its transpose $(\boldsymbol{\beta}'\boldsymbol{X}'\boldsymbol{y})'$, and we have $\boldsymbol{\beta}'\boldsymbol{X}'\boldsymbol{y} = (\boldsymbol{\beta}'\boldsymbol{X}'\boldsymbol{y})' = \boldsymbol{y}'\boldsymbol{X}\boldsymbol{\beta}$.

We minimize $\text{LS}(\boldsymbol{\beta})$ by setting the vector of first derivatives zero and by showing that the matrix of second derivatives is positive definite. Readers unfamiliar with vector and matrix calculus should first inspect Appendix A.8 (p. 635). Applying two rules for the differentiation of vector functions [see Theorem A.33 (1) and (3)], we obtain

$$\frac{\partial\, \text{LS}(\boldsymbol{\beta})}{\partial \boldsymbol{\beta}} = -2\,\boldsymbol{X}'\boldsymbol{y} + 2\,\boldsymbol{X}'\boldsymbol{X}\boldsymbol{\beta}. \tag{3.12}$$

Taking second derivatives results in

$$\frac{\partial^2\, \text{LS}(\boldsymbol{\beta})}{\partial \boldsymbol{\beta}\,\partial \boldsymbol{\beta}'} = \frac{\partial\,(-2\,\boldsymbol{X}'\boldsymbol{y} + 2\,\boldsymbol{X}'\boldsymbol{X}\boldsymbol{\beta})}{\partial \boldsymbol{\beta}'} = 2\,\boldsymbol{X}'\boldsymbol{X}.$$

We thereby used Theorem A.33 (5). According to our assumptions, the columns of the design matrix $\boldsymbol{X}$ are linear independent, i.e., $\text{rk}(\boldsymbol{X}) = k + 1 = p$. Thus, according to Theorem A.30 (p. 634), the matrix $\boldsymbol{X}'\boldsymbol{X}$ is positive definite.

We therefore obtain the least squares estimator $\hat{\boldsymbol{\beta}}$ by setting Eq. (3.12) to zero or equivalently by solving the so-called *normal equations*:

$$\boldsymbol{X}'\boldsymbol{X}\hat{\boldsymbol{\beta}} = \boldsymbol{X}'\boldsymbol{y}. \tag{3.13}$$

Since $\boldsymbol{X}'\boldsymbol{X}$ is positive definite and invertible (see Theorem A.28, p. 634), the normal equations have a unique solution given by the least squares estimator

$$\hat{\boldsymbol{\beta}} = (\boldsymbol{X}'\boldsymbol{X})^{-1}\boldsymbol{X}'\boldsymbol{y}. \tag{3.14}$$

Maximum Likelihood Estimation

We obtained the least squares estimator without any specific distributional assumptions regarding the error term $\boldsymbol{\varepsilon}$. Assuming normally distributed errors, i.e., $\boldsymbol{\varepsilon} \sim \mathrm{N}(\boldsymbol{0}, \sigma^2\boldsymbol{I})$, the maximum likelihood (ML) estimator for the unknown parameters can be computed (see Appendix B.4 for an introduction to the theory of maximum likelihood estimators). We next show that the ML estimator for $\boldsymbol{\beta}$ is equivalent to the least squares estimator.

Assuming normally distributed errors we have $\boldsymbol{y} \sim \mathrm{N}(\boldsymbol{X}\boldsymbol{\beta}, \sigma^2\boldsymbol{I})$, which yields the likelihood

$$L(\boldsymbol{\beta}, \sigma^2) = \frac{1}{(2\pi\sigma^2)^{n/2}} \exp\left(-\frac{1}{2\sigma^2}(\boldsymbol{y} - \boldsymbol{X}\boldsymbol{\beta})'(\boldsymbol{y} - \boldsymbol{X}\boldsymbol{\beta})\right). \tag{3.15}$$

The log-likelihood is thus given by

$$l(\boldsymbol{\beta}, \sigma^2) = -\frac{n}{2}\log(2\pi) - \frac{n}{2}\log(\sigma^2) - \frac{1}{2\sigma^2}(\boldsymbol{y} - \boldsymbol{X}\boldsymbol{\beta})'(\boldsymbol{y} - \boldsymbol{X}\boldsymbol{\beta}). \tag{3.16}$$

When maximizing the log-likelihood with respect to $\boldsymbol{\beta}$, we can ignore the first two terms of the sum in Eq. (3.16) because they are independent of $\boldsymbol{\beta}$. Maximizing $-\frac{1}{2\sigma^2}(\boldsymbol{y} - \boldsymbol{X}\boldsymbol{\beta})'(\boldsymbol{y} - \boldsymbol{X}\boldsymbol{\beta})$ is equivalent to minimizing $(\boldsymbol{y} - \boldsymbol{X}\boldsymbol{\beta})'(\boldsymbol{y} - \boldsymbol{X}\boldsymbol{\beta})$, which is the least squares criterion (3.10). The maximum likelihood estimator of $\boldsymbol{\beta}$ is therefore identical to the least squares estimator (3.14).

Predicted Values and Residuals

Based on the least squares estimator $\hat{\boldsymbol{\beta}} = (\boldsymbol{X}'\boldsymbol{X})^{-1}\boldsymbol{X}'\boldsymbol{y}$ for $\boldsymbol{\beta}$, we can estimate the (conditional) mean of $\boldsymbol{y}$ by

$$\widehat{\mathrm{E}(\boldsymbol{y})} = \hat{\boldsymbol{y}} = \boldsymbol{X}\hat{\boldsymbol{\beta}}.$$

Substituting the least squares estimator further results in

$$\hat{\boldsymbol{y}} = \boldsymbol{X}(\boldsymbol{X}'\boldsymbol{X})^{-1}\boldsymbol{X}'\boldsymbol{y} = \boldsymbol{H}\boldsymbol{y},$$

with the $n \times n$-matrix

$$\boldsymbol{H} = \boldsymbol{X}(\boldsymbol{X}'\boldsymbol{X})^{-1}\boldsymbol{X}'.$$

3.5 Properties of the Hat Matrix

The hat matrix $\boldsymbol{H} = \boldsymbol{X}(\boldsymbol{X}'\boldsymbol{X})^{-1}\boldsymbol{X}'$ has the following properties:
1. $\boldsymbol{H}$ is symmetric.
2. $\boldsymbol{H}$ is idempotent. For the definition of idempotent matrices, see Appendix A, Definition A.12.
3. $\text{rk}(\boldsymbol{H}) = \text{tr}(\boldsymbol{H}) = p$, i.e., the trace is equal to the rank.
4. $\frac{1}{n} \leq h_{ii} \leq \frac{1}{r}$, where r represents the number of rows in $\boldsymbol{X}$ with different $\boldsymbol{x}_i$. If the rows are all different, then $\frac{1}{n} \leq h_{ii} \leq 1$.
5. The matrix $\boldsymbol{I} - \boldsymbol{H}$ is also symmetric and idempotent, with $\text{rk}(\boldsymbol{I} - \boldsymbol{H}) = n - p$.

The matrix $\boldsymbol{H}$ is also called the *prediction matrix* or *hat matrix* . Some easily verifiable properties of the hat matrix are summarized in Box 3.5.

With the help of the hat matrix $\boldsymbol{H}$, it is also possible to express the residuals $\hat{\varepsilon}_i = y_i - \hat{y}_i$ in matrix notion. We have

$$\hat{\boldsymbol{\varepsilon}} = \boldsymbol{y} - \hat{\boldsymbol{y}} = \boldsymbol{y} - \boldsymbol{H}\boldsymbol{y} = (\boldsymbol{I} - \boldsymbol{H})\boldsymbol{y}.$$

3.2.2 Estimation of the Error Variance

Maximum Likelihood Estimation

It is rather natural to estimate the variance σ^2 using maximum likelihood. We already determined the likelihood $L(\boldsymbol{\beta}, \sigma^2)$ and the log-likelihood $l(\boldsymbol{\beta}, \sigma^2)$ for the linear model; see Eqs. (3.15) and (3.16). Differentiation of the log-likelihood (3.16) with respect to σ^2, and setting to zero, provides

$$\frac{\partial l(\boldsymbol{\beta}, \sigma^2)}{\partial \sigma^2} = -\frac{n}{2\sigma^2} + \frac{1}{2\sigma^4}(\boldsymbol{y} - \boldsymbol{X}\boldsymbol{\beta})'(\boldsymbol{y} - \boldsymbol{X}\boldsymbol{\beta}) \stackrel{!}{=} 0.$$

Substituting the ML or least squares estimator $\hat{\boldsymbol{\beta}}$ for $\boldsymbol{\beta}$ results in

$$-\frac{n}{2\sigma^2} + \frac{1}{2\sigma^4}(\boldsymbol{y} - \boldsymbol{X}\hat{\boldsymbol{\beta}})'(\boldsymbol{y} - \boldsymbol{X}\hat{\boldsymbol{\beta}}) = -\frac{n}{2\sigma^2} + \frac{1}{2\sigma^4}(\boldsymbol{y} - \hat{\boldsymbol{y}})'(\boldsymbol{y} - \hat{\boldsymbol{y}}) = -\frac{n}{2\sigma^2} + \frac{1}{2\sigma^4}\hat{\boldsymbol{\varepsilon}}'\hat{\boldsymbol{\varepsilon}} \stackrel{!}{=} 0,$$

which yields

$$\hat{\sigma}^2_{ML} = \frac{\hat{\boldsymbol{\varepsilon}}'\hat{\boldsymbol{\varepsilon}}}{n}.$$

3.6 Parameter Estimation in the Classical Linear Model

Estimation of β

In the classical linear model, the estimator

$$\hat{\boldsymbol{\beta}} = (\boldsymbol{X}'\boldsymbol{X})^{-1}\boldsymbol{X}'\boldsymbol{y}$$

minimizes the least squares criterion

$$\text{LS}(\boldsymbol{\beta}) = \sum_{i=1}^{n} (y_i - \boldsymbol{x}_i'\boldsymbol{\beta})^2.$$

Under the assumption of normally distributed errors, the least squares estimator is also the ML estimator for $\boldsymbol{\beta}$.

Estimation of σ^2

The estimator

$$\hat{\sigma}^2 = \frac{1}{n-p}\hat{\boldsymbol{\varepsilon}}'\hat{\boldsymbol{\varepsilon}}$$

is unbiased and can be characterized as the REML estimator for σ^2.

However, this estimator for σ^2 is rarely used. The mean of the sum of squared residuals is

$$\text{E}(\hat{\boldsymbol{\varepsilon}}'\hat{\boldsymbol{\varepsilon}}) = (n-p)\cdot\sigma^2 \tag{3.17}$$

and thus

$$\text{E}(\hat{\sigma}^2_{ML}) = \frac{n-p}{n}\sigma^2.$$

This implies that the ML estimator for σ^2 is biased. A proof for Eq. (3.17) is provided in Sect. 3.5.2 on p. 168.

Restricted Maximum Likelihood Estimation

Using Eq. (3.17), we easily get an unbiased estimator $\hat{\sigma}^2$ for σ^2. We obtain

$$\hat{\sigma}^2 = \frac{1}{n-p}\hat{\boldsymbol{\varepsilon}}'\hat{\boldsymbol{\varepsilon}}, \tag{3.18}$$

which is a commonly used estimator for σ^2. Estimator (3.18) also possesses the interesting feature of being the *restricted maximum likelihood estimator* (REML). It can be shown that Eq. (3.18) maximizes the *marginal likelihood*

$$L(\sigma^2) = \int L(\boldsymbol{\beta}, \sigma^2)\, d\boldsymbol{\beta},$$

which is obtained by integration over the vector of regression coefficients $\boldsymbol{\beta}$. In general, the REML estimator for variance parameters is less biased than the ML estimator and thus is generally preferred. In the present case the REML estimator is even unbiased, in contrast to the ML estimator. A further application of the REML principle can be found when estimating linear mixed models; see Chap. 7.

3.2.3 Properties of the Estimators

We will now discuss the properties of the estimators in the classical linear model. First, we will discuss the geometric properties of the least squares estimator and then move to its statistical properties for finite and infinite samples. We will also discuss properties of the residuals.

Geometric Properties of Least Squares

From a geometric perspective, the (conditional) mean $\boldsymbol{\mu} = \mathrm{E}(\boldsymbol{y}) = \boldsymbol{X}\boldsymbol{\beta}$ of the linear model $\boldsymbol{y} = \boldsymbol{X}\boldsymbol{\beta} + \boldsymbol{\varepsilon}$ is an n-dimensional vector in $\mathbb{R}^n$. Since $\boldsymbol{\mu}$ is a linear combination of the columns of the design matrix $\boldsymbol{X}$, $\boldsymbol{\mu}$ must be an element of the column space of $\boldsymbol{X}$. See Definition A.15 in Appendix A for the definition of the column space of a matrix. The column space of $\boldsymbol{X}$ is spanned by the columns $\mathbf{1}, \boldsymbol{x}^1, \ldots, \boldsymbol{x}^k$, which are assumed to be linearly independent ($\mathrm{rk}(\boldsymbol{X}) = k + 1 = p$). Thus the columns of $\boldsymbol{X}$ define a p-dimensional vector space.

Geometrically, the response vector $\boldsymbol{y}$ is also a vector in $\mathbb{R}^n$. However, $\boldsymbol{y}$ is usually (similar to $\boldsymbol{\varepsilon}$) *not* an element of the column space of $\boldsymbol{X}$. We illustrate the scenario with the help of the very simple model

$$\boldsymbol{y} = \begin{pmatrix} y_1 \\ y_2 \end{pmatrix} = \boldsymbol{X}\boldsymbol{\beta} + \boldsymbol{\varepsilon} = \begin{pmatrix} 1 \\ 1 \end{pmatrix} \beta_0 + \begin{pmatrix} \varepsilon_1 \\ \varepsilon_2 \end{pmatrix},$$

having only two observations and one parameter β_0. Since the design matrix only consists of the column vector $\boldsymbol{x}^0 = (1, 1)'$, the column space of $\boldsymbol{X}$ consists of all points in $\mathbb{R}^2$ that are on the line defined by the origin and the point $(1, 1)$; see Fig. 3.14.

For example, if we observe the vector $\boldsymbol{y} = (2, 3)'$, we obtain the value $\hat{\beta}_0 = 2.5$ as the least squares estimator for β_0. The resulting predicted values are $\hat{\boldsymbol{y}} = (2.5, 2.5)'$ (see Fig. 3.14). Since the method of least squares minimizes $(\boldsymbol{y} - \boldsymbol{X}\boldsymbol{\beta})'(\boldsymbol{y} - \boldsymbol{X}\boldsymbol{\beta}) = \boldsymbol{\varepsilon}'\boldsymbol{\varepsilon}$, the *Euclidean distance* between $\boldsymbol{y}$ and $\boldsymbol{X}\boldsymbol{\beta}$ is minimized. Figure 3.14 shows this distance is minimized if $\hat{\beta}_0$ is chosen such that the line connecting $\boldsymbol{y}$ with $\hat{\boldsymbol{y}}$ is orthogonal to $\hat{\boldsymbol{y}}$. The connecting line is in fact the residual vector $\hat{\boldsymbol{\varepsilon}}$. This in turn implies that the residuals $\hat{\boldsymbol{\varepsilon}}$ and the predicted values $\hat{\boldsymbol{y}}$ are orthogonal, i.e., $\hat{\boldsymbol{\varepsilon}}'\hat{\boldsymbol{y}} = 0$. Notice that $\boldsymbol{x}^0$ and $\hat{\boldsymbol{\varepsilon}}$ are also orthogonal to each other.

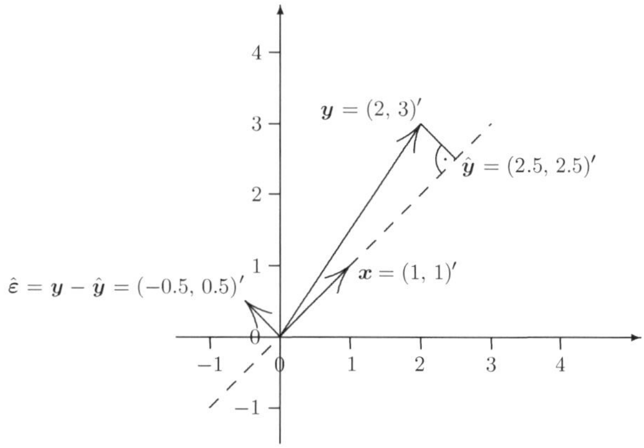

Fig. 3.14 Visualization of the geometric properties of the least squares estimator

We can generalize these observations for arbitrary linear models: The method of least squares yields parameter estimates $\hat{\boldsymbol{\beta}}$ such that the residuals $\hat{\boldsymbol{\varepsilon}}$ and the predicted values $\hat{\boldsymbol{y}}$ are orthogonal to each other. This can be easily proved using properties of the hat matrix $\boldsymbol{H}$. Using $\hat{\boldsymbol{y}} = \boldsymbol{H}\boldsymbol{y}$ and $\hat{\boldsymbol{\varepsilon}} = (\boldsymbol{I} - \boldsymbol{H})\boldsymbol{y}$ (see p. 108), we obtain

$$\hat{\boldsymbol{y}}'\hat{\boldsymbol{\varepsilon}} = \boldsymbol{y}'\boldsymbol{H}\,(\boldsymbol{I} - \boldsymbol{H})\boldsymbol{y} = \boldsymbol{y}'\boldsymbol{H}\boldsymbol{y} - \boldsymbol{y}'\boldsymbol{H}\boldsymbol{H}\boldsymbol{y} = \boldsymbol{y}'\boldsymbol{H}\boldsymbol{y} - \boldsymbol{y}'\boldsymbol{H}\boldsymbol{y} = 0.$$

Moreover, all columns of the design matrix are orthogonal to the residuals, i.e., $(\boldsymbol{x}^j)'\hat{\boldsymbol{\varepsilon}} = 0,\ j = 0, \ldots, k$, or $\boldsymbol{X}'\hat{\boldsymbol{\varepsilon}} = \boldsymbol{0}$. For a proof we take again advantage of properties of the hat matrix:

$$\boldsymbol{X}'\hat{\boldsymbol{\varepsilon}} = \boldsymbol{X}'(\boldsymbol{I} - \boldsymbol{H})\boldsymbol{y} = \boldsymbol{X}'\boldsymbol{y} - \boldsymbol{X}'\boldsymbol{H}\boldsymbol{y} = \boldsymbol{X}'\boldsymbol{y} - \boldsymbol{X}'\boldsymbol{X}(\boldsymbol{X}'\boldsymbol{X})^{-1}\boldsymbol{X}'\boldsymbol{y} = \boldsymbol{0}.$$

The fact that the columns of the design matrix and the residuals are orthogonal provides some more interesting implications; see properties 3–5 in Box 3.7. A proof can be found in Sect. 3.5.2, on p. 169.

Example 3.9 Orthogonal Design

The fact that the residuals and the columns of the design matrix are orthogonal can be used for constructing an orthogonal design matrix. An orthogonal design matrix implies uncorrelated covariates. Orthogonalizing the design matrix is useful, e.g., when constructing orthogonal polynomials to model nonlinear relationships (see Example 3.5, p. 90).

Let $\boldsymbol{X}$ be the design matrix with columns $\boldsymbol{x}^j$. The goal is to transform the columns $\boldsymbol{x}^j$ such that the resulting columns $\tilde{\boldsymbol{x}}^j$ of the transformed design matrix $\tilde{\boldsymbol{X}}$ are orthogonal. To achieve this we use for $j = 1, \ldots, k$ the transformations

$$\tilde{\boldsymbol{x}}^j = \boldsymbol{x}^j - \tilde{\boldsymbol{X}}_j(\tilde{\boldsymbol{X}}_j'\tilde{\boldsymbol{X}}_j)^{-1}\tilde{\boldsymbol{X}}_j'\boldsymbol{x}^j,$$

where matrix $\tilde{\boldsymbol{X}}_j$ contains the first $j - 1$ transformed vectors. We do not transform the first column of the design matrix $\boldsymbol{X}$, associated with the intercept. The transformed vector $\tilde{\boldsymbol{x}}_j$ can be viewed as the residual vector of a regression with $\boldsymbol{x}^j$ as the response variable and the j transformed vectors $\tilde{\boldsymbol{x}}^0 = \boldsymbol{1}, \tilde{\boldsymbol{x}}^1, \ldots, \tilde{\boldsymbol{x}}^{j-1}$ as the covariates. Due to the orthogonality

3.7 Geometric Properties of the Least Squares Estimator

The method of least squares has the following geometric properties:

1. The predicted values $\hat{\boldsymbol{y}}$ are orthogonal to the residuals $\hat{\boldsymbol{\varepsilon}}$, i.e., $\hat{\boldsymbol{y}}'\hat{\boldsymbol{\varepsilon}} = 0$.
2. The columns $\boldsymbol{x}^j$ of $\boldsymbol{X}$ are orthogonal to the residuals $\hat{\boldsymbol{\varepsilon}}$, i.e., $(\boldsymbol{x}^j)'\hat{\boldsymbol{\varepsilon}} = 0$ or $\boldsymbol{X}'\hat{\boldsymbol{\varepsilon}} = \boldsymbol{0}$.
3. The average of the residuals is zero, i.e.,

$$\sum_{i=1}^{n} \hat{\varepsilon}_i = 0 \quad \text{or} \quad \frac{1}{n}\sum_{i=1}^{n} \hat{\varepsilon}_i = 0.$$

4. The average of the predicted values $\hat{y}_i$ is equal to the average of the observed response y_i, i.e.,

$$\frac{1}{n}\sum_{i=1}^{n} \hat{y}_i = \bar{y}.$$

5. The regression hyperplane runs through the average of the data, i.e.,

$$\bar{y} = \hat{\beta}_0 + \hat{\beta}_1 \bar{x}_1 + \cdots + \hat{\beta}_k \bar{x}_k.$$

of the residuals, $\tilde{\boldsymbol{x}}^j$ is orthogonal to the columns of $\tilde{\boldsymbol{X}}_j$. This implies that $\tilde{\boldsymbol{x}}^j$ is orthogonal to all $j-1$ previously constructed variables $\tilde{\boldsymbol{x}}^1, \ldots, \tilde{\boldsymbol{x}}^{j-1}$. Notice that the first transformed variable $\tilde{\boldsymbol{x}}^1$ results from a simple centering around the column mean value of $\boldsymbol{x}^1$. In linear algebra, this method is also known as Gram–Schmidt orthogonalization.

△

Analysis of Variance and Coefficient of Determination

Using the geometric properties of the least squares estimator, we can derive a fundamental analysis of variance formula for the empirical variance of observed responses y_i. This allows us to define the *coefficient of determination* or the proportion of total variance that is explained by the regression model. The coefficient of determination is closely related to the empirical correlation coefficient and can be used as a goodness-of-fit measure (among many others).

In Sect. 3.5.2 (p. 169), we prove the following decomposition formula:

$$\sum_{i=1}^{n} (y_i - \bar{y})^2 = \sum_{i=1}^{n} (\hat{y}_i - \bar{y})^2 + \sum_{i=1}^{n} \hat{\varepsilon}_i^2. \tag{3.19}$$

Division by n (or $n-1$) on both sides leads to the analysis of variance formula:

$$s_y^2 = s_{\hat{y}}^2 + s_{\hat{\varepsilon}}^2,$$

where s_y^2, $s_{\hat{y}}^2$, and $s_{\hat{\varepsilon}}^2$ are the empirical variances of the observed response, the predicted values, and the residuals. Thus we obtain an additive decomposition of the empirical variance of the response into the empirical variance of the predicted values and the residuals.

Based on the analysis of variance formula, we can define the coefficient of determination R^2 as a goodness-of-fit measure. It is defined as

$$R^2 = \frac{\sum_{i=1}^{n} (\hat{y}_i - \bar{y})^2}{\sum_{i=1}^{n} (y_i - \bar{y})^2} = 1 - \frac{\sum_{i=1}^{n} \hat{\varepsilon}_i^2}{\sum_{i=1}^{n} (y_i - \bar{y})^2}.$$

As a consequence of Eq. (3.19), $0 \leq R^2 \leq 1$.

Using the analysis of variance formula, we can interpret the coefficient of determination in the following way: The closer R^2 is to 1, the smaller the residual sum of squares $\sum \hat{\varepsilon}_i^2$ and thus the better the fit to the data. The extreme case of $R^2 = 1$ implies $\sum \hat{\varepsilon}_i^2 = 0$, i.e., all residuals are zero and the fit to the data is perfect. If R^2 is close to 0, the sum of squared residuals is relatively large and the model fit is poor. The limit $R^2 = 0$ implies $\sum(\hat{y}_i - \bar{y})^2 = 0$, and hence $\hat{y}_i = \bar{y}$ for all i. That is the prediction of y_i is always equal to the mean $\bar{y}$ of the response variable and is thus *independent* of the explanatory variables. Consequently, the covariates do not have any explanatory power with regard to the mean of y. However, it is important to note that the model can be misspecified, e.g., there could be a nonlinear relationship for one of the covariates. In such a case, covariates can hold uncovered explanatory power, even if the coefficient of determination is close to zero.

In the special case of a simple regression model

$$y = \beta_0 + \beta_1 x + \varepsilon,$$

it can be shown that

$$R^2 = r_{xy}^2,$$

where r_{xy}^2 is the squared empirical correlation coefficient between $\boldsymbol{x}$ and $\boldsymbol{y}$. This property of the coefficient of determination is responsible for the notation R^2. In the simple regression model, as well as in the multiple regression model, we can interpret the coefficient of determination as the squared empirical correlation coefficient between the observations $\boldsymbol{y}$ and the predicted values $\hat{\boldsymbol{y}}$. Formally we have

$$R^2 = r_{y\hat{y}}^2.$$

Finally, a warning is in order. It is quite common to choose models with preferably high coefficients of determination. However, we must be careful when

Table 3.1 Munich rent index: comparison of different models for the relationship between net rent per square meter (*rentsqm*) and area (*area*)

Model	Equation	R^2
M1	$\widehat{rentsqm} = 9.47 - 0.035 \cdot area$	0.116
M2	$\widehat{rentsqm} = 4.73 + 140.18 \cdot 1/area$	0.154
M3	$\widehat{rentsqm} = 11.83 - 0.106 \cdot area + 0.00047 \cdot area^2$	0.143
M4	$\widehat{rentsqm} = 14.28 - 0.22 \cdot area + 0.0020 \cdot area^2 - 0.000006 \cdot area^3$	0.150

comparing different models using the coefficients of determination. Model comparison using R^2 is only meaningful if the following three conditions are fulfilled:

- Every model must use the same response variable y, e.g., it is not possible to compare models with an original response y to that of a transformed response, e.g., $\log(y)$.
- Every model must have the same number of parameters.
- Every model should include an intercept β_0.

Thus, in general, we *cannot* use the coefficient of determination for model comparison. It can be shown that the coefficient of determination cannot decrease if an additional explanatory variable is added to the model. The following example will illustrate the difficulty of using R^2 for model comparison.

Example 3.10 Munich Rent Index—Comparison of Models Using R^2

In Sect. 3.1.3 (p. 87), we examined several models to investigate the relationship between net rent per square meter and living area. Table 3.1 contains the estimated models and their corresponding coefficients of determination. First, we find that every coefficient of determination is relatively small. One reason is the strong variability in the data; see the scatter plot in Fig. 3.8 on p. 88. Another reason is that many important covariates of the original rent index are missing in the models of this illustrative example. The final model that was chosen for the official rent index for 1999 contained more than 20 explanatory variables, which led to a coefficient of determination of 0.49.

If we compare model M1 and M2, we find that M2 has a higher coefficient of determination. Since both models have the same number of parameters, a comparison based on the coefficient of determination is appropriate, and Model M2 should be preferred. Note, however, that we came to the same conclusion with residual analysis; see Fig. 3.8 (p. 88). The models M1, M3, and M4 are nested, which means that M4 contains both M3 and M1, and M3 contains M1 as a special case. As a consequence M1 *must* have the smallest R^2 value, M3 has the second smallest R^2 value, and M4 has the highest R^2 value. A comparison on the basis of the coefficient of determination does not make sense here, as the number of parameters differs across models. Nevertheless, we point out that a model with a higher number of regression parameters does not necessarily imply a higher coefficient of determination. This is the case, e.g., if the models have different explanatory variables (non-nested models). A comparison of models M2 and M3 demonstrates such a comparison: Even though M3 has three regression parameters and M2 only has two, M2 has a higher coefficient of determination. In such a case, M2 is preferred over M3, as it contains less parameters with a higher coefficient of determination. The coefficient of determination only increases automatically with an increase of parameters for *nested models*, i.e., if more complex models contain the other models as special cases, as is the case with the models M1, M3, and M4.

△

3.8 The Coefficient of Determination

Definition

$$R^2 = \frac{\sum_{i=1}^{n} (\hat{y}_i - \bar{y})^2}{\sum_{i=1}^{n} (y_i - \bar{y})^2} = 1 - \frac{\sum_{i=1}^{n} \hat{\varepsilon}_i^2}{\sum_{i=1}^{n} (y_i - \bar{y})^2}.$$

Interpretation

The closer the coefficient of determination is to 1, the smaller is the residual sum of squares, and the better the fit is to the data. If $R^2 = 1$, all residuals are zero with perfect fit to the data.

Properties

1. In the simple linear regression model $y_i = \beta_0 + \beta_1 x_i + \varepsilon_i$, the coefficient of determination corresponds to the squared correlation coefficient (according to Bravais–Pearson), implying $R^2 = r_{xy}^2$.
2. In the multiple regression model, the coefficient of determination can be understood as a squared correlation coefficient between the observations $\boldsymbol{y}$ and the predicted values $\hat{\boldsymbol{y}}$, implying $R^2 = r_{y\hat{y}}^2$.
3. Let the vector $\boldsymbol{x}$ of the explanatory variables be partitioned into two subvectors $\boldsymbol{x}_1$ and $\boldsymbol{x}_2$. Consider the full model M1 $y_i = \beta_0 + \boldsymbol{\beta}_1' \boldsymbol{x}_{i1} + \boldsymbol{\beta}_2' \boldsymbol{x}_{i2} + \varepsilon_i$ and the submodel M2 $y_i = \beta_0 + \boldsymbol{\beta}_1' \boldsymbol{x}_{i1} + \varepsilon_i$. Then

$$R_{M1}^2 \geq R_{M2}^2,$$

   i.e., for nested models, the coefficient of determination of the submodel is always less than or equal to the coefficient of determination of the full model.
4. Comparing the coefficient of determination across models is meaningful for a common response variable and if the models contain the same number of parameters and an intercept.

In Sect. 3.4, we will investigate other model choice criteria that are appropriate to compare models containing a varying number of parameters.

Statistical Properties Without Specific Distributional Assumptions

We will now investigate the statistical properties of the least squares estimator. Thereby we will admit an arbitrary distribution for the error term. In particular, we do not assume that the errors necessarily follow a normal distribution. Note, however, that the following properties are only true if the model is otherwise *correctly specified*; see Sect. 3.4 for the consequences of model misspecification.

Expectation and Bias

For the expectation of the least squares estimator, we have

$$\mathrm{E}(\hat{\boldsymbol{\beta}}) = \mathrm{E}\left((\boldsymbol{X}'\boldsymbol{X})^{-1}\boldsymbol{X}'\boldsymbol{y}\right) = (\boldsymbol{X}'\boldsymbol{X})^{-1}\boldsymbol{X}'\mathrm{E}(\boldsymbol{y}) = (\boldsymbol{X}'\boldsymbol{X})^{-1}\boldsymbol{X}'\boldsymbol{X}\boldsymbol{\beta} = \boldsymbol{\beta}.$$

Thus the least squares estimator is unbiased for $\boldsymbol{\beta}$.

Covariance Matrix

Using Theorem B.2.5 (p. 647), the covariance matrix of the least squares estimator is given by

$$\begin{aligned}\mathrm{Cov}\,(\hat{\boldsymbol{\beta}}) &= \mathrm{Cov}\left((\boldsymbol{X}'\boldsymbol{X})^{-1}\boldsymbol{X}'\boldsymbol{y}\right) = (\boldsymbol{X}'\boldsymbol{X})^{-1}\boldsymbol{X}'\mathrm{Cov}(\boldsymbol{y})((\boldsymbol{X}'\boldsymbol{X})^{-1}\boldsymbol{X}')' \\ &= \sigma^2\,(\boldsymbol{X}'\boldsymbol{X})^{-1}\boldsymbol{X}'\boldsymbol{X}\,(\boldsymbol{X}'\boldsymbol{X})^{-1} = \sigma^2\,(\boldsymbol{X}'\boldsymbol{X})^{-1}.\end{aligned}$$

For the variances of the estimated regression coefficients $\hat{\beta}_j$, there exists a more easily interpretable formula (see, e.g., Wooldridge, 2006):

$$\mathrm{Var}(\hat{\beta}_j) = \frac{\sigma^2}{(1 - R_j^2)\sum_{i=1}^{n}(x_{ij} - \bar{x}_j)^2}.$$

We define R_j^2 here as the coefficient of determination for the regression between x_j, as the response variable, and all of the other explanatory variables (except x_j). We immediately see which components determine the precision of the estimates for the regression coefficients:

- The smaller the model variance σ^2, the smaller the variance of $\hat{\beta}_j$ and thus the more accurate the estimation.
- The smaller the linear dependence between x_j and the other explanatory variables (measured through R_j^2), the smaller is the variance of $\hat{\beta}_j$. The variance, $\mathrm{Var}(\hat{\beta}_j)$, is minimized for $R_j^2 = 0$, i.e., when the covariates are uncorrelated. *Orthogonal designs* guarantee uncorrelated regressors and are typically used in experimental design situations to maximize the precision of estimators. On the other hand, when some of the covariates are highly correlated, the estimators can be very imprecise. In the extreme case of $R_j^2 \to 1$ the variances explode towards infinity. We will explain this *collinearity* problem in detail in Sect. 3.4.
- The larger the variability of covariate x_j around its average, the smaller is the variance of $\hat{\beta}_j$.

The true covariance of $\hat{\boldsymbol{\beta}}$ cannot be calculated in applications, as the error variance σ^2 is unknown. Rather, we have to estimate $\mathrm{Cov}(\hat{\boldsymbol{\beta}})$, by replacing σ^2 with its estimate $\hat{\sigma}^2 = \hat{\boldsymbol{\varepsilon}}'\hat{\boldsymbol{\varepsilon}}/(n-p)$:

$$\widehat{\text{Cov}(\hat{\boldsymbol{\beta}})} = \hat{\sigma}^2 \, (\boldsymbol{X}'\boldsymbol{X})^{-1} = \frac{1}{n-p} \hat{\boldsymbol{\varepsilon}}'\hat{\boldsymbol{\varepsilon}}(\boldsymbol{X}'\boldsymbol{X})^{-1}.$$

The diagonal elements of this matrix represent the estimated variances of the least squares estimator $\hat{\beta}_j$. The square root of the diagonal elements are the estimated standard errors, which we abbreviate as se_j from this point forward:

$$\text{se}_j = \widehat{\text{Var}(\hat{\beta}_j)}^{1/2}, \quad j = 0, 1, \ldots, k.$$

The estimated covariance matrix, especially the standard errors se_j, are used for statistical testing and confidence intervals for the regression coefficients; see Sect. 3.3.

Comparison with Linear Estimators

In the following, we want to compare the least squares estimator with the more general class of *linear estimators*. A linear estimator $\hat{\boldsymbol{\beta}}^L$ takes the form of

$$\hat{\boldsymbol{\beta}}^L = \boldsymbol{a} + \boldsymbol{A}\,\boldsymbol{y},$$

where $\boldsymbol{a} = (a_0, \ldots, a_k)'$ is a $p \times 1$-vector, and $\boldsymbol{A} = (a_{ij})$ is a matrix of dimension $p \times n$. Thus, the components β_j of $\boldsymbol{\beta}$ are estimated through a linear combination of the response observations y_i:

$$\hat{\beta}_j^L = a_j + a_{j1}y_1 + \cdots + a_{jn}y_n \qquad j = 0, \ldots, k.$$

Obviously, the least squares estimator $\hat{\boldsymbol{\beta}}$ is a special case of a linear estimator with $\boldsymbol{a} = \boldsymbol{0}$ and $\boldsymbol{A} = (\boldsymbol{X}'\boldsymbol{X})^{-1}\boldsymbol{X}'$. We can derive the expectation and the covariance matrix of linear estimators in a similar way as we did with the least squares estimator. We have

$$\text{E}(\hat{\boldsymbol{\beta}}^L) = \boldsymbol{a} + \boldsymbol{A}\boldsymbol{X}\boldsymbol{\beta} \qquad \text{Cov}\,(\hat{\boldsymbol{\beta}}^L) = \sigma^2\,\boldsymbol{A}\boldsymbol{A}'.$$

Linear estimators are, thus, not necessarily unbiased. A comparison of the least squares estimator with the subclass of *linear unbiased estimators* $\hat{\boldsymbol{\beta}}^L$ shows that the least squares estimator has minimal variances, i.e.,

$$\text{Var}(\hat{\beta}_j^L) \geq \text{Var}(\hat{\beta}_j), \quad j = 0, \ldots, k.$$

This property also holds for any linear combination

$$b_0\beta_0 + b_1\beta_1 + \ldots + b_k\beta_k = \boldsymbol{b}'\boldsymbol{\beta}$$

of $\boldsymbol{\beta}$, i.e.,

$$\mathrm{Var}(\boldsymbol{b}'\hat{\boldsymbol{\beta}}^L) \geq \mathrm{Var}(\boldsymbol{b}'\hat{\boldsymbol{\beta}}).$$

These properties of the least squares estimator are known as Gauß–Markov Theorem. A proof can be found in this chapter's appendix, on p. 170. The Gauß–Markov Theorem can also be used to obtain an "optimal" prediction for a new (future) observation y_0 with a given covariate vector $\boldsymbol{x}_0$. We use the conditional mean for prediction:

$$\mathrm{E}(y_0\,|\boldsymbol{x}_0) = \boldsymbol{x}_0{}'\boldsymbol{\beta}.$$

An optimal estimator of the expectation (in terms of the Gauß–Markov Theorem) is then given by

$$\hat{y}_0 = \boldsymbol{x}_0{}'\hat{\boldsymbol{\beta}}.$$

Statistical Properties Under Normality Assumption

Thus far, the derived statistical properties of the least squares estimator have been obtained without assuming any specific distribution for the error term $\boldsymbol{\varepsilon}$. Assuming a normal distribution for the errors, i.e., $\boldsymbol{\varepsilon} \sim \mathrm{N}(\boldsymbol{0}, \sigma^2\boldsymbol{I})$, additional properties of the least squares estimator can be derived. They will be especially useful when constructing hypothesis tests and confidence intervals for $\boldsymbol{\beta}$.

Normality of the errors results in $\boldsymbol{y} \sim \mathrm{N}(\boldsymbol{X}\boldsymbol{\beta}, \sigma^2\boldsymbol{I})$. It follows that the least squares estimator $\hat{\boldsymbol{\beta}} = (\boldsymbol{X}'\boldsymbol{X})^{-1}\boldsymbol{X}'\boldsymbol{y}$ will also follow a normal distribution, as $\hat{\boldsymbol{\beta}}$ is a linear transformation of $\boldsymbol{y}$ (refer to Theorem B.5 on p. 649). Using the previously derived results for the expectation and the covariance matrix of the least squares estimator, we immediately have

$$\hat{\boldsymbol{\beta}} \sim \mathrm{N}(\boldsymbol{\beta}, \sigma^2(\boldsymbol{X}'\boldsymbol{X})^{-1}).$$

Moreover, Theorem B.8.1 (p. 651) shows that the distribution for the distance (weighted by the inverse covariance matrix) between the least squares estimator $\hat{\boldsymbol{\beta}}$ and the true $\boldsymbol{\beta}$ follows a χ^2-distribution with p degrees of freedom:

$$\frac{(\hat{\boldsymbol{\beta}} - \boldsymbol{\beta})'(\boldsymbol{X}'\boldsymbol{X})(\hat{\boldsymbol{\beta}} - \boldsymbol{\beta})}{\sigma^2} \sim \chi^2_p.$$

This property is useful when deriving the distribution of a test statistic in hypothesis testing.

Asymptotic Properties of the Least Squares Estimator

The distributional properties of the least squares estimator are the basis for tests and confidence intervals that we will discuss in the next chapter. In order to have exact tests and intervals, the assumption of normally distributed errors is needed. However, some of the properties are still approximately valid, as long as the sample size n tends to infinity or is sufficiently large. For clarification, we index the model with the number of observations n:

3.9 Statistical Properties of the Least Squares Estimator

Without Specific Distributional Assumptions

1. *Expectation:* $\mathrm{E}(\hat{\boldsymbol{\beta}}) = \boldsymbol{\beta}$, implying that the least squares estimator is unbiased.
2. *Covariance Matrix:* $\mathrm{Cov}(\hat{\boldsymbol{\beta}}) = \sigma^2 (\boldsymbol{X}'\boldsymbol{X})^{-1}$. The diagonal elements can be expressed as

$$\mathrm{Var}(\hat{\beta}_j) = \frac{\sigma^2}{(1 - R_j^2) \sum_{i=1}^{n} (x_{ij} - \overline{x}_j)^2},$$

where R_j^2 is the coefficient of determination of the regression between x_j as the response variable and the remaining columns of $\boldsymbol{X}$ as regressors. An estimator for the covariance matrix is given by

$$\widehat{\mathrm{Cov}}(\hat{\boldsymbol{\beta}}) = \hat{\sigma}^2 (\boldsymbol{X}'\boldsymbol{X})^{-1} = \frac{1}{n-p} \hat{\boldsymbol{\varepsilon}}'\hat{\boldsymbol{\varepsilon}} (\boldsymbol{X}'\boldsymbol{X})^{-1}.$$

3. *Gauß–Markov Theorem:* Among all linear and unbiased estimators $\hat{\boldsymbol{\beta}}^L$, the least squares estimator has minimal variances, implying

$$\mathrm{Var}(\hat{\beta}_j) \le \mathrm{Var}(\hat{\beta}_j^L), \quad j = 0, \ldots, k.$$

For any linear combination $\boldsymbol{b}'\boldsymbol{\beta}$ it holds

$$\mathrm{Var}(\boldsymbol{b}'\hat{\boldsymbol{\beta}}) \le \mathrm{Var}(\boldsymbol{b}'\hat{\boldsymbol{\beta}}^L).$$

With Normality Assumption

1. *Distribution of the response variable:*

$$\boldsymbol{y} \sim \mathrm{N}(\boldsymbol{X}\boldsymbol{\beta}, \sigma^2 \boldsymbol{I}).$$

2. *Distribution of the least squares estimator:*

$$\hat{\boldsymbol{\beta}} \sim \mathrm{N}(\boldsymbol{\beta}, \sigma^2 (\boldsymbol{X}'\boldsymbol{X})^{-1}).$$

3. *Distribution of weighted distance:*

$$\frac{(\hat{\boldsymbol{\beta}} - \boldsymbol{\beta})'(\boldsymbol{X}'\boldsymbol{X})(\hat{\boldsymbol{\beta}} - \boldsymbol{\beta})}{\sigma^2} \sim \chi_p^2.$$

$$\boldsymbol{y}_n = \boldsymbol{X}_n\boldsymbol{\beta} + \boldsymbol{\varepsilon}_n, \quad \mathrm{E}(\boldsymbol{\varepsilon}_n) = \mathbf{0}, \quad \mathrm{Cov}(\boldsymbol{\varepsilon}_n) = \sigma^2\boldsymbol{I}_n.$$

Similarly, we index the least squares estimator $\hat{\boldsymbol{\beta}}_n$ and the variance estimator $\hat{\sigma}_n^2$ with n. To obtain valid asymptotic results, we need to go beyond the assumptions 1–3 stated in Box 3.1 (p. 76). Further assumptions are needed regarding the limiting behavior of the design matrix $\boldsymbol{X}_n$ and with it the sequence $\boldsymbol{x}_1, \ldots, \boldsymbol{x}_n, \ldots$ of the design vectors. A standard assumption is that the matrix $\boldsymbol{X}_n'\boldsymbol{X}_n$ averaged over n converges to a limiting positive definite matrix $\boldsymbol{V}$, i.e.,

$$\lim_{n\to\infty} \frac{1}{n}\boldsymbol{X}_n'\boldsymbol{X}_n = \boldsymbol{V}, \quad \boldsymbol{V} \text{ positive definite.} \tag{3.20}$$

In this case we have the following asymptotic results:

3.10 Asymptotic Properties of the Least Squares Estimator

1. The least squares estimator $\hat{\boldsymbol{\beta}}_n$ for $\boldsymbol{\beta}$ and the ML or REML estimator $\hat{\sigma}_n^2$ for the variance σ^2 are consistent.
2. The least squares estimator asymptotically follows a normal distribution, specifically

$$\sqrt{n}(\hat{\boldsymbol{\beta}}_n - \boldsymbol{\beta}) \xrightarrow{d} \mathrm{N}(\mathbf{0}, \sigma^2\boldsymbol{V}^{-1}).$$

   That is the difference $\hat{\boldsymbol{\beta}}_n - \boldsymbol{\beta}$ normalized with $\sqrt{n}$ converges in distribution to the normal distribution on the right-hand side.

We use these asymptotic results for a sufficiently large sample size n as follows. First, $\hat{\boldsymbol{\beta}}_n$ has an approximately normal distribution

$$\hat{\boldsymbol{\beta}}_n \overset{a}{\sim} \mathrm{N}(\boldsymbol{\beta}, \sigma^2\boldsymbol{V}^{-1}/n).$$

If we replace σ^2 with the consistent estimator $\hat{\sigma}_n^2$ and $\boldsymbol{V}$ with the approximation $\boldsymbol{V} \overset{a}{\sim} 1/n\boldsymbol{X}_n'\boldsymbol{X}_n$, we have

$$\hat{\boldsymbol{\beta}}_n \overset{a}{\sim} \mathrm{N}(\boldsymbol{\beta}, \hat{\sigma}_n^2(\boldsymbol{X}_n'\boldsymbol{X}_n)^{-1}).$$

This implies that, with sufficiently large sample size and provided that Eq. (3.20) holds, the least squares estimator has the same approximate normal distribution, regardless of the normal assumption for $\boldsymbol{\varepsilon}$. Assumption (3.20) is particularly ensured if the observed covariate vectors $\boldsymbol{x}_i$, $i = 1, \ldots, n$, are independent and identically distributed realizations of stochastic covariates $\boldsymbol{x} = (1, x_1, \ldots, x_k)'$, i.e., if the observations $(y_i, \boldsymbol{x}_i)$ form a random sample from $(y, \boldsymbol{x})$. This condition is met for many empirical studies, e.g., in our applications on the Munich rent index and on malnutrition in developing countries. In such cases, the law of large numbers implies

$$\frac{1}{n}\boldsymbol{X}_n'\boldsymbol{X}_n = \frac{1}{n}\sum_{i=1}^{n}\boldsymbol{x}_i\boldsymbol{x}_i' \rightarrow \mathrm{E}(\boldsymbol{x}\boldsymbol{x}') =: \boldsymbol{V}.$$

However, the condition is typically violated for regressors following a trend. A simple example is a linear trend $x_i = i$, i.e.,

$$y_i = \beta \cdot i + \varepsilon_i, \quad i = 1, \ldots, n.$$

For simplicity the intercept β_0 is assumed to be zero. We have

$$\frac{1}{n}\boldsymbol{X}_n'\boldsymbol{X}_n = \frac{1}{n}\sum_{i=1}^{n}x_i^2 = \frac{1}{n}(1 + \cdots + i^2 + \cdots + n^2) \rightarrow \infty,$$

resulting in a violation of assumption (3.20).

In fact, consistency can be derived under the following more general condition:

$$(\boldsymbol{X}_n'\boldsymbol{X}_n)^{-1} \rightarrow \boldsymbol{0}. \tag{3.21}$$

Informally, this implies that the covariate information grows as the sample size increases. It can be shown that Eq. (3.21) is a necessary and sufficient condition for consistency of the least squares estimator and its variance estimator; see Lai, Robins, and Wei (1979) and Drygas (1976).

For asymptotic normality, we additionally need the condition

$$\max_{i=1,\ldots,n} \boldsymbol{x}_i'(\boldsymbol{X}_n'\boldsymbol{X}_n)^{-1}\boldsymbol{x}_i \rightarrow 0 \quad \text{for} \quad n \rightarrow \infty. \tag{3.22}$$

In short, this implies that the influence of each observation $\boldsymbol{x}_i$ is negligible relative to the entire information $\boldsymbol{X}_n'\boldsymbol{X}_n = \sum_{i=1}^{n}\boldsymbol{x}_i\boldsymbol{x}_i'$. The central limit theorem (specifically in the Lindeberg–Feller form) then applies. Provided that the conditions (3.21) and (3.22) are satisfied, the normal approximation still holds:

$$\hat{\boldsymbol{\beta}}_n \overset{a}{\sim} \mathrm{N}(\boldsymbol{\beta}, \hat{\sigma}_n^2(\boldsymbol{X}_n'\boldsymbol{X}_n)^{-1}),$$

which is important for practical use.

Example 3.11 Simple Linear Regression

For the simple linear regression model (excluding the intercept β_0)

$$y_i = \beta x_i + \varepsilon_i,$$

the following can be easily verified:

1. For a linear trend $x_i = i$, conditions (3.21) and (3.22) are satisfied.
2. For $x_i = 1/i$, both Eqs. (3.21) and (3.22) are violated, which implies that the least squares estimator is neither consistent nor asymptotically normal. The reason is that the

sequence of regressor values $x_i = 1/i$ converges too fast towards zero and, thus, does not provide increasing information with increasing sample size.

3. For $x_i = 1/\sqrt{i}$, conditions (3.21) and (3.22) are met. Despite that $x_i \to 0$, enough covariate information is provided as $n \to \infty$.

△

Statistical Properties of Residuals

We close this section with the examination of the statistical properties of the residuals $\hat{\varepsilon}_i = y_i - \boldsymbol{x}_i'\hat{\boldsymbol{\beta}}$. In terms of the hat matrix $\boldsymbol{H} = \boldsymbol{X}(\boldsymbol{X}'\boldsymbol{X})^{-1}\boldsymbol{X}'$, the residuals can be expressed as

$$\hat{\boldsymbol{\varepsilon}} = (\boldsymbol{I} - \boldsymbol{H})\boldsymbol{y} = \boldsymbol{y} - \boldsymbol{X}(\boldsymbol{X}'\boldsymbol{X})^{-1}\boldsymbol{X}'\boldsymbol{y};$$

see p. 107.

Thus, we obtain the expectation of the residuals as

$$\mathrm{E}(\hat{\boldsymbol{\varepsilon}}) = \mathrm{E}(\boldsymbol{y}) - \boldsymbol{X}(\boldsymbol{X}'\boldsymbol{X})^{-1}\boldsymbol{X}'\mathrm{E}(\boldsymbol{y}) = \boldsymbol{X}\boldsymbol{\beta} - \boldsymbol{X}(\boldsymbol{X}'\boldsymbol{X})^{-1}\boldsymbol{X}'\boldsymbol{X}\boldsymbol{\beta} = \boldsymbol{0},$$

as well as their covariance matrix

$$\mathrm{Cov}(\hat{\boldsymbol{\varepsilon}}) = \mathrm{Cov}((\boldsymbol{I} - \boldsymbol{H})\boldsymbol{y}) = (\boldsymbol{I} - \boldsymbol{H})\sigma^2\boldsymbol{I}(\boldsymbol{I} - \boldsymbol{H})' = \sigma^2(\boldsymbol{I} - \boldsymbol{H}).$$

In deriving the covariance matrix, we used Theorem B.2.5 (p. 647) in Appendix B in combination with the fact that the matrix $\boldsymbol{I} - \boldsymbol{H}$ is symmetric and idempotent. Specifically for the variances of the residuals, we have

$$\mathrm{Var}(\hat{\varepsilon}_i) = \sigma^2(1 - h_{ii}),$$

where h_{ii} is the ith diagonal element of the hat matrix. We state the following:

- Similar to the error terms, the residuals have mean zero.
- In contrast to the error terms, the residuals are *not* uncorrelated.
- In contrast to the error terms, the residuals have *heteroscedastic variances*. Since $\frac{1}{n} \leq h_{ii} \leq 1$ (see p. 108), the variance of the ith residual approaches zero as h_{ii} approaches one.

If we additionally assume normally distributed errors, we are able to derive the distribution of the residuals. We then have

$$\hat{\boldsymbol{\varepsilon}} \sim \mathrm{N}(\boldsymbol{0}, \sigma^2(\boldsymbol{I} - \boldsymbol{H})). \tag{3.23}$$

Due to $\mathrm{rk}(\boldsymbol{H}) = p \leq n$, this is a singular normal distribution; see also Sect. B.3.2 (p. 650) in Appendix B.

Using Eq. (3.23), we can also make statements about the residual sum of squares. In Sect. 3.5.2 (p. 171), we show

$$\frac{\hat{\boldsymbol{\varepsilon}}'\hat{\boldsymbol{\varepsilon}}}{\sigma^2} \sim \chi^2_{n-p},$$

3.11 Statistical Properties of Residuals

Without Specific Distributional Assumptions

1. *Expectation:* $\mathrm{E}(\hat{\boldsymbol{\varepsilon}}) = \mathbf{0}$, i.e., the residuals have mean zero.
2. *Variance:*
$$\mathrm{Var}(\hat{\varepsilon}_i) = \sigma^2(1 - h_{ii}),$$
which means that the residuals (in contrast to the errors ε_i) have heteroscedastic variances.
3. *Covariance Matrix:*
$$\mathrm{Cov}(\hat{\boldsymbol{\varepsilon}}) = \sigma^2(\boldsymbol{I} - \boldsymbol{H}) = \sigma^2(\boldsymbol{I} - \boldsymbol{X}(\boldsymbol{X}'\boldsymbol{X})^{-1}\boldsymbol{X}'),$$
which implies that the residuals (in contrast to the errors) are not uncorrelated.

With Normal Assumption

4. *Distribution of Residuals:*
$$\hat{\boldsymbol{\varepsilon}} \sim \mathrm{N}(\mathbf{0}, \sigma^2(\boldsymbol{I} - \boldsymbol{H})).$$
5. *Distribution of the Residual Sum of Squares:*
$$\frac{\hat{\boldsymbol{\varepsilon}}'\hat{\boldsymbol{\varepsilon}}}{\sigma^2} = (n-p)\frac{\hat{\sigma}^2}{\sigma^2} \sim \chi^2_{n-p}.$$
6. *Independence:* The residual sum of squares $\hat{\boldsymbol{\varepsilon}}'\hat{\boldsymbol{\varepsilon}}$ and the least squares estimator $\hat{\boldsymbol{\beta}}$ are independent.

which is equivalent to
$$(n-p)\frac{\hat{\sigma}^2}{\sigma^2} \sim \chi^2_{n-p}.$$

Moreover, it can be shown that the residual sum of squares and the least squares estimator are independent (again see Sect. 3.5.2 p. 171 for a proof). Both statements are necessary for the derivation of hypothesis tests regarding the regression coefficients.

Standardized and Studentized Residuals

In practice, the residuals are often used to confirm model assumptions in a linear model. However, the residuals are not always adequate for this purpose. As we have seen, the residuals are neither homoscedastic nor uncorrelated. In most cases, the correlation is negligible, but this is not the case with the heteroscedasticity of the residuals. Thus, the verification of the assumption of homoscedastic errors is problematic, as heteroscedastic residuals are in general not an indicator for heteroscedastic errors. An obvious solution of the heteroscedasticity problem is standardization. Dividing through the estimated standard deviation of the residuals, we obtain the *standardized residuals*

$$r_i = \frac{\hat{\varepsilon}_i}{\hat{\sigma}\sqrt{1-h_{ii}}}. \tag{3.24}$$

Provided that the model assumptions are correct, the standardized residuals are homoscedastic. Thus, the analysis of standardized residuals helps to assess whether or not the assumption of homoscedastic variances is violated. We often plot the standardized residuals against the predicted values or the values of covariates; see Sect. 3.4.4 on model diagnosis for examples.

Since the residuals are normally distributed and $(n-p)\hat{\sigma}^2/\sigma^2$ follows a χ^2_{n-p}-distribution, it is tempting to assume the standardized residuals follow a t-distribution; see Definition B.13 in Appendix B. This conclusion is not correct, as $\hat{\varepsilon}_i$ is part of $\hat{\sigma}$, thus numerator and denominator in Eq. (3.24) are not stochastically independent. However, it is possible to bypass the problem of dependence. To do so, we define "leave-one-out" estimators, which are based on all observations excluding the ith observation. We can then define residuals, which are based on these "leave-one-out" estimators, and show that they follow a valid t-distribution.

Define $\boldsymbol{X}_{(i)}$ and $\boldsymbol{y}_{(i)}$ as the design matrix and response vector, respectively, from which the ith row is removed. The corresponding least squares estimator $\hat{\boldsymbol{\beta}}_{(i)}$, which is based on all observations except the ith one, is then given by

$$\hat{\boldsymbol{\beta}}_{(i)} = (\boldsymbol{X}'_{(i)}\boldsymbol{X}_{(i)})^{-1}\boldsymbol{X}'_{(i)}\boldsymbol{y}_{(i)}.$$

Using $\hat{\boldsymbol{\beta}}_{(i)}$, we obtain predicted values $\hat{y}_{(i)} = \boldsymbol{x}'_i\hat{\boldsymbol{\beta}}_{(i)}$ and thus residuals

$$\hat{\varepsilon}_{(i)} = y_i - \hat{y}_{(i)} = y_i - \boldsymbol{x}'_i(\boldsymbol{X}'_{(i)}\boldsymbol{X}_{(i)})^{-1}\boldsymbol{X}'_{(i)}\boldsymbol{y}_{(i)}$$

for the ith observation. Simple calculations show

$$\hat{\varepsilon}_{(i)} \sim \mathrm{N}(0, \sigma^2(1 + \boldsymbol{x}'_i(\boldsymbol{X}'_{(i)}\boldsymbol{X}_{(i)})^{-1}\boldsymbol{x}_i))$$

or

$$\frac{\hat{\varepsilon}_{(i)}}{\sigma(1 + \boldsymbol{x}'_i(\boldsymbol{X}'_{(i)}\boldsymbol{X}_{(i)})^{-1}\boldsymbol{x}_i)^{1/2}} \sim \mathrm{N}(0, 1).$$

According to property 5 of the preceding Box 3.11, it follows

$$(n-p-1)\frac{\hat{\sigma}^2_{(i)}}{\sigma^2} \sim \chi^2_{n-p-1},$$

where

$$\hat{\sigma}^2_{(i)} = \frac{1}{n-p-1}\left[\left(y_1 - \mathbf{x}_1'\hat{\boldsymbol{\beta}}_{(i)}\right)^2 + \ldots + \left(y_{i-1} - \mathbf{x}_{i-1}'\hat{\boldsymbol{\beta}}_{(i)}\right)^2 + \left(y_{i+1} - \mathbf{x}_{i+1}'\hat{\boldsymbol{\beta}}_{(i)}\right)^2 + \ldots + \left(y_n - \mathbf{x}_n'\hat{\boldsymbol{\beta}}_{(i)}\right)^2\right]$$

is an estimator for σ^2, which is not based on the ith observation. Now we can use Definition B.13 of the t-distribution (p. 644) and obtain the *studentized residuals*

$$r_i^* = \frac{\hat{\varepsilon}_{(i)}}{\hat{\sigma}_{(i)}(1 + \mathbf{x}_i'(\mathbf{X}_{(i)}'\mathbf{X}_{(i)})^{-1}\mathbf{x}_i)^{1/2}} \sim \mathrm{t}_{n-p-1}.$$

A crucial requirement for this distributional statement is that $\hat{\varepsilon}_{(i)}$ and $\hat{\sigma}_{(i)}$ are independent, which holds because we did not use the ith observation y_i when calculating $\hat{\sigma}_{(i)}$. Another requirement is that the model is correctly specified. Knowledge of the distribution of the studentized residuals can be used for model diagnosis to identify observations which are not in agreement with the estimated model; see Sect. 3.4.4 for details.

The standardized and studentized residuals help to circumvent the problem of heteroscedastic residuals. However there remains the problem of correlation among the residuals. The literature does offer suggestions on how to define uncorrelated residuals. As mentioned, in practice the correlation of standardized and studentized residuals in correctly specified models can be neglected. Thus we do not pursue this issue further.

Box 3.12 summarizes the various definitions of residuals. The box also shows (without proof) alternative definitions of studentized residuals. In particular, we find that the studentized residuals can be computed from the standardized residuals. The beauty of this result is that it is unnecessary to recompute the least squares estimator every time we delete one observation: all that we need is the full model least squares estimator and the diagonal elements of the hat matrix. Finally, we recap the definition of partial residuals as used in Example 3.5 (p. 90).

3.3 Hypothesis Testing and Confidence Intervals

This section describes statistical tests for hypotheses regarding the unknown regression parameters $\boldsymbol{\beta}$. Due to the duality between statistical tests and confidence intervals, we are also able to construct confidence intervals for the regression parameters $\boldsymbol{\beta}$. A requirement for the construction of (exact) tests and confidence intervals is the assumption of normally distributed errors. Therefore, we first assume independently and identically normally distributed errors, i.e., $\varepsilon_i \sim \mathrm{N}(0, \sigma^2)$. However, tests and confidence intervals are relatively robust to mild departures from the normality assumption. Moreover, in Sect. 3.3.1 we will show that the tests and confidence intervals, derived under the assumption of normality, remain valid for large sample size even with non-normal errors.

3.12 Overview of Residuals

Ordinary Residuals

The residuals are given by

$$\hat{\varepsilon}_i = y_i - \hat{y}_i = y_i - \boldsymbol{x}_i'\hat{\boldsymbol{\beta}} \quad i = 1, \ldots, n.$$

Standardized Residuals

The standardized residuals are defined by

$$r_i = \frac{\hat{\varepsilon}_i}{\hat{\sigma}\sqrt{1 - h_{ii}}},$$

where h_{ii} is the ith diagonal element of the hat matrix.

Studentized Residuals

The studentized residuals are defined by

$$r_i^* = \frac{\hat{\varepsilon}_{(i)}}{\hat{\sigma}_{(i)}(1 + \boldsymbol{x}_i'(\boldsymbol{X}_{(i)}'\boldsymbol{X}_{(i)})^{-1}\boldsymbol{x}_i)^{1/2}} = \frac{\hat{\varepsilon}_i}{\hat{\sigma}_{(i)}\sqrt{1 - h_{ii}}} = r_i \left(\frac{n - p - 1}{n - p - r_i^2}\right)^{1/2}.$$

The studentized residuals are used to verify model assumptions and to discover outliers (see Sect. 3.4.4).

Partial Residuals

The partial residuals regarding covariate x_j are defined by

$$\hat{\varepsilon}_{x_j,i} = y_i - \hat{\beta}_0 - \ldots - \hat{\beta}_{j-1}x_{i,j-1} - \hat{\beta}_{j+1}x_{i,j+1} - \ldots - \hat{\beta}_k x_{ik} = \hat{\varepsilon}_i + \hat{\beta}_j x_{ij}.$$

In the partial residuals $\hat{\varepsilon}_{x_j,i}$, all covariate effects with the exception of the one associated with x_j are removed. Hence, they are very useful for exploring whether the influence of x_j is modeled correctly (see Sect. 3.4.4).

Example 3.12 Munich Rent Index—Hypothesis Testing

We revisit the data from the Munich rent index to illustrate hypothesis testing. We use the data for the 1999 rent index, in combination with the follow-up data from 2001; see Example 3.7 (p. 100). Consider the regression model

$$\begin{aligned} \mathit{rentsqm}_i &= \beta_0 + \beta_1\,\mathit{areainvc}_i + \beta_2\,\mathit{yearco}_i + \beta_3\,\mathit{yearco2}_i \\ &\quad + \beta_4\,\mathit{yearco3}_i + \beta_5\,\mathit{nkitchen} + \beta_6\,\mathit{pkitchen} + \beta_7\,\mathit{year01} + \varepsilon_i, \end{aligned} \tag{3.25}$$

where *areainvc* is the transformation $1/area$ centered around zero and *yearo*, *yearo2*, *yearo3* constitute third-order orthogonal polynomials (see Example 3.5 on p. 90) for the year of construction. The dummy variable *year01* is an indicator for the follow-up period, i.e., $year01 = 1$ if an observation is from the follow-up and $year01 = 0$ otherwise. The following estimated model results:

$$\begin{aligned}\widehat{rentsqm}_i &= 6.94 + 124.34\, areainvc_i + 0.73\, yearco_i + 0.44\, yearco2_i \\ &\quad -0.01\, yearco3_i + 1.04\, nkitchen + 1.31\, pkitchen - 0.19\, year01.\end{aligned}$$

We find that the average net rent per square meter in the year 2001 decreases about 0.19 Euro, in comparison to the net rent in 1999. First of all, this change is relatively small. Moreover a decrease from 1999 to 2001 is surprising. It raises the question if such a net rent decrease can be extrapolated to the entire population or if it is merely a random phenomenon of the observed random sample. Thus we wish to test if the regression parameter β_7 is significantly different from zero. This can be achieved by testing the hypothesis

$$H_0 : \beta_7 = 0 \qquad \text{against} \qquad H_1 : \beta_7 \neq 0,$$

using an appropriate statistical test. Of course, we also want to test whether or not other variables should be included in the regression model. The test for significance of the variable *kitchen* with the three categories, "substandard kitchen" (reference category), "standard kitchen" (dummy variable *nkitchen*), and "premium kitchen" (dummy variable *pkitchen*), is more complicated since it involves a categorical variable at three levels. In this case, we have to test the hypothesis

$$H_0 : \begin{pmatrix}\beta_5\\ \beta_6\end{pmatrix} = \begin{pmatrix}0\\ 0\end{pmatrix} \qquad \text{against} \qquad H_1 : \begin{pmatrix}\beta_5\\ \beta_6\end{pmatrix} \neq \begin{pmatrix}0\\ 0\end{pmatrix}.$$

Generally, a rent index should be as simple as possible. In statistical terms, this means that the model should be sparse and avoid overparameterization. Having this in mind, we may ask whether it is really necessary to differentiate between standard and premium kitchens. The estimated regression coefficients for standard kitchens are not very different from premium ones. The corresponding statistical hypotheses are given by

$$H_0 : \beta_5 = \beta_6 \qquad \text{against} \qquad H_1 : \beta_5 \neq \beta_6$$

or equivalently

$$H_0 : \beta_5 - \beta_6 = 0 \qquad \text{against} \qquad H_1 : \beta_5 - \beta_6 \neq 0.$$

△

The discussed problems are exemplary for the most common statistical hypotheses in linear models:

1. *Test of significance*:

$$H_0 : \beta_j = 0 \qquad \text{against} \qquad H_1 : \beta_j \neq 0.$$

2. *Composite test of a subvector* $\boldsymbol{\beta}_1 = (\beta_1, \ldots, \beta_r)'$:

$$H_0 : \boldsymbol{\beta}_1 = \mathbf{0} \qquad \text{against} \qquad H_1 : \boldsymbol{\beta}_1 \neq \mathbf{0}.$$

3. *Test of equality*:

$$H_0 : \beta_j - \beta_r = 0 \qquad \text{against} \qquad H_1 : \beta_j - \beta_r \neq 0.$$

We can treat these three test problems as special cases of tests for *general linear hypotheses*

$$H_0 : \boldsymbol{C\beta} = \boldsymbol{d} \qquad \text{against} \qquad H_1 : \boldsymbol{C\beta} \neq \boldsymbol{d}, \tag{3.26}$$

where $\boldsymbol{C}$ is an $r \times p$-matrix with $\text{rk}(\boldsymbol{C}) = r \leq p$. This means that under H_0 a total of r linear-independent conditions apply. When testing the significance of an influential variable, $\boldsymbol{d} = 0$ is a scalar and $\boldsymbol{C}$ is a $1 \times p$-matrix given by $\boldsymbol{C} = (0, \ldots, 0, 1, 0, \ldots, 0)$, where the one is at the $(j+1)$-th column of the matrix. In the case of testing the first r components, we obtain the r-dimensional vector $\boldsymbol{d} = \boldsymbol{0}$ and the $r \times p$-matrix

$$\boldsymbol{C} = \begin{pmatrix} 0 & 1 & 0 & \cdots & 0 & 0 & \cdots & 0 \\ 0 & 0 & 1 & \cdots & 0 & 0 & \cdots & 0 \\ \vdots & & & \ddots & & 0 & \cdots & 0 \\ 0 & 0 & 0 & \cdots & 1 & 0 & \cdots & 0 \end{pmatrix}.$$

Lastly, when testing equality of two coefficients, we obtain the scalar $\boldsymbol{d} = 0$ and the $1 \times p$-matrix $\boldsymbol{C} = (0, \ldots, 1, \ldots, -1, \ldots, 0)$, where the one is at the $(j+1)$-th position and the minus one at the $(r+1)$-th position.

In the next section, we will develop a test for general linear hypotheses (3.26), which will contain the test problems 1–3 given above as special cases.

3.3.1 Exact F-Test

We first assume Gaussian errors to derive an exact test. In order to develop an appropriate test statistic for the general test problem (3.26), we proceed as follows:

1. Compute the residual sum of squares $\text{SSE} = \hat{\boldsymbol{\varepsilon}}'\hat{\boldsymbol{\varepsilon}}$ for the full model.
2. Compute the residual sum of squares $\text{SSE}_{H_0} = \hat{\boldsymbol{\varepsilon}}'_{H_0}\hat{\boldsymbol{\varepsilon}}_{H_0}$ for the model under the null hypothesis, i.e., under the restriction $\boldsymbol{C\beta} = \boldsymbol{d}$.
3. Calculate the statistic
$$\frac{\Delta\text{SSE}}{\text{SSE}} = \frac{\text{SSE}_{H_0} - \text{SSE}}{\text{SSE}}, \tag{3.27}$$
which is the relative distance in residual sum of squares between the restricted model and the full model.

To interpret the test statistic, we first note that the fit to the data under the restriction is at most equal to the unrestricted fit. Figure 3.15 gives an illustration in the case of a linear model $y = \beta x + \varepsilon$, with only one parameter β. This figure shows the residual sum of squares $\text{LS}(\beta)$ depending on β. The least squares estimator $\hat{\beta} = 1.78$ is additionally marked. If the restriction $0 \leq \beta \leq 1$ applies, the parameter estimates can only be chosen within the two vertical lines. In this limited parameter

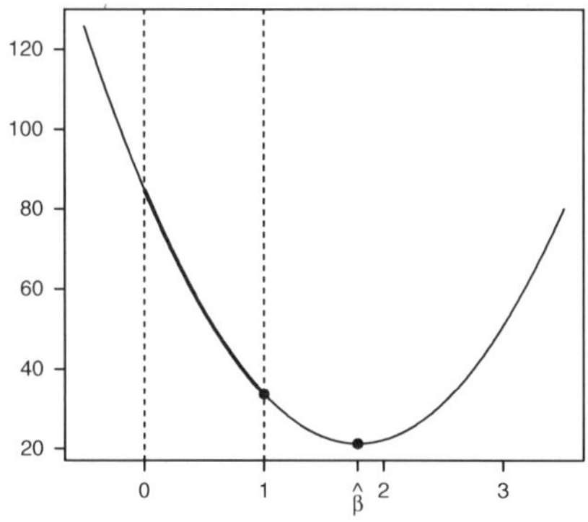

Fig. 3.15 Illustration of the difference in goodness of fit between the unconstrained least squares estimator and the estimator under the constraint $0 \leq \beta \leq 1$. The (unconstrained) least squares estimator is labeled as $\hat{\beta}$. For the constrained solution, we have $\hat{\beta} = 1$

space, the residual sum of squares reaches its minimum for $\hat{\beta} = 1$. In summary, the difference $\text{SSE}_{H_0} - \text{SSE}$ is always greater or equal to zero, since the fit to the data under the restriction $\boldsymbol{C\beta} = \boldsymbol{d}$ can be, at best, as good as with the unconstrained least squares estimator. A formal proof for $\Delta\text{SSE} \geq 0$ will be given in the appendix of this chapter on p. 172.

The above illustration also shows the main idea behind the statistic (3.27). The smaller the difference between SSE_{H_0} and SSE, the closer the two minima are, shown in Fig. 3.15, and the more likely it is that we will retain the null hypothesis. On the other hand, the larger the difference, the more likely it is that we will reject the null hypotheses. The test statistic actually used is

$$F = \frac{\frac{1}{r}\Delta\text{SSE}}{\frac{1}{n-p}\text{SSE}} = \frac{n-p}{r}\frac{\Delta\text{SSE}}{\text{SSE}}, \tag{3.28}$$

where r represents the number of (linear independent) restrictions, or the number of rows in $\boldsymbol{C}$. The additional constant factor $\frac{n-p}{r}$ is not important for interpretation. It ensures that the distribution of the test statistic under the null hypothesis is a known distribution.

In order to derive the distribution of the test statistic under H_0, we proceed as follows:

1. *Determine the least squares estimator under H_0*

   In Sect. 3.5.2 (p. 172), we derive the least squares estimator $\hat{\boldsymbol{\beta}}^R$ under H_0, i.e., under the restriction $\boldsymbol{C\beta} = \boldsymbol{d}$. We obtain

$$\hat{\boldsymbol{\beta}}^R = \hat{\boldsymbol{\beta}} - (\boldsymbol{X}'\boldsymbol{X})^{-1}\boldsymbol{C}'(\boldsymbol{C}\,(\boldsymbol{X}'\boldsymbol{X})^{-1}\boldsymbol{C}')^{-1}(\boldsymbol{C}\,\hat{\boldsymbol{\beta}} - \boldsymbol{d}).$$

2. *Determine the difference in residual sum of squares*

   In Sect. 3.5.2, we derive the difference ΔSSE in the residual sum of squares, given by

$$\Delta \text{SSE} = (C\hat{\beta} - d)' \left(C(X'X)^{-1}C'\right)^{-1} (C\hat{\beta} - d).$$

3. *Stochastic properties of the difference in residual sum of squares*

   When determining the distribution of the test statistic under H_0, we need the following stochastic properties of ΔSSE:

   (a) $\text{E}(\Delta\text{SSE}) = r\sigma^2 + (C\beta - d)' \left(C(X'X)^{-1}C'\right)^{-1} (C\beta - d)$ (whether the restriction is fulfilled or not).

   (b) Under H_0, we have $\frac{1}{\sigma^2} \cdot \Delta\text{SSE} \sim \chi^2_r$.

   (c) ΔSSE and SSE are stochastically independent.

   Proofs of these statements are again found in Sect. 3.5.2.

4. *Distribution of the test statistic*

   Using the stochastic properties in 3, we can now derive the distribution of the test statistic under the null hypothesis: According to property 3(b), we obtain under H_0:

$$\frac{1}{\sigma^2}\Delta\text{SSE} \sim \chi^2_r.$$

   Furthermore, we have

$$\frac{1}{\sigma^2}\text{SSE} \sim \chi^2_{n-p};$$

   refer to Box 3.11 on p. 123. Finally, ΔSSE and SSE are stochastically independent according to property 3(c). Thus, under H_0, the test statistic (3.28) follows an F-distribution with r and $n-p$ degrees of freedom, i.e., $F \sim \text{F}_{r,n-p}$. This result follows from the definition of the F-distribution; see Definition B.14 (p. 645) in Appendix B.1.

This leads us to the following test: Let α be the significance level. We then reject the null hypothesis if the test statistic is larger than the $(1-\alpha)$-quantile of the corresponding F-distribution, i.e.,

$$F > F_{r,n-p}(1-\alpha).$$

Connection to the Wald Test

We next demonstrate an interesting connection to the Wald test. See Sect. B.4.4 (p. 662) in Appendix B for the general background and the idea of the Wald test. The steps to derive the distribution of the test statistic for the F-test (see above) yield the relationship

$$\Delta \text{SSE} = (C\hat{\beta} - d)' \left(C(X'X)^{-1}C'\right)^{-1} (C\hat{\beta} - d).$$

With $\text{SSE} = \hat{\varepsilon}'\hat{\varepsilon} = (n-p)\hat{\sigma}^2$, we have

$$\begin{aligned} F &= \frac{(C\hat{\beta} - d)' \left(\hat{\sigma}^2 C(X'X)^{-1}C'\right)^{-1} (C\hat{\beta} - d)}{r} \\ &= \frac{(C\hat{\beta} - d)'\widehat{\text{Cov}(C\hat{\beta})}^{-1}(C\hat{\beta} - d)}{r}. \end{aligned}$$

This representation offers another interesting interpretation of the test statistic. Clearly, F compares the difference between the estimator $\boldsymbol{C}\hat{\boldsymbol{\beta}}$ and the hypothesized value $\boldsymbol{d}$, weighted by the inverse of the estimated covariance matrix

$$\widehat{\text{Cov}(\boldsymbol{C}\hat{\boldsymbol{\beta}})} = \hat{\sigma}^2 \boldsymbol{C}(\boldsymbol{X}'\boldsymbol{X})^{-1}\boldsymbol{C}'$$

of $\boldsymbol{C}\hat{\boldsymbol{\beta}}$. Note that $\hat{\sigma}^2$ denotes the estimator of σ^2 based on the unconstrained model. The Wald test has an analogous construction, and we have the relationship

$$W = rF$$

between the Wald W and the F statistics. Lastly, we note that the F-test can also be equivalently derived as a likelihood ratio test.

F-Tests for Some Specific Test Problems

In the following, we examine the F-test statistic in detail for some specific test problems:

1. *Test of significance* (t-*test*):

$$H_0 : \beta_j = 0 \qquad \text{against} \qquad H_1 : \beta_j \neq 0 \qquad j = 1, \ldots, k.$$

In this special case, it can be shown that

$$F = \frac{\hat{\beta}_j^2}{\widehat{\text{Var}(\hat{\beta}_j)}} \sim \text{F}_{1,n-p}.$$

Equivalently, the test can be based on the "t-statistic,"

$$t_j = \frac{\hat{\beta}_j}{\text{se}_j}, \qquad (3.29)$$

where $\text{se}_j = \widehat{\text{Var}(\hat{\beta}_j)}^{1/2}$ denotes the estimated standard deviation or standard error of $\hat{\beta}_j$. We can view the original F-test statistic as the square of t_j. According to Definition B.14 in Appendix B.1 t_j is t-distributed with $n - p$ degrees of freedom. We obtain the critical value for the rejection region of the null hypothesis as the $(1 - \alpha/2)$-quantile of the t-distribution with $n - p$ degrees of freedom. Thus, we reject the null hypothesis, if

$$|t_j| > t_{1-\alpha/2}(n - p).$$

We can test the slightly more general hypotheses

$$H_0 : \beta_j = d_j \qquad \text{against} \qquad H_1 : \beta_j \neq d_j \qquad j = 1, \ldots, k$$

in a similar way, by using the modified test statistic

$$t_j = \frac{\beta_j - d_j}{\text{se}_j}.$$

2. *Composite test of an r-dimensional subvector $\boldsymbol{\beta}_1 = (\beta_1, \ldots, \beta_r)'$:*

$$H_0 : \boldsymbol{\beta}_1 = \mathbf{0} \qquad \text{against} \qquad H_1 : \boldsymbol{\beta}_1 \neq \mathbf{0}.$$

In this case, we obtain the test statistic

$$F = \frac{1}{r} \hat{\boldsymbol{\beta}}_1' \widehat{\text{Cov}(\hat{\boldsymbol{\beta}}_1)}^{-1} \hat{\boldsymbol{\beta}}_1 \sim \text{F}_{r,n-p}. \tag{3.30}$$

The estimated covariance matrix of the subvector $\hat{\boldsymbol{\beta}}_1$ consists of the corresponding elements of the full covariance matrix $\hat{\sigma}^2(\boldsymbol{X}'\boldsymbol{X})^{-1}$.
3. *Test for significance of regression:*
We want to test if there is a linear relationship between the response and *any* of the regressors, i.e.,

$$H_0 : \beta_1 = \beta_2 = \cdots = \beta_k = 0.$$

Note, that the alternative hypothesis does not imply that *all* variables are influential. Rather it simply states that at least one of the k covariates is influential. Under H_0 the least squares estimator simplifies to $\hat{\beta}_0 = \bar{y}$. Consequentially, we obtain

$$\text{SSE}_{H_0} = \sum_{i=1}^{n} (y_i - \bar{y})^2,$$

for the residual sum of squares under the null hypothesis. For the difference in the residual sum of squares between the null model and the full model we then have

$$\Delta\text{SSE} = \text{SSE}_{H_0} - \text{SSE} = \sum_{i=1}^{n} (\hat{y}_i - \bar{y})^2,$$

where we applied the analysis of variance formula (3.19) on p. 112. This yields

$$\begin{aligned} F &= \frac{n-p}{k} \frac{\sum (\hat{y}_i - \bar{y})^2}{\sum \hat{\varepsilon}_i^2} \\ &= \frac{n-p}{k} \frac{\sum (\hat{y}_i - \bar{y})^2}{\sum (y_i - \bar{y})^2 - \sum (\hat{y}_i - \bar{y})^2} \end{aligned}$$

Table 3.2 Munich rent index: output for the regression model (3.25) on p. 126

Variable	Coefficient	Standard error	t-statistic	p-value	95 % Confidence-interval	
intercept	6.936	0.037	184.50	<0.001	6.862	7.009
areainv	124.341	4.445	27.97	<0.001	115.625	133.057
yearco	0.729	0.030	24.29	<0.001	0.670	0.788
yearco2	0.436	0.030	14.53	<0.001	0.377	0.495
yearco3	−0.011	0.029	−0.38	0.701	−0.068	0.046
nkitchen	1.044	0.101	10.28	<0.001	0.845	1.243
pkitchen	1.306	0.152	8.57	<0.001	1.007	1.604
year01	−0.192	0.062	−3.08	0.002	−0.314	−0.069

$$= \frac{n-p}{k} \frac{\sum(\hat{y}_i - \bar{y})^2 / \sum(y_i - \bar{y})^2}{1 - \sum(\hat{y}_i - \bar{y})^2 / \sum(y_i - \bar{y})^2}$$

$$= \frac{n-p}{k} \frac{R^2}{1 - R^2}.$$

This result can be interpreted as follows: For a small coefficient of determination R^2 it is more likely to retain the null hypothesis "no linear relationship" (because F is small) than when the coefficient of determination is close to one (as F is then comparably large).

Example 3.13 Munich Rent Index—Standard Output and Hypothesis Tests

At this point, we are able to understand standard output for regression problems provided by software packages. Table 3.2 shows estimation results for the regression model (3.25) of Example 3.12 on p. 126. The table consists of six columns. From the left to the right we find the variable names, the estimated regression coefficients $\hat{\beta}_j$, the standard errors of the coefficients se_j, the test statistics t_j of the tests for $H_0 : \beta_j = 0$ against $H_1 : \beta_j \neq 0$, the corresponding p-values, as well as the respective 95 % confidence intervals (see Sect. 3.3.2 on p. 136). According to Eq. (3.29), the t-statistic is the ratio between the estimated regression coefficient (second column) and the standard error (third column). As usual, the p-value states the minimal significance level α such that the null hypothesis $H_0 : \beta_j = 0$ can be rejected.

We first notice that the decrease of the average net rent by 0.19 Euro in the year 2001 is significant when comparing it to the year of 1999. We can reject the hypothesis $H_0 : \beta_7 = 0$ for every significance level $\alpha > 0.002$. As a matter of fact, there appeared to be a small market break in the housing market during the survey period of the rent index recording. Thus, the estimated decline is explainable.

In order to analyze the significance of the kitchen effect, we test the hypotheses

$$H_0 : \begin{pmatrix} \beta_5 \\ \beta_6 \end{pmatrix} = \begin{pmatrix} 0 \\ 0 \end{pmatrix} \quad \text{versus} \quad H_1 : \begin{pmatrix} \beta_5 \\ \beta_6 \end{pmatrix} \neq \begin{pmatrix} 0 \\ 0 \end{pmatrix}.$$

To compute the test statistic (3.30), we need the estimated covariance matrix $\widehat{\text{Cov}(\hat{\boldsymbol{\beta}})}$ of $\hat{\boldsymbol{\beta}}$, which is given by

$$\begin{pmatrix} 0.00141321 & & & & & & & \\ 0.00732468 & 19.765577 & & & & & & \\ 0.00007468 & -0.01912079 & 0.00090096 & & & & & \\ 0.000076 & 0.02088758 & 0.00001069 & 0.00090025 & & & & \\ -0.00003344 & 0.00206132 & 0.00000174 & 0.00000641 & 0.00085407 & & & \\ -0.00109763 & -0.09199374 & -0.00038591 & -0.00039643 & 0.00009397 & 0.01031718 & & \\ -0.00113005 & 0.0237961 & -0.00062755 & -0.00072908 & -0.00023632 & 0.0014074 & 0.02321696 & \\ -0.00127122 & 0.0025953 & -0.00003465 & -0.00002294 & 0.00010375 & 0.00005632 & 0.00018081 & 0.0038842 \end{pmatrix}.$$

Due to symmetry in the covariance matrix, only the elements below the main diagonal are provided. The test statistic can now be computed as

$$F = \frac{1}{2} \begin{pmatrix} 1.044 \\ 1.306 \end{pmatrix}' \begin{pmatrix} 0.01031718 & 0.0014074 \\ 0.0014074 & 0.02321696 \end{pmatrix}^{-1} \begin{pmatrix} 1.044 \\ 1.306 \end{pmatrix} = 82.22.$$

With a significance level of $\alpha = 0.05$, the corresponding quantile of the F-distribution is given by

$$F_{2,4559-8}(0.95) = 3.00.$$

Since $F = 82.22 > 3.00 = F_{2,4551}(0.95)$, we can reject the null hypothesis at this level. The quality of the kitchen, thus, has a significant influence on the average net rent. Note that this test does not necessarily imply that both regression coefficients are different from zero for the two *kitchen* dummy variables. The null hypothesis is rejected when at least one coefficient significantly differs from zero. However, taking a look at the p-values in Table 3.2 for the variables *nkitchen* and *pkitchen*, we do find that both regression coefficients are significant. The practical implementation of the test is not as intricate as demonstrated here. Most of the statistics software packages provide simple and fast routines for linear tests. In `STATA`, for example, the command `test nkitchen pkitchen` would have produced the desired result.

We conclude this example by answering the question whether a distinction between a standard and a premium kitchen is necessary. We want to test the hypothesis

$$H_0 : \beta_5 - \beta_6 = 0 \quad \text{against} \quad H_1 : \beta_5 - \beta_6 \neq 0.$$

We obtain the test statistic $F = 2.23$. For $\alpha = 0.05$, the corresponding quantile of the F-distribution is given by $F_{1,4551}(0.95) = 3.84$. Since $F = 2.23 < 3.84 = F_{1,4551}(0.95)$, we cannot reject the null hypothesis. Hence, the difference between a standard and premium kitchen is not significant. This test has a p-value of $p = 0.13$.

△

Asymptotic Properties of the F-Test

The applicability of the F-test, considered thus far, requires normally distributed errors. Respecting the connection to the Wald test, we can show that the test can even be applied if the errors are not normally distributed. As already shown, we have the relationship $W = rF$. Moreover, $W \overset{a}{\sim} \chi^2_r$ holds; see Sect. B.4.4 in Appendix B. Thus the $F_{r,n-p}$-distribution converges in distribution with $n \to \infty$ to a χ^2_r / r-distribution, and we are able to use critical values or p-values of the F-test (for large sample size) even in the case of non-normal errors.

3.13 Testing Linear Hypotheses

Hypotheses

1. General linear hypothesis:

$$H_0 : \boldsymbol{C}\boldsymbol{\beta} = \boldsymbol{d} \qquad \text{against} \qquad H_0 : \boldsymbol{C}\boldsymbol{\beta} \neq \boldsymbol{d}$$

   where $\boldsymbol{C}$ is a $r \times p$-matrix with $\text{rk}(\boldsymbol{C}) = r \leq p$ (r linear independent restrictions).
2. Test of significance (t-test):

$$H_0 : \beta_j = 0 \qquad \text{against} \qquad H_1 : \beta_j \neq 0$$

3. Composite test of a subvector:

$$H_0 : \boldsymbol{\beta}_1 = \boldsymbol{0} \qquad \text{against} \qquad H_1 : \boldsymbol{\beta}_1 \neq \boldsymbol{0}$$

4. Test for significance of regression:

$$H_0 : \beta_1 = \beta_2 = \cdots = \beta_k = 0 \text{ against}$$
$$H_1 : \beta_j \neq 0 \text{ for at least one } j \in \{1, \ldots, k\}$$

Test Statistics

Assuming normal errors we obtain under H_0:

1. $F = 1/r\,(\boldsymbol{C}\hat{\boldsymbol{\beta}} - \boldsymbol{d})' \left(\hat{\sigma}^2 \boldsymbol{C}(\boldsymbol{X}'\boldsymbol{X})^{-1}\boldsymbol{C}'\right)^{-1} (\boldsymbol{C}\hat{\boldsymbol{\beta}} - \boldsymbol{d}) \sim \text{F}_{r,n-p}$
2. $t_j = \dfrac{\hat{\beta}_j}{\text{se}_j} \sim \text{t}_{n-p}$
3. $F = \frac{1}{r}(\hat{\boldsymbol{\beta}}_1)'\widehat{\text{Cov}(\hat{\boldsymbol{\beta}}_1)}^{-1}(\hat{\boldsymbol{\beta}}_1) \sim \text{F}_{r,n-p}$
4. $F = \dfrac{n-p}{k}\dfrac{R^2}{1-R^2} \sim \text{F}_{k,n-p}$

Critical Values

Reject H_0 in the case of:

1. $F > F_{r,n-p}(1-\alpha)$
2. $|t| > t_{n-p}(1-\alpha/2)$
3. $F > F_{r,n-p}(1-\alpha)$
4. $F > F_{k,n-p}(1-\alpha)$

The tests are relatively robust against moderate departures from normality. In addition, the tests can be applied for large sample size, even with non-normal errors.

3.3.2 Confidence Regions and Prediction Intervals

Confidence Intervals and Ellipsoids for Regression Coefficients

The duality between two-sided tests and confidence intervals or confidence regions allows us to construct a confidence interval for a single parameter β_j, $j = 0, \dots, k$ or a confidence ellipsoid for a subvector $\boldsymbol{\beta}_1$ of $\boldsymbol{\beta}$. To construct a confidence interval for β_j under normality, we use the test statistic $t_j = (\hat{\beta}_j - d_j)/\text{se}_j$ corresponding to the test $H_0 : \beta_j = d_j$. Recall that we reject the null hypothesis when $|t_j| > t_{n-p}(1 - \alpha/2)$. The test is constructed such that the probability of rejecting H_0 when H_0 is true equals α. Therefore, under H_0 we have

$$\mathrm{P}(|t_j| > t_{n-p}(1 - \alpha/2)) = \alpha.$$

Thus the probability of not rejecting H_0 (given H_0 is true) is provided by

$$\mathrm{P}(|t_j| < t_{n-p}(1 - \alpha/2)) = \mathrm{P}(|(\hat{\beta}_j - \beta_j)/\text{se}_j| < t_{n-p}(1 - \alpha/2)) = 1 - \alpha.$$

This is equivalent to

$$\mathrm{P}(\hat{\beta}_j - t_{n-p}(1 - \alpha/2) \cdot \text{se}_j < \beta_j < \hat{\beta}_j + t_{n-p}(1 - \alpha/2) \cdot \text{se}_j) = 1 - \alpha,$$

and we obtain

$$[\hat{\beta}_j - t_{n-p}(1 - \alpha/2) \cdot \text{se}_j, \hat{\beta}_j + t_{n-p}(1 - \alpha/2) \cdot \text{se}_j]$$

as a $(1 - \alpha)$-confidence interval for β_j.

In a similar way, we can create a $(1 - \alpha)$-confidence region for a r-dimensional subvector $\boldsymbol{\beta}_1$ of $\boldsymbol{\beta}$; see Box 3.14.

Example 3.14 Munich Rent Index—Confidence Intervals

We illustrate the construction of confidence intervals using the regression model taken from the last example; see p. 133. A 95 % confidence interval for the regression coefficient β_7, associated with the time dummy variable, can be calculated as

$$-0.192 \pm 1.96 \cdot 0.062 = [-0.313, -0.070].$$

Here we used $\text{se}_7 = 0.062$ (see Table 3.2) and $t_{n-p}(1 - \alpha/2) = t_{4551}(0.975) = 1.96$. The minor differences found between this confidence interval and the one found in Table 3.2 result from rounding error.

△

Prediction Intervals

In Sect. 3.2.3 on p. 118, we derived an optimal prediction $\hat{y}_0 = \boldsymbol{x}_0'\hat{\boldsymbol{\beta}}$ for a new (future) observation at location $\boldsymbol{x}_0$. Specifically, $\hat{y}_0$ is an estimator for the (conditional) mean $\mathrm{E}(y_0) = \boldsymbol{x}_0'\boldsymbol{\beta} = \mu_0$ of the future observation y_0. In addition to the point estimate, interval estimation for μ_0 is often of interest and is easy to construct. Since $\hat{\boldsymbol{\beta}} \sim \mathrm{N}(\boldsymbol{\beta}, \sigma^2(\boldsymbol{X}'\boldsymbol{X})^{-1})$, it follows for the linear combination

3.14 Confidence Regions and Prediction Intervals

Provided that we have (at least approximately) normally distributed errors or a large sample size, we obtain the following confidence intervals or regions and prediction intervals:

Confidence Interval for β_j

A confidence interval for β_j with level $1-\alpha$ is given by

$$[\hat{\beta}_j - t_{n-p}(1-\alpha/2)\cdot \mathrm{se}_j, \hat{\beta}_j + t_{n-p}(1-\alpha/2)\cdot \mathrm{se}_j].$$

Confidence Ellipsoid for Subvector $\boldsymbol{\beta}_1$

A confidence ellipsoid for $\boldsymbol{\beta}_1 = (\beta_1, \ldots, \beta_r)'$ with level $1-\alpha$ is given by

$$\left\{ \boldsymbol{\beta}_1 : \frac{1}{r}(\hat{\boldsymbol{\beta}}_1 - \boldsymbol{\beta}_1)'\widehat{\mathrm{Cov}(\hat{\boldsymbol{\beta}}_1)}^{-1}(\hat{\boldsymbol{\beta}}_1 - \boldsymbol{\beta}_1) \le F_{r,n-p}(1-\alpha) \right\}.$$

Confidence Interval for μ_0

A confidence interval for $\mu_0 = \mathrm{E}(y_0)$ of a future observation y_0 at location $\boldsymbol{x}_0$ with level $1-\alpha$ is given by

$$\boldsymbol{x}_0'\hat{\boldsymbol{\beta}} \pm t_{n-p}(1-\alpha/2)\hat{\sigma}(\boldsymbol{x}_0'(\boldsymbol{X}'\boldsymbol{X})^{-1}\boldsymbol{x}_0)^{1/2}.$$

Prediction Interval

A prediction interval for a future observation y_0 at location $\boldsymbol{x}_0$ with level $1-\alpha$ is given by

$$\boldsymbol{x}_0'\hat{\boldsymbol{\beta}} \pm t_{n-p}(1-\alpha/2)\hat{\sigma}(1 + \boldsymbol{x}_0'(\boldsymbol{X}'\boldsymbol{X})^{-1}\boldsymbol{x}_0)^{1/2}.$$

$$\boldsymbol{x}_0'\hat{\boldsymbol{\beta}} \sim \mathrm{N}(\boldsymbol{x}_0'\boldsymbol{\beta}, \sigma^2\boldsymbol{x}_0'(\boldsymbol{X}'\boldsymbol{X})^{-1}\boldsymbol{x}_0).$$

Standardizing yields

$$\frac{\boldsymbol{x}_0'\hat{\boldsymbol{\beta}} - \mu_0}{\sigma(\boldsymbol{x}_0'(\boldsymbol{X}'\boldsymbol{X})^{-1}\boldsymbol{x}_0)^{1/2}} \sim \mathrm{N}(0,1).$$

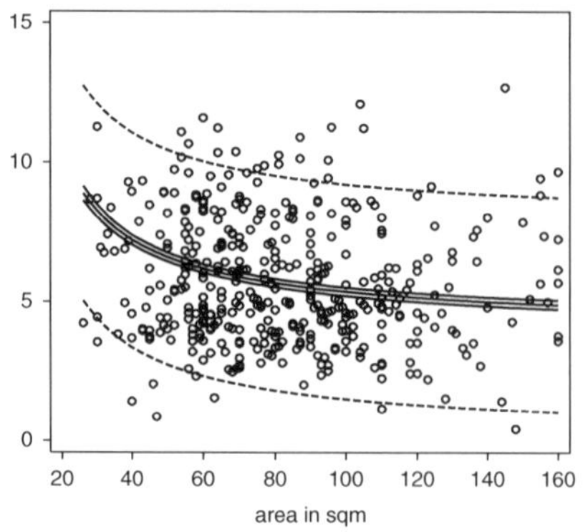

Fig. 3.16 Munich rent index: estimated rent per square meter depending on the living area including 95 % confidence interval (*solid lines*) and 95 % prediction interval (*dashed lines*). The values of the remaining covariates have been set to *yearc* = 1918, *nkitchen* = 0, *gkitchen* = 0, and *year01* = 0. Additionally included are the observations available for this covariate pattern

If we substitute σ^2 with the estimator $\hat{\sigma}^2$, the resulting expression follows a t-distribution with $n - p$ degrees of freedom, and we have

$$P\left(-t_{n-p}(1-\alpha/2) \leq \frac{\boldsymbol{x}_0'\hat{\boldsymbol{\beta}} - \mu_0}{\hat{\sigma}(\boldsymbol{x}_0'(\boldsymbol{X}'\boldsymbol{X})^{-1}\boldsymbol{x}_0)^{1/2}} \leq t_{n-p}(1-\alpha/2)\right) = 1 - \alpha.$$

Consequently,

$$[\boldsymbol{x}_0'\hat{\boldsymbol{\beta}} - t_{n-p}(1-\alpha/2)\hat{\sigma}(\boldsymbol{x}_0'(\boldsymbol{X}'\boldsymbol{X})^{-1}\boldsymbol{x}_0)^{1/2}, \boldsymbol{x}_0'\hat{\boldsymbol{\beta}} + t_{n-p}(1-\alpha/2)\hat{\sigma}(\boldsymbol{x}_0'(\boldsymbol{X}'\boldsymbol{X})^{-1}\boldsymbol{x}_0)^{1/2}]$$

is a $(1-\alpha)$-confidence interval for μ_0.

In many cases, one is also interested in determining an interval, which contains the future observation y_0 with high probability. In relation to the rent index, this means that a prospective tenant wants to know the range of reasonable rent values for a specific apartment. Thus, we are looking for a prediction interval for the future observation y_0. For this purpose we look at the prediction error $\hat{\varepsilon}_0 = y_0 - \boldsymbol{x}_0'\hat{\boldsymbol{\beta}}$. We have

$$\hat{\varepsilon}_0 \sim \mathrm{N}(0, \sigma^2 + \sigma^2\boldsymbol{x}_0'(\boldsymbol{X}'\boldsymbol{X})^{-1}\boldsymbol{x}_0).$$

Standardizing and replacing $\hat{\sigma}^2$ for σ^2 yields

$$\frac{y_0 - \boldsymbol{x}_0'\hat{\boldsymbol{\beta}}}{\hat{\sigma}(1 + \boldsymbol{x}_0'(\boldsymbol{X}'\boldsymbol{X})^{-1}\boldsymbol{x}_0)^{1/2}} \sim \mathrm{t}_{n-p}$$

and therefore

$$P\left(-t_{n-p}(1-\alpha/2) \leq \frac{y_0 - \boldsymbol{x}_0'\hat{\boldsymbol{\beta}}}{\hat{\sigma}(1 + \boldsymbol{x}_0'(\boldsymbol{X}'\boldsymbol{X})^{-1}\boldsymbol{x}_0)^{1/2}} \leq t_{n-p}(1-\alpha/2)\right) = 1 - \alpha.$$

Thus, with $(1-\alpha)$-confidence, we find the future observation at $\boldsymbol{x}_0$ to be within the prediction interval

$$\boldsymbol{x}_0'\hat{\boldsymbol{\beta}} \pm t_{n-p}(1-\alpha/2)\hat{\sigma}(1+\boldsymbol{x}_0'(\boldsymbol{X}'\boldsymbol{X})^{-1}\boldsymbol{x}_0)^{1/2}.$$

Per construction, the prediction interval is always wider than the corresponding confidence interval for μ_0. In applications with high error variance, the interval can be considerably wider; see the following Example 3.15 on the rent index.

Even though at first the two intervals appear to be similar, they are in fact very different. In the first case, we constructed a confidence interval for $E(y_0) = \mu_0$. This implies that the random interval overlaps the (fixed and constant) mean $\mathrm{E}(y_0)$ with probability $1-\alpha$. In the second case, we rather constructed an interval which is very likely (more precisely with probability $1-\alpha$) to contain the (random) future observation y_0.

Example 3.15 Munich Rent Index—Prediction Intervals

We again use the model from Example 3.13 (p. 133). Figure 3.16 shows the estimated rent per square meter for apartments constructed in the year 1918, with a standard kitchen (*nkitchen* = 0, *pkitchen* = 0) and time of assessment in 1999, depending on the living area. The 95 % confidence and prediction intervals are also included: individual confidence intervals were connected with lines leading to the bands shown. Due to the high variability of the rent index data, prediction intervals are much wider than confidence intervals for $E(y_0)$.

△

3.4 Model Choice and Variable Selection

In many applications, a large (potentially enormous) number of candidate predictor variables are available, and we face the challenge and decision as to which of these variables to include in the regression model. As a typical example, take the Munich rent index data with approximately 200 variables that are collected through a questionnaire. The following are two naive (but often practiced) approaches to the model selection problem:

- Strategy 1: Estimate the most complex model which includes all available covariates.
- Strategy 2: First, estimate a model with all variables. Then, remove all insignificant variables from the model.

We discuss and illustrate each of these approaches using two simulated data sets.

Example 3.16 Polynomial Regression

We investigate the simulated data illustrated in Fig. 3.17a. The scatter plot suggests polynomial modeling of the relationship between y and x resulting in the regression model

$$y_i = \beta_0 + \beta_1 x_i + \beta_2 x_i^2 + \cdots + \beta_l x_i^l + \varepsilon_i.$$

As with any polynomial regression, we have to choose the polynomial degree l. Panels (c–e) in Fig. 3.17 show the estimated relationship for $l = 1$ (regression line), $l = 2$, and $l = 5$. Apparently, a simple regression line, i.e., a first-order polynomial, does not fit the

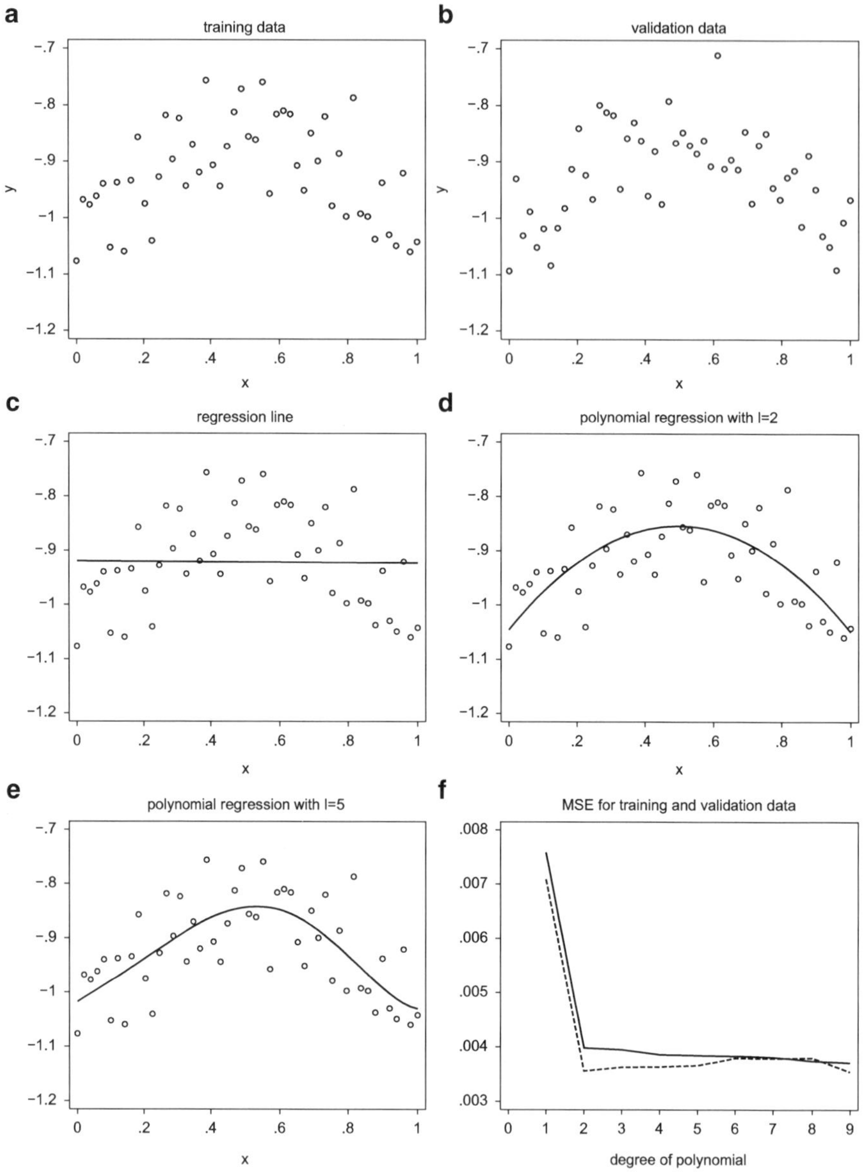

Fig. 3.17 Simulated training data y_i [panel (**a**)] and validation data y_i^* [panel (**b**)] based on 50 design points $x_i, i = 1, \ldots, 50$. The true model used for simulation is $y_i = -1 + 0.3x_i + 0.4x_i^2 - 0.8x_i^3 + \varepsilon_i$ with $\varepsilon_i \sim \text{N}(0, 0.07^2)$. Panels (**c–e**) show estimated polynomials of degree $l = 1, 2, 5$ based on the training set. Panel (**f**) displays the mean squared error MSE(l) of the fitted values in relation to the polynomial degree (*solid line*). The *dashed line* shows MSE(l), if the estimated polynomials are used to predict the validation data y_i^*

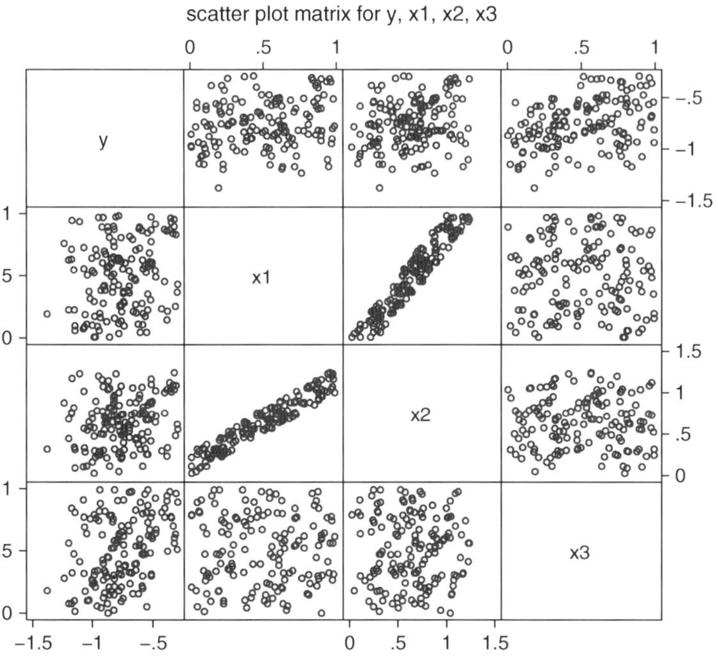

Fig. 3.18 Scatter plot matrix for the variables y, x_1, x_2, and x_3

data well. With polynomial degrees $l > 2$ onward, a satisfactory fit to the data appears to be guaranteed. Figure 3.17f additionally displays the mean squared error

$$\mathrm{MSE}(l) = \frac{1}{50}\sum_{i=1}^{50}(y_i - \hat{y}_i(l))^2$$

of the fitted models depending on the order of the polynomial (continuous line). Clearly, MSE(l) decreases monotonically with increased l. This suggests that the fit to the data is better with larger polynomial order. This finding appears to confirm the first strategy described above, namely to include as many regressors as possible into the model.

In a next step, we investigate how well the fitted models predict new observations that have been simulated according to the same model. Figure 3.17b shows additionally simulated observations for every design point x_i, $i = 1, \ldots, 50$. We refer to this data set as the validation sample, whereas we refer to the first data set (used for estimation) as the training set. Figure 3.17f shows the mean squared error of $\hat{y}_i^*$ for the data y_i^* (dashed line) in the validation set. Apparently, the fit to the new data is initially getting better with an increase of the polynomial order. However, from the polynomial order $l = 3$ onward, the fit is getting worse. We recognize the following: The more complex the model, the better is the fit to the data that were used for estimation. However, with new data resulting from the same data generating process, models that are too complex can cause a poorer fit. △

Example 3.17 Correlated Covariates

Consider the $n = 150$ observations $(y_i, x_{i1}, x_{i2}, x_{i3})$, $i = 1, \cdots, 150$, in the scatter plot matrix in Fig. 3.18. The data were generated as follows: The variables x_1 and x_3 are

Table 3.3 Results for the model based on covariates x_1, x_2, and x_3

Variable	Coefficient	Standard error	t-value	p-value	95 % Confidence interval	
intercept	−0.970	0.047	−20.46	<0.001	−1.064	−0.877
x_1	0.146	0.187	0.78	0.436	−0.224	0.516
x_2	0.027	0.177	0.15	0.880	−0.323	0.377
x_3	0.227	0.052	4.32	<0.001	0.123	0.331

Table 3.4 Results for the correctly specified model based on covariates x_1 and x_3

Variable	Coefficient	Standard error	t-value	p-value	95 % Confidence interval	
intercept	−0.967	0.039	−24.91	<0.001	−1.042	−0.889
x_1	0.173	0.055	3.17	0.002	0.065	0.281
x_3	0.226	0.052	4.33	<0.001	0.123	0.330

independent and uniformly distributed on [0,1]. The variable x_2 is defined as $x_2 = x_1 + u$, where u is also uniformly distributed on [0,1]. Thus, the variables x_1 and x_2 are highly correlated. Finally, the response variable y is simulated according to the model

$$y \mid x_1, x_2, x_3 \sim \mathrm{N}(-1 + 0.3x_1 + 0.2x_3, 0.2^2).$$

The conditional mean of y is thus dependent on x_1 and x_3, but not on x_2. In the following, we assume, however, that we do not know the true model (as is typically the case in practice). At first, we estimate a regression model with all available covariates x_1, x_2, and x_3, and we obtain the results provided in Table 3.3. Clearly, x_1 and x_2 are nonsignificant. If we followed strategy 2, i.e., if we eliminate the nonsignificant variables from the model, we would eliminate not only the nonrelevant covariate x_2, but also the relevant variable x_1.
If we instead estimate a correctly specified model with true predictor variables x_1 and x_3, we obtain the results shown in Table 3.4. When having a correct model specification, not only is x_3 significant but so is the previously insignificant variable x_1. We conclude: If we first consider all variables and then eliminate the insignificant variables from the model, it is possible that also important variables will be eliminated. The main reason for such unfortunate model estimation circumstances is the existing correlation among covariates. △

3.4.1 Effect of Model Specification on Bias, Variance, and Prediction Quality

We now strengthen the new insights of the previous examples with more theoretical considerations. In particular, we focus on the following questions:

1. *Irrelevant Variables:* What can be said about the bias and the variance of the least squares estimator, in the case that we include irrelevant variables in the model?
2. *Missing Variables:* What can be said about the bias and the variance of the least squares estimator, if we omit relevant variables in the model?
3. *Prediction Quality:* What effect does the model specification, more specifically the selected variables in the model, have on prediction?

Effect of Model Specification on Bias and Variance of the Least Squares Estimator

Consider a partition of the available explanatory variables $\boldsymbol{x} = (x_0, x_1, \ldots, x_k)'$ with $x_0 \equiv 1$ into the subsets $\boldsymbol{x}_1 = (x_0, x_1, \ldots, x_{k_1})'$ and $\boldsymbol{x}_2 = (x_{k_1+1}, \ldots, x_k)'$. We look at the two models

$$\boldsymbol{y} = \boldsymbol{X}\boldsymbol{\beta} + \boldsymbol{\varepsilon} = \boldsymbol{X}_1\boldsymbol{\beta}_1 + \boldsymbol{X}_2\boldsymbol{\beta}_2 + \boldsymbol{\varepsilon}$$

and

$$\boldsymbol{y} = \boldsymbol{X}_1\boldsymbol{\beta}_1 + \boldsymbol{u}.$$

The first model uses all available variables. The second model uses only the subset $\boldsymbol{x}_1$. We obtain the least squares estimators

$$\hat{\boldsymbol{\beta}} = (\boldsymbol{X}'\boldsymbol{X})^{-1}\boldsymbol{X}'\boldsymbol{y} \quad \text{and} \quad \tilde{\boldsymbol{\beta}}_1 = (\boldsymbol{X}_1'\boldsymbol{X}_1)^{-1}\boldsymbol{X}_1'\boldsymbol{y}$$

respectively. For the estimator $\tilde{\boldsymbol{\beta}}_1$ of the submodel, we obtain

$$\begin{aligned} \mathrm{E}(\tilde{\boldsymbol{\beta}}_1) &= \mathrm{E}((\boldsymbol{X}_1'\boldsymbol{X}_1)^{-1}\boldsymbol{X}_1'\boldsymbol{y}) \\ &= (\boldsymbol{X}_1'\boldsymbol{X}_1)^{-1}\boldsymbol{X}_1'\mathrm{E}(\boldsymbol{X}_1\boldsymbol{\beta}_1 + \boldsymbol{X}_2\boldsymbol{\beta}_2 + \boldsymbol{\varepsilon}) \\ &= (\boldsymbol{X}_1'\boldsymbol{X}_1)^{-1}\boldsymbol{X}_1'(\boldsymbol{X}_1\boldsymbol{\beta}_1 + \boldsymbol{X}_2\boldsymbol{\beta}_2) \\ &= \boldsymbol{\beta}_1 + (\boldsymbol{X}_1'\boldsymbol{X}_1)^{-1}\boldsymbol{X}_1'\boldsymbol{X}_2\boldsymbol{\beta}_2 \end{aligned}$$

and

$$\begin{aligned} \mathrm{Cov}(\tilde{\boldsymbol{\beta}}_1) &= \mathrm{Cov}((\boldsymbol{X}_1'\boldsymbol{X}_1)^{-1}\boldsymbol{X}_1'\boldsymbol{y}) = (\boldsymbol{X}_1'\boldsymbol{X}_1)^{-1}\boldsymbol{X}_1'\sigma^2\boldsymbol{I}\boldsymbol{X}_1(\boldsymbol{X}_1'\boldsymbol{X}_1)^{-1} \\ &= \sigma^2(\boldsymbol{X}_1'\boldsymbol{X}_1)^{-1}. \end{aligned}$$

for the mean and covariance matrix. The statistical properties of the estimator $\hat{\boldsymbol{\beta}}$ in the full model have already been derived in Sect. 3.2.3; see Box 3.9 on p. 119.

We now investigate the following two situations:

- *Missing Variables:* Even though the complete model $\boldsymbol{y} = \boldsymbol{X}\boldsymbol{\beta} + \boldsymbol{\varepsilon}$ is correct, we mistakenly estimate the reduced model $\boldsymbol{y} = \boldsymbol{X}_1\boldsymbol{\beta}_1 + \boldsymbol{u}$. In this case we neglect the relevant variables $\boldsymbol{x}_2$.
- *Irrelevant Variables:* Even though the reduced model $\boldsymbol{y} = \boldsymbol{X}_1\boldsymbol{\beta}_1 + \boldsymbol{u}$ is correct, we mistakenly estimate the full model $\boldsymbol{y} = \boldsymbol{X}\boldsymbol{\beta} + \boldsymbol{\varepsilon}$. In this case, we included irrelevant variables in the model. The variables in $\boldsymbol{x}_2$ are redundant.

In the first case of missing variables the following applies:

- $\boldsymbol{\beta}_1$ is biased. An exception is the case when $\boldsymbol{X}_1'\boldsymbol{X}_2 = \boldsymbol{0}$, i.e., every variable in $\boldsymbol{X}_1$ is uncorrelated to every variable in $\boldsymbol{X}_2$.
- It can be shown that the difference $\mathrm{Cov}(\hat{\boldsymbol{\beta}}_1) - \mathrm{Cov}(\tilde{\boldsymbol{\beta}}_1)$ of covariance matrices is positive semi-definite. This implies that the components of the estimator $\tilde{\boldsymbol{\beta}}_1$ based on the submodel $\boldsymbol{y} = \boldsymbol{X}_1\boldsymbol{\beta}_1 + \boldsymbol{u}$ show a smaller variance than the corresponding

components of the estimator $\hat{\boldsymbol{\beta}}_1$ based on the correct model $\boldsymbol{y} = \boldsymbol{X}\boldsymbol{\beta} + \boldsymbol{\varepsilon}$. Thus we have $\text{Var}(\hat{\beta}_j) \geq \text{Var}(\tilde{\beta}_j)$.

- Moreover, it can be shown that situations exist, in which the components in $\tilde{\boldsymbol{\beta}}_1$ based on the misspecified submodel actually show a smaller MSE than the components in $\hat{\boldsymbol{\beta}}_1$, which are based on the full model, i.e., $\text{MSE}(\hat{\beta}_j) \geq \text{MSE}(\tilde{\beta}_j)$. Hence, it is possible that a sparse model with unconsidered covariates has better statistical properties than the correctly specified full model.

In the second case of irrelevant variables, we have:

- Even though irrelevant variables were considered, $\hat{\boldsymbol{\beta}}$ is unbiased. Of course, the estimator $\hat{\boldsymbol{\beta}}_1$ based on the true model is also unbiased.
- It can be shown that the estimators for the components in $\boldsymbol{\beta}_1$ based on $\hat{\boldsymbol{\beta}}$ have larger variance than based on $\hat{\boldsymbol{\beta}}_1$. Thus, we have $\text{Var}(\hat{\beta}_j) \geq \text{Var}(\tilde{\beta}_j)$. If the estimated model contains irrelevant variables, then the precision of the estimators decreases.

Analogous statements can be made about $\hat{\boldsymbol{y}}$.

We can reach the following conclusion: Preferably, the specified model should not contain irrelevant covariates. Moreover, we should aim for a sparse model so that bias and variance, and thus MSE, are small.

Consequences of the Model Specification on Prediction Quality

Next we take a look at prediction quality in linear models. Thereby, we do not necessarily assume that the model is correctly specified. Our considerations will become useful when constructing model choice criteria for the comparison of different models.

We assume independent observations $y_i, i = 1, \ldots, n$, with expectation $\text{E}(y_i) = \mu_i$ and variance $\text{Var}(y_i) = \sigma^2$. The variables $x_0 = 1, x_1, \ldots, x_k$ are available as potential regressors. In the following we assume that a subset of the available variables will be used for estimation. The specified model is defined by the subset $M \subset \{0, 1, 2, \ldots, k\}$ of included covariates with corresponding design matrix $\boldsymbol{X}_M$. For the least squares estimator we obtain

$$\hat{\boldsymbol{\beta}}_M = (\boldsymbol{X}'_M \boldsymbol{X}_M)^{-1} \boldsymbol{X}'_M \boldsymbol{y}.$$

An estimator $\hat{\boldsymbol{y}}_M$ for the vector $\boldsymbol{\mu}$ of means $\mu_i = \text{E}(y_i)$ is given by

$$\hat{\boldsymbol{y}}_M = \boldsymbol{X}_M \hat{\boldsymbol{\beta}}_M.$$

We can view the estimator $\hat{y}_{iM}$ also as a prediction for future observations $y_{n+i} = \mu_i + \varepsilon_{n+i}$, $i = 1, \ldots, n$, with given covariates $x_{i1}, \ldots, x_{ik}$. In the following, we derive a formula for the sum of the expected squared prediction errors $\sum \text{E}(y_{n+i} - \hat{y}_{iM})^2$. To do so, we need the following, easily verifiable, properties of $\hat{\boldsymbol{y}}_M$:

1. Expectation:

$$\text{E}(\hat{\boldsymbol{y}}_M) = \boldsymbol{X}_M (\boldsymbol{X}'_M \boldsymbol{X}_M)^{-1} \boldsymbol{X}'_M \text{E}(\boldsymbol{y}).$$

2. Covariance matrix:

$$\operatorname{Cov}(\hat{\boldsymbol{y}}_M) = \sigma^2 \boldsymbol{X}_M (\boldsymbol{X}'_M \boldsymbol{X}_M)^{-1} \boldsymbol{X}'_M.$$

3. Sum of the variances:

$$\sum_{i=1}^{n} \operatorname{Var}(\hat{y}_{iM}) = \sigma^2 \operatorname{tr}(\boldsymbol{X}_M (\boldsymbol{X}'_M \boldsymbol{X}_M)^{-1} \boldsymbol{X}'_M) = |M|\,\sigma^2,$$

   where $|M|$ represents the cardinal number of M, i.e., the number of the covariates included in the model. The sum of the variances increases as more covariates are included in the model.
4. Sum of the mean squared errors (SMSE):

$$\begin{aligned}
\text{SMSE} &= \sum_{i=1}^{n} \operatorname{E}(\hat{y}_{iM} - \mu_i)^2 \\
&= \sum_{i=1}^{n} \operatorname{E}\left((\hat{y}_{iM} - \mu_{iM}) + (\mu_{iM} - \mu_i)\right)^2 \\
&= \sum_{i=1}^{n} \operatorname{Var}(\hat{y}_{iM}) + 2\sum_{i=1}^{n} \operatorname{E}\left((\hat{y}_{iM} - \mu_{iM})(\mu_{iM} - \mu_i)\right) + \sum_{i=1}^{n} (\mu_{iM} - \mu_i)^2 \\
&= |M|\,\sigma^2 + \sum_{i=1}^{n} (\mu_{iM} - \mu_i)^2.
\end{aligned}$$

   Here we used $\mu_{iM} = \operatorname{E}(\hat{y}_{iM})$ as an abbreviation for the expectation of the estimator $\hat{y}_{iM}$.
   These properties provide us with the expected squared prediction error:

$$\begin{aligned}
\text{SPSE} &= \sum_{i=1}^{n} \operatorname{E}(y_{n+i} - \hat{y}_{iM})^2 \\
&= \sum_{i=1}^{n} \operatorname{E}((y_{n+i} - \mu_i) - (\hat{y}_{iM} - \mu_i))^2 \\
&= \sum_{i=1}^{n} (\operatorname{E}(y_{n+i} - \mu_i)^2 - 2\operatorname{E}((y_{n+i} - \mu_i)(\hat{y}_{iM} - \mu_i)) + \operatorname{E}(\hat{y}_{iM} - \mu_i)^2) \\
&= \sum_{i=1}^{n} \operatorname{E}(y_{n+i} - \mu_i)^2 + \sum_{i=1}^{n} \operatorname{E}(\hat{y}_{iM} - \mu_i)^2 \\
&= n\,\sigma^2 + \text{SMSE} \\
&= n\,\sigma^2 + |M|\,\sigma^2 + \sum_{i=1}^{n} (\mu_{iM} - \mu_i)^2.
\end{aligned}$$

Note that in line 3 of the above derivation, the expectation for the cross product term can be written as the product of expectations due to the independence of $\hat{y}_{iM}$ and y_{n+i}. This way the entire term becomes zero. Thus, the expected squared prediction error can be decomposed into three additive terms:

- *Irreducible Prediction Error:* The first term $n\,\sigma^2$ depends on the error variance. Hence, it cannot be reduced, even by sophisticated inference techniques. This term is therefore referred to as the irreducible prediction error.
- *Variance:* The second term consists of the sum of variances $\text{Var}(\hat{y}_{iM})$ of the estimators $\hat{y}_{iM}$. This term can be manipulated through model choice. It becomes smaller as fewer variables are included in the model.
- *Squared Bias:* The last term $\sum(\mu_{iM} - \mu_i)^2$ can be seen as a bias term. It consists of the squared bias of the estimator $\hat{y}_{iM}$ for the expectation μ_i. This term can also be manipulated through model choice and becomes smaller as more variables are included in the model.

The decomposition of the expected prediction error into an irreducible error, a variance term, and a squared bias term is not limited to linear models but rather a *fundamental property of prediction* in all statistical models.

The formula for SPSE shows a classical bias–variance trade-off. The more complex the model, the smaller the squared bias and the greater the variance. On the contrary, simpler models show a greater squared bias and in return for that a smaller variance. This bias–variance trade-off is not only characteristic for linear models, but for all statistical models; see for example the discussion of the bias–variance trade-off with regard to smoothing techniques in Sect. 8.1.8.

3.4.2 Model Choice Criteria

The theoretical considerations regarding prediction quality in the last section, especially the fundamental formula for the expected squared prediction error, help to develop tools for model choice, in particular variable selection. A possible approach is to *minimize* the expected squared prediction error: We include those covariates in the model, which minimize SPSE. Unfortunately, SPSE is not directly accessible, since μ_i and σ^2 are unknown. Thus, we need to estimate SPSE. Two strategies are possible:

1. *Estimate SPSE using new and independent data*

   If in fact additional observations y_{n+i} are available, we are able to estimate $\text{SPSE} = \sum \text{E}(y_{n+i} - \hat{y}_{iM})^2$ simply by

$$\widehat{\text{SPSE}} = \sum_{i=1}^{n} (y_{n+i} - \hat{y}_{iM})^2.$$

   In practice, it is usually not possible to use this approach, as it rarely happens that additional observations are collected. An alternative procedure is the following:

   - Randomly split the data into two parts, i.e., a test and a validation sample.

- Use the test data set to estimate the specified model, i.e., for the estimation of regression coefficients $\boldsymbol{\beta}$ and the expectations μ_i.
- Use the validation set to assess the goodness of fit, i.e., for the estimation of SPSE.

This approach is possible if the available data set is large enough. If the data set is too small, partitioning the data would cause loss of accuracy in estimation of the regression coefficients. How large the sample size has to be in order to proceed in a reasonable manner depends on the problem. As far as we are aware, there is not any rule of thumb or general advice in the literature. This is the reason why we often use the second strategy.

2. *Estimate SPSE using existing data*

A naive estimator for SPSE would be the use of the squared sum of residuals $\sum(y_i - \hat{y}_{iM})^2$. Note that this sum *underestimates* on average the expected prediction error, as it can be shown that

$$\mathrm{E}\left(\sum_{i=1}^{n}(y_i - \hat{y}_{iM})^2\right) = \mathrm{SPSE} - 2|M|\sigma^2.$$

Thus a better estimate for SPSE is given by

$$\widehat{\mathrm{SPSE}} = \sum_{i=1}^{n}(y_i - \hat{y}_{iM})^2 + 2|M|\hat{\sigma}^2.$$

Accordingly, we choose a model that minimizes $\widehat{\mathrm{SPSE}}$. In doing so we have to keep in mind that we always use the same estimator for $\hat{\sigma}^2$. Preferably, this estimator should be based on the full model with all available variables, in order to keep the bias in $\hat{\sigma}^2$ small. The criterion $\widehat{\mathrm{SPSE}}$ has the typical structure of many model choice criteria. It consists of two terms: The first term, the sum of squared residuals, measures the goodness of fit and becomes smaller the more complex the model becomes. The second term $2|M|\hat{\sigma}^2$ measures model complexity and becomes smaller as models become simpler.

We next present some of the most widely used criteria for model choice in linear models. Their derivation follows ideas similar to that of $\widehat{\mathrm{SPSE}}$.

The Corrected Coefficient of Determination

In Sect. 3.2.3, we defined the coefficient of determination R^2 as a measure for the goodness of fit to the data. When comparing different models, the use of the coefficient of determination is limited, since the coefficient of determination will always increase (never decrease) with the addition of a new covariate into the model. The *corrected coefficient of determination* adjusts for this problem, by including a correction term for the number of parameters. The corrected coefficient of determination is defined by

$$\bar{R}^2 = 1 - \frac{n-1}{n-p}(1 - R^2).$$

The corrected coefficient of determination is very popular and is provided by default with all statistical program packages. At this point, we advice against its usage, since the "penalty" for newly included covariates appears to be too small. It can be shown that $\bar{R}^2$ already increases when a variable with a t-value greater than 1 is included in the model, implying we would include variables with a p-value of about 0.3.

Mallow's C_p

Mallow's C_p ("complexity parameter") relies directly on the ideas presented above and is defined by

$$C_p = \frac{\sum_{i=1}^{n}(y_i - \hat{y}_{iM})^2}{\hat{\sigma}^2} - n + 2|M|.$$

C_p can be understood as an estimate of SMSE$/\sigma^2$. Thus minimizing C_p produces the same optimal model as minimizing $\widehat{\text{SPSE}}$.

Akaike Information Criterion

The Akaike information criterion (AIC) is one of the most widely used criteria for model choice within the scope of likelihood-based inference; see Appendix B, p. 664. In general, AIC is defined by

$$\text{AIC} = -2 \cdot l(\hat{\boldsymbol{\beta}}_M, \hat{\sigma}^2) + 2\,(|M| + 1),$$

where $l(\hat{\boldsymbol{\beta}}_M, \hat{\sigma}^2)$ is the maximum value of the log-likelihood, i.e., when the ML estimators $\hat{\boldsymbol{\beta}}_M$ and $\hat{\sigma}^2$ are inserted into the log-likelihood. Smaller values of the AIC correspond to a better model fit. Note that the total number of parameter is $|M| + 1$ since the error variance σ^2 is also counted as a parameter. In a linear model with Gaussian errors, we obtain

$$\begin{aligned} -2\,l(\hat{\boldsymbol{\beta}}_M, \hat{\sigma}^2) &= n\,\log(\hat{\sigma}^2) + \tfrac{1}{\hat{\sigma}^2}(\boldsymbol{y} - \boldsymbol{X}_M\hat{\boldsymbol{\beta}}_M)'(\boldsymbol{y} - \boldsymbol{X}_M\hat{\boldsymbol{\beta}}_M) \\ &= n\,\log(\hat{\sigma}^2) + \frac{n\hat{\sigma}^2}{\hat{\sigma}^2} \\ &= n\,\log(\hat{\sigma}^2) + n, \end{aligned}$$

and thus

$$\text{AIC} = n \cdot \log(\hat{\sigma}^2) + 2\,(|M| + 1),$$

when ignoring the constant n.

Note that the ML estimator $\hat{\sigma}^2 = \hat{\boldsymbol{\varepsilon}}'\hat{\boldsymbol{\varepsilon}}/n$ is considered in AIC and not the usual unbiased variance estimator.

Cross Validation

Cross validation imitates partitioning of the data into a test set for parameter estimation and a validation set to assess predictive quality. Cross validation is based on the following general principle:

- Partition the data set into r subsets $1, \ldots, r$, of similar size.
- Start with the first data set as validation set and use the combined remaining $r-1$ data sets for parameter estimation. Based on the estimates, obtain predictions for the validation set and determine the sum of the squared prediction errors.
- Cycle through the partitions using the second, third, up to the rth data set as validation sample and all other data sets for estimation. Determine the sum of squared prediction errors.
- Use the model with the smallest sum of squared prediction errors, where the final parameter estimates reflect all data. The partition into test and validation samples serves only to estimate the expected squared prediction error.

An important special case is the so-called "leave-one-out" cross validation, which uses all observations with the exception of *one* for the estimation of model parameters. We use this "leave-one-out" estimator to predict the deleted observation and to determine the squared prediction error. If we denote the "leave-one-out" estimator with $\hat{y}_{iM}^{-i}$, we obtain the cross validation score

$$\text{CV} = \frac{1}{n}\sum_{i=1}^{n}(y_i - \hat{y}_{iM}^{-i})^2.$$

At first glance, computation of the cross validation score CV appears to be quite expensive, since the least squares estimator has to be calculated n times. It can be shown, however, that the cross validation score can be computed with the help of the original estimators $\hat{y}_{iM}$ that are based on all data. We obtain

$$\text{CV} = \frac{1}{n}\sum_{i=1}^{n}\left(\frac{y_i - \hat{y}_{iM}}{1 - h_{iiM}}\right)^2,$$

where h_{iiM} are the diagonal elements of the hat matrix $\boldsymbol{H}_M = \boldsymbol{X}_M(\boldsymbol{X}'_M\boldsymbol{X}_M)^{-1}\boldsymbol{X}'_M$.

Bayesian Information Criterion

The Bayesian information criterion (BIC) is generally defined by

$$\text{BIC} = -2 \cdot l(\hat{\boldsymbol{\beta}}_M, \hat{\sigma}^2) + \log(n)\,(|M| + 1);$$

see in Appendix B, p. 676. The BIC multiplied by $1/2$ is also known as Schwartz criterion. Assuming Gaussian errors, we obtain

$$\text{BIC} = n \cdot \log(\hat{\sigma}^2) + \log(n)\,(|M| + 1).$$

The form of the BIC is similar to that of the AIC, and again smaller values indicate a better model fit. Note, however, that the BIC and AIC are motivated in a very

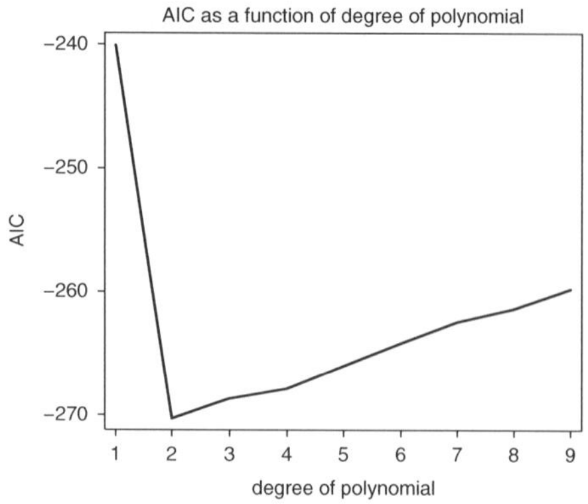

Fig. 3.19 AIC as a function of the polynomial degree for the simulated data of Example 3.16 (p. 139)

different way. From a practical point of view, the main difference is that the BIC penalizes complex models much more than the AIC. Thus, the resulting "best" models are typically more parsimonious when using the BIC rather than the AIC.

Example 3.18 Polynomial Regression—Model Choice with AIC

Figure 3.19 plots AIC for the simulated data from Example 3.16 (p. 139) as a function of the polynomial degree. AIC obtains a minimum for $l = 2$ resulting in a reasonable model, even though we do not obtain the polynomial order of the true model with $l = 3$. △

3.4.3 Practical Use of Model Choice Criteria

We can use the various model choice criteria to select the most promising models from candidate models. We recommend the following approach:

- On the basis of scientific knowledge, perhaps gained from previous research, we obtain a preselection of potential models. The models can differ in the number of variables but also in model type (e.g., linear versus nonlinear). The total number of potential models should be as small as possible.
- All potential models can now be assessed with the aid of one of the various model choice criteria. The summary of the results should not be restricted to the "best" model. As a rule, there are a number of competitive models having approximately equal model fit, differing only in small aspects from each other. These differences cause some uncertainty regarding the conclusions.

Unfortunately, this method is not always practical, since the number of regressor variables and modeling variants can be very large in many applications. In short, the calculation of all models is often impossible. In this case, we can use the following partially heuristic methods:

- *Complete Model Selection (All-Subset-Selection)*: In case that the number of covariates is smaller than about 40, we can determine the best model (in the sense of a model choice criterion) with the "leaps and bounds" algorithm introduced by Furnival and Wilson (1974). The algorithm returns the optimal model thereby avoiding the computation of all models. An implementation can be found in the software SAS and also in the R package `leaps`.

- *Forward Selection:* Based on a starting model, forward selection includes one additional variable in every iteration of the algorithm. The variable which offers the greatest reduction of a preselected model choice criteria (C_p, AIC, CV, BIC) is chosen. The algorithm terminates if no further reduction is possible.
- *Backward Elimination:* Backward elimination starts with the full model containing all potential covariates. Subsequently, in every iteration, the covariate which provides the greatest reduction of the model choice criteria (C_p, AIC, CV, BIC) is eliminated from the model. The algorithm terminates if no further reduction is possible.
- *Stepwise Selection:* Stepwise selection is a combination of forward selection and backward elimination. In every iteration of the algorithm, the inclusion and the deletion of a variable are both possible.

Generally speaking, forward, backward, and stepwise selection do not offer the best model in the sense of the model choice criteria. However, typically a model that is close to the best model is selected.

Unfortunately, the available software for model choice is often problematic. A main problem is that dummy variables of multi-categorical variables are, in general, neither jointly included nor removed from the model. Instead it can happen that only a subset of the dummies of a categorical variable appears in the model. This causes an indirect change of the reference category and the interpretation of the regression coefficients. Moreover, other difficulties can occur, e.g., methods for hierarchical terms are frequently missing. For example, if we use polynomials for modeling nonlinear covariate effects, it can happen that the selected model has a squared and cubic term, but not a linear one. Often it is impossible to exclude certain terms from selection and force them into the model.

The listed procedures should not be confounded with an algorithm proposed by Efroymson in the 1960s, even though the approach is similar. In contrast to what has been proposed above, the Efroymson algorithm includes or excludes those variables in/from the model, which have the highest or lowest t-value.

The procedure terminates when no variable that potentially needs to be included has a p-value of less than a previously fixed maximal p-value (e.g., 0.05) and when no variable that needs to be excluded has a p-value greater than a minimal p-value (e.g., 0.1). This automatic procedure, which is implemented in all major statistical software packages, is often viewed as obscure among statisticians due to the following two reasons:

- Forward, backward, and stepwise selection usually provide different results. This also happens when using a global model choice criterion such as AIC. We can, however, compare the different selected models with the help of the global model choice criterion. When using the Efroymson approach, discrimination between the different models is impossible.
- The repetitive use of the t-test statistic, to assess whether or not a regression coefficient is different from zero, suggests exact tests. However, the t-test statistic

does not follow a t-distribution under the null hypothesis, since during the selection process we do not test an arbitrary variable but rather the variable with the *maximal* t-value.

Example 3.19 Prices of Used Cars—Model Choice

We illustrate the approaches for model choice using data from the sales price of pre-owned VW Golf automobiles. Our goal is to model the relationship between the sales price in 1,000 Euro (variable *price*) and the five explanatory variables "age of the car in months" (*age*), "kilometer reading in 1,000 km" (*kilometer*), "number of months until the next appointment with the Technical Inspection Agency" (*TIA*), "ABS brake yes/no" (*extras1*), and "sunroof yes/no" (*extras2*). Examinations of this kind play an important role in the context of hedonic price indices. In contrast to common price indices, hedonic price indices take quality differences of a product into consideration. This is especially of importance when dealing with product groups that undergo fast technological changes.

Figure 3.20 provides scatter plots and box plots to get a first impression about the relationships between the response variable and the various explanatory variables. The plots suggest the following effects: We can assume a linear or monotonically decreasing nonlinear effect for the variables *age* and *kilometer*, which could be appropriately modeled using (orthogonal) polynomials of degree three or less. The variable *TIA* appears to either have no effect or a very weak linear effect on the average sales price. Cars with ABS (*extras1*) seem to be slightly more expensive than cars without ABS; the effect, however, remains arguable. We can attest to no difference in the average sales price for models with or without sunroof (*extras2*). All in all, there seems to be a relationship with age and the kilometer reading. The effects of the remaining variables appear doubtful.

Based on these considerations, we first examine eight regression models (see Table 3.5), which do not differ in the modeling of the variables *age* and *kilometer*. For the remaining three regressors, all possible model combinations will be tested. Using the AIC criterion, we obtain the first model in Table 3.5 as the preliminary best model. Figure 3.21 displays the AIC values for the eight models under consideration. In addition, the AIC for a ninth model based on automatic variable selection is provided; see below.

Since only five explanatory variables are available, we can even determine the AIC best model with the help of the "leaps and bounds" algorithm. This model attains an AIC value of 389.35. It differs from the current "best model," in that it only makes use of polynomials of second degree (not third) for the variables *age* and *kilometer*. Figure 3.21 shows that the obtained AIC for this ninth model is considerably smaller than the best AIC value of all the models that we examined so far. Table 3.6 shows the estimation output for model 9. Figure 3.22 provides additional visualization of the estimated effects for age and kilometer reading. Table 3.6 reveals that all regression coefficients differ significantly from zero at the 5 % level, and only one quadratic term is not significant at the 1 % level. Note that a model, which is selected according to a global goodness-of-fit criterion, can contain nonsignificant coefficients. This is only remarkable at first glance. All selection criteria are global measures, which, for example, minimize the prediction error. Statistical tests can rather be seen as local measures, which in addition are constructed asymmetrically. This means that the type I error is controlled and kept small so that we favor the null hypothesis over the alternative hypothesis.

△

We finally point out that there are alternative ways for variable selection than the optimization of a model choice criteria such as AIC or the cross validation score. In fact there is vital current research in variable selection for linear models. In this book we will present some of the most interesting approaches in Chap. 4. More specifically, we will discuss:

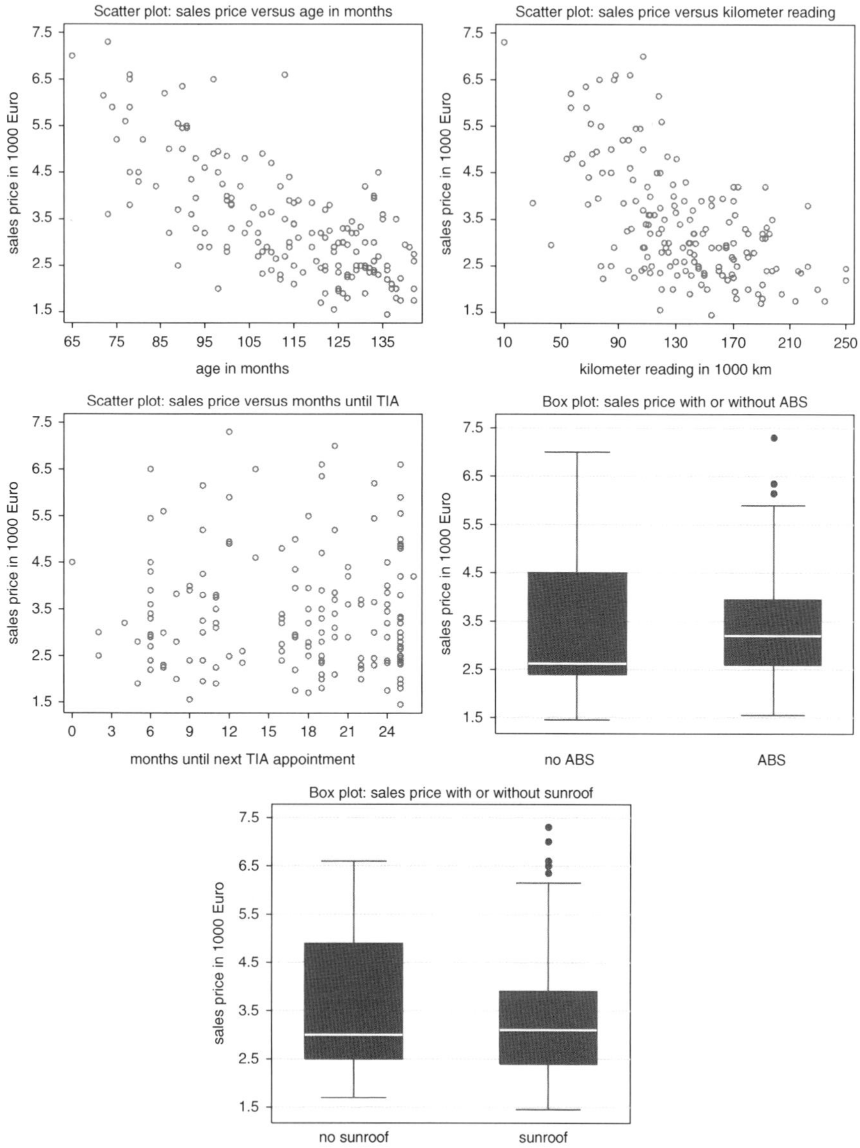

Fig. 3.20 Prices of used cars: scatter plots between sales prices and the continuous covariates *age*, *kilometer*, and *TIA*. Box plots of sales prices stratified according to the values of the binary variables *extras1* and *extras2*

- *Regularization techniques for the least squares estimator, in particular the LASSO:* Regularization of the least squares estimator allows to set some of the regression coefficients zero, so that the corresponding covariates are essentially removed from the model; see Sect. 4.2 for details.

Table 3.5 Prices of used cars: potential models

Model	*kilometer* degree 3	*age* degree 3	*extras1* linear	*extras2*	*TIA*	AIC
1	x	x				393.234 (1)
2	x	x	x			394.566 (2)
3	x	x		x		395.119 (4)
4	x	x			x	394.973 (3)
5	x	x	x	x		396.481 (6)
6	x	x	x		x	396.143 (5)
7	x	x		x	x	396.881 (7)
8	x	x	x	x	x	398.085 (8)

The values in brackets indicate the rank of the models according to AIC

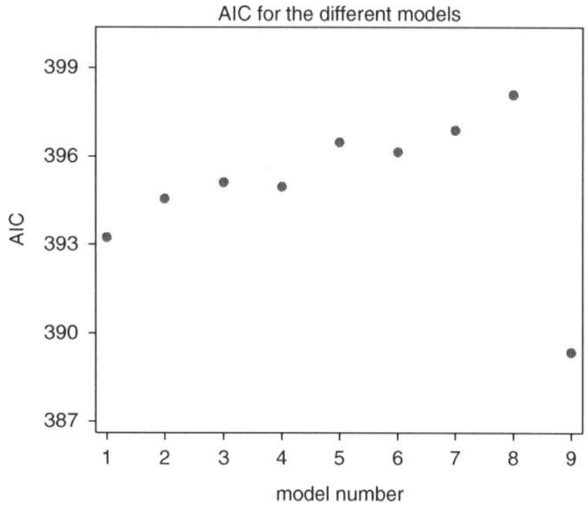

Fig. 3.21 Prices of used cars: AIC values for the potential models

Table 3.6 Prices of used cars: estimation results for the best model according to AIC

Variable	Coefficient	Standard error	t-value	p-value	95 % Confidence interval	
intercept	3.397	0.056	60.220	<0.001	3.285	3.508
ageop1	−0.705	0.061	−11.470	<0.001	−0.826	−0.584
ageop2	0.187	0.057	3.270	0.001	0.074	0.300
kilometerop1	−0.439	0.061	−7.170	<0.001	−0.560	−0.318
kilometerop2	0.141	0.057	2.460	0.015	0.028	0.254

- *Boosting:* Similar to the LASSO, boosting is able to simultaneously estimate the regression coefficients and select relevant variables; see Sect. 4.3.
- *Bayesian variable selection:* The last decades have seen a large number of practical Bayesian approaches for variable selection. We will present one approach with corresponding software in detail and give an overview over alternative approaches; see Sect. 4.4, in particular Sect. 4.4.3.

We will apply the alternative techniques for variable selection in our case study on prices of used cars and compare the results to those obtained by optimizing AIC in the previous example; refer to Examples 4.11 (LASSO), 4.12 (boosting), and 4.15, 4.16 (Bayesian variable selection).

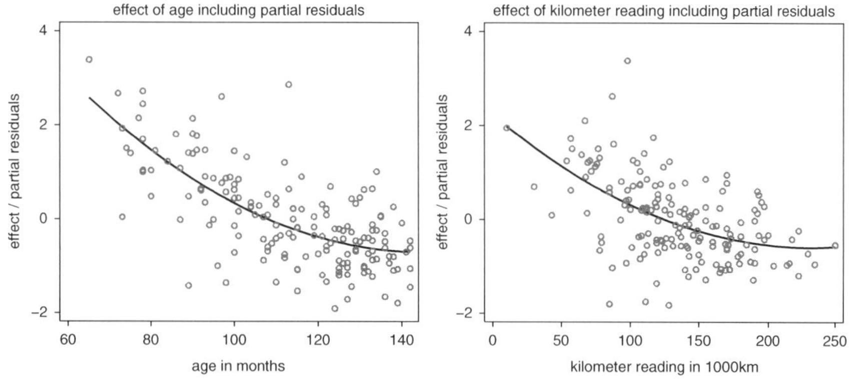

Fig. 3.22 Prices of used cars: model 9 based on all-subset selection, effects of age, and kilometer reading including partial residuals

3.4.4 Model Diagnosis

Having selected a working "best" model, the next step is to assess the validity of the model using diagnostic tools, in search of an improved model. In principle, model diagnosis pursues the following goals:

- *Examination of the model assumptions:* No model is correct. Nevertheless, the assumptions on which a model is based should be met at least approximatively. In the classical linear model, we have to evaluate the assumptions of homoscedastic, uncorrelated, and possibly normally distributed errors, as well as the linearity of the predictors.
- *Outlier detection:* In some situations, the results are heavily influenced by a few "unusual" observations. These observations should be identified and their effect on the estimates quantified.
- *Collinearity analysis:* Theoretical results have shown that highly correlated covariates can have a strong influence on the stability of the estimates; see pages 115ff. or Box 3.9 on p. 119. Thus it is important to examine the correlation structure of the covariates in the model.

In the following sections and Sect. 4.1, we describe a selection of tools useful for model diagnosis.

Examination of Model Assumptions

Homoscedasticity: The examination of the assumption of homoscedastic error variances will be outlined in Sect. 4.1.3 (p. 182). Important tools include residual plots and tests for heteroscedasticity. When heteroscedasticity occurs, transformation of the response variable or the use of weighted regression are possible remedies; see Sect. 4.1.3 in the next chapter.

Uncorrelated Errors: We can detect correlated errors through residual plots over time or through the use of statistical tests; see Sect. 4.1.4 (p. 191) of the

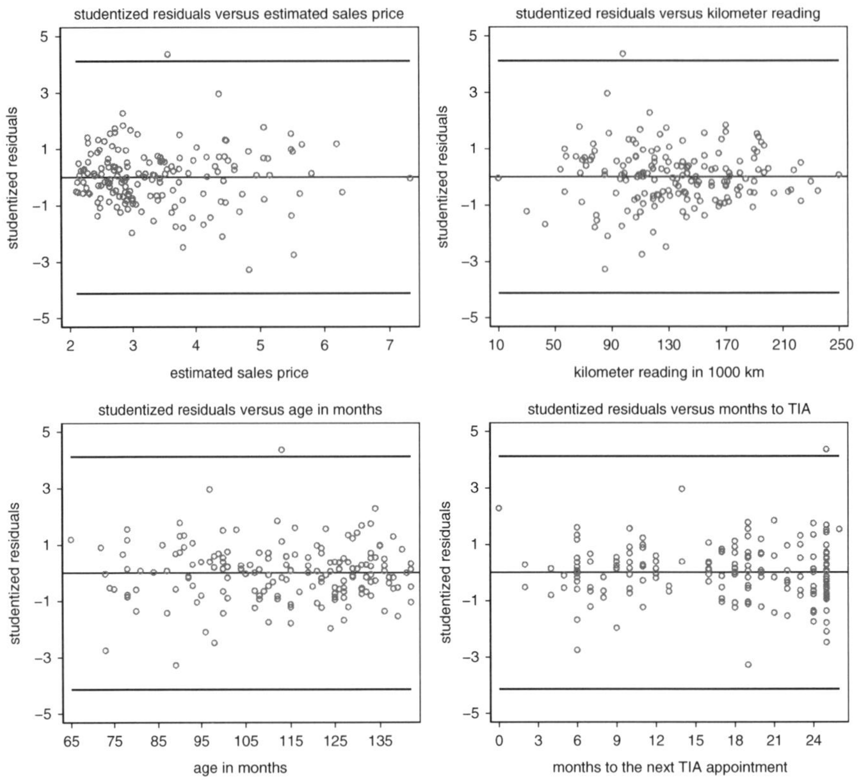

Fig. 3.23 Prices of used cars: studentized residuals against $\widehat{price}$ (*top left*) and against the continuous covariates. The *horizontal lines* mark the critical values of the test for outliers to be discussed in Sect. 3.4.4 (p. 160)

next chapter. Since correlated errors indicate misspecification (especially through omitted explanatory variables and unnoticed nonlinearity), we should first examine whether or not model specification can be improved. If not, we can use one of the estimation procedures discussed in Sect. 4.1.4 (p. 191).

Linearity: We can eventually discover undetected nonlinear effects of covariates with the help of scatter plots of (standardized or studentized) residuals as a function of estimated values. Moreover, plots of partial residuals $\hat{\varepsilon}_{x_j,i} = \hat{\varepsilon}_i + \hat{\beta}_j x_{ij}$ against covariate x_{ij} can be useful.

Assumption of Normality: We can test distributional assumptions with the quantile–quantile plot (Q–Q plot). In Q–Q plots, the empirical quantiles are compared to the quantiles of the theoretical distribution. If the data follows the distribution, the points should closely follow the 45° bisecting line.

Example 3.20 Price of Used Cars—Model Diagnosis

Figure 3.23 plots the studentized residuals versus the estimated price and the continuous covariates for the model estimated in Example 3.19. There is no evidence of any additional

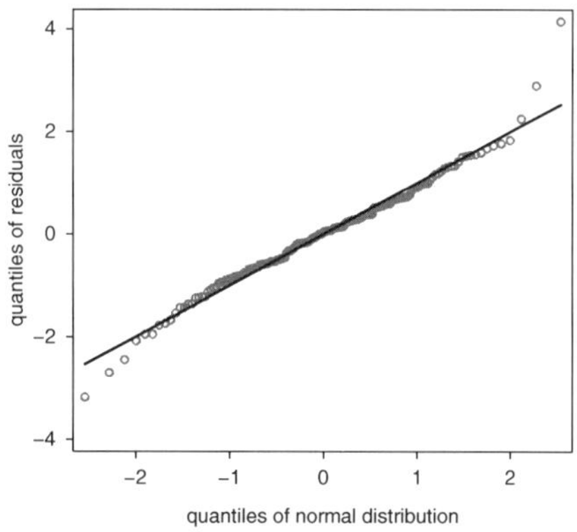

Fig. 3.24 Prices of used cars: Q–Q plot of standardized residuals

nonlinear effects. Nevertheless, evidence of (mild) heteroscedastic errors exists. The variability of studentized residuals decreases as kilometer reading increases. Also note the increasing variability of residuals for VW Golf models that have 18 months or more until their next TIA checkup. It appears that the variable TIA does not affect the conditional mean of sales prices but rather the conditional variance. Figure 3.24 shows the Q–Q plot for the standardized residuals. Significant deviations from the normal distribution are visible only in the extreme left and right tails.

△

Collinearity Analysis

We first illustrate the goals of collinearity analysis with the most extreme situation. Consider the model

$$y = \beta_0 + \beta_1 x_1 + \beta_2 x_2 + \beta_3 x_3 + \varepsilon,$$

with covariates x_1, x_2, and x_3. We first assume that there is perfect linear dependence $x_1 = a\, x_2$ between x_1 and x_2. The variable x_1 could represent, for example, weight in kilogram and x_2 weight in tons, that is, x_1 and x_2 only differ in the measure unit. In this extreme case, we cannot uniquely estimate the regression coefficients β_1 and β_2 belonging to x_1 and x_2, since

$$y = \beta_0 + \beta_1 x_1 + \beta_2 x_2 + \beta_3 x_3 + \varepsilon = \beta_0 + (\beta_1 a + \beta_2)\, x_2 + \beta_3 x_3 + \varepsilon.$$

In practice, the covariates typically do not show such an exact linear dependence, but they can be highly correlated, which is called (multi)collinearity. As we have seen in Sect. 3.2.3 on p. 116, highly correlated covariates cause imprecise estimation of the regression parameters, as again seen through the variance formula for $\hat{\beta}_j$,

$$\mathrm{Var}(\hat{\beta}_j) = \frac{\sigma^2}{(1 - R_j^2) \sum_{i=1}^{n} (x_{ij} - \overline{x}_j)^2}.$$

The greater the linear dependence of a covariate x_j with the other explanatory variables (measured through the coefficient of determination R_j^2), the greater is the variance $\text{Var}(\hat{\beta}_j)$. In the extreme case of $R_j^2 \to 1$, the variance explodes towards infinity. The variance formula also provides a diagnostic tool for measuring the degree of collinearity, i.e., the *variance inflation factor*:

$$\text{VIF}_j = \frac{1}{1 - R_j^2}.$$

The variance inflation factor quantifies the factor increase of the variance of $\hat{\beta}_j$ due to the linear dependence of x_j with the other regressors. The greater the correlation between x_j and the other covariates, the greater is R_j^2, and thus the greater is VIF_j. As a general benchmark, we say that a serious collinearity problem exists when the variance inflation factor is greater than 10, i.e., $\text{VIF}_j > 10$.

Example 3.21 Prices of Used Cars—Collinearity

For the regression model of Example 3.19, the variance inflation factors range between 1.03 and 1.19. In other words, the variances of the estimated coefficients increase, at most, by factor 1.19, when compared to an "ideal" (orthogonal) setting. Thus, there is no evidence for a serious collinearity problem.

△

The literature offers a variety of proposals on how to deal with strong collinearity:

- *Omission of some affected covariates:* This method is probably both the most practiced and the most criticized in literature. However, in many cases there are different measures for the same purpose. Consider, for instance, the various measures for the nutritional status of the mother in our data example on malnutrition in Zambia. The original data set offers the body mass index, as well as the Rohrer Index as measures of nutritional status. Although constructed differently, these two variables are highly correlated, with a correlation coefficient greater than 0.95. In this case, as well as in other similar cases, we are forced to decide which index to use and which to exclude from further analysis.
- *Construct a single combined (easily interpretable) variable from the variables in question:* Of particular interest are linear combinations of the variables or simple summaries, such as the minimum or maximum of the variables. Recall the supermarket scanner data, which examines different marketing measures and relates them to the sale of a particular coffee brand (see Example 3.2 on p. 86). When modeling the sales of a coffee brand, in addition to its own price, the prices of other competitors are important covariates. For some competitors, several brands exist, which only differ from each other by package size. The prices for these brands often change at the same time and proportional to package size. Thus, strong collinearity results if we include all brands as separate effects into the model. We can avoid the collinearity problem by computing, for example, the average price (a specific linear combination) of the competitive brands, which is then included in the model in place of the individual brands. Another possibility is to use a simple summary, e.g., the lowest price (minimum).

- *Ridge Regression:* The ridge estimator is an alternative to the least squares estimator and is defined as

$$\hat{\boldsymbol{\beta}} = (\boldsymbol{X}'\boldsymbol{X} + \lambda \boldsymbol{I})^{-1}\boldsymbol{X}'\boldsymbol{y},$$

  where $\lambda \geq 0$ is an appropriately chosen tuning-parameter. The ridge estimator, unlike the least squares estimator, is biased. The literature suggests the use of the ridge estimator, especially in the presence of collinearity problems. The addition of $\lambda \boldsymbol{I}$ to $\boldsymbol{X}'\boldsymbol{X}$ leads to a regularized and invertible matrix, even if $\boldsymbol{X}'\boldsymbol{X}$ is near singular due to severe collinearity. Thus, the components of the ridge estimator often have a smaller MSE when compared to the corresponding components of the least squares estimator. The least squares estimator is only optimal in the class of linear and unbiased estimators. If we also allow biased estimators, it is possible that other estimators are better in terms of the MSE than the least squares estimators. Even though the ridge estimator has been highly researched and elaborated on in many publications, until recently it was rarely used in practice. One reason is that even when using the ridge estimator, separating the covariate effects is impossible with severe collinearity. However, ridge and related estimators have recently received increasing attention in order to cope with situations that have a huge number of possible regressors in combination with a relatively small sample size, i.e., large p, small n problems. Such problems occur frequently in biostatistics when analyzing gene expression and related data. More details on ridge and related estimators are given in Sect. 4.2.

  Estimators with a similar structure to that of the ridge estimator are also widely used in nonparametric regression; see Chap. 8.

- *Principal component regression:* This variant does not directly include all available covariates in the model but rather appropriate linear combinations of covariates. Throughout, continuous covariates are assumed. The basis of principal component regression is the spectral analysis of the matrix $\boldsymbol{X}'\boldsymbol{X}$ (see Theorem A.25 on p. 633), i.e.,

$$\boldsymbol{X}'\boldsymbol{X} = \boldsymbol{P}\mathrm{diag}(\lambda_1, \ldots, \lambda_p)\boldsymbol{P}'.$$

  The matrix $\boldsymbol{P}$ consists of the eigenvectors $\boldsymbol{p}_j$ or the principal components of $\boldsymbol{X}'\boldsymbol{X}$. The first principal component $\boldsymbol{p}_1$ has the characteristic that the linear combination $z_1 = \boldsymbol{X}\boldsymbol{p}_1$ has the largest variance among all linear combinations of the columns of $\boldsymbol{X}$. The second principal component $\boldsymbol{p}_2$ determines the linear combination $z_2 = \boldsymbol{X}\boldsymbol{p}_2$, which has the second highest variance among all linear combinations that are also orthogonal to z_1. Accordingly, the linear combination $z_j = \boldsymbol{X}\boldsymbol{p}_j$ of the jth principal component has the highest variance relative to the linear combinations which are orthogonal to all of the first $j-1$ principle components. Instead of the original regressor variables, we use the derived linear combinations $z_1, z_2, \ldots$ as covariates in the principal component regression. Typically only the first $q \leq p$ of the p linear combinations are considered and sufficient for modeling. The remaining ones are neglected due to a lack of interpretability or due to low variance.

An application of principal component regression arises in the context of modeling determinants of malnutrition. Here, a number of variables describing the economical situation of the household is available. In most cases it does not make sense to include all economical covariates in the model. Based on a principal component analysis, easily interpretable linear combinations as measurements for the economical status are included instead. In almost all cases, only the linear combination with the highest variance based on the first principle component is used.

Outlier and Influence Analysis

In addition to checking model assumptions and collinearity analysis, another important aspect of model diagnosis is to investigate the influence of individual observations on the estimators. We will see that *outliers* are of particular importance, which we next examine in more detail.

Outliers

There is no exact definition for the term outlier. In this book, outliers are observations which do not adhere to the fitted model. We (informally) define the ith observation as an outlier if the (conditional) mean of y_i is not $\mathrm{E}(y_i \mid \boldsymbol{x}_i) = \boldsymbol{x}_i'\boldsymbol{\beta}$, as is postulated by the model, but rather shifted by the value d, i.e., $\mathrm{E}(y_i \mid \boldsymbol{x}_i) = \boldsymbol{x}_i'\boldsymbol{\beta} + d$.

We illustrate the outlier problem using the simulated data presented in Fig. 3.25. Panel (a) displays simulated observations (y_i, x_i) together with an estimated regression line (solid line). The solid point in the center of the range of values of x appears to be a severe outlier. If we estimate a new regression line, this time without considering the outlier, the resulting estimate (dotted line) only slightly changes. The estimated standard errors and, thus, also the confidence intervals however do change dramatically. Using all observations, we obtain 0.241 as estimated standard error for $\hat{\beta}_1$, whereas without the outlier, we obtain the value 0.069. Accordingly, we obtain the confidence interval [1.53, 2.50] for β_1 based on all observations and [1.84, 2.11] without considering the outlier. Even though in this case the outlier only slightly changes the estimated regression line, statistical inference is influenced considerably. In comparison to panel (a), panel (c) shows a slightly modified situation, where in this case, the outlier is not in the center but rather on the boundary of the range of values. As such, the outlier has now an effect on both the estimated standard errors *and* the estimated regression line. If multiple outliers appear as in panel (e), the impact is even greater.

Summarizing, we can state: Outliers can have a significant effect on estimation and inference. Further, the influence becomes more significant on the boundary of the covariate range and with multiple outliers. Thus, the following questions arise: First, how can we detect outliers? Secondly, what should we do when outliers are detected? The literature offers a multitude of possible answers to both questions. We first address the question how to detect outliers.

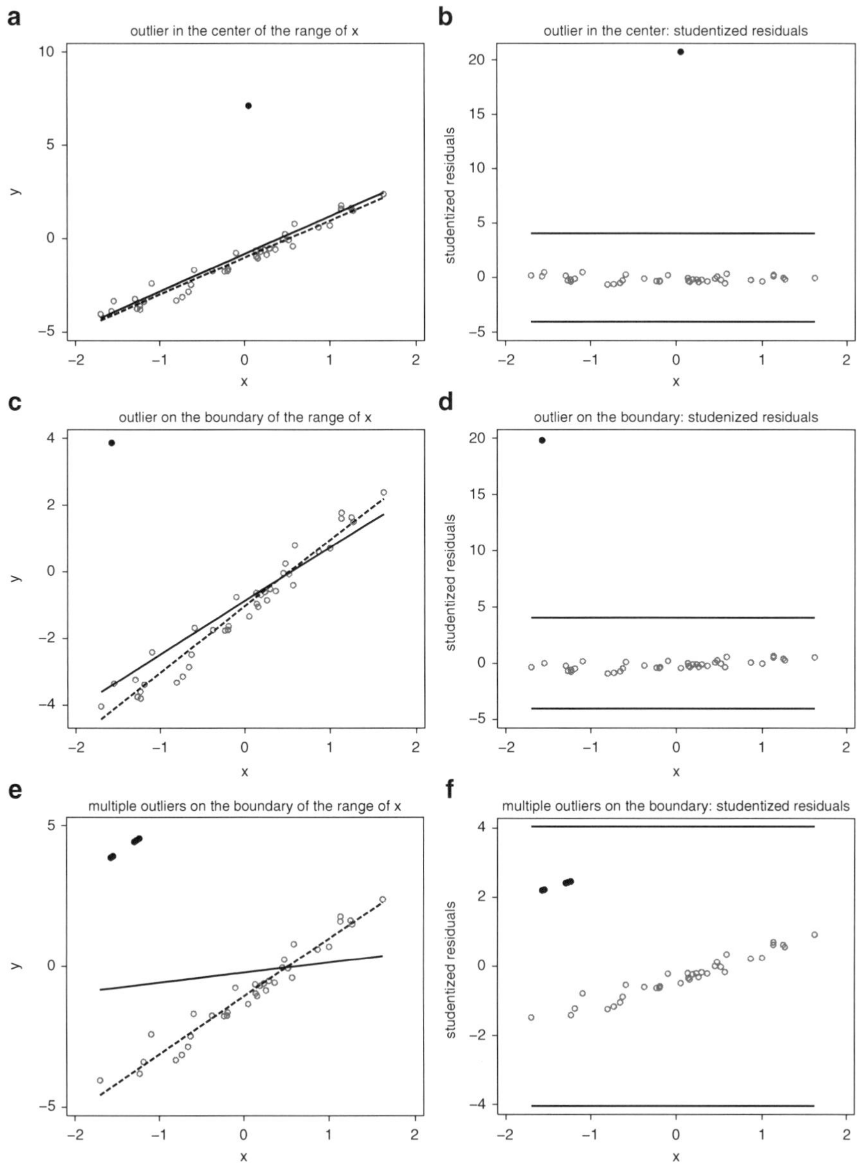

Fig. 3.25 Illustration of the outlier problem: The *left panels* show scatter plots between y and x together with the estimated regression line (*solid line*) and estimated regression line if the outliers are not considered (*dashed line*). The *right panels* display studentized residuals against x. Outliers are marked by *solid points*

A good starting point to detect outliers seems to be the search for large residuals. The discovery of outliers is, however, complicated by the fact that they can have a considerable effect on estimation. Note that estimates are "pulled" towards the outliers, such that the corresponding residuals can be relatively small, especially at the boundary of the covariate range. Therefore, if we wish to examine whether or not the ith observation is an outlier, it should be *excluded from estimation*.

In Sect. 3.10 on p. 123, we defined the "leave-one-out" residuals:

$$\hat{\varepsilon}_{(i)} = y_i - \hat{y}_{(i)} = y_i - \boldsymbol{x}_i'(\boldsymbol{X}_{(i)}'\boldsymbol{X}_{(i)})^{-1}\boldsymbol{X}_{(i)}'\boldsymbol{y}_{(i)}.$$

These residuals are based on estimates which do not consider the ith observation. With standardization, we obtained the studentized residuals:

$$r_i^* = \frac{\hat{\varepsilon}_{(i)}}{\hat{\sigma}_{(i)}(1 + \boldsymbol{x}_i'(\boldsymbol{X}_{(i)}'\boldsymbol{X}_{(i)})^{-1}\boldsymbol{x}_i)^{1/2}},$$

which follow a t-distribution with $n - p - 1$ degree of freedom when the model is correctly specified. Observations with large studentized residuals are thus potential outlier candidates. Since we know the distribution of studentized residuals in a correctly specified model, we can formally test whether or not the ith observation y_i is an outlier. In order to do so, we have to compare the ith studentized residual r_i^* with the $\alpha/2$ or the $1 - \alpha/2$ quantile of the t-distribution with $n - p - 1$ degree of freedom, for a given significance level α. If the residual is smaller than the $\alpha/2$ or larger than the $1 - \alpha/2$ quantile, then it is an outlier candidate.

In most cases, we do not want to test a single observation, but we want to test all observations *simultaneously*. In order to obtain a correct overall significance level α, we have to adjust the significance level for each observation. A straightforward approach is the Bonferroni correction, which uses the level α/n in each test. The Bonferroni correction is relatively conservative, which means that in general (especially with a large sample size) we tend not to discover all outliers even when they exist.

Figure 3.25b, d, f shows studentized residuals as a function of x for the simulated data sets. The plotted horizontal lines mark the 0.5 % and 99.5 % quantiles of the t-distribution with $n - p - 1 = 40 - 2 - 1 = 37$ degrees of freedom. Observations that are below or above the two lines are potential outlier candidates. It appears that the two individual outliers are identified, one in the center and the other on the boundary of the covariate range. The multiple outliers in panel (e) are, however, not identified. In general, it can be very difficult to discover multiple outliers. In this case, the visualization of the studentized residuals in panel (f) does offer us some help to notice the unusual observations.

Example 3.22 Prices of Used Cars—Outliers

Figure 3.23 shows the studentized residuals as a function of the fitted values and the covariates in the regression model derived from Example 3.19 (p. 152). The 0.5 % and 99.5 % quantiles of the t-distribution with $n-p-1 = 172-6-1 = 165$ degrees of freedom (horizontal lines) are also plotted. Based on the overall significance level of $\alpha = 0.01$, we obtain one observation above the line with a studentized residual of $r_i^* = 4.36$.

△

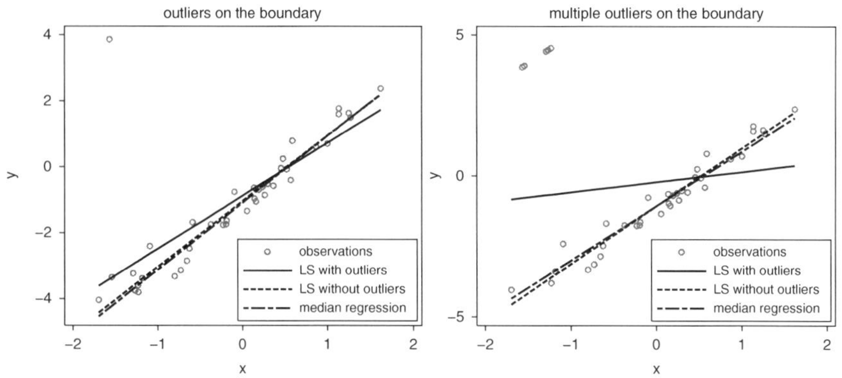

Fig. 3.26 Robust regression: estimated regression line based on various estimation techniques (least squares estimator based on all observations, least squares estimator without considering the outliers and median regression). *Left*: outlier on the border of the covariate range. *Right*: multiple outliers on the border of the covariate range

In practice, outlier candidates are often excluded from the analysis without further comment. This procedure is often not the best alternative. Instead, we recommend the following:

- *Exclusion of data errors:* Outliers can indicate data errors, for example, incorrect data coding. This source of error should first be eliminated.
- *Search for an explanation of outliers:* In some applications, the outliers of a model are the most interesting observations. A case in point is the satellite Nimbus 7 operated by the NASA to provide information about the atmosphere. In the year 1985, the satellite had been in use for several years and scientists observed an extremely large decrease of the ozone concentration in the atmosphere, the ozone hole was discovered. In this case, an explanation for the unusual observations existed.
- *Description of the differences in results:* Instead of excluding outliers from the analysis, it is possible to estimate the model with and without outliers. In this way, we can describe the emerged model differences and we do not lose the outlier information.
- *Robust regression:* Since the 1960s, robust methods have become increasingly more important in statistics. These methods are less sensitive to outliers, i.e., estimation results are not (or only slightly) influenced by outliers. The first robust method is median regression proposed by Boscovich already in 1760 (see Sect. 3.2.1). Median regression minimizes the sum of the absolute deviations instead of the sum of the squared deviations. Such an approach is much more robust against outliers than the least squares method; see Fig. 3.26, which compares several estimation methods in the presence of one or multiple outliers at the boundary of the covariate range. The least squares results are only insensitive to the outliers, in the cases that they are excluded from estimation. When using all observations, the estimated least squares regression lines are "pulled" towards the

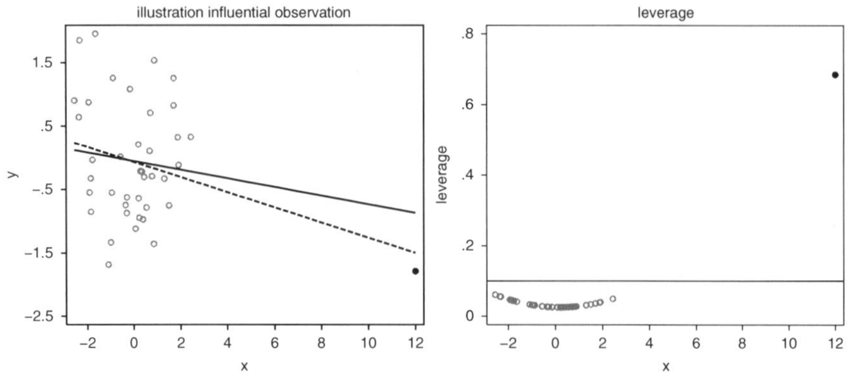

Fig. 3.27 The *left panel* shows a scatter plot between y and x including an estimated regression line (*dashed line*) and an estimated regression line if the influential observation is not considered (*solid line*). The *right panel* shows the leverage depending on x

outliers. However, outliers do not affect the median regression, even when they are not excluded from estimation. Median regression then has the advantage that we do not need to detect potential outliers, since estimation is unaffected. The use of median regression does become problematic when we have a relatively large amount of outliers. In such cases, we have to use other specialized methods, which, however, would go beyond the scope of this book; see, for example, Rousseeuw and Leroy (2003). Note that median regression is a special case of more general *quantile regression* which has recently become quite popular. More on quantile regression can be found in Chap. 10.

We may wonder why the least squares method is still used more often than the method of median regression, especially in light of the fact that median regression is much less sensitive to outliers. The answer is that the least squares estimator has much better estimation properties when no outliers exist. When the data follow a normal distribution, median regression needs about 50 % more data in order to obtain the same estimation precision compared to the least squares method. Nevertheless, within the scope of diagnostics, we also recommend to use robust methods (when available) to examine the sensitivity or robustness of the obtained least squares estimators.

Influence Analysis

Influence analysis focuses on measuring the impact of specific observations on estimation. The main objective is to find observations which have a large influence on $\hat{\boldsymbol{\beta}}$ and, thus, $\hat{\mathbf{y}}$. Estimation can drastically change, when such observations are removed. In many cases, influential observations are outliers, as informally defined above. In extreme cases, a single observation can drive the entire estimated relationship or the elimination of a certain observation destroys the relationship. Figure 3.27 (left panel) illustrates such a case using simulated data.

The literature offers numerous suggestions on how to detect highly influential observations including the following:

- *Leverage*: The diagonal elements h_{ii} of the hat matrix $\boldsymbol{H} = \boldsymbol{X}(\boldsymbol{X}'\boldsymbol{X})^{-1}\boldsymbol{X}'$ measure the leverage of the ith data point. The leverage ranges from $1/n$ to 1; see Box 3.5 on p. 108. A large leverage (close to 1) implies two things. First, $\mathrm{Var}(\hat{\varepsilon}_i) = \sigma^2(1 - h_{ii})$ is small, or even near zero when leverage is close to one. In the case of a single covariate, the regression line nearly passes through the point (y_i, x_i), regardless of any other observation. Thus an observation with high leverage has a considerable influence on the estimation results. These considerations can easily be extended to observations with more covariates, where now, in the extreme case, the regression hyperplane nearly intersects the point $(y_i, \boldsymbol{x}_i)$.

  Secondly, large leverage values indicate some unusual covariates values $\boldsymbol{x}_i$. This property is best seen in the case of only one covariate x_i as then the leverage is given by

$$h_{ii} = \frac{1}{n} + \frac{(x_i - \bar{x})^2}{\sum_j (x_j - \bar{x})^2}.$$

  The leverage h_{ii} increases, the further the observation x_i is from the average $\bar{x}$. In this sense, leverage values can also be understood as outliers in x-direction. In general, we have

$$h_{ii} = \frac{1}{n} + (\boldsymbol{x}_i - \bar{\boldsymbol{x}})'(\tilde{\boldsymbol{X}}'\tilde{\boldsymbol{X}})^{-1}(\boldsymbol{x}_i - \bar{\boldsymbol{x}})$$

  using the design matrix of centered covariates

$$\tilde{\boldsymbol{X}} = \begin{pmatrix} (x_{11} - \bar{x}_1) & \cdots & (x_{1k} - \bar{x}_k) \\ (x_{21} - \bar{x}_1) & \cdots & (x_{2k} - \bar{x}_k) \\ \vdots & \cdots & \vdots \\ (x_{n1} - \bar{x}_1) & \cdots & (x_{nk} - \bar{x}_k) \end{pmatrix}.$$

  Ideally, the leverages h_{ii} are uniformly distributed. As a rule of thumb, observations with $h_{ii} > 2p/n$ (twice the average) should be examined more closely. We should keep in mind, however, that large leverages do not necessarily lead to problems.

- *Cook's Distance:* Denote by $\hat{\boldsymbol{y}}_{(i)}$ the estimator for $\mathrm{E}(\boldsymbol{y})$ that uses all observations with exception of the ith observation. An obvious measure for the difference between $\hat{\boldsymbol{y}}$ based on all observations and $\hat{\boldsymbol{y}}_{(i)}$ is the Euclidean distance between the estimators. More precisely, Cook's distance is defined by

$$D_i = \frac{(\hat{\boldsymbol{y}}_{(i)} - \hat{\boldsymbol{y}})'(\hat{\boldsymbol{y}}_{(i)} - \hat{\boldsymbol{y}})}{p \cdot \hat{\sigma}^2},$$

where the numerator is standardized by the estimated variance $\hat{\sigma}^2$. As a rule of thumb, observations with $D_i > 0.5$ are worthy of attention, and observations with $D_i > 1$ should always be examined.

For our simulated data, the right panel of Fig. 3.27 shows the leverages h_{ii} as a function of x_i. The horizontal line marks the rule of thumb $h_{ii} > 2p/n = 2{\cdot}2/40 = 0.1$ for influential observations. The most influential observation is with $x = 12$ having $h_{ii} = 0.6$ and is the only leverage larger than the rule of thumb. Cook's distance with $D_i = 2.55$ is extremely large for this observation, indicating extreme influence of the observation. It is also the only observation with $D_i > 0.5$.

Example 3.23 Prices of Used Cars—Influence Analysis

Cook's distances are all low for the regression model outlined in Example 3.19 (p. 152). The largest Cook's distance is 0.12, and no observation has a D larger than 0.5. Approximately ten observations have leverages larger than the cutoff $h_{ii} > 2p/n = 0.058$. For the most part, these are observations with a rather small ($kilometer < 50$) or large kilometer reading ($kilometer > 230$). The largest leverage is $h_{ii} = 0.23$. It is therefore reasonable to examine the effect on estimation by removing the observations associated with extremely low or large kilometer readings and refitting; see the next Example 3.24. △

Alternative Modeling Approaches Due to Model Diagnosis

Model diagnosis often provides evidence for alternative modeling approaches to be investigated. When estimating alternative models, the main goal is, on the one hand, to find even more adequate models and, on the other hand, to evaluate model stability. In most cases, we do not find a single model that clearly outperforms all the others. On the contrary, we typically find a few competing models with almost equal fit. It is important to extract the common features of these models, as well as refer to their differences. Typically, the models will be different only with respect to some aspects which then cause uncertainty regarding our conclusions.

The following alternative modeling variants are possible:

- Residual plots can particularly provide useful information about *alternative modeling of the effect of one or more covariates.*
- We can *reestimate the model after removing conspicuous observations* having extremely large residuals, leverages, and/or Cook's distance.
- We can reestimate the model using *alternative approaches*, especially robust regression.
- Sometimes model diagnosis provides evidence for transforming the response variable. Especially if the response is strictly positive with a large range, then a logarithmic transformation of the response might be useful.
- If there is evidence of heteroscedastic errors, an alternative is weighted regression, as developed in Sect. 4.1.3 of the next chapter.

Example 3.24 Prices of Used Cars—Alternative Models

The regression model of Example 3.19 (p. 152) has undergone detailed regression diagnostics, as provided in the Examples 3.20–3.23. We conclude the following:

- There is no evidence of misspecification of continuous covariate effects; see Fig. 3.22 (p. 155).
- Figure 3.23 suggests moderately heteroscedastic errors.

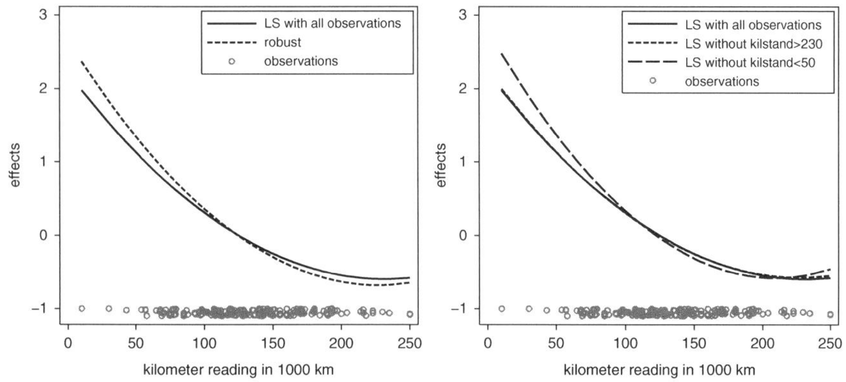

Fig. 3.28 Prices of used cars: effect of the kilometer reading for different modeling alternatives

- The influence analysis suggests unstable effects for Golf models with a kilometer reading smaller than 50 or larger than 230.

Hence, the following alternative model specifications were tested:

- *Weighted regression to cope with heteroscedastic errors:* Since heteroscedastic errors will be treated in Sect. 4.1.3 of Chap. 4, we defer the reader to Example 4.5 for details. For the moment, it suffices to know that the difference between the classical linear model and the heteroscedastic model is marginal. Example 4.5 elaborates this point further.
- *Logarithmic transformation of the response variable price:* Due to the positive and skewed response, we also estimated models with the log price. When using transformed responses, we have to keep in mind that model choice and model diagnoses have to be conducted again. For example, it is possible that outliers in the model with an untransformed response disappear in a model with a log-response, or vice versa. With the transformed response, we obtain a somewhat more parsimonious model, than with an untransformed one, since we can now model the effect of the kilometer reading linearly. Moreover, the model is more stable regarding outliers and influential observations. However, the main conclusions remain the same.
- *Model estimation after removing observations with $kilometer < 50$ or $kilometer > 230$, as well as robust regression:* Figure 3.28 shows the impact of different specifications on the effect of the kilometer reading. All models show a monotonically decreasing effect. We observe some uncertainty in the effect only in the domain $kilometer < 50$.

Based on our results, we can now draw the following final conclusions:

- We find weakly nonlinear and monotonically decreasing effects for age and the kilometer reading.
- The other explanatory variables do not have a significant effect (at most a very minor one) on sale's price.
- The models are relatively stable regarding outliers and influential observations. For the untransformed price response, the effect of the kilometer reading for $kilometer < 50$ is somewhat unstable.
- Any apparent heteroscedasticity can be neglected; see Example 4.5 for more details. △

3.5 Bibliographic Notes and Proofs

3.5.1 Bibliographic Notes

Since linear models play a very prominent role in statistics, a vast amount of literature exists. Faraway (2004), Rawlings, Pantula, and Dickey (2001), and Weisberg (2005) are among some of the textbooks that we recommend.

The linear model also finds a strong representation in econometrics textbooks, often with a different focus. In particular, exploratory and graphical tools for model diagnosis and specification are rare. Rather, statistical tests for model diagnostics and specification do play an important role. A standard textbook is Wooldridge (2006). A somewhat more challenging presentation is the detailed overview provided by Greene (2000); its fifth edition appeared in 2008. We recommend the second edition (2000), since the subsequent editions appear to be somewhat confusing and excessive in coverage.

Some specific topics, which were only lightly touched upon in this chapter are outlined in monographs. We recommend Rousseeuw and Leroy (2003) for robust regression (but see also Chap. 10) and Belsley, Kuh, and Welsch (2003) for identification of outliers and influential observations. A standard book for variable selection in linear models is Miller (2002). We also did not examine models with a multivariate response. These models are described, for example, in Anderson (2003). Further elaboration on SUR models (seemingly unrelated regression) and simultaneous equation systems can be found in econometric textbooks, for example, Greene (2000).

3.5.2 Proofs

Derivation of the Expected Sum of Squared Residuals (p. 109)

Applying Theorem B.2 (p. 647) and using the properties of the matrix $\boldsymbol{I} - \boldsymbol{H}$ (see Box 3.5 on p. 108), we obtain the expected sum of squared residuals as follows:

$$\begin{aligned}
\mathrm{E}(\hat{\boldsymbol{\varepsilon}}'\hat{\boldsymbol{\varepsilon}}) &= \mathrm{E}(\boldsymbol{y}'(\boldsymbol{I}-\boldsymbol{H})\boldsymbol{y}) \\
&= \mathrm{tr}((\boldsymbol{I}-\boldsymbol{H})\sigma^2\boldsymbol{I}) + \boldsymbol{\beta}'\boldsymbol{X}'(\boldsymbol{I}-\boldsymbol{H})\boldsymbol{X}\boldsymbol{\beta} \\
&= \sigma^2(n-p) + \boldsymbol{\beta}'\boldsymbol{X}'(\boldsymbol{I}-\boldsymbol{X}(\boldsymbol{X}'\boldsymbol{X})^{-1}\boldsymbol{X}')\boldsymbol{X}\boldsymbol{\beta} \\
&= \sigma^2(n-p) + \boldsymbol{\beta}'\boldsymbol{X}'\boldsymbol{X}\boldsymbol{\beta} - \boldsymbol{\beta}'\boldsymbol{X}'\boldsymbol{X}(\boldsymbol{X}'\boldsymbol{X})^{-1}\boldsymbol{X}'\boldsymbol{X}\boldsymbol{\beta} \\
&= \sigma^2(n-p) + \boldsymbol{\beta}'\boldsymbol{X}'\boldsymbol{X}\boldsymbol{\beta} - \boldsymbol{\beta}'\boldsymbol{X}'\boldsymbol{X}\boldsymbol{\beta} \\
&= \sigma^2(n-p). \qquad \square
\end{aligned}$$

Proof of Properties 3–5 in Box 3.7 on p. 112

3. Because the first column $\boldsymbol{x}^0$ of the design matrix is the vector $\mathbf{1} = (1, \ldots, 1)'$, property 3 follows from the orthogonality property 2:

$$0 = (\boldsymbol{x}^0)' \,\hat{\boldsymbol{\varepsilon}} = \mathbf{1}' \,\hat{\boldsymbol{\varepsilon}} = \sum_{i=1}^{n} \hat{\varepsilon}_i.$$

4. Property 3 implies

$$\sum_{i=1}^{n} \hat{y}_i = \sum_{i=1}^{n} (y_i - \hat{\varepsilon}_i) = \sum_{i=1}^{n} y_i - \sum_{i=1}^{n} \hat{\varepsilon}_i = \sum_{i=1}^{n} y_i.$$

5. We have

$$\bar{y} = \frac{1}{n} \sum_{i=1}^{n} y_i = \frac{1}{n} \sum_{i=1}^{n} (\hat{y}_i + \hat{\varepsilon}_i) = \frac{1}{n} \sum_{i=1}^{n} (\hat{\beta}_0 + \hat{\beta}_1 x_{i1} + \cdots + \hat{\beta}_k x_{ik} + \hat{\varepsilon}_i)$$

$$= \frac{1}{n} \sum_{i=1}^{n} (\hat{\beta}_0 + \hat{\beta}_1 x_{i1} + \cdots + \hat{\beta}_k x_{ik}) = \hat{\beta}_0 + \hat{\beta}_1 \bar{x}_1 + \cdots + \hat{\beta}_k \bar{x}_k.$$

Note that the proof of these properties essentially relies on the fact that the model contains an intercept term. Therefore, these properties are only valid in models with an intercept. □

Proof of the Variance Decomposition Formula (3.19) on p. 112

For the derivation of the formula for variance decomposition we make use of the special $n \times n$-matrix

$$\boldsymbol{C} = \boldsymbol{I} - \frac{1}{n}\mathbf{1}\mathbf{1}'.$$

Clearly, $\boldsymbol{C}$ is symmetric and idempotent. The matrix has several remarkable properties. For example, multiplying $\boldsymbol{C}$ with a vector $\boldsymbol{a} \in \mathbb{R}^n$, we obtain

$$\boldsymbol{C}\boldsymbol{a} = \begin{pmatrix} a_1 - \bar{a} \\ \vdots \\ a_n - \bar{a} \end{pmatrix}. \tag{3.31}$$

Thus, multiplication with $\boldsymbol{C}$ centers the vector $\boldsymbol{a}$. For the quadratic form $\boldsymbol{a}'\boldsymbol{C}\boldsymbol{a}$, we obtain

$$\boldsymbol{a}'\boldsymbol{C}\boldsymbol{a} = \sum_{i=1}^{n} (a_i - \bar{a})^2. \tag{3.32}$$

Further properties of C are given in Definition A.12 (p. 625). For deriving the variance decomposition formula, however, we only need properties (3.31) and (3.32).

We first multiply the identity $\boldsymbol{y} = \hat{\boldsymbol{y}} + \hat{\boldsymbol{\varepsilon}}$ with $\boldsymbol{C}$ from the left and obtain

$$\boldsymbol{C}\boldsymbol{y} = \boldsymbol{C}\hat{\boldsymbol{y}} + \boldsymbol{C}\hat{\boldsymbol{\varepsilon}}.$$

Using Eq. (3.31) and property 3 on p. 112 gives $\boldsymbol{C}\hat{\boldsymbol{\varepsilon}} = \hat{\boldsymbol{\varepsilon}}$. This implies

$$\boldsymbol{C}\boldsymbol{y} = \boldsymbol{C}\hat{\boldsymbol{y}} + \hat{\boldsymbol{\varepsilon}} \quad \text{and} \quad \boldsymbol{y}'\boldsymbol{C} = \hat{\boldsymbol{y}}'\boldsymbol{C} + \hat{\boldsymbol{\varepsilon}}'.$$

Thus, we get

$$\begin{aligned} \boldsymbol{y}'\boldsymbol{C}\boldsymbol{C}\boldsymbol{y} &= (\hat{\boldsymbol{y}}'\boldsymbol{C} + \hat{\boldsymbol{\varepsilon}}')(\boldsymbol{C}\hat{\boldsymbol{y}} + \hat{\boldsymbol{\varepsilon}}) \\ &= \hat{\boldsymbol{y}}'\boldsymbol{C}\boldsymbol{C}\hat{\boldsymbol{y}} + \hat{\boldsymbol{y}}'\boldsymbol{C}\hat{\boldsymbol{\varepsilon}} + \hat{\boldsymbol{\varepsilon}}'\boldsymbol{C}\hat{\boldsymbol{y}} + \hat{\boldsymbol{\varepsilon}}'\hat{\boldsymbol{\varepsilon}} \qquad (3.33) \\ &= \hat{\boldsymbol{y}}'\boldsymbol{C}\hat{\boldsymbol{y}} + \hat{\boldsymbol{y}}'\hat{\boldsymbol{\varepsilon}} + \hat{\boldsymbol{\varepsilon}}'\hat{\boldsymbol{y}} + \hat{\boldsymbol{\varepsilon}}'\hat{\boldsymbol{\varepsilon}}. \end{aligned}$$

From Eq. (3.33), the left-hand side of Eq. (3.32) is $\boldsymbol{y}'\boldsymbol{C}\boldsymbol{C}\boldsymbol{y} = \boldsymbol{y}'\boldsymbol{C}\boldsymbol{y} = \sum(y_i - \bar{y})^2$. Using $\bar{\hat{y}} = \bar{y}$ (see property 4 on p. 112), it follows that $\hat{\boldsymbol{y}}'\boldsymbol{C}\hat{\boldsymbol{y}} = \sum(\hat{y}_i - \bar{y})^2$. Property 1 on p. 112 implies $\hat{\boldsymbol{y}}'\hat{\varepsilon} = \hat{\varepsilon}'\hat{\boldsymbol{y}} = 0$ and we obtain

$$\sum_{i=1}^{n}(y_i - \bar{y})^2 = \sum_{i=1}^{n}(\hat{y}_i - \bar{y})^2 + \sum_{i=1}^{n}\hat{\varepsilon}_i^2. \qquad \square$$

Proof of the Gauß–Markov Theorem on p. 119

We first consider conditions under which linear estimators are unbiased. For an unbiased linear estimator, the equation $\mathrm{E}(\hat{\boldsymbol{\beta}}^L) = \boldsymbol{a} + \boldsymbol{A}\boldsymbol{X}\boldsymbol{\beta} = \boldsymbol{\beta}$ must hold for *all* $\boldsymbol{\beta} \in \mathbb{R}^p$. Setting $\boldsymbol{\beta} = \boldsymbol{0}$, we obtain $\boldsymbol{a} = \boldsymbol{0}$ as a necessary condition for $\hat{\boldsymbol{\beta}}^L$ being unbiased. Rearranging $\boldsymbol{A}\boldsymbol{X}\boldsymbol{\beta} = \boldsymbol{\beta}$ gives $(\boldsymbol{A}\boldsymbol{X} - \boldsymbol{I}_p)\boldsymbol{\beta} = \boldsymbol{0}$, implying $\boldsymbol{A}\boldsymbol{X} = \boldsymbol{I}_p$ as a further condition for $\hat{\boldsymbol{\beta}}^L$ being unbiased. From $\mathrm{rk}(\boldsymbol{A}\boldsymbol{X}) = \min(\mathrm{rk}(\boldsymbol{X}), \mathrm{rk}(\boldsymbol{A})) = \mathrm{rk}(\boldsymbol{I}_p) = p$, we also obtain $\mathrm{rk}(\boldsymbol{A}) = p$.

Now, without loss of generality, let $\boldsymbol{A}$ be of the form $\boldsymbol{A} = (\boldsymbol{X}'\boldsymbol{X})^{-1}\boldsymbol{X}' + \boldsymbol{B}$. Insertion into the condition $\boldsymbol{I}_p = \boldsymbol{A}\boldsymbol{X}$ yields

$$\boldsymbol{I}_p = \boldsymbol{A}\boldsymbol{X} = (\boldsymbol{X}'\boldsymbol{X})^{-1}\boldsymbol{X}'\boldsymbol{X} + \boldsymbol{B}\boldsymbol{X} = \boldsymbol{I}_p + \boldsymbol{B}\boldsymbol{X}$$

resulting in the condition

$$\boldsymbol{B}\boldsymbol{X} = \boldsymbol{0}.$$

This implies

$$\begin{aligned}\mathrm{Cov}(\hat{\boldsymbol{\beta}}^L) &= \sigma^2\,\boldsymbol{A}\,\boldsymbol{A}' \\ &= \sigma^2\left\{(\boldsymbol{X}'\boldsymbol{X})^{-1}\boldsymbol{X}' + \boldsymbol{B}\right\}\left\{\boldsymbol{X}(\boldsymbol{X}'\boldsymbol{X})^{-1} + \boldsymbol{B}'\right\} \\ &= \sigma^2\Big\{(\boldsymbol{X}'\boldsymbol{X})^{-1}\boldsymbol{X}'\boldsymbol{X}(\boldsymbol{X}'\boldsymbol{X})^{-1} + (\boldsymbol{X}'\boldsymbol{X})^{-1}\boldsymbol{X}'\boldsymbol{B}' \\ &\qquad + \boldsymbol{B}\boldsymbol{X}(\boldsymbol{X}'\boldsymbol{X})^{-1} + \boldsymbol{B}\,\boldsymbol{B}'\Big\} \\ &= \sigma^2\,(\boldsymbol{X}'\boldsymbol{X})^{-1} + \sigma^2\,\boldsymbol{B}\,\boldsymbol{B}' \\ &= \mathrm{Cov}(\hat{\boldsymbol{\beta}}) + \sigma^2\,\boldsymbol{B}\,\boldsymbol{B}'\end{aligned}$$

for the covariance matrix of $\boldsymbol{\beta}^L$. From Theorem A.30 (p. 634), $\boldsymbol{B}\,\boldsymbol{B}'$ is nonnegative definite. Rearranging results in

$$\mathrm{Cov}(\hat{\boldsymbol{\beta}}^L) - \mathrm{Cov}(\hat{\boldsymbol{\beta}}) = \sigma^2\boldsymbol{B}\,\boldsymbol{B}' \geq \boldsymbol{0}.$$

Based on this, we can now derive the optimality properties of the estimator. The variances of $\boldsymbol{b}'\hat{\boldsymbol{\beta}}^L$ and $\boldsymbol{b}'\hat{\boldsymbol{\beta}}$ are given by

$$\mathrm{Var}(\boldsymbol{b}'\hat{\boldsymbol{\beta}}^L) = \boldsymbol{b}'\mathrm{Cov}(\hat{\boldsymbol{\beta}}^L)\boldsymbol{b} \quad \text{and} \quad \mathrm{Var}(\boldsymbol{b}'\hat{\boldsymbol{\beta}}) = \boldsymbol{b}'\mathrm{Cov}(\hat{\boldsymbol{\beta}})\boldsymbol{b}.$$

Since the difference of covariance matrices is nonnegative definite, the difference of variances is nonnegative for any vector $\boldsymbol{b}$ (see Definition A.27 (p. 633) of definite matrices). This proves the proposition

$$\mathrm{Var}(\boldsymbol{b}'\hat{\boldsymbol{\beta}}^L) \geq \mathrm{Var}(\boldsymbol{b}'\hat{\boldsymbol{\beta}})$$

for vector $\boldsymbol{b}$. In particular, we can set $\boldsymbol{b} = (0, \ldots, 1, \ldots, 0)'$ with 1 at position $j+1$ for $j = 0, \ldots, k$. Then we obtain

$$\mathrm{Var}(\hat{\beta}_j^L) = \mathrm{Var}(\boldsymbol{b}'\hat{\boldsymbol{\beta}}^L) \geq \mathrm{Var}(\boldsymbol{b}'\hat{\boldsymbol{\beta}}) = \mathrm{Var}(\hat{\beta}_j). \qquad \square$$

Proof of Propositions 5 and 6 on p. 123

5. The proof makes use of the matrix

$$\boldsymbol{Q} = \boldsymbol{I} - \boldsymbol{H} = \boldsymbol{I} - \boldsymbol{X}(\boldsymbol{X}'\boldsymbol{X})^{-1}\boldsymbol{X}',$$

where $\boldsymbol{H}$ is the hat matrix from Sect. 3.2.1 (p. 107). The matrix $\boldsymbol{Q}$ is symmetric and idempotent with rank $\mathrm{rk}(\boldsymbol{Q}) = n - p$. Multiplication with the design matrix $\boldsymbol{X}$ gives

$$\boldsymbol{Q}\boldsymbol{X} = \boldsymbol{X} - \boldsymbol{X}(\boldsymbol{X}'\boldsymbol{X})^{-1}\boldsymbol{X}'\boldsymbol{X} = \boldsymbol{0}.$$

From this, and because $\boldsymbol{Q}$ is symmetric and idempotent, we obtain

$$\hat{\boldsymbol{\varepsilon}}'\hat{\boldsymbol{\varepsilon}} = \boldsymbol{y}'\boldsymbol{Q}\boldsymbol{y} = (\boldsymbol{X}\boldsymbol{\beta}+\boldsymbol{\varepsilon})'\boldsymbol{Q}\boldsymbol{Q}(\boldsymbol{X}\boldsymbol{\beta}+\boldsymbol{\varepsilon}) = (\boldsymbol{\beta}'\boldsymbol{X}'\boldsymbol{Q}+\boldsymbol{\varepsilon}'\boldsymbol{Q})(\boldsymbol{Q}\boldsymbol{X}\boldsymbol{\beta}+\boldsymbol{Q}\boldsymbol{\varepsilon}) = \boldsymbol{\varepsilon}'\boldsymbol{Q}\boldsymbol{\varepsilon}.$$

The proposition now follows from $\boldsymbol{\varepsilon}/\sigma \sim \mathrm{N}(\boldsymbol{0}, \boldsymbol{I})$ and from Theorem B.8 (2) on p. 651.

6. To prove the proposition, we show that $\frac{1}{\sigma}(\hat{\boldsymbol{\beta}} - \boldsymbol{\beta})$ and $\frac{1}{\sigma^2}\hat{\boldsymbol{\varepsilon}}'\hat{\boldsymbol{\varepsilon}}$ are independent. For this we apply Theorem B.8 (2) on p. 651 with $\boldsymbol{R} = \boldsymbol{Q}$ and $\boldsymbol{B} = (\boldsymbol{X}'\boldsymbol{X})^{-1}\boldsymbol{X}'$. The condition $\boldsymbol{B}\boldsymbol{R} = \boldsymbol{0}$ holds since

$$(\boldsymbol{X}'\boldsymbol{X})^{-1}\boldsymbol{X}'\boldsymbol{Q} = \boldsymbol{0}.$$

Since $\boldsymbol{\varepsilon}/\sigma \sim \mathrm{N}(\boldsymbol{0}, \boldsymbol{I})$, the theorem implies that

$$\frac{\boldsymbol{\varepsilon}}{\sigma}'\boldsymbol{Q}\frac{\boldsymbol{\varepsilon}}{\sigma} = \frac{1}{\sigma^2}\hat{\boldsymbol{\varepsilon}}'\hat{\boldsymbol{\varepsilon}} \quad \text{and} \quad (\boldsymbol{X}'\boldsymbol{X})^{-1}\boldsymbol{X}'\boldsymbol{\varepsilon}/\sigma$$

are independent. The proposition finally follows from

$$\begin{aligned}\frac{1}{\sigma}(\hat{\boldsymbol{\beta}} - \boldsymbol{\beta}) &= \frac{1}{\sigma}\left\{(\boldsymbol{X}'\boldsymbol{X})^{-1}\boldsymbol{X}'\boldsymbol{y} - \boldsymbol{\beta}\right\} \\ &= \frac{1}{\sigma}\left\{(\boldsymbol{X}'\boldsymbol{X})^{-1}\boldsymbol{X}'(\boldsymbol{X}\boldsymbol{\beta} + \boldsymbol{\varepsilon}) - \boldsymbol{\beta}\right\} \\ &= \frac{1}{\sigma}(\boldsymbol{X}'\boldsymbol{X})^{-1}\boldsymbol{X}'\boldsymbol{\varepsilon} \\ &= (\boldsymbol{X}'\boldsymbol{X})^{-1}\boldsymbol{X}'\frac{\boldsymbol{\varepsilon}}{\sigma}.\end{aligned}$$

□

Distribution of the Test Statistic Under Linear Null Hypothesis H_0 (p. 128)

Computation of the Least Squares Estimator Under H_0

To minimize the residual sum of squares under the null hypothesis $\boldsymbol{C}\boldsymbol{\beta} = \boldsymbol{d}$, we apply the *Lagrangeapproach*

$$\begin{aligned}\mathrm{LSR}(\boldsymbol{\beta}; \boldsymbol{\lambda}) &= \mathrm{LS}(\boldsymbol{\beta}) - 2\boldsymbol{\lambda}'(\boldsymbol{C}\boldsymbol{\beta} - \boldsymbol{d}) \\ &= \boldsymbol{y}'\boldsymbol{y} - 2\,\boldsymbol{y}'\boldsymbol{X}\boldsymbol{\beta} + \boldsymbol{\beta}'\boldsymbol{X}'\boldsymbol{X}\boldsymbol{\beta} - 2\,\boldsymbol{\lambda}'\boldsymbol{C}\boldsymbol{\beta} + 2\,\boldsymbol{\lambda}'\boldsymbol{d},\end{aligned}$$

where $\boldsymbol{\lambda}$ is a column vector of Lagrangian multipliers of dimension r. Using Theorems A.33.1 and A.33.3 (p. 636) on differentiation of matrix functions, we obtain

$$\frac{\partial \operatorname{LSR}(\boldsymbol{\beta};\boldsymbol{\lambda})}{\partial \boldsymbol{\beta}} = -2\,\boldsymbol{X}'\boldsymbol{y} + 2\,\boldsymbol{X}'\boldsymbol{X}\boldsymbol{\beta} - 2\,\boldsymbol{C}'\boldsymbol{\lambda}$$
$$\frac{\partial \operatorname{LSR}(\boldsymbol{\beta};\boldsymbol{\lambda})}{\partial \boldsymbol{\lambda}} = -2\,\boldsymbol{C}\boldsymbol{\beta} + 2\,\boldsymbol{d}.$$

Setting these derivatives to zero gives

$$\boldsymbol{X}'\boldsymbol{X}\,\boldsymbol{\beta} - \boldsymbol{X}'\boldsymbol{y} = \boldsymbol{C}'\boldsymbol{\lambda}$$
$$\boldsymbol{C}\,\boldsymbol{\beta} = \boldsymbol{d}.$$

We solve these equations first for $\boldsymbol{\lambda}$ and then for $\boldsymbol{\beta}$. Multiplying the first equation from the left with $(\boldsymbol{X}'\boldsymbol{X})^{-1}$ gives

$$\boldsymbol{\beta} - \hat{\boldsymbol{\beta}} = (\boldsymbol{X}'\boldsymbol{X})^{-1}\boldsymbol{C}'\boldsymbol{\lambda}.$$

Multiplying this equation from the left with the matrix $\boldsymbol{C}$ yields

$$\boldsymbol{C}\,\boldsymbol{\beta} - \boldsymbol{C}\,\hat{\boldsymbol{\beta}} = \boldsymbol{C}\,(\boldsymbol{X}'\boldsymbol{X})^{-1}\boldsymbol{C}'\boldsymbol{\lambda}.$$

Inserting the second equation $\boldsymbol{C}\,\boldsymbol{\beta} = \boldsymbol{d}$, we obtain

$$\boldsymbol{d} - \boldsymbol{C}\,\hat{\boldsymbol{\beta}} = \boldsymbol{C}\,(\boldsymbol{X}'\boldsymbol{X})^{-1}\boldsymbol{C}'\boldsymbol{\lambda},$$

and

$$\boldsymbol{\lambda} = (\boldsymbol{C}\,(\boldsymbol{X}'\boldsymbol{X})^{-1}\boldsymbol{C}')^{-1}(\boldsymbol{d} - \boldsymbol{C}\,\hat{\boldsymbol{\beta}}).$$

In the last step, we have used the fact that $\boldsymbol{C}\,(\boldsymbol{X}'\boldsymbol{X})^{-1}\boldsymbol{C}'$ is positive definite [Theorem A.29.2 (p. 634)] and, therefore, invertible.

Inserting $\boldsymbol{\lambda}$ into the first equation gives

$$\boldsymbol{X}'\boldsymbol{X}\,\boldsymbol{\beta} - \boldsymbol{X}'\boldsymbol{y} = \boldsymbol{C}'(\boldsymbol{C}\,(\boldsymbol{X}'\boldsymbol{X})^{-1}\boldsymbol{C}')^{-1}(\boldsymbol{d} - \boldsymbol{C}\,\hat{\boldsymbol{\beta}}),$$

and finally, the (restricted) least squares estimator

$$\hat{\boldsymbol{\beta}}^R = \hat{\boldsymbol{\beta}} - (\boldsymbol{X}'\boldsymbol{X})^{-1}\boldsymbol{C}'(\boldsymbol{C}\,(\boldsymbol{X}'\boldsymbol{X})^{-1}\boldsymbol{C}')^{-1}(\boldsymbol{C}\,\hat{\boldsymbol{\beta}} - \boldsymbol{d}).$$

Derivation of the Difference of Residual Sum of Squares

We first denote the restricted least squares estimator $\hat{\boldsymbol{\beta}}^R$ as

$$\hat{\boldsymbol{\beta}}^R = \hat{\boldsymbol{\beta}} - \Delta_{H_0},$$

where Δ_{H_0} is defined through

$$\Delta_{H_0} = (\boldsymbol{X}'\boldsymbol{X})^{-1}\boldsymbol{C}'\left(\boldsymbol{C}(\boldsymbol{X}'\boldsymbol{X})^{-1}\boldsymbol{C}'\right)^{-1}(\boldsymbol{C}\hat{\boldsymbol{\beta}} - \boldsymbol{d}).$$

For the values $\hat{\boldsymbol{y}}_{H_0}$ fitted under the restriction of the null hypothesis we then obtain

$$\hat{\boldsymbol{y}}_{H_0} = \boldsymbol{X}\hat{\boldsymbol{\beta}}^R = \boldsymbol{X}(\hat{\boldsymbol{\beta}} - \boldsymbol{\Delta}_{H_0}) = \boldsymbol{X}\hat{\boldsymbol{\beta}} - \boldsymbol{X}\boldsymbol{\Delta}_{H_0} = \hat{\boldsymbol{y}} - \boldsymbol{X}\boldsymbol{\Delta}_{H_0}$$

and

$$\hat{\boldsymbol{\varepsilon}}_{H_0} = \boldsymbol{y} - \hat{\boldsymbol{y}}_{H_0} = \boldsymbol{y} - \hat{\boldsymbol{y}} + \boldsymbol{X}\boldsymbol{\Delta}_{H_0} = \hat{\boldsymbol{\varepsilon}} + \boldsymbol{X}\boldsymbol{\Delta}_{H_0}$$

for the residuals under the null hypothesis. Then the residual sum of squares SSE_{H_0} under H_0 can be written in the form

$$\begin{aligned}
\text{SSE}_{H_0} &= \hat{\boldsymbol{\varepsilon}}'_{H_0}\hat{\boldsymbol{\varepsilon}}_{H_0} \\
&= (\hat{\boldsymbol{\varepsilon}} + \boldsymbol{X}\boldsymbol{\Delta}_{H_0})'(\hat{\boldsymbol{\varepsilon}} + \boldsymbol{X}\boldsymbol{\Delta}_{H_0}) \\
&= \hat{\boldsymbol{\varepsilon}}'\hat{\boldsymbol{\varepsilon}} + \hat{\boldsymbol{\varepsilon}}'\boldsymbol{X}\boldsymbol{\Delta}_{H_0} + \boldsymbol{\Delta}'_{H_0}\boldsymbol{X}'\hat{\boldsymbol{\varepsilon}} + \boldsymbol{\Delta}'_{H_0}\boldsymbol{X}'\boldsymbol{X}\boldsymbol{\Delta}_{H_0} \\
&= \hat{\boldsymbol{\varepsilon}}'\hat{\boldsymbol{\varepsilon}} + \boldsymbol{\Delta}'_{H_0}\boldsymbol{X}'\boldsymbol{X}\boldsymbol{\Delta}_{H_0}.
\end{aligned}$$

The last equality holds because the residuals of the full model and the columns of the design matrix are orthogonal, i.e., $\hat{\boldsymbol{\varepsilon}}'\boldsymbol{X} = \boldsymbol{0}$ (see Box 3.7 on p. 112).

The matrix $\boldsymbol{X}'\boldsymbol{X}$ is positive definite, implying that $\boldsymbol{\Delta}'_{H_0}\boldsymbol{X}'\boldsymbol{X}\boldsymbol{\Delta}_{H_0}$ is nonnegative. Therefore, the residual sum of squares SSE_{H_0} under H_0 is always greater than or equal to the residual sum of squares SSE in the unrestricted model; see p. 128.

For the difference ΔSSE of residual sum of squares we finally obtain

$$\begin{aligned}
\Delta\text{SSE} &= \hat{\boldsymbol{\varepsilon}}'\hat{\boldsymbol{\varepsilon}} + \boldsymbol{\Delta}'_{H_0}\boldsymbol{X}'\boldsymbol{X}\boldsymbol{\Delta}_{H_0} - \hat{\boldsymbol{\varepsilon}}'\hat{\boldsymbol{\varepsilon}} \\
&= \left\{(\boldsymbol{X}'\boldsymbol{X})^{-1}\boldsymbol{C}'\left(\boldsymbol{C}(\boldsymbol{X}'\boldsymbol{X})^{-1}\boldsymbol{C}'\right)^{-1}(\boldsymbol{C}\hat{\boldsymbol{\beta}} - \boldsymbol{d})\right\}'\boldsymbol{X}'\boldsymbol{X} \\
&\quad \cdot\left\{(\boldsymbol{X}'\boldsymbol{X})^{-1}\boldsymbol{C}'\left(\boldsymbol{C}(\boldsymbol{X}'\boldsymbol{X})^{-1}\boldsymbol{C}'\right)^{-1}(\boldsymbol{C}\hat{\boldsymbol{\beta}} - \boldsymbol{d})\right\} \\
&= (\boldsymbol{C}\hat{\boldsymbol{\beta}} - \boldsymbol{d})'\left(\boldsymbol{C}(\boldsymbol{X}'\boldsymbol{X})^{-1}\boldsymbol{C}'\right)^{-1} \\
&\quad \cdot\boldsymbol{C}(\boldsymbol{X}'\boldsymbol{X})^{-1}\boldsymbol{C}'\left(\boldsymbol{C}(\boldsymbol{X}'\boldsymbol{X})^{-1}\boldsymbol{C}'\right)^{-1}(\boldsymbol{C}\hat{\boldsymbol{\beta}} - \boldsymbol{d}) \\
&= (\boldsymbol{C}\hat{\boldsymbol{\beta}} - \boldsymbol{d})'\left(\boldsymbol{C}(\boldsymbol{X}'\boldsymbol{X})^{-1}\boldsymbol{C}'\right)^{-1}(\boldsymbol{C}\hat{\boldsymbol{\beta}} - \boldsymbol{d}).
\end{aligned}$$

Stochastic Properties of the Difference of Residual Sum of Squares

1. To prove this property, we apply Theorem B.2.8 (p. 647) on the expectation of quadratic forms. We have

$$\text{E}(\boldsymbol{C}\hat{\boldsymbol{\beta}} - \boldsymbol{d}) = \boldsymbol{C}\boldsymbol{\beta} - \boldsymbol{d}$$

and

$$\text{Cov}(\boldsymbol{C}\hat{\boldsymbol{\beta}} - \boldsymbol{d}) = \sigma^2\boldsymbol{C}(\boldsymbol{X}'\boldsymbol{X})^{-1}\boldsymbol{C}'.$$

Setting $\boldsymbol{Z} = \boldsymbol{C}\hat{\boldsymbol{\beta}} - \boldsymbol{d}$ and $\boldsymbol{A} = \left(\boldsymbol{C}(\boldsymbol{X}'\boldsymbol{X})^{-1}\boldsymbol{C}'\right)^{-1}$ in Theorem B.2.8, we obtain

$$\begin{aligned}
\mathrm{E}(\Delta\mathrm{SSE}) &= \mathrm{E}\left\{(\boldsymbol{C}\hat{\boldsymbol{\beta}} - \boldsymbol{d})'\left(\boldsymbol{C}(\boldsymbol{X}'\boldsymbol{X})^{-1}\boldsymbol{C}'\right)^{-1}(\boldsymbol{C}\hat{\boldsymbol{\beta}} - \boldsymbol{d})\right\} \\
&= \mathrm{tr}\left\{\sigma^2\left(\boldsymbol{C}(\boldsymbol{X}'\boldsymbol{X})^{-1}\boldsymbol{C}'\right)^{-1}\boldsymbol{C}(\boldsymbol{X}'\boldsymbol{X})^{-1}\boldsymbol{C}'\right\} \\
&\quad +(\boldsymbol{C}\boldsymbol{\beta} - \boldsymbol{d})'\left(\boldsymbol{C}(\boldsymbol{X}'\boldsymbol{X})^{-1}\boldsymbol{C}'\right)^{-1}(\boldsymbol{C}\boldsymbol{\beta} - \boldsymbol{d}) \\
&= \mathrm{tr}(\sigma^2\boldsymbol{I}_r) + (\boldsymbol{C}\boldsymbol{\beta} - \boldsymbol{d})'\left(\boldsymbol{C}(\boldsymbol{X}'\boldsymbol{X})^{-1}\boldsymbol{C}'\right)^{-1}(\boldsymbol{C}\boldsymbol{\beta} - \boldsymbol{d}) \\
&= r\sigma^2 + (\boldsymbol{C}\boldsymbol{\beta} - \boldsymbol{d})'\left(\boldsymbol{C}(\boldsymbol{X}'\boldsymbol{X})^{-1}\boldsymbol{C}'\right)^{-1}(\boldsymbol{C}\boldsymbol{\beta} - \boldsymbol{d}).
\end{aligned}$$

2. The proposition follows from Theorem B.8.1 (p. 651). Defining the random vector $\boldsymbol{Z} = \boldsymbol{C}\hat{\boldsymbol{\beta}}$, we obtain

$$\mathrm{E}(\boldsymbol{Z}) = \boldsymbol{C}\boldsymbol{\beta} = \boldsymbol{d}$$

and

$$\mathrm{Cov}(\boldsymbol{Z}) = \sigma^2\boldsymbol{C}(\boldsymbol{X}'\boldsymbol{X})^{-1}\boldsymbol{C}'$$

under H_0. Since $\hat{\boldsymbol{\beta}}$ is normally distributed, it follows that

$$\boldsymbol{Z} \sim \mathrm{N}(\boldsymbol{d}, \sigma^2\boldsymbol{C}(\boldsymbol{X}'\boldsymbol{X})^{-1}\boldsymbol{C}').$$

The proposition follows directly by applying Theorem B.8.1 to the random vector $\boldsymbol{Z}$.

3. The difference of residual sum of squares ΔSSE only depends on the LS estimator $\hat{\boldsymbol{\beta}}$. The proposition now follows directly from the independence of $\hat{\boldsymbol{\varepsilon}}'\hat{\boldsymbol{\varepsilon}}$ and $\hat{\boldsymbol{\beta}}$; see Box 3.11 on p. 123. □

Extensions of the Classical Linear Model 4

This chapter discusses several extensions of the classical linear model. We first describe in Sect. 4.1 the general linear model and its applications. This model allows for correlated errors and heteroscedastic variances of the errors. Section 4.2 discusses several techniques to regularize the least squares estimator. Such a regularization may be useful in cases where the design matrix is highly collinear or even rank deficient. Moreover, regularization techniques allow for built-in variable selection. Section 4.4 describes Bayesian linear models as an alternative to the frequentist linear model framework. In modern statistics, Bayesian approaches have become increasingly more important and widely used. Hence, Sect. 4.4 (in combination with Appendix B.5) serves as a basis for the Bayesian approaches to be described in chapters to follow. Moreover, we present powerful Bayesian techniques for model choice, in particular variable selection.

4.1 The General Linear Model

Thus far, we have addressed the classical linear model $\boldsymbol{y} = \boldsymbol{X}\boldsymbol{\beta} + \boldsymbol{\varepsilon}$ with uncorrelated and homoscedastic errors, i.e., $\mathrm{Cov}(\boldsymbol{\varepsilon}) = \sigma^2 \boldsymbol{I}$. Least squares estimation based on these assumptions is referred to as *ordinary least squares*. The discussion of the model has shown that the assumption of uncorrelated and homoscedastic errors is not always satisfied. In this section, we want to extend the class of linear models such that correlated and heteroscedastic errors are possible. The resulting model is called the *general linear model*. The classical linear model considered so far is an important special case. In fact, many inference problems in the general linear model can be traced back to the classical linear model.

4.1.1 Model Definition

In the general linear model, we replace

$$\mathrm{Cov}(\boldsymbol{\varepsilon}) = \sigma^2 \boldsymbol{I}$$

L. Fahrmeir et al., *Regression*, DOI 10.1007/978-3-642-34333-9_4,

by the more general assumption

$$\text{Cov}(\boldsymbol{\varepsilon}) = \sigma^2 \boldsymbol{W}^{-1},$$

where $\boldsymbol{W}$ is a positive definite matrix. In the special case of heteroscedastic and (still) uncorrelated errors, we have

$$\boldsymbol{W} = \text{diag}(w_1, \ldots, w_n).$$

The heteroscedastic error variances are then given by $\text{Var}(\varepsilon_i) = \sigma_i^2 = \sigma^2/w_i$.

When introducing a more general model class, more complicated inference techniques are usually needed than in the simpler special case. Thus, it is worth investigating whether or not the use of the more general model is really necessary. We therefore study the consequences of using the comparably simple inference techniques of the classical linear model if the true model is the general linear model, i.e., $\text{Cov}(\boldsymbol{\varepsilon}) = \sigma^2 \boldsymbol{W}^{-1}$ and not $\text{Cov}(\boldsymbol{\varepsilon}) = \sigma^2 \boldsymbol{I}$. We first examine the consequences of using the ordinary least squares estimator $\hat{\boldsymbol{\beta}} = (\boldsymbol{X}'\boldsymbol{X})^{-1}\boldsymbol{X}'\boldsymbol{y}$ within the general linear model. Analogous to the derivations on p. 115 for the classical linear model, we obtain

$$\text{E}(\hat{\boldsymbol{\beta}}) = \boldsymbol{\beta} \qquad \text{Cov}(\hat{\boldsymbol{\beta}}) = \sigma^2(\boldsymbol{X}'\boldsymbol{X})^{-1}\boldsymbol{X}'\boldsymbol{W}^{-1}\boldsymbol{X}(\boldsymbol{X}'\boldsymbol{X})^{-1}. \qquad (4.1)$$

Hence, ordinary least squares developed for the classical linear model is still unbiased for $\boldsymbol{\beta}$ within the general linear model setting. However, the covariance matrix in the general linear model does not correspond with the one found for the classical model, $\sigma^2(\boldsymbol{X}'\boldsymbol{X})^{-1}$. Thus, all derivations that are based on the covariance matrix of $\hat{\boldsymbol{\beta}}$ are wrong. In particular, we obtain incorrect variances and standard errors for the estimated regression coefficients, and thus incorrect tests and confidence intervals.

In the following section, we discuss several alternatives to obtain improved estimators within the general linear model. Section 4.1.2 develops a *weighted least squares estimator* (WLS) as a generalization of the ordinary least squares estimator. The WLS estimator shares many of the desirable properties as found with the ordinary least squares estimator in the classical setting. When using the WLS estimator, the matrix $\boldsymbol{W}$ must be known, which in practice is usually not the case. Therefore Sects. 4.1.3 and 4.1.4 address inference in the case of unknown $\boldsymbol{W}$. We restrict ourselves to the two important special cases of heteroscedastic and autocorrelated errors.

4.1.2 Weighted Least Squares

We now develop an estimator that bypasses the above mentioned problems when using the ordinary least squares estimator in the general linear model. The idea is to transform the response variable, design matrix, and errors such that the

4.1 The General Linear Model

The model

$$y = X\beta + \varepsilon$$

is called general linear model, if the following assumptions hold:

1. $\mathrm{E}(\varepsilon) = \mathbf{0}$.
2. $\mathrm{Cov}(\varepsilon) = \mathrm{E}(\varepsilon\varepsilon') = \sigma^2 W^{-1}$, where W is a known positive definite matrix.
3. The design matrix X has full column rank, i.e., $\mathrm{rk}(X) = k + 1 = p$.

We refer to general normal regression when the additional assumption

4. $\varepsilon \sim \mathrm{N}(\mathbf{0}, \sigma^2 W^{-1})$

applies.

transformed variables follow a classical linear model. For illustration, we first look at a model with uncorrelated and heteroscedastic errors, i.e., $\mathrm{Cov}(\varepsilon) = \sigma^2 W^{-1} = \sigma^2 \mathrm{diag}(1/w_1, \ldots, 1/w_n)$. Multiplication of the errors ε_i with $\sqrt{w_i}$ produces the transformed errors $\varepsilon_i^* = \sqrt{w_i}\varepsilon_i$ with constant variances $\mathrm{Var}(\varepsilon_i^*) = \mathrm{Var}(\sqrt{w_i}\varepsilon_i) = \sigma^2$. To make sure that the model is unchanged, we must transform the response variable and all covariates (including the constant) accordingly. We obtain $y_i^* = \sqrt{w_i}y_i$, $x_{i0}^* = \sqrt{w_i}$, $x_{i1}^* = \sqrt{w_i}x_{i1}, \ldots, x_{ik}^* = \sqrt{w_i}x_{ik}$, and, thus, the classical linear model

$$y_i^* = \beta_0 x_{i0}^* + \beta_1 x_{i1}^* + \ldots + \beta_k x_{ik}^* + \varepsilon_i^*$$

with homoscedastic errors ε_i^*. This transformation formally corresponds to multiplication of the model equation $y = X\beta + \varepsilon$, from the left, with the matrix $W^{1/2} = \mathrm{diag}(\sqrt{w_1}, \ldots, \sqrt{w_n})$, yielding

$$W^{1/2}y = W^{1/2}X\beta + W^{1/2}\varepsilon.$$

Using the transformed values $y^* = W^{1/2}y$, $X^* = W^{1/2}X$, and $\varepsilon^* = W^{1/2}\varepsilon$, we finally obtain

$$y^* = X^*\beta + \varepsilon^*. \tag{4.2}$$

The transformation allows us to work in the familiar framework of the classical linear model. For the least squares estimator we get

$$\begin{aligned}\hat{\beta} &= \left(X^{*\prime}X^*\right)^{-1} X^{*\prime}y^* \\ &= \left(X'W^{1/2}W^{1/2}X\right)^{-1} X'W^{1/2}W^{1/2}y \\ &= (X'WX)^{-1}\, X'W\, y,\end{aligned}$$

which is referred to as the *Aitken* or *WLS estimator*. It can be shown that the Aitken estimator minimizes the "weighted" residual sum of squares

$$\mathrm{GLS}(\boldsymbol{\beta}) = (\boldsymbol{y} - \boldsymbol{X}\boldsymbol{\beta})'\boldsymbol{W}(\boldsymbol{y} - \boldsymbol{X}\boldsymbol{\beta}) = \sum_{i=1}^{n} w_i(y_i - \boldsymbol{x}_i'\boldsymbol{\beta})^2.$$

Observations with a higher variance σ^2/w_i have less weight w_i than observations with a smaller variance.

Assuming normal errors, we can also show that the WLS estimator coincides with the ML estimator for $\boldsymbol{\beta}$, i.e., $\hat{\boldsymbol{\beta}}_{ML} = \hat{\boldsymbol{\beta}}$. For the ML estimator for σ^2, we obtain

$$\hat{\sigma}^2_{ML} = \frac{1}{n}(\boldsymbol{y} - \boldsymbol{X}\hat{\boldsymbol{\beta}})'\boldsymbol{W}(\boldsymbol{y} - \boldsymbol{X}\hat{\boldsymbol{\beta}}) = \frac{1}{n}\hat{\boldsymbol{\varepsilon}}'\boldsymbol{W}\hat{\boldsymbol{\varepsilon}}.$$

As in the classical linear model, this estimator is biased. An unbiased estimator is given by

$$\hat{\sigma}^2 = \frac{1}{n-p}\hat{\boldsymbol{\varepsilon}}'\boldsymbol{W}\hat{\boldsymbol{\varepsilon}} = \frac{1}{n-p}\sum_{i=1}^{n} w_i(y_i - \boldsymbol{x}_i'\hat{\boldsymbol{\beta}})^2.$$

This estimator can be seen as a restricted maximum likelihood estimator; see Sect. 3.2.2 (p. 108). All derivations and proofs are completely analogous to the classical linear model.

The approach exemplified for heteroscedastic errors can be extended to the case of a general covariance matrix $\sigma^2\boldsymbol{W}^{-1}$. When doing so, we use a "square root" $\boldsymbol{W}^{1/2}$ of $\boldsymbol{W}$ with $\boldsymbol{W}^{1/2}(\boldsymbol{W}^{1/2})' = \boldsymbol{W}$. The matrix $\boldsymbol{W}^{1/2}$ is not uniquely determined. It can be obtained using, e.g., the spectral decomposition

$$\boldsymbol{W} = \boldsymbol{P}\,\mathrm{diag}(\lambda_1, \ldots, \lambda_n)\,\boldsymbol{P}'$$

of $\boldsymbol{W}$ (see result A.25 (p. 633) in Appendix A), resulting in

$$\boldsymbol{W}^{1/2} = \boldsymbol{P}\,\mathrm{diag}(\lambda_1^{1/2}, \ldots, \lambda_n^{1/2})\,\boldsymbol{P}'.$$

We can then transform the response vector, the design matrix, and the errors with $\boldsymbol{W}^{1/2}$. This results in a model of the form (4.2), which is a classical linear model since we have

$$\mathrm{E}(\boldsymbol{\varepsilon}^*) = \mathrm{E}(\boldsymbol{W}^{1/2}\boldsymbol{\varepsilon}) = \boldsymbol{W}^{1/2}\,\mathrm{E}(\boldsymbol{\varepsilon}) = \boldsymbol{0}$$

and

$$\mathrm{Cov}(\boldsymbol{\varepsilon}^*) = \mathrm{E}(\boldsymbol{W}^{1/2}\boldsymbol{\varepsilon}\boldsymbol{\varepsilon}'\boldsymbol{W}^{1/2}) = \sigma^2\boldsymbol{W}^{1/2}\boldsymbol{W}\boldsymbol{W}^{1/2} = \sigma^2\boldsymbol{I}.$$

Finally, the WLS estimator shares the same stochastic properties as ordinary least squares; see the following Box 4.2. The proof for the properties of the weighted estimator is completely analogous to the unweighted estimator. Moreover, the tests

4.2 Estimators in the General Linear Model

Weighted Least Squares or ML Estimator for β

$$\hat{\boldsymbol{\beta}} = \left(\boldsymbol{X}'\boldsymbol{W}\boldsymbol{X}\right)^{-1}\boldsymbol{X}'\boldsymbol{W}\boldsymbol{y}.$$

Properties of the Weighted Least Squares Estimator

1. *Expectation:* $\mathrm{E}(\hat{\boldsymbol{\beta}}) = \boldsymbol{\beta}$, i.e., the WLS estimator is unbiased.
2. *Covariance matrix:* $\mathrm{Cov}(\hat{\boldsymbol{\beta}}) = \sigma^2 \left(\boldsymbol{X}'\boldsymbol{W}\boldsymbol{X}\right)^{-1}$.
3. *Gauß-Markov Theorem:* Among all linear and unbiased estimators $\hat{\boldsymbol{\beta}}^L = \boldsymbol{A}\boldsymbol{y}$, the WLS estimator has minimal variance, i.e.,

$$\mathrm{Var}(\hat{\beta}_j) \leq \mathrm{Var}(\hat{\beta}_j^L), \quad j = 0, \ldots, k.$$

REML Estimator for σ^2

$$\hat{\sigma}^2 = \frac{1}{n-p}\hat{\boldsymbol{\varepsilon}}'\boldsymbol{W}\hat{\boldsymbol{\varepsilon}}.$$

The REML estimator is unbiased.

Note that all properties hold only for known $\boldsymbol{W}$ and correctly specified model.

and confidence intervals of the classical linear model can be easily adapted. We simply have to replace the standard errors of the classical linear model by those of the general linear model. They are computed as usual by inserting $\hat{\sigma}^2$ into the covariance $\mathrm{Cov}(\hat{\boldsymbol{\beta}}) = \sigma^2 \left(\boldsymbol{X}'\boldsymbol{W}\boldsymbol{X}\right)^{-1}$ (see Box 4.2) to obtain $\widehat{\mathrm{Cov}(\hat{\boldsymbol{\beta}})}$. The standard errors are then the square roots of the main diagonal.

A first application of the WLS estimator with known weight matrix $\boldsymbol{W}$ is given for grouped data as follows.

Grouped Data

Thus far we have considered *individual* or *ungrouped data*, which implies that we have an observation $(y_i, \boldsymbol{x}_i)$ for every individual or subject i in a sample of size n. Every response y_i and every covariate vector $\boldsymbol{x}_i = (1, x_{i1}, \ldots, x_{ik})'$ belongs to exactly one subject i:

$$
\begin{array}{c}
\text{Unit } 1 \\ \vdots \\ \text{Unit } i \\ \vdots \\ \text{Unit } n
\end{array}
\begin{bmatrix} y_1 \\ \vdots \\ y_i \\ \vdots \\ y_n \end{bmatrix}
\begin{bmatrix} 1 & x_{11} & \cdots & x_{1k} \\ \vdots & \vdots & & \vdots \\ 1 & x_{i1} & & x_{ik} \\ \vdots & \vdots & & \vdots \\ 1 & x_{n1} & \cdots & x_{nk} \end{bmatrix}
$$

If several covariate vectors or rows of the design matrix are identical, we can *group* the data. After sorting and combining the data, the grouped design matrix now only contains rows with *different* covariate vectors $\boldsymbol{x}_i$. We denote by n_i the number of replicates of $\boldsymbol{x}_i$ within the original sample of individual data. We further denote by $\bar{y}_i$ the arithmetic mean of the respective individual response values that were observed for $\boldsymbol{x}_i$:

$$
\begin{array}{c}
\text{Group } 1 \\ \vdots \\ \text{Group } i \\ \vdots \\ \text{Group } G
\end{array}
\begin{bmatrix} n_1 \\ \vdots \\ n_i \\ \vdots \\ n_G \end{bmatrix}
\begin{bmatrix} \bar{y}_1 \\ \vdots \\ \bar{y}_i \\ \vdots \\ \bar{y}_G \end{bmatrix}
\begin{bmatrix} 1 & x_{11} & \cdots & x_{1k} \\ \vdots & \vdots & & \vdots \\ 1 & x_{i1} & \cdots & x_{ik} \\ \vdots & \vdots & & \vdots \\ 1 & x_{G1} & \cdots & x_{Gk} \end{bmatrix}
$$

Thereby G represents the number of unique covariate vectors in the sample. In many cases G is much smaller than the sample size n, especially with binary or categorical covariates.

Grouped data can be easily handled within the scope of the general linear model by defining $\boldsymbol{y} = (\bar{y}_1, \ldots, \bar{y}_G)'$ and $\text{Cov}(\boldsymbol{\varepsilon}) = \sigma^2 \text{diag}(1/n_1, \ldots, 1/n_G)$, i.e., $\boldsymbol{W} = \text{diag}(n_1, \ldots, n_G)$.

Other applications of the WLS estimator with known weight matrix $\boldsymbol{W}$ will be discussed in Chaps. 5 and 8. In Chap. 5, the WLS estimator will be the basis for parameter estimates in generalized linear models. In Chap. 8, WLS estimation will be of importance in relation to *local smoothers*; see Sect. 8.1.7.

In the following two sections, we consider situations in which the weight matrix $\boldsymbol{W}$ is, to some extent, unknown. More specifically, we will discuss heteroscedastic and autocorrelated errors.

4.1.3 Heteroscedastic Errors

In the linear model with heteroscedastic errors, the covariance matrix of the errors is given by $\text{Cov}(\boldsymbol{\varepsilon}) = \sigma^2 \text{diag}(1/w_1, \ldots, 1/w_n)$. In the literature, especially in the econometrics literature, we find a large variety of estimators and modeling approaches, which are beyond the scope of this book. A good overview can be found in Greene (2000). Here, we discuss a two-stage least squares estimator and we briefly sketch the ML estimator. The drawback of both estimators is that they require knowledge of the specific form of heteroscedasticity. We therefore describe an alternative approach developed by White (1980), which avoids assumptions about the form of heteroscedasticity. Before we present the different estimators, we show how to detect heteroscedastic errors.

Detecting Heteroscedastic Errors

As a starting point for detecting heteroscedastic errors, a classical linear model is typically estimated, followed by a careful residual analysis. In the literature, we generally find two different strategies. The statistics and biometry literature proposes graphical tools, residual plots in particular. The econometrics literature rather develops a variety of statistical tests to discover heteroscedasticity.

Residual Plots

In order to detect heteroscedastic errors, it is useful to plot the residuals against the predicted values $\hat{y}_i$ and the covariates x_{ij}. Covariates *not included in the model* should also be considered. Note also that a plot of the residuals against $\boldsymbol{y}$ (rather than the predicted values $\hat{\boldsymbol{y}}$) is not recommended because the residuals $\hat{\boldsymbol{\varepsilon}}$ depend (by definition) on the response $\boldsymbol{y}$ and a plot would reveal this dependency. The standardized or studentized residuals (see Box 3.12 on p. 126) are preferred over the raw residuals, since the latter are heteroscedastic with $\text{Var}(\hat{\varepsilon}_i) = \sigma^2(1 - h_{ii})$. Consequently, raw residuals are less appropriate to examine heteroscedasticity (see p. 122). In the case of homoscedastic error variances, the standardized or studentized residuals exhibit random fluctuation around zero with a *constant variance*. If this is *not* the case, there is evidence for heteroscedastic variances.

Example 4.1 Munich Rent Index—Diagnostics for Heteroscedastic Errors

We illustrate the detection of heteroscedastic errors with the rent index data. For simplicity we restrict ourselves to a model with the net rent as response variable and the covariates living area and year of construction. In Fig. 1.5 (p. 15), the scatter plots between net rent and living area and year of construction show a linear effect of the living area and a slightly nonlinear effect of the year of construction. We model the effect of the year of construction using an orthogonal polynomial of degree three. Thus, we assume the classical linear model

$$rent_i = \beta_0 + \beta_1\, area_i + \beta_2\, yearco_i + \beta_3\, yearco2_i + \beta_4\, yearco3_i + \varepsilon_i. \qquad (4.3)$$

Estimation results can be found in Table 4.1. Figure 4.1a, b shows the estimated effects of living area and year of construction, including the corresponding partial residuals. Panels (c), (d), and (e) show the studentized residuals as a function of the estimated net rent, the living area, and the year of construction. There is clear evidence of heteroscedastic variances depending on both the living area and the year of construction. △

Testing Heteroscedasticity

Tests of heteroscedasticity are treated extensively in the econometrics literature, as, for example, in Greene (2000) and Judge, Griffith, Hill, Lütkepohl, and Lee (1980).

Table 4.1 Munich rent index: estimation results for the unweighted regression

Variable	Coefficient	Standard error	t-value	p-value	95 % Confidence interval	
intercept	459.437	2.631	174.600	<0.001	454.278	464.597
areao	121.817	2.742	44.430	<0.001	116.441	127.193
yearco	54.336	2.706	20.080	<0.001	49.030	59.642
yearco2	31.484	2.668	11.800	<0.001	26.252	36.715
yearco3	−0.198	2.631	−0.080	0.940	−5.358	4.961

We illustrate heteroscedasticity tests with a test according to Breusch and Pagan (1979). The basis for the test is a multiplicative model for the error variances:

$$\sigma_i^2 = \sigma^2 \cdot h(\alpha_0 + \alpha_1 z_{i1} + \cdots + \alpha_q z_{iq}),$$

where h is a function not depending on the unit index i and $z_1, \ldots, z_q$ are covariates that may influence the variance. The hypothesis of homoscedastic variances is equivalent to $\alpha_1 = \cdots = \alpha_q = 0$. The Breusch–Pagan test is based on the hypothesis

$$H_0 : \alpha_1 = \ldots = \alpha_q = 0 \quad \text{against} \quad H_1 : \alpha_j \neq 0 \text{ for at least one } j.$$

In order to conduct the test, an auxiliary regression between the response variable

$$g_i = \frac{\hat{\varepsilon}_i^2}{\hat{\sigma}_{ML}^2}$$

and the explanatory variables $z_1, \ldots, z_q$ is performed. Here, $\hat{\varepsilon}_i$ are the residuals and $\hat{\sigma}_{ML}^2$ is the ML estimator for σ^2 for the linear model $y = \boldsymbol{x}'\boldsymbol{\beta} + \varepsilon$ with homoscedastic errors. The test statistic is given by

$$T = \frac{1}{2} \sum_{i=1}^{n} (\hat{g}_i - \bar{g})^2.$$

In the case that none of the covariates influence the variance, i.e., heteroscedasticity does not exist, then we have $\hat{g}_i \approx \bar{g}$. The larger T is, i.e., the larger the sum of squared deviations of the estimated values $\hat{g}_i$ from the mean $\bar{g}$, the more evidence in favor of the alternative hypothesis of heteroscedastic variances. Under H_0, the distribution of T is independent of the function h and is asymptotically χ^2-distributed with q degrees of freedom, i.e., $T \overset{a}{\sim} \chi_q^2$. With a significance level of α, we reject the null hypothesis if the observed test statistic is greater than the $(1 - \alpha)$-quantile of the χ_q^2-distribution.

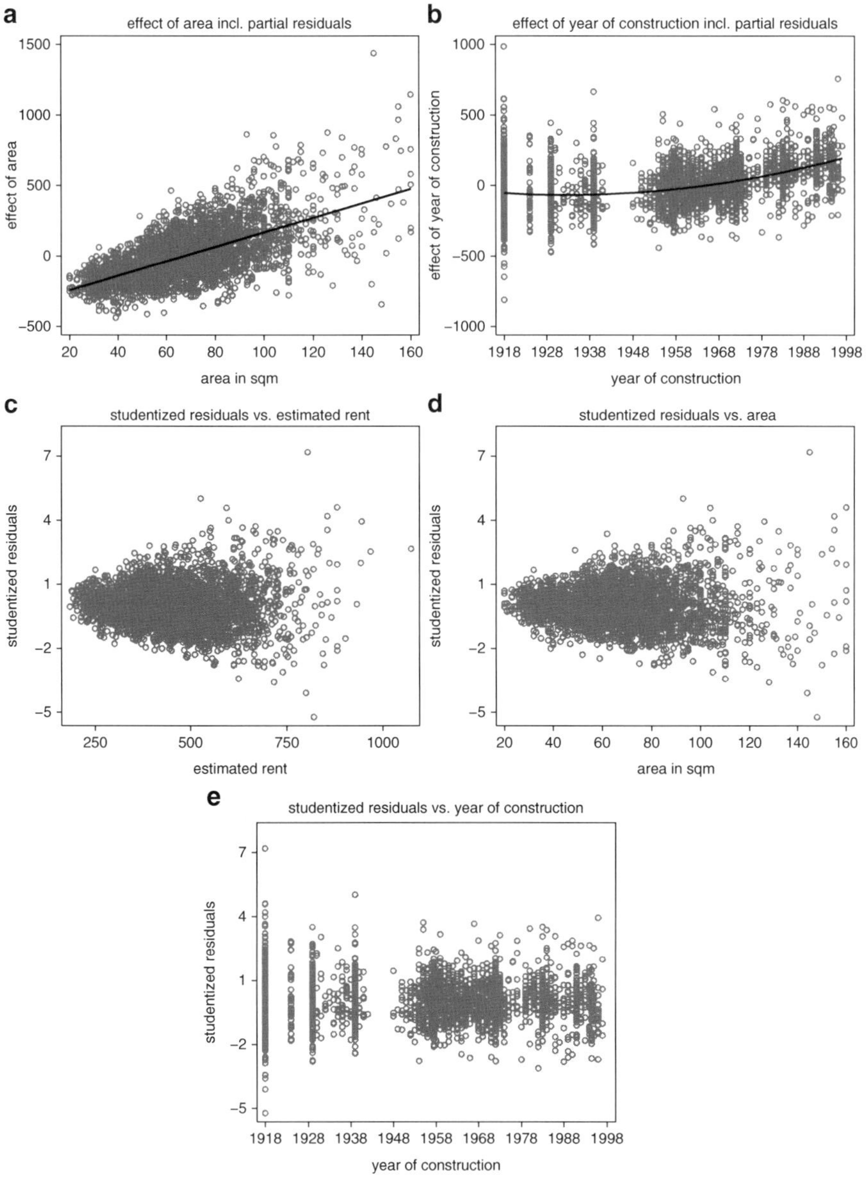

Fig. 4.1 Munich rent index: Panels (**a**) and (**b**) show the effects of living area and year of construction including partial residuals. Panels (**c–e**) display studentized residuals against the estimated net rent, living area, and year of construction

Example 4.2 Munich Rent Index—Breusch–Pagan Test

It is reasonable to assume that the error variances both depend (possibly nonlinearly) on living area and year of construction; see Example 4.1. We therefore assume the variance model

$$\sigma_i^2 = \sigma^2 h(\alpha_0 + \alpha_1 areao_i + \alpha_2 areao2_i + \alpha_3 areao3_i$$
$$+\alpha_4 yearco_i + \alpha_5 yearco2_i + \alpha_6 yearco3_i),$$

where *areao*, *areao2*, and *areao3* are cubic orthogonal polynomials (see Example 3.5 on p. 90) for living area. Based on this model, we obtain $T = 997.164$ as the Breusch–Pagan test statistic. The corresponding p-value is essentially zero so that the Breusch–Pagan test (in addition to the studentized residuals) provides further evidence for heteroscedastic variances. The Breusch–Pagan test has been carried out using function `hettest` of `STATA`.

△

Some concluding (critical) remarks regarding tests on heteroscedasticity are in order. For some of the readers, applying a formal test might seem more exact than the inspection of residual plots. However, in most cases, heteroscedasticity is diagnosed with exploratory techniques, while substantial scientific theory regarding the type and magnitude of heteroscedasticity almost never exists. This is the reason why we face even more uncertainty when modeling error variances in a linear model than when modeling expectations. The validity of statistical tests is extremely dependent on the correctness of models. The Breusch–Pagan test assumes, for example, multiplicative variances with exactly defined covariates. The number of covariates in the variance expression determines the distribution of the test statistic. Hence, tests on heteroscedasticity should be seen as a (heuristic) exploratory tool, similar to residual plots. These tests should by no means be the only device to diagnose heteroscedasticity. Yet, this is suggested by most econometrics textbooks, in which a battery of heteroscedasticity tests are described, while rarely is there a mention of residual plots. Residual plots should always be part of heteroscedasticity analysis, since they are in many cases the only tool to detect the *specific type* of heteroscedasticity or to determine which of the covariates influence the error variances.

Treating Heteroscedastic Variances

When diagnosing heteroscedastic errors, we should react appropriately to prevent incorrect conclusions that may result from ignoring heteroscedasticity. The most widely used approaches are described below.

Variable Transformation

When discussing the assumptions of the linear model, we have encountered models with multiplicative errors (see p. 83). A popular model with multiplicative errors is the exponential model

$$\begin{aligned} y_i &= \exp(\beta_0 + \beta_1 x_{i1} + \ldots + \beta_k x_{ik} + \varepsilon_i) \\ &= \exp(\beta_0)\exp(\beta_1 x_{i1}) \cdot \ldots \cdot \exp(\beta_k x_{ik})\exp(\varepsilon_i). \end{aligned}$$

If the errors are normally distributed $\varepsilon_i \sim \mathrm{N}(0, \sigma^2)$, then $\exp(\varepsilon_i)$ and y_i are log-normally distributed. Using the variance of the log-normal distribution (see p. 641 in Appendix B), we obtain

$$\mathrm{Var}(\exp(\varepsilon_i)) = \exp(\sigma^2) \cdot (\exp(\sigma^2) - 1)$$

and, thus

$$\mathrm{Var}(y_i) = (\exp(\boldsymbol{x}_i'\boldsymbol{\beta}))^2 \exp(\sigma^2) \cdot (\exp(\sigma^2) - 1).$$

This implies that the variances of y_i are heteroscedastic in models with multiplicative errors, even though the variances of the errors are homoscedastic. As mentioned before, it is possible to log-transform the exponential model and obtain the general linear model $\log(y_i) = \boldsymbol{x}_i'\boldsymbol{\beta} + \varepsilon_i$, with homoscedastic variances; see p. 83ff. This result provides us with a simple tool if heteroscedastic variances are diagnosed. In case of a multiplicative model, we can take a simple logarithmic transformation of the response and estimate a classical linear model using the transformed response.

Two-Stage Least Squares

In addition to the regression coefficients $\boldsymbol{\beta}$ and the variance parameter σ^2, the weights w_i are also unknown with heteroscedastic variances. Thus, a reasonable approach is *joint* estimation of all unknown parameters. Since $\mathrm{E}(\varepsilon_i) = 0$, we have $\mathrm{E}(\varepsilon_i^2) = \mathrm{Var}(\varepsilon_i) = \sigma_i^2$, and we can represent ε_i^2 as

$$\varepsilon_i^2 = \sigma_i^2 + v_i,$$

where v_i are the deviations of the squared errors from their expectations. In most cases σ_i^2 depends on one or more covariates. It seems natural to assume

$$\sigma_i^2 = \alpha_0 + \alpha_1 z_{i1} + \ldots + \alpha_q z_{iq} = \boldsymbol{z}_i'\boldsymbol{\alpha},$$

where the vector $\boldsymbol{z}$ consists of all covariates that influence the variance. In many cases, the vector $\boldsymbol{z}$ is identical with the covariate vector $\boldsymbol{x}$. In order to estimate the unknown parameters $\boldsymbol{\alpha}$, we could fit a linear model using the squared errors ε_i^2 as dependent variable and $\boldsymbol{z}_i$ as independent variables. Since the errors are unobserved, they have to be replaced by the residuals $\hat{\varepsilon}_i = y_i - \boldsymbol{x}_i'\hat{\boldsymbol{\beta}}$ resulting from an unweighted regression between y and $\boldsymbol{x}$. As a result, we obtain the following two-stage method:

1. Obtain preliminary estimates $\hat{\boldsymbol{\beta}}$ from an unweighted regression between y and $\boldsymbol{x}$. Compute the residuals $\hat{\varepsilon}_i$.
2. Obtain estimates $\hat{\boldsymbol{\alpha}}$ from an unweighted regression between the squared residuals $\hat{\varepsilon}_i^2$ and the variance explanatory variables $\boldsymbol{z}_i$. Fit a general linear model using the weights

$$\hat{w}_i = \frac{1}{\boldsymbol{z}_i'\hat{\boldsymbol{\alpha}}}.$$

However, this method is not always practical, since we cannot guarantee that $z_i'\hat{\boldsymbol{\alpha}}$ is greater than zero. If $z_i'\hat{\boldsymbol{\alpha}} < 0$, then negative weights $\hat{w}_i$ and negatively estimated variances $\hat{\sigma}_i^2$ would result. Alternatively, we can assume the model

$$\sigma_i^2 = \exp(z_i'\boldsymbol{\alpha})$$

similar to the Breusch–Pagan test. The exponential function ensures that the estimated expected variances are positive. In this model, $\boldsymbol{\alpha}$ can be estimated with the help of the regression

$$\log(\hat{\varepsilon}_i^2) = z_i'\boldsymbol{\alpha} + v_i.$$

The weights for the regression between y and $\boldsymbol{x}$ are then given by

$$\hat{w}_i = \frac{1}{\exp(z_i'\hat{\boldsymbol{\alpha}})}.$$

Example 4.3 Munich Rent Index—Two-Stage Estimation

We illustrate the two-stage least squares approach with the rent index data. Recall that we diagnosed heteroscedastic variances for model (4.3) in Examples 4.1 and 4.2. We obtain the two-stage estimator for this model with the following three steps:

1. *First step: Classical linear model*
   In the first step, we estimate a classical linear model; see the results in Example 4.1. Panels (a) and (b) of Fig. 4.2 show scatter plots between $\log(\hat{\varepsilon}_i^2)$ and living area or year of construction. Both panels indicate that the variances σ_i^2 may depend on both covariates.
2. *Second step: Auxiliary regression*
   To determine the weights $\hat{w}_i$ for the weighted regression, we estimate the regression model
   $$\log(\hat{\varepsilon}_i^2) = \alpha_0 + \alpha_1\, areao_i + \alpha_2\, areao2_i + \alpha_3\, areao3_i$$
   $$+\alpha_4\, yearco_i + \alpha_5\, yearco2_i + \alpha_6\, yearco3_i + v_i$$
   using ordinary least squares. Figure 4.2c, d show the estimated effects of the living area and the year of construction. The variability of the net rent appears to increase monotonically with increased living area. The effect of the year of construction is slightly S-shaped, with a considerably smaller effect than that associated with living area. The interpretation of this effect is that the variability of the net rents for older apartments is higher than for more modern apartments.
3. *Third step: Weighted linear model*
   In the last step we reestimate the regression model (4.3) with the weights
   $$\hat{w}_i = \frac{1}{\exp(\hat{\eta}_i)}$$
   using $\hat{\eta}_i = \hat{\alpha}_0 + \hat{\alpha}_1\, areao_i + \ldots + \hat{\alpha}_6\, yearco3_i$. We obtain the results listed in Table 4.2. The estimated standard errors are based on the estimated covariance matrix
   $$\widehat{\text{Cov}(\hat{\boldsymbol{\beta}})} = \hat{\sigma}^2(\boldsymbol{X}'\text{diag}(\hat{w}_1, \ldots, \hat{w}_n)\boldsymbol{X})^{-1},$$
   which are generally "smaller" than those of the unweighted regression. Using these corrected standard errors, we can perform tests and construct confidence intervals,

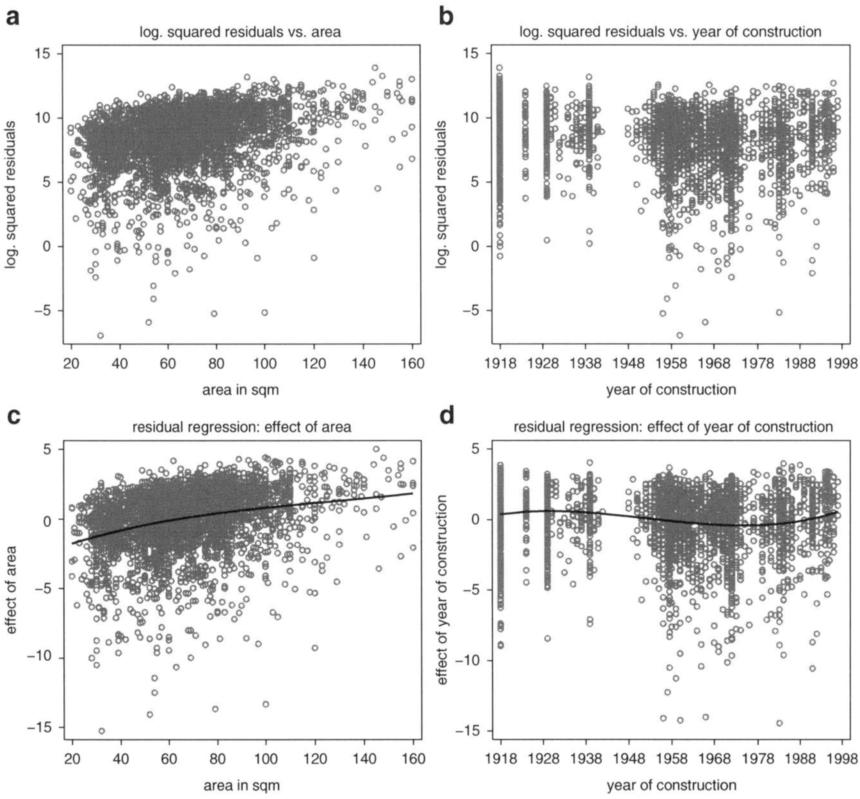

Fig. 4.2 Munich rent index: Panels (**a**) and (**b**) show scatter plots between $\log(\hat{\varepsilon}_i^2)$ and area respectively year of construction. Panels (**c**) and (**d**) display effects of area and year of construction including partial residuals for the variance regression

Table 4.2 Munich rent index: estimation results for the weighted regression

Variable	Coefficient	Standard error	t-value	p-Wert	95 % Confidence interval	
intercept	458.840	2.608	175.900	<0.001	453.726	463.955
areao	117.736	2.381	49.430	<0.001	113.066	122.406
yearco	48.697	2.656	18.330	<0.001	43.488	53.906
yearco2	25.282	2.334	10.830	<0.001	20.706	29.858
yearco3	−1.323	2.381	−0.560	0.578	−5.991	3.344

like the ones that were developed for the classical linear model. In the present case, confidence intervals are somewhat narrower than those based on the unweighted least squares estimator.

△

Simultaneous Estimation

In comparison to a two-stage method, *simultaneous estimation* of $\boldsymbol{\beta}$, $\boldsymbol{\alpha}$, and σ^2 is more favorable, such as full ML estimation or a combination of ML estimation for

$\boldsymbol{\beta}$ and REML estimation for the variance parameters $\boldsymbol{\alpha}$ and σ^2. At this point, we omit a detailed description of these estimation approaches, as they are technically involved and software is limited. Chapter 12 of Greene (2000) describes in detail a maximum likelihood method for a model with multiplicative variances $\sigma_i^2 = \sigma^2 \exp(\boldsymbol{z}_i' \boldsymbol{\alpha})$. An implementation exists, for example, in STATA (function `regh`). The R package `gamlss` allows full maximum likelihood estimation of Gaussian regression models with heteroscedastic variances; see also Sect. 2.9. Both the mean and the standard deviation or variance may depend on covariates. The approach even allows for semiparametric predictors as described in Chaps. 8 and 9. Details can be found in Rigby and Stasinopoulos (2005) and in the documentation to `gamlss`, see the web page `gamlss.org`.

White-Estimators

One of the prerequisites for using the methods discussed thus far is knowledge about the *type of heteroscedasticity*. An alternative approach by White (1980) proposes to *correct* the usual standard errors, confidence intervals, and tests associated with ordinary least squares. The correction is based on a consistent estimate of the covariance matrix (4.1) on p. 178. Under general conditions White (1980) shows that

$$\widehat{\text{Cov}(\hat{\boldsymbol{\beta}})} = (\boldsymbol{X}'\boldsymbol{X})^{-1}\boldsymbol{X}'\text{diag}(\hat{\varepsilon}_1^2, \ldots, \hat{\varepsilon}_n^2)\boldsymbol{X}(\boldsymbol{X}'\boldsymbol{X})^{-1} \tag{4.4}$$

is a consistent estimator for the covariance matrix (4.1) of $\hat{\boldsymbol{\beta}}$. The variances $\sigma_i^2 = \sigma^2/w_i$ in Eq. (4.1) are thus replaced or estimated by the squared residuals $\hat{\sigma}_i^2 = \hat{\varepsilon}_i^2$. Due to its form the matrix (4.4) is referred to as "sandwich matrix." In Sects. 7.3.3 (p. 378) and 8.1.2 (p. 439) we will encounter similar forms for covariance matrices. The estimated robust covariance matrix for $\hat{\boldsymbol{\beta}}$ can be used for computing the F-test statistic when testing general linear hypotheses. In doing so, we obtain asymptotically correct tests and confidence intervals.

The approach has the advantage that it is *not* necessary to know the weight matrix $\boldsymbol{W}$ as is the case for the other approaches. However, there is no free lunch. The price we pay for the generality is less accuracy of estimators through increased variances of the estimators.

Example 4.4 Munich Rent Index—White-Estimator

We again estimate model (4.3) on p. 183. However, this time we use the White covariance matrix (4.4) for the construction of tests and confidence intervals. Using the `regress` function of `STATA` with additional option `robust` we obtain the results listed in Table 4.3. The estimated standard errors tend to be much higher in comparison to the results based on an unadjusted covariance matrix, as well as in comparison to the two-stage estimation method. This should not be surprising: we have less information because we make no specific assumptions regarding the type of heteroscedasticity.

We can conclude as follows: In comparison to the classical linear model, the different variants of heteroscedastic errors provide only little qualitative differences in interpretation. The effect of the living area and the year of construction is significant in all estimated models. Moreover there is evidence that a second degree polynomial for modeling the effect of the year of construction is sufficient. Such relative robustness of results can be observed in

Table 4.3 Munich rent index: estimation results for the unweighted regression with corrected standard errors

Variable	Coefficient	Standard error	t-value	p-value	95 % Confidence interval	
intercept	459.437	2.631	174.600	<0.001	454.277	464.596
areao	121.817	3.681	33.090	<0.001	114.599	129.035
yearco	54.336	3.098	17.530	<0.001	48.260	60.412
yearco2	31.484	2.810	11.200	<0.001	25.974	36.994
yearco3	−0.198	2.919	−0.070	0.946	−5.923	5.526

many applications. Although heteroscedastic variances have been considered for estimation, we often do not find significant changes in the interpretation compared to the classical linear model.

△

Example 4.5 Prices of Used Cars—Weighted Regression

In Chap. 3 we illustrated the practical application of the classical linear model with a detailed case study on prices of used cars in the presence of covariates; see the Examples 3.19 (p. 152), 3.20 (p. 156), 3.21 (p. 158), 3.22 (p. 162), 3.23 (p. 166), and 3.24 (p. 166). Example 3.20 provided some evidence for heteroscedastic errors. We therefore applied the two-stage method outlined on p. 187 for the AIC optimal model with second-degree orthogonal polynomials of *age* and *kilometer*. For the variance equation, we used the covariates *age*, *kilometer*, and *TIA*. We modeled the effects of the age and the kilometer reading using third-degree polynomials. Overall, we found no significant differences relative to unweighted regression.

△

4.1.4 Autocorrelated Errors

In addition to heteroscedastic errors, correlated errors are one of the main reasons why the assumptions of the classical linear model are violated in applications. Section 3.1.2 (p. 80) pointed out that autocorrelated errors may arise if the model is not correctly specified. We mentioned the following reasons for misspecified models:

- *Misspecified covariate effect:* The effect of an explanatory variable is not modeled correctly. The examples in Sect. 3.1.3 could prove helpful in this respect as they discuss various approaches for nonlinear modeling of continuous covariate effects.
- *Omitted variables:* Relevant explanatory variables cannot be observed and consequently cannot be considered in the model. In relation to autocorrelated errors, this problem occurs especially with time series, panel, or longitudinal data, especially if the missing explanatory variables show a temporal trend. In this case, we can use the estimation methods provided in this section in order to achieve more precise estimates and better prediction.

First-Order Autocorrelation

We restrict ourselves to one of the simplest and most frequently used error process, first-order autocorrelation. More specifically, we assume that the errors follow an autoregressive process of first-order (AR(1)) with

$$\varepsilon_i = \rho\,\varepsilon_{i-1} + u_i,$$

where $-1 < \rho < 1$. Regarding the u_i, we assume:
1. $\mathrm{E}(u_i) = 0$
2. $\mathrm{Var}(u_i) = \mathrm{E}(u_i^2) = \sigma_u^2, \quad i = 1, \ldots, n$
3. $\mathrm{Cov}(u_i, u_j) = \mathrm{E}(u_i u_j) = 0, \quad i \neq j$

Related to sample size, we additionally assume that the data are sufficiently historical in nature. Based on these assumptions, a specific linear model with $\mathrm{E}(\boldsymbol{\varepsilon}) = \mathbf{0}$ and with covariance matrix

$$\mathrm{Cov}(\boldsymbol{\varepsilon}) = \sigma^2 \boldsymbol{W}^{-1} = \frac{\sigma_u^2}{1-\rho^2}\begin{pmatrix} 1 & \rho & \rho^2 & \cdots & \rho^{n-1} \\ \rho & 1 & \rho & \cdots & \rho^{n-2} \\ \vdots & \vdots & \vdots & \ddots & \vdots \\ \rho^{n-1} & \rho^{n-2} & \rho^{n-3} & \cdots & 1 \end{pmatrix} \tag{4.5}$$

results; see Sect. 4.5.2 on p. 258 for a derivation.

With the help of the covariance matrix, we are able to calculate the correlation coefficients between the errors ε_i and the errors ε_{i-j} lagged by j periods. We obtain the *autocorrelation function*

$$\mathrm{ACF}(j) = \frac{\mathrm{Cov}(\varepsilon_i, \varepsilon_{i-j})}{\mathrm{Var}(\varepsilon_j)} = \rho^j \quad j = 0, 1, 2, \ldots.$$

An obvious characteristic of first-order autocorrelation is a slow decay of the correlation between ε_i and ε_{i-j} as the lag j increases. If ρ is positive, the correlation decreases geometrically. If ρ is negative, the correlation decreases with alternating signs. This is illustrated in the left column of Fig. 4.3, which shows the autocorrelation function for different values of ρ. These graphical representations are usually referred to as *correlograms*.

In addition to the autocorrelation function, the *partial autocorrelation function* is another tool to characterize correlated errors or, more generally, stochastic processes. The partial autocorrelation $\mathrm{PACF}(j)$ between ε_i and ε_{i-j} is defined as the regression coefficient α_j in the model

$$\varepsilon_i = \alpha_1 \varepsilon_{i-1} + \ldots + \alpha_j \varepsilon_{i-j} + v_i. \tag{4.6}$$

We have $\mathrm{PACF}(1) = \mathrm{ACF}(1)$, independent of the order of the autocorrelation. For first-order autocorrelated errors, we have $\mathrm{PACF}(1) = \rho$ and $\mathrm{PACF}(j) = 0$ for $j > 1$ by definition of the AR(1) process. The right column of Fig. 4.3 shows the

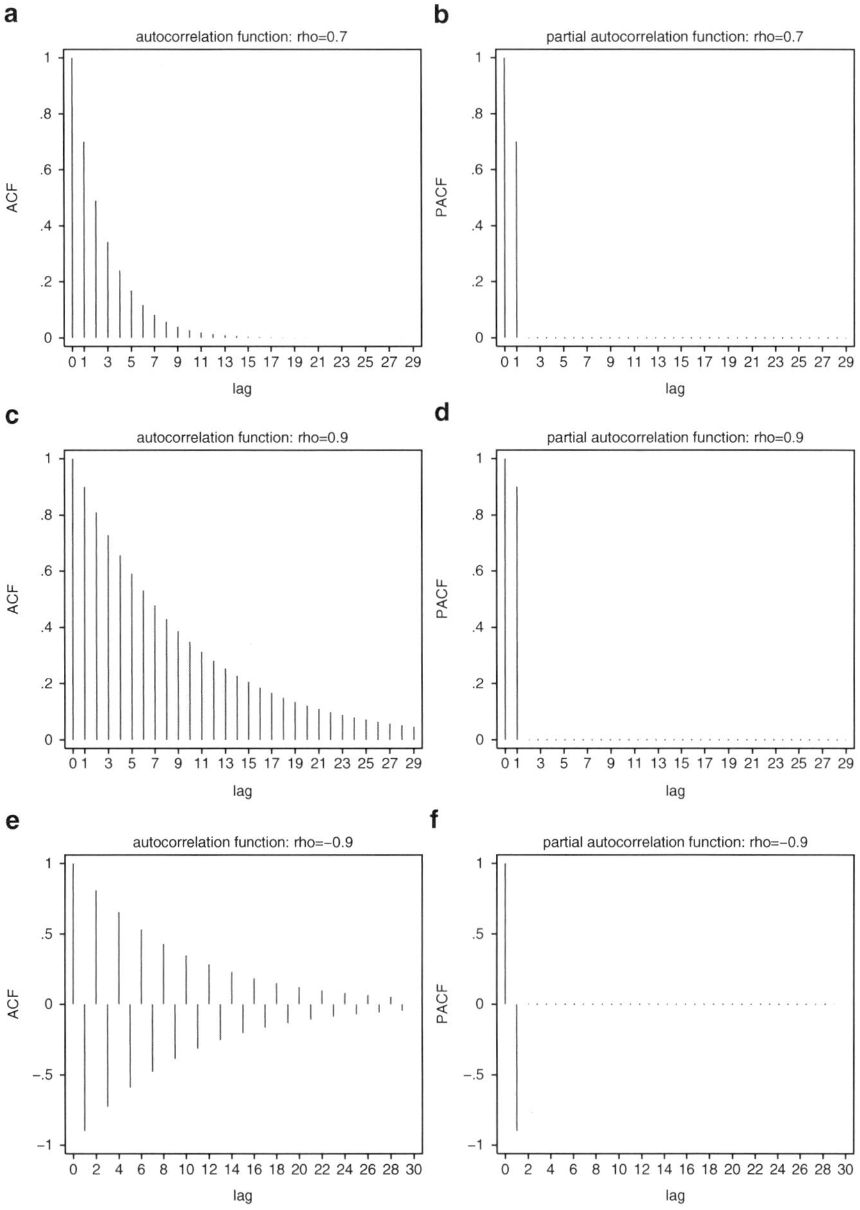

Fig. 4.3 Autocorrelation function and partial autocorrelation function for autocorrelated errors of first order. The functions for $\rho = 0.7, 0.9, -0.9$ are shown

partial autocorrelation function for some AR(1) processes. A characteristic property is the abrupt decline of the partial autocorrelation for $j > 1$. We can interpret the coefficient α_j in Eq. (4.6) as the correlation coefficient between $\varepsilon_i - \alpha_1 \varepsilon_{i-1} - \ldots - \varepsilon_{i-j+1}\alpha_{j-1}$ and ε_{i-j}. The terminology partial correlation coefficient is due to the fact that we do not simply calculate the correlation between ε_i and ε_{i-j}, but rather the correlation between ε_i and ε_{i-j} while controlling the effect of intermediate errors.

Diagnosing Autocorrelated Errors

Similar to the diagnosis of heteroscedastic errors, we first rely on the residuals from a classical linear model. The following diagnostic tools work well in practice:

Graphical Illustration of the Residuals Over Time

A straightforward tool for diagnosing correlated errors are residual plots, more specifically scatter plots of residuals or studentized residuals over time. If a positive (negative) residual is followed in tendency by a positive (negative) residual the errors are positively autocorrelated. Likewise, if in tendency residuals with alternating signs are observed, we have negative autocorrelation.

Empirical Autocorrelation Function

Other useful instruments for diagnosing autocorrelation are the empirical autocorrelation and the partial autocorrelation functions along with their display in correlograms. The empirical autocorrelation function is an estimate for ACF(j) and is given by

$$\widehat{\text{ACF}}(j) = \frac{\widehat{\text{Cov}}(\varepsilon_i, \varepsilon_{i-j})}{\widehat{\text{Var}}(\varepsilon_i)} \quad \text{with} \quad \widehat{\text{Cov}}(\varepsilon_i, \varepsilon_{i-j}) = \frac{1}{n} \sum_{i=j+1}^{n} \hat{\varepsilon}_i \hat{\varepsilon}_{i-j}.$$

Basically, the estimates are empirical correlation coefficients between the residuals and the lagged residuals by j periods.

The partial autocorrelations are obtained by repeatedly estimating the regression model (4.6) for $j = 1, 2, 3, \ldots$. This yields $\widehat{\text{PACF}}(j) = \hat{\alpha}_j$, thereby substituting the errors ε_i in Eq. (4.6) with $\hat{\varepsilon}_i$. By inspecting both the empirical correlogram and the partial correlogram, we obtain information about existing autocorrelation. If all empirical autocorrelations and partial autocorrelations are near zero, we can reasonably assume uncorrelated errors. If the empirical autocorrelations are similar to the theoretical (partial) autocorrelation function of AR(1)-errors, then we have evidence for the existence of first-order autocorrelation. If the correlograms do not follow the typical form of an AR(1)-process, then there is evidence of a more complex correlation structure. Such error structures go beyond the scope of this book and are examined in the literature on statistical time series analysis; see, e.g., Hamilton (1994).

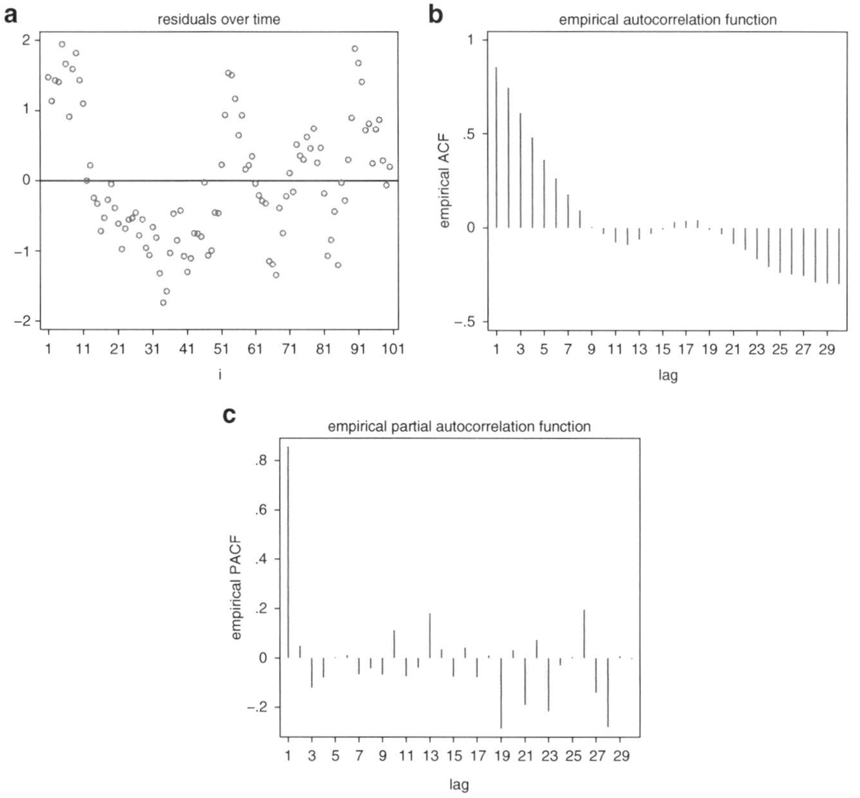

Fig. 4.4 Residuals over time (**a**), empirical autocorrelation function (**b**), and empirical partial autocorrelation function (**c**) based on simulated data with first-order autocorrelation

Example 4.6 Simulated Data—Graphical Diagnosis of Autocorrelated Errors

For illustration purposes, we examine the simulated regression model with positive autocorrelation, as shown in Fig. 3.3 (p. 81). We simulated the model $y_i = -1 + 2x_i + \varepsilon_i$ with $\varepsilon_i = 0.9\varepsilon_{i-1} + u_i$. To show how autocorrelated errors are diagnosed using residual plots and correlograms, we assume a classical linear model for the data. We obtain the estimated model $\hat{y}_i = -1.08 + 2.13\,x_i$. Figure 4.4 shows the residuals over time and the empirical (partial) autocorrelation function. Note that the residual plot in panel (a) differs slightly from the errors over time as presented in Fig. 3.3b, since the residuals are only estimates of the true errors. The plots display almost perfect first-order autocorrelation. The residuals are highly correlated, the empirical autocorrelation decreases geometrically, and the partial autocorrelations are almost zero for $j > 1$.

△

Tests for Autocorrelation—Durbin-Watson Test

In addition to graphical techniques, we can also use statistical tests to uncover autocorrelation. In this context, a test for serial correlation developed by Durbin

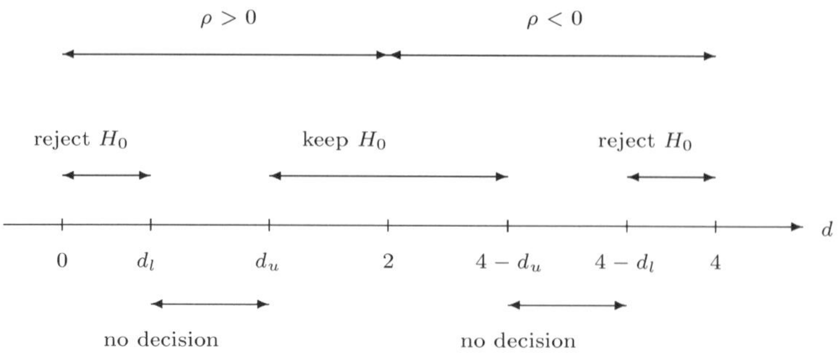

Fig. 4.5 Acceptance and rejection ranges for the Durbin–Watson test

and Watson (1950, 1951, 1971) is widely used. The Durbin–Watson test considers the hypothesis

$$H_0 : \rho = 0 \quad \text{versus} \quad H_1 : \rho \neq 0$$

and is based on the test statistic

$$d = \frac{\sum_{i=2}^{n} (\hat{\varepsilon}_i - \hat{\varepsilon}_{i-1})^2}{\sum_{i=1}^{n} \hat{\varepsilon}_i^2}.$$

The rationale behind the test statistic is as follows: For large sample size n we obtain

$$d = \frac{\sum_{i=2}^{n} \hat{\varepsilon}_i^2 + \sum_{i=2}^{n} \hat{\varepsilon}_{i-1}^2 - 2\sum_{i=2}^{n} \hat{\varepsilon}_i \hat{\varepsilon}_{i-1}}{\sum_{i=1}^{n} \hat{\varepsilon}_i^2} \approx 1 + 1 - 2\hat{\rho} = 2(1 - \hat{\rho}).$$

Since $-1 < \hat{\rho} < 1$, we have $0 < d < 4$. If d is close to 2, then $\hat{\rho} \approx 0$, and we fail to reject the null hypothesis. The closer the test statistic d is to either 0 or 4, the closer is $\hat{\rho}$ to 1 or -1, and we then generally reject the null hypothesis.

The distribution of the test statistic under H_0 is relatively difficult to obtain, since it depends on the design matrix. Hence, a test decision is sometimes difficult. Durbin and Watson partially solved the problem, and were able to provide a decision for certain interval ranges of d. The intervals depend on some lower and upper limits d_l and d_u, which vary for different sample size n and numbers of regressors k. Figure 4.5 provides a graphical presentation of the acceptance and rejection ranges for the Durbin–Watson test.

Nowadays, it has become more routine to obtain the p-values for the Durbin–Watson test numerically using statistical software. An implementation can be found, for instance, in the function `dwtest` of `matlab`. The R function `dwtest` of the package *lmtest* turned out to be erroneous in the following example.

Example 4.7 Simulated Data—Durbin–Watson Test

For the data used in Example 4.6, we obtain the test statistic $d = 0.2619$ and a very small p-value with the matlab function `dwtest`. Thus the Durbin–Watson test, in addition to the previously presented graphical diagnostics, provides clear evidence of first-order autocorrelation.

△

Treating First-Order Autocorrelated Errors

As already pointed out, autocorrelated errors are an indicator for model misspecification. Hence, we should first examine whether or not it is possible to correct the specification problem. Possible improvements can result from the inclusion of other covariates into the model or nonlinear modeling of some continuous covariates.

If these attempts fail to correct for the correlation, we can use estimation procedures for models with autocorrelated errors. In a model with first-order autocorrelated errors, we need estimates for the regression coefficients $\boldsymbol{\beta}$, the variance parameter σ^2, and the correlation coefficient ρ. The statistical literature provides a wealth of methods to estimate these parameters. We first present a two-stage procedure. In the first step, the correlation coefficient is estimated using the ordinary least squares estimates for the regression coefficients. The regression coefficients are then re-estimated with the help of WLS. As an alternative to the two-stage procedure, we briefly describe maximum likelihood estimation. An overview of alternative approaches is provided by Judge et al. (1980).

Two-Stage Estimation

We estimate a model with autocorrelated errors as follows:

1. Obtain a preliminary estimate $\hat{\boldsymbol{\beta}}$ for $\boldsymbol{\beta}$ using the unweighted regression between y and $\boldsymbol{x}$. Denote the resulting residuals with $\hat{\varepsilon}_i$.
2. Obtain an estimate $\hat{\rho}$ of the correlation coefficient through

$$\hat{\rho} = \frac{\sum_{i=2}^{n} \hat{\varepsilon}_i \hat{\varepsilon}_{i-1}}{\sqrt{\sum_{i=2}^{n} \hat{\varepsilon}_i^2}\sqrt{\sum_{i=2}^{n} \hat{\varepsilon}_{i-1}^2}}, \qquad (4.7)$$

   where $\hat{\rho}$ is the empirical correlation coefficient between $\hat{\varepsilon}_i$ and $\hat{\varepsilon}_{i-1}$.
3. Insert $\hat{\rho}$ into the weight matrix $\boldsymbol{W}$ and obtain the estimate $\hat{\boldsymbol{W}}$. Use $\hat{\boldsymbol{W}}$ to re-estimate the regression coefficients using WLS.

To improve the estimates, it is helpful to iterate steps 2 and 3. This method is known as Prais–Winsten estimator. Under quite general conditions, it can be shown that the resulting estimator for $\boldsymbol{\beta}$ is consistent. In addition to the method by Prais and Winsten, there exist numerous modifications; see Greene (2000).

Table 4.4 Prais–Winsten estimator for the simulated data

Variable	Coefficient	Standard error	t-value	p-value	95 % Confidence interval	
intercept	−0.988	0.308	−3.200	0.002	−1.600	−0.376
x	2.095	0.152	13.74	<0.001	1.792	2.397

Table 4.5 ML estimator for the simulated data

Variable	Coefficient	Standard error	t-value	p-value	95 % Confidence interval	
intercept	−0.982	0.307	−3.200	0.001	−1.583	−0.380
x	2.094	0.139	15.000	<0.001	1.821	2.368

Maximum Likelihood Estimation

For maximum likelihood estimation, we need the additional assumption of normally distributed errors. The likelihood is then given by

$$L(\boldsymbol{\beta}, \sigma^2, \rho) = \frac{1}{(2\pi\sigma^2)^{n/2}\,|\boldsymbol{W}^{-1}|^{1/2}} \exp\left(-\frac{1}{2\sigma^2}(\boldsymbol{y} - \boldsymbol{X}\boldsymbol{\beta})'\boldsymbol{W}(\boldsymbol{y} - \boldsymbol{X}\boldsymbol{\beta})\right). \quad (4.8)$$

At first we calculate

$$\boldsymbol{W} = \begin{pmatrix} 1 & -\rho & 0 & \dots & 0 & 0 \\ -\rho & 1+\rho^2 & -\rho & \dots & 0 & 0 \\ 0 & -\rho & 1+\rho^2 & \dots & 0 & 0 \\ \vdots & \vdots & \vdots & \ddots & \vdots & \vdots \\ 0 & 0 & 0 & \dots & 1+\rho^2 & -\rho \\ 0 & 0 & 0 & \dots & -\rho & 1 \end{pmatrix}.$$

and

$$|\boldsymbol{W}^{-1}| = \frac{1}{1-\rho^2}.$$

Then the log-likelihood becomes

$$l(\boldsymbol{\beta}, \sigma^2, \rho) = -\frac{n}{2}\log(\sigma^2) + \frac{1}{2}\log(1-\rho^2) - \frac{1}{2\sigma^2}(\boldsymbol{y} - \boldsymbol{X}\boldsymbol{\beta})'\boldsymbol{W}(\boldsymbol{y} - \boldsymbol{X}\boldsymbol{\beta}),$$

which can not be maximized in closed form. We rather have to compute the ML estimator iteratively. Details are provided in Greene (2000) and in the literature cited therein.

Example 4.8 Simulated Data—ML Estimators

For the simulated data of the preceding examples, Table 4.4 contains the Prais–Winsten estimator and Table 4.5 the maximum likelihood estimator. Both estimators are in close agreement. We used the `STATA` functions `prais` for the Prais–Winsten estimator and `arima` for the ML estimator.

△

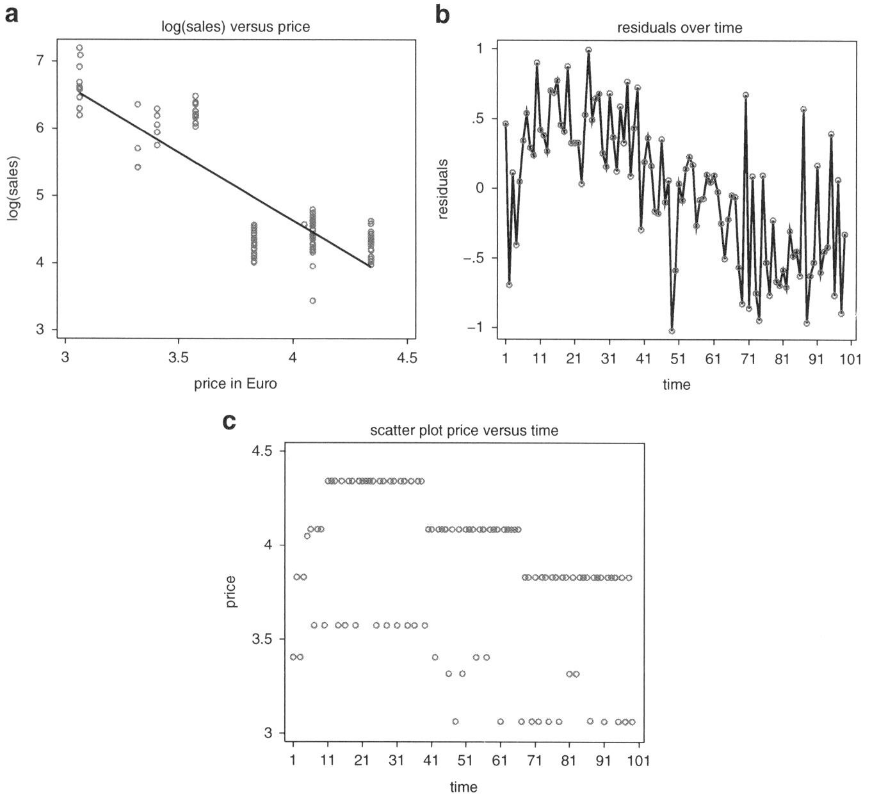

Fig. 4.6 Supermarket scanner data: scatter plot between log-sales and own price including fitted regression line [panel (**a**)]. Panel (**b**) shows the residuals over time. Panel (**c**) displays the price of the coffee brand over time

Example 4.9 Supermarket Scanner Data

For illustration of multiplicative errors, we examined the scanner data in Example 3.2 (p. 86), which are obtained from payments processed at the supermarket check-out counter. The goal is to model the relationship between the weekly sales of a product (in this case a particular coffee brand) and the own price or the price of competing brands (cross price effects); see the scatter plots in Fig. 3.6 (p. 85). In this example, we analyze the data to illustrate the approaches for correlated errors. To simplify matters, we use the data for just one store (Fig. 3.6 on p. 85 contains data of five stores). Due to the multiplicative error structure, we use log-sales rather than original responses. Figure 4.6a shows the scatter plot between log-sales and own price. Additionally, a fitted regression line is included. Panel (b) shows the corresponding residuals over time, which are clearly correlated. The Durbin–Watson test statistic is $d = 1.1751$ (p-value $< 10^{-4}$). The p-value is based on the matlab function `dwtest`. The p-value indicates that the null hypothesis $H_0 : \rho = 0$ can be rejected at the $\alpha = 0.05$ and $\alpha = 0.01$ level. In Fig. 4.7, we additionally examine the empirical autocorrelation and partial autocorrelation function of the residuals. The plots

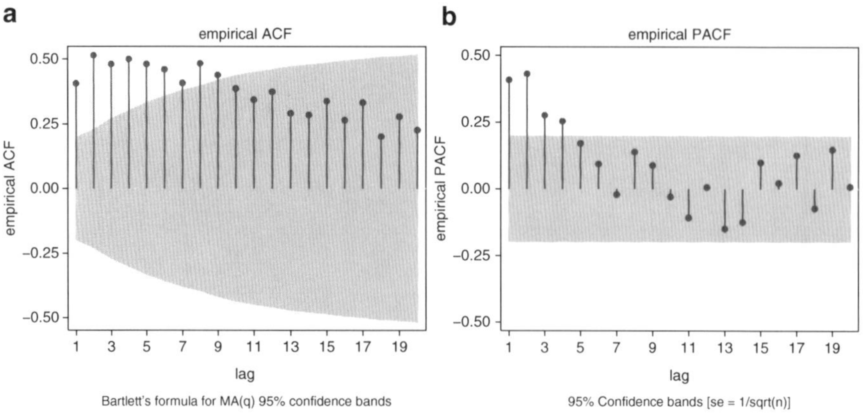

Fig. 4.7 Supermarket scanner data: empirical and partial autocorrelation function of the residuals in the regression between log-sales and price

additionally contain pointwise 95 % confidence intervals (their derivation is omitted at this point). The correlogram provides further evidence for autocorrelated errors, as it decreases slowly with increasing lag. The partial autocorrelation function is relatively small from a lag of 3 onward. Consequently, the correlograms do not point to first-order autocorrelation and a more complex correlation structure should be taken into account (e.g., second-order autocorrelation). We see the benefit here of graphical devices in comparison to formal tests. The Durbin–Watson test provides evidence for autocorrelation, but is not useful to determine the type of correlation.

To correct for autocorrelated errors, we should first examine whether or not an improved model specification eliminates correlations. Estimating a model with autocorrelated errors should be the last resort. In the current example, looking at the prices over time [panel (c) of Fig. 4.6] provides clear evidence how to improve the model. We can identify three time periods with different "regular" sales price of the coffee brand. Initially, the regular price was about 4.35 Euro, then the regular price was further decreased to 4.10 Euro, and finally to 3.8 Euro. During all periods we regularly encounter price reductions to promote the sales of the coffee brand. Clearly, ignoring the temporal development of regular prices is problematic. The marketing literature proposes to replace the price as explanatory variable with the ratio between the actual and the current regular price for the time period; see, for example, Leeflang, Wittink, Wedel, and Naert (2000). We therefore define the new covariate

$$priceq = \frac{\text{actual price}}{\text{regular price}},$$

which takes values between zero and one. If $priceq = 1$, the actual price is equal to the regular price, implying that there was no price reduction during the week. Figure 4.8a shows the scatter plot between log-sales and the ratio resulting from the actual and regular price including the fitted regression line. The residuals over time in panel (b) are now approximately uncorrelated. This is also supported by the Durbin–Watson test with test statistic $d = 1.5946$ (p-value $= 0.04559$). We can conclude that the Durbin–Watson test does not give us any compelling evidence of autocorrelation.

△

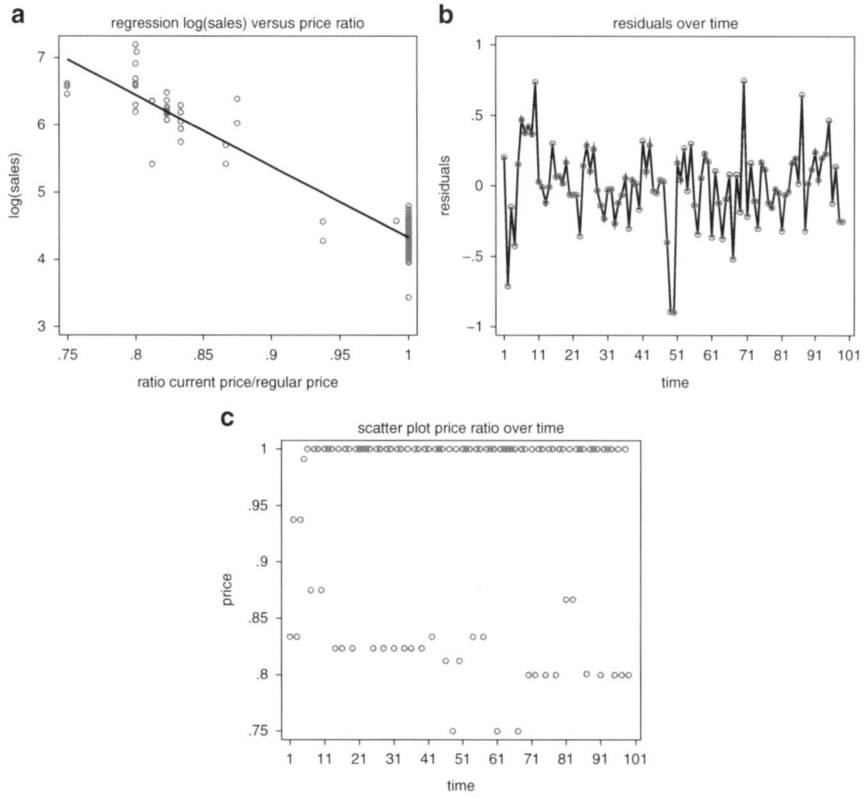

Fig. 4.8 Supermarket scanner data: scatter plot between log-sales and the price ratio actual/regular price including fitted regression line [panel (**a**)]. Panels (**b**) and (**c**) show the residuals and the price ratio over time

4.2 Regularization Techniques

This section deals with another extension of the classical linear model. We present regularization techniques to "regularize" the least squares estimator. This is useful in situations where the least squares estimator is numerically instable leading to (very) high variances. Moreover, regularization techniques provide an alternative to the variable selection procedures discussed in Sect. 3.4.

Throughout this section we assume $\text{Cov}(\boldsymbol{\varepsilon}) = \sigma^2 \boldsymbol{I}$ rather than $\text{Cov}(\boldsymbol{\varepsilon}) = \sigma^2 \boldsymbol{W}^{-1}$ as in the preceding section. However, generalizations of the presented regularization techniques to the general covariance matrix are straightforward.

4.2.1 Statistical Regularization

To compute the least squares estimator in the classical linear model, the system of equations

$$X'X\boldsymbol{\beta} = X'y \tag{4.9}$$

has to be solved with respect to $\boldsymbol{\beta}$. Thus far, we have assumed that X has full rank p so that a unique solution exists. However, when X contains columns that are close to collinear or if the number of regression effects p grows large, computing the solution will become numerically instable, even if the coefficients are still identifiable in theory. Moreover, modern applications arising, for example, in genetics often involve situations where the number of covariates exceeds (sometimes by orders of magnitude) the number of observations. In the statistical community, such problems are commonly referred to as "small n, large p" problems. For example, in studies of gene expression, the expression level of thousands of genes may be measured for only a few dozen individuals. In such situations, regularization techniques are commonly applied to obtain estimates of regression coefficients even in situations where the coefficient matrix $X'X$ in Eq. (4.9) is close to or exactly singular.

The notion of regularization summarizes approaches that make a problem "look nicer than it actually is" by imposing specific restrictions on the set of admissible solutions. In fact, this is closely related to the concept of Bayesian models where prior knowledge on the regression coefficients is incorporated (see Sect. 4.4). Regularization is typically implemented by considering a penalized least squares objective function

$$\text{PLS}(\boldsymbol{\beta}) = (y - X\boldsymbol{\beta})'(y - X\boldsymbol{\beta}) + \lambda \cdot \text{pen}(\boldsymbol{\beta}) \to \min_{\boldsymbol{\beta}},$$

where $\text{pen}(\boldsymbol{\beta})$ is a penalty term measuring the complexity of the vector of regression coefficients and $\lambda \geq 0$ is a smoothing parameter governing the impact of the penalty. In principle, $\text{pen}(\boldsymbol{\beta})$ will be constructed so that it is large when many of the entries in $\boldsymbol{\beta}$ are large. Another related principle would be to consider sparsity penalties that, for example, count the number of nonzero entries in $\boldsymbol{\beta}$ and therefore enforce sparseness of the resulting coefficient vector. The smoothing parameter then determines the trade-off between fidelity to the data as measured by the least squares criterion (λ small) and the impact of the penalty (λ large).

Using mathematical theory for optimization under constraints, one can show that obtaining the penalized least squares estimator

$$\hat{\boldsymbol{\beta}}_{\text{PLS}} = \arg\min_{\boldsymbol{\beta}} \left[(y - X\boldsymbol{\beta})'(y - X\boldsymbol{\beta}) + \lambda \cdot \text{pen}(\boldsymbol{\beta})\right]$$

is equivalent to seeking the solution

$$\hat{\boldsymbol{\beta}}_{\text{PLS}} = \arg\min_{\boldsymbol{\beta}} (\boldsymbol{y} - \boldsymbol{X}\boldsymbol{\beta})'(\boldsymbol{y} - \boldsymbol{X}\boldsymbol{\beta})$$

subject to the constraint

$$\text{pen}(\boldsymbol{\beta}) \leq t,$$

where t is a constant related to the smoothing parameter λ in a one-to-one relationship. A derivation of this well-known fact in optimization theory can be found, e.g., in Sydsaeter, Hammond, Seierstad, and Strom (2005).

4.2.2 Ridge Regression

In Sect. 3.4.4 (p. 159), we already introduced the first statistical regularization approach known as ridge regression. It corresponds to the classical approach of Tikhonov regularization achieved by adding the (squared) L_2-norm of the solution to an optimality criterion. In case of linear regression, the resulting penalty is simply given by the sum of the squared coefficients so that

$$\text{pen}(\boldsymbol{\beta}) = \sum_{j=0}^{k} \beta_j^2 = \boldsymbol{\beta}'\boldsymbol{\beta}$$

and

$$\text{PLS}(\boldsymbol{\beta}) = (\boldsymbol{y} - \boldsymbol{X}\boldsymbol{\beta})'(\boldsymbol{y} - \boldsymbol{X}\boldsymbol{\beta}) + \lambda\boldsymbol{\beta}'\boldsymbol{\beta}.$$

Proceeding in analogy to the derivation of the least squares estimator in Sect. 3.2.1 (p. 105), we first take the derivative of $\text{PLS}(\boldsymbol{\beta})$ with respect to $\boldsymbol{\beta}$ and obtain

$$\begin{aligned} \frac{\partial}{\partial \boldsymbol{\beta}} \text{PLS}(\boldsymbol{\beta}) &= \frac{\partial}{\partial \boldsymbol{\beta}} \left(\boldsymbol{y}'\boldsymbol{y} - 2\boldsymbol{y}'\boldsymbol{X}\boldsymbol{\beta} + \boldsymbol{\beta}'\boldsymbol{X}'\boldsymbol{X}\boldsymbol{\beta} + \lambda\boldsymbol{\beta}'\boldsymbol{\beta}\right) \\ &= -2\,\boldsymbol{X}'\boldsymbol{y} + 2\,\boldsymbol{X}'\boldsymbol{X}\boldsymbol{\beta} + 2\lambda\boldsymbol{\beta}. \end{aligned}$$

Setting the derivative equal to zero and solving for $\boldsymbol{\beta}$ yields the ridge regularized penalized least squares estimator

$$\hat{\boldsymbol{\beta}}_{\text{PLS}} = (\boldsymbol{X}'\boldsymbol{X} + \lambda\boldsymbol{I}_p)^{-1}\boldsymbol{X}'\boldsymbol{y},$$

i.e., $\hat{\boldsymbol{\beta}}_{\text{PLS}}$ only differs from $\hat{\boldsymbol{\beta}}_{\text{LS}}$ by the additional term $\lambda\boldsymbol{I}_p$ that arises from the penalty. For values of the smoothing parameter close to zero, the impact of this additional term practically vanishes and we end up with a solution that is close to the usual least squares estimator. However, if λ is large, $\boldsymbol{X}'\boldsymbol{X} + \lambda\boldsymbol{I}_p$ will be invertible, even if $\boldsymbol{X}'\boldsymbol{X}$ does not have full rank. Moreover, the solution $\hat{\boldsymbol{\beta}}_{\text{PLS}}$ will be shrunken towards zero. This is most easily seen from the penalized least squares criterion: When the smoothing parameter λ grows large, the penalty term will completely determine the optimization problem so that the optimal solution has to minimize the penalty term, i.e., $\text{pen}(\boldsymbol{\beta}) = \boldsymbol{\beta}'\boldsymbol{\beta}$. This will clearly be the case for $\boldsymbol{\beta} = \boldsymbol{0}$.

Typically, penalization of the intercept is not desired in ridge regression so that β_0 should be excluded from the penalty term. This can be achieved in two different ways. A first, popular approach is to center all covariates and the responses so that $\bar{y} = 0$ and $\bar{\boldsymbol{x}} = \boldsymbol{0}$ which automatically results in $\hat{\beta}_0 = 0$. This implies that the intercept can simply be dropped from the model and therefore is also not penalized. A second approach is to modify the penalty to

$$\text{pen}(\boldsymbol{\beta}) = \sum_{j=1}^{k} \beta_j^2 = \boldsymbol{\beta}' \boldsymbol{K} \boldsymbol{\beta},$$

where $\boldsymbol{K} = \text{diag}(0, 1, \ldots, 1) = \text{blockdiag}(0, \boldsymbol{I}_k)$, i.e., we introduce a *penalty matrix* that excludes the intercept but remains the identity matrix for the rest of the coefficient vector. The ridge regression estimate is then given by

$$\hat{\boldsymbol{\beta}}_{\text{PLS}} = (\boldsymbol{X}'\boldsymbol{X} + \lambda \boldsymbol{K})^{-1} \boldsymbol{X}'\boldsymbol{y},$$

i.e., it has exactly the same structure as before but replaces the identity matrix $\boldsymbol{I}_p$ with the penalty matrix $\boldsymbol{K}$. The second approach has the advantage that it can still be used in more complex models when ridge regression shall be combined, for example, with nonlinear regression as introduced in Chap. 8. We therefore follow the second approach in the rest of this section.

The optimality properties derived for the least squares estimator included unbiasedness, i.e.,

$$\text{E}(\hat{\boldsymbol{\beta}}_{\text{LS}}) = \text{E}((\boldsymbol{X}'\boldsymbol{X})^{-1}\boldsymbol{X}'\boldsymbol{y}) = (\boldsymbol{X}'\boldsymbol{X})^{-1}\boldsymbol{X}'\boldsymbol{X}\boldsymbol{\beta} = \boldsymbol{\beta},$$

and the covariance matrix of the least squares estimator is given by

$$\text{Cov}(\hat{\boldsymbol{\beta}}_{\text{LS}}) = \sigma^2 (\boldsymbol{X}'\boldsymbol{X})^{-1}.$$

In contrast, the penalized least squares estimator is biased since

$$\text{E}(\hat{\boldsymbol{\beta}}_{\text{PLS}}) = \text{E}((\boldsymbol{X}'\boldsymbol{X} + \lambda \boldsymbol{K})^{-1}\boldsymbol{X}'\boldsymbol{y}) = (\boldsymbol{X}'\boldsymbol{X} + \lambda \boldsymbol{K})^{-1}\boldsymbol{X}'\boldsymbol{X}\boldsymbol{\beta},$$

as the matrices $(\boldsymbol{X}'\boldsymbol{X} + \lambda \boldsymbol{K})^{-1}$ and $\boldsymbol{X}'\boldsymbol{X}$ do not cancel, unless $\lambda = 0$. Most of the coefficients in $\boldsymbol{\beta}_{\text{PLS}}$ will typically be shrunken towards zero as compared to the coefficients in $\boldsymbol{\beta}_{\text{LS}}$, i.e.,

$$|\hat{\beta}_{j,\text{PLS}}| \leq |\hat{\beta}_{j,\text{LS}}|, \quad j = 1, \ldots, k.$$

This relation holds exactly for orthogonal designs, where $\boldsymbol{X}'\boldsymbol{X}$ is a diagonal matrix, and as such all coefficients can be estimated independently (but may be violated for some coefficients in more general settings).

To motivate the shrinkage effect of ridge regression, note first that $\lambda \boldsymbol{K} = (\boldsymbol{X}'\boldsymbol{X} + \lambda \boldsymbol{K}) - \boldsymbol{X}'\boldsymbol{X}$ is positive semidefinite (for $\lambda > 0$), i.e., "$(\boldsymbol{X}'\boldsymbol{X} + \lambda \boldsymbol{K})$ is larger than (or equal to) $\boldsymbol{X}'\boldsymbol{X}$" in a matrix sense, since a nonnegative amount $\lambda \boldsymbol{K}$ is added to $\boldsymbol{X}'\boldsymbol{X}$. Multiplying two positive semidefinite matrices again yields a positive semidefinite matrix. Therefore, multiplying $(\boldsymbol{X}'\boldsymbol{X} + \lambda \boldsymbol{K})^{-1}$ with $(\boldsymbol{X}'\boldsymbol{X} + \lambda \boldsymbol{K}) - \boldsymbol{X}'\boldsymbol{X}$ yields positive semidefinite $\boldsymbol{I}_p - (\boldsymbol{X}'\boldsymbol{X} + \lambda \boldsymbol{K})^{-1}\boldsymbol{X}'\boldsymbol{X}$. In particular, this implies that the diagonal elements of $(\boldsymbol{X}'\boldsymbol{X} + \lambda \boldsymbol{K})^{-1}\boldsymbol{X}'\boldsymbol{X}$ are smaller than (or equal to) one because the diagonal elements of a positive semidefinite matrix are always greater than (or equal to) zero (see Theorem A.28. 3 of Appendix A.7). Re-expressing the representation of the penalized least squares estimator as

$$\hat{\boldsymbol{\beta}}_{\text{PLS}} = (\boldsymbol{X}'\boldsymbol{X} + \lambda \boldsymbol{K})^{-1}\boldsymbol{X}'\boldsymbol{X}(\boldsymbol{X}'\boldsymbol{X})^{-1}\boldsymbol{X}'\boldsymbol{y} = (\boldsymbol{X}'\boldsymbol{X} + \lambda \boldsymbol{K})^{-1}\boldsymbol{X}'\boldsymbol{X}\hat{\boldsymbol{\beta}}_{\text{LS}},$$

indicates that we can obtain $\hat{\boldsymbol{\beta}}_{\text{PLS}}$ from $\hat{\boldsymbol{\beta}}_{\text{LS}}$ through premultiplication with a matrix that is "smaller than $\boldsymbol{I}_p$." For the special case of an orthogonal design matrix, both $\boldsymbol{X}'\boldsymbol{X}$ and $\boldsymbol{X}'\boldsymbol{X} + \lambda \boldsymbol{K}$ (and therefore also $(\boldsymbol{X}'\boldsymbol{X} + \lambda \boldsymbol{K})^{-1}$ are diagonal matrices so that we can write

$$\hat{\beta}_{j,\text{PLS}} = \frac{(\boldsymbol{x}^j)'\boldsymbol{x}^j}{(\boldsymbol{x}^j)'\boldsymbol{x}^j + \lambda}\hat{\beta}_{j,\text{LS}}, \quad j = 1, \dots, k,$$

where $\boldsymbol{x}^j$ denotes the jth column of $\boldsymbol{X}$. Since the factor $(\boldsymbol{x}^j)'\boldsymbol{x}^j/((\boldsymbol{x}^j)'\boldsymbol{x}^j + \lambda)$ is smaller than one, we immediately obtain the shrinkage effect of ridge regression estimates.

The covariance matrix for the ridge regression estimator is given by

$$\text{Cov}(\hat{\boldsymbol{\beta}}_{\text{PLS}}) = \sigma^2(\boldsymbol{X}'\boldsymbol{X} + \lambda \boldsymbol{K})^{-1}\boldsymbol{X}'\boldsymbol{X}(\boldsymbol{X}'\boldsymbol{X} + \lambda \boldsymbol{K})^{-1}.$$

Proceeding analogously, one can show that

$$\text{Cov}(\hat{\boldsymbol{\beta}}_{\text{LS}}) - \text{Cov}(\hat{\boldsymbol{\beta}}_{\text{PLS}})$$

is positive definite (for $\lambda > 0$), i.e., the covariance matrix of the penalized least squares estimator is smaller than the covariance matrix of the least squares estimator. In particular, this implies that

$$\text{Var}(\hat{\beta}_{j,\text{PLS}}) < \text{Var}(\hat{\beta}_{j,\text{LS}}), \quad j = 1, \dots, k,$$

since for a positive definite matrix the diagonal elements are positive.

In summary, we find that the penalized least squares estimator is biased, but has a smaller variance than the least squares estimator. When combining bias and covariance to the mean squared error one hopes to achieve a smaller MSE when choosing an appropriate smoothing parameter λ. In particular, if $\boldsymbol{X}$ is close to collinear or if X is high-dimensional, the variance usually dominates the mean squared error and thus regularization is desirable. Often, ridge regression is also considered as a possibility to further shrink small coefficients towards zero. In particular, this may give a clearer indication of which variables are actually required in the model so that variable selection can be achieved.

The smoothing parameter is typically determined using r-fold cross validation; see Sect. 3.4, p. 149. Therefore, the data set is split into r parts of approximately equal size. Estimation is performed for a given grid of smoothing parameters on $r - 1$ parts of the data set, while prediction based on these estimates is obtained for the last remaining part (holdout sample). Cycling through the different splits an average prediction error can be obtained for a given smoothing parameter. The "optimal" smoothing parameter can be defined as the one that minimizes the average prediction error on the hold-out samples.

Finally, the scaling of covariates is important when applying a regularization approach. The penalty formed of squared regression coefficients assumes that all coefficients can be compared in their absolute value. However, the scaling of covariates has immediate impact on the interpretation of these absolute values. For example, the coefficient associated with a covariate measuring a distance will be scaled by a factor of 1,000 when the variable is measured in meters instead of kilometers. Hence it is important to make all variables comparable in their scaling before applying a penalized least squares approach. A common solution is to transform all covariates to have zero mean and unit standard deviation, i.e., to standardize all covariates.

Example 4.10 Price of Used Cars—Ridge Regression

We illustrate the impact of ridge regularization using the data on prices of used cars introduced in Example 3.19 (p. 152). Instead of performing a formal variable selection, we will use a relatively complex model comprising all covariates and also higher-order polynomials. Often such a model can yield rather large uncertainty in the coefficient estimates. As a remedy, we apply ridge regularization that effectively reduces the variance while introducing some bias in the coefficient estimates. Estimation is based on the R package `penalized` and in particular function `optL2`.

For the variables representing the age of the car in months (*age*) and kilometer reading in 1,000 km (*kilometer*), we consider orthogonal polynomials of degree three (see Example 3.5 on p. 90) to allow for potential nonlinearity while the remaining covariates, corresponding to months until the next appointment with the technical inspection agency (*TIA*), ABS brake (*extras1*), and sunroof (*extras2*), are included with linear effects. The second column in Table 4.6 represents the least squares estimates obtained for this model specification. Note that our analyses in Example 3.19 have shown that some of the variables may not actually be needed to adequately predict the prices of used cars. As a consequence, regularization in terms of a ridge penalty may be applied with the aim to shrink such redundant parameters towards zero.

Figure 4.9 illustrates the impact of the ridge penalty for a variety of smoothing parameter values λ. On the left-hand side, the graph starts with a rather large value of the smoothing parameter yielding strong shrinkage towards zero. When moving to the right (corresponding to a decrease in the smoothing parameter), the coefficient estimates approach the least squares estimate for $\lambda \to 0$. The surprising scaling of the x-axis with large values on the left and small values on the right is motivated by the fact that many graphics represent coefficient paths as functions of the constraint parameter t instead of the smoothing

Table 4.6 Prices of used cars: estimated coefficients for different models

Variable	LS	Ridge $\lambda = 6.43$	Ridge $\lambda \approx 250$	LASSO $\lambda = 10.34$
intercept	3.585	3.580	3.569	3.421
ageop1	−0.709	−0.672	−0.594	−0.682
ageop2	0.172	0.164	0.146	0.150
ageop3	0.016	0.015	0.013	–
kilometerop1	−0.437	−0.425	−0.393	−0.412
kilometerop2	0.142	0.138	0.128	0.110
kilometerop3	0.009	0.010	0.010	–
TIA	−0.005	−0.005	−0.004	–
extras1	−0.114	−0.104	−0.086	−0.036
extras2	−0.031	−0.042	−0.060	–

Second column: least squares estimate; third column: ridge regression with estimated smoothing parameter based on cross validation; fourth column: ridge regression with fixed, large smoothing parameter; fifth column: LASSO with estimated smoothing parameter based on cross validation

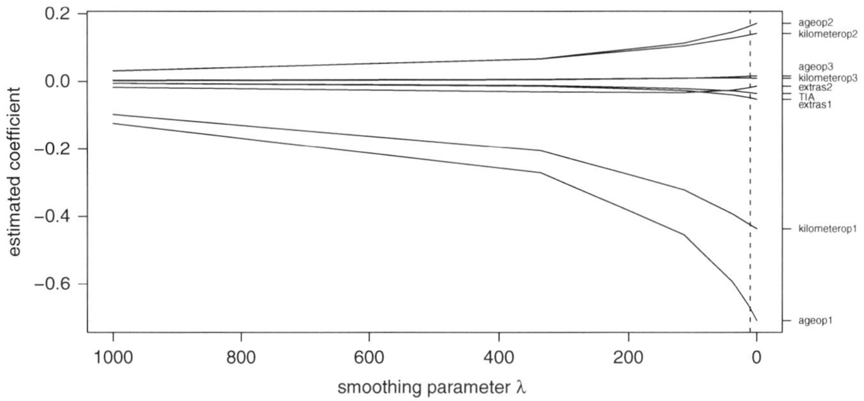

Fig. 4.9 Prices of used cars: estimated ridge regression coefficients as a function of the smoothing parameter λ. The *dashed vertical line* indicates the optimal smoothing parameter determined by tenfold cross validation. All effects refer to standardized versions of the covariates

parameter. In order to comply with the resulting visual impression of coefficient paths evolving from zero to the least squares estimate, we chose a reverse scaling with respect to the smoothing parameter. Note that Fig. 4.9 shows standardized coefficients, i.e., all covariates have been standardized to zero mean and unit variance prior to the analysis to make them comparable in their magnitude. In contrast, Table 4.6 contains estimates transformed back to the original scaling. In fact, the scaling only impacts the categorical covariates since the orthonormal polynomials constructed for continuous covariates are already standardized.

To obtain an optimal amount of shrinkage, the smoothing parameter λ was determined based on tenfold cross validation (the optimal smoothing parameter is indicated as a dashed line in Fig. 4.9). Clearly, only a very small amount of smoothness is needed, so that estimates remain almost unchanged after applying ridge regularization. The third column of Table 4.6 contains the corresponding precise numerical values while the fourth column contains estimates achieved with a rather large smoothing parameter. In this latter case, the impact of the penalty can be seen much more clearly with shrinkage towards zero. Note, however, that single coefficients may actually also increase after applying a penalty as can be seen for the coefficient of *extras2*. While this may seem to be surprising at first sight, it is a result of (negative) correlations in the covariates. If a certain effect is reduced, the effect of a closely related variable may increase. For very large values of the smoothing parameter, however, all coefficient estimates will eventually approach a limiting value of zero. We will provide a geometric explanation later in this section.

△

4.2.3 Least Absolute Shrinkage and Selection Operator

While ridge regression enables estimation of regression coefficients in high-dimensional covariate settings or with design matrices that are close to collinear, it does not yield a sparse solution: all estimated regression coefficients will still be different from zero (with probability one). For interpretational purposes, it would, however, be desirable not only to shrink small coefficients towards zero but to have the possibility to set some effects exactly to zero. In particular, this would allow us to combine model estimation with variable selection in one single model estimation step. Such an approach can be obtained by replacing the penalty of squared regression coefficients with the penalty of absolute values yielding

$$\text{pen}(\boldsymbol{\beta}) = \sum_{j=1}^{k} |\beta_j|$$

and

$$\hat{\boldsymbol{\beta}}_{\text{LASSO}} = \arg\min_{\boldsymbol{\beta}} (\boldsymbol{y} - \boldsymbol{X}\boldsymbol{\beta})'(\boldsymbol{y} - \boldsymbol{X}\boldsymbol{\beta}) + \lambda \sum_{j=1}^{k} |\beta_j|, \tag{4.10}$$

where again we left the intercept unpenalized. As for ridge regression, the penalized least squares criterion balances between fit to the data as measured by the least squares criterion and regularized solutions as determined by the penalty. The trade-off between these two goals is governed by the smoothing parameter λ. Since $\hat{\boldsymbol{\beta}}_{\text{LASSO}}$ is defined in terms of an absolute value penalty and allows to select covariates in a variable selection type fashion (as we will see in the following), it is referred to as the least absolute shrinkage and selection operator (LASSO, Tibshirani, 1996).

The difference between ridge regression and the LASSO can be seen from the different form of the penalty functionals depicted in Fig. 4.10. Ridge regression imposes a quadratic penalty that has very strong impact on large coefficient values but small penalty for values close to zero (compare also the coefficient path depicted in Fig. 4.9). In contrast, the absolute value penalty for the LASSO increases at a slower rate for large coefficient values, but moves away from zero faster for coefficients that are close to zero. Consequently, we expect the desirable behavior that small coefficients will be more strongly shrunken towards zero, while larger coefficients will be less affected by the penalty.

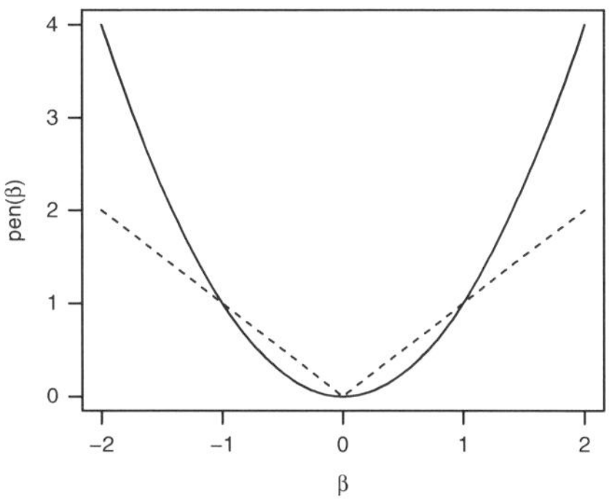

Fig. 4.10 Penalties for ridge regression (*dashed line*) and the LASSO (*solid line*)

In contrast to ridge regression, no closed-form solution for the LASSO-regularized estimate is available. While the penalized least squares criterion (4.10) is not differentiable, because of the inclusion of the absolute value penalty, one can still obtain "estimation equations" similar to the normal equations in the classical linear model. Omitting their derivation they are given by

$$2\boldsymbol{X}'\boldsymbol{X}\boldsymbol{\beta} + 2\boldsymbol{X}'\boldsymbol{y} + \lambda \sum_{j=1}^{k} \text{sign}(\beta_j) = \mathbf{0}.$$

However, no explicit solution can be computed because of the sign function that indicates that the LASSO estimate, in contrast to the ridge estimator, is nonlinear in the data. Stated differently, it cannot be expressed as a linear estimator $\hat{\boldsymbol{\beta}} = \boldsymbol{A}\boldsymbol{y}$ with a $p \times n$-matrix $\boldsymbol{A}$ (compare also Sect. 3.2.3, p. 117). This also implies that statistical properties of $\hat{\boldsymbol{\beta}}_{\text{LASSO}}$ are more difficult to derive than in case of ridge regression. Still, these are in principle analogously to ridge regression, i.e., $\hat{\boldsymbol{\beta}}_{\text{LASSO}}$ is biased, but has smaller variance than the least squares estimate. A comparison between ridge and LASSO is more complicated and there will be no general answer on which of the two is preferable in terms of mean squared error.

The LASSO criterion (4.10) can equivalently be rewritten as

$$\hat{\boldsymbol{\beta}}_{\text{PLS}} = \arg\min_{\boldsymbol{\beta}} (\boldsymbol{y} - \boldsymbol{X}\boldsymbol{\beta})'(\boldsymbol{y} - \boldsymbol{X}\boldsymbol{\beta})$$

subject to the constraint

$$\sum_{j=1}^{k} |\beta_j| \leq t,$$

where again t and the smoothing parameter λ are connected in a one-to-one relationship; see, e.g., Sydsaeter et al. (2005) in Sect. 3.5 for a proof of the equivalence. For interpretational purposes and graphical visualization, one often

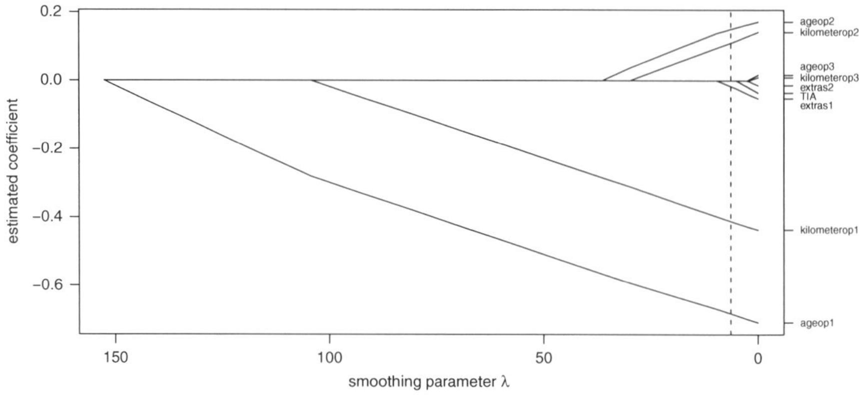

Fig. 4.11 Prices of used cars: estimated LASSO-regularized coefficients as a function of the smoothing parameter λ. The *dashed vertical line* indicates the optimal smoothing parameter determined by tenfold cross validation. All effects refer to standardized versions of the covariates

varies t between the two extremal values 0 and t_{LS}, corresponding to extreme and no regularization, where

$$t_{\text{LS}} = \sum_{j=1}^{k} |\hat{\beta}_{\text{LS},j}|.$$

Estimation of the LASSO relies on numerical optimization approaches. In the original proposal of the LASSO (Tibshirani, 1996), a quadratic programming approach has been applied to optimize the penalized least squares criterion subject to the inequality constraint on the sum of absolute values. Most current implementations use least angle regression (LARS) to determine the LASSO estimates (Efron, Hastie, Johnstone, & Tibshirani, 2004). The main advantage of LARS is that it allows to determine the complete coefficient paths in one estimation run. We used an alternative approach developed in Goeman (2010) and implemented in the R package `penalized` to compute both LASSO-regularized estimates and ridge regression estimates. This package also implements the determination of optimal smoothing parameters based on cross validation.

Example 4.11 Price of Used Cars—LASSO

We repeat the analysis conducted in Example 4.10, now replacing the ridge penalty with the LASSO penalty (estimated via function `optL1` also contained in the R package `penalized`). The resulting coefficient paths are shown in Fig. 4.11. The first striking difference compared to the results achieved with ridge regression is that for large values of the smoothing parameter several coefficients are actually set to zero. At a sequence of threshold values, one covariate at a time enters the model until all coefficients are nonzero for values of the smoothing parameter approaching zero and finally achieve the values of the least squares estimate. Moreover, the paths are all linear in the smoothing parameter and only change their slope once a new variable enters the model.

More specifically, several of the covariates associated with small least squares estimates enter the model very late, with rather small values of the smoothing parameter. When determining a data-driven, optimal value of the smoothing parameter through tenfold cross validation, these coefficients actually drop out of the model so that a sparse solution is obtained. In fact, the resulting model is quite close to the model chosen by AIC in Example 3.19 with quadratic effects of age and kilometer reading and no effects of *TIA* and *extras2*. With AIC, *extras1* was also deleted from the model which is not the case with LASSO regularization. Note, however, that the estimated coefficient is very small especially compared to the least squares estimate (Table 4.6).

When comparing the cross validation criterion for the least squares estimate, ridge regression, and LASSO-regularized estimation, we obtain values of −202.519 (least squares), −204.733 (ridge), and −200.746 (LASSO). The best fit is obtained with the LASSO, while ridge regression actually leads to a larger cross validation criterion in comparison to least squares. Therefore, in this particular example, it is not per se important to regularize estimates, but we have to use a specific form of regularization to achieve an improvement in terms of model fit. This may be explained by the specific properties of the regularization approaches in combination with the settings in our example. While ridge regression shrinks large, important coefficients more strongly than the LASSO, we have the reverse behavior for small coefficients (again see Fig. 4.10). To obtain good prediction for the price of used cars, the type of regularization imposed by the LASSO seems to be more plausible. Note also that we have only a moderate number of covariates compared to the sample size and therefore the need for regularization may not be so great in this example. △

4.2.4 Geometric Properties of Regularized Estimates

As we have seen in the application, LASSO regularization yields estimated coefficient vectors such that some of the parameters are estimated to be exactly zero. To understand this behavior, we have to investigate the geometric properties of penalized least squares estimation. For illustration purposes, we will consider a bivariate coefficient vector $\boldsymbol{\beta} = (\beta_1, \beta_2)'$, but all results are easily generalized to the multivariable setup. Note that we have not included an intercept but consider standardized covariates and a centered response.

First, the least squares criterion $\mathrm{LS}(\boldsymbol{\beta})$ can be rewritten as

$$\begin{aligned}\mathrm{LS}(\boldsymbol{\beta}) &= (\boldsymbol{\beta} - \hat{\boldsymbol{\beta}})'\boldsymbol{X}'\boldsymbol{X}(\boldsymbol{\beta} - \hat{\boldsymbol{\beta}}) + \boldsymbol{y}'(\boldsymbol{I}_n - \boldsymbol{X}(\boldsymbol{X}'\boldsymbol{X})^{-1}\boldsymbol{X}')\boldsymbol{y} \\ &= (\boldsymbol{\beta} - \hat{\boldsymbol{\beta}})'\boldsymbol{X}'\boldsymbol{X}(\boldsymbol{\beta} - \hat{\boldsymbol{\beta}}) + \hat{\boldsymbol{\varepsilon}}'\hat{\boldsymbol{\varepsilon}}\end{aligned} \tag{4.11}$$

and is therefore (up to an additive constant) equivalent to the quadratic form

$$(\boldsymbol{\beta} - \hat{\boldsymbol{\beta}})'\boldsymbol{X}'\boldsymbol{X}(\boldsymbol{\beta} - \hat{\boldsymbol{\beta}})$$

in $\boldsymbol{\beta}$. A proof can be found in the appendix of this chapter on p. 259.

Since Eq. (4.11) defines a quadratic form in $\boldsymbol{\beta}$, the contour lines of $\mathrm{LS}(\boldsymbol{\beta})$, i.e., the values of $\boldsymbol{\beta}$ resulting from solving $\mathrm{LS}(\boldsymbol{\beta}) = c$ for a constant c, are ellipses with the specific shape determined by the matrix $\boldsymbol{X}'\boldsymbol{X}$. On the other hand, for two dimensions, the constraint

$$|\beta_1| + |\beta_2| = t$$

4.3 Regularized Estimation

Penalized Least Squares Criteria

Regularized estimation in the linear model relies on penalized least squares criteria

$$\mathrm{PLS}(\boldsymbol{\beta}) = (\boldsymbol{y} - \boldsymbol{X}\boldsymbol{\beta})'(\boldsymbol{y} - \boldsymbol{X}\boldsymbol{\beta}) + \lambda \cdot \mathrm{pen}(\boldsymbol{\beta})$$

with smoothing parameter $\lambda \geq 0$ and $\mathrm{pen}(\boldsymbol{\beta})$ penalizing model complexity.

Ridge Regression

For ridge regression, the penalty is given by the sum of squared coefficients, i.e.,

$$\mathrm{pen}(\boldsymbol{\beta}) = \sum_{j=1}^{k} \beta_j^2 = \boldsymbol{\beta}'\boldsymbol{K}\boldsymbol{\beta},$$

with penalty matrix $\boldsymbol{K} = \mathrm{diag}(0, 1, \ldots, 1)$. The resulting penalized least squares estimate is

$$\hat{\boldsymbol{\beta}}_{\mathrm{PLS}} = (\boldsymbol{X}'\boldsymbol{X} + \lambda\boldsymbol{K})^{-1}\boldsymbol{X}'\boldsymbol{y}.$$

LASSO

For the LASSO, the penalty is given by the sum of absolute coefficients, i.e.,

$$\mathrm{pen}(\boldsymbol{\beta}) = \sum_{j=1}^{k} |\beta_j|.$$

The resulting estimate is not available in closed form and has to be determined numerically (for example based on quadratic programming).

Choice of the Smoothing Parameter

The optimal smoothing parameter λ can be determined based on r-fold cross validation.

Software

R package `penalized`.

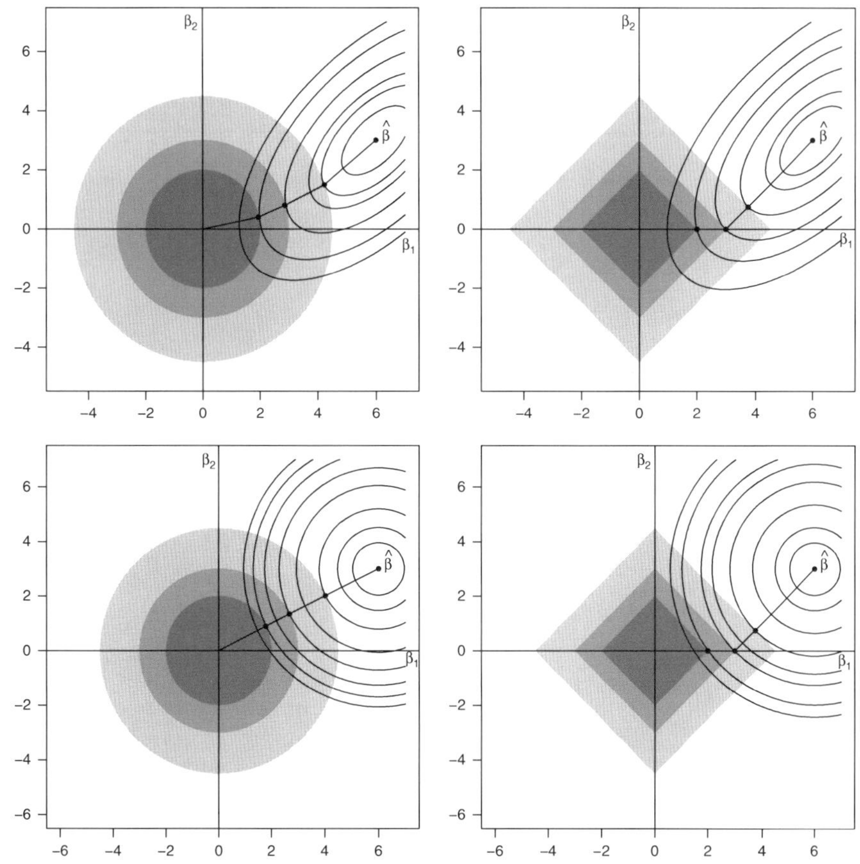

Fig. 4.12 Geometrical interpretation of the penalized least squares criteria for ridge regression (*left panel*) and LASSO (*right panel*). In the *upper panel*, a non-diagonal matrix $\boldsymbol{X}'\boldsymbol{X}$ is considered while the *lower panel* corresponds to an orthonormal design with $\boldsymbol{X}'\boldsymbol{X} = \boldsymbol{I}_2$

defines diamond-shaped contour lines with side lengths $\sqrt{2}t$. Therefore the LASSO-regularized estimate for a given t is the contact point between the two geometrical regions defined by the constraint and the least squares criterion. If this contact point is located in one of the corners of the diamond, some of the coefficients will be estimated to be zero.

This geometric definition of LASSO-regularized regression estimates is visualized in Fig. 4.12 for artificial data and a hypothetical value of $\hat{\boldsymbol{\beta}} = (6, 3)'$ for the least squares estimate. The contour lines of the least squares criterion are centered around this least squares estimate since it is the corresponding unique minimum. In the upper panel of Fig. 4.12, usual data with correlated design vectors have been considered so that the contour lines of the least squares criterion are ellipsoids. In comparison, the lower panel considers the special case of orthonormal design

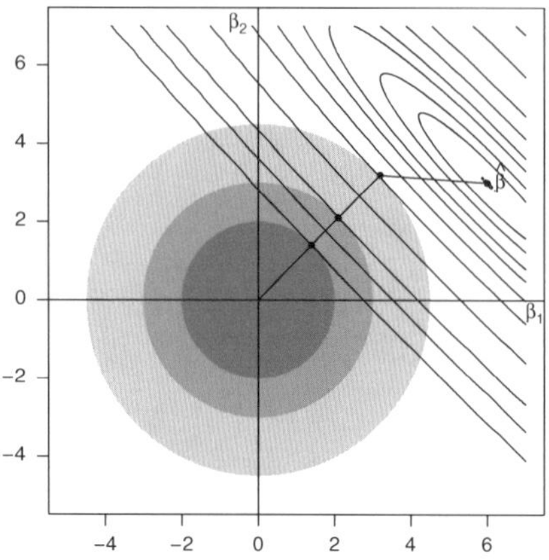

Fig. 4.13 Ridge regression estimates in case of a design matrix with large negative correlation

matrix with $\boldsymbol{X}'\boldsymbol{X} = \boldsymbol{I}_2$ so that the contour lines are exact circles. The contours of the penalty are depicted as shaded areas centered about the minimum of the penalty functional given by $(0, 0)'$. Depending on the value of the smoothing parameter, a specific contact point of the contour lines of the fit and penalty criterion is chosen that determines the regularized estimated coefficients. If the smoothing parameter is large enough, the contact point will be forced to lie on one (or several) of the coefficient axes, and therefore one (or more) coefficients are shrunk to zero.

For ridge regression, the contour lines are circular, defined by the constraint

$$\beta_1^2 + \beta_2^2 = t.$$

When considering candidate contact points within the geometric regions that are defined by the constrained least squares criterion, no possibility arises to set coefficients equal to zero, as illustrated in the left panel of Fig. 4.12. In contrast to the LASSO, an increasing value for the ridge constraint parameter, t, provides a smooth transition of the coefficients towards zero.

Figure 4.13 provides a geometric explanation for the surprising behavior that we observed in Example 4.10, i.e., where ridge regression in fact yielded an increased, instead of shrunken estimates, for some coefficients. We previously argued that negative correlations between covariates are the reason for this observation. In Fig. 4.13, we have generated artificial data with a rather high negative correlation, which is indicated by the negative orientation of the ellipsoids representing the least squares criterion. Seeking the contact point between the outer contour line for the penalty and the least squares criterion yields a shrunken estimate for β_1 but a somewhat increased estimate for β_2 due to the extreme orientation of the least squares contours. When increasing the smoothing parameter (i.e., moving towards the inner contours of the penalty), we again find the expected shrinkage effect and finally approach zero for both regression coefficients.

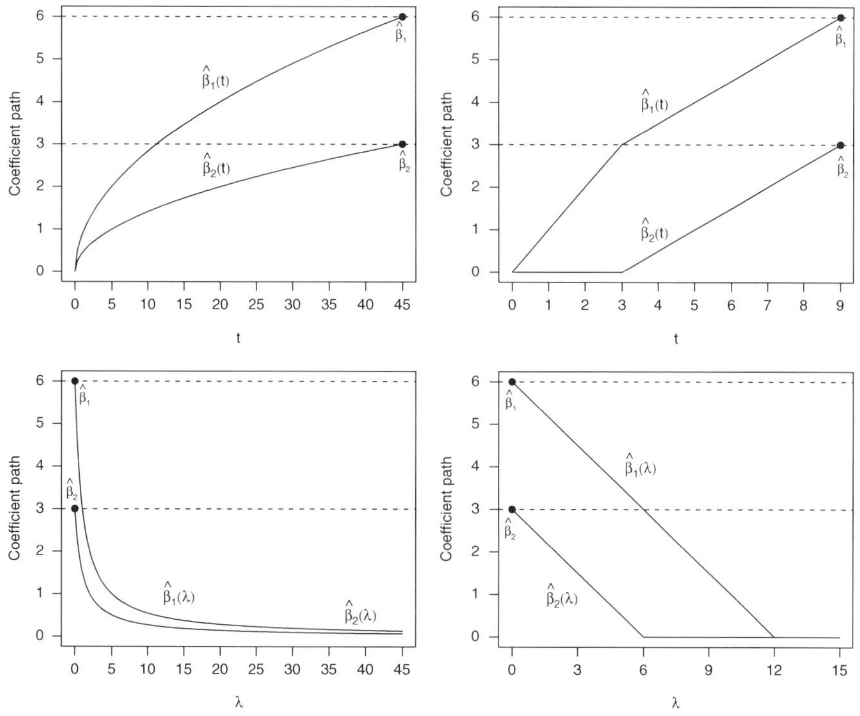

Fig. 4.14 Coefficient paths for ridge regression (*left panel*) and the LASSO (*right panel*) as functions of either the complexity parameter t (*upper panel*) or the smoothing parameter λ (*lower panel*). The *dashed lines* indicate the values of the least squares estimates

Additional insights can be gained when considering situations with orthonormal design matrices, i.e., with $\boldsymbol{X}'\boldsymbol{X} = \boldsymbol{I}$. In this case, the contour lines of the least squares criterion are circular and explicit formulae for both ridge and LASSO-regularized estimates can be derived either in terms of the smoothing parameter λ or the constraint parameter t. For the LASSO, we obtain

$$\hat{\beta}_{\text{LASSO},j}(\lambda) = \text{sign}(\hat{\beta}_{\text{LS},j}) \left[|\hat{\beta}_{\text{LS},j}| - \frac{\lambda}{2} \right]_+ ,$$

where $[x]_+ = \max\{0, x\}$. For ridge regression, we have

$$\hat{\beta}_{\text{PLS},j}(\lambda) = \frac{1}{1+\lambda} \hat{\beta}_{\text{LS},j} .$$

A comparison of these expressions again reveals the distinct, yet different, behavior of ridge regression and the LASSO (also see the graphical representation in Fig. 4.14). For the LASSO, the restriction to positive parts via the function $[x]_+$ allows to set coefficients to zero if they are smaller than 0.5λ. Moreover, the LASSO-regularized estimates move away from zero at a constant slope until a

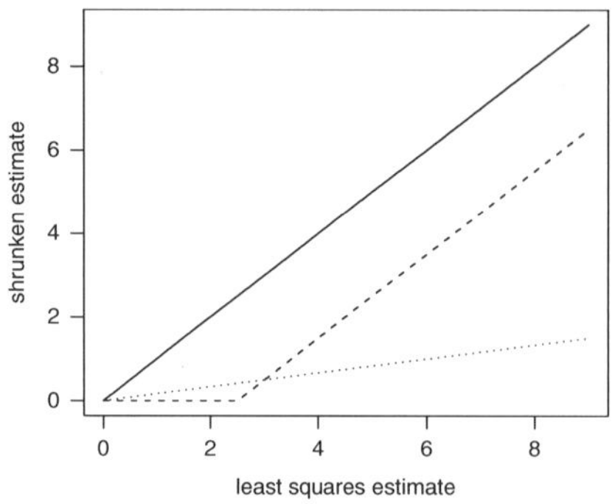

Fig. 4.15 Ridge (*dotted line*) and LASSO (*dashed line*) estimates as a function of the least squares estimate in case of an orthonormal design and for smoothing parameter $\lambda = 5$. The *solid line* represents the unregularized least squares estimate

further nonzero regression coefficient enters the model (top right panel of Fig. 4.14). In contrast, estimates from ridge regression are shrunken towards zero for a large penalty, but actually never reach zero. Specifically, they are simply shrunken by the factor $1/(1+\lambda)$. However, note again that this only holds in case of orthonormal design matrices.

Figure 4.15 illustrates this shrinkage behavior from a different perspective, which shows a regularized estimate as a function of the least squares estimate for a fixed value of the smoothing parameter λ ($\lambda = 5$ in this case). While the LASSO-regularized estimate equals zero for a small coefficient (smaller than 0.5λ), the ridge regression estimate is simply a scaled version of the least squares estimate with the amount of scaling determined by λ.

4.2.5 Partial Regularization

Thus far, we have exclusively dealt with the situation that all coefficients of a regression model (except the intercept) should be subject to regularization. However, it is also possible to penalize only parts of the coefficient vector, yielding, for example,

$$(\boldsymbol{y} - \boldsymbol{X}_1\boldsymbol{\beta}_1 - \boldsymbol{X}_2\boldsymbol{\beta}_2)'(\boldsymbol{y} - \boldsymbol{X}_1\boldsymbol{\beta}_1 - \boldsymbol{X}_2\boldsymbol{\beta}_2) + \lambda\,\mathrm{pen}(\boldsymbol{\beta}_2),$$

where $\boldsymbol{X} = (\boldsymbol{X}_1, \boldsymbol{X}_2)$ is a partition of the design matrix and $\boldsymbol{\beta} = (\boldsymbol{\beta}_1', \boldsymbol{\beta}_2')'$ analogously partitions the coefficients. The penalty now only applies to the part of the model defined in $\boldsymbol{X}_2\boldsymbol{\beta}_2$, while the remaining covariates in $\boldsymbol{X}_1$ are left untouched. Such a distinction can, for example, be relevant when including clinical covariates and gene expression levels in a simultaneous regression model. While the former will typically be included due to prior knowledge and are of a much smaller number, gene expression often contains diverse, heterogeneous, and high-dimensional information. Therefore it is sensible to regularize only the regression coefficients of the latter, while coefficients associated with information such as age or disease status remain unpenalized.

4.3 Boosting Linear Regression Models

In this section, we will describe an alternative regularization approach called boosting where regularization is implicitly achieved through early stopping of an iterative stepwise algorithm. Boosting turns out to be a rather general and versatile regularization approach since estimation problems are described in terms of a loss function and, as a consequence, boosting can also be applied beyond the linear model, e.g., in robust regression or generalized linear models for non-Gaussian responses. Moreover, the implicit regularization performed with boosting allows for automatic model choice and variable selection, similar to the LASSO. A disadvantage of boosting, similar to ridge regression and the LASSO, is that standard inference tools like standard errors or confidence intervals are not easily available. We introduce the general ideas of boosting applied to classical linear models and further present the general framework at the end of the section.

4.3.1 Basic Principles

Consider the linear model

$$y_i = \eta_i + \varepsilon_i = \boldsymbol{x}_i'\boldsymbol{\beta} + \varepsilon_i, \quad i = 1, \dots, n,$$

with the standard predictor $\eta_i = \boldsymbol{x}_i'\boldsymbol{\beta}$. Similar to the cases of ridge regression and the LASSO, we are interested in situations where the ordinary least squares estimate is not optimal. This especially may be the case when the number of regression coefficients is large, when the design matrix is close to collinear, or when we are interested in determining a suitable subset of covariates. In particular in the last case, using either too many or using too few covariates may induce suboptimal predictions (as discussed in Sect. 3.4), and regularized estimation is often a useful alternative to least squares estimation.

The principle idea of boosting can be explained along the following basic algorithm:

1. Choose an initial estimate $\hat{\boldsymbol{\beta}}^{(0)}$ and set $t = 1$.
2. Compute the current residuals $\boldsymbol{u} = \boldsymbol{y} - \boldsymbol{X}\hat{\boldsymbol{\beta}}^{(t-1)}$ and the corresponding least squares estimate $\hat{\boldsymbol{b}} = (\boldsymbol{X}'\boldsymbol{X})^{-1}\boldsymbol{X}'\boldsymbol{u}$. Perform the update

$$\hat{\boldsymbol{\beta}}^{(t)} = \hat{\boldsymbol{\beta}}^{(t-1)} + \nu\hat{\boldsymbol{b}}$$

with $0 < \nu < 1$ and set $t = t + 1$.

3. Iterate step 2 for a fixed number of iterations m_{stop}.

Starting from the initial value $\hat{\boldsymbol{\beta}}^{(0)}$, the algorithm iteratively proceeds towards the final estimate by updating $\hat{\boldsymbol{\beta}}^{(t-1)}$ with *small portions* of the least squares estimate obtained from current residuals. Instead of making large steps towards the least squares estimate, we multiply $\hat{\boldsymbol{b}}$ with the step length factor ν to implicitly implement regularization since—when stopping early enough—the multiplication with ν yields a proportional shrinkage of the estimates analogous to ridge regression. Usually ν will be chosen to be rather small, for example, $\nu = 0.1$ or even $\nu = 0.01$.

In boosting terminology, the least squares estimate used in step 2. of the algorithm is a *base-learning procedure* that provides a means of fitting the model of interest. Often this single estimation step will be replaced with componentwise fits in the following. We then actually fit a sequence of base-learning procedures (for example corresponding to the different covariates in the linear model) and only update the best-fitting one. Yet the application of the step length factor induces regularization.

We now consider a somewhat different interpretation of boosting that is useful to generalize the approach in particular in terms of the available class of base-learning procedures.

When starting with initial guesses $\hat{\boldsymbol{\beta}}^{(0)}$, we can compute the lack of fit information associated with this starting values as the corresponding derivative of the least squares criterion, i.e.,

$$\frac{\partial}{\partial \boldsymbol{\beta}} \text{LS}(\boldsymbol{\beta})\bigg|_{\boldsymbol{\beta}=\hat{\boldsymbol{\beta}}^{(0)}} = -2\boldsymbol{X}'\left(\boldsymbol{y} - \boldsymbol{X}\hat{\boldsymbol{\beta}}^{(0)}\right).$$

This derivative indicates the direction towards the least squares estimate and would be zero if the initial value is already the least squares estimate since it then coincides with the normal equations. In fact, here the minimizer of the least squares criterion could be obtained in one step by setting the derivative to zero and solving with respect to the regression coefficients. However, to achieve regularization, we only do small steps towards the solution so that

$$\hat{\boldsymbol{\beta}}^{(1)} = \hat{\boldsymbol{\beta}}^{(0)} + \nu\hat{\boldsymbol{b}},$$

where $\hat{\boldsymbol{b}} = (\boldsymbol{X}'\boldsymbol{X})^{-1}\boldsymbol{X}'\boldsymbol{u}$ is the least squares estimate obtained from the residuals $\boldsymbol{u} = \boldsymbol{y} - \boldsymbol{X}\hat{\boldsymbol{\beta}}^{(0)}$.

4.3.2 Componentwise Boosting

After introducing the basic principles, we will next move towards more interesting versions of boosting algorithms. Figure 4.16 shows how the boosting approach moves from the initial guess [$(0, 0)'$ in this case] towards the minimum of the least squares criterion. If we use $\hat{\boldsymbol{b}} = (\boldsymbol{X}'\boldsymbol{X})^{-1}\boldsymbol{X}'\boldsymbol{u}$ to update the coefficient vector, we always move along the direction of the steepest descent directly towards the

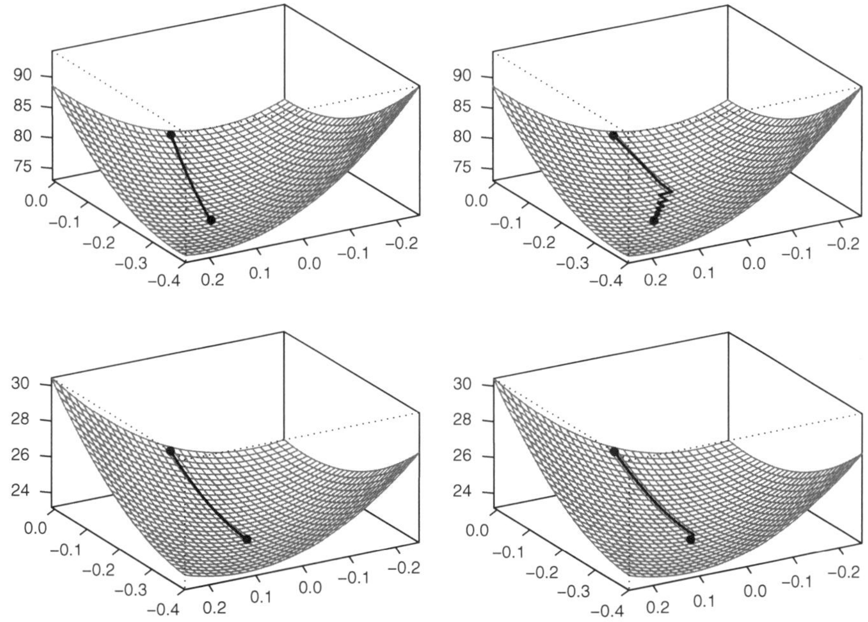

Fig. 4.16 Surface plots of the least squares criterion together with the coefficient path from the starting value $(0, 0)'$ towards the least squares estimate. The *left panels* show direct paths along the steepest descent; the *right panels* restrict the directions to be parallel to the coefficient axes

least squares estimate. In the upper left panel of Fig. 4.16, this is indicated for a least squares estimate where both components are quite different from zero. In the bottom left panel, in contrast, one of the coefficients is estimated to be rather close to zero. In this case, the path along steepest descent is almost parallel to one of the coefficient axes.

The basic idea of componentwise boosting now is to restrict updates in the iterative fitting process to directions that are *exactly parallel* to the coefficient axes. This is implemented as follows: We first compute least squares estimates for all covariates separately, i.e., estimate the models $u_i = b_j x_{ij} + \varepsilon_{ij}$ and obtain coefficient estimates

$$\hat{b}_j = ((\boldsymbol{x}^j)'\boldsymbol{x}^j)^{-1}(\boldsymbol{x}^j)'\boldsymbol{u} = \frac{\sum_{i=1}^{n}(x_{ij} - \bar{x}_j)u_i}{\sum_{i=1}^{n}(x_{ij} - \bar{x}_j)^2}, \quad j = 0, \ldots, k$$

based on the current residuals $\boldsymbol{u}$ where $\boldsymbol{x}^j = (x_{1j}, \ldots, x_{nj})'$ is the column of $\boldsymbol{X}$ corresponding to the jth covariate and $\bar{x}_j$ is the average of the covariate values. We then do not update all coefficients, but rather only the one that leads to the largest reduction in the least squares criterion, i.e., we determine

$$j^* = \arg\min_{j=0,\dots,k} \sum_{i=1}^{n} (u_i - x_{ij}\hat{b}_j)^2,$$

and then update the coefficients using

$$\hat{\beta}_{j^*}^{(1)} = \hat{\beta}_{j^*}^{(0)} + \nu \hat{b}_{j^*},$$
$$\hat{\beta}_{j}^{(1)} = \hat{\beta}_{j}^{(0)}, \quad j \neq j^*.$$

Such an approach is illustrated in the right panel of Fig. 4.16. The upper right panel shows that in the beginning the algorithm moves along one of the coefficient axes for a longer time until a certain decrease in the least squares criterion is achieved. From this, it then alternates between steps in the two different directions. For the lower right panel this implies that the algorithm moves along one direction for many steps (as long as the corresponding covariate is "more informative" about the variability in the response). Finally at the very end of the algorithm, a step is made towards the coefficient that is close to zero.

This finding enables to implement implicit regularization or shrinkage of the estimated coefficients by stopping the algorithm not at the minimum of the least squares criterion, but after a fixed number of iterations. In this case of early stopping, covariates associated with small least squares estimates will drop out of the model, while all other least squares estimates will be shrunken towards the initial guesses (which typically will be $\boldsymbol{\beta}^{(0)} = \mathbf{0}$). In principle, this achieves a similar shrinkage as induced by the LASSO presented in the last section. Recall in the case of an orthonormal design, LASSO-regularized estimates were given by

$$\hat{\beta}_{\text{LASSO},j}(\lambda) = \text{sign}(\hat{\beta}_{\text{LS},j}) \left[|\hat{\beta}_{\text{LS},j}| - \frac{\lambda}{2} \right]_+ ;$$

also see Fig. 4.15. The quantity that plays a similar role to the smoothing parameter λ for the LASSO, in case of boosting, is the number of iterations. In particular, a small number of iterations will induce large regularization and will yield a sparse model with only a few coefficients differing from zero. On the other hand, choosing a large number of iterations will yield estimates that are close to the least squares estimate. As with the smoothing parameter, the optimal number of iterations will typically be determined via r-fold cross validation to minimize prediction error.

Example 4.12 Price of Used Cars—Boosting

To illustrate the application of componentwise boosting, we again consider the data set on prices of used cars with the same model specification as in Example 4.10 (p. 206). Our computations are based on a generic boosting implementation for a wide range of regression models available in the R package `mboost` (function `glmboost`). Figure 4.17 shows the resulting coefficient path as a function of the number of boosting iterations. While the upper plot contains the paths for a large number of iterations, the lower panel shows the path restricted to the optimal number of iterations (determined by tenfold cross validation).

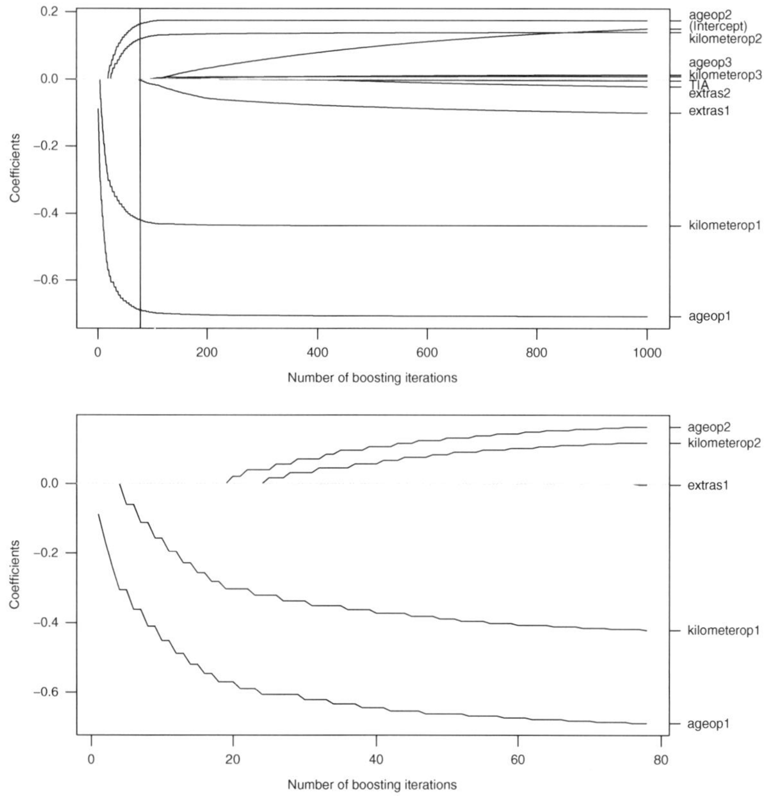

Fig. 4.17 Prices of used cars: estimated regression coefficients as a function of the boosting iterations. The *upper panel* shows the path for 1,000 boosting iterations; the *horizontal line* indicates the optimal stopping iteration determined by tenfold cross validation. The *lower panel* shows the path only until this optimal iteration

Figure 4.18 illustrates the process of determining the optimal number of iterations. For each of the cross validation folds, the prediction error is shown as a function of the boosting iterations (in grey). Averaging over the folds gives the cross validation score shown as a black line. During the first iteration the score strongly decreases but reaches a plateau after about 50 iterations. In fact, the cross validation score reaches a minimum at 78 iterations and then starts to increase again (although very slowly in our example). This indicates that the resulting fit will be rather insensitive to the choice of the boosting iterations. However, since we are interested in a sparse model with not too many nonzero coefficients, a small value will usually be preferable.

Similar to the LASSO, some variables drop out of the model when using the optimal number of boosting iterations, in particular the cubic effects of the age of the car and kilometer reading as well as the effects of *TIA* and *extras2*. These effects are also estimated to be rather small in a least squares approach, as shown in Table 4.7, and directly visible from the overplotting of several variable names that are associated with small effects in the upper panel of Fig. 4.17. Table 4.7 also contains the boosting estimates obtained after 1,000 iterations and for the optimal number of boosting iterations ($m_{\text{stop}} = 78$). While the

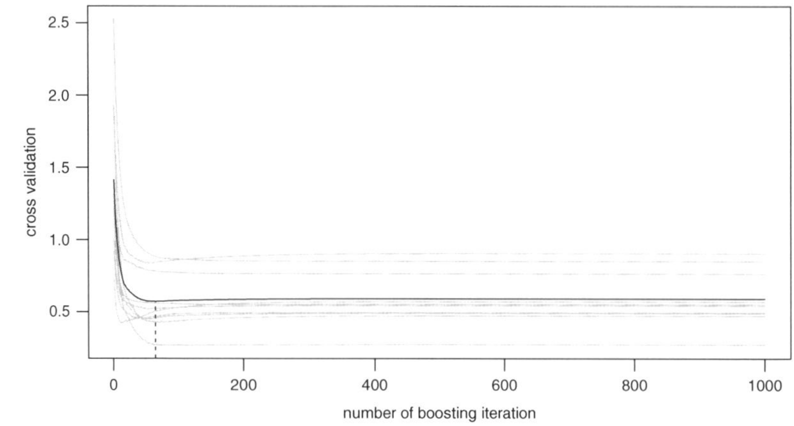

Fig. 4.18 Prices of used cars: cross validation criterion (*black line*) as a function of the boosting iterations. The *grey lines* indicate the prediction error obtained within a single cross validation fold. The *dashed vertical line* shows the optimal boosting iteration, i.e., the minimal cross validation criterion

Table 4.7 Prices of used cars: estimated coefficients for different models

Variable	LS	$m_{\text{stop}} = 1,000$	$m_{\text{stop}} = 78$
intercept	3.585	3.547	3.397
ageop1	−0.709	−0.707	−0.689
ageop2	0.172	0.175	0.163
ageop3	0.016	0.014	–
kilometerop1	−0.437	−0.436	−0.421
kilometerop2	0.142	0.140	0.118
kilometerop3	0.009	0.009	–
TIA	−0.005	−0.004	–
extras1	−0.114	−0.100	−0.003
extras2	−0.031	−0.021	–

Second column: least squares estimate, third column: boosting with large stopping iteration, fourth column: boosting with optimal stopping iteration determined by tenfold cross validation

former are rather close to the least squares estimates (and would be even closer for a further increase in the iterations), the latter are considerably shrunken towards zero.

Overall the results are again very close to the AIC optimal model of Example 3.19 and the LASSO in Example 4.11.

△

4.3.3 Generic Componentwise Boosting

To enable the generalization of boosting to more general model classes, it is advantageous to reformulate it in a somewhat more general context. Instead of relating estimation of a linear model to the determination of regression coefficients, the aim of fitting a linear model can also be interpreted as minimizing the sum of squared errors

$$\mathrm{LS}(\boldsymbol{\eta}) = \sum_{i=1}^{n}(y_i - \eta_i)^2$$

with respect to the predictor vector $\boldsymbol{\eta} = (\eta_1, \ldots, \eta_n)$. If a candidate predictor value $\hat{\boldsymbol{\eta}}^{(0)}$ is given, the corresponding lack of fit can then be evaluated with $\mathrm{LS}(\hat{\boldsymbol{\eta}}^{(0)})$, but more detailed information is contained in the unit-specific derivatives

$$u_i = \frac{\partial}{\partial \eta_i} \mathrm{LS}(\boldsymbol{\eta})\Bigg|_{\eta_i = \hat{\eta}_i^{(0)}} = 2\left(y_i - \hat{\eta}_i^{(0)}\right).$$

For a perfect fit, all these derivatives will be zero while large derivatives point towards observations where the fit could be substantially improved. In fact, the derivatives are basically the residuals obtained by plugging in the candidate predictor (multiplied with a factor of 2).

Now, an improved predictor $\hat{\boldsymbol{\eta}}^{(1)}$ is obtained by fitting a linear model to the derivatives u_i, which correspond to the residuals. This can be conceptualized by multiplying the vector of derivatives $\boldsymbol{u} = (u_1, \ldots, u_n)'$ with the hat matrix

$$\boldsymbol{H} = \boldsymbol{X}(\boldsymbol{X}'\boldsymbol{X})^{-1}\boldsymbol{X}'$$

of the linear model. Similar as in the basic algorithm discussed before, we do not update the predictor $\hat{\boldsymbol{\eta}}^{(0)}$ with the full fit obtained by applying the hat matrix but only with a fraction $0 < \nu < 1$, i.e.,

$$\hat{\boldsymbol{\eta}}^{(1)} = \hat{\boldsymbol{\eta}}^{(0)} + \nu \boldsymbol{H}\boldsymbol{u}.$$

Based on the updated predictor $\hat{\boldsymbol{\eta}}^{(1)}$, we can obtain updated derivatives, fit the linear model again, and iterate this process until finally ending up with the least squares fit for the given linear model. The advantage of the general formulation in terms of predictors instead of regression coefficients is that we can replace the least squares base-learners with any fitting routine that provides predictions for the linear model. In particular, we do not have to rely on parametric models but can instead consider regression trees or semiparametric regression techniques as described in Chap. 8.

Figure 4.19 illustrates predictor-based boosting for the model $y_i = 2x_i + \varepsilon_i$, ε_i i.i.d. $\mathrm{N}(0, 0.25)$. As starting predictor, we chose $\hat{\boldsymbol{\eta}}^{(0)} = \boldsymbol{0}$; an alternative popular choice would be the average response value, i.e., $\hat{\boldsymbol{\eta}}^{(0)} = \bar{y}\boldsymbol{1}_n$. The top left plot of Fig. 4.19 shows the data set together with the starting predictor. The top right plot shows the resulting residuals which in the first iteration are the original data. The middle row shows similar information obtained in the fourth iteration when applying a step length factor of $\nu = 0.1$ (with one intermediate fit shown in addition). The bottom row shows the results after 60 iterations. The estimated function has approximately reached the least squares fit and the residuals do no longer show any specific structure.

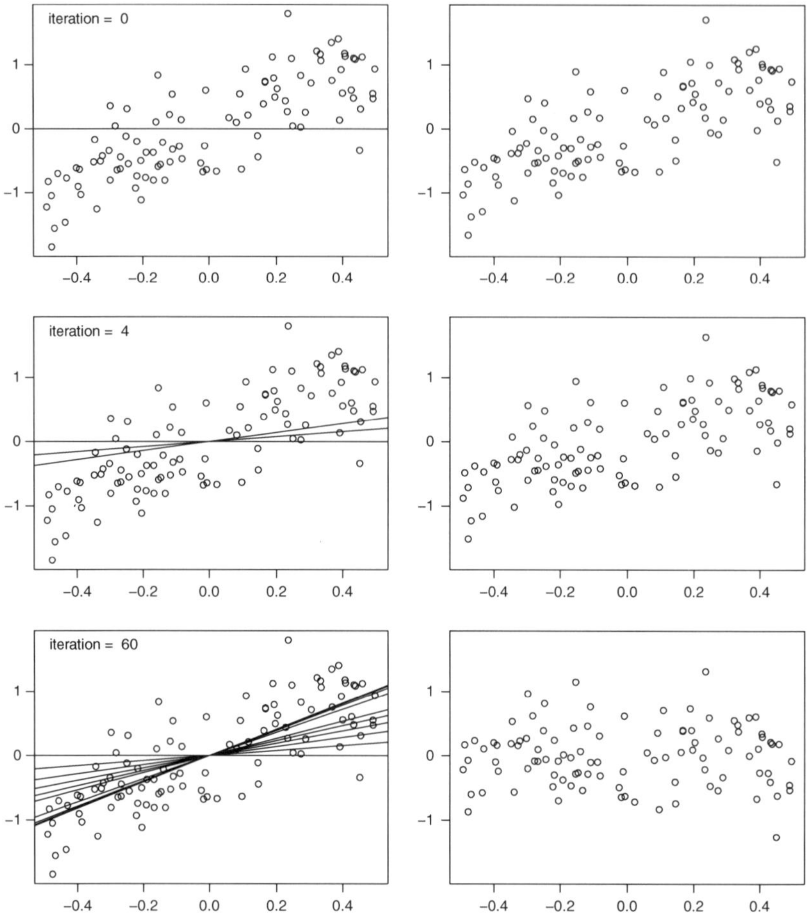

Fig. 4.19 Iterative update of model fits with boosting. The *left panel* shows the current fit (along with some intermediate *regression lines*); the *right panel* shows the corresponding residuals

If boosting only implements an alternative method for regularized estimation in linear models, why should we be interested in using it? There are actually a number of advantages:

- It is possible to replace the least squares criterion $\mathrm{LS}(\boldsymbol{\eta})$ with more general loss functions $\rho(\boldsymbol{\eta})$ such as the absolute value loss

$$\rho(\boldsymbol{\eta}) = \sum_{i=1}^{n} |y_i - \eta_i|$$

  corresponding to median regression. Therefore, boosting does not only allow to fit linear models with regularization, but also more general types of models. For example, boosting can also be used in generalized linear models for non-

Gaussian models as will be discussed in the following chapter when the negative log-likelihood is used as the loss function; see Sect. 5.7.
- We can also use more complex model components and replace the least squares base-learners with more general learning approaches such as regression trees or penalized splines. We will return to such more complex model structures in Chap. 9.
- Boosting combines regularization with implicit variable selection. This again is of particular interest in more complex regression models comprising, for example, nonlinear effects where the LASSO can no longer be used (at least not without suitable modifications).
- Boosting can also be used in situations with high-dimensional covariate vectors, where p may even exceed the sample size n. Since we only fit one linear effect at a time, the implicit regularization achieved by early stopping also allows to fit models for p large compared to n.

Finally, we want to further clarify the role of the step length factor ν in the boosting algorithm. When using a large step length, the boosting algorithm would make large steps towards the least squares estimates in each iteration. While at first glance this may seem to be desirable since it would ensure fast convergence, it actually could be problematic for two reasons: On the one hand, the regularization property of boosting would work very unevenly on the different coefficients while small steps allow to regularize all effects with a comparable strength. On the other hand, in case of correlated covariates, boosting would only be able to pick one covariate, make a large step towards the least squares estimate of the covariate, and would never include the remaining, correlated covariates. Making smaller steps in each iteration avoids this overshooting in specific directions and also allows the simultaneous inclusion of several correlated variables in the final model.

4.4 Bayesian Linear Models

This section covers linear models from a Bayesian point of view. Readers who are unfamiliar with the basic concepts of Bayesian inference should first consult Sect. B.5 in Appendix B. One advantage of the Bayesian approach, in relation to classical inference, is the possibility to consider *prior information* on the parameters in the model. Bayesian models are particularly useful to *regularize* regression problems where data information is limited. We will see, that some of the frequentist regularization techniques in Sect. 4.2, such as ridge regression and the LASSO, have a Bayesian interpretation. Another advantage of the Bayesian approach is that model choice and variable selection can be an integral part of the estimation process.

The starting point of this section is the classical linear model equipped with the conjugate normal-inverse gamma prior. This standard model described in all textbooks on Bayesian analysis will be treated in Sect. 4.4.1. Section 4.4.2 discusses Bayesian regularization techniques as a counterpart to the frequentist techniques encountered in Sect. 4.2. In Sects. 4.4.3 and 4.4.4, we discuss model choice and variable selection from a Bayesian point of view.

4.4 Generic Componentwise Boosting for Linear Models

Model

Given are observations for the model

$$y_i = \boldsymbol{x}_i \boldsymbol{\beta} + \varepsilon_i = \sum_{j=0}^{k} x_{ij}\beta_j + \varepsilon_i$$

with $x_{i0} = 1$ for the intercept.

Algorithm

1. Initialize the regression coefficients, e.g., as

$$\beta_0^{(0)} = \bar{y} \quad \text{and} \quad \beta_j^{(0)} = 0, \; j = 1, \ldots, k,$$

choose a number of iterations m_{stop} and set $t = 0$.

2. Increase t by 1. Compute the negative gradients ("residuals")

$$u_i = -\frac{\partial}{\partial \eta}\rho(y_i, \eta)\big|_{\eta=\hat{\eta}_i^{(t-1)}}, \; i = 1, \ldots, n,$$

where ρ is a general loss function, e.g., the least squares criterion, describing the estimation problem.

3. Fit separate linear models for all covariates, i.e., obtain

$$\hat{b}_j = ((\boldsymbol{x}^j)'\boldsymbol{x}^j)^{-1}(\boldsymbol{x}^j)'\boldsymbol{u}, \; j = 0, \ldots, k$$

and determine the best-fitting variable via

$$j^* = \underset{j=0,\ldots,k}{\arg\min} \sum_{i=1}^{n} (u_i - x_{ij}\hat{b}_j)^2.$$

4. Update the coefficients and the predictor:

$$\eta_i^{(t)} = \eta_i^{(t-1)} + \nu \cdot x_{ij^*}\hat{b}_{j^*}$$

$$\hat{\beta}_{j^*}^{(1)} = \hat{\beta}_{j^*}^{(0)} + \nu \hat{b}_{j^*}$$

$$\hat{\beta}_j^{(1)} = \hat{\beta}_j^{(0)}, \quad j \neq j^*.$$

5. Iterate steps 2–4 until $t = m_{\text{stop}}$.

An optimal value of m_{stop} can be determined via cross validation.

Software

R packages `mboost` and `gbm`.

4.4.1 Standard Conjugate Analysis

Our starting point is the classical linear model $\boldsymbol{y} = \boldsymbol{X}\boldsymbol{\beta} + \boldsymbol{\varepsilon}$. In contrast to classical inference, the Bayesian approach considers the unknown parameters $\boldsymbol{\beta}$ and σ^2 as random variables. Thus, the distribution of the response $\boldsymbol{y}$ can be understood as conditional on the parameters $\boldsymbol{\beta}$ and σ^2, and we obtain the observation model

$$\boldsymbol{y} \mid \boldsymbol{\beta}, \sigma^2 \sim \mathrm{N}(\boldsymbol{X}\boldsymbol{\beta}, \sigma^2 \boldsymbol{I}).$$

We next introduce the classical conjugate prior distribution in Bayesian linear models.

Normal-Inverse Gamma Prior

The standard conjugate prior for linear models described in all introductory textbooks on Bayesian inference is obtained by assuming a multivariate normal prior for the regression coefficients

$$\boldsymbol{\beta} \mid \sigma^2 \sim \mathrm{N}(\boldsymbol{m}, \sigma^2 \boldsymbol{M}),$$

with known expectation $\boldsymbol{m}$ and covariance matrix $\boldsymbol{M}$, e.g., $\boldsymbol{m} = \boldsymbol{0}$ and $\boldsymbol{M} = \boldsymbol{I}$. A normal distribution seems a natural choice as the distribution of the estimated regression coefficients in the classical linear model is (approximately) multivariate normal. For σ^2, we specify an inverse gamma distribution with hyperparameters a and b, i.e.,

$$\sigma^2 \sim \mathrm{IG}(a, b). \tag{4.12}$$

To shed light on the specific form of the inverse gamma prior for σ^2, Fig. 4.20 shows the prior density for various choices of a and b. Of particular interest is the case $a = b$ and both values approaching zero. Then the distribution of $\log \sigma^2$ tends to a uniform distribution as can be shown analytically through the change in variables theorem; see Theorem B.1 of Appendix B.1. This is why small values for a and b are identified with a weakly informative or noninformative prior. We will further elaborate on this point below when we discuss noninformative priors in the linear model.

The joint prior for $\boldsymbol{\beta}$ and σ^2 is a normal-inverse gamma distribution with density

$$\begin{aligned} p(\boldsymbol{\beta}, \sigma^2) &= p(\boldsymbol{\beta} \mid \sigma^2)\, p(\sigma^2) \\ &= \frac{1}{(2\pi)^{\frac{p}{2}} (\sigma^2)^{\frac{p}{2}} |\boldsymbol{M}|^{\frac{1}{2}}} \exp\left(-\frac{1}{2\sigma^2}(\boldsymbol{\beta} - \boldsymbol{m})' \boldsymbol{M}^{-1} (\boldsymbol{\beta} - \boldsymbol{m})\right) \\ &\quad \frac{b^a}{\Gamma(a)} \frac{1}{(\sigma^2)^{a+1}} \exp\left(-\frac{b}{\sigma^2}\right) \end{aligned} \tag{4.13}$$

and parameters $\boldsymbol{m}$, $\boldsymbol{M}$, a, and b; see also Definition B.3.5 in Appendix B.3. We write $\boldsymbol{\beta}, \sigma^2 \sim \mathrm{NIG}(\boldsymbol{m}, \boldsymbol{M}, a, b)$. Ignoring all factors in Eq. (4.13) that are independent of $\boldsymbol{\beta}$ and σ^2, we obtain

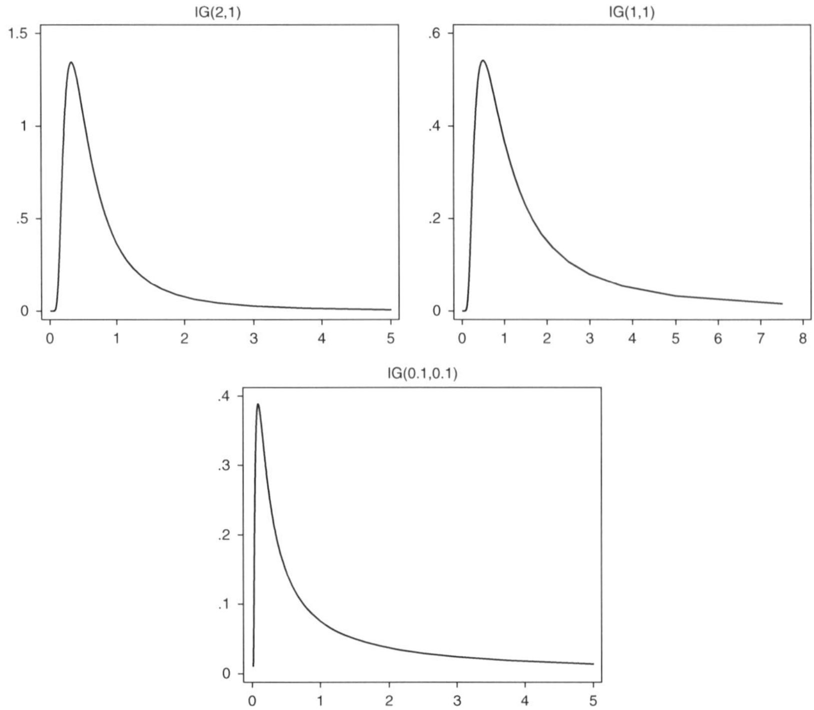

Fig. 4.20 Inverse gamma prior density for σ^2 for various values of the hyperparameters a and b

$$p(\boldsymbol{\beta}, \sigma^2) \propto \frac{1}{(\sigma^2)^{\frac{p}{2}+a+1}} \exp\left(-\frac{1}{2\sigma^2}(\boldsymbol{\beta} - \boldsymbol{m})'\boldsymbol{M}^{-1}(\boldsymbol{\beta} - \boldsymbol{m}) - \frac{b}{\sigma^2}\right) \qquad (4.14)$$

for the density of the normal-inverse gamma prior.

We now derive some useful properties of the general NIG($\boldsymbol{m}, \boldsymbol{M}, a, b$) prior. For convenience, these are summarized in Definition B.3.5 (p. 652) of Appendix B. Readers who are not interested in the technical details can safely skip this more difficult part.

From the properties of the multivariate normal distribution and the inverse gamma distribution (see Definition B.12 in Appendix B), it follows that

$$\mathrm{E}(\boldsymbol{\beta} \mid \sigma^2) = \boldsymbol{m} \qquad \mathrm{Cov}(\boldsymbol{\beta} \mid \sigma^2) = \sigma^2 \boldsymbol{M}$$

and

$$\mathrm{E}(\sigma^2) = \frac{b}{a-1} \quad \text{if a} > 1 \qquad \mathrm{Var}(\sigma^2) = \frac{b^2}{(a-1)^2(a-2)} \quad \text{if } a > 2.$$

We next derive the unconditional prior distribution of $\boldsymbol{\beta}$. Since the conditional mean of $\boldsymbol{\beta}$ is independent of σ^2, we obtain the unconditional mean

$$\mathrm{E}(\boldsymbol{\beta}) = \mathrm{E}(\mathrm{E}(\boldsymbol{\beta} \mid \sigma^2)) = \boldsymbol{m},$$

using the law of iterated expectations (see Theorem B.2. 9 in Appendix B.2). For the unconditional covariance matrix, we have

$$\mathrm{Cov}(\boldsymbol{\beta}) = \mathrm{E}(\mathrm{Cov}(\boldsymbol{\beta} \mid \sigma^2)) + \mathrm{Cov}(\mathrm{E}(\boldsymbol{\beta} \mid \sigma^2)) = \mathrm{E}(\sigma^2)\boldsymbol{M} = \frac{b}{a-1}\boldsymbol{M};$$

see Theorem B.2. 10 in Appendix B.2.

In order to derive the specific distribution of $\boldsymbol{\beta}$, it is useful to determine the conditional distribution of $\sigma^2 \mid \boldsymbol{\beta}$ first. The density of this distribution is proportional to the joint distribution (4.14) of $\boldsymbol{\beta}, \sigma^2$, which yields

$$\begin{aligned} p(\sigma^2 \mid \boldsymbol{\beta}) &\propto p(\boldsymbol{\beta}, \sigma^2) \\ &\propto \frac{1}{(\sigma^2)^{a+\frac{p}{2}+1}} \exp\left(-\frac{1}{\sigma^2}\left(\tfrac{1}{2}(\boldsymbol{\beta}-\boldsymbol{m})'\boldsymbol{M}^{-1}(\boldsymbol{\beta}-\boldsymbol{m}) + b\right)\right). \end{aligned} \tag{4.15}$$

This is the form of an inverse gamma distribution, more specifically

$$\sigma^2 \mid \boldsymbol{\beta} \sim \mathrm{IG}\left(a + \frac{p}{2}, b + \frac{1}{2}(\boldsymbol{\beta}-\boldsymbol{m})'\boldsymbol{M}^{-1}(\boldsymbol{\beta}-\boldsymbol{m})\right).$$

Hence the normalizing constant in Eq. (4.15) is given by

$$c = \frac{(b + \frac{1}{2}(\boldsymbol{\beta}-\boldsymbol{m})'\boldsymbol{M}^{-1}(\boldsymbol{\beta}-\boldsymbol{m}))^{a+\frac{p}{2}}}{\Gamma\left(a+\frac{p}{2}\right)}.$$

Turning again our attention to the unconditional distribution of $\boldsymbol{\beta}$, we can integrate Eq. (4.14) with respect to σ^2. Since the integral of the density of $\sigma^2 \mid \boldsymbol{\beta}$ must equal one, it follows that

$$\begin{aligned} &\int \frac{1}{(\sigma^2)^{a+\frac{p}{2}+1}} \exp\left(-\frac{1}{2\sigma^2}(\boldsymbol{\beta}-\boldsymbol{m})'\boldsymbol{M}^{-1}(\boldsymbol{\beta}-\boldsymbol{m})\right) \exp(-\frac{b}{\sigma^2})\, d\sigma^2 \\ &= \int c^{-1} p(\sigma^2 \mid \boldsymbol{\beta}) d\sigma^2 \\ &= (b + \frac{1}{2}(\boldsymbol{\beta}-\boldsymbol{m})'\boldsymbol{M}^{-1}(\boldsymbol{\beta}-\boldsymbol{m}))^{-(a+\frac{p}{2})}\Gamma\left(a+\frac{p}{2}\right). \end{aligned}$$

We now have

$$p(\boldsymbol{\beta}) \propto \left(b + \tfrac{1}{2}(\boldsymbol{\beta} - \boldsymbol{m})'\boldsymbol{M}^{-1}(\boldsymbol{\beta} - \boldsymbol{m})\right)^{-(a+\frac{p}{2})} \Gamma\left(a + \tfrac{p}{2}\right)$$
$$\propto \left(1 + \tfrac{1}{2a}(\boldsymbol{\beta} - \boldsymbol{m})'(\tfrac{b}{a}\boldsymbol{M})^{-1}(\boldsymbol{\beta} - \boldsymbol{m})\right)^{-(a+\frac{p}{2})}.$$

This is the density of a multivariate t-distribution with $2a$ degrees of freedom, having location parameter $\boldsymbol{m}$, and dispersion matrix $b/a\boldsymbol{M}$, i.e., $\boldsymbol{\beta} \sim \mathrm{t}(2a, \boldsymbol{m}, b/a\boldsymbol{M})$. Details of the multivariate t-distribution can be found in Appendix B.3.4 (p. 651).

Eliciting the NIG$(\boldsymbol{m}, \boldsymbol{M}, a, b)$ prior is a difficult task in practice as there are four parameters to choose (including the prior covariance matrix). Full specification of $\boldsymbol{m}$ and $\boldsymbol{M}$ is typically only possible if there is prior knowledge, e.g., in the form of past results. One then could use the posterior parameters of the previous analysis as the prior values for the new analysis; see Example 4.13 (p. 235) below.

Zellner's g-Prior

In the absence of valid prior knowledge, the choice of the prior parameters is more difficult, if not impossible. For the prior mean a possible choice is $\boldsymbol{m} = (m_1, 0, \ldots, 0)'$ where m_1 is a prior guess for the overall level. Assuming zero means for the other components of $\boldsymbol{\beta}$ reflects the assumption that a priori the effect of the covariates is centered around zero. More difficult is the specification of the prior correlation structure. A widely used choice is Zellner's g-prior (Zellner, 1986)

$$\boldsymbol{\beta} \,|\, \sigma^2 \sim \mathrm{N}(\boldsymbol{m}, \sigma^2(g\boldsymbol{X}'\boldsymbol{X})^{-1}),$$

where $g > 0$ is a hyperparameter. This is a special case of the NIG$(\boldsymbol{m}, \boldsymbol{M}, a, b)$ prior with $\boldsymbol{M} = (g\boldsymbol{X}'\boldsymbol{X})^{-1}$. The prior assumes that the prior precision $\boldsymbol{M}^{-1}$ is a fraction of the precision $\boldsymbol{X}'\boldsymbol{X}$ in the data. Choices for g are discussed, e.g., in Fernandez, Ley, and Steel (2001), and include the following:

- The choice $g = 1/n$ corresponds to assigning roughly the same information as is contained in one observation.
- The choice $g = 1/k^2$ (where k is the number of regressors) is suggested by the risk inflation criterion of Foster and George (1994). They consider the predictive risk function $R(\boldsymbol{\beta}, \hat{\boldsymbol{\beta}}) = \mathrm{E}((\mathbf{X}\hat{\boldsymbol{\beta}} - \mathbf{X}\boldsymbol{\beta})'(\mathbf{X}\hat{\boldsymbol{\beta}} - \mathbf{X}\boldsymbol{\beta}))$, i.e., the expected squared difference between the true, but unknown, mean $\boldsymbol{X}\boldsymbol{\beta}$ and the estimate $\boldsymbol{X}\hat{\boldsymbol{\beta}}$. The risk inflation criterion is defined as

$$\mathrm{RIC} = \sup_{\boldsymbol{\beta}} \frac{R\left(\boldsymbol{\beta}, \hat{\boldsymbol{\beta}}_{working}\right)}{R\left(\boldsymbol{\beta}, \hat{\boldsymbol{\beta}}_{true}\right)},$$

  where $\hat{\boldsymbol{\beta}}_{true}$ is the least squares estimator obtained if we estimate the correctly specified model and $\hat{\boldsymbol{\beta}}_{working}$ is the least squares estimator of a working model that is not necessarily correct (the typical situation in practice). Thus the risk inflation criterion is the maximum of the ratio between the risk of the working model and the risk of the correct model. George and Foster (2000) showed that in a model with Zellner's g-prior and known variance σ^2, the selection of the model with highest posterior probability is equivalent to maximizing the RIC provided that we choose $g = 1/k^2$.

- A compromise between the two choices is given by $g = 1/\max\{n, k^2\}$. This choice is proposed in Fernandez et al. (2001) based on theoretical reasoning and simulation results.

Noninformative Prior

An alternative to Zellner's g-prior is to use fully noninformative priors. As pointed out in Appendix B.5.1, the construction of noninformative priors is a quite delicate task. The widely accepted noninformative prior in the linear model is given by

$$p(\boldsymbol{\beta}, \sigma^2) \propto \frac{1}{\sigma^2}, \tag{4.16}$$

which is the reference prior that maximizes the expected Kullback–Leibler distance of the posterior distribution relative to the prior. Informally, the reference prior can be characterized as the distribution that maximizes the influence of the data on the posterior. Since the density (4.16) cannot be normalized such that it integrates to one, it is an improper prior. The prior can be expressed as the product between a uniform prior $p(\boldsymbol{\beta}) \propto 1$ for $\boldsymbol{\beta}$ and the prior $p(\sigma^2) \propto 1/\sigma^2$ for σ^2 so that $\boldsymbol{\beta}$ and σ^2 are a priori stochastically independent. Note that the prior for σ^2 is equivalent to a uniform prior for $\log(\sigma^2)$. Technically, we can identify the noninformative prior (4.16) with the conjugate $\text{NIG}(\boldsymbol{m}, \boldsymbol{M}, a, b)$ prior by setting $\boldsymbol{m} = \boldsymbol{0}$, $\boldsymbol{M}^{-1} = \boldsymbol{0}$, $a = -p$, and $b = 0$. This is useful for posterior analysis because we can treat the noninformative case within the standard prior. Note, however, that we have to be very careful when proceeding this way. When dealing with improper priors, it is important to check whether the resulting posterior is truly proper. For the noninformative prior (4.16) this is indeed the case.

Another approach to define a noninformative prior is described in O'Hagan (1994). Here we start with the marginal $\text{IG}(a, b)$ distribution for σ^2. If $a \to 0$ and $b \to 0$ tend to zero we obtain

$$p(\sigma^2) \propto \frac{1}{\sigma^2}.$$

The prior corresponds to a uniform distribution for $\log \sigma^2$. For the joint $NIG(\boldsymbol{m}, \boldsymbol{M}, a, b)$ prior, we then have

$$p(\boldsymbol{\beta}, \sigma^2) \propto \frac{1}{(\sigma^2)^{\frac{p}{2}+1}} \exp\left(-\frac{1}{2\sigma^2}(\boldsymbol{\beta} - \boldsymbol{m})'\boldsymbol{M}^{-1}(\boldsymbol{\beta} - \boldsymbol{m})\right).$$

If rather $\boldsymbol{M}^{-1} = \boldsymbol{0}$, we arrive at the alternative noninformative prior

$$p(\boldsymbol{\beta}, \sigma^2) \propto \sigma^{-(p+2)}.$$

This can be shown to be Jeffreys' prior. Although Jeffreys' prior is usually not used in multiparameter settings, our derivation justifies the widely used choice of a and b as equal and near zero as a weakly informative choice for the prior of σ^2 (and more generally variance parameters).

Posterior Analysis

Bayesian inference is based on properties of the posterior distribution, i.e., on the conditional distribution of the unknown parameters $\boldsymbol{\beta}$ and σ^2 given the data $\boldsymbol{y}$. The density of the posterior distribution is proportional to the product of the likelihood and the prior distribution. Hence, we obtain

$$\begin{aligned} p(\boldsymbol{\beta}, \sigma^2 \mid \boldsymbol{y}) &\propto L(\boldsymbol{\beta}, \sigma^2)\, p(\boldsymbol{\beta} \mid \sigma^2)\, p(\sigma^2) \\ &\propto \frac{1}{(\sigma^2)^{\frac{n}{2}}} \exp\left(-\frac{1}{2\sigma^2}(\boldsymbol{y} - \boldsymbol{X}\boldsymbol{\beta})'(\boldsymbol{y} - \boldsymbol{X}\boldsymbol{\beta})\right) \\ &\quad \frac{1}{(\sigma^2)^{\frac{p}{2}}} \exp\left(-\frac{1}{2\sigma^2}(\boldsymbol{\beta} - \boldsymbol{m})'\boldsymbol{M}^{-1}(\boldsymbol{\beta} - \boldsymbol{m})\right) \\ &\quad \frac{1}{(\sigma^2)^{a+1}} \exp\left(-\frac{b}{\sigma^2}\right). \end{aligned} \tag{4.17}$$

Technically, we can determine the properties of the posterior distribution mainly in two ways. On the one hand, we can derive the posterior distribution and its properties analytically. Alternatively, we can draw a random sample from the posterior distribution based on Markov chain Monte Carlo (MCMC) simulation methods. In most cases, an analytical derivation of the posterior distribution is impossible and only the time-consuming sampling-based approach is possible. The linear model is one of a few examples, in which the posterior distribution is analytically tractable, at least for the standard NIG($\boldsymbol{m}, \boldsymbol{M}, a, b$) prior. We can show that the posterior distribution, like the prior distribution, is a normal-inverse gamma distribution. The parameters $\tilde{\boldsymbol{m}}$, $\tilde{\boldsymbol{M}}$, $\tilde{a}$, and $\tilde{b}$ of the distribution can be found in Box 4.5; see Sect. 4.5.2, p. 260, for a derivation. Additionally, the box contains conditional and marginal posteriors of the parameters as well as other quantities of interest. They can be obtained directly from the properties of the normal-inverse gamma distribution derived on p. 228 and summarized in Definition B.3.5 (p. 652) of Appendix B.

Of particular interest is the posterior mean

$$\hat{\boldsymbol{\beta}}_B = \mathrm{E}(\boldsymbol{\beta} \mid \boldsymbol{y}) = \tilde{\boldsymbol{m}} = (\boldsymbol{X}'\boldsymbol{X} + \boldsymbol{M}^{-1})^{-1}(\boldsymbol{M}^{-1}\boldsymbol{m} + \boldsymbol{X}'\boldsymbol{y})$$

4.5 Bayesian Linear Model with Conjugate Prior

Observation Model and Prior Distribution

1. *Observation model:* $y \mid \boldsymbol{\beta}, \sigma^2 \sim \mathrm{N}(\boldsymbol{X}\boldsymbol{\beta}, \sigma^2 \boldsymbol{I})$.
2. *Prior distribution:* $\boldsymbol{\beta} \mid \sigma^2 \sim \mathrm{N}(\boldsymbol{m}, \sigma^2 \boldsymbol{M})$ and $\sigma^2 \sim \mathrm{IG}(a, b)$.

Posterior

$\boldsymbol{\beta}, \sigma^2 \mid \boldsymbol{y} \sim \mathrm{NIG}(\tilde{\boldsymbol{m}}, \tilde{\boldsymbol{M}}, \tilde{a}, \tilde{b})$ with parameters

$$\tilde{\boldsymbol{M}} = (\boldsymbol{X}'\boldsymbol{X} + \boldsymbol{M}^{-1})^{-1} \qquad \tilde{\boldsymbol{m}} = \tilde{\boldsymbol{M}}(\boldsymbol{M}^{-1}\boldsymbol{m} + \boldsymbol{X}'\boldsymbol{y}),$$

and

$$\tilde{a} = a + \frac{n}{2} \qquad \tilde{b} = b + \frac{1}{2}\left(\boldsymbol{y}'\boldsymbol{y} + \boldsymbol{m}'\boldsymbol{M}^{-1}\boldsymbol{m} - \tilde{\boldsymbol{m}}'\tilde{\boldsymbol{M}}^{-1}\tilde{\boldsymbol{m}}\right).$$

- The conditional posterior distribution of $\boldsymbol{\beta}$ given σ^2 is $\boldsymbol{\beta} \mid \sigma^2, \boldsymbol{y} \sim \mathrm{N}(\tilde{\boldsymbol{m}}, \sigma^2 \tilde{\boldsymbol{M}})$.
- The marginal posterior of $\boldsymbol{\beta}$ is $\boldsymbol{\beta} \mid \boldsymbol{y} \sim \mathrm{t}(2\tilde{a}, \tilde{\boldsymbol{m}}, \tilde{b}/\tilde{a}\tilde{\boldsymbol{M}})$.

Posterior with Noninformative Prior

In case of a noninformative prior with $\boldsymbol{m} = \boldsymbol{0}$, $\boldsymbol{M}^{-1} = \boldsymbol{0}$, $a = -p$, and $b = 0$, we obtain

$$\tilde{\boldsymbol{M}} = (\boldsymbol{X}'\boldsymbol{X})^{-1} \qquad \tilde{\boldsymbol{m}} = (\boldsymbol{X}'\boldsymbol{X})^{-1}\boldsymbol{X}'\boldsymbol{y} = \hat{\boldsymbol{\beta}}_{\mathrm{LS}}$$

and

$$\tilde{a} = -p + \frac{n}{2} \qquad \tilde{b} = \frac{1}{2}\left(\boldsymbol{y}'\boldsymbol{y} - \hat{\boldsymbol{\beta}}_{\mathrm{LS}}'\boldsymbol{X}'\boldsymbol{X}\hat{\boldsymbol{\beta}}_{\mathrm{LS}}\right).$$

- The conditional posterior distribution of $\boldsymbol{\beta}$ given σ^2 is $\boldsymbol{\beta} \mid \sigma^2, \boldsymbol{y} \sim \mathrm{N}(\hat{\boldsymbol{\beta}}_{\mathrm{LS}}, \sigma^2(\boldsymbol{X}'\boldsymbol{X})^{-1})$.
- The marginal posterior of $\boldsymbol{\beta}$ is $\boldsymbol{\beta} \mid \boldsymbol{y} \sim \mathrm{t}(2\tilde{a}, \hat{\boldsymbol{\beta}}_{\mathrm{LS}}, \tilde{b}/\tilde{a}\tilde{\boldsymbol{M}})$.

Posterior Mean

$$\hat{\boldsymbol{\beta}}_B = \mathrm{E}(\boldsymbol{\beta} \mid \boldsymbol{y}) = \tilde{\boldsymbol{m}} = (\boldsymbol{X}'\boldsymbol{X} + \boldsymbol{M}^{-1})^{-1}(\boldsymbol{M}^{-1}\boldsymbol{m} + \boldsymbol{X}'\boldsymbol{y}).$$

For a noninformative prior the posterior mean coincides with the least squares estimator.

Software

- Functions `bayesLMRef` and `bayesLMConjugate` of the R package `spBayes`.
- Software package `BayesX` for noninformative priors only (see also the R interface `R2BayesX`).
- Function `zlm` of the R package `BMS` (Zellner's g-prior only).

as a point estimate of $\boldsymbol{\beta}$. Using the matrix $\boldsymbol{A} = (\boldsymbol{X}'\boldsymbol{X} + \boldsymbol{M}^{-1})^{-1}\boldsymbol{X}'\boldsymbol{X}$, we can write the Bayes estimator as a weighted average of the prior expectation $\boldsymbol{m}$ and the least squares estimator $\hat{\boldsymbol{\beta}}$:

$$\hat{\boldsymbol{\beta}}_B = (\boldsymbol{I} - \boldsymbol{A})\boldsymbol{m} + \boldsymbol{A}\hat{\boldsymbol{\beta}}.$$

To interpret the Bayes estimator, note that the diagonal elements of $\boldsymbol{M}$ contain (up to the factor σ^2) the prior variances of $\boldsymbol{\beta}$. The greater the diagonal elements of $\boldsymbol{M}$ (i.e., the variances of $\boldsymbol{\beta}$), the smaller are the elements of $\boldsymbol{M}^{-1}$. In the limit $\boldsymbol{M}^{-1} \to \boldsymbol{0}$ the matrix $\boldsymbol{A}$ approaches the identity matrix and $\hat{\boldsymbol{\beta}}_B$ the ordinary least squares estimator. On the contrary, small elements in $\boldsymbol{M}$ (corresponding to small variances of $\boldsymbol{\beta}$) imply that the matrix $\boldsymbol{A}$ approaches the zero matrix and $\boldsymbol{I} - \boldsymbol{A}$ the identity matrix. The Bayes estimator is then identical with the prior mean $\boldsymbol{m}$. This gives us the following interpretation of $\hat{\boldsymbol{\beta}}_B$: The smaller the prior information about $\boldsymbol{\beta}$, i.e., the greater the diagonal elements of $\boldsymbol{M}$, the closer is $\hat{\boldsymbol{\beta}}_B$ to the least squares estimator. The larger the prior information, i.e., the smaller the diagonal elements of $\boldsymbol{M}$, the more the prior mean $\boldsymbol{m}$ dominates $\hat{\boldsymbol{\beta}}_B$. For the noninformative prior with $\boldsymbol{m} = \boldsymbol{0}$ and $\boldsymbol{M}^{-1} = \boldsymbol{0}$ the Bayes estimator coincides with the least squares estimator, i.e.,

$$\hat{\boldsymbol{\beta}}_B = (\boldsymbol{X}'\boldsymbol{X})^{-1}\boldsymbol{X}'\boldsymbol{y} = \hat{\boldsymbol{\beta}}_{\text{LS}}.$$

Full Conditional Densities and MCMC Inference

Since the linear model is often a building block for more complex models where analytical posterior analysis is impossible, we additionally discuss simulation-based inference using MCMC methods. We develop a Gibbs sampler (see Appendix B.5.3 for an introduction), that consecutively draws random numbers from the full conditional distributions of $\boldsymbol{\beta}$ and σ^2.

We start by deriving the full conditional of $\boldsymbol{\beta}$. The density is proportional to the density (4.17) of the posterior distribution, and we can disregard all factors which are independent of $\boldsymbol{\beta}$. In the appendix of this chapter on p. 211, we derive the identity

$$(\boldsymbol{y} - \boldsymbol{X}\boldsymbol{\beta})'(\boldsymbol{y} - \boldsymbol{X}\boldsymbol{\beta}) = (\boldsymbol{\beta} - \hat{\boldsymbol{\beta}})'\boldsymbol{X}'\boldsymbol{X}(\boldsymbol{\beta} - \hat{\boldsymbol{\beta}}) + \boldsymbol{y}'(\boldsymbol{I}_n - \boldsymbol{X}(\boldsymbol{X}'\boldsymbol{X})^{-1}\boldsymbol{X}')\boldsymbol{y} \quad (4.18)$$

for the least squares criterion. Since the second summand does not depend on $\boldsymbol{\beta}$, we obtain

$$\begin{aligned} p(\boldsymbol{\beta} \mid \cdot) &\propto \exp\left(-\tfrac{1}{2\sigma^2}(\boldsymbol{y} - \boldsymbol{X}\boldsymbol{\beta})'(\boldsymbol{y} - \boldsymbol{X}\boldsymbol{\beta})\right) \exp\left(-\tfrac{1}{2\sigma^2}(\boldsymbol{\beta} - \boldsymbol{m})'\boldsymbol{M}^{-1}(\boldsymbol{\beta} - \boldsymbol{m})\right) \\ &\propto \exp\left(-\tfrac{1}{2\sigma^2}(\boldsymbol{\beta} - \hat{\boldsymbol{\beta}})'\boldsymbol{X}'\boldsymbol{X}(\boldsymbol{\beta} - \hat{\boldsymbol{\beta}})\right) \exp\left(-\tfrac{1}{2\sigma^2}(\boldsymbol{\beta} - \boldsymbol{m})'\boldsymbol{M}^{-1}(\boldsymbol{\beta} - \boldsymbol{m})\right) \end{aligned}$$

for the full conditional of $\boldsymbol{\beta}$. This is the form of a multivariate normal distribution; see Appendix B, Theorem B.4. According to the theorem, the covariance matrix is given by

$$\Sigma_\beta = \left(\frac{1}{\sigma^2} X'X + \frac{1}{\sigma^2} M^{-1}\right)^{-1}. \tag{4.19}$$

For the mean μ_β, we obtain

$$\mu_\beta = \Sigma_\beta \left(\frac{1}{\sigma^2} X'y + \frac{1}{\sigma^2} M^{-1} m\right). \tag{4.20}$$

In summary, we have $\beta \mid \cdot \sim \mathrm{N}\left(\mu_\beta, \Sigma_\beta\right)$.

Similarly, the full conditional distribution of σ^2 can be derived. We have

$$\begin{aligned} p(\sigma^2 \mid \cdot) &\propto \frac{1}{(\sigma^2)^{\frac{n}{2}}} \exp(-\tfrac{1}{2\sigma^2}(y - X\beta)'(y - X\beta)) \\ &\quad \frac{1}{(\sigma^2)^{\frac{p}{2}+a+1}} \exp\left(-\tfrac{1}{2\sigma^2}(\beta - m)'M^{-1}(\beta - m) - \tfrac{b}{\sigma^2}\right) \\ &= \frac{1}{(\sigma^2)^{a+\frac{n}{2}+\frac{p}{2}+1}} \\ &\quad \exp\left(-\tfrac{1}{\sigma^2}\left(b + \tfrac{1}{2}(y - X\beta)'(y - X\beta) + \tfrac{1}{2}(\beta - m)'M^{-1}(\beta - m)\right)\right). \end{aligned}$$

This is the form of an inverse gamma distribution with parameters

$$a' = a + \frac{n}{2} + \frac{p}{2} \tag{4.21}$$

and

$$b' = b + \frac{1}{2}(y - X\beta)'(y - X\beta) + \frac{1}{2}(\beta - m)'M^{-1}(\beta - m). \tag{4.22}$$

Summarizing, we obtain the following Gibbs sampler:

1. Define initial values $\beta^{(0)}$ and $(\sigma^2)^{(0)}$. Set $t = 1$.
2. Sample $\beta^{(t)}$ by drawing from the Gaussian full conditional with covariance matrix (4.19) and mean (4.20). Replace in (4.19) and (4.20) σ^2 by the current state of the chain $(\sigma^2)^{(t-1)}$.
3. Sample $(\sigma^2)^{(t)}$ by drawing from the inverse gamma full conditional with parameters a' and b' given by Eqs. (4.21) and (4.22). Replace in Eqs. (4.21) and (4.22) β by the current state of the chain $\beta^{(t)}$.
4. Stop if $t = T$, otherwise set $t = t + 1$ and go to 2.

Example 4.13 Munich Rent Index—Quality of Kitchen

In Example 3.7 (p. 100), we were confronted with the problem to update the Munich rent index 1999 with new data collected in 2001. Recall that we constructed a two period model with an interaction between the quality of kitchen and the data collection period. In this example, we follow a Bayesian approach to cope with the update problem. For illustration, we again use model (3.7) of Example 3.5 (p. 90) augmented by the kitchen dummy variables. Hence, the model consists of the transformed living area $1/area$, a cubic polynomial for year of construction, and the two kitchen dummies *nkitchen* ("normal kitchen") and *pkitchen* ("premium kitchen"). We first develop a Bayesian version of the

Table 4.8 Munich rent index: estimation results based on the noninformative prior (4.16) for the parameters

Variable	Coefficient	Standard deviation	2.5 % Quantile	97.5 % Quantile
invarea	122.5,417	5.5,877	−111.5,955	−133.7,277
yearc	−0.0861	0.0351	−0.1,549	−0.0174
yearc2	0.0015	0.0007	0.0002	0.0028
yearc3	0.0000	0.0000	0.0000	0.0000
nkitchen	0.9274	0.1258	0.6770	1.1840
pkitchen	1.1022	0.1873	−0.7410	1.4718

Results are based on the data collected in 1999

model based solely on the data for 1999. At that time we have no prior information regarding the unknown parameters. We therefore use the noninformative prior (4.16). Although the model could in principle be estimated analytically using the results of Box 4.5, we used the Gibbs sampler outlined on p. 234 for inference. The reason is that the available software packages usually do not support the analytical solution. Using the function `bayesLMRef` of the R package `spBayes`, we obtained the results of Table 4.8. Up to sampling imprecision, the posterior mean is identical to the least squares estimator (as suggested by the analytically derived posterior). The posterior standard deviations and the quantiles are also very close to the respective least squares standard errors and the 95 % confidence intervals. Note that the analytical posterior mean (and mode) coincides exactly with the least squares estimator, while the posterior standard deviation and quantiles are slightly different from their least squares counterpart.

Turning our attention to the new data collected in 2001 for the rent index update, it seems natural to use the posterior values obtained with the data from 1999 as prior information for the new analysis. That is we estimate a Bayesian linear model using the data collected in 2001 together with a NIG$(\boldsymbol{m}, \boldsymbol{M}, a, b)$ prior with parameters derived from the posterior obtained with the 1999 data. Clearly, $\boldsymbol{m}$ should be the empirical mean vector of the sampled regression coefficients in the MCMC sampler for the 1999 data. Since the prior covariance matrix is given by $\sigma^2\boldsymbol{M}$, we set $\boldsymbol{M} = 1/\hat{\sigma}^2\boldsymbol{S}$, where $\boldsymbol{S}$ is the empirical covariance matrix of the MCMC samples for the 1999 data and $\hat{\sigma}^2$ the empirical mean of the samples for σ^2. To obtain prior values for a and b, we note that the mean and variance of the IG(a, b) prior for σ^2 are given by

$$\begin{aligned} \mathrm{E}(\sigma^2) &= \frac{b}{a-1}, \\ \mathrm{Var}(\sigma^2) &= \frac{b^2}{(a-1)^2(a-2)} = \mathrm{E}(\sigma^2)^2\frac{1}{a-2}. \end{aligned}$$

Solving for a and b yields

$$\begin{aligned} a &= \frac{\mathrm{E}(\sigma^2) + 2\mathrm{Var}(\sigma^2)}{\mathrm{Var}(\sigma^2)}, \\ b &= (a-1)\mathrm{E}(\sigma^2). \end{aligned}$$

Based on the analysis for the 1999 data, we can now replace $\mathrm{E}(\sigma^2)$ and $\mathrm{Var}(\sigma^2)$ by their posterior estimates $\hat{\sigma}^2$ and $s^2_{\hat{\sigma}^2}$ to obtain prior values for a and b.

Using these values for the NIG$(\boldsymbol{m}, \boldsymbol{M}, a, b)$ prior, we arrive at the results given in Table 4.9. The table is obtained from the function `bayesLMConjugate` in the R package `spBayes`. For comparison, we additionally included the least squares estimates for the

Table 4.9 Munich rent index: comparison of the Bayes estimate with informative prior and the least squares estimate for the data collected in 2001

	LS 2001		Bayes 2001	
Variable	Coeff.	Std	Coeff.	Std
invarea	125.8373	7.2360	124.3540	3.3618
yearc	−0.0335	0.0480	−0.0631	0.0208
yearc2	0.0004	0.0009	0.0010	0.0004
yearc3	0.0000	0.0000	0.0000	0.0000
nkitchen	1.2944	0.1701	1.0327	0.0795
pkitchen	1.7935	0.2714	1.2910	0.1193

2001 data. In particular for the kitchen dummies, the least squares estimates for the 2001 data differ considerably from those for the 1999 data (see Table 4.8). The Bayes estimator is a compromise between the least squares results for the 1999 and the 2001 data. Although the new data based on 2001 have an impact on the estimates for *nkitchen* and *pkitchen*, the Bayes estimator is closer to the least squares estimate for the 1999 data. This is a clear result of the prior which pulls the posterior mean to a certain extent towards the prior mean, which is identical to the 1999 least squares estimate. We also observe that the posterior standard deviations of the regression coefficients are considerably lower than the least squares standard errors. This is again a result of the use of additional prior information. △

4.4.2 Regularization Priors

In recent years, a number of "regularization priors" have been proposed in the literature. The main idea is to define priors that have a similar effect as the penalties in the penalized least squares approaches of Sect. 4.2. In this section we will develop some of the most widely used Bayesian regularization priors. As this topic is subject to extensive current research, our presentation cannot be exhaustive (but see Sect. 4.5.1 for references). We start with a Bayesian version of ridge regression, followed by Bayesian LASSO.

Throughout this section, the observation model is a classical linear model given by

$$\boldsymbol{y} \mid \boldsymbol{\beta}, \sigma^2 \sim \mathrm{N}(\beta_0 \mathbf{1} + \tilde{\boldsymbol{X}}\tilde{\boldsymbol{\beta}}, \sigma^2 \boldsymbol{I}),$$

where $\tilde{\boldsymbol{X}}$ is the $n \times k$-design matrix excluding the column of ones for the intercept and $\tilde{\boldsymbol{\beta}}$ is the corresponding vector of regression coefficients excluding β_0. Since the intercept is not subject to regularization, we assume a noninformative (diffuse) prior, i.e.,

$$p(\beta_0) \propto \text{const}.$$

We also assume that the intercept β_0 is independent of the other regression coefficients $\tilde{\boldsymbol{\beta}}$. For the variance parameter σ^2 we specify the usual inverse gamma prior with hyperparameters a and b, i.e., $\sigma^2 \sim \mathrm{IG}(a, b)$.

Bayesian Ridge Regression

Ridge regression minimizes the penalized least squares criterion

$$\mathrm{PLS}(\boldsymbol{\beta}) = (\boldsymbol{y} - \boldsymbol{X}\boldsymbol{\beta})'(\boldsymbol{y} - \boldsymbol{X}\boldsymbol{\beta}) + \lambda\tilde{\boldsymbol{\beta}}'\tilde{\boldsymbol{\beta}}; \tag{4.23}$$

see Sect. 4.2.2. Usually the intercept is not penalized so that here the penalty is restricted to $\tilde{\boldsymbol{\beta}} = (\beta_1, \ldots, \beta_k)'$. The penalty shrinks the parameters towards zero in order to reduce the variance of the least squares estimator at the cost of a (typically small) bias. The amount of penalization is governed by the parameter λ. Small values for λ correspond to negligible penalization, whereas large values lead to strong penalization. Using a particular prior for the regression coefficients, we obtain a Bayesian version of ridge regression. More specifically, we assume a priori independent regression coefficients β_j, $j = 1, \ldots, k$, and set

$$\tilde{\boldsymbol{\beta}} \mid \tau^2 \sim \mathrm{N}(\boldsymbol{0}, \tau^2\boldsymbol{I}). \tag{4.24}$$

Since the prior for the intercept is improper, the joint prior for $\boldsymbol{\beta} = (\beta_0, \tilde{\boldsymbol{\beta}})'$ is also improper. It can be identified with a singular multivariate normal distribution (see Appendix B.3.2) with mean $\boldsymbol{0}$ and precision matrix $\boldsymbol{K} = 1/\tau^2 \mathrm{diag}(0, 1, \ldots, 1)$.

We will show in the appendix of this chapter on p. 261 that maximizing the corresponding posterior with respect to $\boldsymbol{\beta}$ is equivalent to minimizing the penalized least squares criterion (4.23) for fixed $\lambda = \sigma^2/\tau^2$.

While in the frequentist approach to ridge regression, the penalty parameter λ is estimated outside of the optimization criterion, e.g., via cross validation, the Bayesian approach allows for simultaneous inference for the regression coefficients and the amount of penalization measured through τ^2. This is obtained by defining an additional prior for τ^2. A convenient and flexible choice is an inverse gamma distribution $\tau^2 \sim \mathrm{IG}(a_{\tau^2}, b_{\tau^2})$, similar to the conjugate prior for σ^2 outlined above. The advantage of this specification is that the full conditional for τ^2 is again an inverse gamma distribution allowing for straightforward simulation-based MCMC inference; see p. 240 below.

Introducing an additional prior for τ^2, however, changes the interpretation of the prior. This is illustrated in Fig. 4.21 (left panel) which displays the log-prior density, for a single parameter β_j, conditional on τ^2 and the marginal log-prior with the variance parameter τ^2 integrated out. The marginal log-prior is quite different to the Gaussian conditional log-prior. It shows a distinct peak at zero with sharp declines. To derive the marginal prior for $\tilde{\boldsymbol{\beta}}$ formally, we note that the joint prior for $\tilde{\boldsymbol{\beta}}, \tau^2$ is a $\mathrm{NIG}(\boldsymbol{0}, \boldsymbol{I}, a_{\tau^2}, b_{\tau^2})$ distribution. The marginal prior for $\tilde{\boldsymbol{\beta}}$ is then a multivariate t-distribution with $2a_{\tau^2}$ degrees of freedom, location parameter $\boldsymbol{0}$, and dispersion matrix $b_{\tau^2}/a_{\tau^2}\boldsymbol{I}$; see Property 5 in Appendix B.3.5. Note that the diagonal dispersion matrix implies that the regression coefficients are marginally uncorrelated. However, they are not stochastically independent as would be the case if the marginal distribution would be multivariate Gaussian rather than a multivariate t-distribution. In a multivariate t-distribution, a diagonal dispersion matrix implies uncorrelated but *not* stochastically independent components.

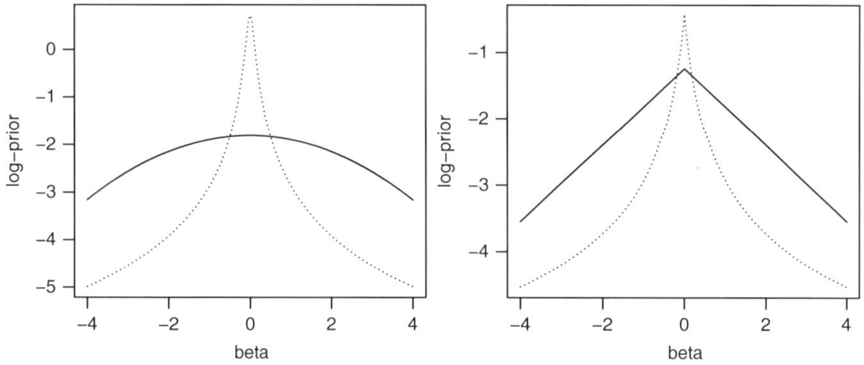

Fig. 4.21 Conditional (*solid lines*) and marginal (*dashed lines*) log-priors for the ridge (*left panel*) and the LASSO prior (*right panel*). In case of the ridge prior, the plots are based on $a = 0.28$ and $b = 0.005$ for the inverse gamma prior. In case of the LASSO, the plots correspond to $a = 0.08$ and $b = 0.001$. With these choices, roughly 90 % of the probability mass are contained in the interval $[-4, 4]$. The hyperparameters are chosen such that the differences between conditional and marginal distributions and between ridge and LASSO are best visible

Bayesian LASSO

The LASSO replaces the quadratic penalty of ridge regression by the sum of absolute values leading to the penalized least squares criterion

$$\text{PLS}(\boldsymbol{\beta}) = (\boldsymbol{y} - \boldsymbol{X}\boldsymbol{\beta})'(\boldsymbol{y} - \boldsymbol{X}\boldsymbol{\beta}) + \lambda \sum_{j=1}^{k} |\boldsymbol{\beta}_j|. \tag{4.25}$$

Similar to ridge regression, we define a prior for the regression coefficients such that the corresponding posterior mode is obtained by minimizing (4.25). We again assume (conditional) independence among the regression coefficients and arrive at the prior

$$\tilde{\boldsymbol{\beta}} \mid \tau_1^2, \ldots, \tau_k^2 \sim \text{N}(\boldsymbol{0}, \text{diag}(\tau_1^2, \ldots, \tau_k^2)), \tag{4.26}$$

where now each regression coefficient β_j has its own variance τ_j^2. The joint prior of $\boldsymbol{\beta} = (\beta_0, \tilde{\boldsymbol{\beta}}')'$ can again be identified with a singular multivariate normal distribution with mean $\boldsymbol{0}$ and precision matrix $\boldsymbol{K} = \text{diag}(0, 1/\tau_1^2, \ldots, 1/\tau_k^2)$.

The variance parameters are assumed mutually independent with priors

$$\tau_j^2 \mid \sim \text{Expo}(0.5\lambda^2);$$

see Definition B.10 of Appendix B.1 for the exponential distribution. The marginal distribution for β_j obtained by integrating over τ_j^2 is a Laplace distribution with location parameter 0 and scale parameter $1/\lambda$, i.e., $p(\beta_j) \propto \exp(-\lambda|\beta_j|)$; see Definition B.8 in Appendix B.1 for the Laplace distribution. Based on this prior specification it can be shown that the posterior mode for $\boldsymbol{\beta}$ with fixed penalty parameter λ corresponds to minimizing the penalized least squares criterion (4.25). The proof is left to the reader as an exercise.

Similar to Bayesian ridge regression, we can assign a hyperprior for λ that allows for simultaneous estimation of the regression coefficients and the amount of penalization. Since the precision of the regression coefficients is given by $\text{Var}(\beta_j)^{-1} = 2\lambda^2$, we assign a gamma distribution to λ^2, i.e., $\lambda^2 \sim \text{G}(a_\lambda, b_\lambda)$.

Summarizing, the joint prior for $\boldsymbol{\beta}$, τ_j^2, $j = 0, \ldots, k$, and λ factors as

$$p(\boldsymbol{\beta}, \tau_1^2, \ldots, \tau_k^2, \lambda) = p(\beta_0)\, p(\tilde{\boldsymbol{\beta}} \mid \tau_1^2, \ldots, \tau_k^2)\, p(\tau_1^2 \mid \lambda) \cdot \ldots \cdot p(\tau_k^2 \mid \lambda)\, p(\lambda^2).$$

A summary of all prior assumptions for the Bayesian LASSO (and ridge regression) can be found in Box 4.6.

We finally compare the LASSO prior with the ridge prior. The right panel of Fig. 4.21 shows the log-prior $\log p(\beta_j \mid \lambda)$, for a single parameter β_j, conditional on the parameter λ together with the marginal log-prior with λ integrated out. As stated, the conditional prior is a Laplace distribution and therefore quite different from the Gaussian conditional log-prior in case of ridge regression (see the left panel of the figure). Somewhat surprisingly, the marginal log-priors appear to be similar although the LASSO prior still has heavier tails than the ridge prior. This is the reason why the Bayesian variants of ridge regression and the LASSO behave often very similar in empirical studies; see Example 4.14.

Posterior Inference

In case of regularization priors the posterior is analytically intractable because of the various variance hyperparameters involved. Therefore inference relies on MCMC simulation techniques. In both cases, a Gibbs sampler can be derived to subsequently draw from the full conditionals of the parameters. Box 4.7 summarizes the resulting algorithms. For the LASSO, the derivation is given in the appendix of this chapter on p. 262. For Bayesian ridge regression the derivation is in complete analogy and left to the reader as an exercise.

Example 4.14 Prices of Used Cars—Bayesian Ridge and LASSO

This example compares the Bayesian variants of ridge regression and the LASSO with their classical counterparts described in Sect. 4.2. Using the software package BayesX, we obtained the estimates displayed in Table 4.10 together with their 95 % credible intervals.

The Bayesian ridge estimates behave quite similar to classical ridge regression. Both variants provide similar results which are also very close to the unpenalized least squares estimate for these data. On the other hand, both LASSO variants show pronounced differences. Most striking is that the Bayesian LASSO does not allow removal of a covariate from the model, as is possible with the classical LASSO. The reason is that the Bayesian LASSO point estimator is the posterior mean or median (rather than the posterior mode) estimated via MCMC. Since the posterior for the regression coefficients

4.6 Regularization Priors

Observation Model

$$y \mid \boldsymbol{\beta}, \sigma^2 \sim \mathrm{N}(\beta_0 \mathbf{1} + \tilde{\boldsymbol{X}}\tilde{\boldsymbol{\beta}}, \sigma^2 \boldsymbol{I})$$

Common Prior Assumptions

$$p(\beta_0) \propto \text{const}$$

$$\sigma^2 \sim \mathrm{IG}(a, b)$$

Ridge Prior

$$\tilde{\boldsymbol{\beta}} \mid \tau^2 \sim \mathrm{N}(\mathbf{0}, \tau^2 \boldsymbol{I})$$

$$\tau^2 \sim \mathrm{IG}(a_{\tau^2}, b_{\tau^2})$$

LASSO Prior

$$\tilde{\boldsymbol{\beta}} \mid \tau_1^2, \ldots, \tau_k^2 \sim \mathrm{N}(\mathbf{0}, \mathrm{diag}(\tau_1^2, \ldots, \tau_k^2))$$

$$\tau_j^2 \mid \lambda \overset{iid}{\sim} \mathrm{Expo}(0.5\lambda^2)$$

$$\lambda^2 \sim \mathrm{G}(a_\lambda, b_\lambda)$$

Software

Software package `BayesX`.

is typically skewed, the posterior mean and median will not coincide with the posterior mode, and, as a consequence, will always be different from zero. This is a distinct disadvantage of the Bayesian LASSO, as the main attraction of the classical LASSO is lost: the ability to perform variable selection. On the other hand, the sampling-based approach provides richer information regarding the posterior, such as posterior standard deviations and quantiles. A possible way to identify redundant covariates with the Bayesian LASSO (as well as Bayesian ridge regression) is to dismiss all covariates which have Bayesian credible intervals covering zero. Bayesian credible intervals are easily constructed from the corresponding posterior quantiles. For instance, the 95 % credible intervals displayed in Table 4.10 are composed of the 2.5 % quantile as the lower bound and the 97.5 % quantile as the upper bound. Based on the 95 % credible intervals only the variables *ageop1*,

4.7 Regularization Priors—Gibbs Sampler

1. *Initialization:*
   - Define initial values $\boldsymbol{\beta}^{(0)}$, $(\sigma^2)^{(0)}$, and $(\tau^2)^{(0)}$ (ridge), $(\tau_1^2)^{(0)}, \dots, (\tau_k^2)^{(0)}$, $\lambda^{(0)}$ (LASSO).
   - Set $t = 1$ and specify the number of iterations T.
2. *Sample $\boldsymbol{\beta}$:* Draw $\boldsymbol{\beta}^{(t)} \mid \cdot \sim \mathrm{N}(\boldsymbol{\mu}_\beta, \Sigma_\beta)$ with $\boldsymbol{\mu}_\beta$ and Σ_β given by

$$\boldsymbol{\Sigma}_\beta = \left(\frac{1}{\sigma^2}\boldsymbol{X}'\boldsymbol{X} + \boldsymbol{K}\right)^{-1} \qquad \boldsymbol{\mu}_\beta = \frac{1}{\sigma^2}\boldsymbol{\Sigma}_\beta \boldsymbol{X}'\boldsymbol{y}.$$

   Here $\boldsymbol{K} = 1/\tau^2 \cdot \mathrm{diag}(0, 1, \dots, 1)$ in case of ridge regression and $\boldsymbol{K} = (0, 1/\tau_1^2, \dots, 1/\tau_k^2)$ in case of LASSO.
3. *Sample σ^2:* Sample $(\sigma^2)^{(t)}$ from the full conditional of σ^2 which is inverse gamma with parameters

$$a^{new} = a + \frac{n}{2}, \qquad b^{new} = b + \frac{1}{2}(\boldsymbol{y} - \boldsymbol{X}\boldsymbol{\beta})'(\boldsymbol{y} - \boldsymbol{X}\boldsymbol{\beta}).$$

4. *Sample variance parameters:*
   - For the ridge prior draw $(\tau^2)^{(t)} \mid \cdot$ from an inverse gamma distribution with parameters

$$a^{new} = a + \frac{k}{2}, \qquad b^{new} = b + \frac{1}{2}\tilde{\boldsymbol{\beta}}'\tilde{\boldsymbol{\beta}}.$$

   - For the LASSO prior, sample

$$(1/\tau_j^2)^{(t)} \mid \cdot \sim \mathrm{InvGauss}\left(\tfrac{|\lambda|}{|\beta_j|}, \lambda^2\right),$$

$$(\lambda^2)^{(t)} \mid \cdot \quad \sim \mathrm{G}\left(a + k, b + \tfrac{1}{2}\sum_{j=1}^{k} \tau_j^2\right).$$

5. Stop if $t = T$, otherwise set $t = t + 1$ and proceed with step 2.

ageop2, *kilometerop1*, and *kilometerop2* are "relevant", while the others are redundant (see Table 4.10).

Finally we note that the Bayesian LASSO appears to induce less shrinkage than the classical LASSO. In fact, Bayesian ridge regression and LASSO show quite similar shrinkage behavior. Indeed, Table 4.10 shows that both regularization variants, albeit conceptually different, produce almost identical posterior estimates. This is in agreement with our theoretical findings on p. 239; see in particular Fig. 4.21.

△

Table 4.10 Prices of used cars: posterior mean and 95 % credible intervals for Bayesian ridge regression and LASSO

	Bayesian ridge		Bayesian LASSO		Ridge	LASSO	LS
Variable	Coeff.	95 % CI	Coeff.	95 % CI	Coeff.	Coeff.	Coeff.
ageop1	−0.682	(−0.802,−0.559)	−0.694	(−0.813, −0.563)	−0.672	−0.682	−0.709
ageop2	0.165	(0.052,0.280)	0.163	(0.052,0.270)	0.164	0.150	0.172
ageop3	0.014	(−0.089,0.128)	0.011	(-0.099,0.113)	0.015	–	0.016
kilometerop1	−0.428	(−0.541, −0.308)	−0.424	(−0.545,−0.303)	−0.425	−0.412	−0.437
kilometerop2	0.140	(0.028,0.252)	0.126	(0.012,0.244)	0.138	0.110	0.142
kilometerop3	0.013	(−0.103,0.125)	0.009	(−0.085,0.108)	0.010	–	0.009
TIA	−0.005	(−0.021,0.011)	−0.004	(−0.020,0.012)	−0.005	–	−0.005
extras1	−0.093	(−0.332,0.152)	−0.075	(−0.306,0.124)	−0.104	−0.036	−0.114
extras2	−0.030	(−0.257,0.211)	−0.022	(−0.240, 0.200)	−0.042	–	−0.031

For comparison the last three columns contain results for classical ridge and LASSO regression as well as the least squares estimator

4.4.3 Classical Bayesian Model Choice (and Beyond)

The classical approach to Bayesian model choice is to compare the models under consideration through their posterior model probabilities (PMP); see Appendix B.5.4 for a general introduction to Bayesian model choice. In this section, we apply this framework to linear models.

Suppose we are given a number of potential covariates and there is uncertainty as to which of the covariates should enter the model. For k possible regressors there are 2^k different models when the intercept β_0 is always included. Denote by M_r, $r = 1, \dots, 2^k$, the different models. More specifically, M_r is given by

$$\boldsymbol{y} \mid \boldsymbol{\beta}, \sigma^2, M_r = \boldsymbol{y} \mid \beta_0, \tilde{\boldsymbol{\beta}}_r, \sigma^2, M_r \sim \mathrm{N}(\beta_0 \mathbf{1} + \tilde{\boldsymbol{X}}_r \tilde{\boldsymbol{\beta}}_r, \sigma^2 \boldsymbol{I}),$$

where the $k_r \times n$-design matrix $\tilde{\boldsymbol{X}}_r$ consists of all k_r covariates included in M_r and $\tilde{\boldsymbol{\beta}}_r$ is the corresponding vector of regression coefficients. As usual, the vector $\boldsymbol{\beta}$ is the full vector of regression coefficients (including the intercept). Those components of $\boldsymbol{\beta}$ not contained in $\tilde{\boldsymbol{\beta}}_r$ are zero in model M_r.

We assign the normal-inverse gamma prior discussed in Sect. 4.4.1 to $\beta_0, \tilde{\boldsymbol{\beta}}_r$, and the error variance σ^2, i.e., $\beta_0, \tilde{\boldsymbol{\beta}}_r \mid \sigma^2, M_r \sim \mathrm{N}(\boldsymbol{m}_r, \sigma^2 \boldsymbol{M}_r)$, and $\sigma^2 \mid M_r = \sigma^2 \sim \mathrm{IG}(a, b)$. This results in the prior distribution

$$p(\beta_0, \tilde{\boldsymbol{\beta}}_r, \sigma^2 \mid M_r) = p(\beta_0, \tilde{\boldsymbol{\beta}}_r \mid \sigma^2, M_r)\, p(\sigma^2) = \mathrm{N}(\boldsymbol{m}_r, \sigma^2 \boldsymbol{M}_r) \cdot \mathrm{IG}(a, b).$$

For completeness, we combine the zero components of $\boldsymbol{\beta}$ in the $(k - k_r)$-dimensional vector $\tilde{\boldsymbol{\beta}}_{-r}$ with a Dirac prior at $(0, \dots, 0)'$

$$p(\tilde{\boldsymbol{\beta}}_{-r} \mid \beta_0, \tilde{\boldsymbol{\beta}}_r, \sigma^2, M_r) = p(\tilde{\boldsymbol{\beta}}_{-r} \mid M_r) = \mathrm{Dirac}(0, \dots, 0).$$

The Dirac prior concentrates all probability mass onto the point $(0,\ldots,0)'$, i.e., $\mathrm{P}(\tilde{\boldsymbol{\beta}}_{-r} = (0,\ldots,0)' \mid M_r) = 1$ and zero otherwise. Thus the prior for $\boldsymbol{\beta}$ and σ^2 under model M_r factors into

$$p(\boldsymbol{\beta},\sigma^2 \mid M_r) = p(\tilde{\boldsymbol{\beta}}_{-r} \mid \beta_0, \tilde{\boldsymbol{\beta}}_r, \sigma^2, M_r)\, p(\beta_0, \tilde{\boldsymbol{\beta}}_r, \sigma^2 \mid M_r).$$

To compare models using their posterior probabilities, we additionally have to assign prior probabilities to each model M_r. A popular prior is

$$p(M_r) = \theta^{k_r}(1-\theta)^{k-k_r}, \tag{4.27}$$

i.e., every possible covariate enters the model independently and with inclusion probability $\theta \in (0,1)$. The "natural" choice $\theta = 1/2$ results in a uniform prior $p(M_r) = 1/2^k$, i.e., each model M_r has the same prior probability.

Prior (4.27) with inclusion probability θ implies a certain prior distribution on the size of the models, denoted by S. Let δ_j, $j = 1,\ldots,k$, be inclusion indicators with $\delta_j = 1$ if covariate x_j is included in the model and 0 otherwise. Then δ_j has a Bernoulli distribution, i.e., $\delta_j \sim \mathrm{B}(1,\theta)$, and the model size S has a binomial distribution with

$$S = \sum_{j=1}^{k} \delta_j \sim \mathrm{B}(k,\theta),$$

resulting in a prior mean model size of $\mathrm{E}(S) = \theta \cdot k$ and variance $\mathrm{Var}(S) = \theta \cdot (1-\theta)\cdot k$ (see Definition B.1 in Appendix B.1). In light of these results, the choice $\theta = 1/2$ seems less natural as suggested at first sight. In particular for a large number k of possible predictors, the prior expected model size appears to be much higher than one would typically expect in applications. For instance, for $k = 50$ potential covariates, the prior expected model size of $\mathrm{E}(S) = 25$ seems to be far too high for most applications.

A convenient way to elicit the model prior (4.27) is to specify the prior mean model size $\mathrm{E}(S)$ and then to set $\theta = \mathrm{E}(S)/k$. For instance, if we have $k = 20$ potential regressors and assume a priori a model size of $\mathrm{E}(S) = 5$, then we must set $\theta = 5/20 = 1/4$.

Based on our prior assumptions, the posterior probability for model M_r is given by

$$p(M_r \mid \boldsymbol{y}) = \frac{p(\boldsymbol{y} \mid M_r)\, p(M_r)}{\sum_{h=1}^{2^k} p(\boldsymbol{y} \mid M_h)\, p(M_h)}, \tag{4.28}$$

where $p(\boldsymbol{y} \mid M_r)$ is the marginal likelihood obtained as

$$p(\boldsymbol{y} \mid M_r) = \int p(\boldsymbol{y} \mid \beta_0, \tilde{\boldsymbol{\beta}}_r, \sigma^2, M_r)\, p(\beta_0, \tilde{\boldsymbol{\beta}}_r, \sigma^2 \mid M_r)\, d\beta_0\, d\tilde{\boldsymbol{\beta}}_r\, d\sigma^2.$$

It can be shown that $p(\boldsymbol{y} \mid M_r)$ is multivariate t-distributed with $2a$ degrees of freedom, location parameter $\boldsymbol{X}_r \boldsymbol{m}_r$, and dispersion matrix $\boldsymbol{I} + \boldsymbol{X}_r \boldsymbol{M}_r \boldsymbol{X}_r'$, where $\boldsymbol{X}_r = (\boldsymbol{1}\, \tilde{\boldsymbol{X}}_r)$; see p. 264 in the appendix of this chapter for the derivation. The Bayes factor for two competing models M_r and M_s is obtained as

$$\mathrm{BF}_{rs} = \frac{p(\boldsymbol{y} \mid M_r)}{p(\boldsymbol{y} \mid M_s)}.$$

With $\tilde{\boldsymbol{M}}_r = (\boldsymbol{X}_r'\boldsymbol{X}_r + \boldsymbol{M}_r^{-1})^{-1}$, it results in

$$\mathrm{BF}_{rs} = \left(\frac{|\tilde{\boldsymbol{M}}_r||\boldsymbol{M}_s|}{|\tilde{\boldsymbol{M}}_s||\boldsymbol{M}_r|}\right)^{1/2} \left(\frac{2a + (\boldsymbol{y} - \boldsymbol{X}_s\boldsymbol{m}_s)'(\boldsymbol{I} - \boldsymbol{X}_s\tilde{\boldsymbol{M}}_s\boldsymbol{X}_s')(\boldsymbol{y} - \boldsymbol{X}_s\boldsymbol{m}_s)}{2a + (\boldsymbol{y} - \boldsymbol{X}_r\boldsymbol{m}_r)'(\boldsymbol{I} - \boldsymbol{X}_r\tilde{\boldsymbol{M}}_r\boldsymbol{X}_r')(\boldsymbol{y} - \boldsymbol{X}_r\boldsymbol{m}_r)}\right)^{a+n/2}; \tag{4.29}$$

see again p. 264 for a derivation.

Once the posterior probabilities $p(M_r \mid \boldsymbol{y})$ are computed for every model M_r under consideration, there are several ways to summarize the results. If the primary focus is on selecting one single (preferably sparse) model, often the model M_* with *highest posterior probability* is taken and inference for the regression coefficients is based on the posterior $p(\boldsymbol{\beta}_*, \sigma^2 \mid \boldsymbol{y}, M_*)$ conditional on model M_*. If model selection is done by minimizing the BIC, as in Sect. 3.4.2 of the previous chapter, we exactly follow this strategy. However, Barbieri and Berger (2004) point out that often the model M_* with highest posterior probability is not optimal in terms of prediction. They show that the optimal predictive model is often the *median probability model*. This model consists of those covariates with posterior probability of 1/2 and higher for being in the model.

Both approaches, however, ignore model uncertainty. In many applications, there are a number of models which are close in terms of posterior probabilities. If this is the case, inference for $\boldsymbol{\beta}$ (or any other quantity of interest) is better conducted by *model averaging* where the models are weighted by their posterior probability. More specifically, the posterior is given by

$$p(\boldsymbol{\beta}, \sigma^2 \mid \boldsymbol{y}) = \sum_{r=1}^{2^k} p(\boldsymbol{\beta}, \sigma^2 \mid \boldsymbol{y}, M_r)\, p(M_r \mid \boldsymbol{y}), \tag{4.30}$$

where $p(\boldsymbol{\beta}, \sigma^2 \mid \boldsymbol{y}, M_r)$ is the conditional posterior under model M_r and $p(M_r \mid \boldsymbol{y})$ is the corresponding posterior model probability given in Eq. (4.28). While models with high posterior probability make important contributions to the posterior, those with negligible posterior probability will contribute only very little information. If most models coincide in their posterior assessment of specific subvectors of $\boldsymbol{\beta}$, this assessment will also carry over to the model-averaged estimate.

While the computation of the posteriors $p(\boldsymbol{\beta}, \sigma^2 \mid \boldsymbol{y}, M_r)$ and $p(M_r \mid \boldsymbol{y})$ required to obtain Eq. (4.30) is straightforward for a particular model M_r, it may be prohibitive for *all models*. The problem is that the number of possible models grows exponentially with k. For $k \leq 25$ regressors, enumeration of all models under consideration is usually possible in the available software packages. If k exceeds 25, more sophisticated algorithms are necessary. The model space then can be explored via MCMC simulation techniques. In doing so, we usually do not visit all models but those models with relatively high posterior probabilities. One such Monte Carlo approach is the MC^3 algorithm of Madigan and York (1995); see also Fernandez et al. (2001). The MC^3 algorithm for exploring model space works as described in Box 4.8.

As already mentioned, eliciting the conjugate prior is difficult in practice as it contains four hyperparameters: the prior mean $\boldsymbol{m}_r$, the prior covariance $\boldsymbol{M}_r$ for the regression coefficients, as well as the inverse gamma parameters a and b of the variance prior. Therefore in applications and software packages a modified prior structure often is assumed that requires to specify less hyperparameters. Fully noninformative priors are not possible if variable selection should be an integral part of the analysis as then the Bayes factor is not determined; see Appendix B.5.4. A widely used choice is to assume a noninformative prior for the intercept, Zellner's g-prior for the regression coefficients in combination with a zero prior mean, and a noninformative prior for the variance. This leads to the prior structure

$$\begin{aligned} p(\beta_0) &\propto \text{const}, \\ p(\tilde{\boldsymbol{\beta}}_r \mid \sigma^2, M_r) &\sim \mathrm{N}(\boldsymbol{0}, \sigma^2 (g \tilde{\boldsymbol{X}}_r' \tilde{\boldsymbol{X}}_r)^{-1}), \\ p(\sigma^2) &\propto \tfrac{1}{\sigma^2}. \end{aligned} \tag{4.31}$$

This prior depends only on one hyperparameter which reduces initialization of the prior to the choice of the factor g. An alternative to Zellner's g-prior is to assume a priori independence among the regression coefficients leading to $p(\tilde{\boldsymbol{\beta}}_r \mid \sigma^2, M_r) \sim \mathrm{N}(\boldsymbol{0}, \sigma^2 \boldsymbol{I})$. These priors take us outside the conjugate model framework because of the improper priors for the intercept and the overall variance. Although the prior is partially improper, posterior analysis including model choice is possible because the intercept and the variance are the same for all models, so that Bayes factors are uniquely determined. Moreover, the MC^3 algorithm for exploring the model space can be easily adapted. Using Zellner's g-prior, the Bayes factor required to compute the acceptance rate (4.33) is given by

$$\mathrm{BF}_{rs} = \left(\frac{g}{g+1}\right)^{\frac{k_r - k_s}{2}} \left(\frac{1 + g - R_s^2}{1 + g - R_r^2}\right)^{\frac{n-1}{2}}, \tag{4.32}$$

4.8 MC^3 Algorithm

Algorithm

1. Choose a start model M_{r_0} with $r_0 \in \{1, \ldots, 2^k\}$ and the number T of iterations. Set $t = 1$.
2. Propose a new model M_{r*} randomly from the neighborhood models of the current model $M_{r_{t-1}}$. The neighborhood of $M_{r_{t-1}}$ consists of all models that are obtained by adding or deleting one variable. Accept the newly proposed model with acceptance probability

$$\begin{aligned}\alpha(M_{r*} \mid M_{r_{t-1}}) &= \min\left\{\frac{p(\boldsymbol{y} \mid M_{r*})\, p(M_{r*})}{p(\boldsymbol{y} \mid M_{r_{t-1}})\, p(M_{r_{t-1}})}, 1\right\} \\ &= \min\left\{\mathrm{BF}_{r*,r_{t-1}} \frac{p(M_{r*})}{p(M_{r_{t-1}})}, 1\right\}\end{aligned} \tag{4.33}$$

   as the current model M_{r_t}. Otherwise set $M_{r_t} = M_{r_{t-1}}$.
3. Update the full parameter vector $\boldsymbol{\beta}$ by sampling from the full conditional of the model parameter vector $\boldsymbol{\beta}_{r_t}$ (see the derivation on p. 234) and setting all other parameters in $\boldsymbol{\beta}$ to zero.
4. Update the error variance σ^2 by sampling from the full conditional (see again the derivation on p. 234).
5. Stop if $t = T$, otherwise set $t = t + 1$ and go to 2.

Software

- R package `BMS`.
- Other R packages for Bayesian variable selection (partly based on other methodology than described here) are `BAS` and `BMA`.

where R_r^2 and R_s^2 are the coefficients of determination of models M_r and M_s. Sampling the regression parameters $\boldsymbol{\beta}$ and σ^2 is done by a Gibbs sampler that consecutively samples from the full conditionals of $\boldsymbol{\beta}_r = (\beta_0, \tilde{\boldsymbol{\beta}}_r)'$ and σ^2. Since the prior for β_0 is improper, joint updating β_0 and $\tilde{\boldsymbol{\beta}}_r$ is not straightforward. The joint prior for β_0 and $\tilde{\boldsymbol{\beta}}_r$ can be written as

$$p(\beta_0, \tilde{\boldsymbol{\beta}}_r \mid \sigma^2) \propto \exp\left(\frac{1}{2\sigma^2}\boldsymbol{\beta}_r' \boldsymbol{K} \boldsymbol{\beta}_r\right),$$

where the precision matrix $\boldsymbol{K}$ is given by

$$\boldsymbol{K} = \begin{pmatrix} 0 & \boldsymbol{0} \\ \boldsymbol{0} & g\tilde{\boldsymbol{X}}_r'\tilde{\boldsymbol{X}}_r \end{pmatrix}.$$

Now the full conditional is expressed as

$$
\begin{aligned}
p(\boldsymbol{\beta}_r \mid \cdot) &\propto \exp\left(-\tfrac{1}{2\sigma^2}(\boldsymbol{y} - \boldsymbol{X}_r\boldsymbol{\beta}_r)'(\boldsymbol{y} - \boldsymbol{X}_r\boldsymbol{\beta}_r)\right) \exp\left(-\tfrac{1}{2}\boldsymbol{\beta}_r'\boldsymbol{K}\boldsymbol{\beta}_r\right) \\
&\propto \exp\left(-\tfrac{1}{2\sigma^2}(\boldsymbol{\beta}_r - \hat{\boldsymbol{\beta}}_r)'\boldsymbol{X}_r'\boldsymbol{X}_r(\boldsymbol{\beta}_r - \hat{\boldsymbol{\beta}}_r)\right) \exp\left(-\tfrac{1}{2}\boldsymbol{\beta}_r'\boldsymbol{K}\boldsymbol{\beta}_r\right),
\end{aligned}
$$

where we have again used identity (4.18). Applying Theorem B.4 (p. 649) of Appendix B.3 with $\boldsymbol{A} = \frac{1}{\sigma^2}\boldsymbol{X}_r'\boldsymbol{X}_r$, $\boldsymbol{a} = \hat{\boldsymbol{\beta}}_r$, $\boldsymbol{B} = \boldsymbol{K}$, and $\boldsymbol{b} = \boldsymbol{0}$ shows that the full conditional is multivariate Gaussian with covariance matrix $\boldsymbol{\Sigma}_{\beta_r}$ and mean $\boldsymbol{\mu}_{\beta_r}$ given by

$$
\boldsymbol{\Sigma}_{\beta_r} = \left(\frac{1}{\sigma^2}\boldsymbol{X}_r'\boldsymbol{X}_r + \boldsymbol{K}\right)^{-1} \qquad \boldsymbol{\mu}_{\beta_r} = \frac{1}{\sigma^2}\boldsymbol{\Sigma}_{\beta_r}\boldsymbol{X}_r'\boldsymbol{y}.
$$

The full conditional of σ^2 is given by

$$
\begin{aligned}
p(\sigma^2 \mid \cdot) \propto & \frac{1}{(\sigma^2)^{n/2}} \exp\left(-\tfrac{1}{2\sigma^2}(\boldsymbol{y} - \boldsymbol{X}_r\boldsymbol{\beta}_r)'(\boldsymbol{y} - \boldsymbol{X}_r\boldsymbol{\beta}_r)\right) \\
& \frac{1}{(\sigma^2)^{k_r/2}} \exp\left(-\tfrac{1}{2\sigma^2}\tilde{\boldsymbol{\beta}}_r' g \tilde{\boldsymbol{X}}_r'\tilde{\boldsymbol{X}}_r\tilde{\boldsymbol{\beta}}_r\right) \frac{1}{\sigma^2},
\end{aligned}
$$

which can be identified with an inverse gamma distribution with parameters

$$
a = \frac{n + k_r}{2} \qquad b = \frac{1}{2}\left((\boldsymbol{y} - \boldsymbol{X}_r\boldsymbol{\beta}_r)'(\boldsymbol{y} - \boldsymbol{X}_r\boldsymbol{\beta}_r) + \tilde{\boldsymbol{\beta}}_r' g \tilde{\boldsymbol{X}}_r'\tilde{\boldsymbol{X}}_r\tilde{\boldsymbol{\beta}}_r\right).
$$

Assuming $p(\tilde{\boldsymbol{\beta}}_r \mid \sigma^2, M_r) \sim \mathrm{N}(\boldsymbol{0}, \sigma^2\boldsymbol{I})$ instead of Zellner's g-prior, the Bayes factor and the Gibbs sampler are similar, details are left to the reader as an exercise.

Example 4.15 Prices of Used Cars—Bayesian Model Averaging (1)

We illustrate the Bayesian approach for model choice using the data on the price of used cars. As in Example 3.19 (p. 152), we assume possibly nonlinear effects for the variables *age* and *kilometer* modeled through the orthogonal cubic polynomials *ageop1*, *ageop2*, *ageop3* and *kilop1*, *kilop2*, *kilop3*; see Example 3.5 on p. 90 for orthogonal polynomials. Together with the regressors *TIA*, *extras1*, and *extras2* we have $k = 9$ potential covariates. We used the package BMS of R for the analysis; see Zeugner (2010) for a tutorial. We started with a uniform prior for the models, i.e., $\theta = 1/2$ in Eq. (4.27), and $g = 1/n$ for Zellner's g-prior. Table 4.11 provides a summary of the preliminary results. From left to right, the columns correspond to the variable names, the posterior inclusion probabilities (PIP), the posterior estimates for the regression coefficients together with their standard deviations, the probability of a positive sign for the respective coefficient, and (for comparison) the least squares estimates. The PIP is the ratio between the number of visited models that include a particular covariate and the total number of visited models. Similarly, the probability of a positive sign reflects the ratio between models with positive sign for a covariate and the total number of models.

For the covariates *ageop1*, *ageop2*, *kilop1*, and *kilop2* with high inclusion probabilities (> 0.5), the Bayesian estimator averaged over the models is quite close to the ordinary least squares estimator. For the remaining covariates, the inclusion probabilities are very

Table 4.11 Prices of used cars: posterior inclusion probabilities (PIP), model averaged estimated coefficients and standard deviations, probabilities of positive sign, and for comparison the ordinary least squares estimates

Variable	PIP	Mean	Std. dev.	Cond. pos sign	LS
ageop1	1.00	−0.7065	0.0621	0	−0.7085
ageop2	0.94	0.1823	0.0721	1	0.1716
ageop3	0.07	0.0009	0.0156	1	0.0162
kilop1	1.00	−0.4345	0.0616	0	−0.4366
kilop2	0.61	0.0872	0.0827	1	0.1417
kilop3	0.07	0.0011	0.0160	1	0.0090
TIA	0.08	−0.0003	0.0025	0	−0.0051
extras1	0.11	−0.0127	0.0565	0	−0.1135
extras2	0.07	−0.0029	0.0374	0	−0.0315

The results are based on a uniform prior for the models ($\theta = 0.5$) and $g = 1/n$ for Zellner's g

Table 4.12 Prices of used cars: top five models T1–T5 with highest posterior probabilities

Variable	T1	T2	T3	T4	T5
ageop1	+	+	+	+	+
ageop2	+	+	+	+	+
ageop3	−	−	−	−	−
kilop1	+	+	+	+	+
kilop2	+	−	+	+	+
kilop3	−	−	−	−	−
TIA	−	−	−	+	−
extras1	−	−	+	−	−
extras2	−	−	−	−	+
PMP	0.375	0.250	0.040	0.032	0.030

The results are based on a uniform prior for the models ($\theta = 0.5$) and $g = 1/n$ for Zellner's g. A plus (minus) sign indicates that the variable is included in (excluded from) the model. The last row displays the posterior model probabilities (PMP).

low and the model averaged estimates are shrunk to zero compared to least squares. As can be seen from Table 4.12, the covariates with high inclusion probability also build the two models with by far the highest posterior probabilities. The top model, with posterior model probability of 0.37, consists exactly of the four covariates with inclusion probabilities higher than 0.5. In this case, the median probability model, that consists of those covariates with posterior probability of 1/2 and higher for being in the model, coincides with the model with highest posterior probability. The second model, with posterior probability 0.25, additionally excludes the variable *kilop2*. All other models have comparably low posterior probabilities. Note that the model with highest posterior probability corresponds exactly to the best AIC model of Example 3.19. It is also close to the model obtained with the LASSO in Example 4.11, which additionally contains covariate *extras1*.

We conclude the example by an investigation of the sensitivity of results on prior assumptions, particularly on the prior expected model size $\mathrm{E}(S)$. The upper panel of Fig. 4.22 compares the PIP for the three models with $\theta = 1/2$, $\theta = 2/9$, and $\theta = 8/9$ corresponding to expected model sizes $\mathrm{E}(S) = 4.5$, $\mathrm{E}(S) = 2$, and $\mathrm{E}(S) = 8$. Clearly, the results are quite sensitive to the choice of θ and $\mathrm{E}(S)$. Except for *ageop1*, *ageop2*,

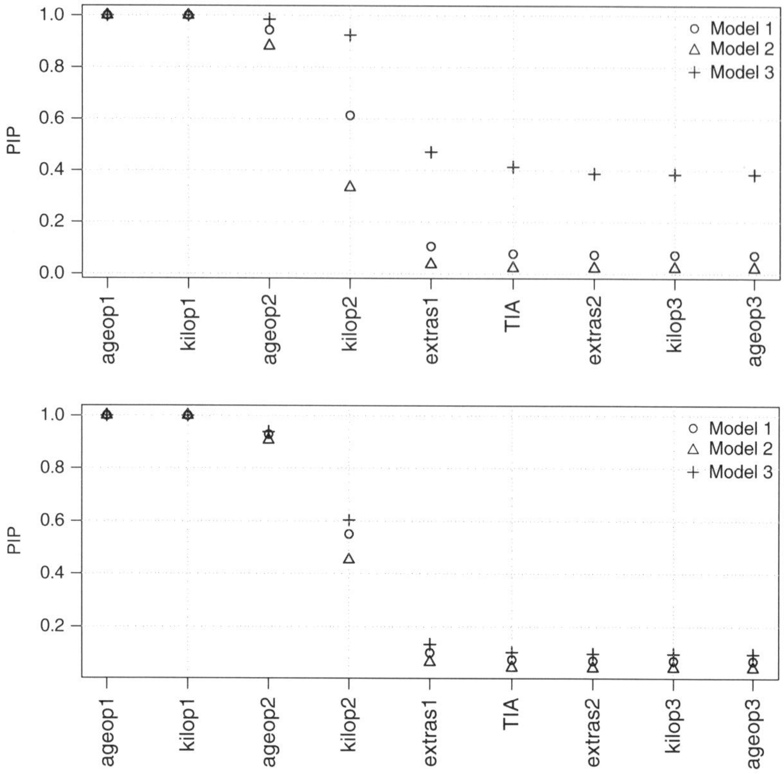

Fig. 4.22 Prices of used cars: comparison of posterior inclusion probabilities with $\theta = 1/2$ (corresponds to expected model size $\mathrm{E}(S) = 4.5$), $\theta = 2/9$ ($\mathrm{E}(S) = 2$), and $\theta = 8/9$ ($\mathrm{E}(S) = 8$). The *upper panel* corresponds to fixed θ. The *lower panel* corresponds to a beta hyperprior for θ

and *kilop1*, the PIP differ considerably, e.g., for *kilop2* from 0.33 ($\mathrm{E}(S) = 2$) over 0.61 ($\mathrm{E}(S) = 4.5$) to 0.92 ($\mathrm{E}(S) = 8$). We will explain the reasons of this undesirable behavior below.

△

The example shows that the results can be strongly affected by the prior choice for θ and the corresponding prior on the model size. Indeed, if θ is chosen as $\theta = \mathrm{E}(S)/k$, then the induced prior for the model size S places relatively small probability mass on model sizes that are moderately far away from the mean $\mathrm{E}(S)$. This is illustrated with Fig. 4.23 which shows, for $k = 10$ (left column) and $k = 40$ (right column) regressors, some priors for model size based on different choices for θ (dashed lines). In particular, with asymmetric prior expected model size $\mathrm{E}(S)$ and for $k = 40$, a broad range of possible model sizes has virtually no prior probability mass. As a result, estimates are often quite sensitive to the choice of the prior expected model size.

A remedy to reduce the dependence of results on the prior choice is to introduce a hyperprior in a further stage of the hierarchy. In our case, a flexible and

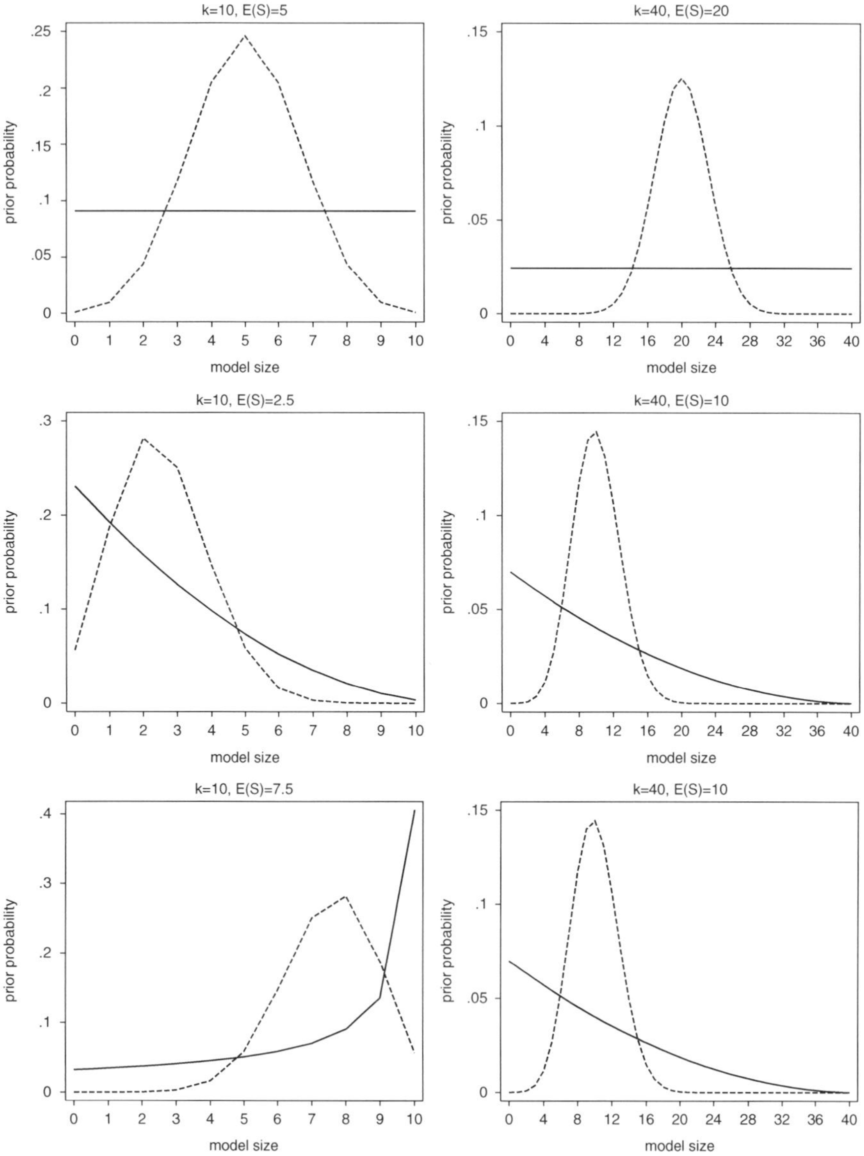

Fig. 4.23 Priors for model size. The *left column* shows various priors for model size with $k = 10$ potential regressors; the *right column* corresponds to $k = 40$. The *solid lines* correspond to a Beta(a, b) hyperprior for θ. The *dashed lines* correspond to fixed θ

convenient choice for θ is to assume a beta distribution with hyperparameters a and b (Ley & Steel, 2009), i.e., $\theta \sim \text{Beta}(a, b)$. For the beta distribution see also Definition B.2 in Appendix B.1.

Since $S \mid \theta \sim \mathrm{B}(k, \theta)$, the marginal distribution for S is beta-binomial (see Definition B.3 in Appendix B.1), i.e., $S \sim \mathrm{BetaB}(k, a, b)$ with prior model size distribution

$$\mathrm{P}(S = s) = \frac{\Gamma(a+b)}{\Gamma(a)\Gamma(b)\Gamma(a+b+k)} \binom{k}{s} \Gamma(a+s)\Gamma(b+k-s),$$

and mean and variance given by

$$\begin{aligned} \mathrm{E}(S) &= \frac{a}{a+b}k, \\ \mathrm{Var}(S) &= \frac{ab(a+b+k)}{(a+b)^2(a+b+1)}k. \end{aligned} \tag{4.34}$$

To ease initiation of the prior, we fix $a = 1$, which still allows for very flexible priors. To choose b, we rearrange $\mathrm{E}(S)$ in Eq. (4.34) to obtain $b = (k - \mathrm{E}(S))/\mathrm{E}(S)$. Hence, similar to fixed θ, we can choose the prior expected model size to fully specify the prior. As can be seen from Fig. 4.23, the (marginal) prior for model size S shows much more variability compared to fixed θ so that all possible model sizes have positive probability mass.

Using a $\mathrm{Beta}(a, b)$ hyperprior requires a slight modification of the MC^3 algorithm as the prior model odds in Eq. (4.33) has to be adapted. The marginal prior for model M_r is obtained by integrating over θ

$$p(M_r) = \frac{\Gamma(a+b)}{\Gamma(a)\Gamma(b)} \frac{\Gamma(a+k_r)\Gamma(b+k-k_r)}{\Gamma(a+b+k)}; \tag{4.35}$$

see the appendix of this chapter, p. 266, for a derivation.

Using formula (4.35) the prior odds between two models M_r and M_s required for Eq. (4.33) in the MC^3 algorithm are easily computed as

$$\frac{p(M_r)}{p(M_s)} = \frac{\Gamma(a+k_r)\Gamma(b+k-k_r)}{\Gamma(a+k_s)\Gamma(b+k-k_s)}.$$

Example 4.16 Prices of Used Cars—Bayesian Model Averaging (2)

We rerun the three regressions of Example 4.15, again with prior expected model size $\mathrm{E}(S) = 2, 4.5, 8$, but now with a beta hyperprior for θ. The bottom panel of Fig. 4.22 compares the PIP for the nine possible covariates. The probabilities are now quite close to each other. An exception are the PIP for *kilop2*, where there is some inclusion uncertainty. For two of the three model priors (with $\mathrm{E}(S) = 4.5$ and $\mathrm{E}(S) = 8$) the best model with highest posterior probability consists of the covariates *ageop1*, *ageop2*, *kilop1*, and *kilop2*. The model that excludes *kilop2* is the second best model. It is also the best model for prior expected model size $\mathrm{E}(S) = 2$. All other models have very low posterior probabilities.

In summary, we are quite certain that only two covariates, the age of the car and the kilometer reading, are relevant predictors for the price. A second-order polynomial is sufficient to model the nonlinear effects of the two covariates. There is some uncertainty whether a linear effect for kilometer reading is sufficient. Note also that the quadratic effect of kilometer reading is already close to linearity; see Fig. 3.22 (p. 155).

△

We finally note that there are also approaches to specify an additional hyperprior for the factor g. Details can be found in Liang, Paulo, Molina, Clyde, and Berger (2008). The approach is also implemented in the R package BMS.

4.4.4 Spike and Slab Priors

We again consider the Bayesian linear variable selection model of the previous section; see Box 4.9 for a summary. For simplicity, we restrict ourselves to a priori independent regression coefficients, i.e., $p(\boldsymbol{\beta}_r \,|\, \sigma^2, M_r) \sim \text{N}(\mathbf{0}, \sigma^2 \boldsymbol{I})$.

We now present an equivalent model formulation that provides further insight and gives rise to interesting modifications. In the last section, we specified the prior $p(M_r \,|\, \theta) = \theta^{k_r} \cdot (1-\theta)^{k-k_r}$ for the models M_r together with a Beta(a_θ, b_θ) prior for θ. Each model M_r can also be identified with a specific vector $\boldsymbol{\delta}_r$ of the inclusion indicators $\boldsymbol{\delta} = (\delta_1, \ldots, \delta_k)'$. Recall that the jth component δ_j of $\boldsymbol{\delta}$ defines whether or not the jth covariate x_j is included in the model. The model prior can now be specified equivalently in terms of the inclusion indicator $\boldsymbol{\delta}$ by assuming

$$\delta_j \,|\, \theta \overset{i.i.d.}{\sim} \text{B}(1, \theta),$$

i.e., $\text{P}(\delta_j = 1) = \theta$ and $\text{P}(\delta_j) = 1 - \theta$. Hence, we assume a priori that a particular covariate x_j enters the model with inclusion probability θ. Since posterior analysis is based on MCMC simulations, we can interpret the relative frequency of $\delta_j = 1$ in the MCMC samples as the posterior probability that x_j enters the model.

The priors for the regression coefficients β_j, $j = 1, \ldots, k$, can now be defined in dependence of the inclusion vector $\boldsymbol{\delta}$ by assuming the mixture prior

$$p(\beta_j \,|\, \delta_j, \sigma^2) = (1 - \delta_j)\text{Dirac}(0) + \delta_j \ \text{N}(0, \sigma^2). \tag{4.36}$$

The mixture prior has the following interpretation: If $\delta_j = 0$ the first component in Eq. (4.36) is "activated" and the corresponding regression coefficient is zero with probability one. In other words, covariate x_j is removed from the model. We call this first part of the mixture prior the *spike component*. For $\delta_j = 1$, we have the usual normal prior for β_j. This part of the mixture prior is called the *slab component*.

We summarize the alternative formulation of the model in Box 4.10. This is fully equivalent to the model of the previous Sect. 4.4.3. We presented the alternative formulation in terms of spike and slab priors because it gives rise to various other approaches related to Bayesian variable selection, which may be preferable in some situations. The main disadvantage of the spike and slab mixture prior (4.36) (and the equivalent model of the previous section) is that the MCMC algorithms for Bayesian inference require computation of the marginal likelihood, which is only analytically

4.9 Bayesian Linear Variable Selection Model

Observation Model Conditional on Model M_r

$$y \mid \boldsymbol{\beta}, \sigma^2, M_r \sim \mathrm{N}(\beta_0 \mathbf{1} + \tilde{\boldsymbol{X}}_r \tilde{\boldsymbol{\beta}}_r, \sigma^2 \boldsymbol{I}).$$

Prior Structure Conditional on Model M_r

$$\begin{aligned} p(\boldsymbol{\beta}, \sigma^2 \mid M_r) &= p(\tilde{\boldsymbol{\beta}}_{-r} \mid \beta_0, \tilde{\boldsymbol{\beta}}_r, \sigma^2, M_r)\, p(\beta_0, \tilde{\boldsymbol{\beta}}_r, \sigma^2 \mid M_r) \\ &= p(\tilde{\boldsymbol{\beta}}_{-r} \mid \beta_0, \tilde{\boldsymbol{\beta}}_r, \sigma^2, M_r)\, p(\beta_0 \mid M_r)\, p(\tilde{\boldsymbol{\beta}}_r \mid \sigma^2, M_r)\, p(\sigma^2 \mid M_r), \end{aligned}$$

with

$$\begin{aligned} p(\tilde{\boldsymbol{\beta}}_{-r} \mid \beta_0, \tilde{\boldsymbol{\beta}}_r, \sigma^2, M_r) &= \mathrm{Dirac}(0, \ldots, 0), \\ p(\beta_0 \mid M_r) &= p(\beta_0) \propto \mathrm{const}, \\ p(\tilde{\boldsymbol{\beta}}_r \mid \sigma^2, M_r) &\sim \mathrm{N}(\mathbf{0}, \sigma^2 (g \tilde{\boldsymbol{X}}_r' \tilde{\boldsymbol{X}}_r)^{-1}), \\ p(\sigma^2 \mid M_r) &= p(\sigma^2) \sim \frac{1}{\sigma^2}. \end{aligned}$$

An alternative to Zellner's g-prior is to assume independent regression coefficients and assume $p(\tilde{\boldsymbol{\beta}}_r \mid \sigma^2, M_r) \sim \mathrm{N}(\mathbf{0}, \sigma^2 \boldsymbol{I})$.

Model Prior

Fixed inclusion parameter

$$p(M_r) = \theta^{k_r} (1 - \theta)^{k - k_r}.$$

Stochastic inclusion parameter

$$\begin{aligned} p(M_r \mid \theta) &= \theta^{k_r} (1 - \theta)^{k - k_r} \\ \theta &\sim \mathrm{Beta}(a_\theta, b_\theta). \end{aligned}$$

available in special cases, e.g., the linear model. In many other settings, for instance in the regression models with non-normal responses, discussed in Chaps. 5 and 6, the marginal likelihood is not analytically, and also not often numerically, tractable.

The recent literature has therefore proposed various alternatives. An approach due to George and Mc Culloch (1993) that is conceptually close replaces the mixture prior (4.36) by

$$p(\beta_j \mid \delta_j, \sigma^2) = (1 - \delta_j)\, \mathrm{N}(0, \nu_0 \sigma^2) + \delta_j\ \mathrm{N}(0, \sigma^2), \tag{4.37}$$

4.10 Spike and Slab Priors

Observation Model

$$y \mid \boldsymbol{\beta}, \sigma^2 \sim \mathrm{N}(\beta_0 \mathbf{1} + \tilde{\boldsymbol{\beta}} \tilde{\boldsymbol{X}}, \sigma^2 \boldsymbol{I})$$

Spike and Slab Prior for the Regression Coefficients

Dirac spike

$$p(\beta_j \mid \delta_j, \sigma^2) \overset{iid}{=} (1-\delta_j)\,\mathrm{Dirac}(0) + \delta_j\ \mathrm{N}(0, \sigma^2) \text{ for } j = 1, \ldots, k.$$

Continuous normal spike

$$p(\beta_j \mid \delta_j, \sigma^2) \overset{iid}{=} (1-\delta_j)\ \mathrm{N}(0, \nu_0\, \sigma^2) + \delta_j\ \mathrm{N}(0, \sigma^2), \text{ with } \nu_0 \text{ “small.”}$$

Spike and Slab Prior for the Variances

$$\beta_1, \ldots, \beta_k \mid \tau_1^2, \ldots, \tau_k^2 \sim \mathrm{N}(\mathbf{0}, \mathrm{diag}(\tau_1^2, \ldots, \tau_k^2)).$$

Dirac spike

$$\tau_j^2 \mid \delta_j \overset{iid}{=} (1-\delta_j)\,\mathrm{Dirac}(0) + \delta_j\ \mathrm{IG}(a_{\tau^2}, b_{\tau^2}).$$

Continuous inverse gamma spike

$$\tau_j^2 \mid \delta_j \overset{iid}{=} (1-\delta_j)\ \mathrm{IG}(a_{\tau^2}, \nu_0 b_{\tau^2}) + \delta_j\ \mathrm{IG}(a_{\tau^2}, b_{\tau^2}) \text{ with } \nu_0 \text{ “small.”}$$

Common Prior Assumptions

$$p(\beta_0) \propto \text{const}$$

$$\sigma^2 \sim \mathrm{IG}(a, b)$$

$$\delta_j \mid \theta \overset{iid}{\sim} \mathrm{B}(1, \theta)$$

$$\theta \sim \mathrm{Beta}(a_\theta, b_\theta)$$

Software

- Software package `BayesX` for continuous variance spikes.
- R package `BMS` for Dirac spikes for the regression coefficients.

where the Dirac spike at 0 is replaced by the $N(0, \nu_0\sigma^2)$ distribution. The variance component ν_0 is a hyperparameter and is chosen very small (close to zero), such that the spike part has very small variance. Consequently if $\delta_j = 0$, then the small variance of the spike part, in combination with the zero mean, ensures that the corresponding regression coefficient is essentially zero. Hence, $\delta_j = 0$ still implies that covariate x_j is (essentially) removed from the model. The main advantage of the alternative mixture prior (4.37) is that MCMC inference is simplified considerably since the computation of marginal likelihoods can be avoided.

A different concept due to Ishwaran and Rao (2005) focuses on the prior variance of the regression coefficients. Recall the Bayesian LASSO, where we assigned independent normal distributions to the regression coefficients β_j conditional on the variance parameters τ_j^2, i.e., $\beta_j \mid \tau_j^2 \sim N(0, \tau_j^2)$ or

$$\tilde{\boldsymbol{\beta}} \mid \tau_1^2, \ldots, \tau_k^2 \sim N(\mathbf{0}, \text{diag}(\tau_1^2, \ldots, \tau_k^2)).$$

Spike and slab priors for the variances are based on the observation that the regression coefficient β_j is estimated essentially zero if the corresponding variance parameter τ_j^2 is zero or at least close to zero. The prior mean of zero, in combination with (near) zero variance, then guarantees that β_j cannot deviate far from zero, which in turn implies that the corresponding covariate x_j is essentially removed from the model.

The idea now is to specify a prior for the variances τ_j^2 that puts relatively large probability mass on near zero variance. This can again be obtained using a mixture distribution. One such approach is a mixture of a Dirac spike at 0 and an inverse gamma slab given by

$$\tau_j^2 \mid \delta_j = (1 - \delta_j) \cdot \text{Dirac}(0) + \delta_j \cdot \text{IG}(a_{\tau^2}, b_{\tau^2}),$$

where $\delta_j \in \{0, 1\}$ again is an indicator variable. Alternatively, we can use a mixture of two inverse gamma distributions. More specifically, we obtain

$$\tau_j^2 \mid \delta_j = (1 - \delta_j) \cdot \text{IG}(a_{\tau^2}, \nu_0 b_{\tau^2}) + \delta_j \cdot \text{IG}(a_{\tau^2}, b_{\tau^2}), \tag{4.38}$$

where the hyperparameter ν_0 is chosen to be small. This prior has the following interpretation: The mean, mode, and variance of an $\text{IG}(a, b)$ distribution are given by

$$\begin{aligned} E(X) &= \frac{b}{a-1} \quad a > 1, \\ \text{Mod}(X) &= \frac{b}{a+1}, \\ \text{Var}(X) &= \frac{b^2}{(a-1)^2(a-2)} \quad a > 2. \end{aligned}$$

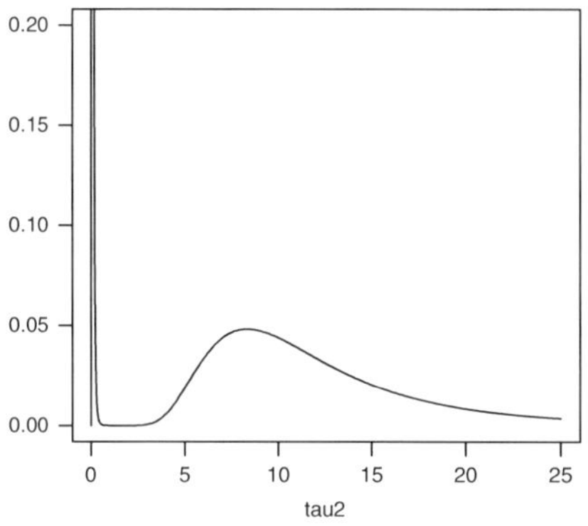

Fig. 4.24 Spike and slab priors: marginal density of the variance parameter τ_j^2 for some hyperparameter combinations

Provided that ν_0 is very small, the first component in Eq. (4.38) has its mode (and mean if it exists) close to zero while the variability of the prior is limited. Thus, the first component makes sure that the marginal prior for τ_j^2 will have a *spike* near zero. The second component in Eq. (4.38), the *slab* part, can be seen as the "usual" variance prior if the corresponding covariate should be included in the model. To shed more light on the distribution of the variance component, Fig. 4.24 shows the marginal density (with δ_j integrated out) of the variance parameter τ_j^2 for some hyperparameter combinations. The characteristic feature is the "spike" at zero.

A summary of the various spike and slab approaches is given in Box 4.10. For illustration, in the appendix of this chapter (p. 266), we derive a Gibbs sampler for the continuous variance spike and slab prior (4.38). The derivation of Gibbs samplers for the other spike and slab variants is similar and left as an exercise to the reader. The great advantage of spike and slab priors for the variances, rather than the regression coefficients, is that updating of the variance and inclusion parameters is independent of the observation model. This means that the concept carries over to more complex models, e.g., the regression models for non-normal responses of Chaps. 5 and 6. If we are able to update the regressions coefficients, which is typically possible even in complex models, the remaining update steps remain unchanged.

4.5 Bibliographic Notes and Proofs

4.5.1 Bibliographic Notes

For modeling correlated errors in Sect. 4.1.4, we only examined autoregressive processes of first order. More complex error structures, especially ARMA models (Autoregressive Moving Average), are treated in greater detail in the literature focusing on time series analysis. Standard books are Brockwell and Davis (2002) and Hamilton (1994).

A standard book for variable selection in linear models is Miller (2002). A good overview on regularization approaches, such as ridge regression and the LASSO, can be found in Hastie, Tibshirani, and Friedman (2009). Alternative regularization approaches based on different penalties are, for example, bridge regression, where general norms of coefficient vectors are used as penalties (Fu, 1998), or the elastic net that is a compromise between ridge regression and LASSO (Zou & Hastie 2005). The latter has the particular advantage to select complete blocks of correlated coefficients instead of only single representatives of such groups. Penalties with a similar aim are defined in the octogonal shrinkage and clustering algorithm for regression (OSCAR, Bondell & Reich, 2008) or with the smoothly clipped absolute deviation (SCAD, Fan & Li, 2001).

Boosting approaches in a regression context, including theoretical properties, are described in Friedman (2001), Bühlmann and Yu (2003), and Bühlmann (2006). Bühlmann and Hothorn (2007) give a general introduction into boosting based on a functional gradient interpretation. A different variant of boosting that relies on likelihood maximization techniques (likelihood-based boosting) is considered, for example, in Tutz and Binder (2006).

Bayesian linear models are covered in all introductory books on Bayesian analysis. Standard references are the textbooks of Gelman, Carlin, Stern, and Rubin (2003) and O'Hagan (1994). Still interesting is the classical book by Box and Tiao (1992). The Bayesian LASSO is due to Park and Casella (2008). An overview over other Bayesian regularization priors is given in Fahrmeir, Kneib, and Konrath (2010). Recent references to the Bayesian linear variable selection model of Box 4.9 are the papers by Fernandez et al. (2001), Ley and Steel (2009), and the many references therein. Spike and slab priors with continuous spike for the regression coefficients are introduced in George and Mc Culloch (1993) and George and Mc Culloch (1997). Spike and slab priors for the variances are proposed in Ishwaran and Rao (2003, 2005). A good overview of spike and slab priors is provided in Malsiner-Walli and Wagner (2011).

4.5.2 Proofs

Autocorrelated Errors of First Order (p. 192)

We first derive the so-called MA(∞) presentation for the errors ε_i. Based on this representation, we then derive means, variances, and covariances of errors.

MA(∞) Presentation

Our goal is to represent ε_i as a weighted infinite sum of the u_i. This is based on the idea that the process started far back in the past. Inserting first $\varepsilon_{i-1} = \rho\, \varepsilon_{i-2} + u_{i-1}$ into $\varepsilon_i = \rho\, \varepsilon_{i-1} + u_i$, we obtain

$$\varepsilon_i = \rho\, (\rho\, \varepsilon_{i-2} + u_{i-1}) + u_i = \rho^2\, \varepsilon_{i-2} + \rho\, u_{i-1} + u_i\,.$$

Inserting $\varepsilon_{i-2} = \rho\, \varepsilon_{i-3} + u_{i-2}$ yields

$$\varepsilon_i = \rho^3\, \varepsilon_{i-3} + \rho^2\, u_{i-2} + \rho\, u_{i-1} + u_i\,.$$

Successive insertion finally results in the presentation

$$\varepsilon_i = \sum_{k=0}^{\infty} \rho^k\, u_{i-k} = u_i + \rho u_{i-1} + \rho^2 u_{i-2} + \ldots. \tag{4.39}$$

Mean of Errors
Using representation (4.39) we obtain

$$\mathrm{E}(\varepsilon_i) = \sum_{k=0}^{\infty} \rho^k\, \mathrm{E}(u_{i-k}) = 0.$$

Covariance of Errors
Using again representation (4.39) as well as $\mathrm{E}(\varepsilon_i) = 0$ and $\mathrm{E}(u_i u_j) = 0$ for $i \neq j$, we obtain

$$\begin{aligned}
\mathrm{Cov}(\varepsilon_i, \varepsilon_{i-j}) &= \mathrm{E}(\varepsilon_i \varepsilon_{i-j}) \\
&= \mathrm{E}\left(\left(u_i + \rho u_{i-1} + \ldots + \rho^j u_{i-j} + \ldots\right)\left(u_{i-j} + \rho u_{i-j-1} + \ldots\right)\right) \\
&= \mathrm{E}(\rho^j u_{i-j}^2) + \mathrm{E}(\rho^{j+2} u_{i-j-1}^2) + \ldots \\
&= \rho^j \sum_{k=0}^{\infty} (\rho^2)^k\, \mathrm{E}\left(u_{i-j-k}^2\right) \\
&= \rho^j \sum_{k=0}^{\infty} (\rho^2)^k\, \sigma_u^2 \\
&= \rho^j \frac{\sigma_u^2}{1-\rho^2}.
\end{aligned}$$

The special case $j = 0$ implies

$$\mathrm{Var}(\varepsilon_i) = \frac{\sigma_u^2}{1-\rho^2}.$$

□

Proof of the Least Squares Identity (p. 211)
We show

$$\begin{aligned}
\mathrm{LS}(\hat{\boldsymbol{\beta}}) &= (\boldsymbol{\beta} - \hat{\boldsymbol{\beta}})' \boldsymbol{X}' \boldsymbol{X} (\boldsymbol{\beta} - \hat{\boldsymbol{\beta}}) + \boldsymbol{y}'(\boldsymbol{I}_n - \boldsymbol{X}(\boldsymbol{X}'\boldsymbol{X})^{-1}\boldsymbol{X}')\boldsymbol{y} \\
&= (\boldsymbol{\beta} - \hat{\boldsymbol{\beta}})' \boldsymbol{X}' \boldsymbol{X} (\boldsymbol{\beta} - \hat{\boldsymbol{\beta}}) + \hat{\boldsymbol{\varepsilon}}'\hat{\boldsymbol{\varepsilon}}.
\end{aligned} \tag{4.40}$$

To verify Eq. (4.40) we first evaluate the products for the least squares criterion to obtain

$$(y - X\beta)'(y - X\beta) = y'y - 2\beta'X'y + \beta'X'X\beta. \tag{4.41}$$

On the other hand, we expand the quadratic form in Eq. (4.40) which gives

$$(\beta - \hat{\beta})'X'X(\beta - \hat{\beta}) = \beta'X'X\beta - 2\beta'X'X\hat{\beta} + \hat{\beta}'X'X\hat{\beta}.$$

Inserting $\hat{\beta} = (X'X)^{-1}X'y$ yields

$$2\beta'X'X\hat{\beta} = 2\beta'X'X(X'X)^{-1}X'y = 2\beta X'y$$

for the second addend and

$$\hat{\beta}'X'X\hat{\beta} = y'X(X'X)^{-1}X'X(X'X)^{-1}X'y = y'X(X'X)^{-1}X'y$$

for the third addend. We thus obtain

$$(\beta - \hat{\beta})'X'X(\beta - \hat{\beta}) = \beta'X'X\beta - 2\beta X'y + y'X(X'X)^{-1}X'y.$$

Inserting this expression in the right-hand side of Eq. (4.40) gives exactly the right-hand side of Eq. (4.41) which proves the least squares identity. The second line of Eq. (4.40) states that

$$y'(I_n - X(X'X)^{-1}X')y = \hat{\varepsilon}'\hat{\varepsilon},$$

which is easily verified by expanding

$$\hat{\varepsilon}'\hat{\varepsilon} = (y - X\hat{\beta})'(y - X\hat{\beta})$$

and inserting $\hat{\beta} = (X'X)^{-1}X'y$.

Derivation of the Posterior in the Linear Model with NIG Prior (p. 227)
We first rewrite the expression

$$(y - X\beta)'(y - X\beta) + (\beta - m)'M^{-1}(\beta - m).$$

Defining

$$\tilde{M} = (X'X + M^{-1})^{-1}$$
$$\tilde{m} = \tilde{M}(M^{-1}m + X'y),$$

we obtain

$$\begin{aligned}
&(\boldsymbol{y} - \boldsymbol{X}\boldsymbol{\beta})'(\boldsymbol{y} - \boldsymbol{X}\boldsymbol{\beta}) + (\boldsymbol{\beta} - \boldsymbol{m})'\boldsymbol{M}^{-1}(\boldsymbol{\beta} - \boldsymbol{m}) \\
&\quad = \boldsymbol{y}'\boldsymbol{y} - 2\boldsymbol{\beta}'\boldsymbol{X}'\boldsymbol{y} + \boldsymbol{\beta}'\boldsymbol{X}'\boldsymbol{X}\boldsymbol{\beta} + \boldsymbol{\beta}'\boldsymbol{M}^{-1}\boldsymbol{\beta} - 2\boldsymbol{\beta}'\boldsymbol{M}^{-1}\boldsymbol{m} + \boldsymbol{m}'\boldsymbol{M}^{-1}\boldsymbol{m} \\
&\quad = \boldsymbol{y}'\boldsymbol{y} + \boldsymbol{\beta}'(\boldsymbol{X}'\boldsymbol{X} + \boldsymbol{M}^{-1})\boldsymbol{\beta} - 2\boldsymbol{\beta}'(\boldsymbol{X}'\boldsymbol{y} + \boldsymbol{M}^{-1}\boldsymbol{m}) + \boldsymbol{m}'\boldsymbol{M}^{-1}\boldsymbol{m} \\
&\quad = \boldsymbol{y}'\boldsymbol{y} + \boldsymbol{\beta}'\tilde{\boldsymbol{M}}^{-1}\boldsymbol{\beta} - 2\boldsymbol{\beta}'\tilde{\boldsymbol{M}}^{-1}\tilde{\boldsymbol{M}}(\boldsymbol{X}'\boldsymbol{y} + \boldsymbol{M}^{-1}\boldsymbol{m}) + \boldsymbol{m}'\boldsymbol{M}^{-1}\boldsymbol{m} \\
&\quad = \boldsymbol{y}'\boldsymbol{y} + \boldsymbol{\beta}'\tilde{\boldsymbol{M}}^{-1}\boldsymbol{\beta} - 2\boldsymbol{\beta}'\tilde{\boldsymbol{M}}^{-1}\tilde{\boldsymbol{m}} + \boldsymbol{m}'\boldsymbol{M}^{-1}\boldsymbol{m} \\
&\quad = \boldsymbol{y}'\boldsymbol{y} + (\boldsymbol{\beta} - \tilde{\boldsymbol{m}})'\tilde{\boldsymbol{M}}^{-1}(\boldsymbol{\beta} - \tilde{\boldsymbol{m}}) - \tilde{\boldsymbol{m}}'\tilde{\boldsymbol{M}}^{-1}\tilde{\boldsymbol{m}} + \boldsymbol{m}'\boldsymbol{M}^{-1}\boldsymbol{m}.
\end{aligned}$$

Defining

$$\tilde{a} = a + \frac{n}{2}$$

and

$$\tilde{b} = b + \frac{1}{2}\left(\boldsymbol{y}'\boldsymbol{y} + \boldsymbol{m}'\boldsymbol{M}^{-1}\boldsymbol{m} - \tilde{\boldsymbol{m}}'\tilde{\boldsymbol{M}}^{-1}\tilde{\boldsymbol{m}}\right),$$

insertion into the posterior distribution (4.17) gives

$$\mathrm{P}(\boldsymbol{\beta}, \sigma^2 \mid \boldsymbol{y}) \propto \tfrac{1}{(\sigma^2)^{\frac{p}{2}}} \exp\left(-\tfrac{1}{2\sigma^2}(\boldsymbol{\beta} - \tilde{\boldsymbol{m}})'\tilde{\boldsymbol{M}}^{-1}(\boldsymbol{\beta} - \tilde{\boldsymbol{m}})\right) \tfrac{1}{(\sigma^2)^{\tilde{a}+1}} \exp\left(-\tfrac{\tilde{b}}{\sigma^2}\right).$$

Comparison with the density of a normal-inverse gamma distribution (compare the density (4.14) on p. 228) shows that the posterior is again a normal-inverse gamma distribution with parameters $\tilde{\boldsymbol{m}}$, $\tilde{\boldsymbol{M}}$, $\tilde{a}$, and $\tilde{b}$.

□

Derivation of the Posterior Mode for Bayesian Ridge Regression (p. 238)
As stated in the text, the derivation is based on fixed σ^2 and τ^2. The posterior for $\boldsymbol{\beta}$ is then given by

$$\begin{aligned}
p(\boldsymbol{\beta} \mid \boldsymbol{y}) &\propto p(\boldsymbol{y} \mid \boldsymbol{\beta})\, p(\beta_0)\, p(\tilde{\boldsymbol{\beta}}) \\
&\propto \exp\left(-\tfrac{1}{2\sigma^2}(\boldsymbol{y} - \boldsymbol{X}\boldsymbol{\beta})'(\boldsymbol{y} - \boldsymbol{X}\boldsymbol{\beta})\right) \exp\left(-\tfrac{1}{2\tau^2}\tilde{\boldsymbol{\beta}}'\tilde{\boldsymbol{\beta}}\right).
\end{aligned}$$

Maximizing the posterior is equivalent to maximizing $\log p(\boldsymbol{\beta} \mid \boldsymbol{y})$ or minimizing $-\log p(\boldsymbol{\beta} \mid \boldsymbol{y})$. This computes to

$$\begin{aligned}
-\log p(\boldsymbol{\beta} \mid \boldsymbol{y}) &= \tfrac{1}{2\sigma^2}(\boldsymbol{y} - \boldsymbol{X}\boldsymbol{\beta})'(\boldsymbol{y} - \boldsymbol{X}\boldsymbol{\beta}) + \tfrac{1}{2\tau^2}\tilde{\boldsymbol{\beta}}'\tilde{\boldsymbol{\beta}} \\
&= \tfrac{1}{2\sigma^2}\left((\boldsymbol{y} - \boldsymbol{X}\boldsymbol{\beta})'(\boldsymbol{y} - \boldsymbol{X}\boldsymbol{\beta}) + \lambda\tilde{\boldsymbol{\beta}}'\tilde{\boldsymbol{\beta}}\right),
\end{aligned}$$

with $\lambda = \sigma^2/\tau^2$. Hence the posterior mode is for given λ obtained by minimizing the penalized least squares criterion

$$PLS(\lambda) = (\boldsymbol{y} - \boldsymbol{X}\boldsymbol{\beta})'(\boldsymbol{y} - \boldsymbol{X}\boldsymbol{\beta}) + \lambda\,\tilde{\boldsymbol{\beta}}'\tilde{\boldsymbol{\beta}}.$$

Derivation of the Gibbs Sampler for the Bayesian LASSO (p. 239)

We develop a Gibbs sampler that subsequently draws from the full conditionals of $\boldsymbol{\beta} = (\beta_0, \tilde{\boldsymbol{\beta}})'$, σ^2, $\tau_1^2, \ldots, \tau_k^2$, and λ^2. All full conditionals are proportional to the posterior which is given by

$$p(\boldsymbol{\beta}, \sigma^2, \tau_1^2, \ldots, \tau_k^2, \lambda \mid \boldsymbol{y}) \propto p(\boldsymbol{y} \mid \boldsymbol{\beta}, \sigma^2)\, p(\beta_0) p(\tilde{\boldsymbol{\beta}} \mid \tau_1^2, \ldots, \tau_k^2) \prod_{j=1}^{k} p(\tau_j^2 \mid \lambda)\, p(\lambda^2).$$

To derive the specific distribution of the full conditionals we can always disregard those factors in the posterior that do not depend on the parameter of current interest.

Full Conditional of $\boldsymbol{\beta}$

To derive the full conditional for $\boldsymbol{\beta}$ we first note that $p(\boldsymbol{\beta} \mid \tau_1^2, \ldots, \tau_k^2) = p(\beta_0)$ $p(\tilde{\boldsymbol{\beta}} \mid \tau_1^2, \ldots, \tau_k^2)$ can be expressed as

$$p(\boldsymbol{\beta} \mid \tau_1^2, \ldots, \tau_k^2) \propto \exp\left(-\frac{1}{2}\boldsymbol{\beta}'\boldsymbol{K}\boldsymbol{\beta}\right),$$

where $\boldsymbol{K} = \mathrm{diag}(0, 1/\tau_1^2, \ldots, 1/\tau_k^2)$. Now the full conditional is given by

$$p(\boldsymbol{\beta} \mid \cdot) \propto \exp\left(-\frac{1}{2\sigma^2}(\boldsymbol{y} - \boldsymbol{X}\boldsymbol{\beta})'(\boldsymbol{y} - \boldsymbol{X}\boldsymbol{\beta})\right) \exp\left(-\frac{1}{2}\boldsymbol{\beta}'\boldsymbol{K}\boldsymbol{\beta}\right).$$

Since the least squares criterion can be written as

$$(\boldsymbol{y} - \boldsymbol{X}\boldsymbol{\beta})'(\boldsymbol{y} - \boldsymbol{X}\boldsymbol{\beta}) = (\boldsymbol{\beta} - \hat{\boldsymbol{\beta}})'\boldsymbol{X}'\boldsymbol{X}(\boldsymbol{\beta} - \hat{\boldsymbol{\beta}}) + \boldsymbol{y}'(\boldsymbol{I}_n - \boldsymbol{X}(\boldsymbol{X}'\boldsymbol{X})^{-1}\boldsymbol{X}')\boldsymbol{y}$$

(see the derivation on p. 211) with the second summand independent of $\boldsymbol{\beta}$, we obtain

$$p(\boldsymbol{\beta} \mid \cdot) \propto \exp\left(-\frac{1}{2\sigma^2}(\boldsymbol{\beta} - \hat{\boldsymbol{\beta}})'\boldsymbol{X}'\boldsymbol{X}(\boldsymbol{\beta} - \hat{\boldsymbol{\beta}})\right) \exp\left(-\frac{1}{2}\boldsymbol{\beta}'\boldsymbol{K}\boldsymbol{\beta}\right).$$

Invoking Theorem B.4 of Appendix B.3 with $\boldsymbol{A} = \frac{1}{\sigma^2}\boldsymbol{X}'\boldsymbol{X}$, $\boldsymbol{a} = \hat{\boldsymbol{\beta}}$, $\boldsymbol{B} = \boldsymbol{K}$, and $\boldsymbol{b} = \boldsymbol{0}$ shows that the full conditional is multivariate Gaussian with covariance matrix $\boldsymbol{\Sigma}_{\beta}$ and mean $\boldsymbol{\mu}_{\beta}$ given by

$$\boldsymbol{\Sigma} = \left(\frac{1}{\sigma^2}\boldsymbol{X}'\boldsymbol{X} + \boldsymbol{K}\right)^{-1} \qquad \boldsymbol{\mu}_{\beta} = \frac{1}{\sigma^2}\boldsymbol{\Sigma}_{\beta}\boldsymbol{X}'\boldsymbol{y}.$$

In summary we have $\boldsymbol{\beta} \mid \cdot \sim \mathrm{N}(\boldsymbol{\mu}_{\beta}, \boldsymbol{\Sigma}_{\beta})$.

Full Conditional of σ^2

The full conditional for σ^2 is given by

$$\begin{aligned} p(\sigma^2 \mid \cdot) &\propto p(\boldsymbol{y} \mid \boldsymbol{\beta}, \sigma^2)\, p(\sigma^2) \\ &\propto \tfrac{1}{(\sigma^2)^{n/2}} \exp\left(-\tfrac{1}{2\sigma^2}(\boldsymbol{y} - \boldsymbol{X\beta})'(\boldsymbol{y} - \boldsymbol{X\beta})\right) \tfrac{1}{(\sigma^2)^{a+1}} \exp\left(-\tfrac{b}{\sigma^2}\right) \\ &= \tfrac{1}{(\sigma^2)^{a+n/2+1}} \exp\left(-\tfrac{1}{\sigma^2}(b + \tfrac{1}{2}(\boldsymbol{y} - \boldsymbol{X\beta})'(\boldsymbol{y} - \boldsymbol{X\beta}))\right), \end{aligned}$$

which can be identified as an inverse gamma distribution with parameters

$$a^{new} = a + \frac{n}{2} \qquad b^{new} = b + \frac{1}{2}(\boldsymbol{y} - \boldsymbol{X\beta})'(\boldsymbol{y} - \boldsymbol{X\beta}).$$

Full Conditional of τ_j^2

Define

$$\mu = \left(\frac{\lambda^2}{\beta_j^2}\right)^{1/2}.$$

The full conditional of τ_j^2, $j = 1, \ldots, k$, is then given by

$$\begin{aligned} p(\tau_j^2 \mid \cdot) &\propto p(\beta_j \mid \tau_j^2)\, p(\tau_j^2 \mid \lambda) \\ &\propto \tfrac{1}{(\tau_j^2)^{1/2}} \exp\left(-\tfrac{1}{2\tau_j^2}\beta_j^2\right) \exp\left(-\tfrac{\lambda^2}{2}\tau_j^2\right) \\ &= \tfrac{1}{(\tau_j^2)^{1/2}} \exp\left(-\tfrac{\beta_j^2}{2\tau_j^2} - \tfrac{\lambda^2\tau_j^2}{2}\right) \\ &\propto \tfrac{1}{(\tau_j^2)^{1/2}} \exp\left(-\tfrac{\beta_j^2}{2\tau_j^2} + \tfrac{\lambda^2}{\mu} - \tfrac{\lambda^2\tau_j^2}{2}\right) \\ &\propto \tfrac{1}{(\tau_j^2)^{1/2}} \exp\left(-\tfrac{\lambda^2\tau_j^2}{2\mu^2}\left(\tfrac{1}{(\tau_j^2)^2} - 2\tfrac{\mu}{\tau_j^2} + \mu^2\right)\right) \\ &= \tfrac{1}{(\tau_j^2)^{1/2}} \exp\left(-\tfrac{\lambda^2\tau_j^2}{2\mu^2}\left(\tfrac{1}{\tau_j^2} - \mu\right)^2\right). \end{aligned}$$

This is not a standard distribution. However, it turns out that the distribution of $\omega_j = 1/\tau_j^2$ is inverse Gaussian with location parameter μ and scale parameter λ^2. Updating of τ_j^2 is then obtained by first sampling $1/\tau_j^2$ from the inverse Gaussian distribution and then inverting the result to obtain τ_j^2.

To derive the distribution of ω_j we apply the change of variables Theorem B.1 of Appendix B.1. With $\omega_j = g(\tau_j^2) = 1/\tau_j^2$, $g^{-1}(\omega_j) = 1/\omega_j$, and $g'(\tau_j^2) = -1/(\tau_j^2)^2$, we obtain

$$p(\omega_j \mid \cdot \propto \left(\frac{1}{\omega_j^3}\right)^{1/2} \exp\left(-\frac{\lambda^2}{2\mu\,\omega_j}(\omega_j - \mu)^2\right),$$

which has the form of the proposed inverse Gaussian distribution.

Full Conditional of λ^2

We update λ by drawing from the full conditional of λ^2. It is given by

$$\begin{aligned} p(\lambda^2 \mid \cdot) &\propto \prod_{j=1}^{k} p(\tau_j^2 \mid \lambda)\, p(\lambda^2) \\ &\propto \left(\frac{\lambda^2}{2}\right)^k \exp\left(-\tfrac{1}{2}\lambda^2 \sum_{j=1}^{k} \tau_j^2\right) (\lambda^2)^{a-1} \exp\left(-b\lambda^2\right) \\ &\propto (\lambda^2)^{a+k-1} \exp\left(-\lambda^2\left(b + \tfrac{1}{2}\sum_{j=1}^{k} \tau_j^2\right)\right), \end{aligned}$$

which can be identified with a gamma distribution. More specifically,

$$\lambda^2 \mid \cdot \sim \mathrm{G}\left(a + k, b + \frac{1}{2}\sum_{j=1}^{k} \tau_j^2\right).$$

Derivation of the Marginal Likelihood and the Bayes Factor in the Linear Model (p. 243)

We derive the marginal likelihood for model M_r given by

$$\boldsymbol{y} \mid \boldsymbol{\beta}, \sigma^2, M_r = \boldsymbol{y} \mid \beta_0, \tilde{\boldsymbol{\beta}}_r, \sigma^2, M_r \sim \mathrm{N}(\beta_0\mathbf{1} + \tilde{\boldsymbol{X}}_r\tilde{\boldsymbol{\beta}}_r, \sigma^2\boldsymbol{I}),$$

with priors $\beta_0, \tilde{\boldsymbol{\beta}}_r \mid \sigma^2, M_r \sim \mathrm{N}(\boldsymbol{m}_r, \sigma^2\boldsymbol{M}_r)$, and $\sigma^2 \mid M_r \sim \mathrm{IG}(a, b)$; see Sect. 4.4.3.

To derive the marginal distribution we write model M_r as

$$\begin{aligned} \boldsymbol{y} &= \boldsymbol{X}_r\boldsymbol{\beta}_r + \boldsymbol{\varepsilon}_1, \quad \boldsymbol{\varepsilon}_1 \sim \mathrm{N}(\mathbf{0}, \sigma^2\boldsymbol{I}), \\ \boldsymbol{\beta}_r &= \boldsymbol{m}_r + \boldsymbol{\varepsilon}_2 \qquad \boldsymbol{\varepsilon}_2, \sim \mathrm{N}(\mathbf{0}, \sigma^2\boldsymbol{M}_r), \end{aligned}$$

where $\boldsymbol{X}_r = (\mathbf{1}\,\tilde{\boldsymbol{X}}_r)$, $\boldsymbol{\beta}_r = (\beta_0, \tilde{\boldsymbol{\beta}}_r)'$, and $\boldsymbol{\varepsilon}_1$ and $\boldsymbol{\varepsilon}_2$ are stochastically independent. Inserting $\boldsymbol{\beta}_r$ into the first equation gives

$$\boldsymbol{y} = \boldsymbol{X}_r(\boldsymbol{m}_r + \boldsymbol{\varepsilon}_2) + \boldsymbol{\varepsilon}_1 = \boldsymbol{X}_r\boldsymbol{m}_r + \boldsymbol{X}_r\boldsymbol{\varepsilon}_2 + \boldsymbol{\varepsilon}_1 \sim \mathrm{N}(\boldsymbol{X}_r\boldsymbol{m}_r, \sigma^2(\boldsymbol{I} + \boldsymbol{X}_r\boldsymbol{M}_r\boldsymbol{X}_r')),$$

which is the conditional distribution of $\boldsymbol{y}$ given σ^2. To derive the covariance matrix $\sigma^2(\boldsymbol{I} + \boldsymbol{X}_r\boldsymbol{M}_r\boldsymbol{X}_r')$, we used properties 5 and 6 of Theorem B.2 in Appendix B.2. Since $\boldsymbol{y} \mid \sigma^2 \sim \mathrm{N}(\boldsymbol{X}_r\boldsymbol{m}_r, \sigma^2(\boldsymbol{I} + \boldsymbol{X}_r\boldsymbol{M}_r\boldsymbol{X}_r'))$ and $\sigma^2 \sim \mathrm{IG}(a,b)$, the joint distribution is $\boldsymbol{y}, \sigma^2 \sim \mathrm{NIG}(\boldsymbol{X}_r\boldsymbol{m}_r, \boldsymbol{I} + \boldsymbol{X}_r\boldsymbol{M}_r\boldsymbol{X}_r', a, b)$. Hence $\boldsymbol{y}$ is marginally multivariate t with $\nu = 2a$ degrees of freedom, location parameter $\boldsymbol{\mu} = \boldsymbol{X}_r\boldsymbol{m}_r$, and dispersion matrix $\boldsymbol{\Sigma} = \boldsymbol{I} + \boldsymbol{X}_r\boldsymbol{M}_r\boldsymbol{X}_r'$; see property 5 in Definition B.3.5 of Appendix B.3 and the derivation on p. 228.

To obtain the Bayes factor we need to compute the inverse $\boldsymbol{\Sigma}^{-1}$ of the dispersion matrix. Using the matrix inversion lemma (Appendix A.3, Theorem A.14) with $\boldsymbol{A} = \boldsymbol{I}$, $\boldsymbol{B} = -\boldsymbol{X}_r$, $\boldsymbol{C} = \boldsymbol{X}_r$, and $\boldsymbol{D} = \boldsymbol{M}_r^{-1}$, we obtain

$$\boldsymbol{\Sigma}^{-1} = (\boldsymbol{I} + \boldsymbol{X}_r\boldsymbol{M}_r\boldsymbol{X}_r')^{-1} = \boldsymbol{I} - \boldsymbol{X}_r(\boldsymbol{X}_r'\boldsymbol{X}_r + \boldsymbol{M}_r^{-1})^{-1}\boldsymbol{X}_r' = \boldsymbol{I} - \boldsymbol{X}_r\tilde{\boldsymbol{M}}_r\boldsymbol{X}_r', \tag{4.42}$$

with

$$\tilde{\boldsymbol{M}}_r = (\boldsymbol{X}_r'\boldsymbol{X}_r + \boldsymbol{M}_r^{-1})^{-1}.$$

The Bayes factor for two competing models M_r and M_s is obtained as

$$\mathrm{BF}_{js} = \frac{p(\boldsymbol{y} \mid M_r)}{p(\boldsymbol{y} \mid M_s)}$$

and results in:

$$\begin{aligned}
\mathrm{BF}_{js} &= \frac{t_{2a}(\boldsymbol{X}_r\boldsymbol{m}_r, \boldsymbol{I} + \boldsymbol{X}_r\boldsymbol{M}_r\boldsymbol{X}_r')}{t_{2a}(\boldsymbol{X}_s\boldsymbol{m}_s, \boldsymbol{I} + \boldsymbol{X}_s\boldsymbol{M}_s\boldsymbol{X}_s')} \\
&= \left(\frac{|\boldsymbol{I} + \boldsymbol{X}_s\boldsymbol{M}_s\boldsymbol{X}_s'|}{|\boldsymbol{I} + \boldsymbol{X}_r\boldsymbol{M}_r\boldsymbol{X}_j'|}\right)^{1/2} \left(\frac{2a + (\boldsymbol{y} - \boldsymbol{X}_s\boldsymbol{m}_s)'(\boldsymbol{I} - \boldsymbol{X}_s\tilde{\boldsymbol{M}}_s\boldsymbol{X}_s')(\boldsymbol{X}_s\boldsymbol{m}_s)}{2a + (\boldsymbol{y} - \boldsymbol{X}_r\boldsymbol{m}_r)'(\boldsymbol{I} - \boldsymbol{X}_r\tilde{\boldsymbol{M}}_r\boldsymbol{X}_r')(\boldsymbol{X}_r\boldsymbol{m}_r)}\right)^{a+n/2}.
\end{aligned}$$

In the literature, the Bayes factor is usually given in a slightly different form. Using Appendix A.4 property 4 of Theorem A.17 and Eq. (4.42) we obtain

$$\left(\frac{|\boldsymbol{I} + \boldsymbol{X}_s\boldsymbol{M}_s\boldsymbol{X}_s'|}{|\boldsymbol{I} + \boldsymbol{X}_r\boldsymbol{M}_r\boldsymbol{X}_r'|}\right)^{1/2} = \left(\frac{\boldsymbol{I} - \boldsymbol{X}_r\tilde{\boldsymbol{M}}_r\boldsymbol{X}_r'}{\boldsymbol{I} - \boldsymbol{X}_s\tilde{\boldsymbol{M}}_s\boldsymbol{X}_s'}\right)^{1/2}.$$

Invoking Sylvester's theorem (Appendix A.4, Theorem A.17, property 6) with $\boldsymbol{A} = \tilde{\boldsymbol{M}}_r^{-1} = \boldsymbol{X}_r'\boldsymbol{X}_r + \boldsymbol{M}_r^{-1}$, $\boldsymbol{B} = \boldsymbol{X}_r'$, and $\boldsymbol{C} = -\boldsymbol{X}_r$ yields

$$\begin{aligned}
&|\tilde{\boldsymbol{M}}_r^{-1} - \boldsymbol{X}_r'\boldsymbol{X}_r| = |\tilde{\boldsymbol{M}}_r^{-1}| \cdot |\boldsymbol{I} - \boldsymbol{X}_r\tilde{\boldsymbol{M}}_j\boldsymbol{X}_r'| \\
\Leftrightarrow\ &|\boldsymbol{M}_r^{-1} + \boldsymbol{X}_r'\boldsymbol{X}_r - \boldsymbol{X}_r'\boldsymbol{X}_r| = \frac{1}{|\tilde{\boldsymbol{M}}_r|}|\boldsymbol{I} - \boldsymbol{X}_r\tilde{\boldsymbol{M}}_r\boldsymbol{X}_r'| \\
\Leftrightarrow\ &\frac{|\tilde{\boldsymbol{M}}_r|}{|\boldsymbol{M}_r|} = |\boldsymbol{I} - \boldsymbol{X}_r\tilde{\boldsymbol{M}}_r\boldsymbol{X}_r'|.
\end{aligned}$$

We finally arrive at the Bayes factor

$$\mathrm{BF}_{rs} = \left(\frac{|\tilde{\boldsymbol{M}}_r||\boldsymbol{M}_s|}{|\tilde{\boldsymbol{M}}_s||\boldsymbol{M}_r|}\right)^{1/2} \left(\frac{2a + (\boldsymbol{y} - \boldsymbol{X}_s\boldsymbol{m}_s)'(\boldsymbol{I} - \boldsymbol{X}_s\tilde{\boldsymbol{M}}_s\boldsymbol{X}_s')(\boldsymbol{y} - \boldsymbol{X}_s\boldsymbol{m}_s)}{2a + (\boldsymbol{y} - \boldsymbol{X}_r\boldsymbol{m}_r)'(\boldsymbol{I} - \boldsymbol{X}_r\tilde{\boldsymbol{M}}_r\boldsymbol{X}_r')(\boldsymbol{y} - \boldsymbol{X}_r\boldsymbol{m}_r)}\right)^{a+n/2}.$$

Derivation of the Marginal Model Prior If a Beta Prior Is Assumed for θ (p. 252)

We obtain

$$\begin{aligned}
p(M_r) &= \int_0^1 p(M_r \mid \theta)\, p(\theta)\, d\theta \\
&= \int_0^1 \theta^{k_r}(1-\theta)^{k-k_r} \frac{\Gamma(a+b)}{\Gamma(a)\Gamma(b)} \theta^{a-1}(1-\theta)^{b-1}\, d\theta \\
&= \frac{\Gamma(a+b)}{\Gamma(a)\Gamma(b)} \int_0^1 \theta^{a+k_r-1}(1-\theta)^{b+k-k_r-1}\, d\theta \\
&= \frac{\Gamma(a+b)}{\Gamma(a)\Gamma(b)} \frac{\Gamma(a+k_r)\Gamma(b+k-k_r)}{\Gamma(a+b+k)} \int_0^1 \mathrm{Beta}(a+k_r, b+k-k_r)\, d\theta \\
&= \frac{\Gamma(a+b)}{\Gamma(a)\Gamma(b)} \frac{\Gamma(a+k_r)\Gamma(b+k-k_r)}{\Gamma(a+b+k)}.
\end{aligned}$$

The integration has been done by expanding the term within the integral sign such that the density of a $\mathrm{Beta}(a + k_r, b + k - k_r)$ distribution results. Then the corresponding integral over the unit interval is one.

Derivation of the Gibbs Sampler for the Continuous Spike and Slab Variance Prior (p. 253)

We develop a Gibbs sampler that consecutively samples from the full conditionals of $\boldsymbol{\beta}$, σ^2, $\tau_1^2, \ldots, \tau_k^2$, $\delta_1, \ldots, \delta_k$, and θ.

The full conditionals of $\boldsymbol{\beta}$ and σ^2 are identical to those of the Bayesian LASSO. It remains to derive the full conditionals of the variances τ_j^2, the indicators δ_j, and the prior inclusion parameter θ.

Full Conditional of τ_j^2

Suppose first that $\delta_j = 1$. We then have

$$\begin{aligned}
p(\tau_j^2 \mid \cdot) &\propto \frac{1}{(\tau_j^2)^{1/2}} \exp\left(-\frac{1}{2\tau_j^2}\beta_j^2\right) \frac{1}{(\tau_j^2)^{a_{\tau^2}+1}} \exp\left(-\frac{b_{\tau^2}}{\tau_j^2}\right) \\
&= \frac{1}{(\tau_j^2)^{a_{\tau^2}+1/2+1}} \exp\left(-\frac{1}{\tau_j^2}\left(b_{\tau^2} + \frac{1}{2}\beta_j^2\right)\right),
\end{aligned}$$

which is the $IG(a_{\tau^2} + 1/2, b_{\tau^2} + 1/2\beta_j^2)$ distribution.

For $\delta_j = 0$, we have to exchange b_{τ^2} by $\nu_0 b$ and arrive at the $\text{IG}(a_{\tau^2} + 1/2, \nu_0 b_{\tau^2} + 1/2\beta_j^2)$ distribution. Summarizing we have

$$\tau_j^2 \,|\, \cdot = (1-\delta_j)\text{IG}(a_{\tau^2} + 1/2, \nu_0 b_{\tau^2} + 1/2\beta_j^2) + \delta_j\ \text{IG}(a_{\tau^2} + 1/2, b_{\tau^2} + 1/2\beta_j^2).$$

Full Conditional of δ_j

For $\delta_j = 1$, we have

$$\text{P}(\delta_j = 1 \,|\, \cdot) \propto \theta \cdot \text{IG}(\tau_j^2, a, b),$$

where $IG(\tau_j^2, a, b)$ denotes the density of the respective inverse gamma distribution evaluated at the current value τ_j^2. For $\delta_j = 0$ the corresponding probability is given as

$$\text{P}(\delta_j = 0 \,|\, \cdot) \propto (1-\theta) \cdot \text{IG}(\tau_j^2, a, \nu_0 b).$$

Thus $\delta_j \,|\, \cdot \sim \text{B}(1, \theta^{new})$ with

$$\theta^{new} = \frac{\theta \cdot \text{IG}(\tau_j^2, a, b)}{\theta \cdot \text{IG}(\tau_j^2, a, b) + (1-\theta) \cdot \text{IG}(\tau_j^2, a, \nu_0 b)}.$$

Full Conditional of θ

Let

$$s = \sum_{j=1}^{k} \delta_j$$

be the number indicators with $\delta_j = 1$. Then the full conditional is given by

$$\begin{aligned} p(\theta \,|\, \cdot) &\propto \prod_{j=1}^{k} p(\delta_j \,|\, \theta) p(\theta) \\ &\propto \theta^s (1-\theta)^{k-s} \theta^{a-1} (1-\theta)^{b-1} \\ &= \theta^{a+s-1} (1-\theta)^{b+k-s-1}, \end{aligned}$$

which is a beta distribution. More specifically,

$$\theta \,|\, \cdot \sim \text{Beta}(a+s, b+k-s).$$

Generalized Linear Models 5

Linear models are well suited for regression analyses when the response variable is continuous and at least approximately normal. In some cases, an appropriate transformation is needed to ensure approximate normality of the response. In addition, the expectation of the response is assumed to be a linear combination of covariates. Again, these covariates may be transformed before being included in the linear predictor. However, in many applications the response is not a continuous variable, but rather binary, categorical, or a count variable as in the following examples:

- Patent opposition (yes/no), see Sect. 2.3 (p. 33).
- Creditworthiness of a client (yes/no).
- Benign or malignant tumor.
- Person is unemployed, part-time employed, or fully employed.
- Tree is very damaged, averagely or lightly damaged, or not damaged at all.
- Number of cases of illness, insurance claims, or problematic credits within a certain time frame.

Moreover, we are not always able to perform a satisfactory regression analysis for certain types of continuous response variables using a linear model. This is the case when dealing with a variable whose distribution is considerably skewed, as, for example, a life span, the amount of damages, or income. Even though data with a skewed distribution can sometimes be transformed into one with an approximately symmetric distribution, it is often advantageous to apply, for example, a gamma regression model to the original response variable.

Within a broad framework, generalized linear models (GLMs) unify many regression approaches with response variables that do not necessarily follow a normal distribution, including, for example, the logit model for binary response variables (Sect. 2.3) as well as the classical linear model with normally distributed errors. GLMs still rely on the assumption that the effect of covariates can be modeled through a linear predictor, similar as in logit and linear models. We start our description of GLMs with regression models for binary responses in Sect. 5.1. Next, Sect. 5.2 describes regression models for count data, especially Poisson regression. Section 5.3 is dedicated to models for nonnegative, continuous

L. Fahrmeir et al., *Regression*, DOI 10.1007/978-3-642-34333-9_5,

responses. Along with the introduction of suitable models, we discuss statistical inference relying on the likelihood principle. Section 5.4 offers a general unified discussion of GLMs and likelihood inference, while Sect. 5.5 outlines quasi-likelihood inference. Section 5.6 considers Bayesian GLMs. Finally, Sect. 5.7 transfers the boosting idea outlined for linear models in Sect. 4.3 to GLMs.

5.1 Binary Regression

5.1.1 Binary Regression Models

As in the previous chapters, we assume that (ungrouped) data on n objects or individuals are given in the form $(y_i, x_{i1}, \ldots, x_{ik})$, $i = 1, \ldots, n$, with the binary response y coded by 0 and 1 and covariates denoted by $x_1, \ldots, x_k$. Similar to linear and logit models in Example 2.8, $x_1, \ldots, x_k$ may have been derived from an appropriate transformation or coding of the original covariates. The main goal of a binary regression analysis is then to model and estimate the effects of the covariates on the (conditional) probability

$$\pi_i = \mathrm{P}(y_i = 1) = \mathrm{E}(y_i),$$

for the outcome $y_i = 1$ and given values of the covariates $x_{i1}, \ldots, x_{ik}$. In this specification, the response variables are assumed to be (conditionally) independent.

We already discussed the disadvantages of the *linear probability model*

$$\pi_i = \beta_0 + \beta_1 x_{i1} + \ldots + \beta_k x_{ik}$$

for binary response variables in Sect. 2.3. In particular, the *linear predictor*

$$\eta_i = \beta_0 + \beta_1 x_{i1} + \ldots + \beta_k x_{ik} = \boldsymbol{x}_i' \boldsymbol{\beta},$$

with $\boldsymbol{\beta} = (\beta_0, \beta_1, \ldots, \beta_k)'$ and $\boldsymbol{x}_i = (1, x_{i1}, \ldots, x_{ik})'$ must lie in the interval $[0, 1]$ for all vectors $\boldsymbol{x}$. This requires restrictions on the parameters $\boldsymbol{\beta}$ that are difficult to handle in the estimation process. Thus, all popular binary regression models combine the probability π_i with the linear predictor η_i through a relation of the form

$$\pi_i = h(\eta_i) = h(\beta_0 + \beta_1 x_{i1} + \ldots + \beta_k x_{ik}), \tag{5.1}$$

where h is a strictly monotonically increasing cumulative distribution function on the real line. This ensures $h(\eta) \in [0, 1]$ and Eq. (5.1) can always be expressed in the form

$$\eta_i = g(\pi_i),$$

with the inverse function $g = h^{-1}$. Within the framework of GLMs, h is called the *response function* and $g = h^{-1}$ is known as the *link function*. Logit and probit models are the most widely used binary regression models.

Logit Model

The logit model presented in Sect. 2.3 results from the choice of the logistic response function

$$\pi = h(\eta) = \frac{\exp(\eta)}{1 + \exp(\eta)} \tag{5.2}$$

or (equivalently) the logit link function

$$g(\pi) = \log\left(\frac{\pi}{1-\pi}\right) = \eta = \beta_0 + \beta_1 x_1 + \ldots + \beta_k x_k. \tag{5.3}$$

This yields a linear model for the logarithmic odds (log-odds) $\log(\pi/(1-\pi))$. Transformation with the exponential function gives

$$\frac{\pi}{1-\pi} = \exp(\beta_0)\exp(\beta_1 x_1) \cdot \ldots \cdot \exp(\beta_k x_k), \tag{5.4}$$

implying that the effects of the covariates affect the odds $\pi/(1-\pi)$ in an exponential-multiplicative form; see Sect. 2.3 for this interpretation. Another interpretation—which is also available for the two models introduced in the following—results from the connection to latent linear models; see p. 274 for details.

Probit Model

For h, we use the standard normal cumulative distribution function Φ, i.e.,

$$\pi = \Phi(\eta) = \Phi(\boldsymbol{x}'\boldsymbol{\beta}). \tag{5.5}$$

A (minor) disadvantage is the required numerical evaluation of Φ in the maximum likelihood estimation of the parameter $\boldsymbol{\beta}$.

Complementary Log–Log Model

The complementary log–log model uses the extreme minimum-value cumulative distribution function

$$h(\eta) = 1 - \exp(-\exp(\eta)) \tag{5.6}$$

as response function, with the inverse

$$g(\pi) = \log(-\log(1-\pi))$$

as link function. In comparison to logit and probit models, this model is useful in more specific applications, for example, when modeling discrete duration times; see, e.g., Fahrmeir and Tutz (2001) for an introduction to discrete time duration models.

Figure 5.1 (left) shows the response functions of the three binary regression models, i.e., the logistic distribution function (5.2), the standard normal distribution function (5.5), and the extreme-value distribution function (5.6).

At first glance, the three models seem very different from each other: Even though the response function of logit and probit models are both symmetric around

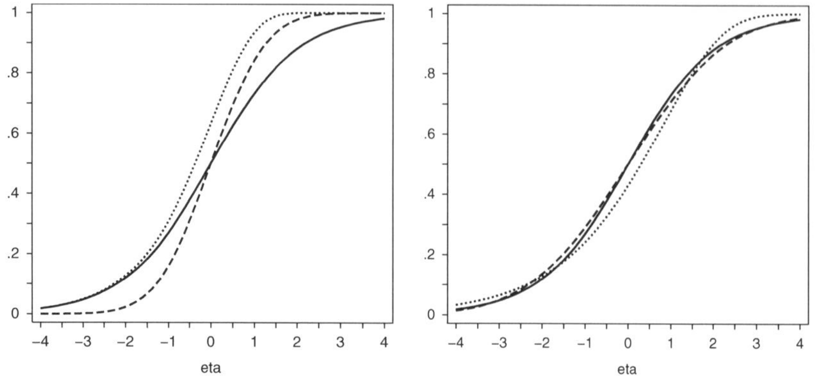

Fig. 5.1 Response functions (*left*) and adjusted response functions (*right*) in binary regression models: logit model (—), probit model (- - -), complementary log–log model (···)

zero, the logistic distribution function approaches 0 or 1 much slower for $\eta \to -\infty$ or $\eta \to +\infty$, respectively. In contrast, the response function of the complementary log–log model is asymmetric, following a similar pattern as the logit response function for small η, but showing a faster approach towards 1 as $\eta \to +\infty$. Thus, statistical analyses involving the three models might be expected to lead to very different results. However, for an adequate comparison of the models, we have to keep in mind that we could have used the more general cumulative distribution function h of a $\mathrm{N}(0, \sigma^2)$ distribution with any choice of variance $\sigma^2 \neq 1$, rather than the standard normal cumulative distribution function of the N(0, 1) distribution that defines the probit model. Standardizing h yields the relation

$$\pi(\eta) = h(\boldsymbol{x}'\boldsymbol{\beta}) = \Phi(\boldsymbol{x}'\boldsymbol{\beta}/\sigma) = \Phi(\boldsymbol{x}'\tilde{\boldsymbol{\beta}}),$$

where $\tilde{\boldsymbol{\beta}} = \boldsymbol{\beta}/\sigma$. Hence, even though the two response functions Φ (with $\sigma^2 = 1$) and h (with $\sigma^2 \neq 1$, e.g. $\sigma^2 = 4$) differ from each other, the resulting model for the probability $\pi(\eta)$ based on $h(\eta)$ with $\eta = \boldsymbol{x}'\boldsymbol{\beta}$ is equivalent to a probit model with the rescaled parameters $\tilde{\boldsymbol{\beta}} = \boldsymbol{\beta}/\sigma$. In this sense, the requirement of $\sigma^2 = 1$ in the probit model is arbitrary and we might just as well have assumed $\sigma^2 = 4$. We also obtain the same equivalence when deriving binary regression models from latent linear models; see p. 274.

For a fair comparison of logit and probit models, we need to put each on equal footing. Since the logistic distribution function has variance $\pi^2/3$ with the circular constant $\pi = 3.141593\ldots$, we need to compare it to a rescaled normal distribution function whose variance is adjusted to $\sigma^2 = \pi^2/3$. Figure 5.1 (right) shows the similarity of the logit and the adjusted probit response function.

Statistical analyses with logit and probit models therefore lead to similar estimated probabilities. The scaling $\tilde{\boldsymbol{\beta}} = \boldsymbol{\beta}/\sigma$ or $\boldsymbol{\beta} = \tilde{\boldsymbol{\beta}}\,\sigma$ will automatically be taken into account in the estimation process. Thus, the estimated coefficients

5.1 Binary Regression Models

Data

The binary response variables y_i are coded 0/1 and are (conditionally) independent given the covariates $x_{i1}, \ldots, x_{ik}$.

Models

The probability $\pi_i = \mathrm{P}(y_i = 1) = \mathrm{E}(y_i)$ and the *linear predictor*

$$\eta_i = \beta_0 + \beta_1 x_{i1} + \ldots + \beta_k x_{ik} = \boldsymbol{x}_i' \boldsymbol{\beta}$$

are connected by the *response function* $h(\eta) \in [0, 1]$ via

$$\pi_i = h(\eta_i).$$

Logit model

$$\pi = \frac{\exp(\eta)}{1 + \exp(\eta)} \quad \Longleftrightarrow \quad \log \frac{\pi}{1 - \pi} = \eta.$$

Probit model

$$\pi = \Phi(\eta) \quad \Longleftrightarrow \quad \Phi^{-1}(\pi) = \eta.$$

Complementary log–log model

$$\pi = 1 - \exp(-\exp(\eta)) \quad \Longleftrightarrow \quad \log(-\log(1 - \pi)) = \eta.$$

$\tilde{\beta}_1, \tilde{\beta}_2, \ldots$ of a logit model differ from the corresponding values $\beta_1, \beta_2, \ldots$ of a probit model (with $\sigma^2 = 1$) approximately by the factor $\sigma = \pi/\sqrt{3} \approx 1.814$, yet the estimated probabilities $\pi(\eta)$ are very similar. Since the ratios $\tilde{\beta}_1/\tilde{\beta}_2 = \beta_1/\beta_2$ etc. are independent of σ, therefore we should not interpret the absolute (estimated) coefficients, but rather the ratios β_1/β_2 etc., as illustrated in Example 5.1 (p. 275).

Similar considerations apply to the comparison with the complementary log–log model. Since the extreme-value distribution has variance $\sigma^2 = \pi^2/6$ and expectation -0.5772, the response function has to be adjusted to the variance $\sigma^2 = \pi^2/3$ and expectation 0 for a comparison with the logistic distribution function. This adjustment does have additional impact on the (estimated) intercept β_0. Figure 5.1 (right) shows the corresponding adjusted response function, which follows a similar form as those of the logit and probit function for small η, but also shows clear differences for larger η. Accordingly, the results of statistical analyses obtained with the complementary log–log model differ more substantially from those obtained by logit or probit models.

Binary Models and Latent Linear Models

Binary regression models can be derived by considering a *latent (unobserved) continuous response variable*, which is connected with the observed binary response via a threshold mechanism. Suppose we are investigating the decision of some individuals $i = 1, \dots, n$ when choosing between two alternatives $y = 0$ and $y = 1$. Typical examples include decision problems, e.g., related to buying a certain product or not. We further assume that individuals assign utilities u_{i0} and u_{i1} to each of the two alternatives. The alternative that maximizes the utility is chosen, i.e., $y_i = 1$ if $u_{i1} > u_{i0}$ and $y_i = 0$ if $u_{i1} \le u_{i0}$.

Now suppose a researcher investigates the choice problem. However, one is not able to observe the latent utilities behind the decision, but rather observes the binary decisions y_i together with a number of explanatory variables $x_{i1}, \dots, x_{ik}$, which may influence the choice between the two alternatives. Assuming that the unobserved utilities can be additively decomposed and follow a linear model, we obtain

$$u_{i1} = \boldsymbol{x}_i'\tilde{\boldsymbol{\beta}}_1 + \tilde{\varepsilon}_{i1},$$

$$u_{i0} = \boldsymbol{x}_i'\tilde{\boldsymbol{\beta}}_0 + \tilde{\varepsilon}_{i0},$$

with $\boldsymbol{x}_i = (1, x_{i1}, \dots, x_{ik})'$. The unknown coefficient vectors $\tilde{\boldsymbol{\beta}}_1$ and $\tilde{\boldsymbol{\beta}}_0$ determine the effect of the explanatory variables on the utilities. The "errors" $\tilde{\varepsilon}_{i1}$ and $\tilde{\varepsilon}_{i0}$ include the effects of unobserved explanatory variables. Equivalently, we may choose to investigate utility differences, then obtaining

$$\tilde{y}_i = u_{i1} - u_{i0} = \boldsymbol{x}_i'(\tilde{\boldsymbol{\beta}}_1 - \tilde{\boldsymbol{\beta}}_0) + \tilde{\varepsilon}_{i1} - \tilde{\varepsilon}_{i0} = \boldsymbol{x}_i'\boldsymbol{\beta} + \varepsilon_i,$$

with $\boldsymbol{\beta} = \tilde{\boldsymbol{\beta}}_1 - \tilde{\boldsymbol{\beta}}_0$ and $\varepsilon_i = \tilde{\varepsilon}_{i1} - \tilde{\varepsilon}_{i0}$. The connection to the observable binary variables y_i is now given by $y_i = 1$ if $\tilde{y}_i = u_{i1} - u_{i0} > 0$ and $y_i = 0$ if $\tilde{y}_i = u_{i1} - u_{i0} \le 0$.

Based on this framework, the binary responses y_i follow a Bernoulli distribution, i.e., $y_i \sim \mathrm{B}(1, \pi_i)$ with

$$\pi_i = \mathrm{P}(y_i = 1) = \mathrm{P}(\tilde{y}_i > 0) = \mathrm{P}(\boldsymbol{x}_i'\boldsymbol{\beta} + \varepsilon_i > 0) = \int I(\boldsymbol{x}_i'\boldsymbol{\beta} + \varepsilon_i > 0) f(\varepsilon_i)\, d\varepsilon_i,$$

where $I(\cdot)$ is the indicator function and f is the probability density of ε_i. We obtain different models depending on the choice of f. Specifically, when ε_i follows a logistic distribution, we obtain the logit model, while for standard normal errors $\varepsilon_i \sim \mathrm{N}(0, 1)$ we have the probit model $\pi_i = \Phi(\boldsymbol{x}_i'\boldsymbol{\beta})$. For $\varepsilon_i \sim \mathrm{N}(0, \sigma^2)$, we have

$$\pi_i = \Phi(\boldsymbol{x}_i'\boldsymbol{\beta}/\sigma) = \Phi(\boldsymbol{x}_i'\tilde{\boldsymbol{\beta}}),$$

through standardization with $\tilde{\boldsymbol{\beta}} = \boldsymbol{\beta}/\sigma$. This implies that regression coefficients $\boldsymbol{\beta}$ of a latent linear regression model can only be identified up to a factor $1/\sigma$. However, the ratio of two coefficients, for example, β_1 and β_2, is identifiable, since $\beta_1/\beta_2 = \tilde{\beta}_1/\tilde{\beta}_2$.

5.2 Interpretation of the Logit Model

Based on the linear predictor

$$\eta_i = \beta_0 + \beta_1 x_{i1} + \ldots + \beta_k x_{ik} = \boldsymbol{x}_i'\boldsymbol{\beta},$$

the odds

$$\frac{\pi_i}{1-\pi_i} = \frac{\mathrm{P}(y_i = 1 \,|\, \boldsymbol{x}_i)}{\mathrm{P}(y_i = 0 \,|\, \boldsymbol{x}_i)}$$

follow the multiplicative model

$$\frac{\mathrm{P}(y_i = 1 \,|\, \boldsymbol{x}_i)}{\mathrm{P}(y_i = 0 \,|\, \boldsymbol{x}_i)} = \exp(\beta_0) \cdot \exp(x_{i1}\beta_1) \cdot \ldots \cdot \exp(x_{ik}\beta_k).$$

If, for example, x_{i1} increases by 1 unit to $x_{i1} + 1$, the following changes apply to the relationship of the odds:

$$\frac{\mathrm{P}(y_i = 1 \,|\, x_{i1}, \ldots)}{\mathrm{P}(y_i = 0 \,|\, x_{i1}, \ldots)} \Big/ \frac{\mathrm{P}(y_i = 1 \,|\, x_{i1} + 1, \ldots)}{\mathrm{P}(y_i = 0 \,|\, x_{i1} + 1, \ldots)} = \exp(\beta_1).$$

$$\beta_1 > 0: \ \mathrm{P}(y_i = 1)/\mathrm{P}(y_i = 0) \text{ increases},$$

$$\beta_1 < 0: \ \mathrm{P}(y_i = 1)/\mathrm{P}(y_i = 0) \text{ decreases},$$

$$\beta_1 = 0: \ \mathrm{P}(y_i = 1)/\mathrm{P}(y_i = 0) \text{ remains unchanged}.$$

Interpretation of Parameters

One of the main reasons for the popularity of the logit model is its interpretation as a linear model for log-odds, as well as a multiplicative model for the odds $\pi/(1-\pi)$, as outlined in Sect. 2.3 and formulae (5.3) and (5.4). The latent linear model is useful to interpret effects in the probit model, since the covariate effects can be interpreted in the usual way with this model formulation (up to a common multiplicative factor). In general, interpretation best proceeds in two steps: For the linear predictor, we interpret the effects in the same way as in the linear model. Then we transform the linear effect for $\eta = \boldsymbol{x}'\boldsymbol{\beta}$ into a nonlinear effect for $\pi = h(\eta)$ with the response function h.

Example 5.1 Patent Opposition—Binary Regression

In Example 2.8 (p. 35), we analyzed the probability of patent opposition using a logit model with linear predictor

⇩

Table 5.1 Patent opposition: estimated regression coefficients for the logit, probit, and complementary log–log model. Adjusted coefficients for the probit and complementary log–log model are also included

Variable	Logit	Probit	Probit (adj.)	Log–Log	Log–Log (adj.)
intercept	201.740	119.204	216.212	164.519	211.744
year	−0.102	−0.060	−0.109	−0.083	−0.106
ncit	0.113	0.068	0.123	0.088	0.113
nclaim	0.026	0.016	0.029	0.021	0.027
ustwin	−0.406	−0.243	−0.441	−0.310	−0.398
patus	−0.526	−0.309	−0.560	−0.439	−0.563
patgsgr	0.196	0.121	0.219	0.154	0.198
ncountry	0.097	0.058	0.105	0.080	0.103

$$\eta_i = \beta_0 + \beta_1 year_i + \beta_2 ncit_i + \beta_3 nclaims_i + \beta_4 ustwin_i$$
$$+\beta_5 patus_i + \beta_6 patgsgr_i + \beta_7 ncountry_i.$$

For an interpretation of the estimated parameters in the logit model compare Example 2.8. For comparison, we now choose a probit model and a complementary log–log model using the same linear predictor and reanalyze the data. Table 5.1 contains parameter estimates for all three models. In order to compare the probit and logit fits, we have to multiply the estimated coefficients of the probit model with the factor $\pi/\sqrt{3} \approx 1.814$, following our previous considerations. For example, we obtain the estimated effect $-0.060 \cdot 1.814 \approx -0.109$ for the covariate *year* compared to -0.102 in the logit model. For the other coefficients, somewhat higher discrepancies occur at some places (see the fourth column in Table 5.1); however, the discrepancies are much smaller than the standard deviations of the estimates. Since, according to the interpretation of binary models, coefficients can only be interpreted up to a factor of $1/\sigma$, the probit and the logit models provide essentially the same results. After rescaling with the factor $\pi/\sqrt{6} \approx 1.283$, we also obtain comparable coefficients for the complementary log–log model, which are close to those of the logit model; see column 6 in Table 5.1.

△

Grouped Data

Thus far, we have assumed *individual data* or *ungrouped data*, which means that one observation $(y_i, \boldsymbol{x}_i)$ is given for each individual or object i in a sample of size n. Every binary, 0/1 coded value y_i of the response variable and every covariate vector $\boldsymbol{x}_i = (x_{i1}, \ldots, x_{ik})$ then belongs to exactly one unit $i = 1, \ldots, n$.

If some covariate vectors (i.e., rows of the design matrix) are identical, the data can be *grouped* as in Sect. 4.1.2 (p. 181). Specifically, after sorting and summarizing the data, the design matrix only contains rows with unique covariate vectors $\boldsymbol{x}_i$. In addition, the number n_i of replications of $\boldsymbol{x}_i$ in the original sample of the individual data and the relative frequencies $\bar{y}_i$ of the corresponding individual binary values of the response variables are given:

$$\begin{array}{c} \text{Group 1} \\ \vdots \\ \text{Group } i \\ \vdots \\ \text{Group } G \end{array} \begin{bmatrix} n_1 \\ \vdots \\ n_i \\ \vdots \\ n_G \end{bmatrix} \begin{bmatrix} \bar{y}_1 \\ \vdots \\ \bar{y}_i \\ \vdots \\ \bar{y}_G \end{bmatrix} \begin{bmatrix} 1 & x_{11} & \cdots & x_{1k} \\ & \vdots & & \vdots \\ 1 & x_{i1} & & x_{ik} \\ & \vdots & & \vdots \\ 1 & x_{G1} & \cdots & x_{Gk} \end{bmatrix}$$

The number of unique covariate vectors in the sample G is often much smaller than the sample size n, especially when covariates are binary or categorial. Rather than *relative* frequencies $\bar{y}_i$, we can also provide the *absolute* frequencies $n_i \bar{y}_i$. Grouped data are then often presented in condensed form in a contingency table, as in the following Example 5.2.

The grouping of individual data decreases computing time, as well as memory requirements, and is also done to ensure data identification protection. Moreover, some inferential methods are only applicable for grouped data, especially when testing the goodness of fit for the model or for model diagnostics; see Sect. 5.1.4 (p. 287).

Individual data y_i are Bernoulli distributed with $P(y_i = 1) = \pi_i$, i.e. $y_i \sim B(1, \pi_i)$. If the response variables y_i are (conditionally) independent, the absolute frequencies $n_i \bar{y}_i$ of grouped data are binomially distributed, i.e.,

$$n_i \bar{y}_i \sim B(n_i, \pi_i),$$

with $E(n_i \bar{y}_i) = n_i \pi_i$, $Var(n_i \bar{y}_i) = n_i \pi_i (1 - \pi_i)$. The relative frequencies then follow a "scaled" binomial distribution

$$\bar{y}_i \sim B(n_i; \pi_i)/n_i,$$

i.e., the range of values of the probability function for relative frequencies is $\{0, 1/n_i, 2/n_i, \ldots, 1\}$, instead of $\{0, 1, 2, \ldots, n_i\}$. The probability function is

$$P(\bar{y}_i = j/n_i) = \binom{n_i}{j} \pi_i^j (1 - \pi_i)^{n_i - j} \qquad j = 0, \ldots, n_i.$$

The mean and the variance are given by

$$E(\bar{y}_i) = \pi_i, \quad Var(\bar{y}_i) = \frac{\pi_i (1 - \pi_i)}{n_i}.$$

For modeling the probability π_i, we can use the same binary regression models as in case of individual data.

Table 5.2 Grouped infection data

	C-section			
	Planned		Not planned	
	Infection		Infection	
	Yes	No	Yes	No
Antibiotics				
Risk factor	1	17	11	87
No risk factor	0	2	0	0
No antibiotics				
Risk factor	28	30	23	3
No risk factor	8	32	0	9

Example 5.2 Caesarean Delivery—Grouped Data

Table 5.2 contains grouped data on infections of mothers after a C-section collected at the clinical center of the University of Munich. The response variable y "infection" is binary with

$$y = \begin{cases} 1 & \text{infection,} \\ 0 & \text{no infection.} \end{cases}$$

After each childbirth the following three binary covariates were collected:

$$NPLAN = \begin{cases} 1 & \text{C-section was not planned,} \\ 0 & \text{planned,} \end{cases}$$

$$RISK = \begin{cases} 1 & \text{risk factors existed,} \\ 0 & \text{no risk factors,} \end{cases}$$

$$ANTIB = \begin{cases} 1 & \text{antibiotics were administered as prophylaxis,} \\ 0 & \text{no antibiotics.} \end{cases}$$

After grouping the individual data of 251 mothers, the data can be represented in the form of a contingency table; see Table 5.2.

If we model the probability for an infection with a logit model

$$\log \frac{\text{P(Infection)}}{\text{P(No Infection)}} = \beta_0 + \beta_1 \, NPLAN + \beta_2 \, RISK + \beta_3 \, ANTIB,$$

we obtain the estimated coefficients

$$\hat{\beta}_0 = -1.89, \quad \hat{\beta}_1 = 1.07, \quad \hat{\beta}_2 = 2.03, \quad \hat{\beta}_3 = -3.25.$$

The multiplicative effect $\exp(\hat{\beta}_2) = 7.6$ implies that the odds of an infection is seven times higher when risk factors are present, for fixed levels of the other two factors. Such an interpretation of course requires that the chosen model without any interaction terms is adequate. We will return to this question in Example 5.3.

If we select a probit model with the same linear predictor, we obtain the estimated coefficients

$$\hat{\beta}_0 = -1.09, \quad \hat{\beta}_1 = 0.61, \quad \hat{\beta}_2 = 1.20, \quad \hat{\beta}_3 = -1.90.$$

Similar to Example 5.1, the absolute values seem to be very different. However, the relative effects, e.g., the ratios $\hat{\beta}_1/\hat{\beta}_2$, are again very similar.

△

Overdispersion

For grouped data, we can estimate the variance within a group via $\bar{y}_i(1-\bar{y}_i)/n_i$, since $\bar{y}_i$ is the ML estimator for π_i based on the data in group i, disregarding the covariate information. In applications, this *empirical* variance is often much larger than the variance $\hat{\pi}_i(1-\hat{\pi}_i)/n_i$ predicted by a binomial regression model with $\hat{\pi}_i = h(\boldsymbol{x}_i'\hat{\boldsymbol{\beta}})$. This phenomenon is called overdispersion, since the data show a higher variability than is presumed by the model. The two main reasons for overdispersion are *unobserved heterogeneity*, which remains unexplained by the observed covariates, and *positive correlations* between the individual binary observations of the response variables, for example, when individual units belong to one *cluster* such as the same household. In either case, the individual binary response variables within a group are then (in most cases positively) correlated. The sum of binary responses is then no longer binomially distributed and has a larger variance according to the variance formula for correlated variables; see in Appendix B.2 Theorem B.2.4. This situation occurs in Sect. 5.2 for Poisson distributed response variables, where a data example of overdispersion is presented.

The easiest way to address the increased variability is through the introduction of a multiplicative overdispersion parameter $\phi > 1$ into the variance formula, i.e., we assume

$$\text{Var}(y_i) = \phi \frac{\pi_i(1-\pi_i)}{n_i}.$$

Estimation of the overdispersion parameter is described in Sect. 5.1.5.

5.1.2 Maximum Likelihood Estimation

The primary goal of statistical inference is the estimation of parameters $\boldsymbol{\beta} = (\beta_0, \beta_1, \ldots, \beta_k)'$ and hypothesis testing for these effects, similar to linear models in Chap. 3. The methodology of this section is based on the likelihood principle: For given data $(y_i, \boldsymbol{x}_i)$, estimation of the parameters relies on the maximization of the likelihood function. Hypotheses regarding the parameters are tested using either likelihood ratio, Wald, or score tests; see Sect. 5.1.3. Appendix B.4.4 provides a general introduction into likelihood-based hypothesis testing.

Due to the (conditional) independence of the response variables, the likelihood $L(\boldsymbol{\beta})$ is given as the product

$$L(\boldsymbol{\beta}) = \prod_{i=1}^{n} f(y_i \mid \boldsymbol{\beta}) \tag{5.7}$$

of the densities of y_i, which depend on the unknown parameter $\boldsymbol{\beta}$ through $\pi_i = \text{E}(y_i) = h(\boldsymbol{x}_i'\boldsymbol{\beta})$. Maximization of $L(\boldsymbol{\beta})$ or the log-likelihood $l(\boldsymbol{\beta}) = \log(L(\boldsymbol{\beta}))$

then yields the ML estimator $\hat{\boldsymbol{\beta}}$. It turns out that the ML estimator has no closed form as for linear models. Instead we rely on iterative methods, in particular Fisher scoring as briefly described in Appendix B.4.2. In order to compute the ML estimator numerically we require the score function $\boldsymbol{s}(\boldsymbol{\beta})$ and the observed or expected Fisher matrix $\boldsymbol{H}(\boldsymbol{\beta})$ or $\boldsymbol{F}(\boldsymbol{\beta})$.

We consider the case of individual data and describe the necessary steps for deriving ML estimates in the binary logit model:

1. Likelihood

For binary response variables $y_i \sim \mathrm{B}(1, \pi_i)$ with $\pi_i = \mathrm{P}(y_i = 1) = \mathrm{E}(y_i) = \mu_i$, the (discrete) density is given by

$$f(y_i \mid \pi_i) = \pi_i^{y_i}(1 - \pi_i)^{1-y_i}.$$

Since $\pi_i = h(\boldsymbol{x}_i'\boldsymbol{\beta})$, the density depends on $\boldsymbol{\beta}$ for given $\boldsymbol{x}_i$, and we can therefore also denote it as $f(y_i \mid \boldsymbol{\beta})$. The density also defines the likelihood contribution $L_i(\boldsymbol{\beta})$ of the ith observation. Due to the (conditional) independence of the responses y_i, the likelihood $L(\boldsymbol{\beta})$ is given by

$$L(\boldsymbol{\beta}) = \prod_{i=1}^{n} L_i(\boldsymbol{\beta}) = \prod_{i=1}^{n} \pi_i^{y_i}(1 - \pi_i)^{1-y_i},$$

i.e., the product of the individual likelihood contributions $L_i(\boldsymbol{\beta})$.

2. Log-likelihood

The log-likelihood results from taking the logarithm of the likelihood yielding

$$l(\boldsymbol{\beta}) = \sum_{i=1}^{n} l_i(\boldsymbol{\beta}) = \sum_{i=1}^{n} \{y_i \log(\pi_i) - y_i \log(1 - \pi_i) + \log(1 - \pi_i)\},$$

with the *log-likelihood contributions*

$$\begin{aligned} l_i(\boldsymbol{\beta}) = \log L_i(\boldsymbol{\beta}) &= y_i \log(\pi_i) - y_i \log(1 - \pi_i) + \log(1 - \pi_i) \\ &= y_i \log\left(\frac{\pi_i}{1 - \pi_i}\right) + \log(1 - \pi_i). \end{aligned}$$

For the logit model, we have

$$\pi_i = \frac{\exp(\boldsymbol{x}_i'\boldsymbol{\beta})}{1 + \exp(\boldsymbol{x}_i'\boldsymbol{\beta})} \quad \text{or} \quad \log\left(\frac{\pi_i}{1 - \pi_i}\right) = \boldsymbol{x}_i'\boldsymbol{\beta} = \eta_i$$

and $(1 - \pi_i) = (1 + \exp(\boldsymbol{x}_i'\boldsymbol{\beta}))^{-1}$. Therefore we obtain

$$l_i(\boldsymbol{\beta}) = y_i(\boldsymbol{x}_i'\boldsymbol{\beta}) - \log(1 + \exp(\boldsymbol{x}_i'\boldsymbol{\beta})) = y_i\eta_i - \log(1 + \exp(\eta_i)).$$

3. Score function
To calculate the ML estimator, defined as the maximizer of the log-likelihood $l(\boldsymbol{\beta})$, we require the score function, i.e., the first derivative of $l(\boldsymbol{\beta})$ with respect to $\boldsymbol{\beta}$:

$$s(\boldsymbol{\beta}) = \frac{\partial l(\boldsymbol{\beta})}{\partial \boldsymbol{\beta}} = \sum_{i=1}^{n} \frac{\partial l_i(\boldsymbol{\beta})}{\partial \boldsymbol{\beta}} = \sum_{i=1}^{n} s_i(\boldsymbol{\beta}).$$

The individual contributions are given by $s_i(\boldsymbol{\beta}) = \partial l_i(\boldsymbol{\beta})/\partial \boldsymbol{\beta}$, or more specifically for logistic regression, using the chain rule,

$$\frac{\partial l_i(\boldsymbol{\beta})}{\partial \boldsymbol{\beta}} = \frac{\partial l_i}{\partial \eta_i}\frac{\partial \eta_i}{\partial \boldsymbol{\beta}} = \left[y_i - \frac{1}{1+\exp(\eta_i)}\exp(\eta_i) \right] \boldsymbol{x}_i,$$

with p-dimensional vector $\partial \eta_i/\partial \boldsymbol{\beta} = \boldsymbol{x}_i$. Further substitution of $\pi_i = \exp(\boldsymbol{x}_i'\boldsymbol{\beta})/(1+\exp(\boldsymbol{x}_i'\boldsymbol{\beta}))$ provides

$$s_i(\boldsymbol{\beta}) = \boldsymbol{x}_i(y_i - \pi_i)$$

and the score function

$$s(\boldsymbol{\beta}) = \sum_{i=1}^{n} \boldsymbol{x}_i(y_i - \pi_i). \tag{5.8}$$

Here, $s(\boldsymbol{\beta})$ depends on $\pi_i = \pi_i(\boldsymbol{\beta}) = h(\boldsymbol{x}_i'\boldsymbol{\beta}) = \exp(\boldsymbol{x}_i'\boldsymbol{\beta})/(1+\exp(\boldsymbol{x}_i'\boldsymbol{\beta}))$ and is therefore nonlinear in $\boldsymbol{\beta}$. From $\mathrm{E}(y_i) = \pi_i$ it follows

$$\mathrm{E}(s(\boldsymbol{\beta})) = \boldsymbol{0}.$$

Equating the score function to zero leads to the *ML equations*

$$s(\hat{\boldsymbol{\beta}}) = \sum_{i=1}^{n} \boldsymbol{x}_i \left(y_i - \frac{\exp(\boldsymbol{x}_i'\hat{\boldsymbol{\beta}})}{1+\exp(\boldsymbol{x}_i'\hat{\boldsymbol{\beta}})} \right) = \boldsymbol{0}. \tag{5.9}$$

This p-dimensional, nonlinear system of equations for $\hat{\boldsymbol{\beta}}$ is usually solved iteratively by the Newton–Raphson or Fisher scoring algorithm; see p. 283.
4. Information matrix
For the estimation of the regression coefficients and the covariance matrix of the ML estimator $\hat{\boldsymbol{\beta}}$, we need the *observed information matrix*

$$\boldsymbol{H}(\boldsymbol{\beta}) = -\frac{\partial^2 l(\boldsymbol{\beta})}{\partial \boldsymbol{\beta}\partial \boldsymbol{\beta}'},$$

with the second derivatives $\partial^2 l(\boldsymbol{\beta})/\partial \beta_j \partial \beta_r$ as elements of the matrix $\partial^2 l(\boldsymbol{\beta})/\partial \boldsymbol{\beta}\partial \boldsymbol{\beta}'$, or the *Fisher matrix (expected information matrix)*

$$F(\boldsymbol{\beta}) = \mathrm{E}\left(-\frac{\partial^2 l(\boldsymbol{\beta})}{\partial\boldsymbol{\beta}\partial\boldsymbol{\beta}'}\right) = \mathrm{Cov}(\boldsymbol{s}(\boldsymbol{\beta})) = \mathrm{E}(\boldsymbol{s}(\boldsymbol{\beta})\boldsymbol{s}'(\boldsymbol{\beta})).$$

The last equality holds since $\mathrm{E}(\boldsymbol{s}(\boldsymbol{\beta})) = \mathbf{0}$. To derive the Fisher matrix note that $F(\boldsymbol{\beta})$ is additive, i.e., $\boldsymbol{F}(\boldsymbol{\beta}) = \sum_{i=1}^{n} \boldsymbol{F}_i(\boldsymbol{\beta})$, where $\boldsymbol{F}_i(\boldsymbol{\beta}) = \mathrm{E}(\boldsymbol{s}_i(\boldsymbol{\beta})\boldsymbol{s}_i(\boldsymbol{\beta})')$ is the contribution of the ith observation. For $\boldsymbol{F}_i(\boldsymbol{\beta})$ we obtain

$$\begin{aligned}
\boldsymbol{F}_i(\boldsymbol{\beta}) &= \mathrm{E}\,(\boldsymbol{s}_i(\boldsymbol{\beta})\boldsymbol{s}_i(\boldsymbol{\beta})') \\
&= \mathrm{E}\left(\boldsymbol{x}_i\boldsymbol{x}_i'(y_i - \pi_i)^2\right) \\
&= \boldsymbol{x}_i\boldsymbol{x}_i'\mathrm{E}\,(y_i - \pi_i)^2 \\
&= \boldsymbol{x}_i\boldsymbol{x}_i'\mathrm{Var}(y_i) \\
&= \boldsymbol{x}_i\boldsymbol{x}_i'\pi_i(1 - \pi_i).
\end{aligned}$$

We finally get

$$\boldsymbol{F}(\boldsymbol{\beta}) = \sum_{i=1}^{n} \boldsymbol{F}_i(\boldsymbol{\beta}) = \sum_{i=1}^{n} \boldsymbol{x}_i\boldsymbol{x}_i'\pi_i(1 - \pi_i).$$

Since $\pi_i = h(\boldsymbol{x}_i'\boldsymbol{\beta})$, the Fisher matrix also depends on $\boldsymbol{\beta}$.

To derive the observed information matrix we use Definition A.29 of Appendix A.8. We obtain $\boldsymbol{H}(\boldsymbol{\beta}) = -\partial^2 l(\boldsymbol{\beta})/\partial\boldsymbol{\beta}\partial\boldsymbol{\beta}' = -\partial\boldsymbol{s}(\boldsymbol{\beta})/\partial\boldsymbol{\beta}'$ through another differentiation of

$$-\boldsymbol{s}(\boldsymbol{\beta}) = \sum_{i=1}^{n} \boldsymbol{x}_i(\pi_i(\boldsymbol{\beta}) - y_i).$$

Using the chain rule, this yields

$$\boldsymbol{H}(\boldsymbol{\beta}) = -\frac{\partial\boldsymbol{s}(\boldsymbol{\beta})}{\partial\boldsymbol{\beta}'} = \sum_{i=1}^{n} \boldsymbol{x}_i \frac{\partial\pi_i(\boldsymbol{\beta})}{\partial\boldsymbol{\beta}'} = \sum_{i=1}^{n} \boldsymbol{x}_i \frac{\partial\eta_i}{\partial\boldsymbol{\beta}'}\frac{\partial\pi_i(\boldsymbol{\beta})}{\partial\eta_i} = \sum_{i=1}^{n} \boldsymbol{x}_i\boldsymbol{x}_i'\pi_i(\boldsymbol{\beta})(1 - \pi_i(\boldsymbol{\beta})).$$

We thereby used

$$\frac{\partial\eta_i}{\partial\boldsymbol{\beta}'} = \left(\frac{\partial\eta_i}{\partial\boldsymbol{\beta}}\right)' = \boldsymbol{x}_i'$$

and

$$\frac{\partial\pi_i(\boldsymbol{\beta})}{\partial\eta_i} = \frac{(1 + \exp(\eta_i))\exp(\eta_i) - \exp(\eta_i)\exp(\eta_i)}{(1 + \exp(\eta_i))^2} = \pi_i(\boldsymbol{\beta})(1 - \pi_i(\boldsymbol{\beta})).$$

The expected and the observed information matrix are, thus, identical for the logit model, i.e., $\boldsymbol{H}(\boldsymbol{\beta}) = \boldsymbol{F}(\boldsymbol{\beta})$. This relationship, however, does not hold for other models, e.g., the *probit* or the *complementary log–log model*. In these models, we usually use the Fisher matrix $\boldsymbol{F}(\boldsymbol{\beta})$, which is typically easier to compute than the observed Fisher matrix $\boldsymbol{H}(\boldsymbol{\beta})$. Its general form will be given in Sect. 5.4.2.

If now instead of individual data with binary response variables $y_i \sim \mathrm{B}(1, \pi_i)$, we rather consider a binomially distributed response $y_i \sim \mathrm{B}(n_i, \pi_i)$ or relative frequencies

$$\bar{y}_i \sim \mathrm{B}(n_i, \pi_i)/n_i, \qquad i = 1, \ldots, n,$$

as, for example, in the case of grouped data, the formulae for $l(\boldsymbol{\beta})$, $\boldsymbol{s}(\boldsymbol{\beta})$, and $\boldsymbol{F}(\boldsymbol{\beta})$ have to be modified appropriately. Analogous arguments than for individual data yield

$$\begin{aligned} l(\boldsymbol{\beta}) &= \sum_{i=1}^{G} \{y_i \log(\pi_i) - y_i \log(1 - \pi_i) + n_i \log(1 - \pi_i)\} \\ \boldsymbol{s}(\boldsymbol{\beta}) &= \sum_{i=1}^{G} \boldsymbol{x}_i (y_i - n_i \pi_i) = \sum_{i=1}^{G} n_i \boldsymbol{x}_i (\bar{y}_i - \pi_i) \\ \boldsymbol{F}(\boldsymbol{\beta}) &= \sum_{i=1}^{G} \boldsymbol{x}_i \boldsymbol{x}_i' n_i \pi_i (1 - \pi_i). \end{aligned}$$

Iterative Calculation of the ML Estimator

Several iterative algorithms that compute the ML estimator as the solution of the ML equation $\boldsymbol{s}(\hat{\boldsymbol{\beta}}) = \mathbf{0}$ can be used for computing $\hat{\boldsymbol{\beta}}$. The most common method is the Fisher scoring algorithm; see Sect. B.4.2 in Appendix B. Given starting values $\hat{\boldsymbol{\beta}}^{(0)}$, e.g., the least squares estimate, the algorithm iteratively performs updates

$$\hat{\boldsymbol{\beta}}^{(t+1)} = \hat{\boldsymbol{\beta}}^{(t)} + \boldsymbol{F}^{-1}(\hat{\boldsymbol{\beta}}^{(t)}) \boldsymbol{s}(\hat{\boldsymbol{\beta}}^{(t)}), \quad t = 0, 1, 2, \ldots. \tag{5.10}$$

Once a convergence criterion is met, for example, $||\hat{\boldsymbol{\beta}}^{(t+1)} - \hat{\boldsymbol{\beta}}^{(t)}||/||\hat{\boldsymbol{\beta}}^{(t)}|| \leq \varepsilon$ (with $|| \cdot ||$ denoting the L_2-norm of a vector), the iterations will be stopped, and $\hat{\boldsymbol{\beta}} \equiv \hat{\boldsymbol{\beta}}^{(t)}$ is the ML estimator. Since $\boldsymbol{F}(\boldsymbol{\beta}) = \boldsymbol{H}(\boldsymbol{\beta})$ in the logit model, the Fisher scoring algorithm corresponds to a Newton method. The iterations Eq. (5.10) can also be expressed in the form of an iteratively weighted least squares estimation; see Sect. 5.4.2 (p. 306).

The Fisher scoring iterations can only converge to the ML solution $\hat{\boldsymbol{\beta}}$ if the Fisher matrix $\boldsymbol{F}(\boldsymbol{\beta})$ is invertible for all $\boldsymbol{\beta}$. As in the linear regression model, this requires that the design matrix $\boldsymbol{X} = (\boldsymbol{x}_1, \ldots, \boldsymbol{x}_n)'$ has full rank p. Then $\boldsymbol{F}(\boldsymbol{\beta})$ is invertible for the types of regression models that we have considered thus far. For example, in case of the logit model, $\boldsymbol{F}(\boldsymbol{\beta}) = \sum_i \boldsymbol{x}_i \boldsymbol{x}_i' \pi_i (1 - \pi_i)$ has full rank because $\boldsymbol{X}'\boldsymbol{X} = \sum_i \boldsymbol{x}_i \boldsymbol{x}_i'$ has full rank p and $\pi_i(1 - \pi_i) > 0$ for all $\boldsymbol{\beta} \in \mathbb{R}^p$. Hence, as in the linear regression model, we will always assume that

$$\mathrm{rk}(\boldsymbol{X}) = p.$$

Typically, the algorithm then converges and stops close to the maximum after only a few iterations.

Nevertheless, it is possible that iterations diverge, i.e., that the successive differences $||\hat{\boldsymbol{\beta}}^{(t+1)} - \hat{\boldsymbol{\beta}}^{(t)}||$ increase instead of converging towards zero. This is the case when the likelihood does not have a maximum for finite $\boldsymbol{\beta}$, i.e., if at least one component in $\hat{\boldsymbol{\beta}}^{(t)}$ diverges to $\pm\infty$, and no finite ML estimator exists. In general, the non-existence of the ML estimator is observed in very unfavorable data configurations, especially when the sample size n is small in comparison to the dimension p.

Even though several authors have elaborated on conditions of the uniqueness and existence of ML estimators, these conditions are, to some extent, very complex. For practical purposes it is, thus, easier to check the convergence or divergence of the iterative method empirically.

Example 5.3 Caesarian Delivery—Binary Regression

In Example 5.2, we chose a main effects model

$$\eta = \beta_0 + \beta_1\, NPLAN + \beta_2\, RISK + \beta_3\, ANTIB$$

for the linear predictor, i.e., a model without interactions between the covariates. If we want to improve the model fit by introducing interaction terms, we observe the following:

If we only include the interaction $NPLAN \cdot ANTIB$, the corresponding estimated coefficient is close to zero. If we include the interactions $RISK \cdot ANTIB$ or $NPLAN \cdot RISK$, we observe the problem of a nonexistent maximum, i.e., the ML estimator diverges. The reason is that we exclusively observed "no infection" for the response variable for both $NPLAN = 0$, $RISK = 0$, $ANTIB = 1$ and $NPLAN = 1$, $RISK = 0$, $ANTIB = 0$. This leads to the divergence towards infinity for the estimated effects of $ANTIB$ and $RISK \cdot ANTIB$ or $NPLAN$ and $NPLAN \cdot RISK$, and a termination before convergence yields exceptionally high estimated interaction effects and standard errors. Depending on the chosen software, the user may receive a warning or not. In any case, very high estimated regression coefficients and/or standard errors may be a sign for non-convergence of the ML estimator.

It is clear that the problem is dependent on the specific data configuration: If we were to move one observation from the two empty cells over to the "infection" category, then the interactions converge and finite ML estimators exist.

△

Comparison of the ML and Least Squares Estimator

In a linear regression model with normally distributed error terms, we have

$$y_i \sim \mathrm{N}(\mu_i = \boldsymbol{x}_i'\boldsymbol{\beta}, \sigma^2).$$

Apart from constant factors, the score function is then given by

$$\boldsymbol{s}(\boldsymbol{\beta}) = \sum_{i=1}^{n} \boldsymbol{x}_i(y_i - \mu_i),$$

where $\mathrm{E}(y_i) = \mu_i = \boldsymbol{x}_i'\boldsymbol{\beta}$ linearly depends on $\boldsymbol{\beta}$. For the logit model, the score function (5.8) follows the same structure, with $\mathrm{E}(y_i) = \pi_i = \mu_i$. However, $\boldsymbol{s}(\boldsymbol{\beta})$ is nonlinear in $\boldsymbol{\beta}$ since $\pi_i = \mu_i = \exp(\boldsymbol{x}_i'\boldsymbol{\beta})/\{1+\exp(\boldsymbol{x}_i'\boldsymbol{\beta})\}$. The ML or least squares system of equations for the linear model has the form

$$\boldsymbol{s}(\hat{\boldsymbol{\beta}}) = \sum_{i=1}^{n} \boldsymbol{x}_i(y_i - \boldsymbol{x}_i'\hat{\boldsymbol{\beta}}) = \boldsymbol{X}'\boldsymbol{y} - \boldsymbol{X}'\boldsymbol{X}\hat{\boldsymbol{\beta}} = \boldsymbol{0},$$

with responses $\boldsymbol{y} = (y_1, \ldots, y_n)'$. If the design matrix $\boldsymbol{X}$ has full rank p, we obtain the estimated regression coefficients as the solution of the system of equations $\boldsymbol{X}'\boldsymbol{X}\hat{\boldsymbol{\beta}} = \boldsymbol{X}'\boldsymbol{y}$ in a single step, yielding

$$\hat{\boldsymbol{\beta}} = (\boldsymbol{X}'\boldsymbol{X})^{-1}\boldsymbol{X}'\boldsymbol{y}.$$

In contrast, the solution to the nonlinear system of equations (5.9) has to be obtained numerically in several iterative steps in the logit model. The (observed and expected) information matrix in the linear model is

$$\boldsymbol{F}(\boldsymbol{\beta}) = \sum_{i=1}^{n} \boldsymbol{x}_i'\boldsymbol{x}_i/\sigma^2 = \frac{1}{\sigma^2}\boldsymbol{X}'\boldsymbol{X}.$$

The structure is again very similar to the one in the logit model, but the information matrix does not depend on $\boldsymbol{\beta}$.

Asymptotic Properties of the ML Estimator

Under relatively weak regularity conditions, one can shows that asymptotically (i.e., for $n \to \infty$), the ML estimator exists, is consistent, and follows a normal distribution. This result does not require that the sample size goes to infinity for each distinct location in the covariate space, but it is sufficient that the total sample size goes to infinity, i.e., $n \to \infty$. Then, for a sufficiently large sample size n, $\hat{\boldsymbol{\beta}}$ has an approximate normal distribution

$$\hat{\boldsymbol{\beta}} \overset{a}{\sim} \mathrm{N}(\boldsymbol{\beta}, \boldsymbol{F}^{-1}(\hat{\boldsymbol{\beta}})),$$

with estimated covariance matrix

$$\widehat{\mathrm{Cov}}(\hat{\boldsymbol{\beta}}) = \boldsymbol{F}^{-1}(\hat{\boldsymbol{\beta}})$$

equal to the inverse Fisher matrix evaluated at the ML estimator $\hat{\boldsymbol{\beta}}$. The diagonal element a_{jj} of the inverse Fisher matrix $\boldsymbol{A} = \boldsymbol{F}^{-1}(\hat{\boldsymbol{\beta}})$ is then an estimator of the variance of the jth component $\hat{\beta}_j$ of $\hat{\boldsymbol{\beta}}$, i.e.,

$$\widehat{\mathrm{Var}}(\hat{\beta}_j) = a_{jj},$$

and $\mathrm{se}_j = \sqrt{a_{jj}}$ is the standard error of $\hat{\beta}_j$ or in other words an estimator for the standard deviation $\sqrt{\mathrm{Var}(\hat{\beta}_j)}$. More details regarding the asymptotic properties of the ML estimator can be found in Fahrmeir and Kaufmann (1985).

5.1.3 Testing Linear Hypotheses

Linear hypotheses have the same form as in linear models:

$$H_0 : \boldsymbol{C}\boldsymbol{\beta} = \boldsymbol{d} \quad \text{versus} \quad H_1 : \boldsymbol{C}\boldsymbol{\beta} \neq \boldsymbol{d},$$

with $\boldsymbol{C}$ having full row rank $r \leq p$. We can use the likelihood ratio, the score and the Wald statistics for testing; see Appendix B.4.4. The *likelihood ratio statistic*

$$lr = -2\{l(\tilde{\boldsymbol{\beta}}) - l(\hat{\boldsymbol{\beta}})\}$$

measures the deviation in log-likelihood between the unrestricted maximum $l(\hat{\boldsymbol{\beta}})$ and that of the restricted maximum $l(\tilde{\boldsymbol{\beta}})$ under H_0, where $\tilde{\boldsymbol{\beta}}$ is the ML estimator under the restriction $\boldsymbol{C}\boldsymbol{\beta} = \boldsymbol{d}$. For the special case

$$H_0 : \boldsymbol{\beta}_1 = \boldsymbol{0} \quad \text{versus} \quad H_1 : \boldsymbol{\beta}_1 \neq \boldsymbol{0}, \tag{5.11}$$

where $\boldsymbol{\beta}_1$ is a subset of $\boldsymbol{\beta}$, we test the significance of the effects belonging to $\boldsymbol{\beta}_1$. The computation of $\tilde{\boldsymbol{\beta}}$ then simply requires ML estimation of the corresponding submodel. The numerical complexity is much greater for general linear hypotheses, since maximization has to be carried out under the constraint $\boldsymbol{C}\boldsymbol{\beta} = \boldsymbol{d}$.

The *Wald statistic*

$$w = (\boldsymbol{C}\hat{\boldsymbol{\beta}} - \boldsymbol{d})'[\boldsymbol{C}\boldsymbol{F}^{-1}(\hat{\boldsymbol{\beta}})\boldsymbol{C}']^{-1}(\boldsymbol{C}\hat{\boldsymbol{\beta}} - \boldsymbol{d})$$

measures the distance between the estimate $\boldsymbol{C}\hat{\boldsymbol{\beta}}$ and the hypothetical value $\boldsymbol{d}$ under H_0, weighted with the (inverse) asymptotic covariance matrix $\boldsymbol{C}\boldsymbol{F}^{-1}(\hat{\boldsymbol{\beta}})\boldsymbol{C}'$ of $\boldsymbol{C}\hat{\boldsymbol{\beta}}$.

The *score statistic*

$$u = \boldsymbol{s}'(\tilde{\boldsymbol{\beta}})\boldsymbol{F}^{-1}(\tilde{\boldsymbol{\beta}})\boldsymbol{s}(\tilde{\boldsymbol{\beta}})$$

measures the distance between $\boldsymbol{0} = \boldsymbol{s}(\hat{\boldsymbol{\beta}})$, i.e., the score function evaluated at the ML estimator $\hat{\boldsymbol{\beta}}$, and $\boldsymbol{s}(\tilde{\boldsymbol{\beta}})$, i.e., the score function evaluated at the restricted ML estimator $\tilde{\boldsymbol{\beta}}$.

Wald tests are mathematically convenient when an estimated model is to be tested against a simplified submodel, since it does not require additional estimation of the submodel. Conversely, the score test is convenient when an estimated model is to be tested against a more complex model alternative.

For the special hypothesis Eq. (5.11), the Wald and score statistic are reduced to

$$w = \hat{\boldsymbol{\beta}}_1'\hat{\boldsymbol{A}}_1^{-1}\hat{\boldsymbol{\beta}}_1$$

and

$$u = \boldsymbol{s}_1(\tilde{\boldsymbol{\beta}}_1)'\tilde{\boldsymbol{A}}_1\boldsymbol{s}_1(\tilde{\boldsymbol{\beta}}_1),$$

where $\boldsymbol{A}_1$ represents the submatrix of $\boldsymbol{A} = \boldsymbol{F}^{-1}$ and $\boldsymbol{s}_1(\tilde{\boldsymbol{\beta}}_1)$ represents the subvector of the score function $\boldsymbol{s}(\tilde{\boldsymbol{\beta}})$ that corresponds to the elements of $\tilde{\boldsymbol{\beta}}_1$. The notation "^" or "~" reflects the respective evaluation at $\hat{\boldsymbol{\beta}}$ or $\tilde{\boldsymbol{\beta}}$.

Under weak regularity conditions, similar to those required for the asymptotic normality of the ML estimators, the three test statistics are asymptotically equivalent under H_0 and approximately follow a χ^2-distribution with r degrees of freedom:

$$lr, w, u \overset{a}{\sim} \chi_r^2.$$

Critical values or p-values are calculated using this asymptotic distribution. For moderate sample sizes, the approximation through the χ^2-distribution is generally sufficient. For a smaller sample size, e.g., $n \leq 50$, the values of the test statistics can, however, differ considerably.

In the special case $H_0 : \beta_j = 0$ versus $H_1 : \beta_j \neq 0$ the Wald statistic equals the squared "t-value"

$$w = t_j^2 = \frac{\hat{\beta}_j^2}{a_{jj}},$$

with a_{jj} as the jth diagonal element of the asymptotic covariance matrix $\boldsymbol{A} = \boldsymbol{F}^{-1}(\hat{\boldsymbol{\beta}})$. Then the test is usually based on t_j which is asymptotically $\mathrm{N}(0, 1)$ distributed. The null hypothesis is then rejected if $|t_j| > z_{1-\alpha/2}$ where $z_{1-\alpha/2}$ is the $(1 - \alpha/2)$-quantile of the $\mathrm{N}(0, 1)$ distribution.

5.1.4 Criteria for Model Fit and Model Choice

Assessing the fit of an estimated model relies on the following idea: When the data have been maximally grouped, we can estimate the group-specific parameter π_i using the mean value $\bar{y}_i$. The use of these mean values as estimators corresponds to the *saturated model*, i.e., the model which contains separate parameters for each group. Thus the saturated model provides the best fit to the data and serves as a benchmark when evaluating the fit of estimated regression models. We now can formally test whether the departure between the estimated model and the saturated model is significant or not. The Pearson statistic and the deviance are the most frequently used goodness-of-fit statistics used for testing such a departure, both requiring that the data have been grouped as much as possible.

The *Pearson statistic* is given by the sum of the squared standardized residuals:

$$\chi^2 = \sum_{i=1}^{G} \frac{(\bar{y}_i - \hat{\pi}_i)^2}{\hat{\pi}_i(1 - \hat{\pi}_i)/n_i},$$

where G represents the number of groups, $\bar{y}_i$ is the relative frequency for group i, $\hat{\pi}_i = h(\boldsymbol{x}_i'\hat{\boldsymbol{\beta}})$ is the probability $\mathrm{P}(y_i = 1)$ estimated by the model, and $\hat{\pi}_i(1-\hat{\pi}_i)/n_i$ is the corresponding estimated variance.

The *deviance* is defined by

$$D = -2\sum_{i=1}^{G}\{l_i(\hat{\pi}_i) - l_i(\bar{y}_i)\},$$

where $l_i(\hat{\pi}_i)$ and $l_i(\bar{y}_i)$ represent the log-likelihood of group i for the estimated and the saturated model, respectively. The Pearson statistic looks similar to conventional chi-square statistics for testing if a random sample comes from a hypothesized discrete distribution: The squared differences between data and estimates are standardized by the variance and then summed up. The deviance compares the

Table 5.3 Patent opposition: estimation results from the logit model

Variable	Coefficient	Standard error	t-value	p-value	95 % Confidence interval	
intercept	201.740	22.321	9.04	<0.001	157.991	245.489
year	−0.102	0.011	−9.10	<0.001	−0.124	−0.080
ncit	0.114	0.022	5.09	<0.001	0.070	0.157
nclaim	0.027	0.006	4.49	<0.001	0.015	0.038
ustwin	−0.403	0.100	−4.03	<0.001	−0.599	−0.207
patus	−0.526	0.113	−4.67	<0.001	−0.747	−0.306
patgsgr	0.197	0.117	1.68	0.094	−0.033	0.427
ncountry	0.098	0.015	6.55	<0.001	0.068	0.127

(maximum of the) log-likelihood of the estimated model to the value of the log-likelihood of the saturated model, i.e., the largest value of the log-likelihood that can be attained. For finite samples, the Pearson and the deviance statistic will differ, but it can be shown that they are asymptotically equivalent for grouped data. If the n_i are sufficiently large for *all* groups $i = 1, \ldots, G$, both statistics are approximately χ^2_{G-p}-distributed, where p represents the number of estimated coefficients. Based on this approximate distribution, we can conduct a formal test for model fit by comparing the observed value of the test statistic to the corresponding quantile of the χ^2_{G-p}-distribution. Larger values in the observed test statistic indicate lack of fit and therefore correspond to larger p-values. For a prespecified significance level α a model is rejected if the $(1-\alpha)$-quantile is exceeded or the p-value is smaller than α. However, if n_i is small (especially if $n_i = 1$ as with ungrouped individual data), conducting such a test can be problematic. In this case, large values of χ^2 or D do not necessarily indicate a poor fit.

As already discussed for the coefficient of determination in linear regression (section "Analysis of Variance and Coefficient of Determination" of Sect. 3.2.3), a model choice strategy that tries to make the goodness-of-fit statistics as small as possible will usually result in an overfit model choice. When comparing models with different predictors and parameters, a compromise should be found between a good model fit obtained with a large number of parameters and model complexity. A well-known compromise is Akaike's information criterion

$$\text{AIC} = -2l(\hat{\boldsymbol{\beta}}) + 2p\,,$$

in which the term $2p$ penalizes complex models with a large number of parameters. When choosing between several models, we prefer those with small AIC values. Rather than the AIC value, one also often considers AIC/n, i.e., the AIC standardized for sample size n. Another alternative is the BIC; see Appendix B.5.4.

Example 5.4 Patent Opposition—Testing and Model Choice

Table 5.3 presents the estimated coefficients for the logit model in Example 5.1 (p. 275), along with the corresponding standard errors, t-values, p-values, and 95 % confidence intervals. For the log-likelihood and the AIC criterion of the estimated model, we have

Table 5.4 Patent opposition: estimation results from the probit model

Variable	Coefficient	Standard error	t-value	p-value	95 % Confidence interval	
intercept	119.204	13.192	9.04	<0.001	93.349	145.060
year	−0.060	0.007	−9.11	<0.001	−0.073	−0.047
ncit	0.068	0.014	5.02	<0.001	0.041	0.094
nclaim	0.016	0.004	4.46	<0.001	0.009	0.023
ustwin	−0.243	0.060	−4.07	<0.001	−0.360	−0.126
patus	−0.309	0.066	−4.72	<0.001	−0.438	−0.181
patgsgr	0.121	0.071	1.71	0.086	−0.017	0.260
ncountry	0.059	0.009	6.51	<0.001	0.041	0.076

Table 5.5 Patent opposition: estimation results of the extended logit model

Variable	Coefficient	Standard error	t-value	p-value	95 % Confidence interval	
intercept	198.131	22.739	8.71	<0.001	153.563	242.699
year	−0.101	0.011	−8.82	<0.001	−0.123	−0.078
ncit	0.113	0.022	5.08	<0.001	0.070	0.157
nclaim	0.026	0.006	4.45	<0.001	0.015	0.038
ustwin	−0.409	0.100	−4.09	<0.001	−0.605	−0.213
patus	−0.539	0.113	−4.77	<0.001	−0.761	−0.318
patgsgr	0.180	0.119	1.52	0.130	−0.053	0.414
ncountry	0.394	0.184	2.14	0.032	0.034	0.754
$ncountry^2$	−0.038	0.024	−1.58	0.113	−0.085	0.009
$ncountry^3$	0.001	0.001	1.50	0.134	−0.000	0.003

$$l(\hat{\boldsymbol{\beta}}) = -1488.560\,, \quad \text{AIC} = 2993.12\,.$$

With a p-value of 0.094, the effect of the variable *patgsgr* is at best marginally significant. If we choose $\alpha = 5\,\%$ as the significance level, the hypothesis $H_0 : \beta_6 = 0$ will not be rejected. This implies that the increased probability of patent objection if the patent comes from Germany, Switzerland, or Great Britain appears nonsignificant.

Table 5.4 contains the corresponding values for the probit model. Even though the estimated coefficients and their standard deviations differ due to the absence of a proper adjustment (see Example 5.1), the t-values and p-values are in good agreement and lead to the same conclusions regarding the significance of the effects. With

$$l(\hat{\boldsymbol{\beta}}) = -1488.407\,, \quad \text{AIC} = 2992.815,$$

we obtain very similar values for the log-likelihood and the AIC criterion. Since the results for the patent data are comparable for the logit and probit model, we only further describe the findings for the logit model.

In order to examine whether or not the effect of the continuous covariate *ncountry* is linear, we included a cubic polynomial

$$\beta_7\, ncountry + \beta_8\, ncountry^2 + \beta_9\, ncountry^3$$

into the linear predictor as in Example 2.8 (p. 35) and estimated this modified logit model. Table 5.5 contains the estimated coefficients, their standard errors, as well as t-values,

Table 5.6 Credit scoring: description of the covariates including summary statistics

Variable	Description	Mean/ frequency in %	Std. dev.	Min/max
acc1	1 = no running account	27.40		
	0 = good or bad running account	72.60		
acc2	1 = good running account	39.40		
	0 = no or bad running account	60.60		
duration	Duration of the credit in months	20.90	12.06	4/72
amount	Credit amount in 1000 Euro,	1.67	1.44	0.13/9.42
moral	Previous payment behavior			
	1 = good	91.10		
	0 = bad	8.90		
intuse	Intended use			
	1 = private	65.70		
	0 = business	34.30		

p-values, and 95 % confidence intervals. The t-values and the p-values corresponding to $ncountry^2$ and $ncountry^3$ already indicate that the more conservative linear model may be sufficient and that the nonlinearity is over-interpreted. The log-likelihood and the AIC criterion for the extended model yields

$$l(\hat{\boldsymbol{\beta}}) = -1487.232\,, \quad \text{AIC} = 2994.463\,.$$

This further confirms that we should rather choose the simpler model with linear modeling of the *ncountry* effect. We can also investigate nonlinearity by testing the hypotheses

$$H_0 : (\beta_8, \beta_9) = (0, 0) \qquad \text{versus} \qquad H_1 : (\beta_8, \beta_9) \neq (0, 0).$$

The likelihood ratio test statistic results in

$$lr = -2\{-1488.56 - (-1487.23)\} = 2.66\,.$$

The 95 % quantile of the (approximate) $\chi^2(2)$-distribution is $\chi^2_{95\,\%}(2) = 5.99$, thus H_0 cannot be rejected, which also follows from the p-value of 0.269. In summary, the assumption of a linear effect of covariate *ncountry* cannot be rejected. The Wald test also leads to the same result.

△

Example 5.5 Credit Scoring—Binary Regression

When issuing credit, banks check the "solvency" or "creditworthiness" of clients, i.e., their ability and willingness to pay back the credit in the specified time frame. To evaluate creditworthiness using statistical methods (credit scoring), characteristics of the borrower are requested that reflect his or her personal and economic situation and thus influence the probability of creditworthiness. Binary regression models are suited for such evaluations since they model the probability of a loan default for given characteristics of the client.

We use a data set on $n = 1{,}000$ private credits issued by a German bank published in Fahrmeir, Hamerle, and Tutz (1996). Every client is associated with a binary response y defined as

$$y = \begin{cases} 1 & \text{client is not creditworthy,} \\ 0 & \text{client is creditworthy.} \end{cases}$$

Among a total of 20 characteristics, we use those described in Table 5.6 as covariates.

Table 5.7 Credit scoring: estimation results for the logit model

Variable	Coefficient	Standard error	t-value	p-value	95 % Confidence interval	
intercept	0.487	0.266	1.83	0.067	−0.034	1.007
acc1	0.618	0.175	3.53	<0.001	0.275	0.960
acc2	−1.338	0.201	−6.65	<0.001	−1.732	−0.944
durationo	0.401	0.093	4.29	<0.001	0.218	0.584
amounto	0.066	0.092	0.72	0.474	−0.115	0.247
moral	−0.986	0.251	−3.93	<0.001	−1.478	−0.494
intuse	−0.426	0.158	−2.69	0.007	−0.736	−0.115

Table 5.8 Credit scoring: results for the extended logit model

Variable	Coefficient	Standard error	t-value	p-value	95 % Confidence interval	
intercept	0.474	0.270	1.75	0.079	−0.055	1.004
acc1	0.618	0.176	3.51	<0.001	0.272	0.963
acc2	−1.337	0.202	−6.61	<0.001	−1.734	−0.941
durationo	0.508	0.100	5.07	<0.001	0.312	0.705
duration2o	−0.173	0.079	−2.20	0.028	−0.327	−0.019
amounto	0.035	0.098	0.36	0.720	−0.155	−0.224
amount2o	0.288	0.097	3.07	0.002	0.104	0.471
moral	−0.995	0.255	−3.90	<0.001	−1.495	−0.495
intuse	−0.404	0.160	−2.52	0.012	−0.718	−0.090

We model the probability $P(y = 1)$ for a weak creditworthiness with a logit model and the linear predictor

$$\eta = \beta_0 + \beta_1\, acc1 + \beta_2\, acc2 + \beta_3\, durationo + \beta_4\, amounto + \beta_5\, moral + \beta_6\, intuse.$$

Since we will later also estimate quadratic orthogonal polynomials (see Example 3.5 on p. 90) for the effects of the continuous covariates *duration* and *amount* we included the linear parts *durationo* and *amounto* of these orthogonal polynomials in the predictor rather than the original covariates. Table 5.7 lists the estimated coefficients, along with their corresponding standard errors, t-values, p-values, and 95 % confidence intervals. The p-value for the effect of *amounto* indicates that the corresponding effect is not significant. The AIC value for this model is 1,043.815.

In a second step of our analysis, we assume a quadratic orthogonal polynomial for the effects of the continuous covariates *duration* and *amount* to detect possible nonlinearity. Table 5.8 contains the corresponding estimated results. All p-values, also those for squared effects, now show significance. Furthermore, the lower AIC value of 1,035.371 indicates an improved model fit.

Figure 5.2 shows the estimated linear effects of credit amount and duration together with the quadratic, nonlinear effects. The "bathtub" shape of the squared effects of the credit amount illustrates that small and large credit increases the risk of not paying back the credit. This effect is missed when reducing the model to linear effects.

We finally apply the likelihood ratio and Wald test in order to test the model having quadratic effects against the submodel with linear effects for credit amount and duration. The likelihood ratio and the Wald statistic yield

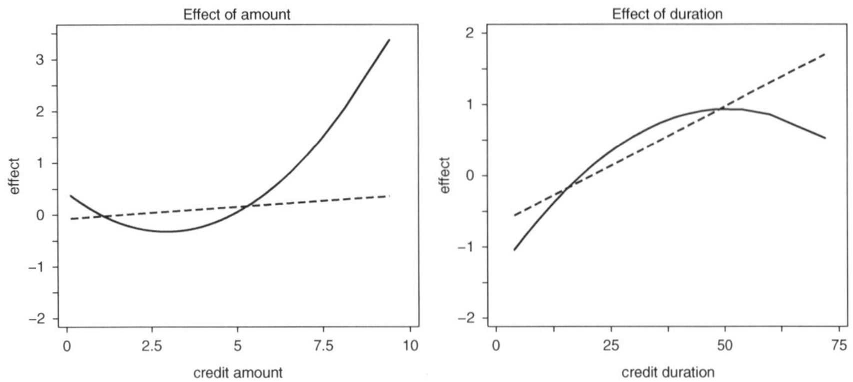

Fig. 5.2 Credit scoring: estimated linear (- - -) and quadratic (—) effects of credit amount and credit duration

$$lr = 12.44\,, \qquad w = 11.47$$

with two degrees of freedom and corresponding p-values of 0.0020 and 0.0032, respectively. Hence, both tests again confirm the specification of the more complex model with quadratic effects.

△

5.1.5 Estimation of the Overdispersion Parameter

As discussed in Sect. 5.1 (p. 279), we may observe overdispersion when working with grouped data. To allow for overdispersion, we assume

$$\mathrm{Var}(y_i) = \phi \frac{\pi_i(1-\pi_i)}{n_i}.$$

The overdispersion parameter ϕ can be estimated as the average Pearson statistic or the average deviance:

$$\hat{\phi}_P = \frac{1}{n-p}\chi^2 \quad \text{or} \quad \hat{\phi}_D = \frac{1}{n-p}D\,.$$

This is analogous to the estimation of the error variance in the linear model, with χ^2 or D replacing the residual sum of squares.

Accordingly, we multiply the estimated covariance matrix with $\hat{\phi}$, i.e., $\widehat{\mathrm{Cov}}(\hat{\boldsymbol{\beta}}) = \hat{\phi}\boldsymbol{F}^{-1}(\hat{\boldsymbol{\beta}})$. Strictly speaking, this approach to treat overdispersion does not correspond to a true likelihood method, but rather to a quasi-likelihood model; see Sect. 5.5.

Since we only need $\pi_i = \mathrm{E}(y_i)$ and $\mathrm{Var}(y_i)$, and not the likelihood itself for the maximum likelihood estimation of $\boldsymbol{\beta}$, both $\boldsymbol{\beta}$ and ϕ can be formally estimated just as if we considered a distribution with scale parameter ϕ, such as a normal or gamma distribution; see Sect. 5.4.2. In fact, the introduction of an overdispersion parameter leads to one of the simplest forms of quasi-likelihood estimation. Even though distributions with variance $\phi\pi_i(1-\pi_i)/n_i$ exist, for example, the beta-binomial distribution, their actual likelihood is not necessary and will also not be used in the estimation process. Other approaches to account for overdispersion are, for example, models with random effects; see Chap. 7. A good additional reference on models with overdispersion is Collett (1991).

5.2 Count Data Regression

Count data are frequently observed when the number of certain events within a fixed time frame or frequencies in a contingency table have to be analyzed. Sometimes, a normal approximation can be sufficient, particularly when the events occur with high frequencies. In situations with only a small number of counts, models for categorial response variables (Chap. 6) can be an alternative. In general, however, discrete distributions recognizing the specific properties of count data are most appropriate. The Poisson distribution is the simplest and most widely used choice, but model modifications and alternatives such as the negative binomial distribution are also used. For details on such extensions, we refer to the specialized literature on count data regression given in the final section of this chapter.

5.2.1 Models for Count Data

Log-Linear Poisson Model

The most widely used model for count data connects the rate $\lambda_i = \mathrm{E}(y_i)$ of the Poisson distribution with the linear predictor $\eta_i = \boldsymbol{x}_i'\boldsymbol{\beta} = \beta_0 + \beta_1 x_{i1} + \ldots + \beta_k x_{ik}$ via

$$\lambda_i = \exp(\eta_i) = \exp(\beta_0)\exp(\beta_1 x_{i1}) \cdot \ldots \cdot \exp(\beta_k x_{ik})$$

or in log-linear form through

$$\log(\lambda_i) = \eta_i = \boldsymbol{x}_i'\boldsymbol{\beta} = \beta_0 + \beta_1 x_{i1} + \ldots + \beta_k x_{ik}. \tag{5.12}$$

The effect of covariates on the rate λ is, thus, exponentially multiplicative similar to the effect on the odds $\pi/(1-\pi)$ in the logit model. The effect on the logarithm of the rate in Eq. (5.12) is linear.

Linear Poisson Model

The direct relationship

$$\lambda_i = \eta_i = \boldsymbol{x}_i'\boldsymbol{\beta}$$

5.3 Log-Linear Poisson Model for Count Data

Data

The response variables y_i take values $\{0, 1, 2, \ldots\}$ and are (conditionally) independent given the covariates $x_{i1}, \ldots, x_{ik}$.

Model Without Overdispersion

$y_i \sim \text{Po}(\lambda_i)$ with

$$\lambda_i = \exp(\boldsymbol{x}_i'\boldsymbol{\beta}) \quad \text{or} \quad \log(\lambda_i) = \boldsymbol{x}_i'\boldsymbol{\beta}.$$

Model with Overdispersion

$$\text{E}(y_i) = \lambda_i = \exp(\boldsymbol{x}_i'\boldsymbol{\beta}), \quad \text{Var}(y_i) = \phi\lambda_i$$

with overdispersion parameter ϕ.

is useful when the covariates have an additive effect on the rate. Since $\boldsymbol{x}_i'\boldsymbol{\beta}$ must not be nonnegative, this usually implies restrictions for the parameter space of $\boldsymbol{\beta}$.

Overdispersion

The assumption of a Poisson distribution for the responses implies

$$\lambda_i = \text{E}(y_i) = \text{Var}(y_i).$$

For similar reasons as in case with binomial data, a significantly higher empirical variance is frequently observed in applications of Poisson regression. For this reason it is often useful to introduce an overdispersion parameter ϕ by assuming

$$\text{Var}(y_i) = \phi\lambda_i.$$

As for binomial data, there are also more complex modeling approaches for count data which take the additional variability into account. One possibility is the use of the negative binomial distribution, which is closely related to the use of random effects models; see Chap. 7.

Example 5.6 Number of Citations from Patents—Poisson Regression

We illustrate the use of regression models for counts with the patent data described in Example 1.3 (p. 8). In contrast to Examples 5.1 and 5.4, we now choose the number of citations for a patent, variable *ncit*, as the response. We also use the complete data set, i.e., patents which either belong to the biotechnology or to the pharmaceutical industry. We incorporate the remaining variables described in Table 1.4 (p. 8) as covariates. As in Sect. 3.1.2 (p. 92), we center the continuous covariates *yearc*, *ncountryc*, and *nclaimsc* around their means and use these centered covariates in the linear predictor. Based on previous descriptive analysis in Example 2.8, we exclude all observations with $nclaims > 60$ and $ncit > 15$ from further analysis.

As a first step, we examine a log-linear model for the rate $\lambda_i = \mathrm{E}(ncit_i)$ with purely linear predictor

$$\log(\lambda_i) = \eta_i = \beta_0 + \beta_1 yearc_i + \beta_2 ncountryc_i + \beta_3 nclaimc_i + \beta_4 biopharm_i + \beta_5 ustwin_i + \beta_6 patus_i + \beta_7 patgsgr_i + \beta_8 opp_i \,.$$

In Example 5.7, we estimate a Poisson model without an overdispersion parameter, as well as a model that includes an overdispersion parameter ϕ in the variance $\mathrm{Var}(ncit_i) = \phi\lambda_i$. To allow for possibly nonlinear effects of the continuous covariates, we further considered polynomial effects for *yearc*, *ncountryc*, and *nclaimsc* and compare the different models using AIC.

△

5.2.2 Estimation and Testing: Likelihood Inference

We again assume that the response variables y_i are (conditionally) independent. The derivations of the likelihood, score function, and the information matrix are analogous to the developments for binary data in Sect. 5.1.

Maximum Likelihood Estimation

For the Poisson distributed response variable, the discrete density (or the likelihood $L_i(\boldsymbol{\beta})$ of the ith observation) is given by

$$f(y_i \mid \boldsymbol{\beta}) = \frac{\lambda_i^{y_i} \exp(-\lambda_i)}{y_i!}, \qquad \mathrm{E}(y_i) = \lambda_i .$$

It depends on $\boldsymbol{\beta}$ through $\lambda_i = \boldsymbol{x}_i'\boldsymbol{\beta}$ in the linear Poisson model and through $\lambda_i = \exp(\boldsymbol{x}_i'\boldsymbol{\beta})$ in the log-linear Poisson model. The ML estimator for the log-linear Poisson model is obtained in the following steps:

1. Log-likelihood

The *log-likelihood* is given by

$$l(\boldsymbol{\beta}) = \sum_{i=1}^{n} l_i(\boldsymbol{\beta}) = \sum_{i=1}^{n} (y_i \log(\lambda_i) - \lambda_i),$$

apart from the additive constant $-n\log(y_i!)$ (that is independent of $\boldsymbol{\beta}$). The Poisson log-linear model with $\log(\lambda_i) = \boldsymbol{x}_i'\boldsymbol{\beta} = \eta_i$ yields

$$l(\boldsymbol{\beta}) = \sum_{i=1}^{n} l_i(\boldsymbol{\beta}) = \sum_{i=1}^{n} y_i(\boldsymbol{x}_i'\boldsymbol{\beta}) - \exp(\boldsymbol{x}_i'\boldsymbol{\beta}) = \sum_{i=1}^{n} (y_i \eta_i - \exp(\eta_i)) .$$

2. Score function
Differentiating according to the chain rule $\partial l_i(\boldsymbol{\beta})/\partial\boldsymbol{\beta} = (\partial l_i/\partial\eta_i)\cdot\partial\eta_i/\partial\boldsymbol{\beta} = \partial l_i/\partial\eta_i\cdot\boldsymbol{x}_i$, we obtain the *score function*

$$s(\boldsymbol{\beta}) = \sum_{i=1}^{n} \boldsymbol{x}_i(y_i - \exp(\eta_i)) = \sum_{i=1}^{n} \boldsymbol{x}_i(y_i - \lambda_i).$$

3. Fisher information
Using the same arguments as in the logit model, we obtain the *Fisher information*

$$\boldsymbol{F}(\boldsymbol{\beta}) = \mathrm{E}(\boldsymbol{s}(\boldsymbol{\beta})\boldsymbol{s}'(\boldsymbol{\beta})) = \sum_{i=1}^{n} \boldsymbol{x}_i\boldsymbol{x}_i'\lambda_i,$$

utilizing $\mathrm{E}(y_i - \lambda_i)^2 = \mathrm{Var}(y_i) = \lambda_i$.

4. Numerical computation
Due to $\lambda_i = \exp(\boldsymbol{x}_i'\boldsymbol{\beta})$, we obtain the nonlinear equation system

$$\boldsymbol{s}(\hat{\boldsymbol{\beta}}) = \boldsymbol{0}$$

for $\hat{\boldsymbol{\beta}}$. The numerical computation of $\hat{\boldsymbol{\beta}}$ is again carried out using Fisher scoring Eq. (5.10), inserting the corresponding expressions for $\boldsymbol{s}(\boldsymbol{\beta})$ and $\boldsymbol{F}(\boldsymbol{\beta})$. Similar to linear and binary regression, we assume

$$\mathrm{rk}(\boldsymbol{X}) = p$$

for the design matrix $\boldsymbol{X} = (\boldsymbol{x}_1, \ldots, \boldsymbol{x}_n)'$. The remarks made in Sect. 5.1.2 also apply for the convergence or divergence of the iterations in the Poisson model.

Under moderate regularity assumptions and for large n (more precisely for $n \to \infty$), we have the asymptotic result

$$\hat{\boldsymbol{\beta}} \overset{a}{\sim} \mathrm{N}(\boldsymbol{\beta}, \boldsymbol{F}^{-1}(\hat{\boldsymbol{\beta}}))$$

with estimated covariance matrix $\widehat{\mathrm{Cov}}(\hat{\boldsymbol{\beta}}) = \boldsymbol{F}^{-1}(\hat{\boldsymbol{\beta}})$.

Testing Linear Hypotheses
We use the same test statistics as in binary regression models for testing linear hypotheses $\boldsymbol{C}\boldsymbol{\beta} = \boldsymbol{d}$, where the appropriate expressions for $l(\boldsymbol{\beta}), \boldsymbol{s}(\boldsymbol{\beta})$, and $\boldsymbol{F}(\boldsymbol{\beta})$ associated with the Poisson model are to be inserted. In addition, the same statements regarding the asymptotic or approximate χ^2-distribution of the test statistics apply.

5.2.3 Criteria for Model Fit and Model Choice

The criteria discussed in Sect. 5.1.2 for binary regression models can be transferred to the Poisson case. Since $\text{Var}(y_i) = \lambda_i$ for the Poisson distribution, we obtain the Pearson statistic

$$\chi^2 = \sum_{i=1}^{G} \frac{(y_i - \hat{\lambda}_i)^2}{\hat{\lambda}_i / n_i}.$$

The Poisson log-likelihood must be inserted into the definition of the deviance and the AIC. Note that by convention $0 \cdot \log(0) = 0$ (for $y_i = 0$).

5.2.4 Estimation of the Overdispersion Parameter

As previously stated, in situations where we allow for possible overdispersion with the assumption $\text{Var}(y_i \mid \boldsymbol{x}_i) = \phi\lambda_i$, the overdispersion parameter ϕ can be estimated as the average Pearson statistic or the average deviance:

$$\hat{\phi}_P = \frac{1}{n-p}\chi^2 \quad \text{or} \quad \hat{\phi}_D = \frac{1}{n-p} D \,.$$

This is analogous to the estimation of the error variance in the linear model, with χ^2 or D replacing the residual sum of squares.

We then have to multiply the estimated covariance matrix with $\hat{\phi}$, i.e., $\widehat{\text{Cov}(\hat{\boldsymbol{\beta}})} = \hat{\phi}\boldsymbol{F}^{-1}(\hat{\boldsymbol{\beta}})$. Strictly speaking, this approach to the estimation of overdispersion does not correspond to a true likelihood method, but rather to a quasi-likelihood model; see Sect. 5.5.

Example 5.7 Number of Citations from Patents—Overdispersion

Table 5.9 shows estimation results for the log-linear Poisson model of Example 5.6 having no overdispersion (i.e., $\phi = 1$) and only linear effects (AIC $= 19{,}092.25$, deviance $= 12{,}085.31$, Pearson-$\chi^2 = 14{,}091.66$).

The p-values indicate significance of all covariates, with the exception of *ustwin*. Since overdispersion is very common with Poisson models, we reanalyze the model by estimating the overdispersion parameter as the mean Pearson statistic or mean deviance. We obtain

$$\hat{\phi}_P = \frac{1}{n-p}\chi^2 = 2.935 \quad \text{resp.} \quad \hat{\phi}_D = \frac{1}{n-p} D = 2.518 \,,$$

with $n = 4{,}809$, $p = 9$. In contrast to the Poisson model, the estimated variance or standard deviation of the estimated regression coefficients needs to be multiplied by $\hat{\phi}$ and $\hat{\phi}^{1/2}$, respectively, while the point estimates are the same as for the pure Poisson model without overdispersion. Table 5.10 lists the results for $\hat{\phi}_P$. In comparison to Table 5.9, we see that the standard errors increase by the factor $\hat{\phi}_P^{1/2} = 1.71$. This adjustment causes an increase of the p-values, such that the effect of variable *patus* is now insignificant, while the analysis without overdispersion resulted in a p-value that was significant.

In order to detect possibly nonlinear effects of the centered continuous covariates *yearc*, *ncountryc*, and *nclaimc*, we construct polynomials of degree three. The centering

Table 5.9 Number of citations from patents: model with linear effects and $\phi = 1$

Variable	Coefficient	Standard error	t-value	p-value	95 % Confidence interval	
intercept	0.158	0.033	4.85	<0.001	0.094	0.222
yearc	−0.072	0.003	−24.17	<0.001	−0.078	−0.066
ncountryc	−0.028	0.004	−6.60	<0.001	−0.036	−0.020
nclaimc	0.018	0.001	14.16	<0.001	0.016	0.021
biopharm	0.239	0.032	7.42	<0.001	0.176	0.302
ustwin	0.002	0.026	0.09	0.926	−0.048	0.053
patus	−0.078	0.027	−2.84	0.005	−0.132	−0.024
patgsgr	−0.198	0.032	−6.24	<0.001	−0.260	−0.136
opp	0.372	0.025	14.81	<0.001	0.322	0.421

Table 5.10 Number of citations from patents: model with linear effects and overdispersion

Variable	Coefficient	Standard error	t-value	p-value	95 % Confidence interval	
intercept	0.158	0.056	2.83	0.005	0.049	0.267
yearc	−0.072	0.005	−14.11	<0.001	−0.082	−0.062
ncountryc	−0.028	0.007	−3.85	<0.001	−0.042	−0.014
nclaimc	0.018	0.002	8.26	<0.001	0.014	0.022
biopharm	0.239	0.055	4.33	<0.001	0.131	0.347
ustwin	0.002	0.044	0.05	0.957	−0.084	0.088
patus	−0.078	0.047	−1.66	0.098	−0.170	0.014
patgsgr	−0.198	0.054	−3.64	<0.001	−0.305	−0.091
opp	0.372	0.043	8.64	<0.001	0.287	0.456

is conducted as described in section "Continuous Covariates" of Sect. 3.1.3. The model obtains AIC $= 18,786.32$, deviance$= 11,767.37$, Pearson-$\chi^2 = 13,815.96$, $\hat{\phi}_D = 2.45$, $\hat{\phi}_P = 2.88$. Compared to the model with only linear effects the fit is considerably improved. Table 5.11 shows the results. The variable *ustwin* remains nonsignificant while *patus* is now weakly significant. The other variables all remain significant (on a level of 5 %) with the exception of some of the polynomial terms. This indicates that lower-order polynomials might be sufficient to model the nonlinearity of the covariate effects. Figure 5.3 compares linear and nonlinear effects of the continuous covariates. We leave the interpretation of the results to the reader; see Example 2.13 (p. 54) on how to interpret the (nonlinear) effects in Poisson regression models.

△

5.3 Models for Nonnegative Continuous Response Variables

The classical linear model

$$y_i = \boldsymbol{x}_i'\boldsymbol{\beta} + \varepsilon_i, \quad \mathrm{E}(\varepsilon_i) = 0, \quad \mathrm{Var}(\varepsilon_i) = \sigma^2$$

Table 5.11 Number of citations from patents: extended model with overdispersion

Variable	Coefficient	Standard error	t-value	p-value	95 % Confidence interval	
intercept	0.17115	0.05558	3.08	0.002	0.06221	0.28009
yearc	−0.09924	0.00966	−10.27	<0.001	−0.11818	−0.08031
yearc^2	−0.00974	0.00226	−4.31	<0.001	−0.01417	−0.00531
yearc^3	−0.00011	0.00030	−0.37	0.715	−0.00070	0.00048
ncountryc	0.01552	0.01322	1.17	0.241	−0.01040	0.04143
ncountryc^2	−0.00213	0.00206	−1.03	0.301	−0.00618	0.00191
ncountryc^3	−0.00157	0.00044	−3.60	<0.001	−0.00243	−0.00071
nclaimc	0.02611	0.00352	7.42	<0.001	0.01922	0.03301
nclaimc^2	−0.00046	0.00036	−1.30	0.195	−0.00116	0.00024
nclaimc^3	0.00000	0.00000	0.18	0.855	−0.00002	0.00002
biopharm	0.15504	0.05564	2.79	0.005	0.04598	0.26410
ustwin	−0.00288	0.04338	−0.07	0.947	−0.08791	0.08215
patus	−0.09502	0.04715	−2.02	0.044	−0.18743	−0.00260
patgsgr	−0.20185	0.05446	−3.71	<0.001	−0.30859	−0.09511
opp	0.37154	0.04253	8.74	<0.001	0.28819	0.45489

is well suited for analyzing regression data when the errors ε_i have (at least approximately) a normal distribution. In this case, the response variables y_i, for given covariate vector $\boldsymbol{x}_i$, are (conditionally) independent and follow a normal distribution with

$$y_i \sim \mathrm{N}(\mu_i, \sigma^2), \qquad \mu_i = \mathrm{E}(y_i) = \boldsymbol{x}_i'\boldsymbol{\beta}.$$

In many applications, the response variable cannot be negative, for example, in case of life times, claim sizes, and costs. Such responses are also usually highly non-normal, often following a (right) skewed distribution.

Lognormal Model

To enable the application of linear models, the response variable y is often transformed logarithmically such that a usual linear model with normal errors can be assumed for $\tilde{y} = \log(y)$, i.e.,

$$\tilde{y}_i = \boldsymbol{x}_i'\boldsymbol{\beta} + \varepsilon_i \quad \text{or} \quad \tilde{y}_i \sim \mathrm{N}(\boldsymbol{x}_i'\boldsymbol{\beta}, \sigma^2).$$

This implies that the original variable y follows a log-normal distribution (see Definition B.6 in Appendix B.1) with

$$\mathrm{E}(y_i) = \exp(\boldsymbol{x}_i'\boldsymbol{\beta} + \sigma^2/2), \qquad \mathrm{Var}(y_i) = \exp(2\boldsymbol{x}_i'\boldsymbol{\beta} + \sigma^2)(\exp(\sigma^2) - 1). \quad (5.13)$$

We can obtain "plug-in" estimators for Eq. (5.13), using the least squares estimates $\hat{\boldsymbol{\beta}}$ and the estimated variance $\hat{\sigma}^2$ for the linear model. When estimating $\hat{\mu}_i = \exp(\hat{\eta}_i) = \exp(\boldsymbol{x}_i\hat{\boldsymbol{\beta}} + \hat{\sigma}^2/2)$, considerable bias can be induced by the nonlinear back-transformation with the exponential function.

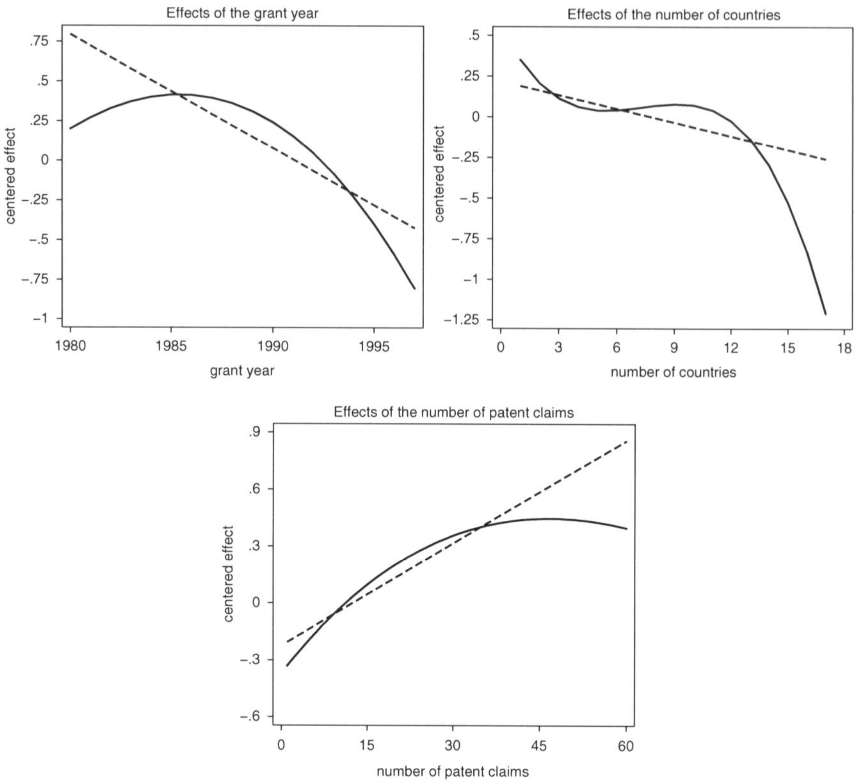

Fig. 5.3 Number of citations from patents: linear (- - -) and nonlinear (—) effects of the continuous covariates *year*, *ncountry*, and *nclaim*

Gamma Regression

To circumvent this difficulty, the assumption of a gamma distribution (see Definition B.9 in Appendix B.1), with expectation $\mathrm{E}(y_i) = \mu_i$ and scale parameter ν for the response variables y_i, can be a valuable alternative. The variance is then given by

$$\mathrm{Var}(y_i) = \sigma_i^2 = \mu_i^2/\nu.$$

For the nonnegative, gamma-distributed response, we have $\mathrm{E}(y_i) = \mu_i > 0$. A direct linear link

$$\mu_i = \boldsymbol{x}_i'\boldsymbol{\beta}$$

is again problematic, since we have to comply with the condition $\boldsymbol{x}_i'\boldsymbol{\beta} > 0$. Thus a *multiplicative exponential model*

$$\mu_i = \exp(\eta_i) = \exp(\boldsymbol{x}_i'\boldsymbol{\beta}) = \exp(\beta_0)\exp(\beta_1 x_{i1}) \cdot \ldots \cdot \exp(\beta_k x_{ik}), \qquad (5.14)$$

with response function $h(\eta) = \exp(\eta)$, is often better suited than the linear link function.

Another possible choice for the response or link function is the *reciprocal*

$$\mu_i = \frac{1}{\eta_i} = \frac{1}{\boldsymbol{x}_i'\boldsymbol{\beta}}.$$

Since $\boldsymbol{x}_i'\boldsymbol{\beta} > 0$ has to be fulfilled, the choice again implies restrictions for $\boldsymbol{\beta}$. Even though the reciprocal response function is the so-called *natural* or *canonical response function* for the gamma distribution (see Sect. 5.4), the multiplicative exponential model (5.14) is usually more adequate for both modeling and interpretation.

5.4 Generalized Linear Models

5.4.1 General Model Definition

The linear model and the regression models for non-normal response variables described in the preceding sections have common properties that can be summarized in a unified framework:

1. The mean $\mu = \mathrm{E}(y)$ of the response y is connected with the linear predictor $\eta = \boldsymbol{x}'\boldsymbol{\beta}$ by a response function h or by a link function $g = h^{-1}$:

$$\mu = h(\eta) \quad \text{or} \quad \eta = g(\mu).$$

2. The distribution of the response variables (normal, binomial, Poisson, and gamma distribution) can be written in the form of a *univariate exponential family*:

5.4 Exponential Family

The density of a univariate exponential family for the response variable y is defined by

$$f(y \mid \theta) = \exp\left(\frac{y\theta - b(\theta)}{\phi} w + c(y, \phi, w)\right).$$

The log-density is given by

$$\log f(y \mid \theta) = \frac{y\theta - b(\theta)}{\phi} w + c(y, \phi, w).$$

The parameter θ is called the *natural* or *canonical* parameter. For the function $b(\theta)$ it is required that $f(y \mid \theta)$ can be normalized and the first and second derivative $b'(\theta)$ and $b''(\theta)$ exist. The second parameter ϕ is a dispersion parameter, while w is a known value (usually a weight).

As a consequence, both the definition of GLMs and the corresponding statistical inference can be presented in a unified framework. More generally, the resulting concepts can also be applied to regression problems with distributions that do not belong to the exponential family. To treat individual data and grouped data simultaneously, we introduce the weight factor w. For individual data, we set $w = 1$, whereas in the case when the response is summarized as a group mean, w is rather set to the corresponding group size. In the case when the sum of the individual responses of group i is selected for the response variable y_i, the weight equals $1/n_i$.

The Bernoulli and Poisson distributions do not include a dispersion parameter, i.e., $\phi = 1$. For the normal distribution, we have $\phi = \sigma^2$. The parameter θ represents the parameter of main interest that is connected to the linear predictor $\eta = \boldsymbol{x}'\boldsymbol{\beta}$, while the parameter ϕ is often considered a "nuisance parameter" of secondary interest. The term $c(y, \phi, a)$ does not depend on θ. It can be shown that

$$\mathrm{E}(y) = \mu = b'(\theta), \quad \mathrm{Var}(y) = \phi\, b''(\theta)/w.$$

Example 5.8 Bernoulli, Poisson, and Normal Distribution

1. *Bernoulli distribution:* A Bernoulli variable has probability mass function or (discrete) density

$$f(y \mid \pi) = \mathrm{P}(Y = y) = \pi^y (1-\pi)^{1-y}, \quad y = 0, 1,$$

where $\mathrm{P}(Y = 1) = \pi = \mathrm{E}(Y) = \mu$ and $\mathrm{Var}(Y) = \pi(1-\pi)$. Taking the logarithm yields

$$\log(f(y \mid \pi)) = y\log(\pi) - y\log(1-\pi) + \log(1-\pi).$$

If we define $\theta = \log(\pi) - \log(1-\pi) = \log(\pi/(1-\pi))$ as the natural parameter and take $\log(1-\pi) = -\log(1+\exp(\theta))$ into account, we obtain the density in the form of an exponential family:

$$f(y \mid \theta) = \exp(y\theta - \log(1+\exp(\theta))),$$

with $b(\theta) = \log(1+\exp(\theta))$, $\phi = 1$ and $c = 0$. Differentiation results in $b'(\theta) = \exp(\theta)/(1+\exp(\theta))$ and $b''(\theta) = \exp(\theta)/(1+\exp(\theta))^2$. Solving $\theta = \log(\pi/(1-\pi))$ for π results in

$$\pi = \exp(\theta)/(1+\exp(\theta)),$$

so that

$$\mathrm{E}(y) = b'(\theta) = \pi, \quad \mathrm{Var}(y) = b''(\theta) = \pi(1-\pi)$$

holds.

2. *Poisson distribution:* A Poisson variable has the (discrete) density

$$f(y \mid \lambda) = \mathrm{P}(Y = y) = \frac{\lambda^y \exp(-\lambda)}{y!}, \quad y = 0, 1, \ldots$$

The logarithm of this density results in

$$\log(f(y \mid \lambda)) = y\log(\lambda) - \lambda - \log(y!).$$

With $\theta = \log(\lambda)$ as the natural parameter, we obtain

$$\log(f(y \mid \theta)) = y\theta - \exp(\theta) - \log(y!).$$

Table 5.12 Univariate exponential families

(a) Density

$$f(y \mid \theta, \phi, w) = \exp\left(\frac{y\theta - b(\theta)}{\phi} w + c(y, \phi, w)\right)$$

(b) Exponential family parameters

Distribution		$\theta(\mu)$	$b(\theta)$	ϕ
Normal	$N(\mu, \sigma^2)$	μ	$\theta^2/2$	σ^2
Bernoulli	$B(1, \pi)$	$\log(\pi/(1-\pi))$	$\log(1+\exp(\theta))$	1
Poisson	$Po(\lambda)$	$\log(\lambda)$	$\exp(\theta)$	1
Gamma	$G(\mu, \nu)$	$-1/\mu$	$-\log(-\theta)$	ν^{-1}
Inverse				
Gaussian	$IG(\mu, \sigma^2)$	$-1/(2\mu^2)$	$-(-2\theta)^{1/2}$	σ^2

(c) Expectation and variance

Distribution	$E(y) = b'(\theta)$	$b''(\theta)$	$\mathrm{Var}(y) = b''(\theta)\phi/w$
Normal	$\mu = \theta$	1	σ^2/w
Bernoulli	$\pi = \frac{\exp(\theta)}{1+\exp(\theta)}$	$\pi(1-\pi)$	$\pi(1-\pi)/w$
Poisson	$\lambda = \exp(\theta)$	λ	λ/w
Gamma	$\mu = -1/\theta$	μ^2	$\mu^2\nu^{-1}/w$
Inverse			
Gaussian	$\mu = (-2\theta)^{-1/2}$	μ^3	$\mu^3\sigma^2/w$

It follows that $b(\theta) = \exp(\theta) = \lambda$, $\phi = 1$ and $c(y, \phi) = -\log(y!)$. With $b'(\theta) = b''(\theta) = \exp(\theta) = \lambda$, we obtain

$$E(y) = \mu = \lambda, \quad \mathrm{Var}(y) = \lambda,$$

i.e., the equality of expectation and variance that is characteristic for Poisson variables.

3. *Normal distribution:* The density of the normal distribution is

$$f(y \mid \mu) = \frac{1}{(2\pi\sigma^2)^{1/2}} \exp\left[-\frac{1}{2\sigma^2}(y-\mu)^2\right],$$

where $\mu = E(y)$ is the parameter of interest and $\sigma^2 = \mathrm{Var}(y)$ is the nuisance parameter. The density can be written in the form of an exponential family

$$f(y \mid \mu) = \exp\left[-\frac{y^2}{2\sigma^2} + \frac{y\mu}{\sigma^2} - \frac{\mu^2}{2\sigma^2} - \frac{1}{2}\log(2\pi\sigma^2)\right]$$

$$= \exp\left[\frac{y\mu - \mu^2/2}{\sigma^2} - \frac{y^2}{2\sigma^2} - \frac{1}{2}\log(2\pi\sigma^2)\right]$$

with $\theta = \mu$, $\phi = \sigma^2$, $b(\theta) = \mu^2/2 = \theta^2/2$ and $c(y, \phi) = -y^2/(2\sigma^2) - 0.5\log(2\pi\sigma^2)$. It follows, as expected, that

$$b'(\theta) = \theta = \mu = E(y), \quad b''(\theta) = 1$$

and thus

$$\mathrm{Var}(y) = \phi b''(\theta) = \sigma^2.$$

△

Similarly, one can derive the properties for the other distributions; see the summary in Table 5.12.

5.5 Generalized Linear Model

Distributional Assumptions

For given covariates $\boldsymbol{x}_i = (1, x_{i1}, \ldots, x_{ik})'$, the response variables are (conditionally) independent and the (conditional) density of y_i belongs to an exponential family with

$$f(y_i \mid \theta_i) = \exp\left(\frac{y_i\theta_i - b(\theta_i)}{\phi} w_i + c(y_i, \phi, w_i)\right).$$

The parameter θ_i is the natural parameter and ϕ is a common dispersion parameter, independent of i. For $\mathrm{E}(y_i) = \mu_i$ and $\mathrm{Var}(y_i)$, we have

$$\mathrm{E}(y_i) = \mu_i = b'(\theta_i), \quad \mathrm{Var}(y_i) = \sigma_i^2 = \phi\, b''(\theta_i)/w_i.$$

The weight parameter w_i is 1 for ungrouped data ($i = 1, \ldots, n$). In the case when the *sum* of the individual responses of group i is selected for the response variable y_i, the weight equals $1/n_i$ for grouped data ($i = 1, \ldots, G$). Note $w_i = n_i$ when the group mean, rather than the sum, is selected.

Structural Assumptions

The (conditional) mean μ_i is connected to the linear predictor $\eta_i = \boldsymbol{x}_i'\boldsymbol{\beta} = \beta_0 + \beta_1 x_{i1} + \ldots + \beta_k x_{ik}$ through

$$\mu_i = h(\eta_i) = h(\boldsymbol{x}_i'\boldsymbol{\beta}) \quad \text{or} \quad \eta_i = g(\mu_i),$$

where

h is a (one-to-one and twice differentiable) *response function* and
g is the *link function*, i.e., the inverse $g = h^{-1}$.

In summary, a specific GLM is completely determined by the type of the exponential family (Gaussian, binomial, Poisson, gamma, inverse Gaussian), the choice of the link or response function, and the definition and selection of covariates.

The choice of an appropriate response or link function is, as presented in the preceding examples, dependent on the type of the response variable. Every exponential family has a unique *canonical* (or *natural*) link function, defined by $\theta_i = \eta_i = \boldsymbol{x}_i'\boldsymbol{\beta}$. According to Table 5.12, the linear model $\mu_i = \eta_i = \boldsymbol{x}_i'\boldsymbol{\beta}$ corresponds to the natural link function for the normal distribution, whereas the logit model is obtained in binary regression models, and the log-linear model results for Poisson models.

5.6 Maximum Likelihood Estimation in GLMs

Definition

The ML estimator $\hat{\boldsymbol{\beta}}$ maximizes the (log-)likelihood and is defined as the solution

$$s(\hat{\boldsymbol{\beta}}) = \mathbf{0}$$

of the score function given by

$$s(\boldsymbol{\beta}) = \sum \boldsymbol{x}_i \frac{h'(\eta_i)}{\sigma_i^2}(y_i - \mu_i) = \boldsymbol{X}'\boldsymbol{D}\boldsymbol{\Sigma}^{-1}(\boldsymbol{y} - \boldsymbol{\mu}),$$

where $\boldsymbol{D} = \text{diag}(h'(\eta_1), \ldots, h'(\eta_n))$, $\boldsymbol{\Sigma} = \text{diag}(\sigma_1^2, \ldots, \sigma_n^2)$ and $\boldsymbol{\mu} = (\mu_1, \ldots, \mu_n)'$ is the vector of expectations $\text{E}(y_i) = \mu_i = h(\eta_i)$.
The Fisher matrix is

$$\boldsymbol{F}(\boldsymbol{\beta}) = \sum \boldsymbol{x}_i \boldsymbol{x}_i' \tilde{w}_i = \boldsymbol{X}'\boldsymbol{W}\boldsymbol{X}$$

where $\boldsymbol{W} = \text{diag}(\tilde{w}_1, \ldots, \tilde{w}_n)$ is the diagonal matrix of working weights

$$\tilde{w}_i = \frac{(h'(\eta_i))^2}{\sigma_i^2}.$$

Numerical Computation

The ML estimator $\hat{\boldsymbol{\beta}}$ is obtained iteratively using Fisher scoring in form of iteratively weighted least squares estimates

$$\hat{\boldsymbol{\beta}}^{(t+1)} = (\boldsymbol{X}'\boldsymbol{W}^{(t)}\boldsymbol{X})^{-1}\boldsymbol{X}'\boldsymbol{W}^{(t)}\tilde{\boldsymbol{y}}^{(t)}, \quad t = 0, 1, 2, \ldots$$

with working weights and observations given in Eqs. (5.17) and (5.16).

For canonical link functions, the log-likelihood is always concave so that the ML estimator is always unique (if it exists). Moreover, it can be shown that the expected and observed information matrix coincide, i.e., $F(\boldsymbol{\beta}) = H(\boldsymbol{\beta})$.

5.4.2 Likelihood Inference

Inference in GLMs is again based on the likelihood principle. Let

$$X = \begin{pmatrix} x'_1 \\ \vdots \\ x'_n \end{pmatrix} = \begin{pmatrix} 1 & x_{11} & \cdots & x_{1k} \\ \vdots & & & \vdots \\ 1 & x_{n1} & \cdots & x_{nk} \end{pmatrix}$$

be the design matrix with $\text{rk}(X) = p$. In Sect. 5.8.2 we derive the log-likelihood $l(\boldsymbol{\beta})$, score function $\boldsymbol{s}(\boldsymbol{\beta})$, and Fisher information $\boldsymbol{F}(\boldsymbol{\beta})$; see Box 5.6 for a summary. Based on the score function and the Fisher information, Sect. 5.8.2 also shows that the ML estimator for $\boldsymbol{\beta}$ can be iteratively obtained as

$$\hat{\boldsymbol{\beta}}^{(t+1)} = \left(\boldsymbol{X}'\boldsymbol{W}^{(t)}\boldsymbol{X}\right)^{-1} \boldsymbol{X}'\boldsymbol{W}^{(t)}\tilde{\boldsymbol{y}}^{(t)}, \quad t = 0, 1, 2, \ldots. \tag{5.15}$$

Here, $\tilde{\boldsymbol{y}}^{(t)} = \left(\tilde{y}_1\left(\hat{\eta}_1^{(t)}\right), \ldots, \tilde{y}_n\left(\hat{\eta}_n^{(t)}\right)\right)'$ is a vector of "working observations" with elements

$$\tilde{y}_i\left(\hat{\eta}_i^{(t)}\right) = \hat{\eta}_i^{(t)} + \frac{\left(y_i - h\left(\hat{\eta}_i^{(t)}\right)\right)}{h'\left(\hat{\eta}_i^{(t)}\right)}, \tag{5.16}$$

where $\hat{\eta}_i^{(t)} = \boldsymbol{x}'_i\hat{\boldsymbol{\beta}}^{(t)}$ is the actual predictor, h is the response function, and $h'(\eta) = \partial h(\eta)/\partial\eta$ is the derivative of h with respect to η. The matrix

$$\boldsymbol{W}^{(t)} = \text{diag}\left(\tilde{w}_1\left(\hat{\eta}_1^{(t)}\right), \ldots, \tilde{w}_n\left(\hat{\eta}_n^{(t)}\right)\right)$$

is a diagonal matrix of the "working weights"

$$\tilde{w}_i\left(\hat{\eta}_i^{(t)}\right) = \frac{\left(h'\left(\hat{\eta}_i^{(t)}\right)\right)^2}{\sigma_i^2\left(\hat{\eta}_i^{(t)}\right)}, \tag{5.17}$$

where $\sigma_i^2\left(\hat{\eta}_i^{(t)}\right)$ is the (conditional) variance $\text{Var}(y_i)$ evaluated at $\eta = \hat{\eta}_i^{(t)}$. The required quantities to compute the weighted least squares estimator can be easily obtained from Table 5.12. A key role in the iterations Eq. (5.15) plays the Fisher matrix $\boldsymbol{F}(\boldsymbol{\beta}) = \boldsymbol{X}'\boldsymbol{W}\boldsymbol{X}$. Since the elements $\tilde{w}_i$ of the diagonal matrix $\boldsymbol{W}$ depend on the covariates $\boldsymbol{x}_i$ and on $\boldsymbol{\beta}$, invertibility of $\boldsymbol{F}(\boldsymbol{\beta})$ in Eq. (5.15) does not follow from the invertibility of $\boldsymbol{X}'\boldsymbol{X}$ (or equivalently the full rank of $\boldsymbol{X}$) in general. However, usually (almost) all of the weights are positive such that $\boldsymbol{F}(\boldsymbol{\beta})$ is invertible, which we assume in the following. Then, according to the stopping criterion, the algorithm

5.7 Asymptotic Properties of the ML Estimator

Let $\hat{\boldsymbol{\beta}}_n$ denote the ML estimator based on a sample of size n. Under regularity conditions, $\hat{\boldsymbol{\beta}}_n$ is consistent and asymptotically normal:

$$\hat{\boldsymbol{\beta}}_n \overset{a}{\sim} \mathrm{N}(\boldsymbol{\beta}, \boldsymbol{F}^{-1}(\boldsymbol{\beta})).$$

This result holds even if the estimator $\boldsymbol{F}(\hat{\boldsymbol{\beta}})$ replaces $\boldsymbol{F}(\boldsymbol{\beta})$.

typically converges close to a maximum after a number of iterative steps. With the natural link function, it can be shown that the achieved maximum is unique. However, this statement does not hold in general and therefore several different starting values should be used to help ensure the global maximum is achieved.

Asymptotic Properties of ML Estimates

As in section "Asymptotic Properties of the Least Squares Estimator" of Sect. 3.2.3 we index the model quantities with the number of observations n. For regressors with compact support, $(\boldsymbol{X}_n'\boldsymbol{X}_n)^{-1} \rightarrow \mathbf{0}$ or $\lambda_{\min}(\boldsymbol{X}_n'\boldsymbol{X}_n) \rightarrow \infty$ are sufficient for asymptotic normality and weak consistency in case of models with canonical link function (where $\lambda_{\min}$ denotes the smallest eigenvalue of $\boldsymbol{X}_n'\boldsymbol{X}_n$). Compare also section "Asymptotic Properties of the Least Squares Estimator" of Sect. 3.2.3 for a brief discussion and some examples of these conditions. For non-canonical link function, stronger conditions on the limiting behavior of $\boldsymbol{X}_n'\boldsymbol{X}_n$ have to be imposed. If, in the case of stochastic regressors, the observations $(y_i, \boldsymbol{x}_i)$ are independent and identically distributed, e.g., $(y, \boldsymbol{x})$, and comply with the assumptions of a general linear model, asymptotic normality follows under mild regularity conditions on the marginal distribution of $\boldsymbol{x}$.

Under the same assumptions, $\hat{\boldsymbol{\beta}}_n$ asymptotically exists with probability 1, i.e.,

$$\lim_{n\to\infty} \mathrm{P}(\hat{\boldsymbol{\beta}}_n \text{ exists}) = 1.$$

Details and general proofs can be found in Fahrmeir and Kaufmann (1985).

The inverse of the Fisher information matrix $\boldsymbol{F}(\boldsymbol{\beta})$, evaluated at the ML estimator $\hat{\boldsymbol{\beta}}$, is the asymptotic or approximate covariance matrix $\boldsymbol{A} = \boldsymbol{F}^{-1}(\hat{\boldsymbol{\beta}})$ of $\hat{\boldsymbol{\beta}}$. The diagonal element a_{jj} is an estimator for the variance $\sigma_j^2 = \mathrm{Var}(\hat{\beta}_j)$ of the jth component and $\sqrt{a_{jj}}$ for the standard deviation σ_j.

Estimation of the Scale or Overdispersion Parameter

Recall that $\mathrm{Var}(y_i) = \phi\, b''(\theta_i)/w_i$ for general exponential families. Denote by $v(\mu_i) = b''(\theta_i)$ the so-called variance function; see Table 5.12 for the specific expression of b''. Note that $b''(\theta_i)$ implicitly depends on μ_i through the relation $b'(\theta_i) = \mu_i$.

Using the variance function the dispersion parameter can then be estimated consistently by

$$\hat{\phi} = \frac{1}{G-p} \sum_{i=1}^{G} \frac{(y_i - \hat{\mu}_i)^2}{v(\hat{\mu}_i)/n_i},$$

where p denotes the number of regression parameters, $\hat{\mu}_i = h(\boldsymbol{x}_i'\hat{\boldsymbol{\beta}})$ is the estimated expectation, $v(\hat{\mu}_i)$ is the estimated variance function, and the data should be grouped as much as possible. We then substitute $\hat{\phi}$ for ϕ in every term containing ϕ, as, for example, in $\boldsymbol{F}(\hat{\boldsymbol{\beta}})$.

Testing Linear Hypotheses

For testing linear hypotheses

$$H_0: \boldsymbol{C}\boldsymbol{\beta} = \boldsymbol{d} \qquad \text{versus} \qquad H_1: \boldsymbol{C}\boldsymbol{\beta} \neq \boldsymbol{d},$$

where $\boldsymbol{C}$ has a full row rank $r \leq p$, we can use the likelihood ratio statistic lr, the Wald statistic w, and the score statistic u as discussed in more detail for binary responses in Sect. 5.1.3; see also Appendix B.4.4 (p. 662) for a general presentation of likelihood-based hypothesis testing. In the corresponding definitions, the specific formulae for the chosen GLM have to be used for $l(\boldsymbol{\beta})$, $s(\boldsymbol{\beta})$, and $F(\boldsymbol{\beta})$. Under conditions similar to those for the asymptotic results on ML estimation, we have $lr, w, s \overset{a}{\sim} \chi^2_r$, allowing for the computation of appropriate critical values and (approximate) p-values.

Criteria for Model Fit and Model Selection

The *Pearson statistic*

$$\chi^2 = \sum_{i=1}^{G} \frac{(y_i - \hat{\mu}_i)^2}{v(\hat{\mu}_i)/w_i}$$

and the *deviance*

$$D = -2\sum_{i=1}^{G}\{l_i(\hat{\mu}_i) - l_i(\bar{y}_i)\}$$

are the two most common global statistics to verify the fit of a model relative to the saturated model. Here, $\hat{\mu}_i$ and $v(\hat{\mu}_i)$ are the estimated expectations and variance functions, respectively, and the ith log-likelihood contribution of the saturated model is $l_i(\bar{y}_i)$, where $\bar{y}_i$ replaces μ_i. This results in the maximum possible value of the log-likelihood. For both model fit statistics, the data should be grouped as much as possible. When n_i is sufficiently large in *all* groups $i = 1, \ldots, G$, both statistics are approximately or asymptotically (for $n \to \infty$) $\phi\chi^2(G-p)$-distributed, where p denotes the number of estimated parameters. In this situation, we can use both statistics for formal testing of model fit, i.e., for comparing the estimated model fit to that of the saturated model. For small n_i, especially for $n_i = 1$, such formal tests

5.8 Testing Linear Hypotheses

Hypotheses

$$H_0 : \boldsymbol{C}\boldsymbol{\beta} = \boldsymbol{d} \qquad \text{versus} \qquad H_1 : \boldsymbol{C}\boldsymbol{\beta} \neq \boldsymbol{d}.$$

Test Statistics

1. *Likelihood ratio statistic:* $lr = -2\{l(\tilde{\boldsymbol{\beta}}) - l(\hat{\boldsymbol{\beta}})\}$
2. *Wald statistic:* $w = (\boldsymbol{C}\hat{\boldsymbol{\beta}} - \boldsymbol{d})'[\boldsymbol{C}\boldsymbol{F}^{-1}(\hat{\boldsymbol{\beta}})\boldsymbol{C}']^{-1}(\boldsymbol{C}\hat{\boldsymbol{\beta}} - \boldsymbol{d})$
3. *Score statistic:* $u = \boldsymbol{s}'(\tilde{\boldsymbol{\beta}})\boldsymbol{F}^{-1}(\tilde{\boldsymbol{\beta}})\boldsymbol{s}(\tilde{\boldsymbol{\beta}})$

where $\tilde{\boldsymbol{\beta}}$ is the ML estimator under H_0.

Test Decision

For large n and under H_0, we have the asymptotic results

$$lr, w, u \overset{a}{\sim} \chi^2_r$$

where r is the (full) row rank of $\boldsymbol{C}$. We reject H_0 when

$$lq, w, u > \chi^2_r(1-\alpha).$$

can be problematic, even with a large sample size n. Large values of χ^2 or D then will not necessarily indicate a poor model fit.

The AIC for model selection is defined generally as

$$\text{AIC} = -2l(\hat{\boldsymbol{\beta}}) + 2p.$$

If the model contains a dispersion parameter ϕ, as is the case with the normal distribution, its maximum likelihood estimator should be substituted into the respective model. Accordingly, the total number of parameters must be increased to $p + 1$.

5.5 Quasi-likelihood Models

For GLMs, the response is assumed to be a member of the exponential family, e.g., the Gaussian, Poisson, or binomial distribution. This distributional assumption, in combination with the mean structure $\text{E}(y) = \mu = h(\boldsymbol{x}'\boldsymbol{\beta})$, implies a specific variance structure $\text{Var}(y) = \phi\, b''(\mu)/w$, where the variance function $v(\mu) = b''(\mu)$ is determined by the exponential family. If the empirical variance does not comply with the estimated variance $\hat{\phi}\, b''(\hat{\mu})/w$, the distribution of the data will be incorrectly specified, i.e., the data do not agree with the chosen distribution from the exponential family.

Quasi-likelihood models allow for a separate specification of the mean and the variance structure. Furthermore, it is not necessary that these specifications correspond to a proper likelihood function. It suffices to specify a correct expectation structure $E(y) = h(\boldsymbol{x}'\boldsymbol{\beta})$, together with a "working" variance structure σ_i^2, and to define parameter estimates as the roots of a quasi-score function or *generalized estimating equation* (GEE) that has the same form as in usual GLMs; see the formula for $\boldsymbol{s}(\boldsymbol{\beta})$ in Box 5.6.

We then start directly with the specification of a *generalized estimating function*

$$\boldsymbol{s}(\boldsymbol{\beta}) = \sum_{i=1}^{n} \boldsymbol{x}_i \frac{h'(\eta_i)}{\sigma_i^2}(y_i - \mu_i). \tag{5.18}$$

Similar as in the score function of Box 5.6 that was obtained as the derivative of the log-likelihood of a GLM, we assume that the expectation $\mu_i = h(\boldsymbol{x}_i'\boldsymbol{\beta})$ of y_i given $\boldsymbol{x}_i$ is correctly specified with

$$E(y_i) = \mu_i = h(\boldsymbol{x}_i'\boldsymbol{\beta}).$$

We then have

$$E(\boldsymbol{s}(\boldsymbol{\beta})) = \sum_{i=1}^{n} \boldsymbol{x}_i \frac{h'(\eta_i)}{\sigma_i^2}(E(y_i) - \mu_i) = \boldsymbol{0},$$

as for a real score function, a property that is crucial for the consistency of parameter estimates.

In contrast, it is not necessary that the specified variance σ_i^2 in Eq. (5.18) equals the true variance $\text{Var}(y_i)$, but it can rather be specified with the help of a given quasi-variance function $v(\mu)$, i.e., $\sigma^2(\mu) = \phi\, v(\mu)/w$. We then call $\sigma^2(\mu) = \phi\, v(\mu)/w$ the *working variance*.

The simplest form of a (working) variance function results from overdispersion in binomial and Poisson models with $w_i = n_i$ and

$$\sigma_i^2(\pi_i) = \phi \frac{\pi_i(1-\pi_i)}{n_i}$$

or

$$\sigma_i^2(\lambda_i) = \phi\lambda_i,$$

respectively. In this case, the quasi-score function (5.18) is identical to the score function of a binomial or Poisson model up to a constant factor $1/\phi$, but it no longer corresponds to the derivative of a log-likelihood function.

The (working) variance function is often parameterized with another parameter θ as

$$\sigma^2(\mu) = \phi v(\mu; \theta).$$

An important special case is

$$v(\mu;\theta) = \mu^{\theta},$$

where we obtain the variance function of the Gaussian, Poisson, gamma, and of the inverse Gaussian distribution, for $\theta = 0, 1, 2, 3$, respectively.

The quasi-ML estimator $\hat{\boldsymbol{\beta}}$ is defined as the root of the quasi-score function, i.e., as the solution to the *generalized estimating equation* (GEE)

$$\boldsymbol{s}(\hat{\boldsymbol{\beta}}) = \boldsymbol{0}.$$

As in case of ML estimation, the solution is obtained iteratively. The quasi-Fisher information matrix $\boldsymbol{F}(\boldsymbol{\beta}) = \mathrm{E}(-\partial \boldsymbol{s}(\boldsymbol{\beta})/\partial \boldsymbol{\beta}')$ becomes

$$\boldsymbol{F}(\boldsymbol{\beta}) = \sum_{i=1}^{n} \boldsymbol{x}_i \boldsymbol{x}_i' \tilde{w}_i,$$

with working variance σ_i^2 included in the working weights $\tilde{w}_i = (h'(\eta_i))^2/\sigma_i^2$. However, $\boldsymbol{F}(\boldsymbol{\beta})$ differs from $\boldsymbol{V}(\boldsymbol{\beta}) = \mathrm{Cov}(\boldsymbol{s}(\boldsymbol{\beta})) = \mathrm{E}(\boldsymbol{s}(\boldsymbol{\beta})\boldsymbol{s}'(\boldsymbol{\beta}))$ in general. In fact, we have

$$\boldsymbol{V}(\boldsymbol{\beta}) = \sum_{i=1}^{n} \boldsymbol{x}_i \boldsymbol{x}_i' \tilde{w}_i \frac{\sigma_{0i}^2}{\sigma_i^2}.$$

Thus only in the case when the working variances equal the true variances σ_{0i}^2 we obtain $\boldsymbol{F}(\boldsymbol{\beta}) = \boldsymbol{V}(\boldsymbol{\beta})$ as in ML estimation.

Under regularity assumptions, quasi-ML estimators are consistent and asymptotically normal

$$\boldsymbol{\beta} \overset{a}{\sim} \mathrm{N}\left(\boldsymbol{\beta}, \hat{\boldsymbol{F}}^{-1} \hat{\boldsymbol{V}} \hat{\boldsymbol{F}}^{-1}\right)$$

with estimates $\hat{\boldsymbol{F}} = \boldsymbol{F}(\hat{\boldsymbol{\beta}})$ and

$$\hat{\boldsymbol{V}} = \sum_{i=1}^{n} \boldsymbol{x}_i \boldsymbol{x}_i' (h'(\hat{\eta}_i))^2 \frac{(y_i - \hat{\mu}_i)^2}{\sigma_i^4(\hat{\boldsymbol{\beta}})}$$

for $\boldsymbol{F}(\boldsymbol{\beta})$ and $\boldsymbol{V}(\boldsymbol{\beta})$. Compared with the asymptotic properties of the ML estimator, only the asymptotic covariance matrix $\mathrm{Cov}(\hat{\boldsymbol{\beta}}) = \hat{\boldsymbol{F}}^{-1}$ has to be corrected to the "sandwich" matrix $\hat{\boldsymbol{A}} = \hat{\boldsymbol{F}}^{-1} \hat{\boldsymbol{V}} \hat{\boldsymbol{F}}^{-1}$. Thus, quasi-likelihood models allow consistent and asymptotically normal estimation of $\boldsymbol{\beta}$ but with some loss of (asymptotic) efficiency. To keep this loss minimal, the working variance structure should be not far off the true variance structure.

5.6 Bayesian Generalized Linear Models

The Bayesian approach for linear models discussed in Sect. 4.4 can, in principle, be applied to GLMs. However, the application is more complicated both mathematically and numerically. Fully Bayesian inference usually requires the use of MCMC simulation techniques that are more complex than the corresponding techniques for linear models. This section gives a brief overview of Bayesian inference in GLMs. We limit the discussion to models without dispersion parameters, specifically focusing on binomial and Poisson models. A more complete discussion of Bayesian GLMs and extensions can be found in Dey, Gosh, and Mallick (2000), as well as in the corresponding sections in Fahrmeir and Tutz (2001).

In Bayesian GLMs, we assume a prior density $p(\boldsymbol{\beta})$ for the parameter vector $\boldsymbol{\beta}$ which is considered a random variable. Similar to Bayesian linear models discussed in Sect. 4.4 we assume a multivariate Gaussian prior, i.e.,

$$\boldsymbol{\beta} \sim \mathrm{N}(\boldsymbol{m}, \boldsymbol{M}), \tag{5.19}$$

where $\boldsymbol{m}$ is the prior mean vector and $\boldsymbol{M}$ the prior covariance matrix. A typical choice is $\boldsymbol{m} = \boldsymbol{0}$ and $\boldsymbol{M} = \boldsymbol{I}$ thereby assuming independence among the regression coefficients. A noninformative prior is obtained by $\boldsymbol{m} = \boldsymbol{0}$ and the limit $\boldsymbol{M}^{-1} = \boldsymbol{0}$. Other choices such as a combination of informative and noninformative priors, Bayesian ridge and LASSO or spike and slab priors, that have been discussed extensively for Bayesian linear models, can be used as well. We restrict the discussion here to the normal prior (5.19) because the only difficulty compared to linear models is inference regarding the regression coefficients. Inference for the hyperparameters is typically based on identical MCMC updating steps as for linear models. The reason is that their full conditionals are independent of the specific observation model. For instance, the Bayesian LASSO assumes

$$\beta_1, \dots, \beta_k \mid \tau_1^2, \dots, \tau_k^2 \sim \mathrm{N}(\boldsymbol{0}, \mathrm{diag}(\tau_1^2, \dots, \tau_k^2)).$$

While updating the regression coefficients in the Bayesian LASSO might be problematic because the full conditional is not Gaussian (see below), updating the variances τ_j^2 proceeds exactly as described in Sect. 4.4.2.

We now discuss the difficulties involved with Bayesian inference for non-Gaussian data. According to Bayes' theorem, inference relies on the posterior density $p(\boldsymbol{\beta} \mid \boldsymbol{y})$, given the (conditionally independent) response variables $\boldsymbol{y} = (y_1, \dots, y_n)'$ and covariates. Suppressing the notational dependence on covariates, this yields

$$p(\boldsymbol{\beta} \mid \boldsymbol{y}) = \frac{L(\boldsymbol{\beta} \mid \boldsymbol{y})\, p(\boldsymbol{\beta})}{\int L(\boldsymbol{\beta} \mid \boldsymbol{y})\, p(\boldsymbol{\beta})\, d\boldsymbol{\beta}} \propto L(\boldsymbol{\beta} \mid \boldsymbol{y})\, p(\boldsymbol{\beta}), \tag{5.20}$$

where

$$L(\boldsymbol{\beta} \mid \boldsymbol{y}) = \prod_{i=1}^{n} f_i(y_i \mid \boldsymbol{\beta})$$

is the likelihood of a given GLM, for example, a binomial logit model or a log-linear Poisson model. The posterior mean is defined as

$$\mathrm{E}(\boldsymbol{\beta} \mid \boldsymbol{y}) = \int \boldsymbol{\beta}\, p(\boldsymbol{\beta} \mid \boldsymbol{y}) d\boldsymbol{\beta}$$

and the corresponding posterior covariance matrix

$$\mathrm{Cov}(\boldsymbol{\beta} \mid \boldsymbol{y}) = \int (\boldsymbol{\beta} - \mathrm{E}(\boldsymbol{\beta} \mid \boldsymbol{y}))(\boldsymbol{\beta} - \mathrm{E}(\boldsymbol{\beta} \mid \boldsymbol{y}))' \, p(\boldsymbol{\beta} \mid \boldsymbol{y}) \, d\boldsymbol{\beta}$$

provides a measure for the precision of the posterior mean. At first glance, it seems straightforward to put Bayesian inference into effect. However, the integrations involved are problematic, as their analytical solution is only possible in a few special cases. Numerical integration methods are applicable, as long as the dimension of $\boldsymbol{\beta}$ remains relatively small (about ≤ 5); extensions to more complex models are described in the following chapters, yet remain widely intractable. Hence, we have two options: First, posteriori mode or MAP (maximum a posteriori) estimation, for which we have to maximize the numerator in Eq. (5.20), or second, fully Bayesian inference with MCMC techniques.

5.6.1 Posterior Mode Estimation

The *posterior mode* $\hat{\boldsymbol{\beta}}_{MAP}$ maximizes the posterior density $p(\boldsymbol{\beta} \mid \boldsymbol{y})$ or the log-posterior

$$\log(p(\boldsymbol{\beta} \mid \boldsymbol{y})) = l(\boldsymbol{\beta}) + \log\, p(\boldsymbol{\beta})\,,$$

where $l(\boldsymbol{\beta})$ is the log-likelihood of the given GLM. For a Gaussian prior

$$\boldsymbol{\beta} \sim \mathrm{N}(\boldsymbol{m}, \boldsymbol{M}), \qquad \boldsymbol{M} \text{ positive definite},$$

we obtain the special case

$$\log(p(\boldsymbol{\beta} \mid \boldsymbol{y})) = l(\boldsymbol{\beta}) - \frac{1}{2}(\boldsymbol{\beta} - \boldsymbol{m})'\boldsymbol{M}^{-1}(\boldsymbol{\beta} - \boldsymbol{m})\,,$$

where terms independent of $\boldsymbol{\beta}$ have been left out. Now $\log(p(\boldsymbol{\beta} \mid \boldsymbol{y}))$ can also be viewed as a *penalized (log-)likelihood*, where the penalty term $(\boldsymbol{\beta} - \boldsymbol{m})'\boldsymbol{M}^{-1}(\boldsymbol{\beta} - \boldsymbol{m})$ penalizes large deviations from the prior mean $\boldsymbol{m}$. This penalization potentially overcomes the problem of non-existence or divergence of the ML estimators. We also refer to the estimator $\hat{\boldsymbol{\beta}}_{MAP}$ as a penalized ML estimator.

For the limiting case $\boldsymbol{M}^{-1} \to \boldsymbol{0}$ of a flat prior

$$p(\boldsymbol{\beta}) \propto const,$$

the penalty disappears, which results in the posterior mode estimator equaling the (unpenalized) ML estimator.

The *ridge estimator* with *shrinkage* parameter $\lambda = 1/(2\tau^2)$ results as a special case with $\boldsymbol{m} = \boldsymbol{0}$ and $\boldsymbol{M} = \tau^2 \mathrm{diag}(0, 1, \ldots, 1)$; see also section "Bayesian Ridge Regression" of Sect. 4.4.2. The penalty then simplifies to

$$\lambda\, \boldsymbol{\beta}'\mathrm{diag}(0, 1, \ldots, 1)\boldsymbol{\beta} = \lambda(\beta_1^2 + \beta_2^2 + \ldots + \beta_k^2)\,,$$

and the parameter λ regularizes shrinkage of the ML estimator $\hat{\boldsymbol{\beta}}$ towards $\mathbf{0}$ and therefore stabilizes the ML estimator in cases of large variability.

Estimation of the posterior mode proceeds analogously to ML estimation. With a Gaussian prior distribution, the score function $\boldsymbol{s}(\boldsymbol{\beta})$ becomes the penalized score function

$$\boldsymbol{s}_p(\boldsymbol{\beta}) = \frac{\partial \log(p(\boldsymbol{\beta} \mid \boldsymbol{y}))}{\partial \boldsymbol{\beta}} = \boldsymbol{s}(\boldsymbol{\beta}) - \boldsymbol{M}^{-1}(\boldsymbol{\beta} - \boldsymbol{m})$$

and the Fisher information matrix $\boldsymbol{F}(\boldsymbol{\beta})$ becomes

$$\boldsymbol{F}_p(\boldsymbol{\beta}) = -\mathrm{E}\left(-\frac{\partial^2 \log(p(\boldsymbol{\beta} \mid \boldsymbol{y}))}{\partial \boldsymbol{\beta} \partial \boldsymbol{\beta}'}\right) = \boldsymbol{F}(\boldsymbol{\beta}) + \boldsymbol{M}^{-1}.$$

Computation is carried out with a modified Fisher scoring algorithm or IWLS algorithm, in which $\boldsymbol{s}_p(\boldsymbol{\beta})$ and $\boldsymbol{F}_p(\boldsymbol{\beta})$ replace $\boldsymbol{s}(\boldsymbol{\beta})$ and $\boldsymbol{F}(\boldsymbol{\beta})$, respectively.

Under regularity assumptions, for $n \to \infty$, $\hat{\boldsymbol{\beta}}_{MAP}$ has an asymptotic (or approximate) normal distribution with

$$\hat{\boldsymbol{\beta}}_{MAP} \overset{a}{\sim} \mathrm{N}\left(\boldsymbol{\beta}, \boldsymbol{F}_p^{-1}(\hat{\boldsymbol{\beta}}_{MAP})\right),$$

and, as a consequence, the posterior mode $\hat{\boldsymbol{\beta}}_{MAP}$ and the (expected) curvature $\boldsymbol{F}_p^{-1}(\hat{\boldsymbol{\beta}}_{MAP})$ are good approximations of the posterior mean $\mathrm{E}(\boldsymbol{\beta} \mid \boldsymbol{y})$ and of the posterior covariance matrix $\mathrm{Cov}(\boldsymbol{\beta} \mid \boldsymbol{y})$, respectively.

5.6.2 Fully Bayesian Inference via MCMC Simulation Techniques

Fully Bayesian inference relies on MCMC techniques (see Appendix B.5) for drawing random numbers from the posterior $p(\boldsymbol{\beta} \mid \boldsymbol{y})$. Posterior means, medians, quantiles, variances, etc. are then approximated with their empirical analogues. For the Gaussian prior $\boldsymbol{\beta} \sim \mathrm{N}(\boldsymbol{m}, \boldsymbol{M})$ and also for the limiting case $\boldsymbol{M}^{-1} \to \mathbf{0}$ of a non-informative prior $p(\boldsymbol{\beta}) \propto const$, we have

$$p(\boldsymbol{\beta} \mid \boldsymbol{y}) \propto \exp\left(l(\boldsymbol{\beta}) - \frac{1}{2}(\boldsymbol{\beta} - \boldsymbol{m})'\boldsymbol{M}^{-1}(\boldsymbol{\beta} - \boldsymbol{m})\right).$$

With the exception of some special cases, no closed analytical form exists for the normalizing constant of this posterior. We therefore use MCMC techniques for drawing samples $\boldsymbol{\beta}^{(t)}, t = 1, \ldots, T$, from $p(\boldsymbol{\beta} \mid \boldsymbol{y})$. Dellaportas and Smith (1993) recommend a Gibbs sampler based on adaptive rejection sampling, which is implemented in the software WinBUGS. We prefer to draw the entire parameter vector $\boldsymbol{\beta}^{(t)}$, in every iteration step t, with a Metropolis–Hastings (MH) algorithm based on IWLS proposals; see Gamerman (1997) and Lenk and DeSarbo (2000). IWLS proposals $q(\boldsymbol{\beta}^* \mid \boldsymbol{\beta}^{(t)})$, for the update $\boldsymbol{\beta}^{(t+1)}$, rely on a normal distribution

5.9 Bayesian GLMs

Posterior Distribution

$$p(\boldsymbol{\beta} \mid \boldsymbol{y}) = \frac{L(\boldsymbol{\beta} \mid \boldsymbol{y}) p(\boldsymbol{\beta})}{\int L(\boldsymbol{\beta} \mid \boldsymbol{y}) p(\boldsymbol{\beta}) d\boldsymbol{\beta}} \propto L(\boldsymbol{\beta} \mid \boldsymbol{y}) p(\boldsymbol{\beta}),$$

where $L(\boldsymbol{\beta} \mid \boldsymbol{y})$ is the likelihood of a GLM and $p(\boldsymbol{\beta})$ is the prior distribution.

Posterior Mode

The *posterior mode* $\hat{\boldsymbol{\beta}}_{MAP}$ maximizes the posterior density $p(\boldsymbol{\beta} \mid \boldsymbol{y})$. With a normal prior distribution

$$\boldsymbol{\beta} \sim \mathrm{N}(\boldsymbol{m}, \boldsymbol{M}),$$

this is equivalent to maximizing the *penalized log-likelihood*

$$\log(p(\boldsymbol{\beta} \mid \boldsymbol{y})) = l(\boldsymbol{\beta}) - \frac{1}{2}(\boldsymbol{\beta} - \boldsymbol{m})' \boldsymbol{M}^{-1} (\boldsymbol{\beta} - \boldsymbol{m}),$$

with $l(\boldsymbol{\beta}) = \log L(\boldsymbol{\beta} \mid \boldsymbol{y})$. The iterative calculation of $\hat{\boldsymbol{\beta}}_{MAP}$ via IWLS relies on the penalized score function and Fisher information matrix

$$\boldsymbol{s}_p(\boldsymbol{\beta}) = \boldsymbol{s}(\boldsymbol{\beta}) - \boldsymbol{M}^{-1}(\boldsymbol{\beta} - \boldsymbol{m}) \qquad \boldsymbol{F}_p(\boldsymbol{\beta}) = \boldsymbol{F}(\boldsymbol{\beta}) + \boldsymbol{M}^{-1}.$$

Fully Bayesian Inference

Fully Bayesian inference is accomplished using an MH algorithm with IWLS proposals for drawing random numbers from the posterior density $p(\boldsymbol{\beta} \mid \boldsymbol{y})$.

Let $\boldsymbol{\beta}^{(t)}$ be the actual state of the Markov chain. We then draw the IWLS proposal $\boldsymbol{\beta}^*$ from a normal distribution density $q(\boldsymbol{\beta}^* | \boldsymbol{\beta}^{(t)})$ with

$$\boldsymbol{\beta}^* \sim \mathrm{N}(\boldsymbol{\mu}^{(t)}, (\boldsymbol{X}' \boldsymbol{W}^{(t)} \boldsymbol{X} + \boldsymbol{M}^{-1})^{-1}).$$

The Fisher matrix $\boldsymbol{F}_p^{(t)} = \boldsymbol{X}' \boldsymbol{W}^{(t)} \boldsymbol{X} + \boldsymbol{M}^{-1}$ is evaluated at the current state $\boldsymbol{\beta}^{(t)}$ and

$$\boldsymbol{\mu}^{(t)} = (\boldsymbol{F}_p^{(t)})^{-1} (\boldsymbol{X}' \boldsymbol{W}^{(t)} \tilde{\boldsymbol{y}}^{(t)} + \boldsymbol{M}^{-1} \boldsymbol{m}),$$

with $\boldsymbol{W}^{(t)} = \boldsymbol{W}(\boldsymbol{\beta}^{(t)})$ and the current working observations $\tilde{\boldsymbol{y}}^{(t)}$ (defined in the same way as for ML estimation). The probability of acceptance is then given by

$$\alpha(\boldsymbol{\beta}^* \mid \boldsymbol{\beta}^{(t)}) = \min \left\{ \frac{L(\boldsymbol{\beta}^*) \, p(\boldsymbol{\beta}^*) \, q(\boldsymbol{\beta}^{(t)} \mid \boldsymbol{\beta}^*)}{L(\boldsymbol{\beta}^{(t)}) \, p(\boldsymbol{\beta}^{(t)}) \, q(\boldsymbol{\beta}^* \mid \boldsymbol{\beta}^{(t)})} \right\},$$

with the likelihood $L(\boldsymbol{\beta})$ of the GLM evaluated at the proposed and current value, $\boldsymbol{\beta}^*$ and $\boldsymbol{\beta}^{(t)}$, respectively.

Table 5.13 Number of citations from patents: Bayesian Poisson model

Variable	Coefficient	Standard deviation	2.5 % Quantile	97.5 % Quantile
intercept	0.156	0.031	0.090	0.218
yearc	−0.072	0.003	−0.077	−0.066
ncountryc	−0.028	0.004	−0.036	−0.020
nclaimc	0.018	0.001	0.015	0.021
biopharm	0.240	0.032	0.180	0.301
ustwin	0.003	0.025	−0.043	0.054
patus	−0.078	0.029	−0.133	−0.019
patgsgr	−0.199	0.032	−0.262	−0.138
opp	0.372	0.025	0.321	0.422

having expectation and covariance matrix that are a (first) approximation of the posterior mode estimator and of the respective covariance matrix. Refer to Box 5.9 for details.

Example 5.9 Number of Citations from Patents—Bayesian Inference

We illustrate Bayesian inference through reanalyzing the log-linear Poisson model of Example 5.7 (p. 297) with a flat prior $p(\boldsymbol{\beta}) \propto const$ for $\boldsymbol{\beta}$. With this choice, the posterior mean obtained from a fully Bayesian model specification and the ML estimator, which is identical to the posterior mode, should not differ too much from each other. Table 5.13 contains the posterior mean estimates, as well as the posterior standard deviations and quantiles, for the Bayesian Poisson model with purely linear effects. Table 5.14 contains the corresponding results obtained with nonlinear effects for the continuous covariates. The results are based on `bayesreg objects` of the software `BayesX`. We find good agreement with our previous results based on ML inference. Note, however, that the existing overdispersion has not been taken into account so that the standard deviations are below the standard errors of Table 5.11 (p. 299).

The fact that the results of ML and Bayes inference differ only slightly from each other in this example provokes the following question: What is the advantage of the comparably computer intensive Bayesian estimator based on MCMC methods? One advantage is that, in addition to point estimates and confidence intervals, we are also able to estimate the entire posterior density $p(\boldsymbol{\beta} \mid \boldsymbol{y})$ based on the sampled random numbers. Figure 5.4 shows kernel density estimates for the posterior densities $p\left(\beta_j \mid \boldsymbol{y}\right)$ of the covariate effects for *ustwin*, *patus*, and *patgsgr*, as well as corresponding normal densities with adjusted expectations and variances. The posterior densities are all close to normality, which should be expected due to the comparably large sample size in this example. In general, Bayesian inference with MCMC is especially important for more complex regression models, if asymptotic normality approximations of likelihood inference are not reliable.

△

5.6.3 MCMC-Based Inference Using Data Augmentation

For a number of response distributions alternative sampling schemes, based on the representation of the models as latent linear models, can be developed. For binary response models, the connection to latent linear models has been pointed out in Sect. 5.1 on p. 274.

Table 5.14 Number of citations from patents: extended Bayesian Poisson model with nonlinear effects

Variable	Coefficient	Standard deviation	2.5 % Quantile	97.5 % Quantile
intercept	0.17022	0.03225	0.10759	0.23645
yearc	−0.09897	0.00556	−0.10975	−0.08807
yearc^2	−0.00973	0.00131	−0.01242	−0.00723
yearc^3	−0.00011	0.00017	−0.00046	0.00022
ncountryc	0.01581	0.00804	−0.00061	0.03318
ncountryc^2	−0.00215	0.00124	−0.00456	0.00025
ncountryc^3	−0.00158	0.00026	−0.00211	−0.00109
nclaimc	0.02609	0.00208	0.02205	0.03016
nclaimc^2	−0.00047	0.00021	−0.00085	−0.00006
nclaimc^3	0.00000	0.00000	−0.00000	−0.00001
biopharm	0.15549	0.03254	0.09443	0.21985
ustwin	−0.00294	0.02458	−0.05460	0.04415
patus	−0.09475	0.02757	−0.14652	−0.04188
patgsgr	−0.20191	0.03207	−0.26613	−0.13890
opp	0.37127	0.02506	0.32115	0.42044

We illustrate the alternative sampling approach for binary probit models. Conditional on the covariates and the parameters, y_i follows a Bernoulli distribution $y_i \sim \mathrm{B}(1, \pi_i)$ with conditional mean $\pi_i = \Phi(\eta_i)$ where Φ is the cumulative distribution function of a standard normal distribution. On p. 274, the probit model was equivalently defined using latent variables

$$\tilde{y}_i = \boldsymbol{x}_i'\boldsymbol{\beta} + \varepsilon_i = \eta_i + \varepsilon_i$$

with normally distributed errors $\varepsilon_i \sim \mathrm{N}(0, 1)$. The connection between the binary responses and the latent variables is $y_i = 1$ if $\tilde{y}_i > 0$, and $y_i = 0$ if $\tilde{y}_i \leq 0$.

The idea is to use the latent variable representation rather than the original formulation for parameter estimation. This approach was first introduced in a paper by Albert and Chib (1993). The main idea is to consider the latent variables as additional parameters in the model, and to base posterior inference on the extended parameter space. Correspondingly, additional sampling steps for updating the $\tilde{y}_i$'s are required. Fortunately, sampling the $\tilde{y}_i$'s is relatively easy and fast because the full conditionals are truncated normal distributions (see Definition B.5 in Appendix B.1). More specifically, $\tilde{y}_i \mid \cdot \sim \mathrm{TN}_{0,\infty}(\eta_i, 1)$ if $y_i = 1$ and $\tilde{y}_i \mid \cdot \sim \mathrm{TN}_{-\infty,0}(\eta_i, 1)$ if $y_i = 0$. Efficient algorithms for drawing random numbers from a truncated normal distribution can be found in Geweke (1991) or Robert (1995) and are implemented in many major statistics packages. The advantage of defining a probit model through the latent variables $\tilde{y}_i$ is that the full conditionals for the regression coefficients $\boldsymbol{\beta}$ are Gaussian with covariance matrix and mean given by

$$\boldsymbol{\Sigma}_\beta = \left(\boldsymbol{X}'\boldsymbol{X} + \boldsymbol{M}^{-1}\right)^{-1}, \qquad \boldsymbol{\mu}_\beta = \boldsymbol{\Sigma}_\beta \left(\boldsymbol{X}'\tilde{\boldsymbol{y}} + \boldsymbol{M}^{-1}\boldsymbol{m}\right). \tag{5.21}$$

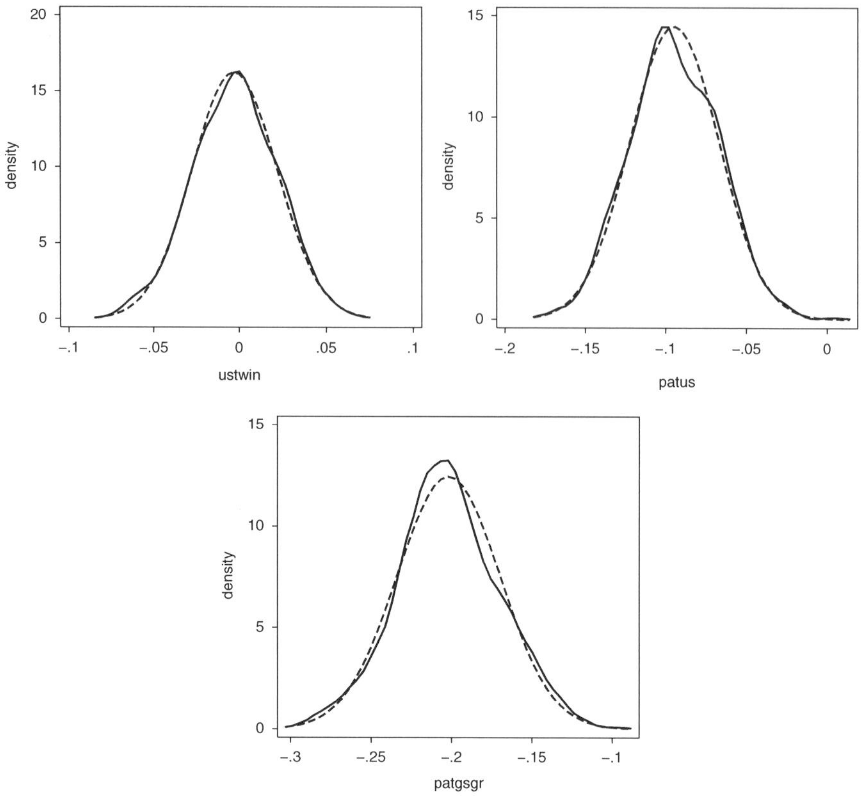

Fig. 5.4 Number of citations from patents: estimated posteriori densities for the effects of *ustwin*, *patus*, and *patgsgr* (*solid line*) together with a normal approximation (*dashed line*)

Hence, we can avoid costly MH steps as is the case with IWLS proposals. Instead, we can resort to the simple Gibbs sampler that was developed for Gaussian responses with slight modifications. Updating of $\boldsymbol{\beta}$ can be done exactly as described in Sect. 4.4.1 (p. 234) using the current values $\tilde{y}_i$ of the latent variables as (pseudo) responses. Another distinct advantage of the Gibbs step is that it works even for high-dimensional parameter vectors, whereas the MH steps with IWLS proposals may break down because acceptance rates typically go down as the parameter dimension increases. The price we pay for the simplicity is the additional update step to draw the latent variables which may be time-consuming in large samples. Then the MH algorithm with IWLS proposals may be faster.

Summarizing, we obtain the following Gibbs sampler:

1. Define initial values $\tilde{\boldsymbol{y}}^{(0)}$ and $\boldsymbol{\beta}^{(0)}$. Set $t = 1$.
2. Sample $\tilde{\boldsymbol{y}}^{(t)}$ by drawing $\tilde{y}_i^{(t)}$, $i = 1, \ldots, n$, from $\mathrm{TN}_{0,\infty}(\eta_i^{(t-1)}, 1)$ if $y_i = 1$ and $\tilde{y}_i \mid \cdot \sim \mathrm{TN}_{-\infty,0}(\eta_i^{(t-1)}, 1)$ if $y_i = 0$.

3. Sample $\boldsymbol{\beta}^{(t)}$ by drawing from the Gaussian full conditional with covariance matrix and mean given in Eq. (5.21) thereby replacing $\tilde{\boldsymbol{y}}$ by the actual state $\tilde{\boldsymbol{y}}^{(t)}$.
4. Stop if $t = T$, otherwise set $t = t + 1$ and go to 2.

We finally note that the data augmentation trick is not limited to binary probit models. Similar algorithms have been developed, e.g., for binary logit models, multi-categorical logit or probit models as outlined in Chap. 6, and Poisson regression. References to the literature are given in Sect. 5.8.

5.7 Boosting Generalized Linear Models

In Sect. 4.3, we introduced a versatile method for obtaining regularized estimates in linear regression with the particular advantage of implicit variable selection (boosting). In fact, the approach can be immediately transferred to the context of GLMs with rather minor modifications. When considering the generic algorithm in Box 4.4 (p. 226), the basic ingredients of a boosting algorithm are:

- The specification of a lack-of-fit criterion via a loss function
- The specification of base learning procedures

A suitable loss function in GLMs is given by the negative log-likelihood such that

$$\rho(\boldsymbol{\eta}) = -l(\boldsymbol{\eta}) = -\sum_{i=1}^{n} \frac{y_i \theta_i - b(\theta_i)}{\phi} w_i .$$

The negative gradients are then still given by

$$u_i = -\frac{\partial}{\partial \eta} \rho(y_i, \eta)\big|_{\eta=\hat{\eta}_i^{(t-1)}}$$

and are, for GLMs, computed as

$$u_i = \frac{h'(\eta_i)^{(t-1)}}{(\sigma_i^2)^{(t-1)}} \left(y_i - \mu_i^{(t-1)}\right).$$

In contrast, no modifications are required for the base learning procedures, and we can still rely on least squares fits applied to the working responses u_i. In summary, boosting can immediately be adapted to generalized response types by providing a suitable loss function. While the negative log-likelihood is a natural choice, different loss functions can in principle be considered. For example, in case of binary regression, the exponential loss

$$\rho(\boldsymbol{\eta}) = \sum_{i=1}^{n} \exp(-y_i \eta_i)$$

(with $y_i \in \{-1, 1\}$ instead of $y_i \in \{0, 1\}$) is sometimes used as an alternative popular in the classification literature; see Friedman, Hastie, and Tibshirani (2000) for details.

5.8 Bibliographic Notes and Proofs

5.8.1 Bibliographic Notes

Nelder and Wedderburn (1972) introduced GLMs as a general class of models for response variables with densities belonging to the exponential family of distributions. Linear, logit, probit, and Poisson models could therefore be subsumed under one conceptual umbrella, leading to important new stimulations for statistical model building, methodological developments, and applications. The book McCullagh and Nelder (1989) gives a detailed outline of GLMs; a more compact introduction can be found in Fahrmeir and Tutz (2001, Chap. 2). Collett (1991) and Tutz (2011, Chaps. 2–5) provide a detailed exposition of binary regression models. These books also give a good overview of methods for model diagnosis that are based on residuals, developed analogously to the linear model of Chap. 3. In the econometrics literature GLMs are usually treated within the field of "microeconometrics." Standard textbooks on microeconometrics are Cameron and Trivedi (2005) and Winkelmann (2010a). Kleiber and Zeileis (2008) discuss econometrics models including GLMs in the software package R.

Several aspects motivated modifications and additions to basic GLMs. For example, the response distribution may be difficult to model with univariate exponential families (as assumed in Sect. 5.4) in some applications. This especially applies to the following regression situation:

- *Regression Models for Count Data:* Although Poisson regression, as illustrated in Sect. 5.2, is the standard model for count data regression, the Poisson distribution is often too simplistic in applications. Cameron and Trivedi (1998) and Winkelmann (2010b) describe enhanced regression modeling for count data. An overview of available count data models in the software package R is given in Zeileis, Kleiber, and Jackman (2008).
- *Life Time (Survival) and Duration Time Models:* Life times, duration times, and waiting times up to a certain event appear in many areas of application. Statistical analyses are then often complicated by incomplete data due to censoring, e.g., when life spans are not terminated until the end of a study period. The Cox model is the most popular regression model for (censored) life times and is closely related to Poisson regression. Standard textbooks on survival and duration time models are Collett (2003), Klein and Moeschberger (2005), Hosmer, Lemeshow, and May (2008), and Therneau and Grambsch (2000).
- *GLMs for Location, Scale, and Shape:* For continuous response variables, we can model the effect of covariates not only on the mean but also on the variance, skewness, or kurtosis; see Sect. 2.9.1 for a brief introduction and Rigby and Stasinopoulos (2005) for more details.
- *Multivariate Response Variables:* If the response $\boldsymbol{y} = (y_1, \dots, y_c)$ consists of several scalar responses $y_1, \dots, y_c$, this yields *multivariate regression*. Anderson (2003) gives an introduction to multivariate linear regression as an extension of the linear regression model. Components $y_1, \dots, y_c$ that do not have a

normal distribution face the difficulty of finding an appropriate joint distribution. *Copula* concepts offer appropriate possibilities (Joe, 1997) especially for continuous components. Other approaches are quasi-likelihood or marginal models (Fahrmeir and Tutz, 2001, Chap. 3), and models with latent variables (Skrondal and Rabe-Hesketh, 2004).

One of the most important extensions of GLMs is the inclusion of nonparametric and semiparametric approaches that allow for flexible modeling of nonlinear covariate effects. The resulting model class, e.g., generalized additive models (GAM), has already been introduced in Chap. 2 and will be discussed in more detail in Chaps. 8 and 9.

In their original definition, GLMs are especially suited for the regression analysis of cross-sectional data. Mixed models (Chap. 7) are a popular tool for the analysis of *clustered* or *longitudinal data*. Depending on the goals of a longitudinal study, *autoregressive* (or *conditional*) models including temporally lagged values of the response variable as additional covariates, or *marginal models* based on quasi-likelihood approaches, can be a reasonable alternative for the analysis; see Diggle, Heagerty, Liang, and Zeger (2002), Fahrmeir and Tutz (2001, Chap. 6), and the additional comments in Sect. 7.8.

In the early 1990s, *Bayesian GLMs* and corresponding extensions have seen a fast development parallel to the spread of MCMC simulation techniques; see Dey et al. (2000) and corresponding sections in the following chapters. IWLS proposals for updating the regression coefficients are due to Gamerman (1997); see also Lenk and DeSarbo (2000) for a slightly modified approach. Estimating Bayesian GLMs using data augmentation similar as described for probit models works for a variety of response distributions; see Holmes and Held (2006) and Frühwirth-Schnatter and Frühwirth (2010) for logit models, Frühwirth-Schnatter and Wagner (2006) and Frühwirth-Schnatter, Frühwirth, Held, and Rue (2009) for Poisson and gamma regression.

GLMs with errors in variables have been developed for data situations where covariates cannot be observed exactly, but only subject to measurement errors. For details on models of this type, we refer to Carroll, Ruppert, Stefanski, and Crainiceanu (2006).

5.8.2 Proofs

Derivation of the ML Estimator in GLMs (Sect. 5.4.2)

The ML estimator in GLMs is derived with the following steps:

1. Log-likelihood

The log-likelihood contribution of an observation $(y_i, \boldsymbol{x}_i)$ (up to an additive constant) is given by

$$l_i(\boldsymbol{\beta}) = \log(f(y_i \mid \boldsymbol{\beta})) = \frac{y_i\theta_i - b(\theta_i)}{\phi} w_i . \tag{5.22}$$

Thereby, the log-likelihood depends on the regression parameters $\boldsymbol{\beta}$ through the natural parameter θ_i of the exponential family via

$$\mu_i = b'(\theta_i) = h(\boldsymbol{x}_i'\boldsymbol{\beta}).$$

Due to the (conditional) independence of y_i,

$$l(\boldsymbol{\beta}) = \sum l_i(\boldsymbol{\beta})$$

is the complete log-likelihood of the sample. To treat individual data ($i = 1, \ldots, n$) and grouped data ($i = 1, \ldots, G$) simultaneously, we omit n or G from the upper limit of the summation signs.

2. Score function

The score function $\boldsymbol{s}(\boldsymbol{\beta}) = \partial l(\boldsymbol{\beta})/\partial \boldsymbol{\beta}$ is obtained by applying the chain rule to the individual score function contributions:

$$\boldsymbol{s}_i(\boldsymbol{\beta}) = \partial l_i(\boldsymbol{\beta})/\partial \boldsymbol{\beta} = \frac{\partial \eta_i}{\partial \boldsymbol{\beta}} \frac{\partial \mu_i}{\partial \eta_i} \frac{\partial \theta_i}{\partial \mu_i} \frac{\partial(y_i\theta_i - b(\theta_i))}{\partial \theta_i} \frac{w_i}{\phi}.$$

The first contribution is simply given by

$$\frac{\partial \eta_i}{\partial \boldsymbol{\beta}} = \boldsymbol{x}_i.$$

The second contribution

$$\frac{\partial \mu_i}{\partial \eta_i} = \frac{\partial h(\eta_i)}{\partial \eta_i} = h'(\eta_i)$$

depends on the response function h and is therefore specific to a given model. In the following we use the shortcut:

$$d_i = h'(\eta_i).$$

The third term is obtained by reversing the nominator and denominator, which yields

$$\frac{\partial \mu_i}{\partial \theta_i} = \frac{\partial b'(\theta_i)}{\partial \theta_i} = b''(\theta_i) = \frac{w_i \operatorname{Var}(y_i)}{\phi} = \frac{w_i \sigma_i^2}{\phi}$$

and therefore

$$\frac{\partial \theta_i}{\partial \mu_i} = \frac{\phi}{w_i \sigma_i^2}.$$

Finally, we have

$$\frac{\partial(y_i\theta_i - b(\theta_i))}{\partial \theta_i} = y_i - b'(\theta_i) = y_i - \mu_i.$$

Putting these pieces together yields the score function as

$$s(\boldsymbol{\beta}) = \sum \boldsymbol{x}_i d_i \frac{\phi}{w_i \sigma_i^2}(y_i - \mu_i)\frac{w_i}{\phi} = \sum \boldsymbol{x}_i \frac{d_i}{\sigma_i^2}(y_i - \mu_i)$$

From $\mathrm{E}(y_i) = \mu_i$, it follows that $\mathrm{E}(\boldsymbol{s}(\boldsymbol{\beta})) = \boldsymbol{0}$ holds.

To express the score function more compactly in matrix notation we define the vectors

$$\boldsymbol{y} = (y_1, \ldots, y_n)', \quad \boldsymbol{\mu} = (\mu_1, \ldots, \mu_n)',$$

and the diagonal matrices

$$\boldsymbol{D} = \mathrm{diag}(d_1, \ldots, d_n), \quad \boldsymbol{\Sigma} = \mathrm{diag}(\sigma_1^2, \ldots, \sigma_n^2).$$

Then we obtain

$$\boldsymbol{s}(\boldsymbol{\beta}) = \boldsymbol{X}'\boldsymbol{D}\boldsymbol{\Sigma}^{-1}(\boldsymbol{y} - \boldsymbol{\mu})$$

where $\boldsymbol{D}$ and $\boldsymbol{\Sigma}$ are both dependent on $\boldsymbol{\beta}$.

3. Information matrix

To derive the Fisher matrix $\boldsymbol{F}(\boldsymbol{\beta}) = \mathrm{E}(\boldsymbol{s}(\boldsymbol{\beta})\boldsymbol{s}'(\boldsymbol{\beta}))$, we note that

$$\boldsymbol{F}(\boldsymbol{\beta}) = \sum \mathrm{E}(\boldsymbol{s}_i(\boldsymbol{\beta})\boldsymbol{s}_i'(\boldsymbol{\beta})).$$

We obtain

$$\begin{aligned}
\mathrm{E}\left(\boldsymbol{s}_i(\boldsymbol{\beta})\boldsymbol{s}_i'(\boldsymbol{\beta})\right) &= \mathrm{E}\left(\boldsymbol{x}_i\boldsymbol{x}_i' \frac{d_i^2}{(\sigma_i^2)^2}(y_i - \mu_i)^2\right) \\
&= \boldsymbol{x}_i\boldsymbol{x}_i' \frac{d_i^2}{(\sigma_i^2)^2}\mathrm{E}(y_i - \mu_i)^2 \\
&= \boldsymbol{x}_i\boldsymbol{x}_i' \frac{d_i^2}{(\sigma_i^2)^2}\mathrm{Var}(y_i) \\
&= \boldsymbol{x}_i\boldsymbol{x}_i' \frac{d_i^2}{\sigma_i^2}.
\end{aligned}$$

This yields

$$\boldsymbol{F}(\boldsymbol{\beta}) = \sum \boldsymbol{x}_i\boldsymbol{x}_i'\tilde{w}_i, \tag{5.23}$$

with the "working weights"

$$\tilde{w}_i = \frac{d_i^2}{\sigma_i^2} = \left(h'(\eta_i)\right)^2 \frac{w_i}{b''(\theta_i)\phi}$$

also depending on $\boldsymbol{\beta}$. In matrix notation the Fisher matrix can be written as

$$\boldsymbol{F}(\boldsymbol{\beta}) = \boldsymbol{X}'\boldsymbol{W}\boldsymbol{X}$$

with the diagonal matrix $\boldsymbol{W} = \text{diag}(\dots, \tilde{w}_i, \dots)$ of working weights. Note that $\boldsymbol{W} = \boldsymbol{D}^2\boldsymbol{\Sigma}^{-1}$.

4. Numerical computation of the ML estimator
Computation of the ML estimator $\hat{\boldsymbol{\beta}}$ is usually based on the Fisher scoring algorithm

$$\hat{\boldsymbol{\beta}}^{(t+1)} = \hat{\boldsymbol{\beta}}^{(t)} + \boldsymbol{F}^{-1}(\hat{\boldsymbol{\beta}}^{(t)})\boldsymbol{s}(\hat{\boldsymbol{\beta}}^{(t)}), \quad t = 0, 1, 2, \dots .$$

Inserting the formulae for $\boldsymbol{s}(\boldsymbol{\beta})$ and $\boldsymbol{F}(\boldsymbol{\beta})$ we obtain

$$\begin{aligned}
\hat{\boldsymbol{\beta}}^{(t+1)} &= \hat{\boldsymbol{\beta}}^{(t)} + (\boldsymbol{X}'\boldsymbol{W}(\hat{\boldsymbol{\beta}}^{(t)})\boldsymbol{X})^{-1}\boldsymbol{X}'\boldsymbol{D}(\hat{\boldsymbol{\beta}}^{(t)})\boldsymbol{\Sigma}(\hat{\boldsymbol{\beta}}^{(t)})^{-1}(\boldsymbol{y} - \boldsymbol{\mu}(\hat{\boldsymbol{\beta}}^{(t)})) \\
&= (\boldsymbol{X}'\boldsymbol{W}(\hat{\boldsymbol{\beta}}^{(t)})\boldsymbol{X})^{-1}\boldsymbol{X}'\boldsymbol{W}(\hat{\boldsymbol{\beta}}^{(t)})\boldsymbol{X}\hat{\boldsymbol{\beta}}^{(t)} \\
&\quad +(\boldsymbol{X}'\boldsymbol{W}(\hat{\boldsymbol{\beta}}^{(t)})\boldsymbol{X})^{-1}\boldsymbol{X}\boldsymbol{W}(\hat{\boldsymbol{\beta}}^{(t)})'\boldsymbol{D}(\hat{\boldsymbol{\beta}}^{(t)})^{-1}(\boldsymbol{y} - \boldsymbol{\mu}(\hat{\boldsymbol{\beta}}^{(t)})) \\
&= (\boldsymbol{X}'\boldsymbol{W}(\hat{\boldsymbol{\beta}}^{(t)})\boldsymbol{X})^{-1}\boldsymbol{X}'\boldsymbol{W}(\hat{\boldsymbol{\beta}}^{(t)})\left[\eta(\hat{\boldsymbol{\beta}}^{(t)}) + \boldsymbol{D}(\hat{\boldsymbol{\beta}}^{(t)})^{-1}(\boldsymbol{y} - \boldsymbol{\mu}(\hat{\boldsymbol{\beta}}^{(t)}))\right].
\end{aligned}$$

Hence the iterations can be expressed as an *iteratively weighted least squares estimator*

$$\hat{\boldsymbol{\beta}}^{(t+1)} = (\boldsymbol{X}'\boldsymbol{W}^{(t)}\boldsymbol{X})^{-1}\boldsymbol{X}'\boldsymbol{W}^{(t)}\tilde{\boldsymbol{y}}^{(t)}, \quad t = 0, 1, 2, \dots$$

where $\tilde{\boldsymbol{y}}^{(t)} = (\dots, \tilde{y}_i(\hat{\boldsymbol{\beta}}^{(t)}), \dots)'$ is a "working response vector" with elements

$$\tilde{y}_i(\hat{\boldsymbol{\beta}}^{(t)}) = \boldsymbol{x}_i'\hat{\boldsymbol{\beta}}^{(t)} + d_i^{-1}(\hat{\boldsymbol{\beta}}^{(t)})(y_i - \hat{\mu}_i(\hat{\boldsymbol{\beta}}^{(t)})),$$

and $\boldsymbol{W}^{(t)}$ is the weight matrix, evaluated at $\boldsymbol{\beta} = \hat{\boldsymbol{\beta}}^{(t)}$. Replacing in $\tilde{w}_i$ and $\tilde{y}_i$ the d_i by $h'(\boldsymbol{x}_i'\hat{\boldsymbol{\beta}}^{(t)})$ and writing the expressions in terms of $\hat{\eta}_i^{(t)} = \boldsymbol{x}_i'\hat{\boldsymbol{\beta}}^{(t)}$ rather than $\hat{\boldsymbol{\beta}}^{(t)}$ we obtain the formulae (5.17) and (5.16) as stated in Sect. 5.4.2.

Categorical Regression Models

6

In Sect. 5.1 we considered binary regression models, that is, regression situations where the response is observed in two categories. In many applications, from social science to medicine, response variables often have more than two categories. For example, consumers may choose between different brands of a product or they may express their opinion about some product in ordered categories ranging from "very satisfied" to "not satisfied at all." Similarly, voters choose between several parties or they assess the quality of candidates in ordered categories. In medicine, we may, for example, not only distinguish between "infection" and "no infection" but also between several types of infection, as in Example 6.1 below. Another application with a categorical response is Example 1.4 (p. 9) on forest health, where the status of trees is assessed in ordered categories, such as "no damage", "medium damage", and "severe damage." This chapter extends regression models and methods for binary responses to the case of categorical responses with more than two categories. Compared to binary regression there is a greater variety of models, depending on the type of response and (underlying) response mechanisms. In particular, we distinguish between models for responses with unordered categories (Sect. 6.2) and ordered categories (Sect. 6.3).

6.1 Introduction

In many applications, the response variable is not binary as in Sect. 5.1, but rather multi-categorical. Such a response variable can be defined as either *ordinal*, i.e., the categories $1, \ldots, c + 1$ of the response can be ordered, or *nominal*, which means that the categories $1, \ldots, c + 1$ are unordered. We distinguish $c + 1$ categories as we will later define c dummy variables representing the response variable and choose the $(c + 1)$th category as the reference.

Example 6.1 Caesarian Delivery—Categorical Response

Table 6.1 shows data on infections after C-sections performed at a clinical center in Munich; see also Example 5.3 (p. 284) of Chap. 5. The response variable has three unordered

L. Fahrmeir et al., *Regression*, DOI 10.1007/978-3-642-34333-9_6,

Table 6.1 Data on infections for 251 C-sections

	C-section					
	Planned			Unplanned		
	Infection			Infection		
	I	II	no	I	II	no
Antibiotics						
Risk factor	0	1	17	4	7	87
No risk factor	0	0	2	0	0	0
No antibiotics						
Risk factor	11	17	30	10	13	3
No risk factor	4	4	32	0	0	9

Table 6.2 Pulmonary function test

		Breathing test		
Age	Smoking status	Normal	Borderline	Abnormal
<40	Nonsmoker	577	27	7
	Former smoker	192	20	3
	Current smoker	682	46	11
40–59	Nonsmoker	164	4	0
	Former smoker	145	15	7
	Current smoker	245	47	27

categories: type I infection, type II infection, and no infection. In addition, information on three covariates is available:

NPLAN : C-section was not planned/planned.

RISK : Presence/absence of risk factors, such as diabetes or obesity.

ANTIB : Antibiotics were prescribed/not prescribed as prophylaxis.

△

Example 6.2 Forest Health Status—Categorical Response

Example 1.4 (p. 9) studies the health status of beech trees measured in terms of the response variable "degree of defoliation" with nine ordered categories (0 %, 12.5 %, ..., 100 % defoliation). We aggregate the response variable into three categories, 1 = no (0 %), 2 = weak (12.5 %–37.5 %), and 3 = severe ($\geq$ 50 %) defoliation, since some original response categories have only a few observations or are completely missing. For every observation in this study, the continuous and categorical covariates that are listed in Table 1.5 (p. 10) have been collected annually along with the response variable. In contrast to Example 6.1, this data set contains ungrouped, individual observations.

△

Example 6.3 Pulmonary Function—Categorical Response

In a study on the impairment of pulmonary function, the age and the smoking behavior of Texan industrial workers were collected as covariates; see Forthofer and Lehnen (1981). The results of a breathing test with categories "normal," "borderline," and "abnormal" are considered as the dependent variable. The resulting data are summarized in the contingency table given in Table 6.2.

△

In all these examples and for the remainder of this chapter, the response variable Y has $c+1$ ordered or unordered categories, i.e., $Y \in \{1,\ldots,c+1\}$. Categorical regression models relate the probabilities

$$\pi_r = \mathrm{P}(Y=r), \qquad r=1,\ldots,c+1,$$

to covariates. To formulate such models, it is useful to represent the response variable by a vector $\boldsymbol{y} = (y_1,\ldots,y_c)'$ of c dummy variables

$$y_r = \begin{cases} 1, & Y=r \\ 0, & \text{otherwise} \end{cases} \qquad r=1,\ldots,c.$$

Here, we have chosen category $c+1$ as the "reference" category so that $y_{c+1} = 1-y_1-\ldots-y_c$. We therefore obtain

$$Y=r \Longleftrightarrow \boldsymbol{y} = (0,\ldots,1,\ldots,0)', \qquad r=1,\ldots,c,$$

with a value of 1 as the rth component of $\boldsymbol{y}$, and

$$\pi_r = \mathrm{P}(Y=r) = \mathrm{P}(y_r=1), \quad r=1,\ldots,c.$$

Correspondingly, for the reference category, we have

$$Y=c+1 \Longleftrightarrow \boldsymbol{y} = (0,\ldots,0,\ldots,0)', \qquad \mathrm{P}(Y=c+1) = 1-\pi_1-\ldots-\pi_c.$$

Multinomial Distribution

The multinomial distribution provides the generalization of the binomial distribution that is necessary for modeling categorical responses. For an individual observation $\boldsymbol{y} = (y_1,\ldots,y_c)'$ with binary 0/1 variables $y_1,\ldots,y_c$ and vector $\boldsymbol{\pi} = (\pi_1,\ldots,\pi_c)'$ of occurrence probabilities, the probability function is given by

$$f(\boldsymbol{y}\mid\boldsymbol{\pi}) = \pi_1^{y_1}\cdot\ldots\cdot\pi_c^{y_c}(1-\pi_1-\ldots-\pi_c)^{1-y_1-\ldots-y_c}.$$

For m independent trials, y_r represents the number of repetitions of category r, $r=1,\ldots,c$. The multinomial probability function of $\boldsymbol{y} = (y_1,\ldots,y_c)'$ is then given by

$$f(\boldsymbol{y}|\boldsymbol{\pi}) = \frac{m!}{y_1!\cdot\ldots\cdot y_c!(m-y_1-\ldots-y_c)!}\pi_1^{y_1}\cdot\ldots\cdot\pi_c^{y_c}(1-\pi_1-\ldots-\pi_c)^{1-y_1-\ldots-y_c}. \tag{6.1}$$

We write

$$\boldsymbol{y} \sim \mathrm{M}(m,\boldsymbol{\pi}),$$

with the parameters m (number of trials) and $\boldsymbol{\pi} = (\pi_1, \ldots, \pi_c)'$ (occurrence probabilities). The first two moments of the multinomial distribution are given by

$$\mathrm{E}(\boldsymbol{y}) = m\boldsymbol{\pi} = \begin{pmatrix} m\pi_1 \\ \vdots \\ m\pi_c \end{pmatrix}, \quad \mathrm{Cov}(\boldsymbol{y}) = m \begin{pmatrix} \pi_1(1-\pi_1) & \cdots & -\pi_1\pi_c \\ \vdots & \ddots & \vdots \\ -\pi_c\pi_1 & \cdots & \pi_c(1-\pi_c) \end{pmatrix}.$$

When using the relative frequencies $\bar{y}_r = y_r/m$ instead of the absolute frequencies y_r, the relative frequency vector $\bar{\boldsymbol{y}} = (\bar{y}_1, \ldots, \bar{y}_c)'$ follows a scaled multinomial distribution, $\bar{\boldsymbol{y}} \sim \mathrm{M}(m, \boldsymbol{\pi})/m$, such that

$$\mathrm{E}(\bar{\boldsymbol{y}}) = \begin{pmatrix} \pi_1 \\ \vdots \\ \pi_c \end{pmatrix}, \quad \mathrm{Cov}(\bar{\boldsymbol{y}}) = \frac{1}{m} \begin{pmatrix} \pi_1(1-\pi_1) & \cdots & -\pi_1\pi_c \\ \vdots & \ddots & \vdots \\ -\pi_c\pi_1 & \cdots & \pi_c(1-\pi_c) \end{pmatrix}.$$

Data

The data structure is analogous to binary and other univariate generalized linear models. For ungrouped individual data, the value of the categorical response variable $Y_i \in \{1, \ldots, c+1\}$ or $\boldsymbol{y}_i = (y_{i1}, \ldots, y_{ic})'$ and the covariate vector $\boldsymbol{x}_i = (1, x_{i1}, \ldots, x_{ik})'$ for each unit i are given:

$$\begin{matrix} \text{Unit } 1 \\ \vdots \\ \text{Unit } i \\ \vdots \\ \text{Unit } n \end{matrix} \begin{pmatrix} \boldsymbol{y}'_1 = (y_{11}, \ldots, y_{1c}) \\ \vdots \\ \boldsymbol{y}'_i = (y_{i1}, \ldots, y_{ic}) \\ \vdots \\ \boldsymbol{y}'_n = (y_{n1}, \ldots, y_{nc}) \end{pmatrix} \begin{pmatrix} 1 & x_{11} & \cdots & x_{1k} \\ \vdots & \vdots & & \vdots \\ 1 & x_{i1} & \cdots & x_{ik} \\ \vdots & \vdots & & \vdots \\ 1 & x_{n1} & \cdots & x_{nk} \end{pmatrix}.$$

With grouped data, we combine observations with identical covariate vectors $\boldsymbol{x}_i$ into a group i of n_i units. Then, absolute frequencies $\boldsymbol{y}_i = (y_{i1}, \ldots, y_{ic})'$ or relative frequencies $\bar{\boldsymbol{y}}_i = (\bar{y}_{i1}, \ldots, \bar{y}_{ic})'$ of observed categories in group $i, i = 1, \ldots, G$ are collected along with the covariates:

$$\begin{matrix} \text{Group } 1 \\ \vdots \\ \text{Group } i \\ \vdots \\ \text{Group } G \end{matrix} \begin{pmatrix} n_1 \\ \vdots \\ n_i \\ \vdots \\ n_G \end{pmatrix} \begin{pmatrix} \bar{\boldsymbol{y}}'_1 = (\bar{y}_{11}, \ldots, \bar{y}_{1c}) \\ \vdots \\ \bar{\boldsymbol{y}}'_i = (\bar{y}_{i1}, \ldots, \bar{y}_{ic}) \\ \vdots \\ \bar{\boldsymbol{y}}'_G = (\bar{y}_{G1}, \ldots, \bar{y}_{Gc}) \end{pmatrix} \begin{pmatrix} 1 & x_{11} & \cdots & x_{1k} \\ \vdots & \vdots & & \vdots \\ 1 & x_{i1} & \cdots & x_{ik} \\ \vdots & \vdots & & \vdots \\ 1 & x_{G1} & \cdots & x_{Gk} \end{pmatrix}.$$

Grouped data are often presented in contingency tables, as in Examples 6.1 and 6.3, indicating absolute frequencies $y_{i1}, \ldots, y_{ic}$ instead of relative frequencies.

Given the covariates $\boldsymbol{x}_i$, individual data $\boldsymbol{y}_i$ follow a multinomial distribution with

$$\boldsymbol{y}_i \sim \mathrm{M}(1, \boldsymbol{\pi}_i), \qquad \boldsymbol{\pi}_i = (\pi_{i1}, \ldots, \pi_{ic})'.$$

For grouped data, the absolute frequencies are multinomial distributed with

$$\boldsymbol{y}_i \sim \mathrm{M}(n_i, \boldsymbol{\pi}_i),$$

whereas the relative frequencies $\bar{\boldsymbol{y}}_i$ follow a scaled multinomial distribution. Categorical regression aims at modeling and estimating the probabilities $\pi_{ir} = \mathrm{P}(Y_i = r) = \mathrm{P}(y_{ir} = 1)$ depending on covariates $\boldsymbol{x}_i$.

In comparison to binary regression models, multinomial regression has a larger variety of modeling possibilities. In the next two sections, we describe some important models for nominal and ordinal response variables. Additional models can be found in Agresti (2002), Fahrmeir and Tutz (2001), and Tutz (2011).

6.2 Models for Unordered Categories

In this section, we assume that the categories $r \in \{1, \ldots, c+1\}$ either do not have an ordered structure or this structure is not modeled in the analyses, which is the case for Example 6.1 and for many applications in biomedicine, social, and economic sciences. Well-known areas of application are, for example, the modeling of decisions, the choice of political party, or the mode of transportation when going to work (e.g., car, train, bike). The categories of the response variable simply represent the different alternatives without requiring any ordered structure. This implies that the response variable is of a nominal scale.

For binary response variables ($c = 1$), we considered the logit model

$$\mathrm{P}(Y_i = 1) = \pi_i = \frac{\exp(\beta_0 + \beta_1 x_{i1} + \ldots + \beta_k x_{ik})}{1 + \exp(\beta_0 + \beta_1 x_{i1} + \ldots + \beta_k x_{ik})} = \frac{\exp(\boldsymbol{x}_i'\boldsymbol{\beta})}{1 + \exp(\boldsymbol{x}_i'\boldsymbol{\beta})},$$

as presented in Sect. 5.1. A direct generalization is the categorical (multinomial) logit model defined in Box 6.1.

It follows from Eq. (6.3) that the linear predictor $\eta_{ir} = \boldsymbol{x}_i'\boldsymbol{\beta}_r$ specifies the (logarithmic) *odds* or *relative risk* between category r and the reference category $c + 1$ in terms of either a log-linear or exponentially multiplicative model. Thus, the interpretation of parameters is analogous to the binary logit model. The only difference is that we do not model the logit for presence ($Y = 1$) in relation to absence ($Y = 0$) but rather the different logits for category r ($Y = r$) relative to the reference category $c + 1$ ($Y = c + 1$).

When interpreting the covariate effects, we have to be aware that a positive regression coefficient β_{rj} does not necessarily imply an increasing probability for category r as x_j increases. It does hold that the odds for category r increase relative

6.1 Multinomial Logit Model

Data

The categorical response variable $Y_i \in \{1, \ldots, c + 1\}$ is measured on a nominal scale. In addition, covariates $\boldsymbol{x}_i$, which are independent of the response category, are given.

Model

The probability of occurrence for category r is specified as

$$P(Y_i = r) = \pi_{ir} = \frac{\exp(\boldsymbol{x}_i'\boldsymbol{\beta}_r)}{1 + \sum_{s=1}^{c} \exp(\boldsymbol{x}_i'\boldsymbol{\beta}_s)} \qquad r = 1, \ldots, c. \tag{6.2}$$

For the reference category, we have

$$\pi_{i,c+1} = 1 - \pi_{i1} - \ldots - \pi_{ic} = \frac{1}{1 + \sum_{s=1}^{c} \exp(\boldsymbol{x}_i'\boldsymbol{\beta}_s)}.$$

An equivalent representation is

$$\log \frac{\pi_{ir}}{\pi_{i,c+1}} = \boldsymbol{x}_i'\boldsymbol{\beta}_r \quad \text{or} \quad \frac{\pi_{ir}}{\pi_{i,c+1}} = \exp(\boldsymbol{x}_i'\boldsymbol{\beta}_r), \qquad r = 1, \ldots, c. \tag{6.3}$$

The parameter vectors $\boldsymbol{\beta}_r = (\beta_{r0}, \beta_{r1}, \ldots, \beta_{rk})'$ and therefore the linear predictor $\eta_{ir} = \boldsymbol{x}_i'\boldsymbol{\beta}_r = \beta_{r0} + x_{i1}\beta_{r1} + \ldots + x_{ik}\beta_{rk}$ are specific for each category, $r = 1, \ldots, c$.

to the reference category. However, this does not imply that the odds for category r relative to all other alternatives also increase. If, for example, category s has a higher regression coefficient than category r, i.e., $\beta_{sj} > \beta_{rj}$, the odds for category s relative to the reference category grows faster than that of category r. Hence, it is possible that the probability for category r decreases despite its positive coefficient β_{rj}. To assess how a changing covariate affects probabilities, it is useful to visualize the probabilities in a plot with all of the remaining covariates kept fixed, e.g., at average values. The most straightforward interpretation is, however, in terms of the log-odds or odds (6.3).

Table 6.3 C-section births: grouped data

		Response		Covariates		
		y_1	y_2	*NPLAN*	*RISK*	*ANTIB*
Group 1	$n_1 = 40$	4	4	0	0	0
Group 2	$n_2 = 58$	11	17	0	1	0
Group 3	$n_3 = 2$	0	0	0	0	1
Group 4	$n_4 = 18$	0	1	0	1	1
Group 5	$n_5 = 9$	0	0	1	0	0
Group 6	$n_6 = 26$	10	13	1	1	0
Group 7	$n_7 = 98$	4	7	1	1	1

Example 6.4 Caesarian Delivery—Categorical Response

The response variable Y "infection" has three categories: "infection type I" ($Y = 1$), "infection type II" ($Y = 2$), and "no infection" ($Y = 3$). With $Y = 3$ as reference category, we code the response variable through $\boldsymbol{y}' = (y_1, y_2)$, with

$$\begin{array}{ll} \boldsymbol{y}' = (1, 0) & \text{infection of type I} \\ \boldsymbol{y}' = (0, 1) & \text{infection of type II} \\ \boldsymbol{y}' = (0, 0) & \text{no infection} \end{array}$$

Similar to Example 5.2 (p. 278), we use the binary covariates *NPLAN*, *RISK*, and *ANTIB*. The grouped data arising from the contingency table of Example 6.1 can be found in Table 6.3.

We consider a main effects multinomial logit model based on (6.3), i.e.,

$$\log \frac{\text{P(infection type } r)}{\text{P(no infection)}} = \beta_{r0} + \beta_{rN} NPLAN + \beta_{rR} RISK + \beta_{rA} ANTIB, \qquad r = 1, 2,$$

or equivalently

$$\frac{\text{P(infection type } r)}{\text{P(no infection)}} = \exp(\beta_{r0}) \exp(\beta_{rN} NPLAN) \exp(\beta_{rR} RISK) \exp(\beta_{rA} ANTIB).$$

As such, $\exp(\beta_{rN})$ is the multiplicative effect of the binary covariate *NPLAN* for category r, when the covariate has the value 1 instead of 0. Alternatively, we can also interpret $\exp(\beta_{rN})$ as the increase in relative risk (odds ratio) if the covariate takes a value 1 relative to 0. If $NPLAN = 0$ is increased to $NPLAN = 1$, we have

$$\exp(\beta_{rN}) = \frac{\text{P(type } r \mid NPLAN = 1, R, A)}{\text{P(no} \mid NPLAN = 1, R, A)} \Big/ \frac{\text{P(type } r \mid NPLAN = 0, R, A)}{\text{P(no} \mid NPLAN = 0, R, A)}.$$

Analogous interpretations hold for the covariates *RISK* and *ANTIB*. Table 6.4 provides the estimated parameters for the given data. The estimates are based on the methods provided in Sect. 6.4.

According to these parameter estimates, antibiotics decrease the relative risk for both types of infections, with a somewhat stronger effect for the type I infections. We also find that the relative risk strongly increases when risk factors exist or when a C-section was not planned. We will analyze whether or not the differences in effects for type I and type II infections are significant in more detail in Sect. 6.4.

Table 6.4 C-section births: estimated coefficients

Type I infection			Type II infection		
	β	$\exp(\beta)$		β	$\exp(\beta)$
intercept	−2.621	0.072	*intercept*	−2.560	0.077
NPLAN	1.174	3.235	*NPLAN*	0.996	2.707
ANTIB	−3.520	0.030	*ANTIB*	−3.087	0.046
RISK	1.829	6.228	*RISK*	2.195	8.980

The results are based on the `STATA` function `mlogit`

Table 6.5 C-section births: estimated probabilities for "no infection," infection type I, and infection type II

NPLAN	*ANTIB*	*RISK*	P(no infection)	P(type I)	P(type II)
0	0	0	0.870	0.063	0.067
1	0	0	0.692	0.163	0.145
0	1	0	0.994	0.002	0.004
0	0	1	0.466	0.211	0.323
1	1	0	0.984	0.007	0.009
1	0	1	0.230	0.337	0.433
0	1	1	0.957	0.013	0.030
1	1	1	0.886	0.038	0.076

Since the estimated parameters cannot easily be transferred to the underlying probabilities, Table 6.5 provides the estimated probabilities for all possible covariate combinations. We see that infections of type I or II are generally unlikely. Exceptions are the cases where risk factors are present *and* antibiotics are not given ($RISK = 1$ and $ANTIB = 0$). Then the probability of "no infection" is below 50%; 0.466 if C-section was planned ($NPLAN = 0$) and 0.230 if C-section was not planned ($NPLAN = 1$). On the other hand, the probabilities for type I and II infections are rather high in these cases.

△

Nominal Models and Latent Utility Models

The multinomial logit model, as well as other categorical regression models, especially probit models, can be motivated and derived with the help of latent utility models. This is not only interesting for alternative interpretations but also useful for statistical inference, especially for Bayesian approaches; see also Sect. 5.6.3.

We assume that each of the $c + 1$ alternatives is associated with a specific utility and that the random utility function of the rth alternative has the form

$$u_r = \tilde{\eta}_r + \varepsilon_r,$$

where $\tilde{\eta}_r$ is a constant (the utility of the rth alternative), and ε_r is a random variable with cumulative distribution function F. According to the principle of maximum utility, we observe alternative r if the associated (unobserved) utility is maximal, i.e.,

$$Y = r \Longleftrightarrow u_r = \max_{s=1,\ldots,c+1} u_s.$$

Assuming independent errors ε_r with extreme maximal-value distribution $F(x) = \exp(-\exp(-x))$ yields

$$\mathrm{P}(Y = r) = \frac{\exp(\tilde{\eta}_r)}{\sum_{s=1}^{c+1} \exp(\tilde{\eta}_s)}, \qquad r = 1, \ldots, c+1;$$

see McFadden (1973). Since only the difference in utility functions is identifiable, we choose a reference category. With category $c + 1$ as the reference category, we obtain

$$\mathrm{P}(Y = r) = \frac{\exp(\tilde{\eta}_r - \tilde{\eta}_{c+1})}{1 + \sum_{s=1}^{c} \exp(\tilde{\eta}_s - \tilde{\eta}_{c+1})} = \frac{\exp(\eta_r)}{1 + \sum_{s=1}^{c} \exp(\eta_s)},$$

with $\eta_r = \tilde{\eta}_r - \tilde{\eta}_{c+1}$.

An extension of this concept results when the utilities do not only depend on the covariates characterizing the statistical units facing the decision but also on the alternatives. Let $\boldsymbol{x}_i$ denote the covariates that have already been included in η_{ir} and $\boldsymbol{w}_{ir}$ the vector of covariates that are specific to the alternative r. If the mode of transportation is considered as the decision problem, the cost or time associated with a specific mode of transportation provides examples of *category-specific covariates*. Since category-specific covariates often also vary between the individual observations, we use a double subscript on $\boldsymbol{w}_{ir}$, where i represents the individual, while r represents the alternative. Assuming independent errors that follow an extreme maximal-value distribution leads to the extended multinomial logit model

$$\pi_{ir} = \frac{\exp(\boldsymbol{x}_i'\boldsymbol{\beta}_r + (\boldsymbol{w}_{ir} - \boldsymbol{w}_{i,c+1})'\boldsymbol{\gamma})}{1 + \sum_{s=1}^{c} \exp(\boldsymbol{x}_i'\boldsymbol{\beta}_s + (\boldsymbol{w}_{is} - \boldsymbol{w}_{i,c+1})'\boldsymbol{\gamma})},$$

$r = 1, \ldots, c$. The model has *category-unspecific* (*global*) covariates $\boldsymbol{x}_i$, which are supplemented with *category-specific* coefficients $\boldsymbol{\beta}_r$, and *category-specific* covariates $\boldsymbol{w}_{ir} - \boldsymbol{w}_{i,c+1}$, which are accompanied by *category-unspecific* (*global*) coefficients $\boldsymbol{\gamma}$. The difference from $\boldsymbol{w}_{ir}$ to $\boldsymbol{w}_{i,c+1}$ of the reference category results from building the difference $\eta_{ir} = \tilde{\eta}_{ir} - \tilde{\eta}_{i,c+1} = \boldsymbol{w}_{ir}'\boldsymbol{\gamma} - \boldsymbol{w}_{i,c+1}'\boldsymbol{\gamma}$.

Other distributional assumptions for the errors in the latent utility model lead to alternative models. If we assume independent but standard normal distributed errors ε_r, we obtain the ("independent") probit model. If we choose a multivariate normal distribution (with a nondiagonal covariance matrix) for the errors $\boldsymbol{\varepsilon}$, we obtain the multivariate probit model.

6.3 Ordinal Models

In this section, we present models that are useful when the response variable is ordinal, i.e., the categories can be ordered. The explicit use of the ordinal scale of Y allows for models with parsimonious parameterization. Ordinal responses have been present in the examples on forest health (Example 6.2), as well as in the example regarding pulmonary function (Example 6.3).

6.3.1 The Cumulative Model

The most widely used ordinal regression model is motivated from a threshold mechanism, similar to the binary regression model in Sect. 5.1 (p. 274). We can derive the model by assuming a latent (unobserved) variable that drives the decision for the observed alternatives. Exceeding a certain threshold on the latent scale results in an observable category of the response variable.

Let u denote the unobserved latent variable. In the examples on forest health or the pulmonary function test, u can be considered the latent damage state of a tree or the lung. The latent variable corresponding to covariate value $\boldsymbol{x}_i$ (excluding the intercept) is assumed to be

$$u_i = -\boldsymbol{x}_i'\boldsymbol{\beta} + \varepsilon_i, \tag{6.4}$$

where $\boldsymbol{\beta}$ is the parameter vector and ε_i is an error variable with cumulative distribution function F. The reason for the minus sign in front of $\boldsymbol{x}_i'\boldsymbol{\beta}$ will become clear later. The link between observed and latent variable is defined by the threshold mechanism

$$Y_i = r \iff \theta_{r-1} < u_i \leq \theta_r, \qquad r = 1, \ldots, c+1, \tag{6.5}$$

where $-\infty = \theta_0 < \theta_1 < \ldots < \theta_{c+1} = \infty$ are the latent, ordered thresholds placed on the latent continuum. Note that the intercept has to be excluded from the predictor $\boldsymbol{x}_i'\boldsymbol{\beta}$ in order to obtain an identifiable model. Otherwise, a shift of the intercept and a corresponding negative shift of the thresholds would lead to an equivalent model.

Category r is observed, if the latent variable falls between the thresholds θ_{r-1} and θ_r. Interpreted as a binary decision, we obtain a response variable $Y_i \leq r$, if $u_i \leq \theta_r$, i.e., when the latent variable remains under the threshold θ_r, and $Y_i > r$ if $u_i > \theta_r$.

Figure 6.1 illustrates the threshold concept. It shows the density of the latent variable $u_i = -\boldsymbol{x}_i'\boldsymbol{\beta} + \varepsilon_i$ for two different realizations of $\boldsymbol{x}_i$. The corresponding densities do have the same form ($\varepsilon_i \sim F$), but they are shifted on the latent continuum. The area between two thresholds corresponds to the probability of a certain category as exemplified by the grey area in Fig. 6.1 for the second category. Obviously there is a strong change in the probability due to the shift

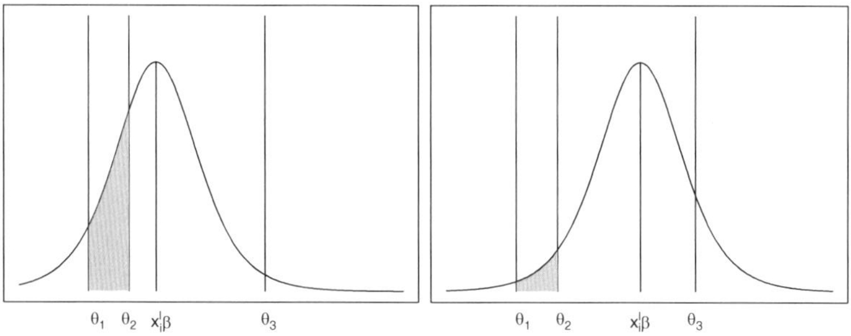

Fig. 6.1 Thresholds and densities of the latent variable

forced through $\boldsymbol{x}_i$. The illustration in Fig. 6.1 is based on a logistic density, but all statements remain unchanged when using a different distribution for the error terms.

The *cumulative model* with distribution function F results from the two assumptions (6.4) and (6.5) through

$$\begin{aligned} \mathrm{P}(Y_i \le r) &= \mathrm{P}(u_i \le \theta_r) \\ &= \mathrm{P}(-\boldsymbol{x}_i'\boldsymbol{\beta} + \varepsilon_i \le \theta_r) \\ &= \mathrm{P}(\varepsilon_i \le \theta_r + \boldsymbol{x}_i'\boldsymbol{\beta}) \\ &= F(\theta_r + \boldsymbol{x}_i'\boldsymbol{\beta}), \qquad r = 1, \dots, c+1. \end{aligned}$$

The name "cumulative model" refers to the specification of cumulative probabilities $\mathrm{P}(Y_i \le r) = \mathrm{P}(Y_i = 1) + \ldots + \mathrm{P}(Y_i = r)$ on the left-hand side of the model equation. The model itself no longer contains the latent variable and can be considered a regression model with regressors $\boldsymbol{x}_i$ and parameters $\theta_1, \dots, \theta_c$ and $\boldsymbol{\beta}$. The occurrence probabilities are given by

$$\mathrm{P}(Y_i = r) = F(\theta_r + \boldsymbol{x}_i'\boldsymbol{\beta}) - F(\theta_{r-1} + \boldsymbol{x}_i'\boldsymbol{\beta}), \quad r = 1, \dots, c+1.$$

Depending on the choice of F, we obtain different models. For example, the *cumulative logit model* results for the logistic distribution function F with

$$\mathrm{P}(Y_i \le r) = \frac{\exp(\theta_r + \boldsymbol{x}_i'\boldsymbol{\beta})}{1 + \exp(\theta_r + \boldsymbol{x}_i'\boldsymbol{\beta})}$$

or equivalently

$$\log \frac{\mathrm{P}(Y_i \le r)}{\mathrm{P}(Y_i > r)} = \theta_r + \boldsymbol{x}_i'\boldsymbol{\beta}.$$

6.2 Cumulative Model

Data

The response variable $Y_i \in \{1, \ldots, c+1\}$ is categorical and measured on an ordinal scale. The covariates $\boldsymbol{x}_i$ do not depend on the response categories.

Model

The response variable Y_i is linked to a latent variable u_i via

$$Y_i = r \quad \Longleftrightarrow \quad \theta_{r-1} < u_i \leq \theta_r, \qquad r = 1, \ldots, c+1,$$

with thresholds $-\infty = \theta_0 < \theta_1 < \ldots < \theta_{c+1} = \infty$. A linear model

$$u_i = -\boldsymbol{x}_i \boldsymbol{\beta} + \boldsymbol{\varepsilon}_i,$$

with error distribution F is assumed for u_i. The *cumulative model* for Y_i is then given by

$$\mathrm{P}(Y_i \leq r) = F(\theta_r + \boldsymbol{x}_i' \boldsymbol{\beta}), \qquad r = 1, \ldots, c+1,$$

or equivalently

$$\mathrm{P}(Y_i = r) = F(\theta_r + \boldsymbol{x}_i' \boldsymbol{\beta}) - F(\theta_{r-1} + \boldsymbol{x}_i' \boldsymbol{\beta}).$$

This model is also called the *proportional odds* model. The term *proportional* refers to the fact that the ratio of the cumulative odds for subpopulations characterized by $\boldsymbol{x}_i$ and $\tilde{\boldsymbol{x}}_i$ is given by

$$\frac{\mathrm{P}(Y_i \leq r \,|\, \boldsymbol{x}_i)/\mathrm{P}(Y_i > r \,|\, \boldsymbol{x}_i)}{\mathrm{P}(Y_i \leq r \,|\, \tilde{\boldsymbol{x}}_i)/\mathrm{P}(Y_i > r \,|\, \tilde{\boldsymbol{x}}_i)} = \exp((\boldsymbol{x}_i - \tilde{\boldsymbol{x}}_i)' \boldsymbol{\beta}). \tag{6.6}$$

The ratio is independent of category r, and therefore the cumulative odds are proportional across all categories.

If we choose F to be the extreme minimum-value distribution $F(x) = 1 - \exp(-\exp(x))$, we obtain the *cumulative extreme value* or *grouped Cox model*

$$\mathrm{P}(Y_i \leq r) = 1 - \exp(-\exp(\theta_r + \boldsymbol{x}_i' \boldsymbol{\beta})), \tag{6.7}$$

or

$$\log(-\log \mathrm{P}(Y_i > r)) = \theta_r + \boldsymbol{x}_i' \boldsymbol{\beta}.$$

The term "grouped Cox" or "proportional hazards model" stems from a property analogous to Eq. (6.6) of an underlying duration time model. With the reparameterization $\tilde{\theta}_r = \log(\exp(\theta_r) - \exp(\theta_{r-1}))$, $r = 1, \ldots, c$, we can equivalently express Eq. (6.7) as

$$P(Y = r \mid Y \geq r) = 1 - \exp(-\exp(\tilde{\theta}_r + \boldsymbol{x}_i'\boldsymbol{\beta})) = F(\tilde{\theta}_r + \boldsymbol{x}_i'\boldsymbol{\beta}). \qquad (6.8)$$

On the left-hand side, we find the probability of the response variable in category r, given that the category r or a larger category is reached. In duration time analysis (when Y represents a discrete time), this conditional probability is referred to as the discrete hazard function. Model (6.8) is the discrete-time analogue of the continuous-time Cox model from duration time analysis.

Example 6.5 Forest Health Status—Cumulative Logit Model

We model the three-categorical response variable representing the health status with a cumulative logit model. The effects of *year* and *age* are approximated by orthogonal cubic polynomials (see Example 3.5 on p. 90), whereas all other continuous covariates from Table 1.5 of Example 1.4 are modeled with linear effects. All binary or categorical covariates are effect-coded with the most common category as reference category (coded as -1). In this analysis, we do not consider the spatial information, which is provided by the location of a tree. The results in Table 6.6 and Fig. 6.2 should therefore be mainly considered as illustrations of a preliminary nature. Estimation is based on the methods outlined in the next section. In Sect. 9.5, we will further analyze the data with a more flexible geoadditive model. The nonlinear effects of age and year shown in Fig. 6.2 are only very rough approximations for the corresponding covariate effects as we will see in Sect. 9.5. There is evidence that younger trees do generally show a lower damage state than middle-aged and older trees. The calendar effect *year* reflects the observed temporal trends. The risk of damage increases starting in 1983, flattens towards the end of the decade and then shows a continuous decrease from 1990 to 2000. In recent years, however, we again find some evidence of an increasing risk. To get an intuitive feel about the variability of the probabilities for the three damage states it is recommended to plot the estimated probabilities in addition to the effect plots. As an example, Fig. 6.3 visualizes the probabilities for "no damage," "weak damage," and "severe damage" depending on the age of the trees. Thereby we kept the effects of the other continuous covariates fixed at their mean value and of the categorical covariates at their mode category; see also Examples 2.8 (p. 35) and 2.13 (p. 54) of Chap. 2.

We only comment on some selected linear effects: A high canopy density, the presence of fertilization, and a moderately dry soil significantly reduce the risk of damage. In contrast, humid soil and a homogeneous deciduous forest increase the probability of damage. We reanalyze the data in Example 9.7 (p. 556) using a flexible geoadditive model, which shows that some effects react sensitively to model choice, especially when spatial effects are included.

△

6.3.2 The Sequential Model

An alternative to the cumulative model is the sequential model, which also explicitly uses the ordinal scaling of the response categories. Due to its construction, the sequential model is useful when the categories of the response variable are obtained

Table 6.6 Forest health status: estimated covariate effects. The results are obtained using function `ologit` of the software `STATA`

Variable	Coefficient	Standard error	t-statistic	p-value	95 % Confidence interval	
θ_0	−5.8994	1.298			−8.4444	−3.3544
θ_1	−1.8424	1.290			−4.3709	0.6861
gradient	0.0004	0.007	0.05	0.959	−0.0130	0.0137
canopyd	−0.0319	0.003	−10.51	<0.001	−0.0378	−0.0259
alt	−0.0004	0.001	−0.26	0.794	−0.0030	0.0023
depth	−0.0206	0.007	−2.98	0.003	−0.0341	−0.0070
ph	−0.8598	0.250	−3.44	0.001	−1.3501	−0.3695
type	−0.3963	0.074	5.35	<0.001	0.2512	0.5414
fert	−0.4453	0.125	−3.56	<0.001	−0.6902	−0.2004
humus0	−0.1256	0.132	−0.95	0.343	−0.3851	0.1340
humus2	0.3100	0.116	2.67	0.007	0.0828	0.5373
humus3	0.1688	0.134	1.26	0.209	−0.0943	0.4319
humus4	−0.2739	0.179	−1.53	0.126	−0.6247	0.0769
watermoisture1	−0.6270	0.150	−4.17	<0.001	−0.9217	−0.3323
watermoisture3	0.4926	0.108	4.58	<0.001	0.2817	0.7036
alkali1	0.3646	0.159	2.30	0.022	0.0536	0.6756
alkali3	−0.5451	0.156	−3.50	<0.001	−0.8500	−0.2401
alkali4	0.9230	0.221	4.18	<0.001	0.4901	1.3559

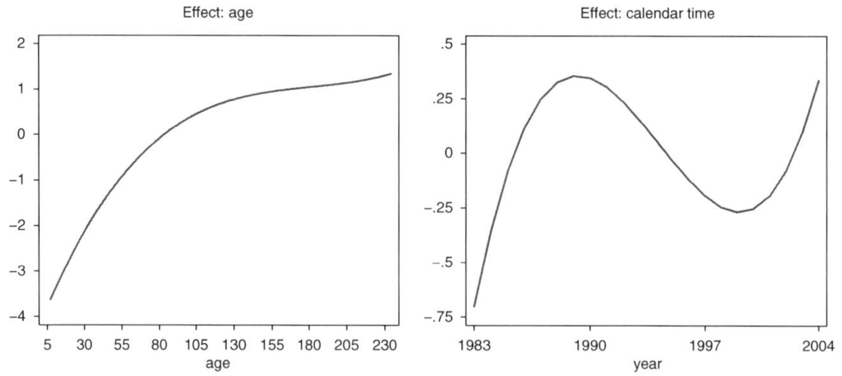

Fig. 6.2 Forest health status: estimated polynomial effects of age and calendar time

sequentially. If we assume that, in Example 6.3, the pulmonary function is originally normal, the abnormal category (for given smoking status and age) can only be reached provided that the borderline category has been reached in between. Another example would be years of unemployment considered as categories of a duration time response. One can only be unemployed for two years provided that one has been unemployed for one year before. The assumptions on which we build

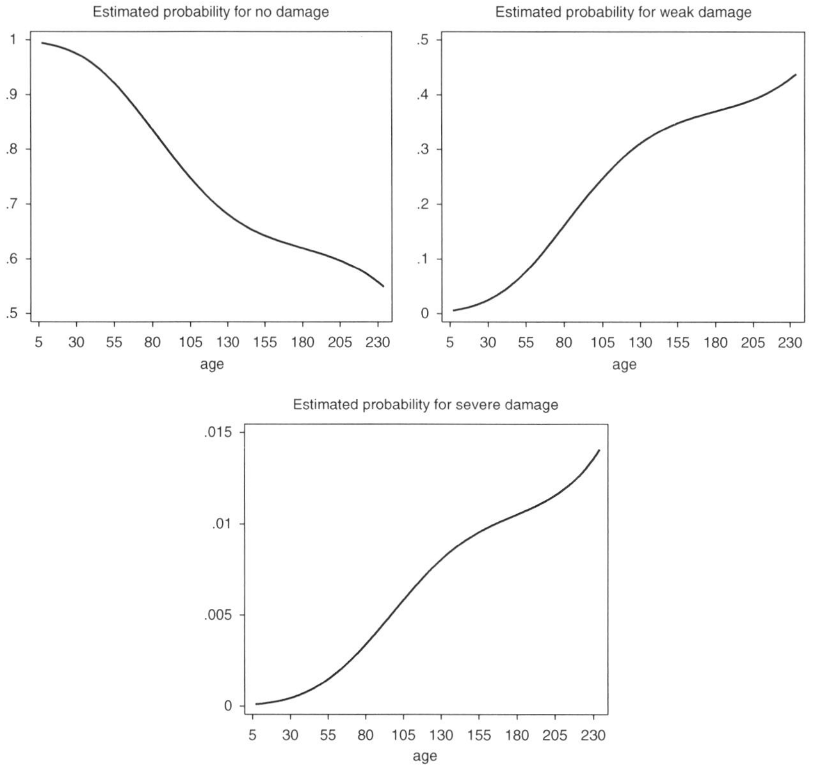

Fig. 6.3 Forest health status: estimated probabilities for "no damage," "weak damage," and "severe damage" depending on age. The remaining continuous covariates are held fixed at their mean values. The categorical covariates are held fixed at their mode category

sequential models are, thus, successively or gradually achievable categories of the variable $Y \in \{1, \ldots, c+1\}$.

The sequential reachability of the categories is modeled explicitly in terms of a sequence of binary transitions. The process starts in $Y_i = 1$ and a binary model of Sect. 5.1 describes the transition to $Y_i > 1$:

$$P(Y_i = 1) = F(\theta_1 + \boldsymbol{x}_i'\boldsymbol{\beta}).$$

If the response variable remains in category 1, the process stops. Otherwise, the following transition is dichotomously modeled with

$$P(Y_i = 2 \mid Y_i \geq 2) = F(\theta_2 + \boldsymbol{x}_i'\boldsymbol{\beta})$$

as the conditional probability for remaining in $Y_i = 2$ and complementary probability $1 - P(Y_i = 2 \mid Y_i \geq 2)$ for the transition to $Y_i > 2$.

Accordingly, the rth step of the sequential mechanism is based on

$$\mathrm{P}(Y_i = r \mid Y_i \geq r) = F(\theta_r + \boldsymbol{x}_i'\boldsymbol{\beta}), \qquad r = 1, \ldots, c. \tag{6.9}$$

The process stops as soon as one of the binary transition equations decides for the process to remain in category r.

It should be pointed out that we are not actually observing individual transitions but only the realized category of the response variable. The sequential mechanism is basically an assumption which helps to formulate the sequential model. Alternatively, model (6.9) can be expressed in terms of marginal probabilities

$$\mathrm{P}(Y_i = r) = F(\theta_r + \boldsymbol{x}_i'\boldsymbol{\beta}) \prod_{s=1}^{r-1} (1 - F(\theta_s + \boldsymbol{x}_i'\boldsymbol{\beta})), \qquad r = 1, \ldots, c,$$

and

$$\mathrm{P}(Y_i = c + 1) = 1 - \sum_{s=1}^{c} \mathrm{P}(Y_i = s)$$

for the reference category.

For the logistic distribution function F, we specifically obtain the *sequential logistic model*; for the extreme minimum-value distribution $F(x) = 1 - \exp(-\exp(x))$, we obtain the *extreme value sequential model*. Possible alternatives are the probit model or other binary regression models. The sequential logit model can be represented through the conditional transition probabilities

$$\log \frac{\mathrm{P}(Y_i = r \mid Y_i \geq r)}{1 - \mathrm{P}(Y_i = r \mid Y_i \geq r)} = \theta_r + \boldsymbol{x}_i'\boldsymbol{\beta}.$$

When the number of categories is small, an extended sequential model with category-specific regression coefficients for some of the covariates can be considered. Instead of the linear term $\theta_r + \boldsymbol{x}_i'\boldsymbol{\beta}$ in Eq. (6.9), we obtain the general linear predictor

$$\theta_r + \boldsymbol{x}_i'\boldsymbol{\beta}_r + \boldsymbol{z}_i'\boldsymbol{\gamma},$$

where $\boldsymbol{x}_i$ and $\boldsymbol{z}_i$ represent groups of variables with category-specific effects $\boldsymbol{\beta}_r$ and global effects $\boldsymbol{\gamma}$, respectively.

Example 6.6 Pulmonary Function—Sequential Model

We specify the influence of age and smoking behavior on the test results of Table 6.2 using a sequential model. Dummy coding is used for the explanatory variables with reference categories *age* $\geq$ 40 and smoking status "current smoker." The results provided in Table 6.7 are based on `remlreg objects` of `BayesX` and are generally as expected. Younger workers of age lower than 40 and nonsmokers are less likely to show a borderline or abnormal breathing test compared to older workers of age larger than 40 and current smokers. The difference between former and current smokers is comparably small (and nonsignificant).

6.3 Sequential Models

Data

The response variable $Y_i \in \{1, \ldots, c+1\}$ is categorical and measured on an ordinal scale. The categories can only be reached sequentially.

Model

The transitions between the categories are modeled using binary regression models. If category $r \in \{1, \ldots, c\}$ has already been reached, the (potential) termination of the process in r is defined by

$$\mathrm{P}(Y_i = r \mid Y_i \geq r) = F(\theta_r + \boldsymbol{x}_i'\boldsymbol{\beta}).$$

Thus, we have

$$\mathrm{P}(Y_i = r) = F(\theta_r + \boldsymbol{x}_i'\boldsymbol{\beta}) \prod_{s=1}^{r-1} (1 - F(\theta_s + \boldsymbol{x}_i'\boldsymbol{\beta})), \qquad r = 1, \ldots, c.$$

If F is the logistic distribution function, we obtain the sequential logit model. If F is the standard normal distribution function, the sequential probit model results. We can also consider category-specific effects of covariates based on the general predictor $\theta_r + \boldsymbol{x}_i'\boldsymbol{\beta}_r + \boldsymbol{z}_i'\boldsymbol{\gamma}$.

Table 6.7 Lung function test: estimated effects for the main effects sequential model

Variable	Coefficient	Standard error	p-value	95 % Confidence interval	
θ_1	1.495	0.120	<0.001	1.260	1.730
θ_2	0.551	0.179	0.002	0.200	0.901
age < 40	−0.742	0.135	<0.001	−1.007	−0.478
non sm.	−0.867	0.175	<0.001	−1.210	0.522
former sm.	−0.188	0.171	0.272	−0.523	0.147

We now additionally include interactions between age and smoking status; see Table 6.8 for the results. The interaction model has a comparably lower AIC of 1575.13 compared to the main effects model which shows an AIC of 1596.59. Quite surprisingly (at least at first sight), the interaction effects are both positive. For instance, nonsmoking workers with age below 40 show a considerably high interaction effect of 2.220 indicating that the risk of borderline and abnormal breathing test is increased. However, it is not justified to interpret the main effects and the interaction effects separately. It is rather necessary to compute the effects of the six combinations of age and smoking behavior. For instance, for the combination *age* < 40 and "nonsmoking," we obtain the effect

$$-1.207 \cdot 1 - 2.519 \cdot 1 - 0.586 \cdot 0 + 2.220 \cdot 1 + 0.832 \cdot 0 = -1.506.$$

Table 6.8 Lung function test: estimated effects for the sequential model with interactions

Variable	Coefficient	Standard error	p-value	95 % Confidence interval	
θ_1	1.249	0.125	<0.001	1.004	1.495
θ_2	0.387	0.181	0.032	0.032	0.742
$age < 40$	−1.207	0.174	<0.001	−1.548	−0.865
non sm.	−2.519	0.521	<0.001	−3.541	−1.496
former sm.	−0.586	0.234	0.012	−1.045	−0.127
$age < 40$, *non*	2.220	0.560	<0.001	−1.122	3.318
$age < 40$, *former*	0.832	0.339	0.014	0.167	1.498

Table 6.9 Lung function test: combined effects in the main effects plus interactions model

Covariate combination	Effect
$age < 40$, *nonsmoker*	$-1.207 - 2.519 + 2.220 = -1.506$
$age < 40$, *former smoker*	$-1.207 - 0.586 + 0.832 = -0.961$
$age < 40$, *current smoker*	−1.207
$age \geq 40$, *nonsmoker*	−2.519
$age \geq 40$, *former smoker*	−0.586
$age \geq 40$, *current smoker*	0

Table 6.10 Lung function test: estimated probabilities for the sequential model with interactions

Covariate combination	P(normal)	P(borderline)	P(abnormal)
$age < 40$, *nonsmoker*	0.940	0.052	0.008
$age < 40$, *former smoker*	0.901	0.078	0.020
$age < 40$, *current smoker*	0.921	0.066	0.013
$age \geq 40$, *nonsmoker*	0.977	0.021	0.001
$age \geq 40$, *former smoker*	0.862	0.100	0.038
$age \geq 40$, *current smoker*	0.778	0.133	0.090

Table 6.9 lists the calculated effects of all covariate combinations and Table 6.10 lists the estimated probabilities. We now clearly see that in both age groups the risk of borderline or abnormal breathing test is lower for nonsmokers and to a lesser extent for former smokers compared to current smokers. Surprisingly, the non-smokers in the age group $age \geq 40$ have an even lower effect than in the group $age < 40$.

The example shows that it can be tedious to interpret models with interaction effects. We have already pointed out in Sect. 3.4 that the interaction between covariates can also be modeled by defining a new variable whose categories consist of *all* possible combinations of age and smoking status. The result of this approach with reference category $age \geq 40$, *current smoker* is given in Table 6.11. We now immediately see the effects of the covariates without any tedious calculations (the differences in the third digit are due to rounding errors).

△

Table 6.11 Lung function test: estimated effects for the sequential model with interactions and alternative dummy coding

Variable	Coefficient	Standard error	p-value	95 % Confidence interval	
θ_1	1.249	0.125	<0.001	1.004	1.495
θ_2	0.387	0.181	0.032	0.032	0.742
$age < 40, non$	−1.505	0.201	<0.001	−1.899	−1.111
$age < 40, former$	−0.960	0.241	<0.001	−1.433	−0.487
$age < 40, current$	−1.207	0.174	<0.001	−1.548	−0.865
$age \geq 40, non$	−2.519	0.521	<0.001	−3.541	−1.496
$age \geq 40, former$	−0.586	0.234	0.013	−1.046	−0.127

6.4 Estimation and Testing: Likelihood Inference

Parameter estimation relies on the same maximum likelihood principles that have been used for binary regression models and more generally with GLMs in Chap. 5. We assume the more general case of grouped data where the responses y_{ir}, $i = 1, \ldots, G$, $r = 1, \ldots, c$ represent the number of repetitions of category r out of n_i independent trials. Since the response variables $\boldsymbol{y}_i = (y_{i1}, \ldots, y_{ic})'$ follow multinomial distributions

$$\boldsymbol{y}_i \sim \mathrm{M}(n_i, \boldsymbol{\pi}_i)$$

and are assumed to be (conditionally) independent, the likelihood function is given by the product

$$L(\boldsymbol{\beta}) = \prod_{i=1}^{G} f(\boldsymbol{y}_i \mid \boldsymbol{\pi}_i)$$

of densities $f(\boldsymbol{y}_i \mid \boldsymbol{\pi}_i)$ of the form (6.1).

The vector of probabilities $\boldsymbol{\pi}_i = (\pi_{i1}, \ldots, \pi_{ic})'$ depends on the vector $\boldsymbol{\beta}$ of all parameters through linear predictors with specific forms depending on the chosen model. Up to an additive constant, the log-likelihood is given by

$$l(\boldsymbol{\beta}) = \sum_{i=1}^{G} (y_{i1} \log \pi_{i1} + \ldots + y_{ic} \log \pi_{ic} + y_{i,c+1} \log \pi_{i,c+1})$$

with $y_{i,c+1} = n_i - y_{i1} - \ldots - y_{ic}$ and $\pi_{i,c+1} = 1 - \pi_{i1} - \ldots - \pi_{ic}$. From this, we can derive the score function $\boldsymbol{s}(\boldsymbol{\beta}) = \partial l(\boldsymbol{\beta})/\partial \boldsymbol{\beta}$ as the vector of first derivatives and the expected information matrix

$$\boldsymbol{F}(\boldsymbol{\beta}) = \mathrm{E}(\boldsymbol{s}(\boldsymbol{\beta})\boldsymbol{s}'(\boldsymbol{\beta})).$$

The ML estimator is computed as the iterative solution of the ML equations

$$\boldsymbol{s}(\hat{\boldsymbol{\beta}}) = \boldsymbol{0}.$$

The score function and the information matrix as well as the iterative computation of the ML estimator are multivariate extensions of the quantities derived for binary regression models and univariate GLMs in Chap. 5; see the following details.

To allow for a unified presentation of all categorical regression models, we introduce some common notation. In every categorical regression model, the occurrence probabilities $\mathrm{P}(Y_i = r) = \mathrm{P}(y_{ir} = 1) = \pi_{ir}$ for category r are connected to linear predictors η_{ir}, $r = 1, \ldots, c$ via response functions h_r, i.e.,

$$\pi_{ir} = h_r(\eta_{i1}, \ldots, \eta_{ic}), \qquad r = 1, \ldots, c.$$

The specification of the response functions and the linear predictors depends on the chosen categorical regression model.

For the multinomial logit model in Box 6.1, we have

$$\pi_{ir} = \frac{\exp(\eta_{ir})}{1 + \sum_{s=1}^{c} \exp(\eta_{is})} = h_r(\eta_{i1}, \ldots, \eta_{ic}),$$

with $\eta_{ir} = \boldsymbol{x}_i' \boldsymbol{\beta}_r = \beta_{r0} + x_{i1}\beta_{r1} + \ldots + x_{ik}\beta_{rk}$. In the extended form with category-specific explanatory variables $\boldsymbol{w}_{ir} - \boldsymbol{w}_{i,c+1}$ and global parameters $\boldsymbol{\gamma}$, the predictor is extended to

$$\eta_{ir} = \boldsymbol{x}_i' \boldsymbol{\beta}_r + (\boldsymbol{w}_{ir} - \boldsymbol{w}_{i,c+1})' \boldsymbol{\gamma}. \tag{6.10}$$

For the ordinal cumulative model in Box 6.2, we have $\pi_{i1} = F(\eta_{i1}) = h_1(\eta_{i1}, \ldots, \eta_{ic})$ and

$$\pi_{ir} = F(\eta_{ir}) - F(\eta_{i,r-1}) = h_r(\eta_{i1}, \ldots, \eta_{ic}), \qquad r = 2, \ldots, c,$$

with

$$\eta_{ir} = \theta_r + x_{i1}\beta_1 + \ldots + x_{ik}\beta_k.$$

Combining the predictors into the vector

$$\boldsymbol{\eta}_i = (\eta_{i1}, \ldots, \eta_{ic})',$$

and the parameters to the parameter vector $\boldsymbol{\beta}$, we can always obtain the form

$$\boldsymbol{\eta}_i = \boldsymbol{X}_i \boldsymbol{\beta},$$

with an appropriately defined design matrix $\boldsymbol{X}_i$.

For the multinomial logit model with the extended predictor (6.10), we have $\boldsymbol{\beta} = (\boldsymbol{\beta}_1, \ldots, \boldsymbol{\beta}_c, \boldsymbol{\gamma})'$ and

$$\boldsymbol{X}_i = \begin{pmatrix} \boldsymbol{x}_i' & & & \boldsymbol{w}_{i1}' - \boldsymbol{w}_{i,c+1}' \\ & \boldsymbol{x}_i' & & \boldsymbol{w}_{i2}' - \boldsymbol{w}_{i,c+1}' \\ & & \ddots & \vdots \\ & & \boldsymbol{x}_i' & \boldsymbol{w}_{ic}' - \boldsymbol{w}_{i,c+1}' \end{pmatrix}.$$

For the ordinal cumulative models, the vector of regression coefficients is $\boldsymbol{\beta} = (\theta_1, \ldots, \theta_c, \beta_1, \ldots, \beta_k)'$ and the design matrix has to be defined as

$$\boldsymbol{X}_i = \begin{pmatrix} 1 & & & \boldsymbol{x}_i' \\ & 1 & & \boldsymbol{x}_i' \\ & & \ddots & \vdots \\ & & 1 & \boldsymbol{x}_i' \end{pmatrix}.$$

If we additionally introduce the c-dimensional response function

$$\boldsymbol{h}(\boldsymbol{\eta}_i) = (h_1(\boldsymbol{\eta}_i), \ldots, h_c(\boldsymbol{\eta}_i))',$$

we can write all categorical regression models for $\boldsymbol{\pi}_i = (\pi_{i1}, \ldots, \pi_{ic})'$ in compact matrix notion as

$$\boldsymbol{\pi}_i = \boldsymbol{h}(\boldsymbol{\eta}_i), \qquad \boldsymbol{\eta}_i = \boldsymbol{X}_i\boldsymbol{\beta}.$$

The score function $\boldsymbol{s}(\boldsymbol{\beta}) = \partial l(\boldsymbol{\beta})/\partial\boldsymbol{\beta}$ of the multinomial distribution does have a structure that is similar to the univariate GLM, presented in Chap. 5. We have

$$\boldsymbol{s}(\boldsymbol{\beta}) = \sum_{i=1}^{G} \boldsymbol{X}_i' \boldsymbol{D}_i \boldsymbol{\Sigma}_i^{-1}(\boldsymbol{y}_i - n_i\boldsymbol{\pi}_i),$$

where $\boldsymbol{D}_i = \partial\boldsymbol{h}(\boldsymbol{\eta}_i)/\partial\boldsymbol{\eta}$ is the $c \times c$-matrix of the partial derivatives, evaluated at $\boldsymbol{\eta}_i = \boldsymbol{X}_i\boldsymbol{\beta}$, and $\boldsymbol{\Sigma}_i$ is the covariance matrix

$$\boldsymbol{\Sigma}_i = n_i \begin{pmatrix} \pi_{i1}(1-\pi_{i1}) & -\pi_{i1}\pi_{i2} & \cdots & -\pi_{i1}\pi_{ic} \\ -\pi_{i2}\pi_{i1} & \pi_{i2}(1-\pi_{i2}) & & \\ \vdots & & \ddots & \vdots \\ -\pi_{ic}\pi_{i1} & \cdots & & \pi_{ic}(1-\pi_{ic}) \end{pmatrix}.$$

Introducing the weight matrix

$$\boldsymbol{W}_i = \boldsymbol{D}_i \boldsymbol{\Sigma}_i^{-1} \boldsymbol{D}_i',$$

(that depends on $\boldsymbol{\beta}$), we can write the Fisher matrix as

$$\boldsymbol{F}(\boldsymbol{\beta}) = \sum_{i=1}^{G} \boldsymbol{X}_i' \boldsymbol{W}_i \boldsymbol{X}_i,$$

which again is the generalization of the Fisher matrix in univariate GLMs.

Combining all observations and probabilities in $\boldsymbol{y} = (\boldsymbol{y}_1', \ldots, \boldsymbol{y}_G')'$ and $\boldsymbol{\pi} = (\boldsymbol{\pi}_1', \ldots, \boldsymbol{\pi}_G')'$, defining the full design matrix

$$\boldsymbol{X} = \begin{pmatrix} \boldsymbol{X}_1 \\ \vdots \\ \boldsymbol{X}_G \end{pmatrix}$$

for all observations, and forming the block-diagonal matrices $\boldsymbol{\Sigma} = \text{blockdiag}(\boldsymbol{\Sigma}_1, \ldots, \boldsymbol{\Sigma}_G)$, $\boldsymbol{D} = \text{blockdiag}(\boldsymbol{D}_1, \ldots, \boldsymbol{D}_G)$, and $\boldsymbol{W} = \text{blockdiag}(\boldsymbol{W}_1, \ldots, \boldsymbol{W}_G)$ finally yields

$$s(\boldsymbol{\beta}) = \boldsymbol{X}'\boldsymbol{D}\boldsymbol{\Sigma}^{-1}(\boldsymbol{y} - n_i\boldsymbol{\pi}), \quad \boldsymbol{F}(\boldsymbol{\beta}) = \boldsymbol{X}'\boldsymbol{W}\boldsymbol{X}.$$

Numerical Computation of the ML Estimator

We can again use the Fisher scoring method discussed in Sect. 5.1.2 for the iterative solution of the ML equations with $(t+1)$-th iteration given by

$$\hat{\boldsymbol{\beta}}^{(t+1)} = \hat{\boldsymbol{\beta}}^{(t)} + \boldsymbol{F}^{-1}(\hat{\boldsymbol{\beta}}^{(t)})\boldsymbol{s}(\hat{\boldsymbol{\beta}}^{(t)}).$$

The iterations can also be expressed as iteratively (re)weighted least squares updates

$$\hat{\boldsymbol{\beta}}^{(t+1)} = (\boldsymbol{X}'\boldsymbol{W}(\hat{\boldsymbol{\beta}}^{(t)})\boldsymbol{X})^{-1}\boldsymbol{X}'\boldsymbol{W}(\hat{\boldsymbol{\beta}}^{(t)})\tilde{\boldsymbol{y}}^{(t)},$$

based on the working observations $\tilde{\boldsymbol{y}}^{(t)} = (\tilde{\boldsymbol{y}}_1(\hat{\boldsymbol{\beta}}^{(t)}), \ldots, \tilde{\boldsymbol{y}}_G(\hat{\boldsymbol{\beta}}^{(t)}))$, with components

$$\tilde{\boldsymbol{y}}_i(\hat{\boldsymbol{\beta}}^{(t)}) = \boldsymbol{X}_i\hat{\boldsymbol{\beta}}^{(t)} + (\boldsymbol{D}_i^{-1}(\hat{\boldsymbol{\beta}}^{(t)}))'(\boldsymbol{y}_i - \boldsymbol{\pi}_i(\hat{\boldsymbol{\beta}}^{(t)})).$$

Asymptotic Properties and Tests of Linear Hypotheses

The asymptotic properties of the ML estimator are analogous to those in the univariate case. Under relatively weak regularity conditions and with increasing sample size $n \to \infty$, the ML estimator $\hat{\boldsymbol{\beta}}$ exists, is consistent, and is asymptotically normal, i.e.,

$$\hat{\boldsymbol{\beta}} \overset{a}{\sim} \mathrm{N}(\boldsymbol{\beta}, \boldsymbol{F}^{-1}(\hat{\boldsymbol{\beta}}));$$

see Fahrmeir and Kaufmann (1985). Tests of linear hypotheses $H_0 : \boldsymbol{C\beta} = \boldsymbol{d}$ can be conducted analogously to the univariate case (Chap. 5) after replacing the log-likelihood, the score function, and the information matrix by their multivariate versions.

Example 6.7 Caesarian Delivery—Hypotheses Testing

We now test whether or not the effects of *NPLAN* and *RISK* differ significantly across both types of infections. Thus we have the hypotheses

$$H_0 : \beta_{1N} = \beta_{2N} \text{ and } \beta_{1R} = \beta_{2R}, \qquad H_1 : \beta_{1N} \neq \beta_{2N} \text{ or } \beta_{1R} \neq \beta_{2R}.$$

This can be written as a special linear hypothesis in the form of $\boldsymbol{C\beta} = \boldsymbol{0}$, where $\boldsymbol{C}$ has rank 2. As for the Wald test statistic, we have $w = 0.33$ which results in a p-value of 0.8465. Thus H_0 cannot be rejected. Therefore, the two effects do not differ significantly across the two categories.

△

6.5 Bibliographic Notes

Analogously to Sect. 5.6, the concepts of *Bayesian inference* can be transferred to categorical regression models. For *fully Bayesian inference*, we can work either with MH algorithms based on IWLS proposals or with data augmentation schemes. IWLS proposals are described in Gamerman (1997) and Lenk and DeSarbo (2000) for univariate responses. Brezger and Lang (2006) propose a scheme for the multinomial logit model. Sampling schemes for categorical probit models based on data augmentation are more complicated than for binary probit models but similar in nature; see Albert and Chib (1993), Chen and Dey (2000), Fahrmeir and Lang (2001), and Imai and van Dyk (2005). Recently similar sampling schemes have been proposed for multinomial logit models; see Holmes and Held (2006) and Frühwirth-Schnatter and Frühwirth (2010).

Categorical regression models are described in full detail in the books by Tutz (2011) and Agresti (2002). A shorter account can be found in Fahrmeir and Tutz (2001, Chap. 3). The derivation of categorical regression models and more general discrete choice models traces back to McFadden, who was awarded the Noble Prize for Economic Sciences for this work; see, for example, McFadden (1984) or Train (2003).

Mixed Models 7

Mixed models extend the predictor $\eta = \boldsymbol{x}'\boldsymbol{\beta}$ of linear, generalized linear, and categorical regression models by incorporating random effects or coefficients in addition to the non-random or "fixed" effects $\boldsymbol{\beta}$. Therefore, mixed models are sometimes also called random effects models, and have become quite popular for analyzing longitudinal data obtained from repeated observations on individuals or objects in longitudinal studies. A closely related situation is the analysis of clustered data, i.e., when observations are obtained from objects selected by subsampling primary sampling units (clusters or groups of objects) in cross-sectional studies. For example, clusters may be defined by hospitals, schools, or firms, where data from (possibly small) subsamples of patients, students, or clients are collected. Generally, clustering may result from any data generating mechanism that induces a cluster structure. In any case, the data consist of n_i repeated observations

$$(y_{i1}, \ldots, y_{ij}, \ldots, y_{in_i}, \boldsymbol{x}_{i1}, \ldots, \boldsymbol{x}_{ij}, \ldots, \boldsymbol{x}_{in_i})$$

of responses and covariates for each individual or cluster $i = 1, \ldots, m$. For longitudinal data, y_{ij} and $\boldsymbol{x}_{ij}$ denote the observed value of the response and the covariate vector, respectively, for individual i at time t_{ij}, $j = 1, \ldots, n_i$, while for clustered data these values are observations for subjects or objects j from cluster i. Mixed models allow estimation of individual- or cluster-specific effects, even in the case of relatively small numbers n_i of repeated individual measurements or sizes n_i of subsamples from clusters. The basic idea is to extend the linear predictor $\eta_{ij} = \boldsymbol{x}'_{ij}\boldsymbol{\beta}$ for observation y_{ij} with fixed population effects $\boldsymbol{\beta}$ to the linear mixed predictor $\eta_{ij} = \boldsymbol{x}'_{ij}\boldsymbol{\beta} + \boldsymbol{u}'_{ij}\boldsymbol{\gamma}_i$. Usually $\boldsymbol{u}'_{ij}$ is a subvector of the covariates, and $\boldsymbol{\gamma}_i$ is a vector of individual- or cluster-specific random effects. The assumption of $\boldsymbol{\gamma}_i$ to be fixed effects as in the standard linear or generalized linear model is often impractical since the number of parameters to be estimated becomes quite large relative to the sample size. On the other hand, the random effects distribution implicitly induces certain regularization properties for the cluster parameters $\boldsymbol{\gamma}_i$. An additional advantage of mixed models is that correlations, induced by repeated observations from individuals or clusters, are taken into account during estimation.

L. Fahrmeir et al., *Regression*, DOI 10.1007/978-3-642-34333-9_7,

In the following, we first describe linear mixed models (LMMs) with (conditionally) Gaussian responses y_{ij}, making the conventional assumption that the random effects are i.i.d. Gaussian variables. We then extend LMMs by allowing correlated Gaussian random effects. This leads to a very broad class of models that are appropriate for analyzing spatial and spatio-temporal data, as well as for Bayesian approaches to non- and semiparametric regression in Chaps. 8 and 9, in particular in Sects. 8.1.9 and 9.6. Statistical inference is described from a frequentist likelihood-oriented as well as a Bayesian perspective.

The second part of chapter extends generalized linear models for non-Gaussian responses to generalized linear mixed models (GLMMs), such as logit or Poisson regression models. Statistical inference for the GLMM is based on similar, but more complicated concepts of the LMM.

7.1 Linear Mixed Models for Longitudinal and Clustered Data

7.1.1 Random Intercept Models

We start with random intercept models which are among the most simple (albeit quite important) mixed models. For notational simplicity we first restrict ourselves to the case of just one covariate x. Let

$$(y_{ij}, x_{ij}), \quad i = 1, \dots, m, \quad j = 1, \dots, n_i$$

denote the values of the response variable y and covariate x observed at times $t_{i1} < \dots < t_{ij} < \dots < t_{in_i}$ for individuals $i = 1, \dots, m$ in a longitudinal study, or for subjects $j = 1, \dots, n_i$ in clusters $i = 1, \dots, m$. Our starting point for modeling the relationship between y and x is the classical linear model

$$y_{ij} = \beta_0 + \beta_1 x_{ij} + \varepsilon_{ij}, \tag{7.1}$$

with i.i.d. errors $\varepsilon_{ij} \sim \mathrm{N}(0, \sigma^2)$. In this model, the fact that we have *repeated measurements* $j = 1, \dots, n_i$ on the *same individual or cluster* i is not taken into account. In particular, we not only assume that the observations y_{ij} and y_{rl}, of different individuals i and r, are (stochastically) independent but also repeated measurements y_{ij} and y_{il} on the same individual or cluster i. A possible graphical way to check this independence assumption is to estimate and plot separate regression lines (or more generally regression curves) for each individual or cluster i. If there is no cluster-specific heterogeneity all regression lines should have similar (not identical due to sampling variability) intercepts and slopes. A typical plot is shown in the left panel of Fig. 7.1. The estimates are based on artificial data drawn from the model $y_{ij} = 1 + x_{ij} + \varepsilon_{ij}$, $i = 1, \dots, 10$, $j = 1, \dots, 20$, for $m = 10$ individuals or clusters, $n_i = 20$ repeated measurements in each cluster, and i.i.d. errors $\varepsilon_{ij} \sim \mathrm{N}(0, 0.1^2)$. Clearly, the estimated cluster-specific regression lines scatter with low variability

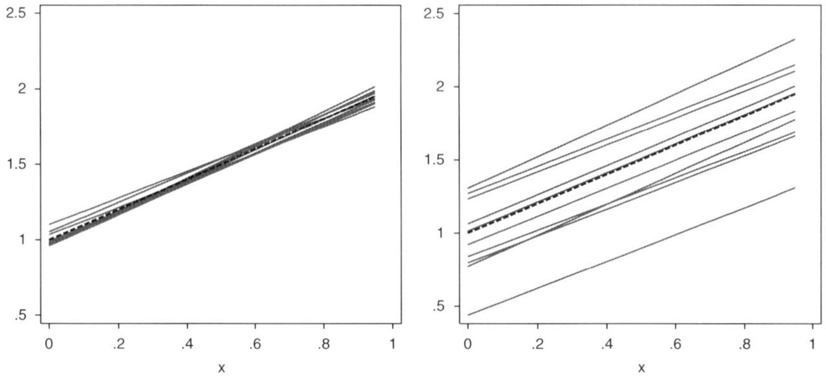

Fig. 7.1 Illustration of random intercept models: both panels show separately estimated regression lines for each cluster. In the *left panel*, there is no cluster-specific random intercept, while in the *right panel* a random intercept is present. The *dashed line* corresponds to the population model

around the true regression line $1 + x$ (dashed line in Fig. 7.1). Hence, there is no reason for assuming cluster-specific heterogeneity and a common regression model for all clusters is sufficient.

The right panel of Fig. 7.1 reveals a different scenario. The estimated cluster-specific regression lines still show a common slope across clusters, but the intercept appears to be different from cluster to cluster. To model this type of cluster-specific heterogeneity we introduce cluster-specific parameters γ_{0i} and obtain

$$y_{ij} = \beta_0 + \beta_1 x_{ij} + \gamma_{0i} + \varepsilon_{ij}, \tag{7.2}$$

where $\varepsilon_{ij} \sim \mathrm{N}(0, \sigma^2)$ are the usual i.i.d. errors of the classical linear model. In Eq. (7.2):

- β_0 is the "fixed" population intercept.
- γ_{0i} is the individual- or cluster-specific (random) deviation from the population intercept β_0.
- $\beta_0 + \gamma_{0i}$ is the (random) intercept for cluster i.
- β_1 is a "fixed" population slope parameter of covariate x that is common across clusters.

Since the individuals or clusters are a random sample from a larger population, the cluster-specific parameters γ_{0i} are assumed to be random with

$$\gamma_{0i} \overset{i.i.d.}{\sim} \mathrm{N}\left(0, \tau_0^2\right). \tag{7.3}$$

We also assume mutual independence between the ε_{ij} and the γ_{0i}. The normal *random effects distribution* in Eq. (7.3) is also sometimes called a *mixture distribution*. The mean can be set to zero because the population mean is already represented by the fixed effect β_0.

The random intercepts $\beta_0 + \gamma_{0i} \sim \mathrm{N}(\beta_0, \tau_0^2)$ may be interpreted as effects of omitted (individual- or cluster-specific) covariates and account for unobserved heterogeneity. Another way to look at the model is to interpret γ_{0i} as an additional error term. The random intercept model appears as a linear regression model with two error terms, where γ_{0i} is then a cluster-level error shared between measurements on the same individual or cluster i and ε_{ij} is the observation error of measurement j in cluster i.

The random intercept model induces a specific correlation or dependence structure on the responses y_{ij}. Given the random intercepts γ_{0i}, the y_{ij} are still conditionally independent with

$$y_{ij} \mid \gamma_{0i} \sim \mathrm{N}(\beta_0 + \beta_1 x_{ij} + \gamma_{0i}, \sigma^2).$$

Marginally, however, repeated measurements y_{ij} for subject or cluster i are correlated with within-subject correlation coefficient

$$\mathrm{Corr}(y_{ij}, y_{il}) = \frac{\tau_0^2}{\tau_0^2 + \sigma^2}, \quad j \neq l; \tag{7.4}$$

see Sect. 7.1.4 for a derivation. Based on the normality assumption of the random effects and the errors, we can further derive the marginal distribution of responses. We have

$$\boldsymbol{y}_i \sim \mathrm{N}(\boldsymbol{X}_i\boldsymbol{\beta}, \sigma^2\boldsymbol{I}_{n_i} + \tau_0^2\boldsymbol{J}_{n_i}), \tag{7.5}$$

where $\boldsymbol{X}_i$ is an $(n_i \times 2)$-design matrix with ones in the first column and the observed x_{ij} in the second column, and $\boldsymbol{J}_{n_i}$ denotes an $(n_i \times n_i)$-matrix of ones. Between two subjects i and r, the observations y_{ij} and y_{rl} are still uncorrelated. The strength of the within-subject correlation Eq. (7.4) depends on the magnitude of the error variances τ_0^2 and σ^2. The higher the random effects variance τ_0^2 relative to the error variance σ^2, the stronger the within-subject correlation. Note also that the within-subject correlation is constant (equicorrelation) from measurement to measurement. This may be questionable for longitudinal data, where we might expect correlations dying off for measurements which are farther apart in time.

The marginal model (7.5) also shows what happens with the estimates for $\boldsymbol{\beta}$ if the classical linear model (7.1) is estimated instead of the random intercept model (7.2). If the assumed correlation structure induced by the random intercept model is correct, estimating a classical linear model means that we mistakenly assume an error covariance matrix $\sigma^2\boldsymbol{I}$ instead of a non-diagonal covariance matrix as imposed in the marginal model (7.5) of the random intercept model. The consequences of using an incorrect covariance matrix have already been established in Sect. 4.1.1 in the context of the general linear model. As stated there, the estimates for the "fixed" regression coefficients $\boldsymbol{\beta}$ are still unbiased. However, the covariance matrix of $\boldsymbol{\beta}$ and all derived quantities in particular standard errors, confidence intervals, and tests are not correct. Note that the standard errors could either be smaller or larger, as in a misspecified classical linear model. A nice description of the consequences of mistakenly specifying a classical linear model is also given in Skrondal and Rabe-Hesketh (2008) in Sect. 3.10.1.

Between- and Within-Cluster Effects

The random intercept model (7.2) considered thus far has an important limitation: The so-called within- and between-cluster effects of x are the same. The within-cluster effect denotes the effect if x changes within the *same individual* at different occasions. The between-cluster effect refers to different x values *between different subjects or clusters*. In either case, a difference of one unit of x in model (7.2) induces a difference of β_1 in expected responses. This equality between the within- and between-cluster effect might be questionable in applications. A possible example is our data on undernutrition in Zambia. Here the individual data on children are nested within the districts of Zambia as the cluster variable. Suppose we are interested in the effect of the households wealth, measured by a wealth index, x say, on the Z-score. Now we focus our attention on two imaginary districts, one comparably rich district i and a second rather poor district l. More precisely, we assume that in district i the population is on average richer than in the second district l, i.e., $\bar{x}_i > \bar{x}_l$ with $\bar{x}_i$, $\bar{x}_l$ being the cluster averages of the wealth index. Suppose first that we compare a child living in the rich district with a child living in the poor district (between district comparison). It is then conceivable that the child living in the rich environment has a higher Z-score, i.e., is better nourished, than the child in the poor environment, even if the individual household wealth is identical. The reason might be that the child profits from the rich environment independent of the individual household situation that might be much less favorable. Economists call this an *external effect*. A possible way to model this between-cluster effect is simply to include the cluster averages $\bar{x}_i$ as an additional covariate into the regression equation. On the other hand, there might be also a within-cluster effect if the individual household wealth is different to that of the average cluster wealth. Children living in households which are wealthier than the average households in the district should have an even higher Z-score (at least on average). This effect may be equal in size to the between-cluster effect but could just as well differ, i.e., being smaller or larger in size. An illustration of different within- and between-cluster effects is shown in Fig. 7.2. The left panel shows different sizes of the within- and between-cluster effects but with identical signs. In the right panel, the signs of the within- and between-cluster effects are opposite of each other.

To deal with possibly different within- and between-cluster effects we can incorporate two covariates derived from x into the predictor: The between-cluster effect can be modeled by the respective cluster means $\bar{x}_i$ as a covariate. The within-cluster effect is modeled by incorporating the individual difference from the cluster mean $x_{ij} - \bar{x}_i$. This yields the extended random intercept model

$$y_{ij} = \beta_0 + \beta_1(x_{ij} - \bar{x}_i) + \beta_2\bar{x}_i + \gamma_{0i} + \varepsilon_{ij}, \tag{7.6}$$

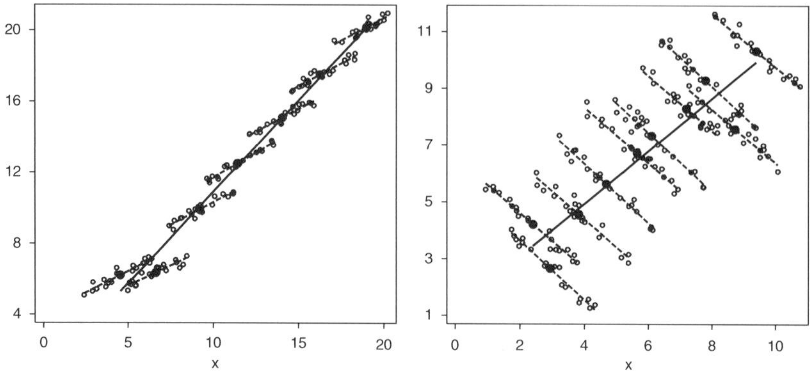

Fig. 7.2 Illustration of within- and between-cluster effects: the between-cluster effects are visualized through the cluster means $\bar{x}_i$ marked by *black dots* and the corresponding linear trends (*solid lines*) which are increasing with x in both panels. The within-cluster effects are illustrated by *dashed lines*

where the coefficient β_2 of the cluster mean represents the between-cluster effect, and the coefficient β_1 of the individual deviation from the cluster mean represents the within-cluster effect. The model collapses to the original random intercept model (7.2) if the within- and between-cluster effects are identical, i.e., if $\beta_1 = \beta_2$. Another interpretation of model (7.6) is obtained by considering $\tilde{\gamma}_{0i} = \beta_2 \bar{x}_i + \gamma_{0i}$ as a random intercept that depends through $\beta_2 \bar{x}_i$ on the covariates.

Alternative Views on the Random Intercept Model

For some applications, the view of the γ_{0i} as random effects can be questionable. This is the case when we do not have the interpretation that clusters are randomly sampled from a larger population. Such a situation arises, for example, when data have been observed on a discrete spatial grid. A typical example is the data for the Munich rent index where the district of each apartment in Munich is given. To account for spatial heterogeneity it might be useful to add a district-specific effect into the predictor; see Example 9.2 of Chap. 9. Since the districts cannot be seen as a random sample of a larger "population" of districts, the interpretation of the district-specific effects as random effects is somewhat artificial. However, there are alternative useful interpretations of the random intercept model which are more suitable for the given situation. In particular, the random effects distribution Eq. (7.3) can be readily understood as the prior for γ_{0i} in a corresponding Bayesian approach. In fact Eq. (7.3) is identical to the Bayesian ridge prior of Sect. 4.4.2. Assuming noninformative priors for the "fixed" effects, i.e., $p(\beta_0) \propto$ const and $p(\beta_1) \propto$ const, the posterior is given by

$$p(\beta_0, \beta_1, \boldsymbol{\gamma} \mid \boldsymbol{y}) \propto L(\beta_0, \beta_1, \boldsymbol{\gamma}) \prod_{i=1}^{m} \frac{1}{\sqrt{\tau_0^2}} \exp\left(-\frac{1}{2\tau_0^2}\gamma_{0i}^2\right),$$

where $\boldsymbol{\gamma} = (\gamma_{01}, \ldots, \gamma_{0m})'$ is the vector of cluster effects and $L(\cdot)$ is the Gaussian likelihood of the conditional model. The variances σ^2 and τ_0^2 are assumed fixed for the moment. Now the posterior mode can be obtained by maximizing the log-posterior resulting in the optimization criterion

$$\text{PLS}(\beta_0, \beta_1, \boldsymbol{\gamma}) = \sum_{i=1}^{m} \sum_{j=1}^{n_i} (y_{ij} - \beta_0 - \beta_1 x_{ij} - \gamma_{0i})^2 + \lambda \sum_{i=1}^{m} \gamma_{0i}^2 \tag{7.7}$$

with $\lambda = \sigma^2/\tau_0^2$. This has the form of a penalized least squares criterion quite similar to ridge regression outlined in Sect. 4.2.2. To understand the nature of the penalization, we consider the particularly simple random intercept model

$$y_{ij} = \beta_0 + \gamma_{0i} + \varepsilon_{ij}, \tag{7.8}$$

without any covariates, and as usual with $\gamma_{0i} \overset{i.i.d.}{\sim} \text{N}(0, \tau_0^2)$, $\varepsilon_{ij} \overset{i.i.d.}{\sim} \text{N}(0, \sigma^2)$. As will be shown in Sect. 7.3.2, the estimator for the γ_{0i} is given by

$$\hat{\gamma}_{0i} = \frac{n_i \tau_0^2}{\sigma^2 + n_i \tau_0^2} \frac{1}{n_i} \sum_{j=1}^{n_i} (y_{ij} - \hat{\beta}_0),$$

where $\hat{\beta}_0 = \bar{y}$, with $\bar{y}$ the overall mean of the responses. The estimator for the cluster mean $\eta_i = \beta_0 + \gamma_{0i}$ is now given by

$$\hat{\eta}_i = \hat{\beta}_0 + \hat{\gamma}_{0i} = \bar{y} + \frac{n_i \tau_0^2}{\sigma^2 + n_i \tau_0^2} \frac{1}{n_i} \sum_{j=1}^{n_i} (y_{ij} - \bar{y}). \tag{7.9}$$

The term $e_i = \frac{1}{n_i} \sum_{j=1}^{n_i} (y_{ij} - \bar{y})$ can be seen as an average residual for individual i, which is a rather natural estimate of γ_{0i}. This is multiplied by the factor

$$\lambda_i = \frac{n_i \tau_0^2}{\sigma^2 + n_i \tau_0^2} < 1,$$

which is sometimes called a shrinkage effect, because the ad hoc estimate e_i for γ_i is shrunken towards the prior mean 0. The larger the n_i, the closer the weight λ_i is to 1 and the smaller the shrinkage. Additional shrinkage is obtained if the error variance σ^2 is large relative to the random effects variance τ_0^2.

It is instructive to contrast the estimator Eq. (7.9) with two extreme modeling strategies:

- *Full ignorance of groups:* On the one extreme we could fully ignore the groups and estimate the model $y_{ij} = \beta_0 + \varepsilon_{ij}$ with fixed overall intercept β_0. This is also called the fully pooled model. Of course the least squares estimator for β_0 (which is then identical to the cluster mean η_i) is given by the overall mean $\bar{y}$ of responses, i.e., $\hat{\eta}_i = \hat{\beta}_0 = \bar{y}$.
- *Full distinction of groups:* The other extreme would be to estimate separate models $y_{ij} = \eta_i + \varepsilon_{ij}$ for each cluster i treating the η_i as fixed parameters without random effects distribution in a model without intercept (to omit collinearity problems). This is also known as fully unpooled estimation. Here, the least squares estimators for the η_i are given by the cluster means $\bar{y}_i$, i.e., $\hat{\eta}_i = \bar{y}_i$.

Now the random effects estimator Eq. (7.9) can be seen as a compromise between the two extreme cases. For large n_i or small σ^2 relative to τ_0^2, the estimator Eq. (7.9) approaches the fully unpooled estimator $\bar{y}_i$. For small n_i or large σ^2 relative to τ_0^2 the fully pooled estimator $\bar{y}$ is approached. In the extreme cases $n_i = 0$ or $\tau_0^2 \to 0$ or $\sigma^2 \to \infty$ the mean $\bar{y}$ is reached as a limit. In the other extreme cases $n_i \to \infty$ or $\tau_0^2 \to \infty$ or $\sigma^2 \to 0$ we reach the cluster mean $\bar{y}_i$ as a limit. An illustration of the shrinkage factor λ_i depending on the cluster size n_i (panel a), the random effects variance τ_0^2 (panel b), and the error variance σ^2 (panel c) can be found in Fig. 7.3.

Key Features of Mixed Models

Although the random intercept model is comparably simple it already reveals the key features and advantages of mixed models:

- Individual- or cluster-specific effects can be introduced to account for specific deviations from the population behavior.
- They allow to correct for unobserved heterogeneity induced by omitted covariates.
- Correlations between observations of the same individual or cluster can be taken into account (at least to some extent). This ensures that inference regarding the regression coefficients is correct in the sense that we obtain correct standard errors, confidence intervals, and tests.
- Estimation is stabilized by assuming a common random effects distribution that acts as a penalty term for the otherwise unpenalized cluster-specific effects.

We finally summarize the various interpretations of the random intercept model:

- A *classical interpretation*, where clusters are a random sample of a larger population and the γ_{0i} are cluster-specific random effects.
- A *marginal interpretation*, where the random effects γ_{0i} induce the general linear model (7.5) with correlated errors for the observed y_{ij}.
- A *Bayesian point of view*, which interprets Eq. (7.3) as an underlying prior.
- A *penalized least squares view*, where the penalized least squares criterion induces a penalty on the cluster-specific effects to regularize the estimated parameters.

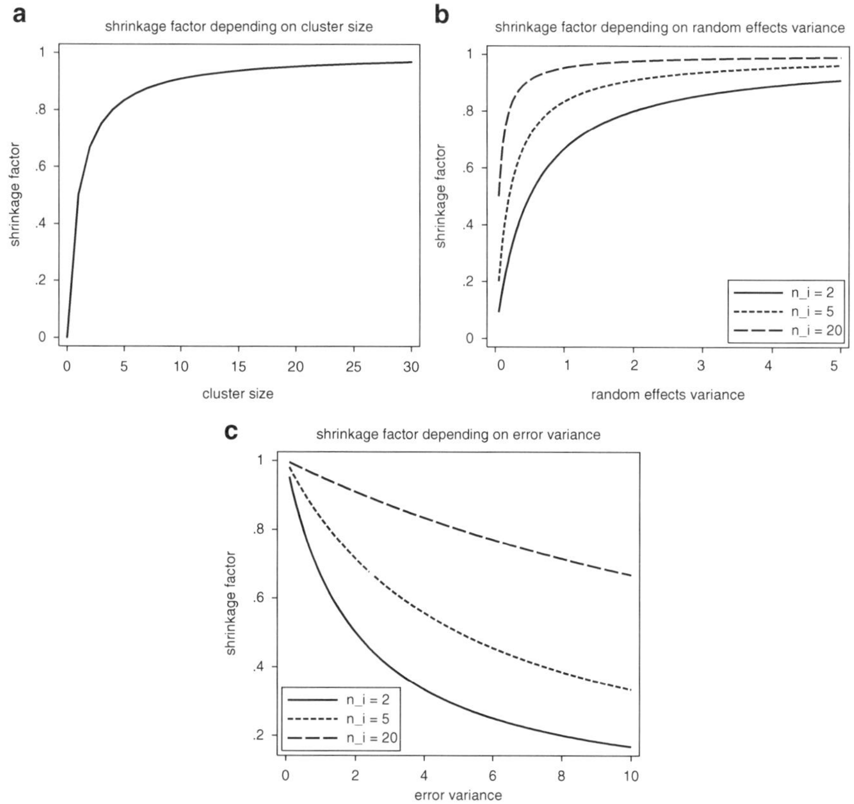

Fig. 7.3 Shrinkage factor in simple random intercept models: Panel (**a**) illustrates the shrinkage factor λ_i depending on the cluster size n_i (for fixed $\sigma^2 = 1$ and $\tau_0^2 = 1$). Panels (**b**) and (**c**) display the shrinkage factor depending on the random effects variance τ_0^2 (for fixed $\sigma^2 = 1$) and the error variance σ^2 (for fixed $\tau_0^2 = 1$), respectively

7.1.2 Random Coefficient or Slope Models

Random intercept models are still based on the assumption that the covariate effect of x is equal in size for each individual or cluster. However, this may be too restrictive as Fig. 7.4 reveals. The left panel shows the separately estimated cluster-specific regression lines for some artificial data. Here, the intercept for the regression lines seems to be identical (or at least close), whereas the slopes are clearly different. More typical is the situation in the right panel of Fig. 7.4, where the regression lines exhibit both cluster-specific intercepts and slopes. To cope with such cluster-specific slopes, we extend the random intercept model (7.2) to obtain

$$y_{ij} = \beta_0 + \beta_1 x_{ij} + \gamma_{0i} + \gamma_{1i} x_{ij} + \varepsilon_{ij}, \tag{7.10}$$

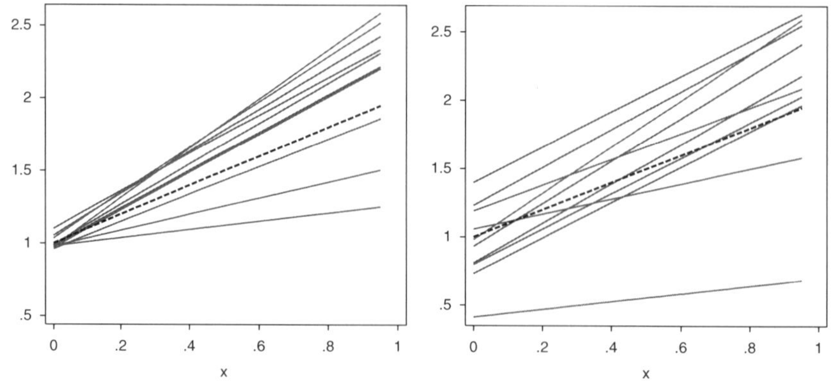

Fig. 7.4 Illustration of random slope models: both panels show separately estimated regression lines for each cluster. In the *left panel*, there is a cluster-specific random slope and a common intercept across clusters. In the *right panel*, both the intercept and the slope are cluster-specific. The *dashed line* corresponds to the population model

where:

- β_1 is the "fixed" population slope of the effect of x.
- γ_{1i} is the individual- or cluster-specific deviation for the slope.
- $\beta_1 x_{ij}$ is the population effect of x.
- $\beta_1 x_{ij} + \gamma_{1i} x_{ij}$ is the cluster-specific effect of x.

The left panel of Fig. 7.4 corresponds to a model with most random intercept parameters γ_{0i} zero or close to zero and random slope parameters γ_{1i} different from zero. In the right panel both random coefficients are mostly found to be away from zero.

For the cluster-specific parameters, we now define the bivariate normal random effects distribution

$$\boldsymbol{\gamma}_i = \begin{pmatrix} \gamma_{0i} \\ \gamma_{1i} \end{pmatrix} \overset{i.i.d.}{\sim} \mathrm{N}\left(\begin{pmatrix} 0 \\ 0 \end{pmatrix}, \begin{pmatrix} \tau_0^2 & \tau_{01} \\ \tau_{10} & \tau_1^2 \end{pmatrix} \right) \tag{7.11}$$

with mean $\mathbf{0} = (0, 0)'$ and covariance matrix

$$\boldsymbol{Q} = \begin{pmatrix} \tau_0^2 & \tau_{01} \\ \tau_{10} & \tau_1^2 \end{pmatrix}.$$

The parameters τ_0^2 and τ_1^2 determine the variability of the cluster-specific intercepts and slopes, respectively. The left panel of Fig. 7.4 corresponds to $\tau_0^2 \approx 0$ and $\tau_1^2 > 0$, while in the right panel, $\tau_0^2 > 0$ also holds.

The covariance $\tau_{01} = \tau_{10}$ can capture correlations between random intercepts and slopes. Such a correlation may be present, for example, when individuals with larger slopes tend to have smaller intercepts, leading to negatively correlated random intercepts and slopes.

Similar to simple random intercepts, the random coefficient model (7.10) induces a certain (marginal) correlation structure on observations y_{ij} and y_{ir} from the same cluster. This correlation structure is, however, more complicated as it depends on the covariate values x_{ij}. First of all, the marginal variances of y_{ij} are heteroscedastic with

$$\text{Var}(y_{ij}) = \tau_0^2 + 2\tau_{01}x_{ij} + \tau_1^2 x_{ij}^2 + \sigma^2;$$

see Sect. 7.1.4 for a derivation. Thus, the marginal variances depend quadratically on the covariate values. The covariance between y_{ij} and y_{ir} can be shown to be

$$\text{Cov}(y_{ij}, y_{ir}) = \tau_0^2 + \tau_{01}x_{ij} + \tau_{01}x_{ir} + \tau_1^2 x_{ij}x_{ir}$$

resulting in an intraclass correlation coefficient

$$\text{Corr}(y_{ij}, y_{ir}) = \frac{\text{Cov}(y_{ij}, y_{ir})}{\text{Var}(y_{ij})^{1/2}\text{Var}(y_{ir})^{1/2}},$$

which depends in a rather complicated way on the observed covariate values and is very difficult to interpret.

We now illustrate the idea of random intercept and slope models with a first example, thereby extending our framework in a natural way to more than one covariate. We reconsider Example 2.9 (p. 39). Another application with cluster data and several random slopes can be found in Sect. 7.7, which presents a case study on sales of orange juice.

Example 7.1 Hormone Therapy with Rats—Linear Mixed Model

To investigate the effect of testosterone on their craniofacial growth, 50 male rats have been randomized to either a control group or one of two treatment groups. Treatment consisted of a low or high dose of the drug Decapeptyl, an inhibitor for testosterone production, and started at the age of 45 days. To measure the growth of the skull of each rat, X-ray pictures were taken every 10 days, with the first observation at the age of 50 days. In this example, height of the skull, defined as the distance (in pixels) between two well-defined points, is considered as the response variable. The individual profiles $\{y_{ij},\ j = 1, \ldots, n_i\}$ for the three groups are shown in Fig. 2.10 (p. 40). Since many rats drop out before the end of the study, the numbers n_i of individual repeated measurements are different; see Table 2.3 (p. 39).

In Example 2.9 (p. 39), we formulated the LMM

$$y_{ij} = \beta_0 + \beta_1 L_i t_{ij} + \beta_2 H_i t_{ij} + \beta_3 C_i t_{ij} + \gamma_{0i} + \gamma_{1i} t_{ij} + \varepsilon_{ij}, \qquad (7.12)$$

with the transformed age

$$t = \log(1 + (age - 45)/10)$$

as the time scale (with $t = 0$ corresponding to begin of treatment), the time points

$$t_{ij} = \log(1 + (age_{ij} - 45)/10)$$

for measurement j of rat i, and the indicator variables

$$L_i = \begin{cases} 1, & \text{rat } i \text{ in low-dose group,} \\ 0, & \text{otherwise,} \end{cases}$$

$$H_i = \begin{cases} 1, & \text{rat } i \text{ in high-dose group,} \\ 0, & \text{otherwise,} \end{cases}$$

$$C_i = \begin{cases} 1, & \text{rat } i \text{ in control group,} \\ 0, & \text{otherwise.} \end{cases}$$

In this model, β_0 is the population intercept of rats before treatment, γ_{0i} are individual-specific deviations from this population intercept, the effects β_1, β_2, and β_3 are the (different) population slopes for the three groups, and γ_{1i} are individual-specific deviations from these slopes. Therefore, $\beta_1 + \gamma_{1i}$, $\beta_2 + \gamma_{1i}$, and $\beta_3 + \gamma_{1i}$ are the individual slopes for rat i, in the low-dose, high-dose, and control group, respectively. Similarly, $\beta_0 + \beta_1 t_{ij} + \gamma_{0i} + \gamma_{1i} t_{ij}$, $\beta_0 + \beta_2 t_{ij} + \gamma_{0i} + \gamma_{1i} t_{ij}$, and $\beta_0 + \beta_3 t_{ij} + \gamma_{0i} + \gamma_{1i} t_{ij}$ are the corresponding linear trends.

In contrast to the "fixed" effects $\boldsymbol{\beta} = (\beta_0, \beta_1, \beta_2, \beta_3)'$, the individual-specific effects $\boldsymbol{\gamma}_i = (\gamma_{0i}, \gamma_{1i})'$ are considered to be "random" because the rats are randomly selected from a population. We make the usual assumption that the random effects are independent and identically distributed Gaussian random variables according to Eq. (7.11) but with the additional restriction $\tau_{01} = \tau_{10} = 0$.

The variances τ_0^2 and τ_1^2 characterize the amount of variability of individual-specific deviations γ_{0i} from the population intercept β_0 and of deviations γ_{1i} from population slopes, respectively. Note that we implicitly assume that deviations from the three population slopes have the same variance.

A closer look at Fig. 2.10 (p. 40) suggests that individual-specific slopes do not differ much from each other, corresponding to the assumption $\gamma_{1i} = 0$ for all i. This leads to the simplified or reduced model

$$y_{ij} = \beta_0 + \beta_1 L_i t_{ij} + \beta_2 H_i t_{ij} + \beta_3 C_i t_{ij} + \gamma_{0i} + \varepsilon_{ij}, \tag{7.13}$$

where only i.i.d. random intercepts $\gamma_{0i} \sim \mathrm{N}(0, \tau_0^2)$ are incorporated, without random slopes.

Table 7.1 contains estimates for the fixed effects and the variance parameters in both models. Estimates are based on methodology outlined in Sect. 7.3 and estimation has been carried out using function `lmer` of the R package `lme4`. Furthermore, Table 7.1 shows the estimated effects of a linear regression model without any random effects.

We first take a look at the random intercept model (7.13). The estimated random effects variance of 3.565 is quite large compared to the overall error variance of 1.445 yielding a comparably high intraclass correlation coefficient of

$$\widehat{\mathrm{Corr}}(y_{ij}, y_{ir}) = \frac{3.565}{3.565 + 1.445} = 0.71.$$

This indicates strong individual-specific heterogeneity. Compared to the linear model without random effects, the estimated population intercepts are in close agreement, whereas estimated treatment effects differ. In particular, the standard errors are higher in the simple linear model, which is in part due to the considerably larger estimate for the error variance σ^2 that has "captured" omitted individual-specific random effects. Comparing the random intercept model (7.13) with the more complex random slope model (7.12), all estimates are

Table 7.1 Hormone therapy with rats: estimates including standard errors for the random slope model, the reduced random intercept model, and a classical linear model without any random effects

		Model (7.12)		Model (7.13)		Linear model	
	Parameter	Estimate	s.e.	Estimate	s.e.	Estimate	s.e.
intercept	β_0	68.606	0.327	68.607	0.331	68.687	0.348
low-dose	β_1	7.505	0.227	7.507	0.225	7.677	0.286
high-dose	β_2	6.875	0.230	6.871	0.228	6.529	0.284
control	β_3	7.318	0.284	7.314	0.281	7.212	0.326
$\text{Var}(\gamma_{0i})$	τ_0^2	3.430		3.565			
$\text{Var}(\gamma_{1i})$	τ_1^2	0.001					
$\text{Var}(\varepsilon_{ij})$	σ^2	1.444		1.445		4.730	

very close. Moreover, the variance for the random slope is very small (0.001), so that the simpler random intercept model may be sufficient in this application. The decision as to whether or not the reduced model (7.13) is appropriate can be also supported by a statistical test; see Sect. 7.3.4.

△

7.1.3 General Model Definition and Matrix Notation

We now define LMMs in the more general matrix notation containing the illustrative models and examples of the previous sections as special cases. Let

$$(y_{ij}, \boldsymbol{x}'_{ij}), \quad i = 1, \ldots, m, \quad j = 1, \ldots, n_i,$$

denote the values of a response variable y and a covariate vector $\boldsymbol{x}$ observed at times $t_{i1} < \ldots < t_{ij} < \ldots < t_{in_i}$ for individuals $i = 1, \ldots, m$ in a longitudinal study (as in Example 7.1) or for subjects $j = 1, \ldots, n_i$ in cluster $i = 1, \ldots, m$.

A LMM is hierarchically defined through the following stages. In the *first stage*, responses y_{ij} are assumed to depend linearly on unknown fixed population effects $\boldsymbol{\beta}$ and on unknown individual- or cluster-specific effects $\boldsymbol{\gamma}_i$ through the *measurement model*

$$y_{ij} = \boldsymbol{x}'_{ij}\boldsymbol{\beta} + \boldsymbol{u}'_{ij}\boldsymbol{\gamma}_i + \varepsilon_{ij}.$$

In this case, $\boldsymbol{x}_{ij} = (1, x_{ij1}, \ldots, x_{ijk})'$, $\boldsymbol{u}_{ij} = (1, u_{ij1}, \ldots, u_{ijq})'$, and ε_{ij} represent design vectors and the error terms, respectively; the latter are assumed to be i.i.d. $\text{N}(0, \sigma^2)$ random variables. The design vectors include original covariates from the given data set or can be constructed from original covariates. The components may be time-dependent, such as the transformed age t in Example 7.1, or may be time-constant. For clustered data, this corresponds to covariates that vary across subjects or only across clusters. Usually, $\boldsymbol{x}_{ij}$ as well as $\boldsymbol{u}_{ij}$ both contain the constant 1, so that the linear predictor includes a fixed population intercept β_0 and a random intercept γ_{0i}. In addition, $\boldsymbol{u}_{ij}$ is typically a subvector of $\boldsymbol{x}_{ij}$ and will not contain additional components. As an example consider the simple random intercept model (7.2), where $\boldsymbol{x}_{ij} = (1, x_{ij})'$ and $\boldsymbol{u}_{ij} = 1$. In the extended random intercept model (7.6), we have $\boldsymbol{x}_{ij} = (1, x_{ij} - \bar{x}_i, \bar{x}_i)'$ and $\boldsymbol{u}_{ij} = 1$, while for the simple random slope model (7.10) we obtain $\boldsymbol{x}_{ij} = (1, x_{ij})'$ and $\boldsymbol{u}_{ij} = (1, x_{ij})'$.

Collecting all individual- or cluster-specific responses y_{ij}, design vectors $\boldsymbol{x}_{ij}, \boldsymbol{u}_{ij}$, and errors ε_{ij}, $j = 1, \ldots, n_i$, into vectors or design matrices

$$\boldsymbol{y}_i = \begin{pmatrix} y_{i1} \\ \vdots \\ y_{ij} \\ \vdots \\ y_{in_i} \end{pmatrix}, \quad \boldsymbol{X}_i = \begin{pmatrix} \boldsymbol{x}'_{i1} \\ \vdots \\ \boldsymbol{x}'_{ij} \\ \vdots \\ \boldsymbol{x}'_{in_i} \end{pmatrix}, \quad \boldsymbol{U}_i = \begin{pmatrix} \boldsymbol{u}'_{i1} \\ \vdots \\ \boldsymbol{u}'_{ij} \\ \vdots \\ \boldsymbol{u}'_{in_i} \end{pmatrix}, \quad \boldsymbol{\varepsilon}_i = \begin{pmatrix} \varepsilon_{i1} \\ \vdots \\ \varepsilon_{ij} \\ \vdots \\ \varepsilon_{in_i} \end{pmatrix}, \tag{7.14}$$

we obtain the measurement model in matrix notation

$$\boldsymbol{y}_i = \boldsymbol{X}_i \boldsymbol{\beta} + \boldsymbol{U}_i \boldsymbol{\gamma}_i + \boldsymbol{\varepsilon}_i \tag{7.15}$$

for individual or cluster $i = 1, \ldots, m$ with $\mathrm{E}(\boldsymbol{\varepsilon}_i) = \boldsymbol{0}$.

In the *second stage*, the model is supplemented through a distributional assumption for the random effects $\boldsymbol{\gamma}_i$, reflecting the idea that the data are generated by drawing individuals or clusters from a population. We assume that the random effects are independent and identically distributed according to a *random effects* or—from a Bayesian perspective—a *prior* distribution. In this book, we make the conventional assumption of a Gaussian random effects distribution

$$\boldsymbol{\gamma}_i \sim \mathrm{N}(\boldsymbol{0}, \boldsymbol{Q}),$$

with unknown $(q+1) \times (q+1)$-covariance matrix $\boldsymbol{Q}$. The covariance matrix may be of a specific structural form. In particular, the special case of (a priori) independent components γ_{il} of $\boldsymbol{\gamma}_i$ corresponds to a diagonal matrix $\boldsymbol{Q} = \mathrm{diag}(\tau_0^2, \ldots, \tau_q^2)$ with elements $\tau_l^2 = \mathrm{Var}(\gamma_{il})$, $l = 0, \ldots, q$. Such models are sometimes called *variance components models* and are among the least complex random effects models.

The assumption of i.i.d. $\mathrm{N}(0, \sigma^2)$ errors implies

$$\boldsymbol{\varepsilon}_i \sim \mathrm{N}(\boldsymbol{0}, \sigma^2 \boldsymbol{I}_{n_i}), \tag{7.16}$$

as well as independence of error vectors $\boldsymbol{\varepsilon}_1, \ldots, \boldsymbol{\varepsilon}_m$. Additionally, error terms and random effects are assumed to be mutually independent. Correlation between repeated observations within individuals or clusters is, however, induced by the common vector $\boldsymbol{\gamma}_i$ of random effects, as is illustrated for simple random intercept and slope models in the previous Sects. 7.1.1 and 7.1.2.

Additional correlation can be specified through

$$\boldsymbol{\varepsilon}_i \sim \mathrm{N}(\mathbf{0}, \boldsymbol{\Sigma}_i)$$

with a non-diagonal $n_i \times n_i$-covariance matrix $\boldsymbol{\Sigma}_i$. Special forms for $\boldsymbol{\Sigma}_i$ result from assumptions about the error process $\{\varepsilon_{ij}\}$. For example, the assumption of autoregressive errors leads to covariance matrices as considered in Sect. 4.1.4. If the errors follow autoregressive processes of first order, i.e.,

$$\varepsilon_{ij} = \rho\, \varepsilon_{i,j-1} + \tilde{\varepsilon}_{ij}, \qquad \tilde{\varepsilon}_{ij} \sim \mathrm{N}(0, \sigma_{\tilde{\varepsilon}}^2), \qquad 0 < \rho < 1,$$

and if observations are made at the same equidistant time points $t_1 < t_2 < \ldots < t_n$, then all covariance matrices $\boldsymbol{\Sigma}_i$ have the form (4.5) (see p. 192). More generally, the correlations $\rho^{|j-j'|}$ in Eq. (4.5) have to be replaced by $\rho^{|t_{ij}-t_{ij'}|}$, $j, j' = 1, \ldots, n_i$. In our examples, we will usually adhere to the stronger assumption (7.16) of i.i.d. errors. A summary of LMMs for longitudinal and clustered data can be found in Box 7.1.

The following example illustrates how the LMM Eq. (7.12) for the rats data is reexpressed in the general form (7.14) and (7.15).

Example 7.2 Hormone Therapy with Rats—Matrix Notation

Measurements are available for rats at the age of $a = 50, 60, \ldots, 110$ days, as long as they remain in the study. After the logarithmic transformation $t = \log(1 + (age - 45)/10)$, time corresponds to the values

$$t_1 = \log(1.5), \quad t_2 = \log(2.5), \quad \ldots, \quad t_7 = \log(7.5)\,.$$

For rat 17 from the low-dose group ($L_{17} = 1, H_{17} = 0, C_{17} = 0$), $n_{17} = 5$ measurements have been made. This results in

$$\boldsymbol{y}_{17} = \begin{pmatrix} y_{17,1} \\ \vdots \\ y_{17,5} \end{pmatrix}, \quad \boldsymbol{X}_{17} = \begin{pmatrix} 1 & \log(1.5) & 0 & 0 \\ \vdots & \vdots & \vdots & \vdots \\ 1 & \log(5.5) & 0 & 0 \end{pmatrix}, \quad \boldsymbol{U}_{17} = \begin{pmatrix} 1 & \log(1.5) \\ \vdots & \vdots \\ 1 & \log(5.5) \end{pmatrix}$$

for model (7.12). For rat 12 from the high-dose group ($L_{12} = 0, H_{12} = 1, C_{12} = 0$) with $n_{12} = 4$ measurements, we obtain

$$\boldsymbol{y}_{12} = \begin{pmatrix} y_{12,1} \\ \vdots \\ y_{12,4} \end{pmatrix}, \quad \boldsymbol{X}_{12} = \begin{pmatrix} 1 & 0 & \log(1.5) & 0 \\ \vdots & \vdots & \vdots & \vdots \\ 1 & 0 & \log(4.5) & 0 \end{pmatrix}, \quad \boldsymbol{U}_{12} = \begin{pmatrix} 1 & \log(1.5) \\ \vdots & \vdots \\ 1 & \log(4.5) \end{pmatrix},$$

and for rat 22 from the control group ($L_{22} = 0, H_{22} = 0, C_{22} = 1$),

$$\boldsymbol{y}_{22} = \begin{pmatrix} y_{22,1} \\ \vdots \\ y_{22,7} \end{pmatrix}, \quad \boldsymbol{X}_{22} = \begin{pmatrix} 1 & 0 & 0 & \log(1.5) \\ \vdots & \vdots & \vdots & \vdots \\ 1 & 0 & 0 & \log(7.5) \end{pmatrix}, \quad \boldsymbol{U}_{22} = \begin{pmatrix} 1 & \log(1.5) \\ \vdots & \vdots \\ 1 & \log(7.5) \end{pmatrix}.$$

7.1 Linear Mixed Model for Longitudinal and Clustered Data

Measurement Model

LMM for longitudinal or clustered data is given by

$$y_{ij} = \boldsymbol{x}'_{ij}\boldsymbol{\beta} + \boldsymbol{u}'_{ij}\boldsymbol{\gamma}_i + \varepsilon_{ij}$$

for individuals or clusters $i = 1, \ldots, m$ observed at occasions

$$t_{i1} < \ldots < t_{ij} < \ldots < t_{in_i},$$

or in matrix notation

$$\boldsymbol{y}_i = \boldsymbol{X}_i\boldsymbol{\beta} + \boldsymbol{U}_i\boldsymbol{\gamma}_i + \boldsymbol{\varepsilon}_i.$$

In this model $\boldsymbol{y}_i$ is the n_i-dimensional vector of responses for individual or cluster i, $\boldsymbol{X}_i$ and $\boldsymbol{U}_i$ are $n_i \times p$- and $n_i \times (q+1)$-dimensional design matrices constructed from known covariates, $\boldsymbol{\beta}$ is the p-dimensional vector of fixed effects, $\boldsymbol{\gamma}_i$ is a $(q+1)$-dimensional vector of individual- or cluster-specific effects, and $\boldsymbol{\varepsilon}_i$ is a n_i-dimensional vector of errors.

Distributional Assumptions

For $\boldsymbol{\gamma}_i$ and $\boldsymbol{\varepsilon}_i$, $i = 1, \ldots, m$, the following distributional assumptions hold:

$$\boldsymbol{\gamma}_i \sim \mathrm{N}(\boldsymbol{0}, \boldsymbol{Q}), \qquad \boldsymbol{\varepsilon}_i \sim \mathrm{N}(\boldsymbol{0}, \sigma^2 \boldsymbol{I}_{n_i}).$$

$\boldsymbol{\gamma}_1, \ldots, \boldsymbol{\gamma}_m, \boldsymbol{\varepsilon}_1, \ldots, \boldsymbol{\varepsilon}_m$ are assumed to be independent.

Software

- Functions `xtreg`, `xtmixed`, and `gllamm` of STATA; see Skrondal and Rabe-Hesketh (2008)
- Package `lme4` of R
- `proc mixed` of SAS
- Software package BayesX; see also the R interface `R2BayesX`

For model (7.13) without random slopes $\gamma_{1i} \cdot t_{ij}$, the second column needs to be deleted in the design matrices $\boldsymbol{U}_{12}$, $\boldsymbol{U}_{17}$, and $\boldsymbol{U}_{22}$.

△

7.1.4 Conditional and Marginal Formulation

The measurement model (7.15) implies the *conditional Gaussian model*

$$\boldsymbol{y}_i \mid \boldsymbol{\gamma}_i \sim \mathrm{N}(\boldsymbol{X}_i\boldsymbol{\beta} + \boldsymbol{U}_i\boldsymbol{\gamma}_i, \sigma^2\boldsymbol{I})$$

for the response vector $\boldsymbol{y}_i$, given the random effect $\boldsymbol{\gamma}_i$. From this conditional perspective, the individual- or cluster-specific effects $\boldsymbol{\gamma}_i$ are interpreted similarly as the usual regression effects, with the difference that they only apply to individual or cluster i. Rewriting the model (7.15) in the form

$$\boldsymbol{y}_i = \boldsymbol{X}_i\boldsymbol{\beta} + \boldsymbol{\varepsilon}_i^*,$$

with errors $\boldsymbol{\varepsilon}_i^* = \boldsymbol{U}_i\boldsymbol{\gamma}_i + \boldsymbol{\varepsilon}_i$, results in a linear Gaussian regression model with correlated errors

$$\boldsymbol{\varepsilon}_i^* \sim \mathrm{N}(\boldsymbol{0}, \boldsymbol{V}_i), \quad \boldsymbol{V}_i = \mathrm{Cov}(\boldsymbol{\varepsilon}_i) + \mathrm{Cov}(\boldsymbol{U}_i\boldsymbol{\gamma}_i) = \sigma^2\boldsymbol{I} + \boldsymbol{U}_i\boldsymbol{Q}\boldsymbol{U}_i'.$$

The first equality for $\boldsymbol{V}_i$ holds because $\boldsymbol{\varepsilon}_i$ and $\boldsymbol{\gamma}_i$ are assumed to be independent, and $\mathrm{Cov}(\boldsymbol{U}_i\boldsymbol{\gamma}_i) = \boldsymbol{U}_i\boldsymbol{Q}\boldsymbol{U}_i'$ follows from Theorem B.2, property 5, in Appendix B.2.

Therefore the corresponding *marginal Gaussian model* for $\boldsymbol{y}_i$ is given by

$$\boldsymbol{y}_i \sim \mathrm{N}(\boldsymbol{X}_i\boldsymbol{\beta}, \sigma^2\boldsymbol{I} + \boldsymbol{U}_i\boldsymbol{Q}\boldsymbol{U}_i').$$

This shows that interpretation of population parameters $\boldsymbol{\beta}$ in the conditional model is the same as in the marginal model. This useful property of LMMs is lost for GLMMs that follow in Sect. 7.5. The non-diagonal covariance matrix $\boldsymbol{V}_i$ induces a correlation structure between repeated measurements of individual or cluster i in the marginal model, even if the errors ε_{ij} in the conditional model are uncorrelated. For the special case of a random intercept model, where $\boldsymbol{U}_i = (1, \ldots, 1)'$ and $\boldsymbol{Q} = \tau_0^2$, we obtain

$$\boldsymbol{U}_i\boldsymbol{Q}\boldsymbol{U}_i' = \tau_0^2 \begin{pmatrix} 1 & \ldots & 1 \\ \vdots & \ddots & \vdots \\ 1 & \ldots & 1 \end{pmatrix} = \tau_0^2 \boldsymbol{J}_{n_i},$$

where $\boldsymbol{J}_{n_i}$ is an $(n_i \times n_i)$-matrix of ones. The covariance matrix $\boldsymbol{V}_i$ reduces to the *equi-covariance matrix* of the marginal form of the random intercept model, which has diagonal elements $\mathrm{Var}(\varepsilon_{ij}^*) = \sigma^2 + \tau_0^2$ and off-diagonal elements $\mathrm{Cov}(\varepsilon_{ij}^*, \varepsilon_{il}^*) = \mathrm{Var}(\gamma_i) = \tau_0^2$, $j \neq l$, implying equal correlations

$$\mathrm{Corr}(\varepsilon_{ij}^*, \varepsilon_{il}^*) = \frac{\tau_0^2}{\sigma^2 + \tau_0^2}, \qquad j \neq l,$$

for errors in cluster i. Errors from different clusters are assumed to be uncorrelated.

7.1.5 Stochastic Covariates

As in linear and generalized linear models, we have to distinguish between deterministic and stochastic covariates. In designed experiments, as in Example 7.1, covariates are typically deterministic. On the other hand, observational studies have mostly stochastic covariates, i.e., both responses $\boldsymbol{y}_i$ and covariates $\boldsymbol{x}_{ij}$, $i = 1, \ldots, m$, $j = 1, \ldots, n_i$, have to be considered as realizations of a random vector $(\boldsymbol{y}, \boldsymbol{x}')$. Our assumptions on errors and random effects in Box 7.1 are then to be understood conditionally on the covariates, as, for example,

$$\boldsymbol{\varepsilon}_i \mid \boldsymbol{X}_i \sim \mathrm{N}(\boldsymbol{0}, \sigma^2 \boldsymbol{I}_{n_i}), \quad \boldsymbol{\gamma}_i \mid \boldsymbol{X}_i \sim \mathrm{N}(\boldsymbol{0}, \boldsymbol{Q}).$$

Since the conditional distributions of $\boldsymbol{\varepsilon}_i$ and $\boldsymbol{\gamma}_i$ do not depend on $\boldsymbol{X}_i$, these two assumptions imply that the errors and random effects are independent from covariates. For errors, this is completely analogous to linear models; see Sect. 3.1. For random effects, this assumption is likely to be violated if we assume that the random effects are surrogates for omitted covariates, as the omitted covariates may well be correlated with the observed covariates. In the econometrics literature, covariates that are correlated with the errors are called endogenous.

We illustrate the problem of correlation between random effects and observed covariates with the simple random intercept model. Following Sect. 3.2.1 of Skrondal and Rabe-Hesketh (2004), we discuss the situation for the case when the correct model is

$$y_{ij} = \beta_0 + \beta_1 x_{ij} + \beta_2 w_i + \gamma_{0i} + \varepsilon_{ij},$$

and the misspecified working model is

$$y_{ij} = \beta_0 + \beta_1 x_{ij} + \tilde{\gamma}_{0i} + \varepsilon_{ij},$$

where the cluster-specific covariate w_i is omitted. Therefore, $\tilde{\gamma}_{0i} = \beta_2 w_i + \gamma_{0i}$. Assume now that w_i and x_{ij} are dependent in form of the regression model

$$w_i = \alpha_0 + \alpha_1 \bar{x}_i + u_i,$$

where $\bar{x}_i$ is the mean of the x_{ij}, $j = 1, \ldots, n_i$. Inserting w_i in the equation for $\tilde{\gamma}_{0i}$ gives

$$\tilde{\gamma}_{0i} = \beta_2 \alpha_0 + \beta_2 \alpha_1 \bar{x}_i + \beta_2 u_i + \gamma_{0i} = \delta_0 + \delta_1 \bar{x}_i + \delta_{0i}$$

after reparameterizing with $\delta_0 = \beta_2 \alpha_0$, $\delta_1 = \beta_2 \alpha_1$, and $\delta_{0i} = \beta_2 u_i + \gamma_{0i}$. Substituting these results into the misspecified model finally yields

$$y_{ij} = \beta_0 + \delta_0 + \delta_1 \bar{x}_i + \beta_1 x_{ij} + \delta_{0i} + \varepsilon_{ij}, \tag{7.17}$$

with the new fixed intercept $\beta_0 + \delta_0$ and the new random intercept δ_{0i}. Therefore, inclusion of $\bar{x}_i$ as a separate covariate retains β_1 as the required effect of x_{ij}, although the covariate w_i is missing. Omitting $\bar{x}_i$ from the model will, however, lead to a biased estimate of β_1 if $\delta_1 = \beta_2\alpha_1 \neq 0$; see also Sect. 3.4 for a discussion of the effect of missing covariates on the bias in regression models. For interpretational purposes it is often better to include the centered covariate $x_{ij} - \bar{x}_i$ rather than x_{ij} into the model. We can then estimate the extended random intercept model (7.6) discussed in Sect. 7.1.1 and interpret the effects of $\bar{x}_i$ and $x_{ij} - \bar{x}_i$ as between- and within-cluster effects, respectively. For models with several random effects, in particular random slopes, analogous results are available; see Snijders and Berkhof (2004).

In the econometrics literature often another approach to cope with the problem of correlation between random effects and observed covariates is preferred. Here, a fixed effects approach is proposed assuming fixed cluster-specific effects γ_{0i} without assuming a random effects distribution. Of course, inclusion of all γ_{0i}, $i = 1, \ldots, m$, into the regression equation is not feasible as then the design matrix is rank deficient. As a remedy we could either omit one of the γ_{0i}'s (the corresponding cluster then serves as the reference category) or omit the global intercept β_0. Omitting the global intercept we arrive at the model

$$y_{ij} = \beta_1 x_{ij} + \gamma_{0i} + \varepsilon_{ij}, \tag{7.18}$$

where the γ_{0i} are treated as fixed parameters and the cluster index has the role of a categorical covariate albeit with a large number of categories. The advantage of such a specification is that we eliminated the correlation of x_{ij} with the random effect simply with the replacement of fixed effects. Since Eq. (7.18) is a classical linear model both the regression parameter β_1 and the cluster dummies γ_{0i} can be estimated unbiasedly using ordinary least squares. However, since the cluster dummies are not regularized as in the random effects approach, they are usually estimated with considerable variance at least if the cluster sizes n_i are small (as is often the case). Moreover, the fixed effects approach uses only within-cluster information. For covariates with equal between- and within-cluster effect, a random effects approach allows more precise estimation of the effect as also the between-cluster variability is used. Finally, covariates at the cluster level are not possible in a fixed effects approach because of perfect linear dependence between the columns in the design matrix corresponding to the cluster dummies and the cluster-specific variables. Nevertheless, the approach can be useful if interest lies solely in estimating the regression coefficient β_1, while the cluster dummies are nuisance parameters. It turns out that the regression coefficient β_1 can be estimated without resorting to the γ_{0i}'s. It can be shown that the least squares estimator for β_1 can be obtained by regressing the centered responses $y_{ij} - \bar{y}_i$ on the centered covariate values $x_{ij} - \bar{x}_i$ in a model without intercept; see p. 410 in Sect. 7.8 for a derivation. This result also carries over to more than one covariate, then all covariates need to be centered by their respective cluster means.

In summary, there are two ways to deal with the problem of random intercepts that are correlated with the available covariates. If interest primarily lies in the fixed effect regression coefficients, we can simply regress the cluster-wise centered responses on the cluster-wise centered covariate values. If we are also interested in estimates for the random effects γ_{0i} and their variance τ_0^2, a random effects approach with the cluster means of the covariates and the deviations from the means as regressors is preferable. This approach also allows to estimate the between- and within-cluster effects.

7.2 General Linear Mixed Models

The LMM $\boldsymbol{y}_i = \boldsymbol{X}_i\boldsymbol{\beta} + \boldsymbol{U}_i\boldsymbol{\gamma}_i + \boldsymbol{\varepsilon}_i, i = 1, \ldots, m$, as well as multilevel extensions and further types of mixed models considered below and in subsequent chapters, can be formulated in a rather compact form by defining the vectors

$$\boldsymbol{y} = \begin{pmatrix} \boldsymbol{y}_1 \\ \vdots \\ \boldsymbol{y}_i \\ \vdots \\ \boldsymbol{y}_m \end{pmatrix}, \quad \boldsymbol{\varepsilon} = \begin{pmatrix} \boldsymbol{\varepsilon}_1 \\ \vdots \\ \boldsymbol{\varepsilon}_i \\ \vdots \\ \boldsymbol{\varepsilon}_m \end{pmatrix}, \quad \boldsymbol{\gamma} = \begin{pmatrix} \boldsymbol{\gamma}_1 \\ \vdots \\ \boldsymbol{\gamma}_i \\ \vdots \\ \boldsymbol{\gamma}_m \end{pmatrix}$$

of all responses, errors, and random effects, respectively, as well as the design matrices

$$\boldsymbol{X} = \begin{pmatrix} \boldsymbol{X}_1 \\ \vdots \\ \boldsymbol{X}_i \\ \vdots \\ \boldsymbol{X}_m \end{pmatrix} \tag{7.19}$$

and

$$\boldsymbol{U} = \text{blockdiag}(\boldsymbol{U}_1, \ldots, \boldsymbol{U}_i, \ldots, \boldsymbol{U}_m) = \begin{pmatrix} \boldsymbol{U}_1 & & & & \boldsymbol{0} \\ & \ddots & & & \\ & & \boldsymbol{U}_i & & \\ & & & \ddots & \\ \boldsymbol{0} & & & & \boldsymbol{U}_m \end{pmatrix}. \tag{7.20}$$

The LMM can then be written as

$$\boldsymbol{y} = \boldsymbol{X}\boldsymbol{\beta} + \boldsymbol{U}\boldsymbol{\gamma} + \boldsymbol{\varepsilon}$$

7.2 General Linear Mixed Model

Model

A general linear mixed model is given by

$$\boldsymbol{y} = \boldsymbol{X}\boldsymbol{\beta} + \boldsymbol{U}\boldsymbol{\gamma} + \boldsymbol{\varepsilon}$$

with

$$\begin{pmatrix} \boldsymbol{\gamma} \\ \boldsymbol{\varepsilon} \end{pmatrix} \sim \mathrm{N}\left(\begin{pmatrix} \mathbf{0} \\ \mathbf{0} \end{pmatrix}, \begin{pmatrix} \boldsymbol{G} & \mathbf{0} \\ \mathbf{0} & \boldsymbol{R} \end{pmatrix} \right).$$

In this model, $\boldsymbol{X}$ and $\boldsymbol{U}$ are design matrices, $\boldsymbol{\beta}$ is a vector of fixed effects, and $\boldsymbol{\gamma}$ is a vector of random effects. The covariance matrices for $\boldsymbol{\gamma}$ and $\boldsymbol{\varepsilon}$ are assumed to be nonsingular, and therefore positive definite, and $\boldsymbol{\gamma}$ and $\boldsymbol{\varepsilon}$ are independent.

Software

Package `lme4` of R and `proc mixed` of SAS.

with $\boldsymbol{\varepsilon} \sim \mathrm{N}(\mathbf{0}, \boldsymbol{R})$, $\boldsymbol{\gamma} \sim \mathrm{N}(\mathbf{0}, \boldsymbol{G})$, and the block diagonal covariance matrices

$$\begin{aligned} \boldsymbol{R} &= \text{blockdiag}(\sigma^2 \boldsymbol{\Sigma}_{n_1}, \ldots, \sigma^2 \boldsymbol{\Sigma}_{n_i}, \ldots, \sigma^2 \boldsymbol{\Sigma}_{n_m}) \\ \boldsymbol{G} &= \text{blockdiag}(\boldsymbol{Q}, \ldots, \boldsymbol{Q}, \ldots, \boldsymbol{Q}). \end{aligned} \tag{7.21}$$

Note that $\boldsymbol{\varepsilon}$ and $\boldsymbol{\gamma}$ are still assumed to be independent. For i.i.d. errors, $\boldsymbol{R}$ simplifies to $\boldsymbol{R} = \sigma^2 \boldsymbol{I}$.

Box 7.2 summarizes LMMs in this compact notation. Such a representation is useful for two main reasons: First, likelihood inference (Sect. 7.3) and Bayesian inference (Sect. 7.4) can be similarly formulated. Secondly, and more importantly, statistical inference remains valid if we allow general design matrices $\boldsymbol{X}$ and $\boldsymbol{U}$ that are not of the special forms (7.19) and (7.20) and further allow more general covariance matrices $\boldsymbol{G}$ and $\boldsymbol{R}$, which are not block-diagonal as in Eq. (7.21). We make primarily use of such general linear mixed models in Chaps. 8 and 9, representing non- and semiparametric regression models as mixed models. To give the reader a flavor of the capabilities of general LMMs, we briefly discuss multilevel models in the remarks that follow.

Remarks

1. *Multilevel models:* Mixed models for longitudinal or clustered data as described in Sect. 7.1 are suitable for modeling and analyzing data structures with one grouping level given by the individual or the cluster to which the observation

belongs. Multilevel models extend mixed models to data structures with more than one grouping level: Each observation may belong to several, possibly nested, groups. The classical example for this type of data structure results from school tests where pupils are grouped within classes and classes are grouped in schools. If we assume nested data, i.e., one pupil is member of exactly one class and each class belongs to exactly one school, then this leads to three-level models such as

$$y_{ijl} = \boldsymbol{x}'_{ijl}\boldsymbol{\beta} + \gamma_i + \gamma_{ij} + \varepsilon_{ijl},$$

where i indexes schools, j indexes classes, l indexes individual pupils within classes, and only random intercepts are considered. Naturally, random slopes can also be included at any level, and it is then of particular interest to compare effects of the same type of covariate on different levels. For example, the covariate "math score" can be constructed both on the class level as an average score and also on the individual level. Such three-level nested models allow for a variance decomposition on the different levels, similar to the presentation in mixed models (where it was limited to only two levels). Although the models have more than one grouping level, they can be written as a general LMM as defined in Box 7.2.

Multilevel models can also be extended to further hierarchical stages, as well as to non-nested models. Such models with more than two hierarchical levels will not be considered in this book, although they are useful in several areas of research, such as meta-analyses in clinical research or models for learning achievements. We recommend introductions provided in, for example, Gelman and Hill (2006), Skrondal and Rabe-Hesketh (2004), Scott, Simonoff, and Marx (2012), and Skrondal and Rabe-Hesketh (2008). The latter is an introduction with STATA but worth reading even if other software is used.

2. *Normality assumption:* As in LMMs for longitudinal and clustered data, the normality assumption is not necessary for all inferential results. However, estimation of unknown parameters in $\boldsymbol{G}$ and $\boldsymbol{R}$ is likelihood-based, thus we include the normality assumption in the definition.
3. *Conditional and marginal formulation:* We distinguish the conditional and the marginal model formulation. The conditional model for $\boldsymbol{y}$ given $\boldsymbol{\gamma}$ is

$$\boldsymbol{y} \,|\, \boldsymbol{\gamma} \sim \mathrm{N}(\boldsymbol{X}\boldsymbol{\beta} + \boldsymbol{U}\boldsymbol{\gamma}, \boldsymbol{R}),$$

together with the random effects distribution

$$\boldsymbol{\gamma} \sim \mathrm{N}(\boldsymbol{0}, \boldsymbol{G}).$$

The marginal model can be obtained by reexpressing the model in the form

$$\boldsymbol{y} = \boldsymbol{X}\boldsymbol{\beta} + \boldsymbol{\varepsilon}^*, \qquad \boldsymbol{\varepsilon}^* = \boldsymbol{U}\boldsymbol{\gamma} + \boldsymbol{\varepsilon}.$$

This results in the general marginal model

$$\boldsymbol{y} \sim \mathrm{N}(\boldsymbol{X}\boldsymbol{\beta}, \boldsymbol{R} + \boldsymbol{U}\boldsymbol{G}\boldsymbol{U}'), \tag{7.22}$$

or $\boldsymbol{y} \sim \mathrm{N}(\boldsymbol{X\beta}, \boldsymbol{V})$ with covariance matrix

$$\boldsymbol{V} = \boldsymbol{R} + \boldsymbol{UGU}'.$$

Note that the marginal model can be derived from the conditional model, but not vice versa. If interest only lies in estimation of fixed effects $\boldsymbol{\beta}$, the marginal model may be used. However, the conditional formulation is needed to estimate random effects $\boldsymbol{\gamma}$.

7.3 Likelihood Inference in LMMs

This section describes in Sects. 7.3.1–7.3.3 likelihood-based estimation of fixed and random effects, as well as of the unknown parameters in $\boldsymbol{G}$ and $\boldsymbol{R}$ in the general LMM of Box 7.2. Section 7.3.4 discusses tests of hypotheses. Bayesian inference is presented in Sect. 7.4.

7.3.1 Known Variance–Covariance Parameters

We first assume that all parameters in $\boldsymbol{R}$ and $\boldsymbol{G}$, and therefore in $\boldsymbol{V} = \boldsymbol{R} + \boldsymbol{UGU}'$, are known.

The unknown parameters can be estimated by simultaneously maximizing the joint log-likelihood of $\boldsymbol{y}, \boldsymbol{\gamma}$ with respect to $\boldsymbol{\beta}$ and $\boldsymbol{\gamma}$. The logarithm of $p(\boldsymbol{y}, \boldsymbol{\gamma}) = p(\boldsymbol{y} \mid \boldsymbol{\gamma})p(\boldsymbol{\gamma})$ yields the log-likelihood (up to an additive constant)

$$-\frac{1}{2}(\boldsymbol{y} - \boldsymbol{X\beta} - \boldsymbol{U\gamma})'\boldsymbol{R}^{-1}(\boldsymbol{y} - \boldsymbol{X\beta} - \boldsymbol{U\gamma}) - \frac{1}{2}\boldsymbol{\gamma}'\boldsymbol{G}^{-1}\boldsymbol{\gamma}.$$

It follows that maximization of the above expression is equivalent to minimizing

$$\mathrm{LS}_{\mathrm{pen}}(\boldsymbol{\beta}, \boldsymbol{\gamma}) = (\boldsymbol{y} - \boldsymbol{X\beta} - \boldsymbol{U\gamma})'\boldsymbol{R}^{-1}(\boldsymbol{y} - \boldsymbol{X\beta} - \boldsymbol{U\gamma}) + \boldsymbol{\gamma}'\boldsymbol{G}^{-1}\boldsymbol{\gamma}. \quad (7.23)$$

The first term corresponds to a (general) *least squares criterion*. Without the second term, $\boldsymbol{\gamma}$ would be estimated exactly as a fixed effect. The second term takes into account that $\boldsymbol{\gamma}$ is random and follows a distribution, i.e., $\boldsymbol{\gamma} \sim \mathrm{N}(\boldsymbol{0}, \boldsymbol{G})$, with positive definite covariance matrix $\boldsymbol{G}$. Additionally, $\boldsymbol{\gamma}'\boldsymbol{G}^{-1}\boldsymbol{\gamma}$ penalizes deviations of $\boldsymbol{\gamma}$ from $\boldsymbol{0}$ and is therefore called a penalty term. The penalty is large if $\boldsymbol{G}$ is "small," and it is small if $\boldsymbol{G}$ is "large." In the limiting case of $\boldsymbol{G}^{-1} \to \boldsymbol{0}$, the penalty term becomes 0, and $\boldsymbol{\gamma}$ is estimated exactly as a fixed effect without any distributional assumptions. For a random intercept model, the penalty Eq. (7.23) reduces to the form (7.7) where the quadratic penalty term shrinks parameters towards zero. See in Sect. 7.1.1 on p. 354f. for a more detailed discussion.

Setting the first derivatives of $\mathrm{LS}_{\mathrm{pen}}(\boldsymbol{\beta}, \boldsymbol{\gamma})$ to zero results in the *mixed model equations*

$$\begin{pmatrix} X'R^{-1}X & X'R^{-1}U \\ U'R^{-1}X & U'R^{-1}U + G^{-1} \end{pmatrix} \begin{pmatrix} \hat{\beta} \\ \hat{\gamma} \end{pmatrix} = \begin{pmatrix} X'R^{-1}y \\ U'R^{-1}y \end{pmatrix}. \tag{7.24}$$

Defining $C = (X, U)$ and the partitioned matrix

$$B = \begin{pmatrix} 0 & 0 \\ 0 & G^{-1} \end{pmatrix},$$

the solution of Eq. (7.24) can be written as

$$\begin{pmatrix} \hat{\beta} \\ \hat{\gamma} \end{pmatrix} = (C'R^{-1}C + B)^{-1}C'R^{-1}y. \tag{7.25}$$

This form also shows the close relationship to ridge estimation; see Sect. 4.2.2 (p. 203). A full derivation of the penalized least squares estimator Eq. (7.25) can be found in Sect. 7.8.2 on p. 412.

Some (tedious) matrix manipulations show that the estimator Eq. (7.25) has the form

$$\hat{\beta} = (X'V^{-1}X)^{-1}X'V^{-1}y \tag{7.26}$$

and

$$\hat{\gamma} = GU'V^{-1}(y - X\hat{\beta}). \tag{7.27}$$

This implies that $\hat{\beta}$ is a weighted least squares estimator with the inverse of the marginal covariance matrix $V = R + UGU'$ of β as the weight matrix.

Omitting the normality assumption, $(\hat{\beta}, \hat{\gamma})$ can also be derived as the best linear unbiased predictor (BLUP); see, for example, McCulloch and Searle (2001).

7.3.2 Unknown Variance–Covariance Parameters

This section describes the maximum likelihood (ML) and restricted maximum likelihood (REML) approach for estimating unknown parameters ϑ in R, G, or V. These estimators are the most widely used, and they are implemented in software such as SAS, STATA, and R. To make dependence on ϑ explicit, we write

$$V = V(\vartheta) = UG(\vartheta)U' + R(\vartheta).$$

Maximum Likelihood Estimation of ϑ

ML estimation of ϑ is based on the likelihood of the marginal model

$$y \sim \mathrm{N}(X\beta, V(\vartheta)).$$

The corresponding log-likelihood, up to additive constants, is

$$\log L(\boldsymbol{\beta}, \boldsymbol{\vartheta}) = l(\boldsymbol{\beta}, \boldsymbol{\vartheta}) = -\frac{1}{2}\left[\log |\boldsymbol{V}(\boldsymbol{\vartheta})| + (\boldsymbol{y} - \boldsymbol{X}\boldsymbol{\beta})'\boldsymbol{V}(\boldsymbol{\vartheta})^{-1}(\boldsymbol{y} - \boldsymbol{X}\boldsymbol{\beta})\right]. \tag{7.28}$$

The maximization of $l(\boldsymbol{\beta}, \boldsymbol{\vartheta})$ with respect to $\boldsymbol{\beta}$ (while holding $\boldsymbol{\vartheta}$ fixed) gives

$$\hat{\boldsymbol{\beta}}(\boldsymbol{\vartheta}) = (\boldsymbol{X}'\boldsymbol{V}(\boldsymbol{\vartheta})^{-1}\boldsymbol{X})^{-1}\boldsymbol{X}'\boldsymbol{V}(\boldsymbol{\vartheta})^{-1}\boldsymbol{y}.$$

Inserting $\hat{\boldsymbol{\beta}}(\boldsymbol{\vartheta})$ in $l(\boldsymbol{\beta}, \boldsymbol{\vartheta})$ results in the *profile log-likelihood*

$$l_P(\boldsymbol{\vartheta}) = -\frac{1}{2}\left[\log |\boldsymbol{V}(\boldsymbol{\vartheta})| + (\boldsymbol{y} - \boldsymbol{X}\hat{\boldsymbol{\beta}}(\boldsymbol{\vartheta}))'\boldsymbol{V}(\boldsymbol{\vartheta})^{-1}(\boldsymbol{y} - \boldsymbol{X}\hat{\boldsymbol{\beta}}(\boldsymbol{\vartheta}))\right].$$

The maximization of $l_P(\boldsymbol{\vartheta})$ with respect to $\boldsymbol{\vartheta}$ provides the ML estimator $\hat{\boldsymbol{\vartheta}}_{\text{ML}}$. Refer to Harville (1977) for explicit formulae and algorithms.

Restricted Maximum Likelihood Estimation of $\boldsymbol{\vartheta}$

Rather than using $l_P(\boldsymbol{\vartheta})$, estimation of $\boldsymbol{\vartheta}$ is often based on the *marginal or restricted log-likelihood*

$$l_R(\boldsymbol{\vartheta}) = \log\left(\int L(\boldsymbol{\beta}, \boldsymbol{\vartheta})d\boldsymbol{\beta}\right),$$

integrating out $\boldsymbol{\beta}$ from the likelihood. This marginal log-likelihood is motivated from an empirical Bayesian perspective, where $\boldsymbol{\beta}$ is assumed to be random with a flat prior $p(\boldsymbol{\beta}) \propto$ const. It can be shown that the restricted log-likelihood is

$$l_R(\boldsymbol{\vartheta}) = l_P(\boldsymbol{\vartheta}) - \frac{1}{2}\log |\boldsymbol{X}'\boldsymbol{V}(\boldsymbol{\vartheta})^{-1}\boldsymbol{X}|, \tag{7.29}$$

and maximization of $l_R(\boldsymbol{\vartheta})$ provides the restricted ML estimator $\hat{\boldsymbol{\vartheta}}_{\text{REML}}$; see Harville (1974) for details.

As discussed in Sect. 3.2.2, the REML estimator

$$\hat{\sigma}^2 = \frac{1}{n-p}\sum_{i=1}^{n}(y_i - \boldsymbol{x}_i'\hat{\boldsymbol{\beta}})^2$$

removes the bias of the ML estimator for the error variance σ^2 in linear models. Just as in this simpler context, reduction of the bias of $\hat{\boldsymbol{\vartheta}}_{\text{ML}}$ is the main reason for preferring $\hat{\boldsymbol{\vartheta}}_{\text{REML}}$ in LMMs as an estimator for $\boldsymbol{\vartheta}$. However, in our more general setting, the REML estimator is generally *not* unbiased. A rather simple example for a biased REML estimator is the random intercept model with unbalanced data where the number of observations n_i is *different* in each cluster. Moreover, there are no general results available to ensure that the MSE is also reduced compared to full maximum likelihood.

The estimators $\hat{\boldsymbol{\vartheta}}_{\text{REML}}$ and $\hat{\boldsymbol{\vartheta}}_{\text{ML}}$ are computed numerically through iterative algorithms, for example, by maximizing $l_R(\boldsymbol{\vartheta})$ or $l_P(\boldsymbol{\vartheta})$ using Newton–Raphson or Fisher scoring algorithms; see Appendix B.4.2.

Finally, plugging in $\hat{\boldsymbol{\vartheta}}$ after convergence provides the estimated covariance matrices

$$\hat{\boldsymbol{R}} = \boldsymbol{R}(\hat{\boldsymbol{\vartheta}}), \quad \hat{\boldsymbol{G}} = \boldsymbol{G}(\hat{\boldsymbol{\vartheta}}), \quad \hat{\boldsymbol{V}} = \boldsymbol{V}(\hat{\boldsymbol{\vartheta}}).$$

Estimation of Fixed and Random Effects

The final estimators $\hat{\boldsymbol{\beta}}$ and $\hat{\boldsymbol{\gamma}}$ of fixed effects $\boldsymbol{\beta}$ and random effects $\boldsymbol{\gamma}$ are obtained using the estimates Eqs. (7.26) and (7.27) after replacing the matrices $\boldsymbol{R}, \boldsymbol{G}$, and $\boldsymbol{V}$ with their estimates $\hat{\boldsymbol{R}}, \hat{\boldsymbol{G}}$, and $\hat{\boldsymbol{V}}$. However, after plugging in $\hat{\boldsymbol{\vartheta}}$, optimality properties are no longer exactly valid.

A summary of likelihood inference in LMMs is provided in Box 7.3.

Random Intercept Model

As a special case, we consider the random intercept model

$$y_{ij} = \boldsymbol{x}'_{ij}\boldsymbol{\beta} + \gamma_{0i} + \varepsilon_{ij}, \quad i = 1, \ldots, m, \quad j = 1, \ldots, n_i,$$

or, equivalently,

$$\boldsymbol{y}_i = \boldsymbol{X}_i\boldsymbol{\beta} + \boldsymbol{U}_i\gamma_{0i} + \boldsymbol{\varepsilon}_i, \quad i = 1, \ldots, m,$$

where $\boldsymbol{\varepsilon}_i \sim \text{N}(\boldsymbol{0}, \sigma^2\boldsymbol{I})$, $\gamma_{0i} \overset{i.i.d.}{\sim} \text{N}(0, \tau_0^2)$, and $\boldsymbol{U}_i = (1, \ldots, 1)'$ is a n_i-dimensional vector of ones.

Now $\boldsymbol{G} = \tau_0^2\boldsymbol{I}_m$, $\boldsymbol{R} = \sigma^2\boldsymbol{I}$, $\boldsymbol{V} = \text{blockdiag}(\boldsymbol{V}_1, \ldots, \boldsymbol{V}_m)$ with

$$\begin{aligned} \boldsymbol{V}_i &= \sigma^2\boldsymbol{I}_{n_i} + \tau_0^2\boldsymbol{J}_{n_i}, \\ \boldsymbol{V}_i^{-1} &= \tfrac{1}{\sigma^2}\left(\boldsymbol{I}_{n_i} - \tfrac{\tau_0^2}{\sigma^2+n_i\tau_0^2}\boldsymbol{J}_{n_i}\right). \end{aligned}$$

The inverse can be verified by direct multiplication. More specifically, the elements on the main diagonal of $\boldsymbol{V}_i^{-1}$ are given by

$$\frac{1}{\sigma^2(\sigma^2 + n_i\tau_0^2)},$$

and the elements above and below the main diagonal are $-\tau_0^2$. Therefore

$$\begin{aligned} \hat{\boldsymbol{\gamma}} &= \hat{\boldsymbol{G}}\boldsymbol{U}'\hat{\boldsymbol{V}}^{-1}(\boldsymbol{y} - \boldsymbol{X}\hat{\boldsymbol{\beta}}) \\ &= \hat{\tau}_0^2\,\text{blockdiag}(\boldsymbol{U}_1, \ldots, \boldsymbol{U}_m)'\,\text{blockdiag}(\hat{\boldsymbol{V}}_1^{-1}, \ldots, \hat{\boldsymbol{V}}_m^{-1})(\boldsymbol{y} - \boldsymbol{X}\hat{\boldsymbol{\beta}}). \end{aligned}$$

7.3 Likelihood Inference in LMMs

Estimation of Variance and Covariance Parameters

We denote $\boldsymbol{\vartheta}$ as the vector of unknown parameters in $\boldsymbol{R} = \boldsymbol{R}(\boldsymbol{\vartheta}), \boldsymbol{G} = \boldsymbol{G}(\boldsymbol{\vartheta})$, and $\boldsymbol{V} = \boldsymbol{V}(\boldsymbol{\vartheta})$. The ML estimator for $\boldsymbol{\vartheta}$ is obtained by maximizing the (profile) log-likelihood

$$l_P(\boldsymbol{\vartheta}) = -\frac{1}{2}\left[\log|\boldsymbol{V}(\boldsymbol{\vartheta})| + (\boldsymbol{y} - \boldsymbol{X}\hat{\boldsymbol{\beta}}(\boldsymbol{\vartheta}))'\boldsymbol{V}(\boldsymbol{\vartheta})^{-1}(\boldsymbol{y} - \boldsymbol{X}\hat{\boldsymbol{\beta}}(\boldsymbol{\vartheta}))\right]$$

with

$$\hat{\boldsymbol{\beta}}(\boldsymbol{\vartheta}) = (\boldsymbol{X}'\boldsymbol{V}(\boldsymbol{\vartheta})^{-1}\boldsymbol{X})^{-1}\boldsymbol{X}'\boldsymbol{V}(\boldsymbol{\vartheta})^{-1}\boldsymbol{y}.$$

The REML (restricted ML) estimator is obtained by maximizing the restricted log-likelihood

$$l_R(\boldsymbol{\vartheta}) = l_P(\boldsymbol{\vartheta}) - \frac{1}{2}\log|\boldsymbol{X}'\boldsymbol{V}(\boldsymbol{\vartheta})^{-1}\boldsymbol{X}|.$$

Estimation of Fixed and Random Effects

Estimators $\hat{\boldsymbol{\beta}}$ and $\hat{\boldsymbol{\gamma}}$ for fixed and random effects are

$$\begin{aligned}\hat{\boldsymbol{\beta}} &= (\boldsymbol{X}'\hat{\boldsymbol{V}}^{-1}\boldsymbol{X})^{-1}\boldsymbol{X}'\hat{\boldsymbol{V}}^{-1}\boldsymbol{y} \\ \hat{\boldsymbol{\gamma}} &= \hat{\boldsymbol{G}}\boldsymbol{U}'\hat{\boldsymbol{V}}^{-1}(\boldsymbol{y} - \boldsymbol{X}\hat{\boldsymbol{\beta}}),\end{aligned}$$

or, equivalently,

$$\begin{pmatrix}\hat{\boldsymbol{\beta}} \\ \hat{\boldsymbol{\gamma}}\end{pmatrix} = (\boldsymbol{C}'\hat{\boldsymbol{R}}^{-1}\boldsymbol{C} + \hat{\boldsymbol{B}})^{-1}\boldsymbol{C}'\hat{\boldsymbol{R}}^{-1}\boldsymbol{y},$$

with $\boldsymbol{C} = (\boldsymbol{X}, \boldsymbol{U})$ and $\hat{\boldsymbol{B}} = \begin{pmatrix}\mathbf{0} & \mathbf{0} \\ \mathbf{0} & \hat{\boldsymbol{G}}^{-1}\end{pmatrix}$. These estimators are also referred to as empirical best linear predictors (EBLP or EBLUP).

Estimated Covariance Matrix

Bayesian Covariance Matrix

$$\widehat{\text{Cov}}\begin{pmatrix}\hat{\boldsymbol{\beta}} \\ \hat{\boldsymbol{\gamma}}\end{pmatrix} = \boldsymbol{A} = (\boldsymbol{C}'\hat{\boldsymbol{R}}^{-1}\boldsymbol{C} + \hat{\boldsymbol{B}})^{-1}$$

Frequentist Covariance Matrix

$$\widehat{\text{Cov}}\left(\begin{pmatrix}\hat{\boldsymbol{\beta}} \\ \hat{\boldsymbol{\gamma}}\end{pmatrix} \mid \boldsymbol{\gamma}\right) = \boldsymbol{A}\boldsymbol{C}'\hat{\boldsymbol{R}}^{-1}\boldsymbol{C}\boldsymbol{A}$$

Table 7.2 Hormone therapy with rats: comparison of estimates based on REML and ML

		REML		ML	
	Parameter	Estimate	s.e.	Estimate	s.e.
intercept	β_0	68.606	0.327	68.607	0.326
low-dose	β_1	7.505	0.227	7.508	0.224
high-dose	β_2	6.875	0.230	6.869	0.226
control	β_3	7.318	0.284	7.315	0.279
$\text{Var}(\gamma_{0i})$	τ_0^2	3.430		3.417	
$\text{Var}(\gamma_{1i})$	τ_1^2	0.001		<0.001	
$\text{Var}(\varepsilon_{ij})$	σ^2	1.444		1.428	

Hence, the estimated random intercepts are

$$\hat{\gamma}_{0i} = \hat{\tau}_0^2 \, (1, \ldots, 1) \, \hat{\boldsymbol{V}}_i^{-1} \left(\boldsymbol{y}_i - \boldsymbol{X}_i \hat{\boldsymbol{\beta}} \right),$$

which can be simplified to

$$\hat{\gamma}_{0i} = \frac{n_i \hat{\tau}_0^2}{\hat{\sigma}^2 + n_i \hat{\tau}_0^2} \left\{ \frac{1}{n_i} \sum_{j=1}^{n_i} \left(y_{ij} - \boldsymbol{x}'_{ij} \hat{\boldsymbol{\beta}} \right) \right\} = \frac{n_i \hat{\tau}_0^2}{\hat{\sigma}^2 + n_i \hat{\tau}_0^2} e_i$$

with the average residual $e_i = \frac{1}{n_i} \sum_{j=1}^{n_i} (y_{ij} - \boldsymbol{x}'_{ij} \hat{\boldsymbol{\beta}})$. An interpretation of this result is already given in Sect. 7.1.1 on p. 354ff.

Example 7.3 Hormone Therapy with Rats—Estimated Effects

The results presented in Example 7.1 (p. 359) are obtained using REML to estimate the variance and covariance parameters of random effects. REML is also the default in most statistical software packages when estimating mixed models. Table 7.2 contrasts the REML estimates with maximum likelihood for the random slope model (7.12). Generally the results are very close. The estimates for the variance components with ML are slightly below those based on REML. This could have been expected as ML is more biased downwards towards zero compared to REML, i.e., it tends to underestimate the variances.

The estimated random effects $\hat{\gamma}_{0i}$ and $\hat{\gamma}_{1i}$ of model (7.12) are best presented graphically. A particularly useful visual representation of the random effects is found in *caterpillar plots*. These plots show the estimated random effects in increasing order together with 95 % confidence intervals; see the next section for the computation of the latter. Figure 7.5 shows such caterpillar plots for model (7.12). Clearly, the estimated random intercept parameters $\hat{\gamma}_{0i}$ show considerable variability indicating large rat-specific deviations from the average height of skull. On the other hand, the estimated random slope parameters $\hat{\gamma}_{1i}$ are all estimated close to zero (note the different scaling of the figures). This is not really a surprise as already the estimated random slope variance is essentially zero; refer to Table 7.2.

Another option to visualize the random effects distribution is to produce (nonparametric) kernel densities computed from the estimates $\hat{\gamma}_{0i}$ and $\hat{\gamma}_{1i}$, $i = 1, \ldots, 50$, of the random effects. Such plots are also helpful to investigate the normality assumption of the random

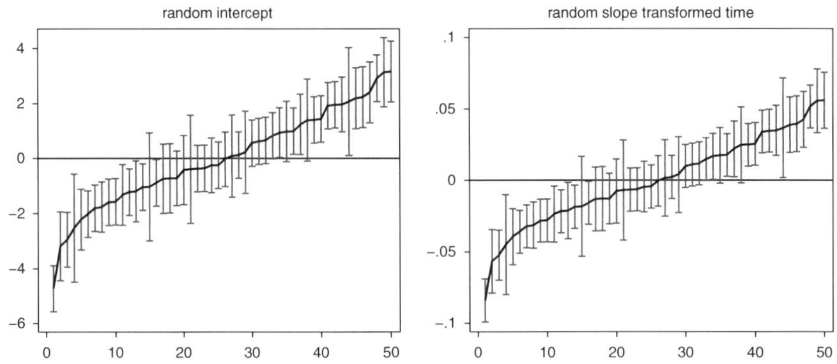

Fig. 7.5 Hormone therapy with rats: caterpillar plots of the estimated random effects $\hat{\gamma}_{0i}$ and $\hat{\gamma}_{1i}$ in the random slope model (7.12)

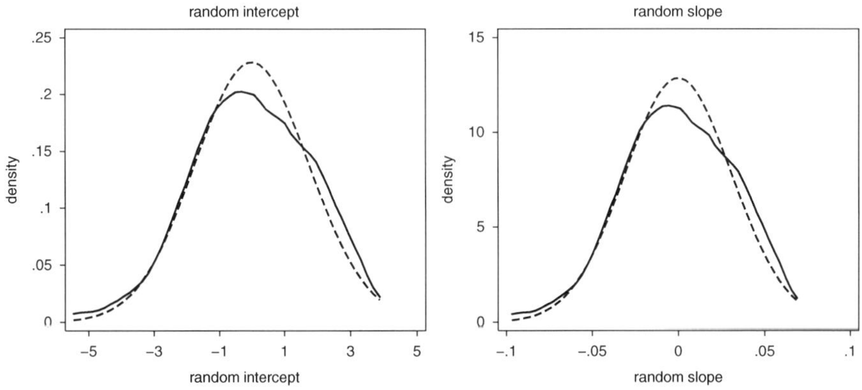

Fig. 7.6 Hormone therapy with rats: kernel density estimators (*solid line*) and normal approximation (*dashed line*) for the random effects

effects (at least to a certain extent). Figure 7.6 shows kernel densities for $\hat{\gamma}_{0i}$ and $\hat{\gamma}_{1i}$. Deviations from the assumed normal distribution are comparably small. Once again, the random slopes are close to zero.

The estimated random effects can also be used to obtain estimates for the individual (rat-specific) trends. These trends are computed by inserting the estimates into the trend lines. For instance, for the low-dose group, we obtain the estimated trends $68.606 + \hat{\gamma}_{0i} + 7.505t_{ij} + \hat{\gamma}_{1i}t_{ij}$. Plots of the individual trends are given in Fig. 7.7 which again reveals considerable rat-specific heterogeneity at least regarding the intercept. The slopes of the individual trends are almost in parallel. This is another indicator that the simple random intercept model (7.13) is sufficient. In fact, if we would superimpose the trends obtained from Eq. (7.13), they would be visually indistinguishable.

△

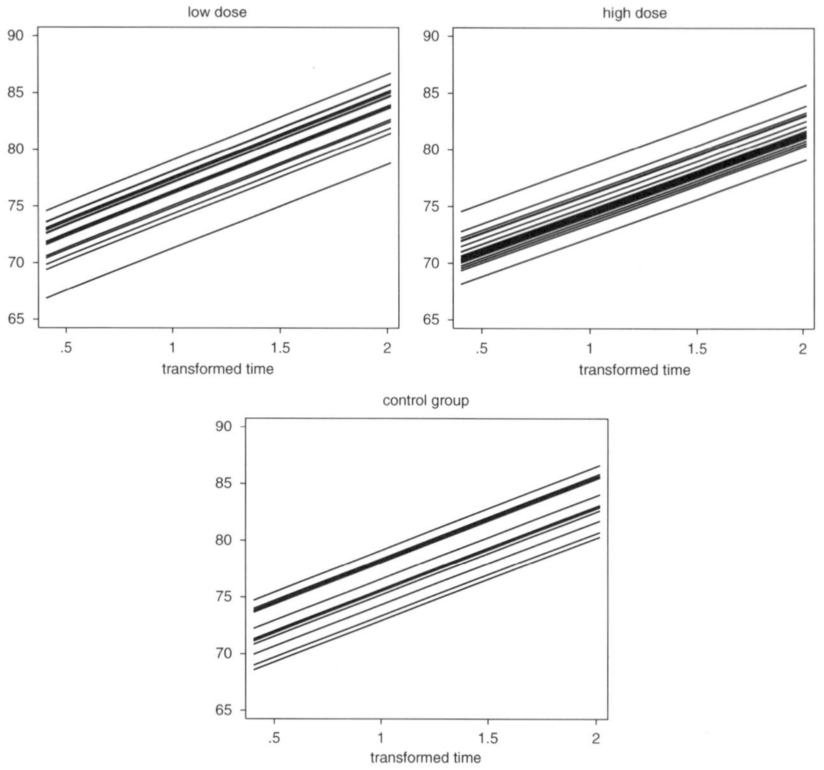

Fig. 7.7 Hormone therapy with rats: individual estimated trends in the random slope model (7.12)

7.3.3 Variability of Fixed and Random Effects Estimators

Bayesian Covariance Matrix

For known variance and covariance parameters, we can derive the covariance matrix

$$\operatorname{Cov}\begin{pmatrix}\hat{\boldsymbol{\beta}}\\ \hat{\boldsymbol{\gamma}}\end{pmatrix} = (\boldsymbol{C}'\boldsymbol{R}^{-1}\boldsymbol{C} + \boldsymbol{B})^{-1} \tag{7.30}$$

from a Bayesian perspective; see Sect. 7.4.1 (p. 384) for a derivation. Plugging in $\hat{\boldsymbol{R}}$ and $\hat{\boldsymbol{B}}$, we obtain the estimated covariance matrix

$$\widehat{\operatorname{Cov}}\begin{pmatrix}\hat{\boldsymbol{\beta}}\\ \hat{\boldsymbol{\gamma}}\end{pmatrix} = \boldsymbol{A} = (\boldsymbol{C}'\hat{\boldsymbol{R}}^{-1}\boldsymbol{C} + \hat{\boldsymbol{B}})^{-1}. \tag{7.31}$$

Note that the additional variability introduced by plugging in estimates $\hat{\boldsymbol{R}}$ and $\hat{\boldsymbol{B}}$ for $\boldsymbol{R}$ and $\boldsymbol{B}$ is not taken into account in Eq. (7.31), implying a tendency to underestimate the true covariance matrix.

To construct confidence intervals or confidence bands for $\hat{\boldsymbol{\beta}}$ and $\hat{\boldsymbol{\gamma}}$, it is assumed that $(\hat{\boldsymbol{\beta}}, \hat{\boldsymbol{\gamma}})$ has an approximate normal distribution for large sample size. For special LMMs, asymptotic results can be derived under appropriate regularity assumptions (e.g., Ruppert et al., 2003). However, generally, rigorous proofs remain open problems. Based on the normality approximation, quantiles and confidence intervals for the components of $\hat{\boldsymbol{\beta}}$ and $\hat{\boldsymbol{\gamma}}$ can be computed.

Frequentist Covariance Matrix

An alternative approximate covariance matrix can be derived by conditioning on $\boldsymbol{\gamma}$. Then $\boldsymbol{y} \mid \boldsymbol{\gamma} \sim \mathrm{N}(\boldsymbol{X\beta} + \boldsymbol{U\gamma}, \boldsymbol{R})$ and $\mathrm{Cov}(\boldsymbol{y} \mid \boldsymbol{\gamma}) = \boldsymbol{R}$. Applying the rules for linear transformations (Appendix B.2, Theorem B.2.5) to Eq. (7.25) yields

$$\mathrm{Cov}\left(\begin{pmatrix}\hat{\boldsymbol{\beta}} \\ \hat{\boldsymbol{\gamma}}\end{pmatrix} \mid \boldsymbol{\gamma}\right) = (\boldsymbol{C}'\boldsymbol{R}^{-1}\boldsymbol{C} + \boldsymbol{B})^{-1}\boldsymbol{C}'\boldsymbol{R}^{-1}\boldsymbol{C}(\boldsymbol{C}'\boldsymbol{R}^{-1}\boldsymbol{C} + \boldsymbol{B})^{-1}. \tag{7.32}$$

This covariance matrix has the form of a "sandwich" matrix, composed of the terms $(\boldsymbol{C}'\boldsymbol{R}^{-1}\boldsymbol{C} + \boldsymbol{B})^{-1}$ and $\boldsymbol{C}'\boldsymbol{R}^{-1}\boldsymbol{C}$ (see also Sects. 4.1.3, p. 190, 8.1.2, p. 439, and Sect. 8.1.9, p. 486). Inserting the estimates $\hat{\boldsymbol{R}}$ and $\hat{\boldsymbol{B}}$ provides the estimated covariance matrix

$$\widehat{\mathrm{Cov}}\left(\begin{pmatrix}\hat{\boldsymbol{\beta}} \\ \hat{\boldsymbol{\gamma}}\end{pmatrix} \mid \boldsymbol{\gamma}\right) = \boldsymbol{A}\boldsymbol{C}'\hat{\boldsymbol{R}}^{-1}\boldsymbol{C}\boldsymbol{A}, \tag{7.33}$$

where $\boldsymbol{A}$ is the Bayesian covariance matrix Eq. (7.31). In comparison to $\boldsymbol{A}$, Eq. (7.33) tends to further underestimate the true covariance matrix because the additional factor $\boldsymbol{C}'\hat{\boldsymbol{R}}^{-1}\boldsymbol{C}\boldsymbol{A}$ is "smaller" than the identity matrix $\boldsymbol{I}$. Therefore, we prefer the Bayesian covariance matrix Eq. (7.31). Motivated by its derivation, Eq. (7.33) is called the frequentist version of the covariance matrix.

Approximate Covariance Matrices for Covariance Parameters

Approximate covariance matrices $\widehat{\mathrm{Cov}}(\hat{\boldsymbol{\vartheta}})$ for the estimators $\hat{\boldsymbol{\vartheta}}_{\mathrm{ML}}$ and $\hat{\boldsymbol{\vartheta}}_{\mathrm{REML}}$ can be obtained from the Fisher information involved in Newton–Raphson or Fisher scoring iterations. However, since the estimators are skewed, particularly for variances, these approximations are relatively imprecise and are less appropriate for constructing confidence intervals or tests.

7.3.4 Testing Hypotheses

Testing Fixed Effects Parameters

Hypotheses on fixed effects are often of primary interest. The simplest case is testing

$$H_0 : \beta_j = d_j \qquad \text{versus} \qquad H_1 : \beta_j \neq d_j$$

for a component β_j of $\boldsymbol{\beta}$.

With $d_j = 0$, this is a test on the significance of the jth covariate, just as in the linear regression model. Assuming an approximate normal distribution for $\hat{\boldsymbol{\beta}}$, a decision can be based on an (approximate) confidence interval for β_j. H_0 is rejected at a significance level α if d_j is outside the corresponding $1-\alpha$ confidence interval. Equivalently, the test statistic

$$t_j = \frac{\hat{\beta}_j - d_j}{\hat{\sigma}_j}$$

can be used, where $\hat{\sigma}_j$ is the square root of the corresponding diagonal element of the approximate covariance matrix Eq. (7.31) of $\hat{\boldsymbol{\beta}}, \hat{\boldsymbol{\gamma}}$. Under normality assumptions on $\boldsymbol{\gamma}$ and $\boldsymbol{\varepsilon}$, the test statistic t_j even has an exact t-distribution (under H_0) in special cases (e.g., Ruppert et al., 2003). In general, it is assumed that t_j is approximately standard normal under the null hypothesis, and H_0 is rejected if $|t_j| > z_{1-\alpha/2}$.

To test linear hypotheses

$$H_0 : \boldsymbol{C}\boldsymbol{\beta} = \boldsymbol{d} \qquad \text{versus} \qquad H_1 : \boldsymbol{C}\boldsymbol{\beta} \neq \boldsymbol{d}$$

one may consider the Wald statistic (see Appendix B.4.4)

$$W = (\boldsymbol{C}\hat{\boldsymbol{\beta}} - \boldsymbol{d})'(\boldsymbol{C}\widehat{\text{Cov}}(\hat{\boldsymbol{\beta}})\boldsymbol{C}')^{-1}(\boldsymbol{C}\hat{\boldsymbol{\beta}} - \boldsymbol{d}),$$

where $\widehat{\text{Cov}}(\hat{\boldsymbol{\beta}}) = (\boldsymbol{X}'\hat{\boldsymbol{V}}^{-1}\boldsymbol{X})^{-1}$ is the estimated covariance matrix of $\hat{\boldsymbol{\beta}}$. However, $(\boldsymbol{X}'\hat{\boldsymbol{V}}^{-1}\boldsymbol{X})^{-1}$ might be a poor estimate of $\text{Cov}(\hat{\boldsymbol{\beta}})$ for small samples, and the asymptotic χ^2-distribution for W is lacking a general rigorous foundation. Therefore Wald tests, or F-tests with $F = W/\text{rk}(\boldsymbol{C})$ as implemented in SAS, STATA, and R, must be used with caution.

Alternatively, tests could be carried out with the likelihood ratio statistic

$$\text{LRT} = 2\{l(\hat{\boldsymbol{\beta}}, \hat{\boldsymbol{\vartheta}}) - l(\tilde{\boldsymbol{\beta}}, \tilde{\boldsymbol{\vartheta}})\}$$

or with the restricted likelihood ratio statistic

$$\text{RLRT} = 2\{l_R(\hat{\boldsymbol{\beta}}, \hat{\boldsymbol{\vartheta}}) - l_R(\tilde{\boldsymbol{\beta}}, \tilde{\boldsymbol{\vartheta}}),\}$$

where $l(\boldsymbol{\beta}, \boldsymbol{\vartheta})$ is the log-likelihood Eq. (7.28) of the marginal model and $l_R(\boldsymbol{\beta}, \boldsymbol{\vartheta}) = l(\boldsymbol{\beta}, \boldsymbol{\vartheta}) - 0.5 \log |\boldsymbol{X}'\boldsymbol{V}(\boldsymbol{\vartheta})^{-1}\boldsymbol{X}|$ is the restricted log-likelihood. Thereby $(\hat{\boldsymbol{\beta}}, \hat{\boldsymbol{\vartheta}})$ are the unrestricted ML estimates and $(\tilde{\boldsymbol{\beta}}, \tilde{\boldsymbol{\vartheta}})$ are the ML estimates under H_0. However, cautionary remarks have to be made for these tests, as outlined in Crainiceanu and Ruppert (2004).

An important example for a general linear hypothesis in the context of LMMs is the test for equality of β_1 and β_2 in the extended random intercept model

$$y_{ij} = \beta_0 + \beta_1(x_{ij} - \bar{x}_i) + \beta_2\bar{x}_i + \gamma_{0i} + \varepsilon_{ij},$$

discussed in Sect. 7.1.1. Equality of β_1 and β_2 means that the between- and within-cluster effects of x are identical. If $H_0 : \beta_1 = \beta_2$ is rejected, there are significantly different within- and between-cluster effects. The test can also be seen as a test for correlation between random effects and observed covariates or endogeneity; see the discussion in Sect. 7.1.5 on stochastic covariates. A rejection of H_0 then provides evidence of correlation between random effects and covariates or endogenous x. In the econometrics literature testing endogeneity is usually performed using the (among econometricians) well-known Hausman specification test; see Hausmann (1978). In the context of the random intercept model the Hausman test is equivalent to the Wald test described above.

Testing Random Effects or Variance Parameters

Testing hypotheses about random effects or random effects variances is more difficult and generally applicable tests are comparably scarce. Consider, for example, the random intercept model

$$y_{ij} = \beta_0 + \beta_1 x_{ij} + \gamma_{0i} + \varepsilon_{ij}, \quad i = 1, \ldots, m, \; j = 1, \ldots, n_i,$$

with $\varepsilon_{ij} \sim \mathrm{N}(0, \sigma^2)$ and $\gamma_{0i} \sim \mathrm{N}(0, \tau_0^2)$. The null hypothesis $H_0 : \tau_0^2 = 0$ is then equivalent to $H_0 : \boldsymbol{\gamma} = (\gamma_{01}, \ldots, \gamma_{0m})' = \mathbf{0}$ and the simple linear regression model

$$y_{ij} = \beta_0 + \beta_1 x_{ij} + \varepsilon_{ij}$$

without the random intercept γ_{0i}.

Extending the random intercept model to a random slope model

$$y_{ij} = \beta_0 + \beta_1 x_{ij} + \gamma_{0i} + \gamma_{1i} x_{ij} + \varepsilon_{ij}$$

with $\gamma_{1i} \sim \mathrm{N}(0, \tau_1^2)$, the null hypothesis $H_0 : \tau_1^2 = 0$ or $H_0 : \boldsymbol{\gamma}_1 = (\gamma_{11}, \ldots, \gamma_{1m})' = \mathbf{0}$ is then equivalent to the simpler random intercept model.

In SAS, an F-test is implemented for testing linear hypotheses about random effects vectors, such as $H_0 : \boldsymbol{\gamma}_0 = \mathbf{0}$ or $H_0 : \boldsymbol{\gamma}_1 = \mathbf{0}$. However, we do not encourage its use because it can be quite conservative and its power is weak. Of course, if a null hypothesis such as $\boldsymbol{\gamma}_0 = \mathbf{0}$ or $\boldsymbol{\gamma}_1 = \mathbf{0}$ is rejected, we can safely include the random effect component. On the other hand, if H_0 is not rejected, there is a considerable risk to falsely omit the random effect component.

As an alternative, (restricted) likelihood ratio tests for testing hypotheses on random effects variances, such as

$$H_0 : \tau_0^2 = 0 \qquad \text{versus} \qquad H_1 : \tau_0^2 > 0 \tag{7.34}$$

or

$$H_0 : \tau_1^2 = 0 \qquad \text{versus} \qquad H_1 : \tau_1^2 > 0 \tag{7.35}$$

should be considered, based on the LRT or, even better, the RLRT statistics. For the random intercept model, the parameters are $\boldsymbol{\beta} = (\beta_0, \beta_1)'$ and $\boldsymbol{\vartheta} = (\sigma^2, \tau_0^2)$, with $\tau_0^2 = 0$ under H_0, while β_0, β_1, and σ^2 are "nuisance parameters." For the random slope model, $\boldsymbol{\vartheta} = (\sigma^2, \tau_0^2, \tau_1^2)$, with $\tau_1^2 = 0$ under H_0, and $\beta_0, \beta_1, \sigma^2$ and τ_0^2 are "nuisance parameters."

However, standard asymptotic theory is violated for tests of this form, because $\tau_0^2 = 0$ or $\tau_1^2 = 0$ is on the boundary of the parameter space. Stram and Lee (1994) show that the asymptotic distribution of LRT and RLRT is a $0.5\chi_0^2 : 0.5\chi_1^2$ mixture distribution, where χ_0^2 denotes a point mass in zero. The latter has to be included because, with a chance of (asymptotically) 50 %, the random effects variance is estimated to be zero under the null hypothesis. Testing the variance of a random effects component based on this asymptotic mixture distribution is implemented, for example, in STATA. However, it has to be applied with caution because of the assumptions made by Stram and Lee (1994). They assume that the response vector $\boldsymbol{y}$ can be divided into subvectors $\boldsymbol{y}_i$, $i = 1, \ldots, m$, which are independent and identically distributed (in the marginal model) under both the null hypothesis and the alternative. Such a decomposition is possible for random intercept models with a balanced design $n_i = n$, i.e., with the same number of observations within each cluster or for each individual. It is not possible, however, for unbalanced designs or for testing Eq. (7.35) in random slope models in the presence of a random intercept. Moreover, it is required that the number m of clusters tends to infinity. This is an appropriate assumption for longitudinal data with a large number m of individuals, but not for clustered data with a small number of clusters, even if the numbers of observations within clusters are large. Giampaoli and Singer (2009) relax the quite restrictive assumptions of Stram and Lee (1994) and prove that the (R)LRT statistic for testing Eq. (7.34) still has an asymptotic $0.5\chi_0^2 : 0.5\chi_1^2$ distribution, without requiring balanced data or large m. They also consider the hypothesis Eq. (7.35) for testing one random slope in the presence of a random intercept. In this case, however, the asymptotic distribution of (R)LRT is a $0.5\chi_1^2 : 0.5\chi_2^2$ mixture. It seems tempting to extend this result to models with more random slope components (as was done in Stram and Lee (1994) under the stronger assumptions discussed above), but no corresponding results are available.

Crainiceanu and Ruppert (2004a) and Crainiceanu, Ruppert, Claeskens, and Wand (2005) have investigated the finite sample and asymptotic distributions of the (R)LRT statistic under various scenarios and found that finite sample deviations from the $0.5\chi_0^2 : 0.5\chi_1^2$ mixture can be considerable.

However, Crainiceanu and Ruppert (2004a) also present representations of the finite sample distribution of (R)LRT tests that allow to construct exact tests, based on simulations, for zero variances in LMMs with one variance component. Scheipl, Greven, and Küchenhoff (2008) compare (R)LRT tests and F-type tests in a massive simulation study, concluding that RLRT tests are superior to F-tests. They include a fast and reliable approximation of the exact RLRT test and an extension to LMMs with several variance components, implemented in the R package `RLRsim`. It is recommended to apply it at least in addition to the tests based on asymptotic χ^2-mixtures.

We summarize as follows: For testing linear hypotheses on fixed effects, the same toolkit of test statistics (t-, F-, Wald, and (restricted) likelihood ratio test statistics) is available as for linear and generalized linear models. Conclusions should be drawn with more caution, however, in particular for more complex models involving many parameters and random effects. The main reason is a lack of rigorous asymptotic justification, caused by correlation of observations induced by random effects.

In comparison, reliable tests to select or deselect random effects components are more scarce. This is not surprising because these tests are faced with non-standard problems, not covered by classical testing theory. For random intercept models, the $0.5\chi_0^2 : 0.5\chi_1^2$ mixture test is a sound asymptotic approximation. In addition, the test based on the exact finite sample distribution in the R package `RLRsim` should be applied. For models with one (scalar) additional random slope, the $0.5\chi_1^2 : 0.5\chi_2^2$ mixture test is still justified asymptotically. In all other cases, with more random slopes, only the test offered in `RLRsim` seems to be a reliable approximation. A good strategy is always to compare results with and without certain random slopes.

We finally note that there are some recent Bayesian approaches for random effects selection; see, e.g., Chen and Dunson (2003) and Frühwirth-Schnatter and Tüchler (2008).

7.4 Bayesian Linear Mixed Models

From a Bayesian perspective, both the population effects $\boldsymbol{\beta}$ and the random effects $\boldsymbol{\gamma}$ in a LMM

$$y = X\boldsymbol{\beta} + U\boldsymbol{\gamma} + \boldsymbol{\varepsilon}$$

are considered to be random variables. The priors for $\boldsymbol{\beta}$ and $\boldsymbol{\gamma}$ are different, however. A *noninformative* prior

$$p(\boldsymbol{\beta}) \propto const$$

corresponds to the frequentist approach, where $\boldsymbol{\beta}$ is assumed to be fixed but (completely) unknown. A more informative prior is, for example, a Gaussian prior

$$\boldsymbol{\beta} \sim \mathrm{N}(\boldsymbol{m}, \boldsymbol{M}),$$

where $\boldsymbol{m}$ and $\boldsymbol{M}$ are assumed to be known. The noninformative, flat prior $p(\boldsymbol{\beta}) \propto const$ is obtained for the limiting case $\boldsymbol{M}^{-1} \to \mathbf{0}$, i.e., when the precision matrix tends to zero. The choice of a Gaussian prior is also motivated by the fact that, as in linear regression models, posterior inference is relatively straightforward. In a Bayesian approach, the random effects distribution

$$\boldsymbol{\gamma} \sim \mathrm{N}(\mathbf{0}, \boldsymbol{G})$$

in the general LMM of Sect. 7.2 is now seen as a prior for the random effects. However, the covariance matrices $\boldsymbol{G}$ and $\boldsymbol{R}$ will usually contain a vector $\boldsymbol{\vartheta}$ of unknown hyperparameters. To make explicit dependence on these hyperparameters, we write

$$\boldsymbol{\gamma} \sim \mathrm{N}(\mathbf{0}, \boldsymbol{G}(\boldsymbol{\vartheta})), \quad \boldsymbol{\varepsilon} \sim \mathrm{N}(\mathbf{0}, \boldsymbol{R}(\boldsymbol{\vartheta})).$$

In addition, $\boldsymbol{\beta}$, $\boldsymbol{\gamma}$, and $\boldsymbol{\varepsilon}$ are assumed to be independent a priori.

Two different concepts exist for Bayesian inference, including the parameter $\boldsymbol{\vartheta}$: In a *fully Bayesian* approach, $\boldsymbol{\vartheta}$ is also considered as a random variable with a hyperprior $p(\boldsymbol{\vartheta})$. In contrast, $\boldsymbol{\vartheta}$ is considered as unknown, but fixed, in an *empirical Bayes* approach and is estimated, for example, through maximization of the (marginal) likelihood. The estimate $\hat{\boldsymbol{\vartheta}}$ is then inserted into $\boldsymbol{G}(\boldsymbol{\vartheta})$ and $\boldsymbol{R}(\boldsymbol{\vartheta})$, and Bayesian inference for $\boldsymbol{\beta}$ and $\boldsymbol{\gamma}$ proceeds as with "known" covariance matrices $\hat{\boldsymbol{G}} = \boldsymbol{G}(\hat{\boldsymbol{\vartheta}})$ and $\hat{\boldsymbol{R}} = \boldsymbol{R}(\hat{\boldsymbol{\vartheta}})$.

7.4.1 Estimation for Known Covariance Structure

We first consider the *posterior distribution* of $\boldsymbol{\beta}$ and $\boldsymbol{\gamma}$ for known (or given) covariance matrices $\boldsymbol{G}$ and $\boldsymbol{R}$, required in both Bayesian approaches. Applying Bayes' theorem, the *joint posterior* is determined through

$$\begin{aligned} p(\boldsymbol{\beta}, \boldsymbol{\gamma} \mid \boldsymbol{y}) &\propto p(\boldsymbol{y} \mid \boldsymbol{\beta}, \boldsymbol{\gamma})\, p(\boldsymbol{\beta})\, p(\boldsymbol{\gamma}) \\ &\propto \exp\left(-\frac{1}{2}(\boldsymbol{y} - \boldsymbol{X}\boldsymbol{\beta} - \boldsymbol{U}\boldsymbol{\gamma})'\boldsymbol{R}^{-1}(\boldsymbol{y} - \boldsymbol{X}\boldsymbol{\beta} - \boldsymbol{U}\boldsymbol{\gamma})\right) \\ &\quad \cdot \exp\left(-\frac{1}{2}(\boldsymbol{\beta} - \boldsymbol{m})'\boldsymbol{M}^{-1}(\boldsymbol{\beta} - \boldsymbol{m})\right) \cdot \exp\left(-\frac{1}{2}\boldsymbol{\gamma}'\boldsymbol{G}^{-1}\boldsymbol{\gamma}\right). \end{aligned}$$

Defining

$$\boldsymbol{C} = (\boldsymbol{X}, \boldsymbol{U}), \quad \boldsymbol{B} = \begin{pmatrix} \boldsymbol{M}^{-1} & \mathbf{0} \\ \mathbf{0} & \boldsymbol{G}^{-1} \end{pmatrix}, \quad \tilde{\boldsymbol{m}} = \begin{pmatrix} \boldsymbol{M}^{-1}\boldsymbol{m} \\ \mathbf{0} \end{pmatrix},$$

algebraic manipulations (similar as for Bayesian linear models in Sect. 4.4) show that the posterior is a multivariate Gaussian distribution with mean vector

$$\begin{pmatrix}\hat{\boldsymbol\beta}\\ \hat{\boldsymbol\gamma}\end{pmatrix} = \mathrm{E}\left(\begin{pmatrix}\boldsymbol\beta\\ \boldsymbol\gamma\end{pmatrix} \mid \boldsymbol y\right) = (\boldsymbol C'\boldsymbol R^{-1}\boldsymbol C + \boldsymbol B)^{-1}(\tilde{\boldsymbol m} + \boldsymbol C'\boldsymbol R^{-1}\boldsymbol y) \tag{7.36}$$

and covariance matrix

$$\mathrm{Cov}\left(\begin{pmatrix}\boldsymbol\beta\\ \boldsymbol\gamma\end{pmatrix} \mid \boldsymbol y\right) = (\boldsymbol C'\boldsymbol R^{-1}\boldsymbol C + \boldsymbol B)^{-1}.$$

For the case of a noninformative prior $p(\boldsymbol\beta) \propto const$, we can simply set $\boldsymbol M^{-1} = \boldsymbol 0$, resulting in

$$\boldsymbol B = \begin{pmatrix}\boldsymbol 0 & \boldsymbol 0\\ \boldsymbol 0 & \boldsymbol G^{-1}\end{pmatrix}, \quad \tilde{\boldsymbol m} = \begin{pmatrix}\boldsymbol 0\\ \boldsymbol 0\end{pmatrix}.$$

The posterior mean then coincides with the BLUP estimator (7.25), while the posterior covariance matrix is the Bayesian covariance matrix (7.30).

7.4.2 Estimation for Unknown Covariance Structure

Empirical Bayes Estimation

In an empirical Bayes approach the covariance parameters $\boldsymbol\vartheta$ are considered as unknown, but fixed, and are estimated by maximizing the (marginal) likelihood

$$p(\boldsymbol y \mid \boldsymbol\vartheta) \propto \int p(\boldsymbol y \mid \boldsymbol\beta, \boldsymbol\gamma, \boldsymbol\vartheta)\, p(\boldsymbol\gamma)\, p(\boldsymbol\beta)\, d\boldsymbol\beta\, d\boldsymbol\gamma \tag{7.37}$$

with respect to $\boldsymbol\vartheta$. For a noninformative prior $p(\boldsymbol\beta) \propto const$, it can be shown that $\log p(\boldsymbol y \mid \boldsymbol\vartheta) = l_R(\boldsymbol\vartheta)$, i.e., the logarithm of the marginal likelihood coincides with the restricted log-likelihood (7.29). Thus, for a noninformative prior for $\boldsymbol\beta$, the empirical Bayes estimator of $\boldsymbol\vartheta$ is identical with the REML estimator $\hat{\boldsymbol\vartheta}_{\text{REML}}$.

The empirical Bayes estimator for $\boldsymbol\beta$ and $\boldsymbol\gamma$ results if the covariance matrices $\boldsymbol G$ and $\boldsymbol R$ in Eq. (7.36) are replaced by their estimates $\hat{\boldsymbol G}$ and $\hat{\boldsymbol R}$. For a noninformative prior for $\boldsymbol\beta$, the empirical Bayes estimator is identical with the empirical best linear predictor $(\hat{\boldsymbol\beta}, \hat{\boldsymbol\gamma})$ of Box 7.3.

Fully Bayesian Estimation

In a fully Bayesian approach, a prior $p(\boldsymbol\vartheta)$ must be specified for the unknown parameters $\boldsymbol\vartheta$ in $\boldsymbol G = \boldsymbol G(\boldsymbol\vartheta)$ and $\boldsymbol R = \boldsymbol R(\boldsymbol\vartheta)$. Inference is then based on the posterior

$$p(\boldsymbol\beta, \boldsymbol\gamma, \boldsymbol\vartheta \mid \boldsymbol y) \propto p(\boldsymbol y \mid \boldsymbol\beta, \boldsymbol\gamma, \boldsymbol\vartheta)\, p(\boldsymbol\beta)\, p(\boldsymbol\gamma \mid \boldsymbol\vartheta)\, p(\boldsymbol\vartheta),$$

with $\boldsymbol\beta, \boldsymbol\gamma$ and $\boldsymbol\vartheta$ assumed to be independent. To ensure that $p(\boldsymbol\beta, \boldsymbol\gamma, \boldsymbol\vartheta \mid \boldsymbol y)$ is a proper posterior density,

$$p(\boldsymbol{y}) = \int p(\boldsymbol{y} \mid \boldsymbol{\beta}, \boldsymbol{\gamma}, \boldsymbol{\vartheta})\, p(\boldsymbol{\beta})\, p(\boldsymbol{\gamma} \mid \boldsymbol{\vartheta})\, p(\boldsymbol{\vartheta})\, d\boldsymbol{\beta}\, d\boldsymbol{\gamma}\, d\boldsymbol{\vartheta} < \infty \quad (7.38)$$

must hold for the marginal density in the denominator of Bayes' theorem. If $p(\boldsymbol{\vartheta})$ is a proper, informative prior, i.e., $\int p(\boldsymbol{\vartheta})\, d\boldsymbol{\vartheta} = 1$, then $p(\boldsymbol{\beta}, \boldsymbol{\gamma}, \boldsymbol{\vartheta} \mid \boldsymbol{y})$ is also proper. For an improper prior with $\int p(\boldsymbol{\vartheta})\, d\boldsymbol{\vartheta} = \infty$, the posterior is not generally guaranteed to be proper.

In general, the integral in Eq. (7.38) cannot be solved analytically, so that the posterior $p(\boldsymbol{\beta}, \boldsymbol{\gamma}, \boldsymbol{\vartheta} \mid \boldsymbol{y})$ is not available in closed form. Therefore, fully Bayesian inference typically relies on MCMC simulation, with the entire vector $(\boldsymbol{\beta}, \boldsymbol{\gamma}, \boldsymbol{\vartheta})$ of parameters partitioned into subvectors. As long as no specific covariance structures are assumed, these subvectors (or blocks) are $\boldsymbol{\beta}, \boldsymbol{\gamma}$, and $\boldsymbol{\vartheta}$. The full conditional distributions of each subvector, given the data $\boldsymbol{y}$ and the rest of parameters, then have to be derived so that random numbers can be drawn sequentially; see Appendix B.5.

Similarly to the classical linear model in Sect. 4.4, it can be shown that the full conditionals for $\boldsymbol{\beta}$ and $\boldsymbol{\gamma}$ are multivariate Gaussian. For $\boldsymbol{\beta}$, we obtain $\boldsymbol{\beta} \mid \cdot \sim \mathrm{N}(\boldsymbol{\mu}_\beta, \boldsymbol{\Sigma}_\beta)$ with

$$\boldsymbol{\mu}_\beta = (\boldsymbol{X}'\boldsymbol{R}(\boldsymbol{\vartheta})^{-1}\boldsymbol{X} + \boldsymbol{M}^{-1})^{-1}(\boldsymbol{M}^{-1}\boldsymbol{m} + \boldsymbol{X}'\boldsymbol{R}(\boldsymbol{\vartheta})^{-1}(\boldsymbol{y} - \boldsymbol{U}\boldsymbol{\gamma})) \quad (7.39)$$

$$\boldsymbol{\Sigma}_\beta = (\boldsymbol{X}'\boldsymbol{R}(\boldsymbol{\vartheta})^{-1}\boldsymbol{X} + \boldsymbol{M}^{-1})^{-1}. \quad (7.40)$$

The full conditional for $\boldsymbol{\gamma}$ is given by $\boldsymbol{\gamma} \mid \cdot \sim \mathrm{N}(\boldsymbol{\mu}_\gamma, \boldsymbol{\Sigma}_\gamma)$ with

$$\boldsymbol{\mu}_\gamma = (\boldsymbol{U}'\boldsymbol{R}(\boldsymbol{\vartheta})^{-1}\boldsymbol{U} + \boldsymbol{G}^{-1})^{-1}(\boldsymbol{U}'\boldsymbol{R}(\boldsymbol{\vartheta})^{-1}(\boldsymbol{y} - \boldsymbol{X}\boldsymbol{\beta}))$$

$$\boldsymbol{\Sigma}_\gamma = (\boldsymbol{U}'\boldsymbol{R}(\boldsymbol{\vartheta})^{-1}\boldsymbol{U} + \boldsymbol{G}^{-1})^{-1}.$$

With a noninformative prior $p(\boldsymbol{\beta}) \propto \text{const}$, we have $\boldsymbol{M}^{-1} = \boldsymbol{0}$ and the conditional mean

$$\boldsymbol{\mu}_\beta = (\boldsymbol{X}'\boldsymbol{R}(\boldsymbol{\vartheta})^{-1}\boldsymbol{X})^{-1}\boldsymbol{X}'\boldsymbol{R}(\boldsymbol{\vartheta})^{-1}(\boldsymbol{y} - \boldsymbol{U}\boldsymbol{\gamma})$$

is the weighted least squares estimator applied to $\boldsymbol{y} - \boldsymbol{U}\boldsymbol{\gamma}$, i.e., the data $\boldsymbol{y}$ adjusted for the random effects $\boldsymbol{U}\boldsymbol{\gamma}$. In analogy, $\boldsymbol{\mu}_\gamma$ is a Bayes estimator with prior $\boldsymbol{\gamma} \sim \mathrm{N}(\boldsymbol{0}, \boldsymbol{G}(\boldsymbol{\vartheta}))$, applied to $\boldsymbol{y} - \boldsymbol{X}\boldsymbol{\beta}$.

The full conditional $p(\boldsymbol{\vartheta} \mid \cdot)$ can be simplified for special LMMs with appropriately specified prior $p(\boldsymbol{\vartheta})$. As an example, we consider the Bayesian LMM for longitudinal or clustered data

$$y_{ij} = \boldsymbol{x}_{ij}'\boldsymbol{\beta} + \boldsymbol{u}_{ij}'\boldsymbol{\gamma}_i + \varepsilon_{ij}, \qquad i = 1, \ldots, m, \; j = 1, \ldots, n_i,$$

with i.i.d. errors ε_{ij}. It follows that $\mathrm{Cov}(\boldsymbol{\varepsilon}_i) = \sigma^2\boldsymbol{I}_{n_i}$, $\boldsymbol{R} = \sigma^2\boldsymbol{I}$, $\mathrm{Cov}(\boldsymbol{\gamma}_i) = \boldsymbol{Q}$, $\boldsymbol{G} = \mathrm{blockdiag}(\boldsymbol{Q}, \ldots, \boldsymbol{Q})$, and

$$X = \begin{pmatrix} X_1 \\ \vdots \\ X_i \\ \vdots \\ X_m \end{pmatrix} \qquad U = \begin{pmatrix} U_1 & & & & \mathbf{0} \\ & \ddots & & & \\ & & U_i & & \\ & & & \ddots & \\ \mathbf{0} & & & & U_m \end{pmatrix},$$

see Sect. 7.2.

The vector $\boldsymbol{\vartheta}$ consists of both σ^2 and the elements of $\boldsymbol{Q}$, without more specific assumptions. The noninformative (but improper) Jeffreys' prior

$$p(\sigma^2) \propto \sigma^{-2}, \quad p(\boldsymbol{Q}) \propto |\boldsymbol{Q}|^{-\frac{q+2}{2}},$$

with $q + 1 = \dim(\boldsymbol{\gamma}_i)$, can lead to improper posteriors such that MCMC techniques also break down. Therefore, a weakly informative inverse gamma prior $\sigma^2 \sim \text{IG}(a, b)$ is commonly proposed with small values a and b. Note that meaningful values for a and b depend on the magnitude of the response observations. We have good experience with $a = b = 0.001$ if the responses are divided by their standard deviation prior to estimation. For $\boldsymbol{Q}$, an inverse Wishart distribution may be specified as a prior, generalizing an inverse gamma distribution to the multivariate case. However, our experience shows, that a full Wishart prior works in practice only for two or three random effects (if at all). To simplify, we further assume that $\boldsymbol{Q}$ is a diagonal matrix with elements τ_r^2, $r = 0, \ldots, q$. The inverse Wishart distribution then factorizes into the product of inverse gamma priors

$$\tau_r^2 \sim \text{IG}(a_r, b_r), \quad r = 0, \ldots, q.$$

We therefore implicitly assume that the components γ_{ir} in $\boldsymbol{\gamma}_i$ are conditionally independent a priori with priors

$$\gamma_{ir} \mid \tau_r^2 \sim \text{N}(0, \tau_r^2), \qquad r = 0, \ldots, q.$$

The full conditionals then become $\sigma^2 \mid \cdot \sim \text{IG}(\tilde{a}, \tilde{b})$ and $\tau_r^2 \mid \cdot \sim \text{IG}(\tilde{a}_r, \tilde{b}_r)$, $r = 0, \ldots, q$, with hyperparameters provided in Box 7.4. Note that the full conditional for $\boldsymbol{\gamma}$ in the general LMM considerably simplifies due to the independence assumptions on the γ_{ir}. In fact, the γ_{ir} can be updated independently by sampling from univariate normal densities. The resulting Gibbs sampler is provided in Box 7.4. Note that this sampler is derived under special prior assumptions, in particular independent inverse gamma priors, for the variances of the random effects priors. The latter assumption is sometimes criticized, and alternatives are suggested; see Gelman (2006). When using inverse gamma priors it is important to investigate how sensitive results are with respect to the values a_r and b_r of the hyperparameters; see the following Example 7.4.

7.4 Gibbs Sampler for Longitudinal and Clustered Data

Model

$$y_{ij} \,|\, \boldsymbol{\beta}, \boldsymbol{\gamma}_i, \sigma^2 \sim \mathrm{N}(\boldsymbol{x}_{ij}'\boldsymbol{\beta} + \boldsymbol{u}_{ij}'\boldsymbol{\gamma}_i, \sigma^2), \qquad i = 1, \ldots, m, \; j = 1, \ldots, n_i.$$

Priors

- $\boldsymbol{\beta} \sim \mathrm{N}(\boldsymbol{m}, \boldsymbol{M})$, where $\boldsymbol{m}$ and $\boldsymbol{M}$ are assumed to be known.
- The random effects parameters γ_{ir} are conditionally independent with $\gamma_{ir} \,|\, \tau_r^2 \sim \mathrm{N}(0, \tau_r^2)$, $i = 1, \ldots, m$, $r = 0, \ldots, q$.
- The variance parameters are independent inverse gamma distributed, i.e., $\sigma^2 \sim \mathrm{IG}(a, b)$ and $\tau_r^2 \sim \mathrm{IG}(a_r, b_r)$, $r = 0, \ldots, q$.

Gibbs Sampler

1. Initialization: Specify initial values $\boldsymbol{\beta}^{(0)}, \boldsymbol{\gamma}_i^{(0)}$, $i = 1, \ldots, m$, $(\sigma^2)^{(0)}$, $(\tau_r^2)^{(0)}$, $r = 0, \ldots, q$ and the number of iterations T. Set $t = 1$.
2. Sample $\boldsymbol{\beta}^{(t)}$ from $\mathrm{N}(\boldsymbol{\mu}_\beta, \boldsymbol{\Sigma}_\beta)$ with

$$\boldsymbol{\Sigma}_\beta = (\frac{1}{\sigma^2}\boldsymbol{X}'\boldsymbol{X} + \boldsymbol{M}^{-1})^{-1}, \qquad \boldsymbol{\mu}_\beta = \boldsymbol{\Sigma}_\beta(\boldsymbol{M}^{-1}\boldsymbol{m} + \frac{1}{\sigma^2}\boldsymbol{X}'(\boldsymbol{y} - \boldsymbol{U}\boldsymbol{\gamma})).$$

   Use $\boldsymbol{\gamma} = \boldsymbol{\gamma}^{(t-1)}, \sigma^2 = (\sigma^2)^{(t-1)}$ in $\boldsymbol{\mu}_\beta$ and $\boldsymbol{\Sigma}_\beta$.
3. For $r = 0, \ldots, q$ sample $\gamma_{ir}^{(t)}$ from $\mathrm{N}(\mu_{\gamma_{ir}}, \sigma_{\gamma_{ir}}^2)$ with

$$\sigma_{\gamma_{ir}}^2 = \sigma^2 \left(\sum_{j=1}^{n_i} u_{ijr}^2 + \sigma^2/\tau_r^2 \right)^{-1}, \qquad \mu_{\gamma_{ir}} = \frac{\sigma_{\gamma_{ir}}^2}{\sigma^2} \sum_{j=1}^{n_i} u_{ijr}(y_{ij} - \boldsymbol{x}_{ij}'\boldsymbol{\beta}).$$

   Use $\boldsymbol{\beta} = \boldsymbol{\beta}^{(t)}, \sigma^2 = (\sigma^2)^{(t-1)}, (\tau_r^2)^{(t-1)}$ in $\mu_{\gamma_{ir}}$ and $\sigma_{\gamma_{ir}}^2$.
4. Sample $(\sigma^2)^{(t)}$ from $\mathrm{IG}(\tilde{a}, \tilde{b})$ with

$$\tilde{a} = a + \frac{1}{2}, \qquad \tilde{b} = b + \frac{1}{2}(\boldsymbol{y} - \boldsymbol{X}\boldsymbol{\beta} - \boldsymbol{U}\boldsymbol{\gamma})'(\boldsymbol{y} - \boldsymbol{X}\boldsymbol{\beta} - \boldsymbol{U}\boldsymbol{\gamma}).$$

   Use $\boldsymbol{\beta} = \boldsymbol{\beta}^{(t)}, \boldsymbol{\gamma} = \boldsymbol{\gamma}^{(t)}$ in $\tilde{a}$ and $\tilde{b}$.
5. For $r = 0, \ldots, q$ sample $(\tau_r^2)^{(t)}$ from $\mathrm{IG}(\tilde{a}_r, \tilde{b}_r)$ with

$$\tilde{a}_r = a_r + \frac{m}{2}, \qquad \tilde{b}_r = b_r + \frac{1}{2}\sum_{i=1}^{m} \gamma_{ir}^2.$$

   Use $\gamma_{ir} = \gamma_{ir}^{(t)}$ in $\tilde{a}_r$ and $\tilde{b}_r$.
6. Stop if $t = T$, otherwise set $t = t + 1$ and go to step 2.

Table 7.3 Hormone therapy with rats: estimated fixed effects in a fully Bayesian approach

Variable	Coefficient	Standard deviation	2.5 % quantile	97.5 % quantile
intercept	68.610	0.343	67.895	69.272
low-dose	7.500	0.227	7.066	7.953
high-dose	6.862	0.235	6.402	7.307
control	7.318	0.284	6.764	7.902

Example 7.4 Hormone Therapy with Rats—Bayesian Linear Mixed Model

We reanalyze the random intercept model (7.13) in Example 7.1, specifying inverse gamma priors for σ^2 and $\tau_0^2 = \text{Var}(\gamma_{0i})$ with $a = a_0 = b = b_0 = 0.001$. Table 7.3 shows estimates corresponding to Table 7.1 and, additionally, 2.5 %- and 97.5 %-quantiles. The results are obtained with the software package BayesX and are based on 102,000 iterations and using every 100th sampled parameter for estimation after the burn in period of 2,000 iterations. The estimates are in good agreement with corresponding ML estimates. The posterior mean of the random effects variance is 3.674 and hence also close to the likelihood-based estimator. A sensitivity analysis regarding the choice of the hyperparameters a_0 and b_0 with alternative choices $a_0 = b_0 = 0.01$ and $a_0 = b_0 = 0.0001$ showed no substantially different results.

△

7.5 Generalized Linear Mixed Models

Similar to linear models, the linear predictor in generalized linear models will be extended by including random effects. This leads to GLMMs, generalizing LMMs to models for non-Gaussian responses that are, for example, binary, discrete, or nonnegative. As expected, statistical inference becomes technically more involved when moving from LMMs to GLMMs.

7.5.1 GLMMs for Longitudinal and Clustered Data

Recall the definitions in Sect. 5.5 (p. 304). The (conditional) density of y_i, given the linear predictor $\eta_i = \boldsymbol{x}_i'\boldsymbol{\beta}$, belongs to an exponential family (e.g., has a binomial, Poisson, or gamma distribution) and the linear predictor is related to the mean $\mu_i = \text{E}(y_i)$ through

$$\mu_i = h(\eta_i) \quad \text{and} \quad \eta_i = g(\mu_i).$$

To define GLMMs, we extend the linear predictor $\eta_i = \boldsymbol{x}_i'\boldsymbol{\beta}$ of a GLM by adding random effects.

The responses y_{ij}, $i = 1, \ldots, m$, $j = 1, \ldots, n_i$, with n_i repeated measurements per individual or cluster, may now be, for example, binary, or count variables. Including individual- or cluster-specific random effects $\boldsymbol{\gamma}_i$, responses y_{ij} given $\boldsymbol{\gamma}_i$ are then (conditionally) independent and could be, e.g., binomial or Poisson distributed. The conditional mean $\mu_{ij} = \mathrm{E}(y_{ij} \,|\, \boldsymbol{\gamma}_i)$ is related to a linear mixed predictor of the same form as for the LMM, i.e.,

$$\eta_{ij} = \boldsymbol{x}'_{ij}\boldsymbol{\beta} + \boldsymbol{u}'_{ij}\boldsymbol{\gamma}_i\,, \qquad i = 1, \ldots, m, \quad j = 1, \ldots, n_i,$$

through the relation $\mu_{ij} = h(\eta_{ij})$, with a suitable response function h.

As in the LMM, $\boldsymbol{u}_{ij}$ is usually a subvector of $\boldsymbol{x}_{ij}$. In particular, the choice $\boldsymbol{u}_{ij} \equiv 1$ defines a random intercept model

$$\eta_{ij} = \boldsymbol{x}'_{ij}\boldsymbol{\beta} + \gamma_{0i} = \beta_0 + \beta_1 x_{ij1} + \ldots + \beta_k x_{ijk} + \gamma_{0i}\,,$$

where γ_{0i} are random deviations from the fixed intercept β_0. For random effects, we make the same assumptions as for LMMs, i.e., they are independent and identically normal with $\boldsymbol{\gamma}_i \sim \mathrm{N}(\boldsymbol{0}, \boldsymbol{Q})$ and independent from covariates.

The assumption of conditionally independent responses, given the random effects, corresponds to the assumption of independent errors in LMMs. This assumption may be relaxed by assuming appropriate dependency structures, e.g., for odds ratios with binary responses. However, such generalized approaches in GLMMs are typically much more complicated compared to introducing correlated residuals in LMMs.

A summary of GLMMs for longitudinal and clustered data is given in Box 7.5. We will now discuss some specific GLMMs.

Mixed Logit and Probit Models

For binary response variables, we obtain mixed logit models, for example,

$$\log \frac{\mathrm{P}(y_{ij} = 1 \,|\, \boldsymbol{\gamma}_i)}{\mathrm{P}(y_{ij} = 0 \,|\, \boldsymbol{\gamma}_i)} = \boldsymbol{x}'_{ij}\boldsymbol{\beta} + \boldsymbol{u}'_{ij}\boldsymbol{\gamma}_i. \tag{7.41}$$

As in Chap. 5, mixed probit models are a possible alternative.

Example 7.5 Speed Dating—Mixed Logit Model

We will illustrate (binary) mixed logit models with data from Fisman et al. (2006) on a speed dating experiment on a sample of a few hundred students in graduate and professional schools at Columbia University. In the speed dating events, the experiment randomly assigned each participant to ten short dates with participants of the opposite sex. For each date, each person rated six attributes (attractive, sincere, intelligent, fun, ambitious, shared interests) of the other person on a 10-point scale and wrote down whether he or she would like to see the other person again. We denote these repeated binary decisions as

$$y_{ij} = \begin{cases} 1 & \text{if person } i \text{ likes to see the } j\text{th partner again} \\ 0 & \text{otherwise,} \end{cases}$$

and analyze the data with a mixed logit model of the form (7.41), where the covariates consist of the ratings, as well as the gender of the person. A similar example, also based on the speed dating data, appears in the exercises to Chap. 14 in Gelman and Hill (2006).

△

7.5 GLMMs for Longitudinal and Clustered Data

Distributional Assumption

Given the random effects $\boldsymbol{\gamma}_i$ and the covariates $\boldsymbol{x}_{ij}, \boldsymbol{u}_{ij}$, the responses y_{ij} are conditionally independent and the conditional density $f(y_{ij} \mid \boldsymbol{\gamma}_i)$ belongs to an exponential family as in GLMs; see Sect. 5.5 (p. 304).

Structural Assumption

The conditional mean $\mu_{ij} = \mathrm{E}(y_{ij} \mid \boldsymbol{\gamma}_i)$ is linked to the linear predictor

$$\eta_{ij} = \boldsymbol{x}_{ij}'\boldsymbol{\beta} + \boldsymbol{u}_i'\boldsymbol{\gamma}_i$$

through

$$\mu_{ij} = h(\eta_{ij}) \quad \text{or} \quad \eta_{ij} = g(\mu_{ij}),$$

where h is the response function and $g = h^{-1}$ the link function.

Distributional Assumption for Random Effects

The random effects $\boldsymbol{\gamma}_i$, $i = 1, \ldots, m$, are independent and identically distributed with

$$\boldsymbol{\gamma}_i \stackrel{i.i.d.}{\sim} \mathrm{N}(\mathbf{0}, \boldsymbol{Q})$$

and positive definite covariance matrix $\boldsymbol{Q}$. An important special case is $\boldsymbol{Q} = \mathrm{diag}(\tau_0^2, \ldots, \tau_q^2)$.

Software

- Function `gllamm` of STATA; see Skrondal and Rabe-Hesketh (2008)
- Package `lme4` of R
- `proc mixed` of SAS
- Software package BayesX (also see the R interface `R2BayesX`)

Mixed Poisson Models

For count responses $y_{ij} \sim \mathrm{Po}(\lambda_{ij} \mid \boldsymbol{\gamma}_i)$ mixed log-linear Poisson models

$$\log(\lambda_{ij}) = \boldsymbol{x}_{ij}'\boldsymbol{\beta} + \boldsymbol{u}_{ij}'\boldsymbol{\gamma}_i,$$

are the most common choice.

While repeated measurements ($n_i > 1$) are usually needed to identify binary random intercept models, this is not necessary for mixed Poisson models. It is possible to specify the (random) rate without repeated observations, i.e.,

$$\lambda_i = \exp(\boldsymbol{x}_i'\boldsymbol{\beta} + \gamma_{0i}), \tag{7.42}$$

with $\gamma_{0i} \sim \mathrm{N}(0, \tau_0^2)$ in a random intercept model. A log-linear random intercept model is also called a Poisson-normal or Poisson-lognormal model because Eq. (7.42) is equivalent to

$$\lambda_i = \nu_i \exp(\boldsymbol{x}_i'\boldsymbol{\beta}),$$

where $\nu_i = \exp(\gamma_{0i})$ has a lognormal distribution. Poisson-(log)normal models can be used to analyze count data with overdispersion; see the following example and the discussion on the implied marginal model in Sect. 7.5.2.

Example 7.6 Number of Citations of Patents—Mixed Poisson Regression

In Example 5.7 (p. 297), log-linear Poisson models, both with and without overdispersion, have been employed for regression analyzes using the response *ncit* (number of citations). There is clear evidence of overdispersion, and log-linear Poisson random intercept models (7.42) provide an alternative for such modeling. For illustrative purposes, we consider a log-linear model with linear effects and extend it to the log-linear Poisson-normal model

$$\begin{aligned}\log(\lambda_i) = \eta_i = \beta_0 &+ \beta_1 yearc_i + \beta_2 ncountryc_i + \beta_3 nclaimsc_i + \beta_4 biopharm_i \\ &+ \beta_5 ustwin_i + \beta_6 patus_i + \beta_7 patgsgr_i + \beta_8 opp_i + \gamma_{0i},\end{aligned}$$

where again *yearc*, *ncountryc*, and *nclaimsc* are the centered versions of the covariates. △

Mixed Models for Categorical Responses

Mixed models can also be formulated for categorical responses through suitable generalization of predictors in categorical regression models (Chap. 6). For example, multinomial logit mixed models for a response Y with unordered categories $(1, \ldots, c+1)$ that include global covariates, but additionally include category-specific fixed effects $\boldsymbol{\beta}_r$ and random effects $\boldsymbol{\gamma}_r$, are defined through

$$\mathrm{P}(Y_{ij} = r \mid \boldsymbol{\gamma}_r) = \frac{\exp(\eta_{ijr})}{1 + \sum_{s=1}^{c} \exp(\eta_{ijs})} \qquad r = 1, \ldots, c,$$

with category-specific predictors

$$\eta_{ijr} = \boldsymbol{x}_{ij}'\boldsymbol{\beta}_r + \boldsymbol{u}_{ij}'\boldsymbol{\gamma}_{ir}.$$

The random effects $\boldsymbol{\gamma}_{i1}, \ldots, \boldsymbol{\gamma}_{i,c}$ are assumed to be i.i.d. Gaussian, $\boldsymbol{\gamma}_{ir} \sim \mathrm{N}(\boldsymbol{0}, \boldsymbol{Q}_r)$.

7.5.2 Conditional and Marginal Models

Conceptually, the marginal density $p(y_{ij})$ can be obtained by integrating the conditional density $p(y_{ij} \mid \boldsymbol{\gamma}_i)$ determined through the specific exponential family, i.e.,

$$p(y_{ij}) = \int p(y_{ij} \mid \boldsymbol{\gamma}_i) p(\boldsymbol{\gamma}_i)\, d\boldsymbol{\gamma}_i .$$

The density $p(\boldsymbol{\gamma}_i)$ of $\boldsymbol{\gamma}_i$ is a $\mathrm{N}(\mathbf{0}, \boldsymbol{Q})$ distribution. Apart from the case of LMMs, where $p(y_{ij} \mid \boldsymbol{\gamma}_i)$ also follows a normal distribution, the integration only can be carried out analytically in rather special cases. This is also one of the main reasons why statistical inference in GLMMs becomes more complex compared to LMMs. Marginal means, variances, and covariances are also only available analytically in special cases.

For the log-linear Poisson random intercept model (7.42) with $n_i = 1$, the marginal means, variances, and covariances can all be determined analytically because

$$\mu_i \mid \gamma_{0i} = \nu_i \exp(\boldsymbol{x}_i' \boldsymbol{\beta}),$$

and $\nu_i = \exp(\gamma_{0i})$ has a lognormal distribution. Thus

$$\mathrm{E}(y_i) = \mathrm{E}(\nu_i \exp(\boldsymbol{x}_i'\boldsymbol{\beta})) = \exp(\tau_0^2/2) \exp(\boldsymbol{x}_i'\boldsymbol{\beta}) = \exp(\tau_0^2/2 + \boldsymbol{x}_i'\boldsymbol{\beta}) .$$

Up to the intercept term, the fixed population effects $\boldsymbol{\beta}$ are identical in the marginal and in the conditional model. Furthermore, it can be shown that

$$\mathrm{Var}(y_i) = \mathrm{E}(y_i)\{1 + \exp(\boldsymbol{x}_i'\boldsymbol{\beta})(\exp(3\tau_0^2/2) - \exp(\tau_0^2/2))\}.$$

Since the second factor is greater than 1, the marginal variance is larger than the marginal mean. Therefore the marginal distribution is not a Poisson distribution, but rather a distribution with overdispersion. It is not possible, however, to derive an explicit form for its density. In addition, it can be shown that

$$\mathrm{Cov}(y_i, y_j) = \exp(\boldsymbol{x}_i'\boldsymbol{\beta} + \boldsymbol{x}_j'\boldsymbol{\beta})\{\exp(\tau_0^2)(\exp(\tau_0^2) - 1)\}.$$

Therefore, y_i and y_j have positive marginal correlations, with specific covariance structure.

In contrast to the log-linear Poisson model, particularly in contrast to the LMM, fixed covariate effects in conditional and marginal models are generally different. For example, in the case of a probit-normal random intercept model

$$\mathrm{E}(y_{ij} \mid \gamma_{0i}) = \mathrm{P}(y_{ij} = 1 \mid \gamma_{0i}) = \Phi(\boldsymbol{x}_{ij}'\boldsymbol{\beta} + \gamma_{0i}), \quad \gamma_{0i} \sim \mathrm{N}(0, \tau_0^2),$$

it can be shown that

$$\mathrm{P}(y_{ij} = 1) = \Phi\left(\frac{\boldsymbol{x}_{ij}'\boldsymbol{\beta}}{\sqrt{\tau_0^2 + 1}}\right) = \Phi(\boldsymbol{x}_{ij}'\boldsymbol{\beta}^*),$$

with $\boldsymbol{\beta}^* = \boldsymbol{\beta}/\sqrt{\tau_0^2 + 1}$. Thus absolute values of fixed effects in the marginal model are shrunken by the factor $1/\sqrt{\tau_0^2 + 1}$. For a logit model with the same predictor, it is not even possible to derive any explicit formula, rather only an approximation based on Taylor expansion; see, for example, Fahrmeir and Tutz (2001, Sect. 7.7).

7.5.3 GLMMs in General Form

Similar to the LMMs in Sect. 7.3, the predictor vector $\boldsymbol{\eta}$ for all observations can be expressed as

$$\boldsymbol{\eta} = \boldsymbol{X}\boldsymbol{\beta} + \boldsymbol{U}\boldsymbol{\gamma},$$

where $\boldsymbol{X}$ and $\boldsymbol{U}$ are appropriate design matrices and $\boldsymbol{\gamma} = (\boldsymbol{\gamma}'_1, \ldots, \boldsymbol{\gamma}'_m)'$ is the vector of all random effects. As for the LMM, we may replace the assumption

$$\boldsymbol{\gamma}_i \overset{i.i.d.}{\sim} \mathrm{N}(\boldsymbol{0}, \boldsymbol{Q}),$$

more generally with $\boldsymbol{\gamma} \sim \mathrm{N}(\boldsymbol{0}, \boldsymbol{G})$, where $\boldsymbol{G}$ is any positive definite covariance matrix. This allows for correlated random effects, as needed in the mixed model representation of spline functions in Chaps. 8 and 9. Furthermore, statistical inference in the next section is formulated using this compact notation for the distribution of $\boldsymbol{\gamma}$.

For GLMMs in general form, we allow $\boldsymbol{X}$ and $\boldsymbol{U}$ to be general design matrices and $\boldsymbol{G}$ may be any positive definite covariance matrix. This is completely analogous to general LMMs in Sect. 7.2. In the following section on inference we will refer ro a specific observation in a general GLMM by the index i, i.e., y_i denotes the ith observation.

7.6 Likelihood and Bayesian Inference in GLMMs

Conceptually, inference in GLMMs is based on the same likelihood or Bayes principles as in LMMs. However, since the conditional likelihood is in general non-Gaussian and the relationship between mean and predictor is nonlinear in GLMMs, important parts of these approaches cannot be carried out analytically but rather have to be replaced by numerical methods or suitable approximations. We focus on an approximate penalized likelihood approach and on MCMC techniques for fully Bayesian inference. Both approaches are strongly related since the penalized likelihood estimator can also be seen as an empirical Bayes estimator. For simplicity, we restrict the presentation to models with scale parameter $\phi = 1$; however, the more general case, such as for gamma-distributed responses, can be dealt with as in the GLM.

7.6.1 Penalized Likelihood and Empirical Bayes Estimation

Known Variance–Covariance Parameters

We first assume that all parameters in $\boldsymbol{G}$ are known.

Similar to Sect. 7.3.1, we consider the joint likelihood

$$L(\boldsymbol{\beta}, \boldsymbol{\gamma}) = p(\boldsymbol{y} \mid \boldsymbol{\beta}, \boldsymbol{\gamma})\, p(\boldsymbol{\gamma})$$

for known $\boldsymbol{G}$, and maximize it simultaneously with respect to $\boldsymbol{\beta}$ and $\boldsymbol{\gamma}$. Since

$$p(\boldsymbol{\beta}, \boldsymbol{\gamma} \mid \boldsymbol{y}) \propto p(\boldsymbol{y} \mid \boldsymbol{\beta}, \boldsymbol{\gamma})\, p(\boldsymbol{\gamma}),$$

this is equivalent to determining the posterior mode estimator $(\hat{\boldsymbol{\beta}}, \hat{\boldsymbol{\gamma}})$ or an *empirical Bayes estimator*. Taking logarithms, we obtain the penalized log-likelihood

$$l_{\text{pen}}(\boldsymbol{\beta}, \boldsymbol{\gamma}) = l(\boldsymbol{\beta}, \boldsymbol{\gamma}) - \frac{1}{2}\boldsymbol{\gamma}'\boldsymbol{G}^{-1}\boldsymbol{\gamma}. \tag{7.43}$$

The log-likelihood $l(\boldsymbol{\beta}, \boldsymbol{\gamma})$ is defined as in a GLM, where only the predictor $\eta_i = \boldsymbol{x}_i'\boldsymbol{\beta}$ for a particular observation i has to be replaced by the extended mixed model predictor $\eta_i = \boldsymbol{x}_i'\boldsymbol{\beta} + \boldsymbol{u}_i'\boldsymbol{\gamma}$. Maximization of $l_{\text{pen}}(\boldsymbol{\beta}, \boldsymbol{\gamma})$ proceeds in complete analogy to simple GLMs as outlined in Sect. 5.4.2 and fully derived in Sect. 5.8.2. With $\boldsymbol{C} = (\boldsymbol{X}\ \boldsymbol{U})$ and the partitioned matrix

$$\boldsymbol{B} = \begin{pmatrix} \boldsymbol{0} & \boldsymbol{0} \\ \boldsymbol{0} & \boldsymbol{G}^{-1} \end{pmatrix},$$

the resulting iteratively weighted least squares estimator

$$\begin{pmatrix} \hat{\boldsymbol{\beta}}^{(t+1)} \\ \hat{\boldsymbol{\gamma}}^{(t+1)} \end{pmatrix} = \left(\boldsymbol{C}'\boldsymbol{W}^{(t)}\boldsymbol{C} + \boldsymbol{B}\right)^{-1} \boldsymbol{C}'\boldsymbol{W}^{(t)}\tilde{\boldsymbol{y}}^{(t)}$$

has the same form as the ML or penalized least squares estimator in the LMM; see formula (7.25) of Sect. 7.3.1. The matrix of working weights $\boldsymbol{W}^{(t)}$ and the working observations $\tilde{y}_i$ are defined in complete analogy to simple GLMs. More specifically, we obtain for a particular observation $\boldsymbol{x}_i, \boldsymbol{u}_i$ in the general GLMM

$$\tilde{y}_i\left(\hat{\eta}_i^{(t)}\right) = \hat{\eta}_i^{(t)} + \frac{\left(y_i - h\left(\hat{\eta}_i^{(t)}\right)\right)}{h'\left(\hat{\eta}_i^{(t)}\right)}, \tag{7.44}$$

where $\hat{\eta}_i^{(t)} = \boldsymbol{x}_i'\hat{\boldsymbol{\beta}}^{(t)} + \boldsymbol{u}_i'\hat{\boldsymbol{\gamma}}^{(t)}$ is the actual predictor, h is the response function, and $h'(\eta) = \partial h(\eta)/\partial\eta$ is the derivative of h with respect to η. The working weight is given by

$$\tilde{w}_i\left(\hat{\eta}_i^{(t)}\right) = \frac{\left(h'\left(\hat{\eta}_i^{(t)}\right)\right)^2}{\sigma_i^2\left(\hat{\eta}_i^{(t)}\right)}, \tag{7.45}$$

where $\sigma_i^2\left(\hat{\eta}_i^{(t)}\right)$ is the (conditional) variance $\text{Var}(y_i)$ evaluated at $\eta = \hat{\eta}_i^{(t)}$. Omitting the iteration index we obtain in matrix notation

$$\begin{aligned} \boldsymbol{W} &= \boldsymbol{D}\boldsymbol{\Sigma}^{-1}\boldsymbol{D}, \\ \tilde{\boldsymbol{y}} &= \boldsymbol{X}\hat{\boldsymbol{\beta}} + \boldsymbol{U}\hat{\boldsymbol{\gamma}} + \boldsymbol{D}^{-1}(\boldsymbol{y} - \boldsymbol{\mu}), \end{aligned} \tag{7.46}$$

with $\boldsymbol{D} = \text{diag}\left(\ldots, \frac{\partial h(\eta_i)}{\partial \eta}, \ldots\right)$ and $\boldsymbol{\Sigma} = \text{diag}(\ldots, \sigma_i^2, \ldots)$.

Unknown Variance–Covariance Parameters

Now assume that the covariance matrix $\boldsymbol{G} = \boldsymbol{G}(\boldsymbol{\vartheta})$ contains unknown parameters $\boldsymbol{\vartheta}$. Estimation of the variance–covariance parameters $\boldsymbol{\vartheta}$ can be accomplished by means of an approximate marginal likelihood. A Laplace approximation argument (Breslow and Clayton, 1993) shows that the conditional log-likelihood can be approximated through a quadratic form:

$$l(\boldsymbol{\beta}, \boldsymbol{\gamma}; \boldsymbol{\vartheta}) = \log\, p(\boldsymbol{y} \mid \boldsymbol{\beta}, \boldsymbol{\gamma}) \approx \frac{1}{2}(\boldsymbol{y} - \boldsymbol{\mu})'\boldsymbol{\Sigma}^{-1}(\boldsymbol{y} - \boldsymbol{\mu}),$$

where $\boldsymbol{\mu}$ and $\boldsymbol{\Sigma}$ depend on $\boldsymbol{\beta}$ and $\boldsymbol{\gamma}$. From the definition Eq. (7.46) of the working observations $\tilde{\boldsymbol{y}}$, we get

$$\boldsymbol{y} - \boldsymbol{\mu} = \boldsymbol{D}(\tilde{\boldsymbol{y}} - \boldsymbol{X}\hat{\boldsymbol{\beta}} - \boldsymbol{U}\hat{\boldsymbol{\gamma}}),$$

which implies (together with $\boldsymbol{W} = \boldsymbol{D}\boldsymbol{\Sigma}^{-1}\boldsymbol{D}$)

$$l(\boldsymbol{\beta}, \boldsymbol{\gamma}; \boldsymbol{\vartheta}) \approx (\tilde{\boldsymbol{y}} - \boldsymbol{X}\hat{\boldsymbol{\beta}} - \boldsymbol{U}\hat{\boldsymbol{\gamma}})'\boldsymbol{W}(\tilde{\boldsymbol{y}} - \boldsymbol{X}\hat{\boldsymbol{\beta}} - \boldsymbol{U}\hat{\boldsymbol{\gamma}}).$$

Therefore, the conditional log-likelihood of a GLMM can be approximated for fixed $\boldsymbol{\vartheta}$ through the log-likelihood of the LMM

$$\tilde{\boldsymbol{y}} \mid \boldsymbol{\gamma} \sim \text{N}(\boldsymbol{X}\boldsymbol{\beta} + \boldsymbol{U}\boldsymbol{\gamma}, \boldsymbol{W}^{-1}).$$

Setting

$$\boldsymbol{V} = \boldsymbol{W} + \boldsymbol{U}\boldsymbol{G}(\boldsymbol{\vartheta})^{-1}\boldsymbol{U}',$$

we obtain the (approximate) *marginal* or *restricted* log-likelihood

$$l_R(\boldsymbol{\vartheta}) = -\frac{1}{2}\log|\boldsymbol{V}(\boldsymbol{\vartheta})| - \frac{1}{2}\log(|\boldsymbol{X}'\boldsymbol{V}(\boldsymbol{\vartheta})^{-1}\boldsymbol{X}|) - \frac{1}{2}(\tilde{\boldsymbol{y}}(\boldsymbol{\vartheta}) - \boldsymbol{X}\hat{\boldsymbol{\beta}})'\boldsymbol{V}(\boldsymbol{\vartheta})^{-1}(\tilde{\boldsymbol{y}}(\boldsymbol{\vartheta}) - \boldsymbol{X}\hat{\boldsymbol{\beta}}),$$

where $\boldsymbol{V}(\boldsymbol{\vartheta})$ and $\tilde{\boldsymbol{y}}(\boldsymbol{\vartheta})$ also depend on $\hat{\boldsymbol{\beta}}$ and $\hat{\boldsymbol{\gamma}}$.

This leads to the estimation algorithm in Box 7.6, which switches between estimation of $\boldsymbol{\beta}$ and $\boldsymbol{\gamma}$ and estimation of $\boldsymbol{\vartheta}$.

The estimated covariance matrix is the inverse of the estimated Fisher information matrix $\hat{\boldsymbol{F}} = \boldsymbol{F}(\hat{\boldsymbol{\beta}}, \hat{\boldsymbol{\gamma}}, \hat{\boldsymbol{\vartheta}})$, i.e.,

$$\widehat{\text{Cov}}\begin{pmatrix}\hat{\boldsymbol{\beta}}\\ \hat{\boldsymbol{\gamma}}\end{pmatrix} \approx \hat{\boldsymbol{F}}^{-1}.$$

Assuming approximate normality, confidence intervals and tests of linear hypotheses can be constructed analogously to those for the LMM. If the number of parameters is large relative to the sample size, we expect increased bias and a degraded normality approximation, in particular for responses with low information, such as binary observations. Additional analyses that use fully Bayesian inference are recommended in such cases.

7.6.2 Fully Bayesian Inference Using MCMC

For fully Bayesian inference, we assume the same priors for $\boldsymbol{\beta}$ and $\boldsymbol{\vartheta}$ as in LMMs, i.e.,

$$\boldsymbol{\beta} \sim \text{N}(\boldsymbol{m}, \boldsymbol{M}),$$

with the limiting case $\boldsymbol{M}^{-1} \to \boldsymbol{0}$ corresponding to a flat prior

$$p(\boldsymbol{\beta}) \propto \text{const}.$$

Usually Bayesian inference using MCMC requires knowledge of the specific structure of the GLMM. Bayesian inference for a general GLMM is usually difficult to accomplish, and there is no guarantee that the sampler works.

We therefore restrict ourselves to the specific GLMM for longitudinal or clustered data,

$$\text{E}(y_{ij} \,|\, \boldsymbol{\gamma}_i) = \mu_{ij} = h(\boldsymbol{x}_{ij}'\boldsymbol{\beta} + \boldsymbol{u}_{ij}'\boldsymbol{\gamma}_i), \quad i = 1, \ldots, m, \quad j = 1, \ldots, n_i,$$

with i.i.d. random effects

$$\boldsymbol{\gamma}_i \sim \text{N}(\boldsymbol{0}, \boldsymbol{Q}).$$

We also make the simplifying assumption $\boldsymbol{Q} = \text{diag}(\tau_0^2, \ldots, \tau_q^2)$ and assume a priori independent diagonal elements with

$$\tau_r^2 \sim \text{IG}(a_r, b_r), \quad r = 0, \ldots, q.$$

Moreover, we only consider GLMMs with scale parameter $\phi = 1$, otherwise an additional prior for ϕ has to be specified.

7.6 Penalized Likelihood/Empirical Bayes Estimation in GLMMs

Score Function and Fisher Information

$$s_{\beta}(\boldsymbol{\beta},\boldsymbol{\gamma},\boldsymbol{\vartheta}) = \frac{\partial l_{\text{pen}}(\boldsymbol{\beta},\boldsymbol{\gamma};\boldsymbol{\vartheta})}{\partial \boldsymbol{\beta}} = \boldsymbol{X}'\boldsymbol{D}\boldsymbol{\Sigma}^{-1}(\boldsymbol{y}-\boldsymbol{\mu}),$$

$$s_{\gamma}(\boldsymbol{\beta},\boldsymbol{\gamma},\boldsymbol{\vartheta}) = \frac{\partial l_{\text{pen}}(\boldsymbol{\beta},\boldsymbol{\gamma};\boldsymbol{\vartheta})}{\partial \boldsymbol{\gamma}} = \boldsymbol{U}'\boldsymbol{D}\boldsymbol{\Sigma}^{-1}(\boldsymbol{y}-\boldsymbol{\mu}) - \boldsymbol{G}(\boldsymbol{\vartheta})^{-1}\boldsymbol{\gamma}.$$

$$F(\boldsymbol{\beta},\boldsymbol{\gamma},\boldsymbol{\vartheta}) = \begin{pmatrix} \boldsymbol{X}'\boldsymbol{W}\boldsymbol{X} & \boldsymbol{X}'\boldsymbol{W}\boldsymbol{U} \\ \boldsymbol{U}'\boldsymbol{W}\boldsymbol{X} & \boldsymbol{U}'\boldsymbol{W}\boldsymbol{U} + \boldsymbol{G}(\boldsymbol{\vartheta})^{-1} \end{pmatrix}$$

with $\boldsymbol{D} = \text{diag}\left(\ldots, \frac{\partial h(\eta_i)}{\partial \eta}, \ldots\right)$, $\boldsymbol{\Sigma} = \text{diag}(\ldots, \sigma_i^2, \ldots)$, and $\boldsymbol{\mu} = (\ldots, h(\eta_i), \ldots)'$.

Numerical Computation

1. Initialize with starting values $\hat{\boldsymbol{\beta}}^{(0)}, \hat{\boldsymbol{\gamma}}^{(0)}, \hat{\boldsymbol{\vartheta}}^{(0)}$. Set $t = 1$.
2. Determine $\hat{\boldsymbol{\beta}}^{(t+1)}$ and $\hat{\boldsymbol{\gamma}}^{(t+1)}$ through

$$\begin{pmatrix} \hat{\boldsymbol{\beta}}^{(t+1)} \\ \hat{\boldsymbol{\gamma}}^{(t+1)} \end{pmatrix} = \left(\boldsymbol{C}'\boldsymbol{W}^{(t)}\boldsymbol{C} + \boldsymbol{B}\right)^{-1} \boldsymbol{C}'\boldsymbol{W}^{(t)}\tilde{\boldsymbol{y}}^{(t)}$$

   with working weights and observations as given in Eqs. (7.45) and (7.44).
3. Compute the (approximate) ML estimator $\hat{\boldsymbol{\vartheta}}^{(t+1)}$ through one iteration step of some numerical algorithm for maximizing $l_R(\boldsymbol{\vartheta})$, for example, a Newton–Raphson step.
4. Terminate the algorithm if convergence (according to some stopping criterion) is reached and obtain $\hat{\boldsymbol{\beta}}, \hat{\boldsymbol{\gamma}}, \hat{\boldsymbol{\vartheta}}$. Otherwise, set $t = t + 1$ and proceed with 2.

Based on this assumptions, we obtain the posterior

$$p(\boldsymbol{\beta}, \boldsymbol{\gamma}_1, \ldots, \boldsymbol{\gamma}_m, \boldsymbol{Q} \mid \boldsymbol{y}) \propto \prod_{i=1}^{m} \prod_{j=1}^{n_i} p(y_{ij} \mid \boldsymbol{\beta}, \boldsymbol{\gamma}_i) p(\boldsymbol{\beta}) \prod_{i=1}^{m} p(\boldsymbol{\gamma}_i \mid \boldsymbol{Q}) p(\boldsymbol{Q}).$$

This implies the *full conditionals*

$$p(\boldsymbol{\beta} \mid \cdot) \propto \prod_{i=1}^{m} \prod_{j=1}^{n_i} p(y_{ij} \mid \boldsymbol{\beta}, \boldsymbol{\gamma}_i) p(\boldsymbol{\beta})$$

$$p(\boldsymbol{\gamma}_i \mid \cdot) \propto \prod_{j=1}^{n_i} p(y_{ij} \mid \boldsymbol{\beta}, \boldsymbol{\gamma}_i) p(\boldsymbol{\gamma}_i \mid \boldsymbol{Q}), \quad i = 1, \ldots, m$$

$$p(\tau_r^2 \mid \cdot) \propto \prod_{i=1}^{m} p(\boldsymbol{\gamma}_{ir} \mid \tau_r^2) p(\tau_r^2), \quad r = 0, \ldots, q.$$

In contrast to the LMM, full conditionals for $\boldsymbol{\beta}$ and $\boldsymbol{\gamma}_i$, $i = 1, \ldots, m$, are no longer available in a known analytical form. Therefore random numbers cannot be directly drawn from these full conditionals, and thus the Metropolis–Hastings (MH) algorithm needs to be employed instead of the Gibbs sampler (as for LMMs in Sect. 7.4). We prefer the automated algorithm of Gamerman (1997) containing IWLS (iteratively weighted least squares) proposals. See Sect. 5.6.2 for the general idea of IWLS proposals.

The full conditionals $p(\tau_r^2 \mid \cdot)$ are again inverse gamma densities

$$\tau_r^2 \mid \cdot \sim \text{IG}(\tilde{a}_r, \tilde{b}_r)$$

with updated parameters

$$\tilde{a}_r = a_r + \frac{m}{2}, \qquad \tilde{b}_r = b_r + \frac{1}{2} \sum_{i=1}^{m} \gamma_{ir}^2.$$

Alternatively the data augmentation schemes outlined in Sect. 5.6.3 for GLMs could be applied in a straightforward way for GLMMs.

Example 7.7 Speed Dating—Mixed Binary Regression

To analyze binary repeated measurement data from the speed dating experiment (Example 7.5), we first reduced the sample by deleting all participants who never wanted or always wanted to see a partner again. Without these "outliers," $m = 390$ persons remained in the sample. For illustration, we only included the two attributes "attractive" and "shared interests," as well as gender, and the interactions between gender and the two attributes as fixed effects into the predictor η_{ij} of a mixed logit model for $P\left(Y_{ij} = 1 \mid \eta_{ij}\right)$, the probability that person i likes to see partner j again. For a random intercept model, the predictor is

$$\begin{aligned}\eta_{ij} = {} & \beta_0 + \beta_1\, gender_i + \beta_2\, attr_{ij} + \beta_3\, shar_{ij} \\ & + \beta_4\, gender_i \cdot attr_{ij} + \beta_5\, gender_{ij} \cdot shar_{ij} + \gamma_{0i}.\end{aligned}$$

Gender is dummy-coded (male = 1, female = 0) and the two attributes are rated on a 10-point scale. The variables *attr* and *shar* are standardized prior to estimation. With such a model we can analyze the effect of gender and the main effects of attractiveness and shared interests on the decision to see a partner again. The interaction terms allow analysis as to whether or not there are gender-specific differences of the effects of the two attributes. The intercept term γ_{0i} captures omitted covariates and induces correlation between repeated

Table 7.4 Speed dating: random intercept and random slope models based on likelihood (L) and Bayesian (B) inference. Posterior estimates and standard errors for fixed effects, (posterior) variance estimates for random effects

Covariate	L: random intercept		L: random slopes	
intercept	−0.6976	(0.118)	−0.7997	(0.126)
gender	−0.1751	(0.166)	0.2823	(0.177)
attr	1.6329	(0.077)	1.9323	(0.105)
shar	1.2546	(0.073)	1.4393	(0.095)
gender · attr	0.4567	(0.116)	0.4630	(0.153)
gender · shar	−0.1933	(0.103)	−0.2334	(0.134)
$\tau^2_{intercept}$	2.55			2.70
τ^2_{gender}	–			0.00
τ^2_{attr}	–			0.69
τ^2_{shar}	–			0.49
Covariate	B: random intercept		B: random slopes	
intercept	−0.6984	(0.121)	−0.8008	(0.128)
gender	0.1737	(0.165)	0.2737	(0.163)
attr	1.6370	(0.082)	1.9520	(0.116)
shar	1.2527	(0.076)	1.4466	(0.104)
gender · attr	0.4501	(0.118)	0.4603	(0.152)
gender · shar	−0.1877	(0.102)	−0.2310	(0.133)
$\tau^2_{intercept}$	2.61			2.77
τ^2_{gender}	–			0.07
τ^2_{attr}	–			0.76
τ^2_{shar}	–			0.56

measurements, at least to some extent. In addition, we also considered a model with random slopes for the main effects, i.e., the predictor is extended to

$$\tilde{\eta}_{ij} = \eta_{ij} + \gamma_{1i}\, gender_i + \gamma_{2i}\, attr_{ij} + \gamma_{3i}\, shar_{ij},$$

and a diagonal matrix $\boldsymbol{Q} = \text{diag}\left(\tau_0^2, \tau_1^2, \tau_2^2, \tau_3^2\right)$ is assumed for $\boldsymbol{\gamma}_i \sim \text{N}(\boldsymbol{0}, \boldsymbol{Q})$.

Table 7.4 contains estimates for the fixed effects and the random effects variances obtained through likelihood and Bayesian MCMC inference. Likelihood-based inference is carried out using function `lmer` of the R package `lme4`. Results for the Bayesian approach are obtained using BayesX with 102,000 iterations, 2,000 burn in iterations, and taking every 100th sampled parameter for inference. The hyperparameters for the inverse gamma variance parameters are $a = b = 0.001$. A sensitivity analysis with the alternatives $a = b = 0.01$ and $a = b = 0.0001$ shows that the results are relatively insensitive regarding the choice of hyperparameters. Only the variance parameter of the random slope for *gender* shows some sensitivity with a posterior mean ranging from 0.045 to 0.12.

As Table 7.4 reveals, the results for both approaches are in very close agreement. Attractiveness and shared interests have significant positive influence on the decision to see a partner again. Whereas the main effect of gender is not significant, the interactions of gender with the attributes reveal significant gender-specific differences. Attractiveness of the (female) partner increases the already positive effect on the decision of males to see her again, while shared interests are less important for males compared to females.

Table 7.5 Number of citations of patents: Poisson-normal model

Variable	Coefficient	Standard error	2.5 % quantile	97.5 % quantile
intercept	−0.347	0.061	−0.466	−0.227
yearc	−0.088	0.006	−0.100	−0.078
ncountryc	−0.036	0.008	−0.052	−0.019
nclaimsc	0.018	0.003	0.013	0.023
biopharm	0.174	0.061	0.051	0.290
ustwin	0.022	0.048	−0.074	0.112
patus	−0.163	0.051	−0.267	−0.063
patgsgr	−0.284	0.058	−0.400	−0.167
opp	0.399	0.045	0.316	0.489

The relatively large variance of the random intercept reflects the fact that there are still many individual-specific covariates or preferences not included in the model. △

Example 7.8 Number of Citations of Patents—Bayesian Linear Mixed Model

We analyze the Poisson-normal random intercept model of Example 7.6 (p. 392) using empirical and fully Bayesian inference. For fully Bayesian inference, we choose a noninformative flat prior for fixed effects and an inverse gamma distribution $\tau_0^2 \sim \text{IG}(a, b)$ with $a = b = 0.001$ for the variance parameter of the random intercepts. Results for empirical and fully Bayesian inference are similar. We therefore restrict the presentation to fully Bayesian inference which has been obtained using `bayesreg objects` of `BayesX`. The estimated fixed effects in Table 7.5, as well as standard errors, are in good agreement with the results in Example 5.7 (Table 5.10, p. 298). The estimated intercept −0.347 cannot be directly compared with the corresponding intercept 0.158 in Table 5.10, i.e., we have to adjust for $\tau_0^2/2$; see the remark on the Poisson-normal model in Sect. 7.5.2. After adjustment, the intercepts are in better agreement. The estimated value of τ_0^2 is $\hat{\tau}_0^2 = 1.17$ (posterior mean). △

7.7 Practical Application of Mixed Models

This section serves as a case study on (generalized) LMMs and how they can be applied in practice. Here, we restrict ourselves to models for longitudinal and cluster data.

7.7.1 General Guidelines and Recommendations

We first present some general guidelines and recommendations on how to apply LMMs or GLMMs in practice:

- *Exploratory analysis of cluster-specific heterogeneity:* Besides the usual inspection of the data through summary statistics and graphical tools (as outlined in Sect. 1.2) some additional graphical utilities may be useful for longitudinal and cluster data. An effective device is to perform a *cluster-wise analysis*.

In its simplest form, one could produce cluster-wise scatter plots of the relationship between the response and the continuous covariates as has been done in Example 2.9 (p. 39, Fig. 2.10) for the rats data. Such plots are also helpful to detect nonlinear relationships between the response and the covariates. Alternatively, one could estimate and visualize separate cluster-specific regression lines (or more generally regression curves), as illustrated in Sect. 7.1.1 (Figs. 7.1 and 7.4). For categorical covariates scatter plots can be replaced by box plots.

Additionally it is helpful to estimate the assumed regression model separately for each cluster (at least for the clusters where there are enough data for a separate analysis). The obtained cluster-specific intercepts and slopes could then be visualized in caterpillar plots together with 95 % confidence intervals. If there is cluster-specific heterogeneity, the cluster-specific intercepts and slopes should show considerable variability. We illustrate this approach in our case study below on sales of orange juice, as also seen in Fig. 7.10.

As mentioned, all cluster-wise analyses naturally require a sufficient amount of data in each cluster. If there are clusters with only a few observations, the cluster-wise analysis must be restricted to those clusters with sufficient data.

- *Model specification:* The specification of random intercepts and slopes is usually guided by subject-matter knowledge and a preceding exploratory analysis. When specifying random coefficient models, one should always think of possibly different between- and within-cluster effects or endogenous (stochastic) covariates; recall the discussions in Sect. 7.1.1 (p. 353) and in Sect. 7.1.5 (p. 366).

  Usually it does not make sense to include random slopes without including a random intercept. It is also typically not sensible to include a random slope without including a corresponding fixed effect. Generally, the specification of random slopes should be as parsimonious as possible. Often convergence problems of estimation algorithms are encountered if too complex models are specified. A nice discussion of the pitfalls in the specification of random coefficient models is given in Sect. 4.10 of Skrondal and Rabe-Hesketh (2008).
- *Model assessment and diagnostics:* In addition to the usual model diagnostic tools discussed in Sect. 3.4.4 (p. 155), tools similar to those used for the exploratory analysis are helpful for model assessment in random coefficients models. In particular, caterpillar plots of the estimated random coefficients help to assess the heterogeneity assumption. It may also be useful to visualize the estimated cluster-specific curves together with partial residuals and kernel densities of random effects to assess the normality assumption (at least to a certain extent). The results of hypothesis tests could be used as an additional source of information for model diagnostics, but should be taken with care; see the discussion in Sect. 7.3.4.

  Since there are a considerable number of competing estimation approaches, it is recommended to estimate the models using these different approaches (ML, REML, fully Bayesian under different choices of hyperparameters) and even different software to assess the stability of estimates and to investigate possible convergence problems (at least when using complex model specifications).

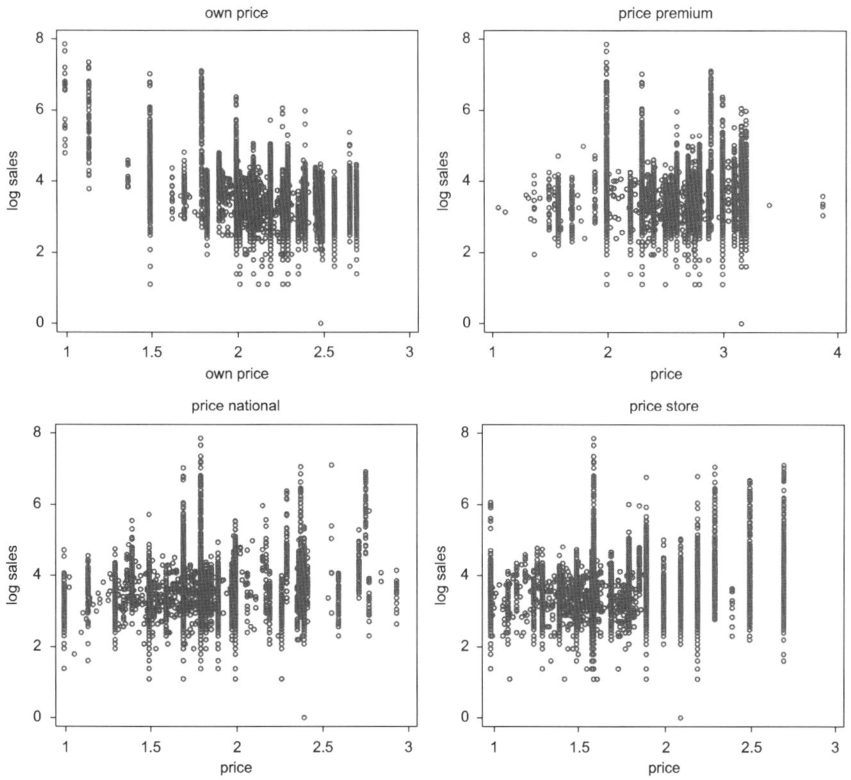

Fig. 7.8 Sales of orange juice: scatter plots between the log number of packages sold and the price of "tree fresh" (*upper left panel*), as well as prices of competitive brands

7.7.2 Case Study on Sales of Orange Juice

Data Description and Exploratory Analysis

We now illustrate the application of LMMs using the supermarket scanner data that were briefly outlined in Example 4.9 (p. 199) in the context of multiplicative errors. The data include weekly unit sales and respective retail prices for different brands of orange juice (premium, national, and store brands) in 81 chains of a store over a time span of 89 weeks. We restrict ourselves to one of the national brands, i.e., the "Tree Fresh" brand. In the following, we model the log sales y_{ij} in store i, $i = 1, \dots, 81$, at time t_{ij} in relation to its own-item price p_{ij} and competitor prices $pc1_{ij}$, $pc2_{ij}$, and $pc3_{ij}$ among the three quality tiers (premium, national, store). To get an intuitive feel for the data, Fig. 7.8 shows scatter plots of log sales against the covariates. We clearly find an inverse relationship between unit sales and

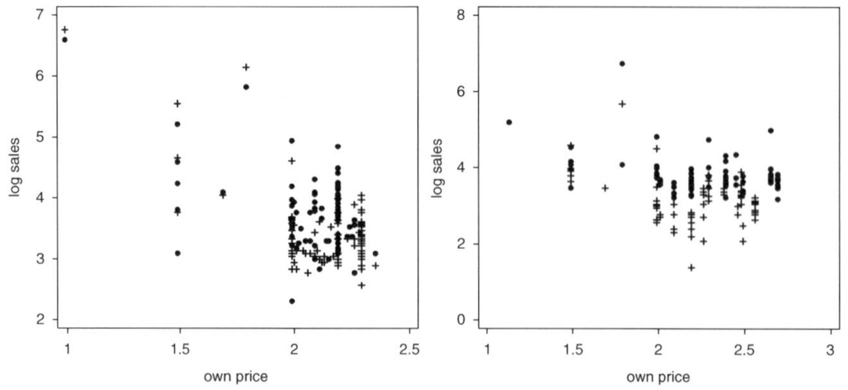

Fig. 7.9 Sales of orange juice: scatter plots between the log number of sold packages and the own price in four different stores. Each plot displays the log sales versus price in two outlets using different symbols (*plus signs and black dots*)

own-item price. The relationship also appears to be nonlinear as the slope decreases for prices higher than values near 2. However, the situation is less clear for the impact of competitive prices on the sales of Tree Fresh, i.e., it is much more difficult to discern any relationships in these scatter plots. Of course, economic theory suggests a positive correlation between sales and prices of competitors, which can be observed at least for the price of the store brand. In order to investigate possible store-specific heterogeneity, it is instructive to plot the log sales price relationship for different stores. Figure 7.9 displays log sales against own-item price for four different stores exhibiting a clear tendency for outlet-specific heterogeneity. It is apparent that the overall sales level and also the slope of the relationship vary across the four stores. The scatter plots further suggest transforming prices to log prices in order to satisfactorily capture the nonlinearity in the relationships. To further check cluster-specific heterogeneity, we estimate the model

$$y_{ij} = \beta_{0i} + \beta_{1i} \cdot logp_{ij} + \beta_{2i} \cdot logpc1_{ij} + \beta_{3i} \cdot logpc2_{ij} + \beta_{4i} \cdot logpc3_{ij} + \varepsilon_{ij} \quad (7.47)$$

separately for each store $i = 1, \ldots, 81$. In this model, $\beta_{1i}, \ldots, \beta_{4i}$ are fixed effects of the log price and the log prices of competitors. All covariates are centered about their mean values to avoid problems with meaningless origins. Figure 7.10 shows caterpillar plots of the estimated cluster-specific coefficients. We clearly see that the intercepts, as well as the regression coefficients, of the own-item price shows considerable cluster-specific heterogeneity. The regression coefficients of the cross-price effects reveal much less heterogeneity.

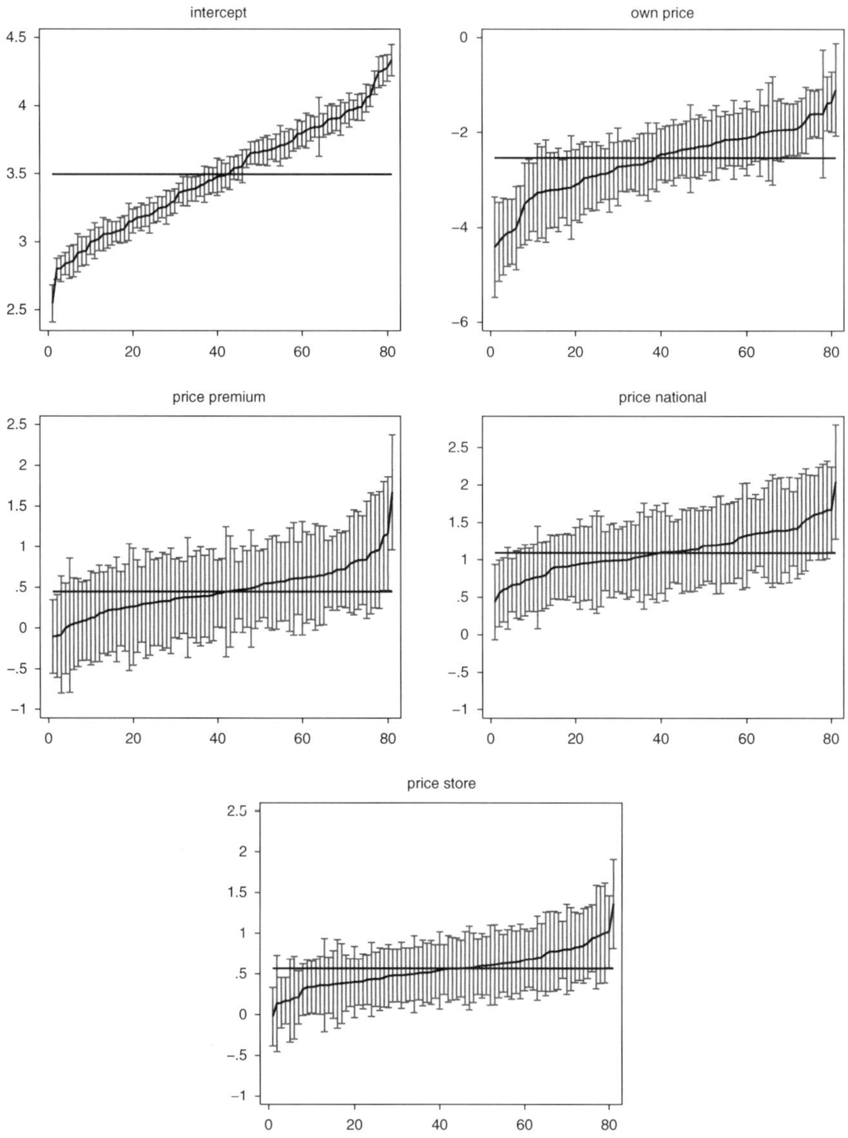

Fig. 7.10 Sales of orange juice: caterpillar plots of the regression coefficients in a cluster-wise analysis of the regression model (7.47). Shown are the estimated regression coefficients and their 95 % confidence intervals. The *solid lines* correspond to the regression coefficients of a global regression model

Table 7.6 Sales of orange juice: estimated fixed effects and variance parameters for the full model (7.48) and a reduced model without random slopes for cross-price effects

		Full		Reduced	
	Parameter	Estimate	s.e.	Estimate	s.e.
intercept	β_0	3.489	0.045	3.490	0.045
logp	β_1	−2.558	0.077	−2.533	0.074
logpc1	β_2	0.454	0.036	0.453	0.033
logpc2	β_3	1.114	0.038	1.107	0.031
logpc3	β_3	0.549	0.029	0.535	0.023
$\text{Var}(\gamma_{0i})$	τ_0^2	0.166		0.165	
$\text{Var}(\gamma_{1i})$	τ_1^2	0.354		0.309	
$\text{Var}(\gamma_{2i})$	τ_2^2	0.020		–	
$\text{Var}(\gamma_{3i})$	τ_3^2	0.041		–	
$\text{Var}(\gamma_{4i})$	τ_4^2	0.023		–	
$\text{Var}(\varepsilon_{ij})$	σ^2	0.186		1.190	

Model Specification and Results

In light of our discussion and exploratory analysis, we estimate the two-level random coefficients model

$$\begin{aligned} y_{ij} = (\beta_0 + \gamma_{0i}) + (\beta_1 + \gamma_{1i}) \cdot logp_{ij} + (\beta_2 + \gamma_{2i}) \cdot logpc1_{ij} \\ +(\beta_3 + \gamma_{3i}) \cdot logpc2_{ij} + (\beta_4 + \gamma_{4i}) \cdot logpc3_{ij} + \varepsilon_{ij}. \end{aligned} \tag{7.48}$$

Here, $\beta_0 + \gamma_{0i}$ are store-specific random intercepts, and $\gamma_{1i}, \ldots, \gamma_{4i}$ are store-specific deviations from the overall slopes of price effects. For the random effects, a fully unspecified covariance matrix is assumed a priori. Estimation has been carried out using function `lmer` of the R package `lme4`.

Table 7.6 shows the estimated fixed effects and variance parameters. The estimated log-linear price effects are in agreement with economic theory. Increasing own prices reduce sales, while increasing prices of competing brands lead to increased sales for "Tree Fresh." The strength of the outlet-specific random variation differs across stores, measured through the random effects variances. While the random intercept (measuring the outlet-specific overall sales level) and the own-price random effects are comparably strong with variances of $\tau_0^2 = 0.166$ and $\tau_1^2 = 0.354$ (in relation to the overall variance of $\sigma^2 = 0.186$), the random effects of cross-price effects are comparably small. Note that a comparison of random effects variances only makes sense if the covariates are measured on the same scale, which is the case for all price covariates.

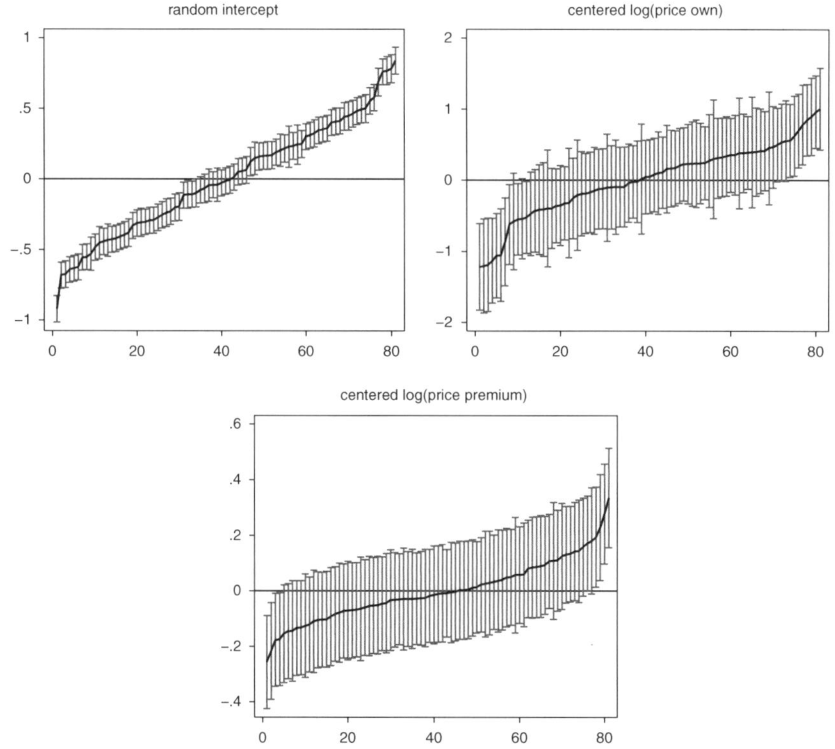

Fig. 7.11 Sales of orange juice: caterpillar plots of some random effects in the full model (7.48). The estimated regression coefficients and their 95 % confidence intervals are shown

Model Diagnostics

Figure 7.11 displays caterpillar plots of the estimated random effects, while Fig. 7.12 shows kernel densities of the random effects. Both figures confirm that the random effects of cross-prices are small compared to the random intercept and the random slope of the own-item price. Another way to get an intuition about the store-specific heterogeneity is to plot the store-specific price curves. This is done for four outlets in Fig. 7.13 showing considerable own-price heterogeneity.

To test the stability of the results, model (7.48) has been reestimated using a fully Bayesian approach based on MCMC methods and the software package `BayesX`. The estimates are built on 102,000 iterations, with 2,000 iterations as burn in period and every 100th sampled parameter used for subsequent inference. We assumed independent inverse gamma priors for the variance parameters $\tau_0^2, \ldots, \tau_4^2$ with hyperparameters $a = b = 0.001$. The results remain stable for the alternative choices $a = b = 0.01$ and $a = b = 0.0001$. Table 7.7 shows the resulting fixed effects and random effects variance parameter estimates. Overall the results are in close agreement with the results of `lme4` given in Table 7.6.

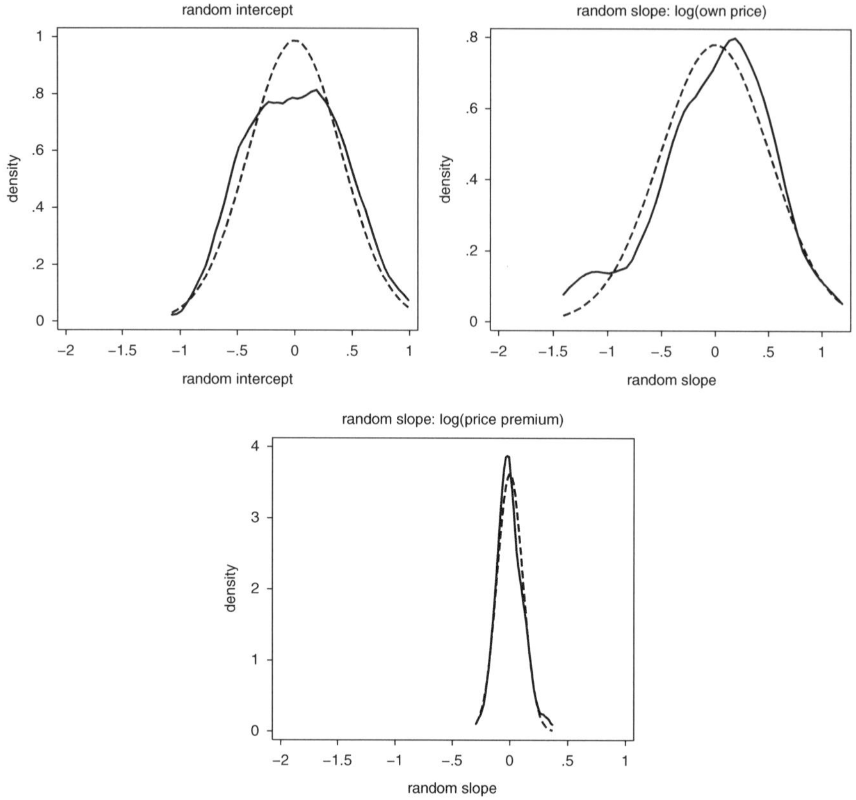

Fig. 7.12 Sales of orange juice: kernel densities of random effects (*solid lines*) together with corresponding normal densities (*dashed lines*) in the full model (7.48)

Since the estimated variances of cross-price random slopes are small, it seems reasonable to simplify the model by removing all or some of the cross-price random effects. Indeed, if we apply the simulation-based restricted likelihood ratio test (function `exactRLRT` of the R package `RLRsim`) for testing $H_0 : \tau_l^2 = 0$, $l = 2, 3, 4$ versus $H_1 : \tau_l^2 \neq 0$, most p-values are comparably large indicating that the null hypotheses cannot be rejected. The only p-value lower than 0.05 is for $logpc3$. Moreover, the DIC in a fully Bayesian approach for a reduced model without cross-price random effects is only 11 points above that of the full model (7.48). Hence, these results are at least supportive that a reduced model without cross-price random slopes may be sufficient for these data. The results for the reduced model are again given in Tables 7.6 and 7.7.

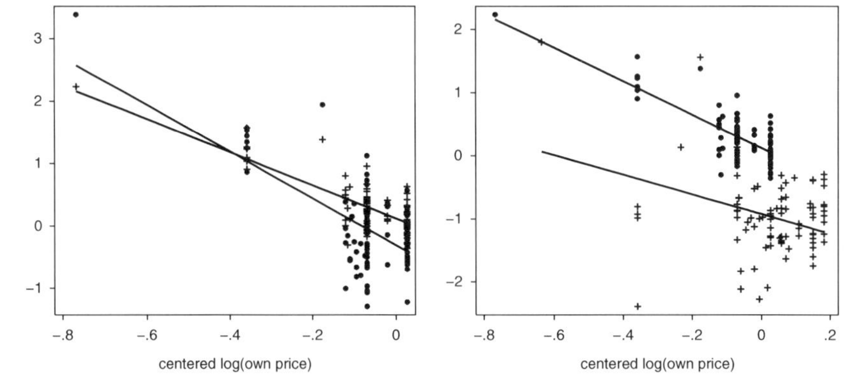

Fig. 7.13 Sales of orange juice: log sales versus log price curves for four different outlets together with corresponding partial residuals in the full model (7.48)

Table 7.7 Sales of orange juice: estimated fixed effects and variance parameters for the full model (7.48) and a reduced model without random slopes for cross-price effects in a fully Bayesian approach

		Full		Reduced	
	Parameter	Estimate	s.e.	Estimate	s.e.
intercept	β_0	3.489	0.046	3.488	0.047
logp	β_1	−2.540	0.076	−2.532	0.074
logpc1	β_2	0.454	0.033	0.453	0.033
logpc2	β_3	1.110	0.034	1.108	0.031
logpc3	β_4	0.542	0.028	0.535	0.023
$\text{Var}(\gamma_{0i})$	τ_0^2	0.171		0.169	
$\text{Var}(\gamma_{1i})$	τ_1^2	0.320		0.317	
$\text{Var}(\gamma_{2i})$	τ_2^2	0.008		–	
$\text{Var}(\gamma_{3i})$	τ_3^2	0.014		–	
$\text{Var}(\gamma_{4i})$	τ_4^2	0.017		–	
$\text{Var}(\varepsilon_{ij})$	σ^2	0.188		1.190	

7.8 Bibliographic Notes and Proofs

7.8.1 Bibliographic Notes

LMMs for longitudinal data are described extensively in Verbeke and Molenberghs (2000), with a focus on likelihood inference and biostatistical applications. Other books including (chapters on) LMMs and GLMMs are Diggle et al. (2002) (also with biostatistical applications), Demidenko (2004), Skrondal and Rabe-Hesketh (2004), and the handbook on longitudinal data edited by Fitzmaurice, Davidian, Verbeke, and Molenberghs (2003). Gelman and Hill (2006) place more emphasis on clustered and, more generally, multilevel data and on social science applications. LMMs from an econometricians point of view are covered in Hsiao (2003). Bayesian LMM and GLMM are described in Fahrmeir and Kneib (2011), Chap. 3, including models with non-Gaussian random effects priors. A nice introduction to LMM for longitudinal and cluster data from a practitioner's point of view is given in Skrondal and Rabe-Hesketh (2008). Although the focus is on analysis with STATA, it is worth reading even if STATA is not used for inference.

Besides mixed models, two other conceptual models exist for the analysis of longitudinal data. *Conditional* or *autoregressive models* extend linear and generalized linear models for cross-sectional data by including effects of response values observed at previous time points $j-1, j-2, \ldots$ into the predictor. The simplest possibility is

$$\eta_{ij} = \boldsymbol{x}'_{ij}\boldsymbol{\beta} + y_{i,j-1}\alpha,$$

i.e., an autoregressive predictor of first order. For statistical inference, $y_{i,j-1}$ is treated formally as an additional covariate.

Marginal models, often in form of generalized estimating equations (*GEE*) approaches, consider multivariate versions of score functions with similar structure as the score functions in Sect. 6.4 for categorical response. However, the true covariance matrix $\boldsymbol{\Sigma}_i$ of the multinomial distribution is replaced through a "working covariance matrix" $\boldsymbol{S}_i$, leading to a multivariate version of the quasi-likelihood approach in Sect. 5.5. Solving the estimating equation $s(\hat{\boldsymbol{\beta}}) = \boldsymbol{0}$ provides the quasi-likelihood or GEE estimator $\hat{\boldsymbol{\beta}}$ in marginal models.

Both classes of models are described, for example, in Diggle et al. (2002) and, in less detail, in Fahrmeir and Tutz (2001).

7.8.2 Proofs

Estimator in the Fixed Effects Model (P. 367)

We first consider the classical linear model

$$\boldsymbol{y} = \boldsymbol{X}\boldsymbol{\beta} + \boldsymbol{\varepsilon} = \boldsymbol{X}_1\boldsymbol{\beta}_1 + \boldsymbol{X}_2\boldsymbol{\beta}_2 + \boldsymbol{\varepsilon}$$

involving two sets of covariates with design matrices $\boldsymbol{X}_1$ and $\boldsymbol{X}_2$. We derive the least squares estimator for $\boldsymbol{\beta}_2$. The normal equations are given by

$$\begin{pmatrix} \boldsymbol{X}'_1\boldsymbol{X}_1 & \boldsymbol{X}'_1\boldsymbol{X}_2 \\ \boldsymbol{X}'_2\boldsymbol{X}_1 & \boldsymbol{X}'_2\boldsymbol{X}_2 \end{pmatrix} \begin{pmatrix} \boldsymbol{\beta}_1 \\ \boldsymbol{\beta}_2 \end{pmatrix} = \begin{pmatrix} \boldsymbol{X}'_1\boldsymbol{X}_1\boldsymbol{\beta}_1 + \boldsymbol{X}'_1\boldsymbol{X}_2\boldsymbol{\beta}_2 \\ \boldsymbol{X}'_2\boldsymbol{X}_1\boldsymbol{\beta}_1 + \boldsymbol{X}'_2\boldsymbol{X}_2\boldsymbol{\beta}_2 \end{pmatrix} = \begin{pmatrix} \boldsymbol{X}'_1\boldsymbol{y} \\ \boldsymbol{X}'_2\boldsymbol{y} \end{pmatrix}.$$

Solving the first set of equations for $\boldsymbol{\beta}_1$ gives the solution

$$\hat{\boldsymbol{\beta}}_1 = (\boldsymbol{X}'_1\boldsymbol{X}_1)^{-1}\boldsymbol{X}'_1\boldsymbol{y} - (\boldsymbol{X}'_1\boldsymbol{X}_1)^{-1}\boldsymbol{X}'_1\boldsymbol{X}_2\hat{\boldsymbol{\beta}}_2 = (\boldsymbol{X}'_1\boldsymbol{X}_1)^{-1}\boldsymbol{X}'_1(\boldsymbol{y} - \boldsymbol{X}_2\hat{\boldsymbol{\beta}}_2). \quad (7.49)$$

Thus, the estimator for $\boldsymbol{\beta}_1$ is obtained by regressing the covariates in $\boldsymbol{X}_1$ on the partial residuals $\boldsymbol{y} - \boldsymbol{X}_2\hat{\boldsymbol{\beta}}_2$. Inserting Eq. (7.49) in the second equation of the normal equations yields

$$\boldsymbol{X}_2'\boldsymbol{X}_1(\boldsymbol{X}_1'\boldsymbol{X}_1)^{-1}\boldsymbol{X}_1'\boldsymbol{y} - \boldsymbol{X}_2'\boldsymbol{X}_1(\boldsymbol{X}_1'\boldsymbol{X}_1)^{-1}\boldsymbol{X}_1'\boldsymbol{X}_2\hat{\boldsymbol{\beta}}_2 + \boldsymbol{X}_2'\boldsymbol{X}_2\hat{\boldsymbol{\beta}}_2 = \boldsymbol{X}_2'\boldsymbol{y}.$$

Rearranging terms then gives

$$\boldsymbol{X}_2'\boldsymbol{X}_2\hat{\boldsymbol{\beta}}_2 - \boldsymbol{X}_2'\boldsymbol{X}_1(\boldsymbol{X}_1'\boldsymbol{X}_1)^{-1}\boldsymbol{X}_1'\boldsymbol{X}_2\hat{\boldsymbol{\beta}}_2 = \boldsymbol{X}_2'(\boldsymbol{I} - \boldsymbol{X}_1(\boldsymbol{X}_1'\boldsymbol{X}_1)^{-1}\boldsymbol{X}_1')\boldsymbol{y}.$$

Solving for $\hat{\boldsymbol{\beta}}_2$ finally produces

$$\begin{aligned}\hat{\boldsymbol{\beta}}_2 &= \left(\boldsymbol{X}_2'(\boldsymbol{I} - \boldsymbol{X}_1(\boldsymbol{X}_1'\boldsymbol{X}_1)^{-1}\boldsymbol{X}_1')\boldsymbol{X}_2\right)^{-1}\left(\boldsymbol{X}_2'(\boldsymbol{I} - \boldsymbol{X}_1(\boldsymbol{X}_1'\boldsymbol{X}_1)^{-1}\boldsymbol{X}_1')\boldsymbol{y}\right)\\ &= \left(\boldsymbol{X}_2'(\boldsymbol{I} - \boldsymbol{H}_1)\boldsymbol{X}_2\right)^{-1}\left(\boldsymbol{X}_2'(\boldsymbol{I} - \boldsymbol{H}_1)\boldsymbol{y}\right),\end{aligned}$$

where $\boldsymbol{H}_1$ is the hat matrix (see Box 3.5 on p. 108) of a regression with the variables in $\boldsymbol{X}_1$ as the regressors. Note that $(\boldsymbol{I} - \boldsymbol{H}_1)\boldsymbol{X}_2$ is a matrix of residuals in regressions of the variables in $\boldsymbol{X}_1$ on the variables in $\boldsymbol{X}_2$. Since both $\boldsymbol{H}_1$ and $\boldsymbol{I} - \boldsymbol{H}_1$ are idempotent, we obtain

$$\hat{\boldsymbol{\beta}}_2 = \left(\tilde{\boldsymbol{X}}_2'\tilde{\boldsymbol{X}}_2\right)^{-1}\tilde{\boldsymbol{X}}_2'\tilde{\boldsymbol{y}},$$

where $\tilde{\boldsymbol{X}}_2 = (\boldsymbol{I} - \boldsymbol{H}_1)\boldsymbol{X}_2$ and $\tilde{\boldsymbol{y}} = (\boldsymbol{I} - \boldsymbol{H}_1)\boldsymbol{y}$. Thus, the estimator $\hat{\boldsymbol{\beta}}_2$ is obtained in two steps. We first regress each column of $\boldsymbol{X}_2$ on the variables in $\boldsymbol{X}_1$. In a second step the set of residuals from the first regression is regressed on the residuals of a regression of $\boldsymbol{X}_1$ on $\boldsymbol{y}$. While this is an important result on its own, we use it to derive the estimator for the regression coefficients in the fixed effects model of p. 367. Consider the model

$$\boldsymbol{y} = \boldsymbol{X}\boldsymbol{\beta} + \boldsymbol{Z}\boldsymbol{\gamma}_0 + \boldsymbol{\varepsilon},$$

where $\boldsymbol{X}$ is the design matrix of the covariates (excluding the intercept) and $\boldsymbol{Z} = (z_1, \ldots, z_m)$ with z_i being a dummy vector indicating the ith unit or cluster. Applying our result on partitioned regression, we obtain

$$\hat{\boldsymbol{\beta}} = (\boldsymbol{X}'(\boldsymbol{I} - \boldsymbol{H}_Z)\boldsymbol{X})^{-1}\boldsymbol{X}'((\boldsymbol{I} - \boldsymbol{H}_Z)\boldsymbol{y}),$$

with $\boldsymbol{H}_Z = \boldsymbol{Z}(\boldsymbol{Z}'\boldsymbol{Z})^{-1}\boldsymbol{Z}'$. This corresponds to a least squares regression using the transformed design matrix $(\boldsymbol{I} - \boldsymbol{H}_Z)\boldsymbol{X}$ and the transformed response vector $(\boldsymbol{I} - \boldsymbol{H}_Z)\boldsymbol{y}$. Due to the specific structure of $\boldsymbol{Z}$, it is easily verified that premultiplication of $(\boldsymbol{I} - \boldsymbol{H}_Z)$ results in cluster-wise centering of the columns of $\boldsymbol{X}$ and of $\boldsymbol{y}$. Thus we obtain $\hat{\boldsymbol{\beta}}$ using a regression of the cluster-wise centered covariates $x_{ijl} - \bar{x}_{il}$ on the cluster-wise centered responses $y_{ij} - \bar{y}_i$.

Derivation of the Penalized Least Squares Estimator in the LMM (Sect. 7.3.1 on p. 371)

We first rearrange the penalized least squares criterion to obtain

$$\begin{aligned}
\mathrm{LS}_{\mathrm{pen}}(\boldsymbol{\beta},\boldsymbol{\gamma}) &= (\boldsymbol{y}-\boldsymbol{X}\boldsymbol{\beta}-\boldsymbol{U}\boldsymbol{\gamma})'\boldsymbol{R}^{-1}(\boldsymbol{y}-\boldsymbol{X}\boldsymbol{\beta}-\boldsymbol{U}\boldsymbol{\gamma})+\boldsymbol{\gamma}'\boldsymbol{G}^{-1}\boldsymbol{\gamma}\\
&= \boldsymbol{y}'\boldsymbol{R}^{-1}\boldsymbol{y}-\boldsymbol{y}'\boldsymbol{R}^{-1}\boldsymbol{X}\boldsymbol{\beta}-\boldsymbol{y}\boldsymbol{R}^{-1}\boldsymbol{U}\boldsymbol{\gamma}\\
&\quad -\boldsymbol{\beta}'\boldsymbol{X}'\boldsymbol{R}^{-1}\boldsymbol{y}+\boldsymbol{\beta}'\boldsymbol{X}'\boldsymbol{R}^{-1}\boldsymbol{X}\boldsymbol{\beta}+\boldsymbol{\beta}'\boldsymbol{X}'\boldsymbol{R}^{-1}\boldsymbol{U}\boldsymbol{\gamma}\\
&\quad -\boldsymbol{\gamma}'\boldsymbol{U}'\boldsymbol{R}^{-1}\boldsymbol{y}+\boldsymbol{\gamma}'\boldsymbol{U}'\boldsymbol{R}^{-1}\boldsymbol{X}\boldsymbol{\beta}+\boldsymbol{\gamma}'\boldsymbol{U}'\boldsymbol{R}^{-1}\boldsymbol{U}\boldsymbol{\gamma}\\
&\quad +\boldsymbol{\gamma}'\boldsymbol{G}^{-1}\boldsymbol{\gamma}\\
&= \boldsymbol{y}'\boldsymbol{R}^{-1}\boldsymbol{y}-2\boldsymbol{y}'\boldsymbol{R}^{-1}\boldsymbol{X}\boldsymbol{\beta}+\boldsymbol{\beta}'\boldsymbol{X}'\boldsymbol{R}^{-1}\boldsymbol{X}\boldsymbol{\beta}+2\boldsymbol{\gamma}'\boldsymbol{U}'\boldsymbol{R}^{-1}\boldsymbol{X}\boldsymbol{\beta}\\
&\quad -2\boldsymbol{y}\boldsymbol{R}^{-1}\boldsymbol{U}\boldsymbol{\gamma}+\boldsymbol{\gamma}'\left(\boldsymbol{U}'\boldsymbol{R}^{-1}\boldsymbol{U}+\boldsymbol{G}^{-}\right)\boldsymbol{\gamma}.
\end{aligned}$$

Applying two rules for the differentiation of vector functions [see Theorem A.33 (1) and (3)], we obtain the derivatives

$$\begin{aligned}
\frac{\partial \mathrm{LS}_{\mathrm{pen}}(\boldsymbol{\beta},\boldsymbol{\gamma})}{\partial\boldsymbol{\beta}} &= -2\boldsymbol{X}'\boldsymbol{R}^{-1}\boldsymbol{y}+2\boldsymbol{X}'\boldsymbol{R}^{-1}\boldsymbol{X}\boldsymbol{\beta}+2\boldsymbol{X}'\boldsymbol{R}^{-1}\boldsymbol{U}\boldsymbol{\gamma},\\
\frac{\partial \mathrm{LS}_{\mathrm{pen}}(\boldsymbol{\beta},\boldsymbol{\gamma})}{\partial\boldsymbol{\gamma}} &= -2\boldsymbol{U}'\boldsymbol{R}^{-1}\boldsymbol{y}+2(\boldsymbol{U}'\boldsymbol{R}^{-1}\boldsymbol{U}+\boldsymbol{G}^{-1})\boldsymbol{\gamma}+2\boldsymbol{U}'\boldsymbol{R}^{-1}\boldsymbol{X}\boldsymbol{\beta}.
\end{aligned}$$

This results in the equation system

$$\begin{pmatrix}\boldsymbol{X}'\boldsymbol{R}^{-1}\boldsymbol{X} & \boldsymbol{X}'\boldsymbol{R}^{-1}\boldsymbol{U}\\ \boldsymbol{U}'\boldsymbol{R}^{-1}\boldsymbol{X} & \boldsymbol{U}'\boldsymbol{R}^{-1}\boldsymbol{U}+\boldsymbol{G}^{-1}\end{pmatrix}\begin{pmatrix}\hat{\boldsymbol{\beta}}\\ \hat{\boldsymbol{\gamma}}\end{pmatrix}=\begin{pmatrix}\boldsymbol{X}'\boldsymbol{R}^{-1}\boldsymbol{y}\\ \boldsymbol{U}'\boldsymbol{R}^{-1}\boldsymbol{y}\end{pmatrix}.$$

Defining $\boldsymbol{C}=(\boldsymbol{X},\boldsymbol{U})$ and the partitioned matrix

$$\boldsymbol{B}=\begin{pmatrix}\boldsymbol{0} & \boldsymbol{0}\\ \boldsymbol{0} & \boldsymbol{G}^{-1}\end{pmatrix},$$

yields

$$(\boldsymbol{C}'\boldsymbol{R}^{-1}\boldsymbol{C}+\boldsymbol{B})\begin{pmatrix}\boldsymbol{\beta}\\ \boldsymbol{\gamma}\end{pmatrix}=\boldsymbol{C}'\boldsymbol{R}^{-1}\boldsymbol{y}.$$

The solution is the penalized least squares estimator (7.25).

Nonparametric Regression 8

The main goal of nonparametric regression is the flexible modeling of effects of continuous covariates on a dependent variable. We have already seen in several practical applications that a purely linear model is not always sufficient. This insufficiency could either result from theoretical considerations about the given application or simply from uncertainty about the specific form of an effect that a covariate has on the response. In Sect. 3.1.3, we considered two possible approaches for modeling the nonlinear effect of a continuous covariate: simple transformations and polynomials. We found that scatter plots of the residuals, in particular partial residual plots, are useful diagnostic devices for identifying nonlinear relations between covariates and response. Even though these methods may be sufficient for simple relationships, they quickly become intractable in situations with a larger number of potential covariates. Moreover, due to the limited number of transformations that are actually used in practice, the functional forms associated with covariate effects are both limited and not very flexible. In this chapter, we will therefore introduce flexible regression techniques that enable the automatic, data-driven estimation of nonlinear effects. We will first concentrate on models for the effect of one single continuous covariate on a response variable, having an approximately normal distribution. The methods developed in this context will be the basis for the bivariate smoothing methods presented in Sect. 8.2, and the more complex additive models of Chap. 9.

To motivate the following discussion, we consider two examples. In the first, we analyze the risk of malnutrition of children in developing countries (Tanzania, in this case), similar to Example 1.2 (p. 5). We consider chronic malnutrition measured in terms of a Z-score as the response variable and examine its dependence on the age of the child. Figure 8.1 shows the corresponding scatter plot for only one of the districts in Tanzania (Ruvuma). We will limit our considerations in this chapter to this single district, since this yields clearer graphical presentations due to the relatively small sample size. However, in general, the results carry over to the entire data set. Figure 8.1 suggests a nonlinear relationship, where young children have a higher Z-score than older children, implying that young children suffer from a lower risk of malnutrition than older ones. At first glance, a simple polynomial or

L. Fahrmeir et al., *Regression*, DOI 10.1007/978-3-642-34333-9_8,

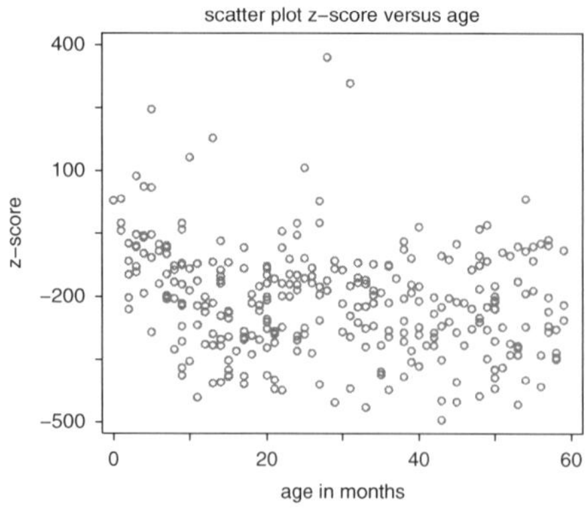

Fig. 8.1 Malnutrition in Tanzania: scatter plot of the Z-score for chronic malnutrition versus the age of the child in months for one of the districts in Tanzania (Ruvuma)

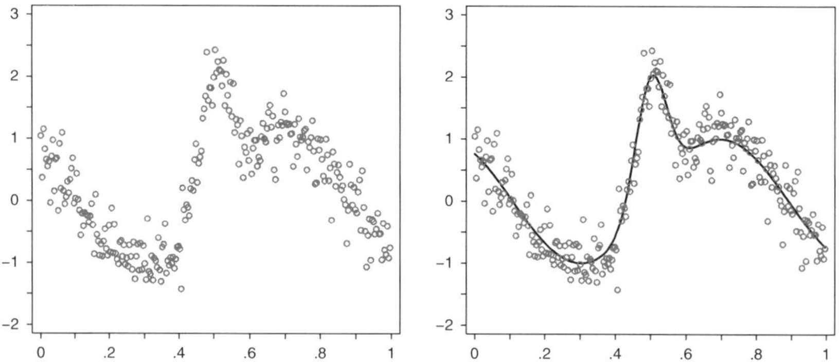

Fig. 8.2 Scatter plot of a simulated data set with nonlinear effect of the covariate: The *right panel* additionally shows the true covariate effect. The data have been simulated according to the model $y = f(x) + \varepsilon$ where $f(x) = \sin(2(4x - 2)) + 2\exp(-(16^2)(x - 0.5)^2)$ and $\varepsilon \sim \mathrm{N}(0, 0.3^2)$

the inverse transformation of age appears adequate for modeling the age effect. In the approaches presented in Sect. 8.1, we will, however, learn that a more flexible model provides additional insights.

As a second example, we use the simulated data set presented in Fig. 8.2. The left panel shows a scatter plot of the simulated data, whereas the right panel also provides the underlying true effect of the covariate. Despite the fact that the scatter plot indicates the structure of the true function fairly well, it is rather difficult to approximate this function through a transformation or in terms of simple polynomials.

8.1 Univariate Smoothing

In this first basic section, we examine several nonparametric regression approaches, which allow flexible modeling of the effect of one continuous covariate on a continuous dependent variable. Such approaches are called *scatter plot smoothers* since the data can be visualized best in a scatter plot, and the goal is to determine a smooth function representing the effect of the covariate. More precisely, we assume that data are given in the form (y_i, z_i), $i = 1, \ldots, n$, where the y_i are observations of the response variable and the z_i represent the corresponding values of the continuous covariate. In the standard univariate nonparametric regression model, we assume that we are able to explain the response variable through a deterministic function of the covariate plus an additive error term:

$$y_i = f(z_i) + \varepsilon_i.$$

Different assumptions regarding the function f then lead to different modeling possibilities. In order to obtain a simpler estimation problem, we often impose qualitative constraints concerning the smoothness of the function f, e.g., regarding its continuity or differentiability. We also should point out that some nonparametric smoothing approaches are, despite their name, in fact purely parametric. In contrast to the regression models we have examined thus far, these models are usually determined by a large number of parameters, so that individual parameters no longer have a meaningful interpretation. Nevertheless, we will continue to use the term nonparametric regression in our presentation, but will also exchangeably use the notion of univariate smoothing.

In nonparametric regression models, we make the same assumptions about the error term as in the classical linear model of Chap. 3: The errors are independent and identically distributed with

$$\mathrm{E}(\varepsilon_i) = 0 \quad \text{and} \quad \mathrm{Var}(\varepsilon_i) = \sigma^2, \quad i = 1, \ldots, n.$$

Similar as in the linear model, it follows

$$\mathrm{E}(y_i) = f(z_i) \quad \text{and} \quad \mathrm{Var}(y_i) = \sigma^2, \quad i = 1, \ldots, n,$$

i.e., the expected value of the response variable is to be modeled through the function f. In some situations, we will additionally assume that the errors follow a normal distribution, especially when constructing confidence intervals for f.

8.1.1 Polynomial Splines

As a first approach for nonparametric regression, we consider *polynomial splines* or *regression splines* which are closely related to the idea of polynomial regression modeling. Therefore, we revisit the polynomial model

$$f(z_i) = \gamma_0 + \gamma_1 z_i + \ldots + \gamma_l z_i^l,$$

8.1 Univariate Smoothing

Data

Measurements (y_i, z_i), $i = 1, \dots, n$, for a continuous response variable y and a continuous covariate z.

Model

$$y_i = f(z_i) + \varepsilon_i$$

with independent and identically distributed errors and

$$\mathrm{E}(\varepsilon_i) = 0 \quad \text{and} \quad \mathrm{Var}(\varepsilon_i) = \sigma^2.$$

In some cases, we additionally assume that the errors are i.i.d. normally distributed, so that

$$\varepsilon_i \sim \mathrm{N}(0, \sigma^2).$$

which models the effect of covariate z on responses y as a polynomial of degree l. This model is within the scope of linear models discussed in Chap. 3 and, thus, we are able to determine the regression coefficients of the polynomials and the complete function f using ordinary least squares. In contrast to Chap. 3, we denote the regression coefficients as γ_j, which will allow us to distinguish the more flexible models discussed in this chapter from linear models.

Based on the simulated data example, Fig. 8.3 shows why a purely polynomial model is often insufficient to estimate nonlinear functions $f(z)$. We considered polynomials up to a degree of $l = 15$ for estimating f. As seen, polynomials are not able to properly reproduce the local maximum at $z = 0.5$, even with (the very high) degree of $l = 11$. For $l = 15$ this maximum is nearly obtained, but the estimated relationship is very wiggly and not quite satisfactory in other regions.

To make the polynomial model more flexible, we next partition the domain of the covariate into intervals and estimate separate polynomials in each interval. In other words, we use several locally defined polynomials instead of one global model. Yet we can still estimate each of the polynomial parts separately using least squares. The left panel in Fig. 8.4 illustrates this approach. Here we split up the domain of z into ten intervals of width 0.1 and fit separate polynomials in each. This actually leads to a more flexible estimate of the function, which reflects the form of the true function. However, the estimated function also exemplifies the major disadvantage of piecewise polynomial models: Since the polynomials are fit separately on the intervals, the polynomial pieces do not lead to an overall smooth function and, in particular, they show different function values at the interval boundaries. Therefore, it would be desirable to impose further smoothness restrictions on the function

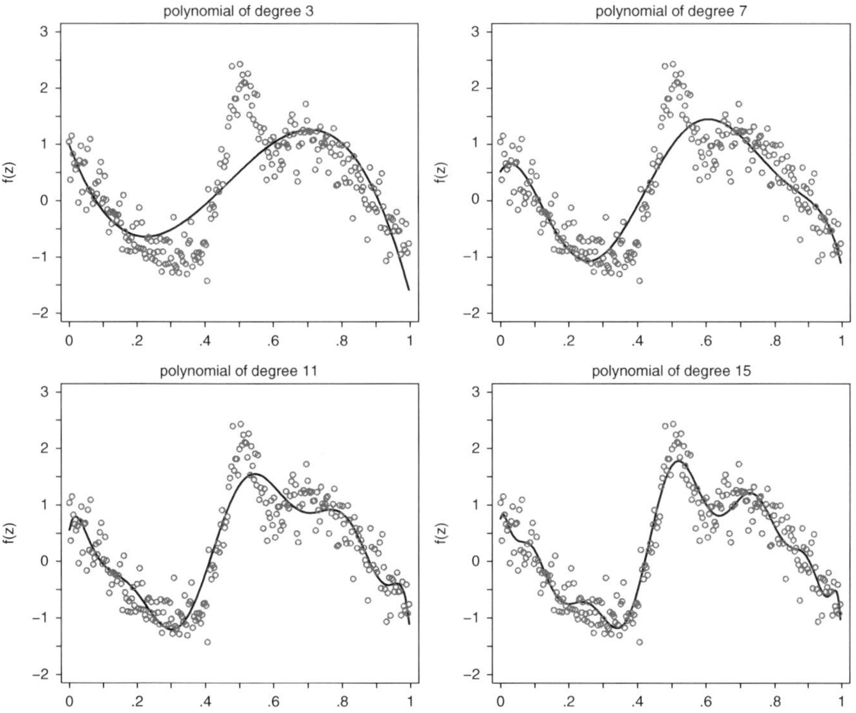

Fig. 8.3 Polynomial regression models for the simulated data set

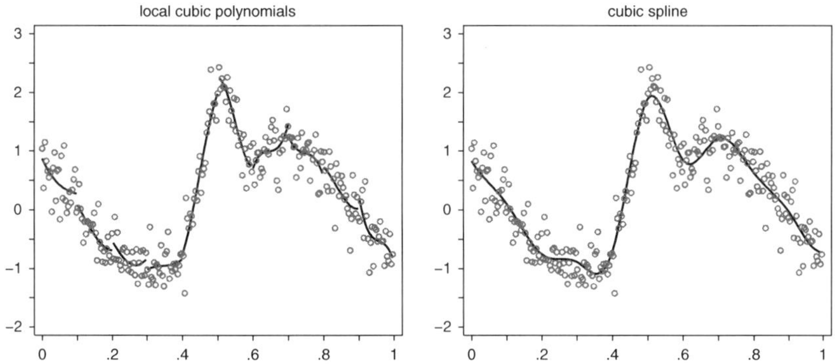

Fig. 8.4 Piecewise polynomial regression (*left*) and polynomial splines (*right*)

at the interval boundaries to obtain a function estimate similar to the one shown in the right panel of Fig. 8.4. Introducing these additional smoothness constraints leads to the class of *polynomial splines*. The main idea still is to define local polynomials on intervals in the domain of the covariate; however, to guarantee sufficient smoothness, we additionally require that the resulting function should be $(l-1)$-times continuously differentiable at the interval boundaries. This yields the following definition:

8.2 Polynomial Splines

A function $f : [a, b] \to \mathbb{R}$ is called a polynomial spline of degree $l \geq 0$ with knots $a = \kappa_1 < \ldots < \kappa_m = b$, if it fulfills the following conditions:

1. $f(z)$ is $(l-1)$-times continuously differentiable. The special case of $l = 1$ corresponds to $f(z)$ being continuous (but not differentiable). We do not state any smoothness requirements for $f(z)$ when $l = 0$.
2. $f(z)$ is a polynomial of degree l on the intervals $[\kappa_j, \kappa_{j+1})$ defined by the knots.

This summarizes our initial considerations, yielding a piecewise polynomial function where the partition of the covariate domain results from the specification of the *knots* $\kappa_1 < \ldots < \kappa_m$. Moreover, the function is assumed to be $(l-1)$-times continuously differentiable to ensure the desired smoothness restrictions at the knots. Figure 8.5 shows some simple polynomial splines with a small number of knots to demonstrate the impact of the chosen spline degree. While the degree l of the spline determines the global smoothness, the diversity of available functions is mainly driven by the number of knots. The more knots we use the higher the number of piecewise polynomials that constitute the polynomial spline. In the following sections, we will discuss the influence of the spline degree and the number (as well as the location) of knots in more detail.

Prior to using polynomial splines in nonparametric regression, we need a representation of the set of polynomial splines for a given degree and knots configuration. This can be achieved with different but equivalent approaches. In the following two sections, we will discuss the two most popular variants: the truncated power series and B-splines.

Polynomial Splines and the Truncated Power Series

Consider the regression model

$$y_i = \gamma_1 + \gamma_2 z_i + \ldots + \gamma_{l+1} z_i^l + \gamma_{l+2}(z_i - \kappa_2)_+^l + \ldots + \gamma_{l+m-1}(z_i - \kappa_{m-1})_+^l + \varepsilon_i$$

with

$$(z - \kappa_j)_+^l = \begin{cases} (z - \kappa_j)^l & z \geq \kappa_j, \\ 0 & \text{otherwise.} \end{cases}$$

The first part of this model is a global polynomial of degree l, as presented at the beginning of this chapter (but with a different way of indexing where the intercept is denoted as γ_1 instead of γ_0 for reasons that will become clearer later on). In contrast, the coefficient of the highest polynomial changes at every single knot $\kappa_2, \ldots, \kappa_{m-1}$.

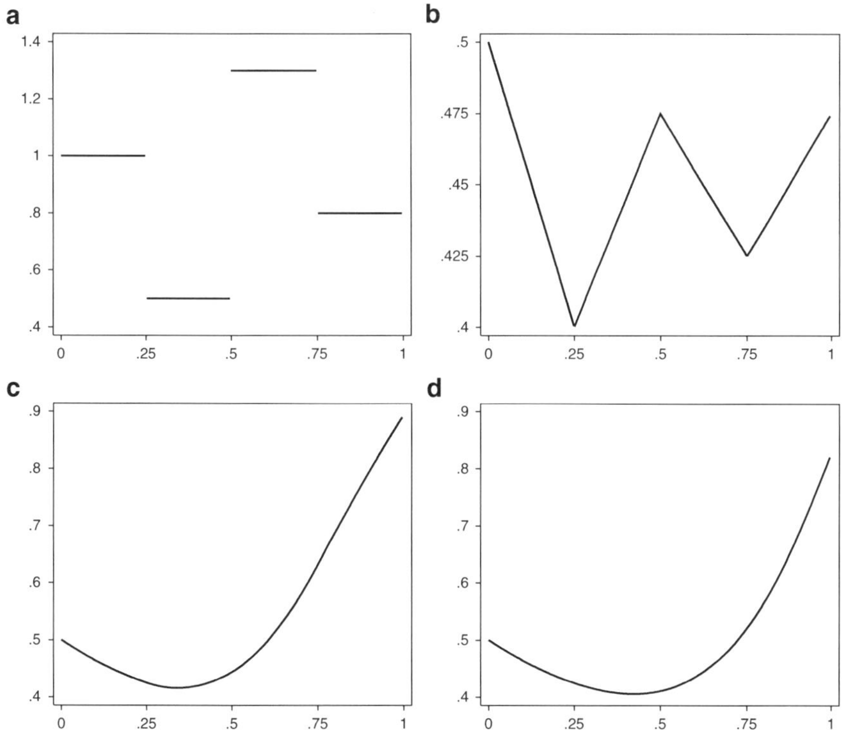

Fig. 8.5 Examples of polynomial splines of degree 0, 1, 2, and 3 with knots $\kappa_1 = 0$, $\kappa_2 = 0.25$, $\kappa_3 = 0.5$, $\kappa_4 = 0.75$, and $\kappa_5 = 1$. (**a**) Spline of degree 0, (**b**) Spline of degree 1, (**c**) Spline of degree 2, (**d**) Spline of degree 3

On the one hand, this specification allows the use of local polynomials in every interval defined by the knots, and on the other hand it fulfills the demand for global smoothness. Figure 8.6 illustrates the concept for a polynomial spline with $l = 1$. Panel (a) shows the functions that define the model, i.e., the global polynomial function of degree $l = 1$ (dashed line) and the additional truncated portions (solid lines). We scale these functions with estimated regression coefficients according to the given data, yielding panel (b). We used equally spaced knots with a distance of 0.1 to define the basis functions.

The horizontal line at $y \approx 0.8$ then corresponds to the global constant γ_1. In the first interval $[0, 0.1)$, we obtain a decreasing function starting from this global level, represented by the slope parameter γ_2. From the knot $\kappa_2 = 0.1$ onward, γ_3 superimposes this slope. In our example, γ_3 is also negative resulting in a somewhat steeper decreasing trend of the function. The positive coefficient γ_4 decreases the negative slope from $\kappa_3 = 0.2$ onward. Nevertheless, we still have a negative trend. When considering the further trend of the function, the coefficients γ_j indicate the change of the slope that occurs at the corresponding knot κ_{j-1}. Due to the special

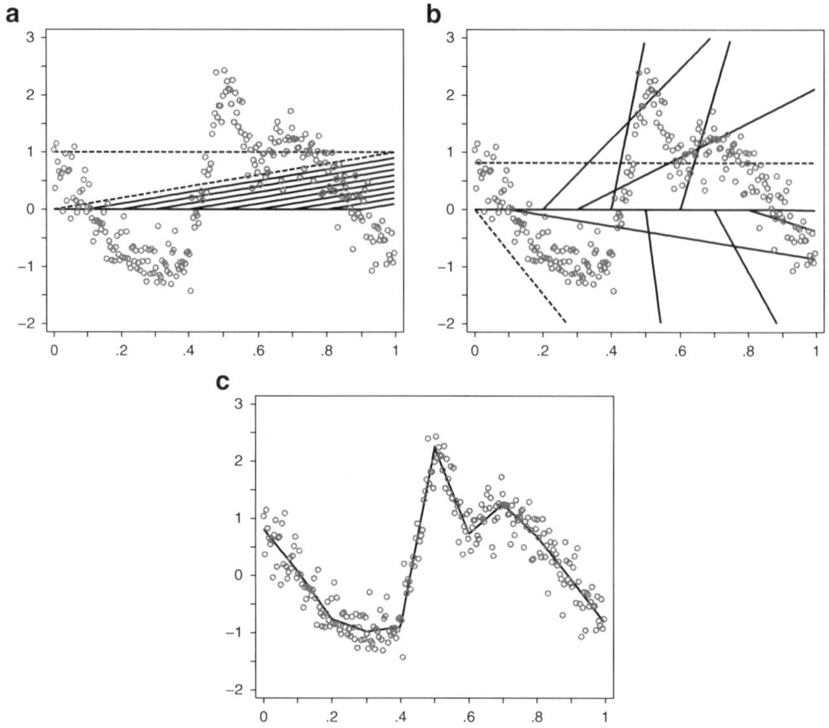

Fig. 8.6 Polynomial spline fit with linear truncated polynomials. (**a**) Basis functions, (**b**) Scaled basis functions (**c**) Sum of scaled basis functions

construction of the truncated function $(z-\kappa_j)_+^l$, we ensure that the change in slope is smooth enough so that the properties of a polynomial spline are preserved. Finally, when adding up all of the scaled functions, we obtain the fit of $f(z)$ illustrated in Fig. 8.6c.

More formally, it can be shown that each polynomial spline of degree l with knots $\kappa_1 < \ldots < \kappa_m$ can be uniquely determined as a linear combination of the $d = m + l - 1$ functions:

$$B_1(z) = 1, \quad B_2(z) = z, \quad \ldots, \quad B_{l+1}(z) = z^l,$$
$$B_{l+2}(z) = (z-\kappa_2)_+^l, \quad \ldots, \quad B_d(z) = (z-\kappa_{m-1})_+^l$$

(see Figs. 8.7 and 8.8, in which these functions are shown for $l = 0$ and $l = 2$ based on a small number of knots). This yields the following representation of the nonparametric regression problem:

$$y_i = f(z_i) + \varepsilon_i = \sum_{j=1}^{d} \gamma_j B_j(z_i) + \varepsilon_i.$$

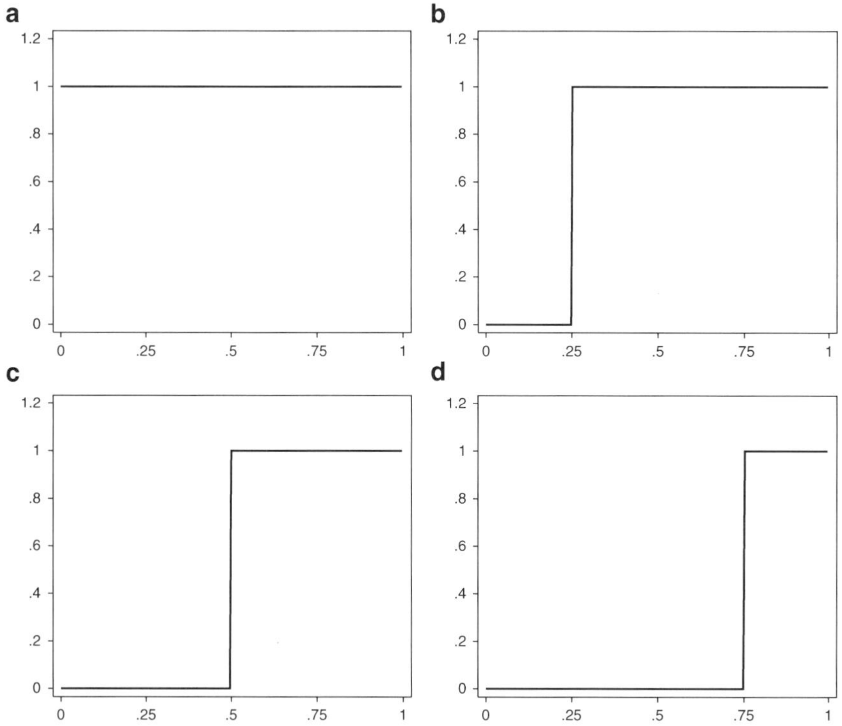

Fig. 8.7 TP basis for splines of degree 0 based on the knots {0,0.25,0.5,0.75,1}. (**a**) Basis function B_1, (**b**) Basis function B_2, (**c**) Basis function B_3, (**d**) Basis function B_4

We call the functions $B_1, \ldots, B_d$ *basis functions*, since we can uniquely represent all polynomial splines by using these functions. To distinguish between various bases for the function space of all splines, we call this basis the *truncated power series basis* (TP basis). One can show that polynomial splines indeed form a d-dimensional vector space so that it is justified to call the TP basis a basis.

Modeling $f(z)$ as a polynomial spline has the advantage that we can still understand the nonparametric regression model as a linear model, but with a possibly large number of parameters. If we define the vectors of the observed response variables $\boldsymbol{y}$ and the errors $\boldsymbol{\varepsilon}$, as well as the design matrix

$$\boldsymbol{Z} = \begin{pmatrix} B_1(z_1) & \ldots & B_d(z_1) \\ \vdots & & \vdots \\ B_1(z_n) & \ldots & B_d(z_n) \end{pmatrix} = \begin{pmatrix} 1 & z_1 & \ldots & z_1^l & (z_1 - \kappa_2)_+^l & \ldots & (z_1 - \kappa_{m-1})_+^l \\ \vdots & & & & & & \vdots \\ 1 & z_n & \ldots & z_n^l & (z_n - \kappa_2)_+^l & \ldots & (z_n - \kappa_{m-1})_+^l \end{pmatrix},$$

we obtain the equation

$$\boldsymbol{y} = \boldsymbol{Z}\boldsymbol{\gamma} + \boldsymbol{\varepsilon},$$

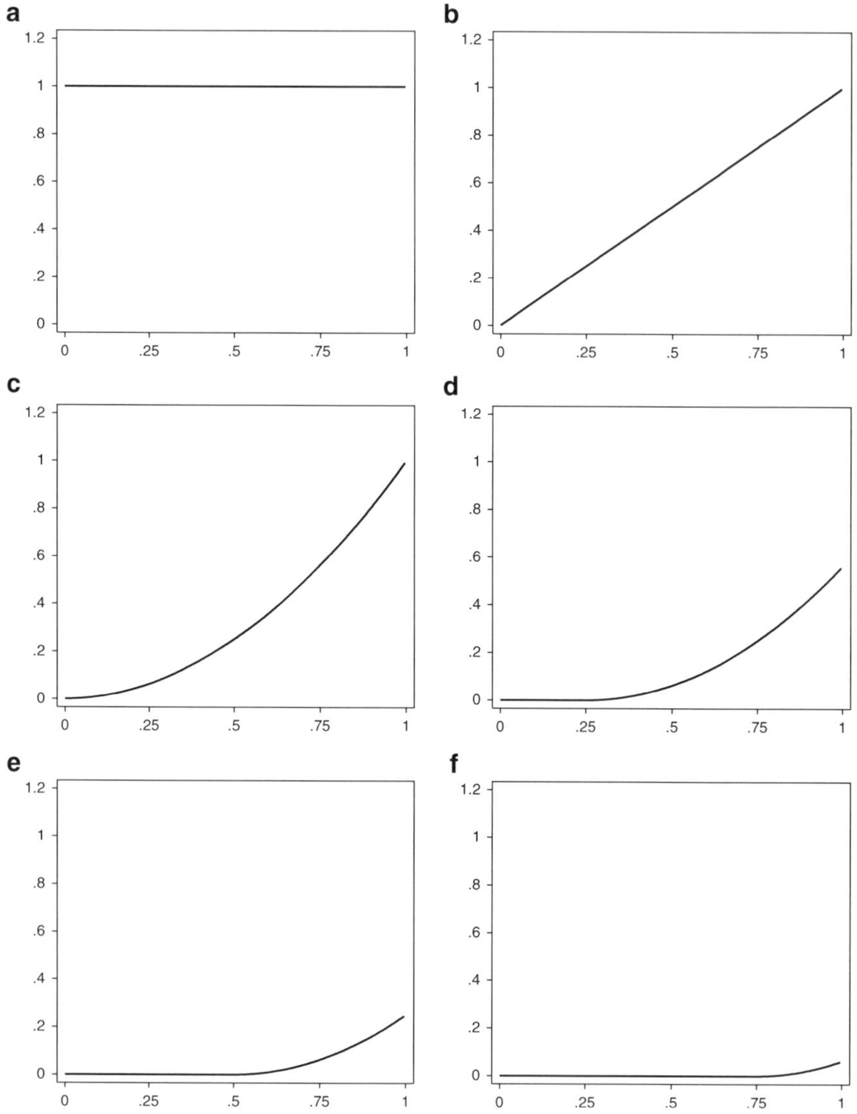

Fig. 8.8 TP basis for splines of degree 2 based on the knots {0,0.25,0.5,0.75,1}. (**a**) Basis function B_1, (**b**) Basis function B_2, (**c**) Basis function B_3, (**d**) Basis function B_4, (**e**) Basis function B_5, (**f**) Basis function B_6

with the coefficient vector $\boldsymbol{\gamma} = (\gamma_1, \dots, \gamma_d)'$. This is indeed a linear model with regression coefficients $\boldsymbol{\gamma}$. The usual least squares estimate is thus

$$\hat{\boldsymbol{\gamma}} = (\boldsymbol{Z}'\boldsymbol{Z})^{-1}\boldsymbol{Z}'\boldsymbol{y}.$$

However, in contrast to the linear models described in Chap. 3, an interpretation of the individual estimated parameters $\hat{\gamma}_j$ is not very informative. Rather, we are interested in the form of the estimated function, which results from the estimated coefficients. In other words, we consider

$$\hat{f}(z) = \boldsymbol{z}'\hat{\boldsymbol{\gamma}}$$

with $\boldsymbol{z} = (B_1(z), \dots, B_d(z))'$ depending on the chosen covariate value z. We can then assess the quality of the model fit in a scatter plot of the data, using the estimated curve. We will return to appropriate measures for assessing the model fit in a subsequent section.

Example 8.1 Malnutrition in Tanzania—Modeling with Polynomial Splines

In our first illustration of nonparametric regression, we compare different specifications for the spline degree and the number of knots using the Tanzania data. Figure 8.9 displays various corresponding fits. Whereas the spline degree mainly determines the overall smoothness of the estimated function, the number of knots influences the flexibility of the estimated curve. For example, with the spline degree $l = 0$, we obtain a piecewise constant function. However, such a fit is inconsistent with the suggestions of a smooth trend for the Z-score with varying age. Increasing the spline degree accordingly, we move from a continuous ($l = 1$) to a continuous and differentiable function ($l = 2$).

For comparison, the numbers of knots are varied (11 and 21). The larger number of knots leads to a "wigglier" estimate for the function, irrespective of the chosen spline degree. It appears that the estimation of the function with 11 knots corresponds better to the theoretical understanding regarding the relationship between age and malnutrition than the one with 21 knots where the additional roughness is difficult to interpret in a meaningful way. It is interesting to note that, in addition to the low risk of malnutrition of young children, we see an unexpected increase in the estimated function after 24 months. This increase actually has an assignable cause due to a change in the reference population to which the Tanzanian children are compared. For young children up to 24 months age, the reference population consists of US American children from white parents of a high socioeconomic status, whereas for older children the reference population is a representative sample from all US American children. This change in reference population is what produces an artificially higher Z-score or a perceived lower risk of malnutrition for the age of 24 months. It is unlikely that we would have been able to identify this effect of the change in the reference population with the use of simple transformations or low-order polynomials.

△

Influence of the Knots

Example 8.1 showed that the choice of the number and the position of the knots, as well as the spline degree, have a definite effect on function estimation. Cubic splines are often used as a default, since they lead to a smooth, twice continuously differentiable function. In contrast, it is much more difficult to provide a "rule of thumb" for the number of knots. As already shown in Example 8.1, a higher

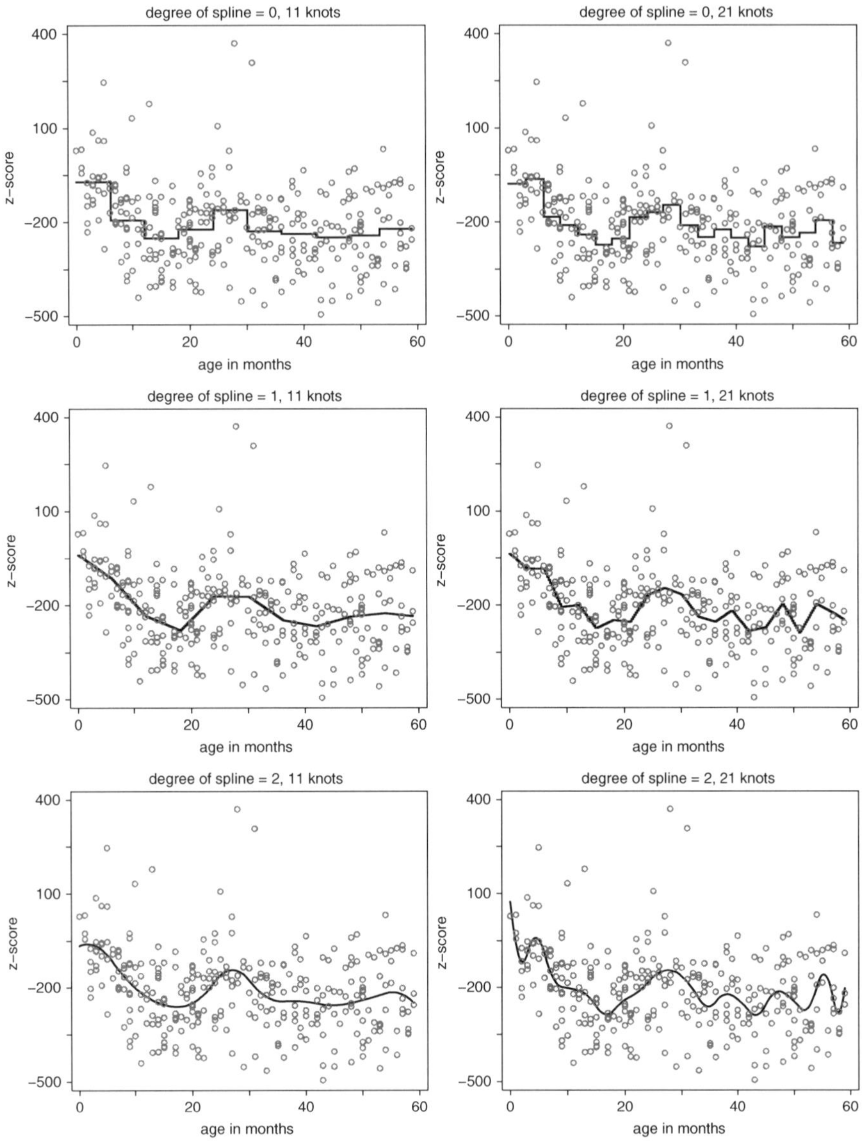

Fig. 8.9 Nonparametric estimates for the age effect based on polynomial splines

(lower) number of knots usually leads to a more (less) flexible estimated function. Figure 8.10 illustrates this effect for simulated data, using a cubic spline while varying the numbers of knots. Since we know the true function for the simulated data, it is easier to assess the quality of the estimates in relation to the number of knots in this case.

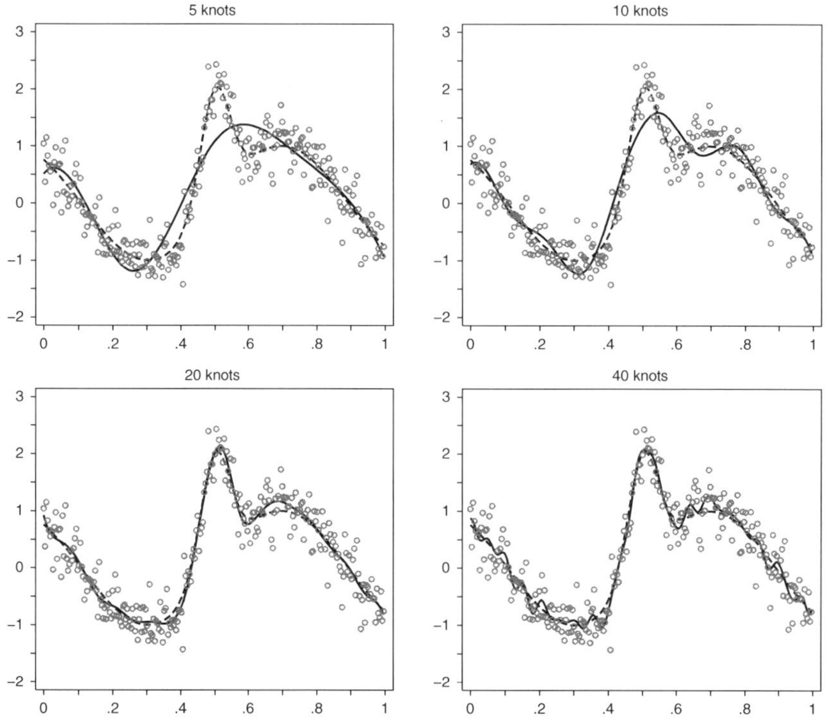

Fig. 8.10 Impact of the number of knots on cubic spline fits: The estimated function is represented as a *solid line* while the true function is superimposed as a *dashed line*

We find results similar to those found in Example 8.1: For a small number of knots, the resulting function estimate is very smooth, and, in our example, does not fit the true function well. In particular, the maximum at $z = 0.5$ is not achieved. We do get more satisfactory results with 20 knots, while further increasing the number of knots produces additional artifacts and results in very rough estimates that are difficult to interpret.

In addition to the number of knots, we also have to choose the distribution or the position of the knots along the covariate axis. The following three approaches are commonly used in practice:

- *Equidistant knots:* The domain $[a, b]$ of z is split into $m - 1$ intervals of width

$$h = \frac{b - a}{m - 1}$$

in order to obtain the knots

$$\kappa_j = a + (j - 1) \cdot h, \quad j = 1, \ldots, m.$$

In all examples considered so far, we have always tacitly assumed equidistant knots.

- *Quantile-based knots:* Use the $(j-1)/(m-1)$-quantiles ($j = 1, \dots, m$) of the observed covariate values $z_1, \dots, z_n$ as knots. By doing so, we place many knots in areas where we have a large number of observations and therefore the distribution of the knots better adapts to the distribution of the explanatory variable.
- *Visual knot choice based on a scatter plot:* Studying the scatter plot of the data allows us to (subjectively) target the placement of the knots either to adapt the knot density to the variability of data or to account for the specific context of the estimation problem.

None of these strategies, however, answers the main question regarding the number of knots. To overcome this problem, we basically have two possibilities: the regularization of the estimation problem through the introduction of a penalty (similar to ridge regression, see Sect. 3.4.4), or the adaptive (i.e., the automatic, data-driven) selection of knots with the help of model choice strategies. Penalization approaches will be the main focus of this book and will be discussed in Sects. 8.1.2–8.1.6. Adaptive methods will be sketched in Sect. 8.1.10. First, we consider an alternative representation of polynomial splines.

B-Splines

Although the TP basis is easy to understand, alternative bases for polynomial splines with numerically favorable properties exist. These alternatives will also be useful in the construction of penalization approaches in the subsequent sections. Due to their construction from truncated polynomials, the calculation of TP basis functions can lead to numerical instabilities for covariates with large values. Moreover, the basis functions of the TP basis are nearly collinear (nearly linear dependent), especially in cases when two knots are very close to each other. Consequently, we will choose the *basic spline* or *B-spline basis* as an alternative basis for polynomial splines.

Prior to giving an exact mathematical definition of B-splines, we first informally motivate the construction of the basis functions. We start with the observations that we made at the beginning of the chapter about the approximation of $f(z)$ with piecewise polynomials. There we realized that additional smoothness conditions had to be imposed on the function $f(z)$. B-spline basis functions are now constructed from piecewise polynomials that are fused smoothly at the knots to achieve the desired smoothness constraints. More specifically, a B-spline basis function consists of $(l+1)$ polynomial pieces of degree l, which are joined in an $(l-1)$-times continuously differentiable way. Figure 8.11 displays the resulting basis function for the spline degrees $l = 0,\ 1,\ 2,\ 3$. All B-spline basis functions are set up based on a given knot configuration. Figure 8.12 displays such complete B-spline bases for equidistant and unevenly distributed knots. Using the complete basis, the function $f(z)$ can again be represented through a linear combination of $d = m + l - 1$ basis functions, i.e.,

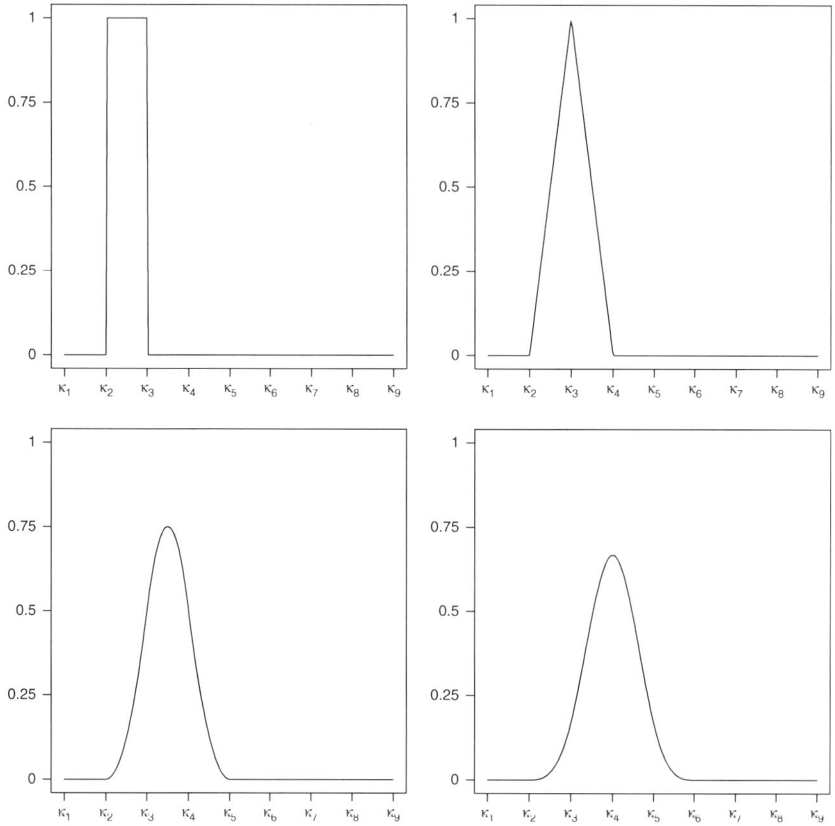

Fig. 8.11 Single B-spline basis functions for degrees $l = 0, 1, 2, 3$ and equidistant knots

$$f(z) = \sum_{j=1}^{d} \gamma_j B_j(z).$$

The main advantage of the B-spline basis is its local definition. In contrast to the truncated polynomials of the TP basis with positive values starting from a certain knot, B-spline basis functions are only positive on an interval based on $l + 2$ knots. Moreover, B-spline basis functions are bounded from above so that the numerical problems of the TP basis do not occur.

From Fig. 8.11, we can immediately deduce the definition for B-splines of order $l = 0$ as

$$B_j^0(z) = I(\kappa_j \leq z < \kappa_{j+1}) = \begin{cases} 1 & \kappa_j \leq z < \kappa_{j+1} \\ 0 & \text{otherwise} \end{cases} \qquad j = 1, \ldots, d-1,$$

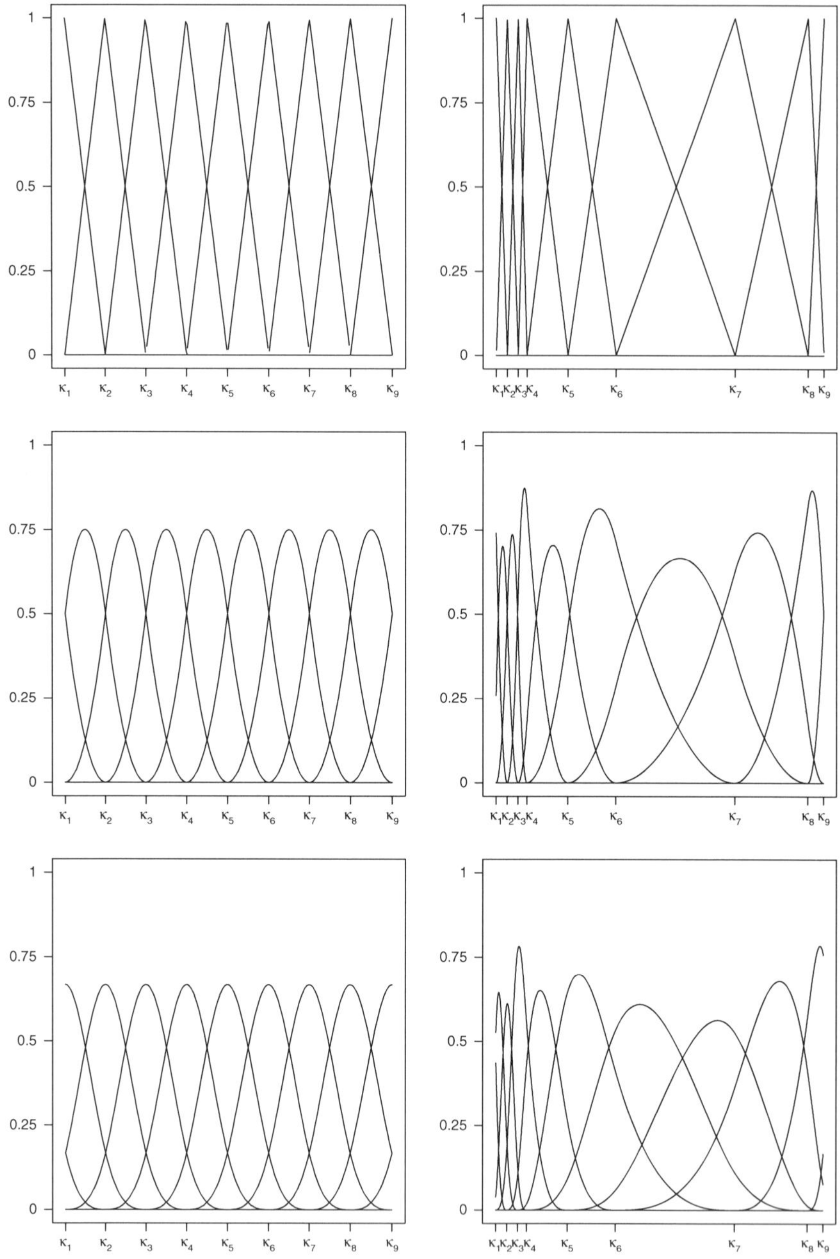

Fig. 8.12 B-spline bases of degree $l = 1, 2, 3$ with equidistant knots (*left panel*) and unevenly distributed knots (*right panel*)

where $I(\cdot)$ denotes the indicator function. In this case, it is also easy to show the equivalence to the TP basis. If we build successive differences of the TP basis functions of degree $l = 0$, we obtain functions that are constant over intervals formed by two adjacent knots, and therefore form a B-spline basis. For B-splines of a higher order, analogous representations based on combinations of piecewise polynomials of degree l can be derived. For example, we obtain the basis functions of degree $l = 1$ as

$$B_j^1(z) = \frac{z - \kappa_{j-1}}{\kappa_j - \kappa_{j-1}} I(\kappa_{j-1} \leq z < \kappa_j) + \frac{\kappa_{j+1} - z}{\kappa_{j+1} - \kappa_j} I(\kappa_j \leq z < \kappa_{j+1}),$$

i.e., each basis function is defined by two linear segments on the intervals $[\kappa_{j-1}, \kappa_j)$ and $[\kappa_j, \kappa_{j+1})$, which are continuously combined at the knot κ_j. In general, higher-order B-splines are defined recursively:

$$B_j^l(z) = \frac{z - \kappa_{j-l}}{\kappa_j - \kappa_{j-l}} B_{j-1}^{l-1}(z) + \frac{\kappa_{j+1} - z}{\kappa_{j+1} - \kappa_{j+1-l}} B_j^{l-1}(z).$$

Applying this formula, for example, for B-splines of order $l = 1$, we obtain exactly the same expression as above. To use the recursive definition of B-splines for the calculation of the basis functions, we need $2l$ *outer knots* outside of the domain $[a, b]$ in addition to the *interior knots* $\kappa_1, \ldots, \kappa_m$. This leads to the expanded knots sequence $\kappa_{1-l}, \kappa_{1-l+1}, \ldots, \kappa_{m+l-1}, \kappa_{m+l}$. The definition of this enlarged number of knots is straightforward for equidistant knots, since we can use the same distance h when defining knots outside of the domain $[a, b]$. This, however, is not possible for unequally distributed knots. Here, it is common to use the distance between the two smallest or largest knots within the covariate domain.

The recursive definition of B-splines makes them appear more complicated than TP-splines; however, this apparent disadvantage is easily balanced with the better numerical properties associated with B-splines. Moreover, most statistical programs (e.g., R or STATA) offer built-in implementations for B-splines so that the user often does not have to deal with such details. For equidistant knots B-splines can also be easily computed from truncated polynomials; see Eilers and Marx (2010) for details.

We next summarize some of the basic characteristics of B-spline basis functions, which can informally be derived from Figs. 8.11 and 8.12:

1. *B-splines form a local basis.* Each basis function is positive only in an interval formed by $l + 2$ adjacent knots. When using equidistant knots, all basis functions have the same form and are only shifted along the z-axis. At any point $z \in [a, b]$, $l + 1$ basis functions are positive.
2. *Unity decomposition.* For every point $z \in [a, b]$, we have:

$$\sum_{j=1}^{d} B_j(z) = 1.$$

3. *Overlapping with 2l adjacent basis functions.* Every basis function (within the domain $[a, b]$) overlaps with exactly $2l$ adjacent basis functions.
4. *Bounded basis functions.* The domain of the individual basis functions is bounded upwards.
5. *Derivatives.* Since the individual basis functions are composed of polynomial parts, it is straightforward to determine derivative formulae. For every single basis function we have

$$\frac{\partial}{\partial z} B_j^l(z) = l \cdot \left(\frac{1}{\kappa_j - \kappa_{j-l}} B_{j-1}^{l-1}(z) - \frac{1}{\kappa_{j+1} - \kappa_{j+1-l}} B_j^{l-1}(z) \right).$$

Therefore, we obtain the derivative for the entire polynomial spline as

$$\frac{\partial}{\partial z} \sum_j \gamma_j B_j^l(z) = l \cdot \sum_j \frac{\gamma_j - \gamma_{j-1}}{\kappa_j - \kappa_{j-l}} B_{j-1}^{l-1}(z). \tag{8.1}$$

As a consequence, we are able to express the derivative of a polynomial spline in terms of the differences of adjacent basis coefficients and B-spline basis functions of one lower degree. Thus, by estimating the coefficients γ_j, we do not only obtain an estimate for the function itself but also for its derivative. Analogously, we can express higher-order derivatives using higher-order differences of the coefficients and basis functions of lower order; see De Boor (2001), p. 115.

Similar to the TP basis, the estimation of a polynomial spline in B-spline representation can be traced back to the estimation of a linear model with a large number of parameters and design matrix

$$\boldsymbol{Z} = \begin{pmatrix} B_1^l(z_1) & \dots & B_d^l(z_1) \\ \vdots & & \vdots \\ B_1^l(z_n) & \dots & B_d^l(z_n) \end{pmatrix}.$$

Due to the properties of the B-spline basis, this design matrix has some special characteristics. It is apparent that the matrix does not contain an explicit intercept term. Since the rows of the design matrix sum to one, due to the unit decomposition of the B-spline basis, the intercept is however implicitly contained in the span of the basis. Specifying an additional intercept would lead to an unidentifiable model. Since the B-spline basis is locally defined, $\boldsymbol{Z}$ has a special structure and mainly consists of zeros. This results in a band matrix structure for $\boldsymbol{Z}'\boldsymbol{Z}$ with bandwidth l. We can use this property to solve the normal equation $\boldsymbol{Z}'\boldsymbol{Z}\boldsymbol{\gamma} = \boldsymbol{Z}'\boldsymbol{y}$ in a numerically efficient way.

Figure 8.13 illustrates the estimation of a B-spline fit for the simulated data example. First, we calculate a complete B-spline basis (in this case of degree 3) for a given number of knots [panel (a)]. The least squares estimate $\hat{\boldsymbol{\gamma}}$ then yields an

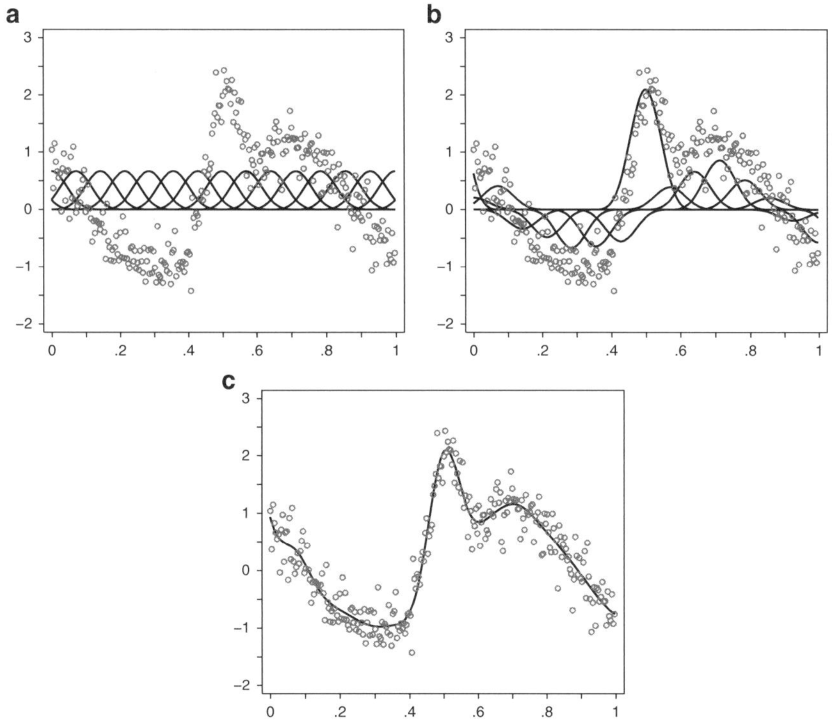

Fig. 8.13 Schematic representation of a nonparametric fit with cubic B-splines. (**a**) B-spline basis (**b**) Scaled B-spline basis (**c**) Sum of scaled B-spline basis functions

amplitude $\hat{\gamma}_j$ for the scaling of every basis function [panel (b)]. When summing the scaled basis functions, we obtain the final estimate [panel (c)].

8.1.2 Penalized Splines (P-Splines)

As we have seen in the previous section, the quality of a nonparametric function estimated by polynomial splines crucially depends on the number of knots. There are basically two strategies to overcome this problem:

- The adaptive choice of knots based on model choice strategies
- The regularization of the estimation problem through the introduction of roughness penalties

We will mostly concentrate on approaches based on penalties and, as mentioned, will only give a brief overview of adaptive methods in Sect. 8.1.10.

The main idea of *penalized splines (P-splines)* can be summarized as follows:

- Approximate the function $f(z)$ with a polynomial spline that uses a generous number of knots (usually about 20–40). This ensures that $f(z)$ can be approximated with enough flexibility to represent even highly complex functions.
- Introduce an additional penalty term that prevents overfitting and minimize a *penalized least squares (PLS) criterion* instead of the usual least squares criterion.

P-Splines Based on a TP Basis

We first consider P-splines based on the TP basis, i.e.,

$$f(z) = \gamma_1 + \gamma_2 z + \ldots + \gamma_{l+1} z^l + \gamma_{l+2}(z - \kappa_2)_+^l + \ldots + \gamma_d (z - \kappa_{m-1})_+^l.$$

This basis consists of two different parts: the first $l + 1$ basis functions, which describe a global polynomial in z, and the truncated powers describing deviations from this polynomial. Estimated functions will be rough when there is a lot of variability in the second part of the model. Therefore, regularized function estimates may be defined by introducing a penalty for the coefficients of the corresponding basis functions. We can create such a penalty, e.g., using the sum of squared coefficients

$$\sum_{j=l+2}^{d} \gamma_j^2,$$

so that large coefficients associated with the truncated powers are penalized. Instead of the usual residual sum of squares that underlies the basis function approaches of the previous section,

$$\text{LS} = \sum_{i=1}^{n} (y_i - f(z_i))^2 = \sum_{i=1}^{n} \left(y_i - \sum_{j=1}^{d} \gamma_j B_j(z_i) \right)^2,$$

we now minimize the penalized residual sum of squares

$$\text{PLS}(\lambda) = \sum_{i=1}^{n} \left(y_i - \sum_{j=1}^{d} \gamma_j B_j(z_i) \right)^2 + \lambda \sum_{j=l+2}^{d} \gamma_j^2.$$

The penalty is constructed to discourage estimated functions that are too rough, thereby preventing overfitting to the data. The *smoothing parameter* $\lambda \geq 0$ controls the influence of the penalty. As $\lambda \to 0$, the effect of the penalty disappears so that the penalized residual sum of squares approximately corresponds to the standard residual sum of squares and we have an estimate for $\boldsymbol{\gamma}$ that is close to the least squares estimate. As $\lambda \to \infty$, the estimation criterion is dominated by the penalty such that $\hat{\gamma}_j = 0$ for $j = l + 2, \ldots, d$, and the estimate for $f(z)$ approaches a polynomial of degree l. Varying the value of the smoothing parameter allows us to choose continuously between these two extremes.

The main advantage of the penalization is that the smoothness of the estimated fit is no longer controlled by the number and the position of knots but rather by one single real-valued parameter, namely the smoothing parameter. Since we use a large number of knots, the exact positioning of these knots is of minor importance. To simplify matters, equidistant knots are often used in practice, but knots based on quantiles are used as well. Figure 8.14 shows that penalized function estimates depend on the number of knots only very moderately (when an optimal value is chosen for the smoothing parameter using one of the methods discussed in Sect. 8.1.9). For a very small number of knots, the estimate may not be flexible enough. However, once a sufficiently large number of knots are used, the estimated function only slightly changes when the number of knots is increased further, provided that the smoothing parameter is adapted appropriately.

All smoothing techniques presented in the following will depend on some kind of smoothing parameter that works similarly as the smoothing parameter of penalized splines. i.e., controls the trade-off between smoothness of the resulting estimate and fidelity to the data. Of course, it will then be important to determine a suitable value for the smoothing parameter from the given data. In Sect. 8.1.9, we will discuss the choice of the smoothing parameter in more detail and will offer different possibilities to estimate it along with the nonparametric function $f(z)$.

P-Splines Based on B-Splines

If we represent $f(z)$ using B-spline basis functions instead of a TP basis, an appropriate penalty is less obvious since the decomposition in a parametric polynomial part and a deviation from the polynomial part does not occur in the basis function definition. In order to create a penalty for B-splines, we therefore take a different approach. To characterize the smoothness of any type of function, the use of (squared) derivatives is appropriate, since these represent measures for the variability of a function. Penalties based on the second derivative, such as

$$\lambda \int (f''(z))^2 dz, \tag{8.2}$$

are particularly attractive since they measure the curvature of a function. This type of penalty forms the basis of the implementation of penalized splines in the R package `mgcv`, see Wood (2006), and can also be used in more general approaches, see Sect. 8.1.3. A simple approximation to the derivative is available for B-splines, which we can use for the construction of specialized penalties. From Eq. (8.1), we know that the first derivative of a B-spline can be written as a function of the first differences of the corresponding coefficient vector. In order to obtain a smooth function and to avoid too large values of the first derivative, we introduce penalties that are based on exactly these differences. In the following, we will limit ourselves to equidistant knots, since they result in very simple expressions for the difference penalty.

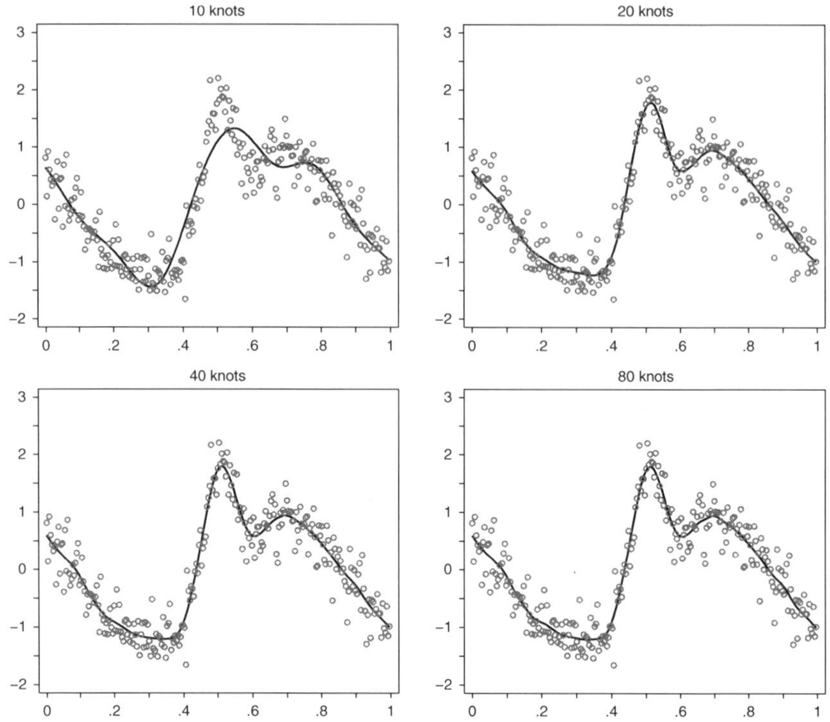

Fig. 8.14 Influence of the number of knots on estimated P-splines

Analogous to first-order differences, we can use differences of a higher order r if we aim at a smooth function in terms of rth-order derivatives. This leads to the penalized residual sum of squares

$$\text{PLS}(\lambda) = \sum_{i=1}^{n} \left(y_i - \sum_{j=1}^{d} \gamma_j B_j(z_i) \right)^2 + \lambda \sum_{j=r+1}^{d} (\Delta^r \gamma_j)^2,$$

where Δ^r denotes rth-order differences, which are recursively defined by

$$\begin{aligned}
\Delta^1 \gamma_j &= \gamma_j - \gamma_{j-1}, \\
\Delta^2 \gamma_j &= \Delta^1 \Delta^1 \gamma_j = \Delta^1 \gamma_j - \Delta^1 \gamma_{j-1} = \gamma_j - 2\gamma_{j-1} + \gamma_{j-2}, \\
&\vdots \\
\Delta^r \gamma_j &= \Delta^{r-1} \gamma_j - \Delta^{r-1} \gamma_{j-1}.
\end{aligned}$$

This type of smoother has been proposed in Eilers and Marx (1996) and is currently one of the most popular smoothers. In order to obtain a better understanding of the impact of the difference penalty on the resulting estimates, we visualize penalized spline estimates for different values of the smoothing parameter, similar as in Fig. 8.13. We will discuss in detail below exactly how these estimates can be obtained. Figure 8.15 shows the results with a second-order difference penalty and cubic B-splines. In most of our applications, second-order differences are chosen, while other orders, e.g., first or third, are possible.

We find that, for a large value of the smoothing parameter ($\lambda \rightarrow \infty$), the function estimate for $f(z)$ is close to linear in case of second-order differences. More generally, as $\lambda \rightarrow \infty$, the fit approaches a polynomial of degree $r - 1$ with r-th order differences, provided that the degree of the spline is at least as large as the order of the differences, i.e., $l \geq r$. For first-order differences, this can easily be shown based on the formula for the first derivative of a B-spline in Eq. (8.1): The first derivative is equal to zero and the corresponding function is a constant if and only if all first-order differences of coefficients are zero. This can be achieved with a large smoothing parameter that places a large weight on the penalty. For penalties with a larger difference order, analogous results can be obtained from formulae for higher-order derivatives.

In comparison to penalized splines that are based on the TP basis, we obtain the additional freedom to separately choose the degree of the limiting polynomial obtained with heavy penalization as $\lambda \rightarrow \infty$ and the degree of the splines to be used for the modeling of $f(z)$. In contrast, the unpenalized polynomial is always of degree l for the TP basis since only the coefficients of the truncated powers are penalized. When using B-splines, we are able to choose the degree of the polynomial that results with heavy penalization, through the choice of the difference order k, which is independent of the degree l of the spline. This provides us with additional flexibility in the modeling process and also with a clear separation of the two different model features.

Penalized Least Squares Estimation

For the derivation of the PLS estimate, it is advantageous to write the penalty in matrix notation. For the TP basis, this is simply achieved via

$$\lambda \sum_{j=l+2}^{d} \gamma_j^2 = \lambda \boldsymbol{\gamma}' \boldsymbol{K} \boldsymbol{\gamma}$$

with the vector $\boldsymbol{\gamma} = (\gamma_1, \ldots, \gamma_d)'$ and the *penalty matrix*

$$\boldsymbol{K} = \text{diag}(\underbrace{0, \ldots, 0}_{(l+1)}, \underbrace{1, \ldots, 1}_{(m-2)}).$$

For the B-spline penalty based on differences, we start by writing the vector of first differences using the difference matrix

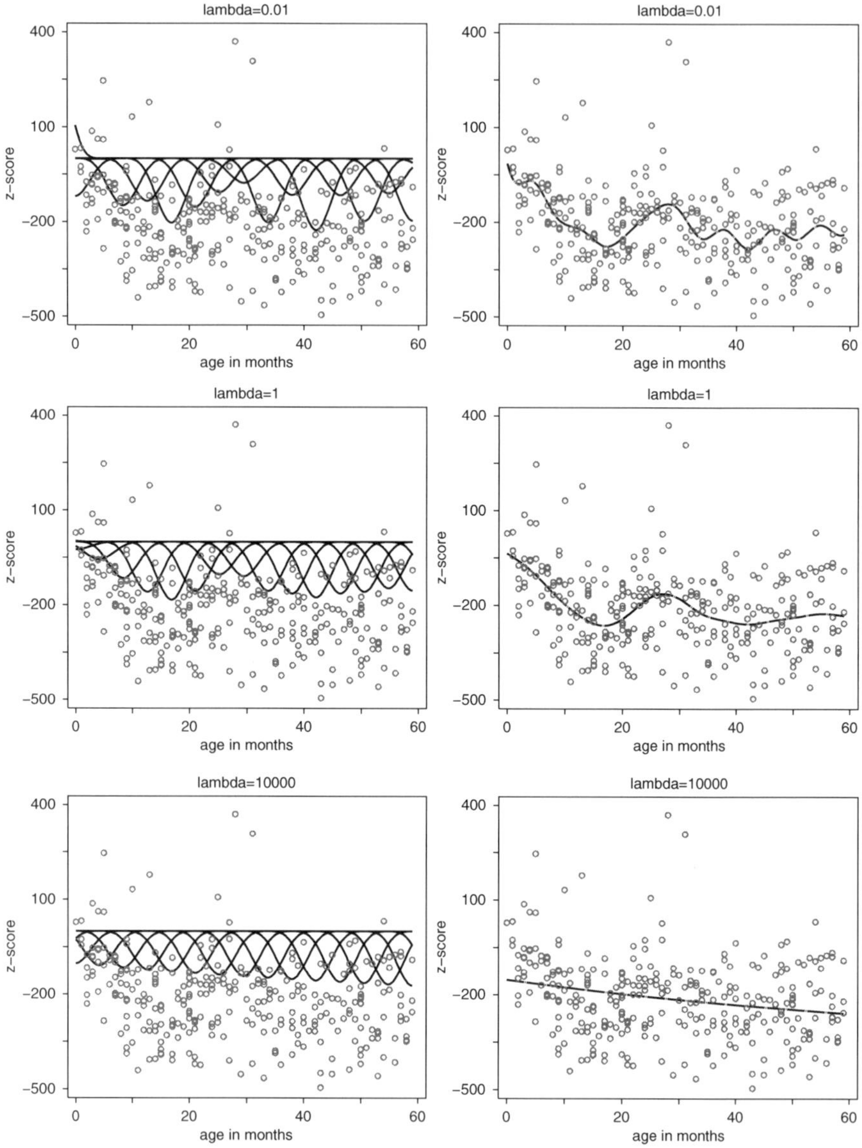

Fig. 8.15 Malnutrition in Tanzania: impact of the smoothing parameter on estimated P-splines with second-order difference penalty

$$D_1 = \begin{pmatrix} -1 & 1 & & & \\ & -1 & 1 & & \\ & & \ddots & \ddots & \\ & & & -1 & 1 \end{pmatrix}$$

of dimension $(d-1) \times d$, yielding

$$\boldsymbol{D}_1\boldsymbol{\gamma} = \begin{pmatrix} \gamma_2 - \gamma_1 \\ \vdots \\ \gamma_d - \gamma_{d-1} \end{pmatrix}.$$

Higher differences can be expressed recursively with the help of difference matrices

$$\boldsymbol{D}_r = \boldsymbol{D}_1 \boldsymbol{D}_{r-1}.$$

For example, with $r = 2$, we obtain the $(d-2) \times d$-difference matrix

$$\boldsymbol{D}_2 = \begin{pmatrix} 1 & -2 & 1 & & & \\ & 1 & -2 & 1 & & \\ & & \ddots & \ddots & \ddots & \\ & & & 1 & -2 & 1 \end{pmatrix}.$$

This yields the penalty

$$\lambda \sum_{j=r+1}^{d} (\Delta^r \gamma_j)^2 = \lambda \boldsymbol{\gamma}' \boldsymbol{D}_r' \boldsymbol{D}_r \boldsymbol{\gamma} = \lambda \boldsymbol{\gamma}' \boldsymbol{K}_r \boldsymbol{\gamma},$$

with first- and second-order difference penalty matrices

$$\boldsymbol{K}_1 = \begin{pmatrix} 1 & -1 & & & \\ -1 & 2 & -1 & & \\ & \ddots & \ddots & \ddots & \\ & & -1 & 2 & -1 \\ & & & -1 & 1 \end{pmatrix}, \qquad \boldsymbol{K}_2 = \begin{pmatrix} 1 & -2 & 1 & & & & \\ -2 & 5 & -4 & 1 & & & \\ 1 & -4 & 6 & -4 & 1 & & \\ & \ddots & \ddots & \ddots & \ddots & \ddots & \\ & & 1 & -4 & 6 & -4 & 1 \\ & & & 1 & -4 & 5 & -2 \\ & & & & 1 & -2 & 1 \end{pmatrix},$$

respectively.

If we define the penalty based on the integral of the squared second derivatives as in Eq. (8.2), we obtain another quadratic penalty

$$\begin{aligned} \int (f''(z))^2 dz &= \int \left(\sum_{j=1}^{d} \gamma_j B_j''(z) \right)^2 dz \\ &= \int \sum_{r=1}^{d} \sum_{j=1}^{d} \gamma_r \gamma_j B_r''(z) B_j''(z) dz \end{aligned}$$

8.3 Penalized Splines

Model

We approximate function f using polynomial splines so that we are able to write the nonparametric regression model as a linear model

$$\boldsymbol{y} = \boldsymbol{Z}\boldsymbol{\gamma} + \boldsymbol{\varepsilon}.$$

Penalized Least Squares Criterion

Rather than using the standard residual sum of squares, we estimate $\boldsymbol{\gamma}$ by minimizing the PLS criterion

$$\text{PLS}(\lambda) = (\boldsymbol{y} - \boldsymbol{Z}\boldsymbol{\gamma})'(\boldsymbol{y} - \boldsymbol{Z}\boldsymbol{\gamma}) + \lambda \boldsymbol{\gamma}'\boldsymbol{K}\boldsymbol{\gamma}.$$

The smoothing parameter $\lambda \geq 0$ controls the compromise between fidelity to the data and smoothness of the resulting function estimate. For splines in a TP basis representation, we penalize the sum of squared coefficients of the truncated powers. For B-splines, we construct the penalty based on the sum of squared differences of neighboring coefficients or based on the integral of the function's squared second derivative.

Penalized Least Squares Estimation

In either case, the PLS estimate has the form

$$\hat{\boldsymbol{\gamma}} = (\boldsymbol{Z}'\boldsymbol{Z} + \lambda \boldsymbol{K})^{-1}\boldsymbol{Z}'\boldsymbol{y}.$$

$$\begin{aligned} &= \sum_{r=1}^{d}\sum_{j=1}^{d} \gamma_r \gamma_j \int B_r''(z) B_j''(z) dz \\ &= \boldsymbol{\gamma}'\boldsymbol{K}\boldsymbol{\gamma}, \end{aligned}$$

with $\boldsymbol{K}[r, j] = \int B_r''(z) B_j''(z) dz$. The entries of the penalty matrix $\boldsymbol{K}$ result from the integrated products of second derivatives of the B-spline basis functions.

In general, we can write the penalized residual sum of squares as

$$\begin{aligned} \text{PLS}(\lambda) &= (\boldsymbol{y} - \boldsymbol{Z}\boldsymbol{\gamma})'(\boldsymbol{y} - \boldsymbol{Z}\boldsymbol{\gamma}) + \lambda \boldsymbol{\gamma}'\boldsymbol{K}\boldsymbol{\gamma} \\ &= \boldsymbol{y}'\boldsymbol{y} - \boldsymbol{y}'\boldsymbol{Z}\boldsymbol{\gamma} - \boldsymbol{\gamma}'\boldsymbol{Z}'\boldsymbol{y} + \boldsymbol{\gamma}'\boldsymbol{Z}'\boldsymbol{Z}\boldsymbol{\gamma} + \lambda \boldsymbol{\gamma}'\boldsymbol{K}\boldsymbol{\gamma} \\ &= \boldsymbol{y}'\boldsymbol{y} - 2\boldsymbol{\gamma}'\boldsymbol{Z}'\boldsymbol{y} + \boldsymbol{\gamma}'(\boldsymbol{Z}'\boldsymbol{Z} + \lambda \boldsymbol{K})\boldsymbol{\gamma} \end{aligned}$$

using an appropriate penalty matrix $\boldsymbol{K}$. Taking the derivative with respect to $\boldsymbol{\gamma}$ (compare Theorem A.33 in Appendix A.8) and setting the derivative equal to zero yields the system of equations

$$-2\boldsymbol{Z}'\boldsymbol{y} + 2(\boldsymbol{Z}'\boldsymbol{Z} + \lambda\boldsymbol{K})\boldsymbol{\gamma} = \boldsymbol{0},$$

and, therefore, the PLS estimate

$$\hat{\boldsymbol{\gamma}} = (\boldsymbol{Z}'\boldsymbol{Z} + \lambda\boldsymbol{K})^{-1}\boldsymbol{Z}'\boldsymbol{y}. \tag{8.3}$$

Consequently, the vector of estimated function evaluations $\hat{\boldsymbol{f}} = (\hat{f}(z_1), \dots, \hat{f}(z_n))'$ is calculated as

$$\hat{\boldsymbol{f}} = \boldsymbol{Z}\hat{\boldsymbol{\gamma}} = \boldsymbol{Z}(\boldsymbol{Z}'\boldsymbol{Z} + \lambda\boldsymbol{K})^{-1}\boldsymbol{Z}'\boldsymbol{y}.$$

Note that the form of the PLS estimate is very similar to the unpenalized least squares estimate, as it only differs by the additional term $\lambda\boldsymbol{K}$. This additional term represents the influence of the penalty within the PLS criterion. For $\lambda = 0$, we again obtain the solution of the unpenalized optimization problem. Based on similar considerations as presented for the least squares estimate in section "Statistical Properties without Specific Distributional Assumptions" of Sect. 3.2.3, we obtain the covariance matrix of the PLS estimate:

$$\text{Cov}(\hat{\boldsymbol{\gamma}}) = \sigma^2(\boldsymbol{Z}'\boldsymbol{Z} + \lambda\boldsymbol{K})^{-1}\boldsymbol{Z}'\boldsymbol{Z}(\boldsymbol{Z}'\boldsymbol{Z} + \lambda\boldsymbol{K})^{-1}.$$

In contrast to the covariance matrix of the least squares estimate, this covariance matrix consists of three parts and has the form of a "sandwich" matrix (known from Sects. 4.1.3, p. 190 and 7.3.3, p. 378). The two additional expressions containing the penalty matrix are responsible for the stabilization of the PLS estimator and, in comparison to the unpenalized least squares estimator, typically lead to smaller variances. However, this achieved efficiency implies that the PLS estimator is no longer unbiased. Provided that the smoothing parameter is chosen appropriately, the PLS estimator generally yields a smaller mean squared error. Thus, from a statistical point of view, the estimator has better properties than the unpenalized least squares estimator.

We find very similar forms when comparing the PLS estimate with that of the posterior expectation

$$\text{E}(\boldsymbol{\gamma} \mid \boldsymbol{y}) = (\boldsymbol{X}'\boldsymbol{X} + \boldsymbol{M}^{-1})^{-1}\boldsymbol{X}'\boldsymbol{y}$$

of a linear model with prior distribution $\boldsymbol{\gamma} \sim \text{N}(\boldsymbol{0}, \sigma^2\boldsymbol{M})$ (derived in Sect. 4.4, pp. 225ff.). We can thus supplement penalized splines with a Bayesian interpretation as detailed below. Additionally, penalized splines are closely related to models containing random effects, as illustrated by a comparison with the estimating equations in Box 7.3 on p. 375. We will use both the Bayesian interpretation and the relation to mixed models in Sect. 8.1.9 when we elaborate on choosing the smoothing parameter λ.

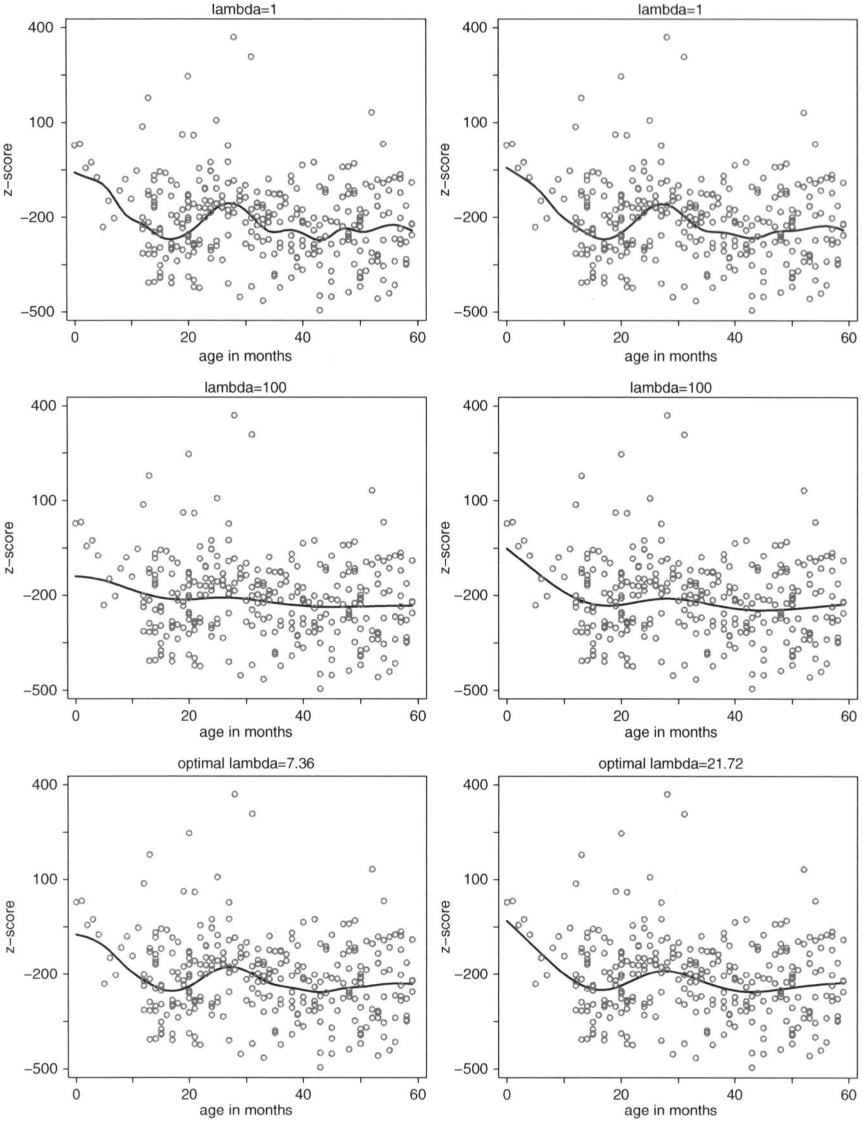

Fig. 8.16 Malnutrition in Tanzania: P-spline estimates based on first (*left panel*) and second (*right panel*) order differences and different smoothing parameters

Example 8.2 Malnutrition in Tanzania: P-Splines

Similar to Example 8.1, we compare several specifications for penalized splines using the data on malnutrition in Tanzania. We focus on P-splines in B-spline representation with 20 inner knots, and examine several smoothing parameter choices as well as penalties that are based on first and second differences. Figure 8.16 shows some of the corresponding results.

In the first row, we can see the estimated function $\hat{f}(z)$, which results from choosing a rather small smoothing parameter. As a consequence, the estimated functions are relatively rough with somewhat smoother results in case of second-order differences. If we choose a larger smoothing parameter (middle row in Fig. 8.16), we obtain very smooth fits. In particular, first-order differences yield an almost constant fit, i.e., the estimated function is close to the limiting case obtained with $\lambda \to \infty$. For second-order differences, we still find some deviation from a linear fit. The bottom row shows estimates resulting from "optimal" smoothing parameters, which were chosen with the help of one of the approaches presented in Sect. 8.1.9. This results in a (at least visually) plausible compromise between fidelity to the data and smoothness. The two aforementioned main features of the relationship between age and malnutrition are easily seen with either penalty: a higher Z-score, which is a lower risk for malnutrition of young children, and a local maximum of the Z-score at the age of approximately 24 months, which is due to the change of the reference population. It would be especially difficult to discover the effect of the change of the reference population without an automatic and data-driven choice of the smoothing parameter.

△

Bayesian P-Splines

Besides their motivation based on a PLS criterion, penalized splines can also be derived in a Bayesian framework. In particular, this allows us to employ Bayesian approaches for the estimation of P-splines including the smoothing parameter; see Sect. 8.1.9. We first focus on penalized splines that are based on B-splines, followed by a brief presentation of Bayesian TP-splines.

For the Bayesian formulation, we start with the same observation model as in the previous section, i.e.,

$$y_i = f(z_i) + \varepsilon_i = \sum_{j=1}^{d} \gamma_j B_j(z_i) + \varepsilon_i,$$

with B-spline basis functions B_j. Instead of imposing a penalty, we will now develop an appropriate prior assumption for $\boldsymbol{\gamma}$ that enforces a smooth function estimation. The stochastic analogue for the difference penalty is *random walks* of order k (RWk). A random walk of first order (RW1) is defined by

$$\gamma_j = \gamma_{j-1} + u_j, \quad u_j \sim \mathrm{N}(0, \tau^2), \quad j = 2, \ldots, d,$$

or equivalently

$$\gamma_j - \gamma_{j-1} = u_j, \quad u_j \sim \mathrm{N}(0, \tau^2), \quad j = 2, \ldots, d,$$

so that a connection to the first-order difference penalty is recognizable. We have to make further assumptions for the prior of the starting value γ_1 and a noninformative prior distribution; $p(\gamma_1) \propto const$ will be our standard option.

Our prior specification corresponds to the conditional distributions

$$\gamma_j \mid \gamma_{j-1}, \ldots, \gamma_1 \sim \mathrm{N}(\gamma_{j-1}, \tau^2). \tag{8.4}$$

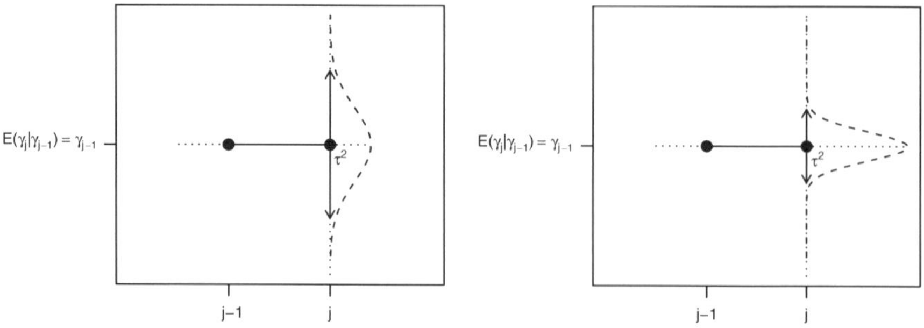

Fig. 8.17 Conditional distribution induced by a first-order random walk with a large (*left panel*) and a small (*right panel*) variance

The RW1 has a special dependence structure such that the conditional distribution of γ_j given all previous values is only dependent on the value lagged by one, i.e., γ_{j-1}. Therefore, the RW1 has the (first-order) *Markov property*.

We next take a closer look at the dependence structure of the parameter vector $\boldsymbol{\gamma}$ imposed by the random walk prior. According to Eq. (8.4), the conditional expectation of γ_j is simply the lagged value γ_{j-1} such that we obtain a constant trend for the expected value. This is visualized in Fig. 8.17, which also shows the effect of the variance τ^2 of the errors u_j. The larger the variance, the larger the possible deviation from the conditional expectation. As we know from the discussion of P-splines in the previous section, a constant value of all B-spline coefficients leads to a constant estimate for the function $f(z)$. This corresponds to the case that the variance of the RW1 is (almost) zero, since only very little deviation between γ_j and γ_{j-1} is allowed in this situation resulting in a (near) constant trend for the sequence $\gamma_1, \ldots, \gamma_d$. In contrast, when having a large variance τ^2, neighboring coefficients are able to deviate from each other, which in turn leads to a rough estimated function. Thus we can interpret the variance parameter τ^2 as an inverse smoothing parameter. We will characterize the exact relationship between λ and τ in more detail below.

Although the definition of a random walk in Eq. (8.4) only determines the conditional distributions $\gamma_j \mid \gamma_{j-1}, \ldots, \gamma_1$, we can actually use these distributions to determine the joint multivariate prior distribution of the complete vector $\boldsymbol{\gamma}$ under an RW1:

$$\begin{aligned} p(\boldsymbol{\gamma} \mid \tau^2) &= \prod_{j=1}^{d} p(\gamma_j \mid \gamma_{j-1}, \ldots, \gamma_1) \\ &= p(\gamma_1) \prod_{j=2}^{d} p(\gamma_j \mid \gamma_{j-1}) \end{aligned}$$

$$\propto \prod_{j=2}^{d} \frac{1}{\sqrt{2\pi\tau^2}} \exp\left(-\frac{1}{2\tau^2}(\gamma_j - \gamma_{j-1})^2\right)$$

$$= \frac{1}{(2\pi\tau^2)^{(d-1)/2}} \exp\left(-\frac{1}{2\tau^2}\sum_{j=2}^{d}(\gamma_j - \gamma_{j-1})^2\right)$$

$$= \frac{1}{(2\pi\tau^2)^{(d-1)/2}} \exp\left(-\frac{1}{2\tau^2}\boldsymbol{\gamma}'\boldsymbol{K}_1\boldsymbol{\gamma}\right).$$

The first step is based on using the rule of total probability to factor the joint density. Due to the Markov property, we then obtain simplified expressions for the conditional distributions, so that we can insert the density of the normal distribution that appears in the RW1 prior. The proportionality sign results from the flat prior for the first regression coefficient γ_1, which we did not specify exactly but only up to a proportionality constant. The joint distribution has the form of a multivariate normal distribution with expectation $\mathbf{0}$ and precision matrix $\boldsymbol{K}_1/\tau^2$, where $\boldsymbol{K}_1 = \boldsymbol{D}_1'\boldsymbol{D}_1$ is defined by the first-order difference matrix $\boldsymbol{D}_1$ as in the previous section. However, this precision matrix does not have full rank, since rows and columns sum to zero. Consequently, we cannot obtain the inverse, and the covariance matrix "$\tau^2\boldsymbol{K}_1^{-1}$" of the prior distribution does not exist. Thus, we cannot normalize the density of $\boldsymbol{\gamma}$, and the integral of the density diverges. Therefore, the joint prior of $\boldsymbol{\gamma}$ is an improper, singular normal distribution (see Appendix B.3.2). More precisely, we call the distribution partially improper since the precision matrix is not of full rank, but $\text{rk}(\boldsymbol{K}_1) = d - 1 > 0$ holds. For a first-order random walk, the deterministic part of the prior distribution (as defined in Appendix B.3.2) corresponds to a constant vector representing the level of function f. With the partially improper prior, we define a flat prior for this level. The stochastic part of the prior represents the deviations from the constant. We will use similar ideas in Sect. 8.1.9 in order to create a representation of penalization approaches as mixed models.

The multivariate prior distribution for the vector $\boldsymbol{\gamma}$ also determines the undirected forms of the conditional distribution (8.4). In this case, we condition on all remaining parameters and not only on the preceding values $\gamma_{j-1}, \ldots, \gamma_1$. For the RW1, we obtain

$$\gamma_j \mid \cdot \sim \text{N}\left(\frac{1}{2}(\gamma_{j-1} + \gamma_{j+1}), \frac{\tau^2}{2}\right), \quad j = 2, \ldots, d-1$$

(with the exception of the boundary parameters γ_1 and γ_d). We again recognize the Markov property (this time, however, in undirected form), according to which the conditional distribution of γ_j, given all remaining parameters, only depends on the two immediate neighbors. More precisely, the conditional expectation is the local average of the two adjacent regression coefficients. A more detailed discussion of the random walk prior that also contains information about the boundary parameters can be found in Lang and Brezger (2004).

Although the joint prior distribution is partially improper, we obtain a proper multivariate normal posterior distribution for $\boldsymbol{\gamma}$. The derivation is similar to the steps to obtain the posterior in Bayesian linear model; see Sect. 4.4. In particular, we obtain the following expressions for the posterior expectation and covariance matrix of $\boldsymbol{\gamma}$:

$$\mathrm{E}(\boldsymbol{\gamma} \mid \boldsymbol{y}, \sigma^2, \tau^2) = \left(\boldsymbol{Z}'\boldsymbol{Z} + \frac{\sigma^2}{\tau^2}\boldsymbol{K}_1\right)^{-1} \boldsymbol{Z}'\boldsymbol{y},$$

i.e., the PLS estimate with smoothing parameter $\lambda = \sigma^2/\tau^2$, and

$$\mathrm{Cov}(\boldsymbol{\gamma} \mid \boldsymbol{y}, \sigma^2, \tau^2) = \sigma^2(\boldsymbol{Z}'\boldsymbol{Z} + \lambda\boldsymbol{K}_1)^{-1}.$$

In contrast to the non-Bayesian results of the last section, we obtain a simpler form for the covariance matrix. The reason is due to the stochastic formulation of the Bayesian model: Whereas the frequentist approach considers the regression coefficients $\boldsymbol{\gamma}$ as fixed and unknown parameters, we now consider the regression coefficients as random.

In the Bayesian formulation, the P-spline smoothing parameter λ is given by the ratio of the error variance and the variance of the RW1. This solidifies our previous conjectures regarding the influence of the variance τ^2 of the random walk and further leads to an interesting interpretation of the smoothing parameter λ: The larger the variance of the prior distribution is relative to the variance of the residuals, the less the estimation will be penalized. Consequently, we always have to interpret the value of τ^2 relative to the variance σ^2 that is associated with the measurement error $\boldsymbol{\varepsilon}$. We can then refer to λ as the noise-to-signal ratio.

For random walks of a higher order, we can derive analogous results. We now briefly consider the special case of the RW2. A RW2 prior for $\boldsymbol{\gamma}$ is defined by

$$\gamma_j = 2\gamma_{j-1} - \gamma_{j-2} + u_j, \quad u_j \sim \mathrm{N}(0, \tau^2), \quad j = 3, \ldots, d,$$

usually in combination with noninformative prior distributions $p(\gamma_1) \propto const$ and $p(\gamma_2) \propto const$ for the initial values. When reformulating the RW2 definition to

$$\gamma_j - 2\gamma_{j-1} + \gamma_{j-2} = u_j, \quad u_j \sim \mathrm{N}(0, \tau^2), \quad j = 3, \ldots, d,$$

we again can identify the similarity to the second-order difference penalty that was presented in the previous section. If we specify the assumption of a RW2 in terms of conditional distributions, we obtain

$$\gamma_j \mid \gamma_{j-1}, \gamma_{j-2}, \ldots, \gamma_1 \sim \mathrm{N}(2\gamma_{j-1} - \gamma_{j-2}, \tau^2)$$

corresponding to a local linear extrapolation of the conditional expectation; see Fig. 8.18. For the joint distribution of the entire vector $\boldsymbol{\gamma}$, we have

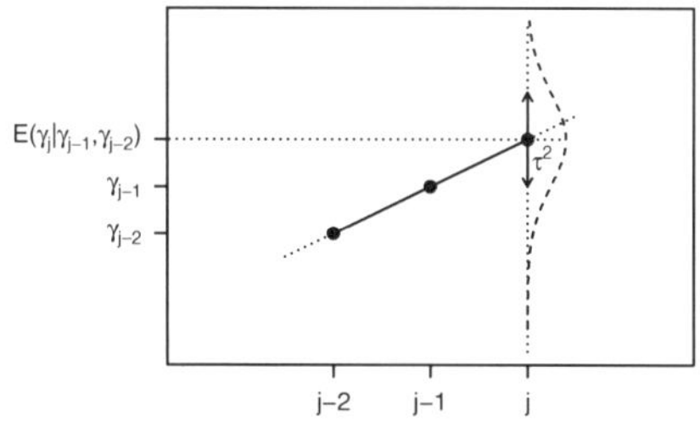

Fig. 8.18 Conditional distribution imposed by a second-order random walk

$$p(\boldsymbol{\gamma} \mid \tau^2) \propto \frac{1}{(\tau^2)^{(d-2)/2}} \exp\left(-\frac{1}{2\tau^2}\boldsymbol{\gamma}'\boldsymbol{K}_2\boldsymbol{\gamma}\right),$$

with $\boldsymbol{K}_2 = \boldsymbol{D}_2'\boldsymbol{D}_2$. Again, the precision matrix is not of full rank and the prior distribution for the RW2 is partially improper with $\text{rk}(\boldsymbol{K}_2) = d - 2$.

Although we can obtain point estimates and covariance matrices for Bayesian P-splines analytically when the smoothing parameter is given, we mainly use the Bayesian formulation to develop a Bayesian inferential framework for penalized splines and the more general penalization approaches in Sect. 8.1.9. This also opens up the possibility to determine the smoothing parameter in a rather simple way. In addition, it is easier to implement the additive extensions, within the scope of Bayesian approaches, discussed in Chap. 9.

P-splines in TP basis representation can also be motivated in a Bayesian framework. In fact, it is even easier to construct the prior distribution, since an i.i.d. prior for the regression coefficients lends itself to the equivalence for the ridge penalty described in the previous section. More precisely, we assume that the coefficients $\gamma_{l+2}, \dots, \gamma_d$ of the truncated polynomials are independently and identically $\text{N}(0, \tau^2)$ distributed. For the coefficients of the global polynomials $\gamma_1, \dots, \gamma_{l+1}$, we consider noninformative prior distributions, i.e., $p(\gamma_j) \propto const$, $j = 1, \dots, l + 1$. Similar to the frequentist penalization approach, the two different types of prior distributions reflect a distinction between parameters to be estimated in a restricted and an unrestricted way. The Bayesian formulation of TP-splines also introduces a connection between nonparametric regression models and mixed models since the distributional assumptions cannot only be interpreted as prior information, but also distinguish between fixed and random effects in a mixed model. Such a connection will also help us to derive an estimation approach for the smoothing parameter.

Similar to P-splines based on B-splines, we can now derive the posterior expectation and covariance matrix for TP-splines. In particular, the PLS estimator is obtained as the posterior expectation.

8.1.3 General Penalization Approaches

Prior to discussing further models of nonparametric regression, we collect some properties of penalized splines that also apply to a number of general penalization approaches. In fact, many of the univariate and bivariate smoothing methods that we discuss in Sects. 8.1 and 8.2 can be included in such a general framework.

Representing polynomial splines with basis functions as in

$$f(z) = \sum_{j=1}^{d} \gamma_j B_j(z) \tag{8.5}$$

resulted in a "large" linear model of the form

$$\boldsymbol{y} = \boldsymbol{Z}\boldsymbol{\gamma} + \boldsymbol{\varepsilon}, \tag{8.6}$$

in which the design matrix was defined by the evaluations of the basis functions. However, in Eq. (8.5), we are of course not limited to TP-splines or B-splines, but we can rather use any type of functions for the approximation of $f(z)$. The choice of the basis functions B_j then determines the class of functions we employ to describe $f(z)$. For example, either the TP or the B-spline basis results in the space of all polynomial splines for the given knot configuration and degree l. In the following sections we will learn about other models of the form (8.6) that are useful in nonparametric regression. The concept of penalized residual sums of squares used to derive P-splines can also be implemented more generally. Assuming that Eq. (8.5) holds, we can always define a suitable penalty via the integral of the squared second derivative (as shown on p. 437):

$$\int (f''(z))^2 dz = \sum_{i=1}^{d}\sum_{j=1}^{d} \gamma_i \gamma_j \int B_i''(z) B_j''(z) dz = \boldsymbol{\gamma}'\boldsymbol{K}\boldsymbol{\gamma}.$$

We again find a quadratic penalty for $\boldsymbol{\gamma}$ where the entries of the penalty matrix $\boldsymbol{K}$ are determined by the second derivatives of the basis functions. As with P-splines, alternative construction mechanisms for penalties may exist for more general types of basis functions, but these often lead to quadratic penalties $\lambda\boldsymbol{\gamma}'\boldsymbol{K}\boldsymbol{\gamma}$ as well.

As an optimization criterion, we always obtain a PLS criterion of the form

$$\text{PLS}(\lambda) = (\boldsymbol{y} - \boldsymbol{Z}\boldsymbol{\gamma})'(\boldsymbol{y} - \boldsymbol{Z}\boldsymbol{\gamma}) + \lambda\boldsymbol{\gamma}'\boldsymbol{K}\boldsymbol{\gamma},$$

differing only in the specification of the design matrix $\boldsymbol{Z}$ and the penalty matrix $\boldsymbol{K}$. Minimization of the PLS criterion yields the PLS estimate

$$\hat{\boldsymbol{\gamma}} = (\boldsymbol{Z}'\boldsymbol{Z} + \lambda\boldsymbol{K})^{-1}\boldsymbol{Z}'\boldsymbol{y}.$$

8.4 General Penalization Approaches

Model

The function f is represented through a large linear model, such that

$$y = Z\gamma + \varepsilon.$$

Often this is achieved by approximating the function $f(z)$ using basis functions, i.e.,

$$f(z) = \sum_{j=1}^{d} \gamma_j B_j(z).$$

Penalization

To regularize estimation, we assume a quadratic penalty of the form

$$\lambda\gamma' K\gamma$$

or the normal prior distribution

$$p(\gamma \mid \tau^2) \propto \left(\frac{1}{2\pi\tau^2}\right)^{\mathrm{rk}(K)/2} \exp\left(-\frac{1}{2\tau^2}\gamma' K\gamma\right)$$

for γ. This leads to the PLS criterion

$$(y - Z\gamma)'(y - Z\gamma) + \lambda\gamma' K\gamma$$

and the PLS estimate

$$\hat{\gamma} = (Z'Z + \lambda K)^{-1} Z' y.$$

We can also consider penalty approaches with quadratic penalties in a Bayesian context. Penalties of the form $\lambda\gamma' K\gamma$ correspond to the assumption of a multivariate normal prior with density

$$p(\gamma \mid \tau^2) \propto \left(\frac{1}{2\pi\tau^2}\right)^{\mathrm{rk}(K)/2} \exp\left(-\frac{1}{2\tau^2}\gamma' K\gamma\right)$$

for the coefficient vector $\boldsymbol{\gamma}$. However, we should keep in mind that the penalty matrix $\boldsymbol{K}$ may not be of full rank. In such cases, the density cannot be normalized and is only defined upon proportionality.

Two of the approaches that we will discuss in Sect. 8.1.9 to determine the optimal smoothing parameter will be explicitly tailored to models using general penalty approaches.

At this point, we also want to briefly discuss the possibility to model non-normally distributed response variables within the nonparametric regression framework (which will be discussed in more detail in Chap. 9). Transferring the ideas of nonparametric regression to response variables from exponential families, as considered in Chap. 5, we obtain models of the form

$$E(y_i \mid \eta_i) = h(\eta_i), \qquad \eta_i = f(z_i),$$

for the expectation of the response variables. As in Chap. 5, $h(\cdot)$ defines a known response function, e.g., the logistic cumulative distribution function in the logit model or the exponential function associated with log-linear Poisson models. If we use basis functions for approximating the function f, we obtain the vector of linear predictors as $\boldsymbol{\eta} = \boldsymbol{Z}\boldsymbol{\gamma}$. Instead of imposing a penalty on the least squares criterion as we did in nonparametric regression for normally distributed responses, we now consider the penalized log-likelihood criterion

$$l_{pen}(\boldsymbol{\gamma}) = l(\boldsymbol{\gamma}) - \frac{\lambda}{2}\boldsymbol{\gamma}'\boldsymbol{K}\boldsymbol{\gamma} \tag{8.7}$$

for the estimation of the regression coefficients. We denote by $l(\boldsymbol{\gamma})$ the usual log-likelihood function of a generalized linear model; see Sect. 5.4.1. The smoothing parameter has to be divided by 2 to make the likelihood-based approach equivalent to the PLS approach discussed so far. In fact, Eq. (8.7) is equivalent to the PLS criterion in case of normally distributed responses.

In order to determine the penalized ML estimate, we amend the Fisher scoring approach, discussed in Sect. 5.4.2, with the penalty. In particular, the estimated $\hat{\boldsymbol{\gamma}}$ can be determined using iteratively weighted PLS estimation; see Chap. 9 for details. To keep our discussion of further modeling alternatives in nonparametric regression models as simple as possible, we will restrict ourselves to normally distributed responses in this chapter but will return to the more general setting in Chap. 9.

8.1.4 Smoothing Splines

When working with penalized splines, we limit ourselves, right from the start, to a function space spanned by the spline basis functions. However, we can also work with more general function spaces, as we will demonstrate with the example of *smoothing splines*. Here we only assume that the function $f(z)$ is twice continuously differentiable, so that we can use the PLS criterion

$$\sum_{i=1}^{n}(y_i - f(z_i))^2 + \lambda \int (f''(z))^2 dz \tag{8.8}$$

to determine an estimate of f. In this setting, we define the penalty to be the integrated squared second derivative, which yields a measure for the overall curvature of the function. Surprisingly, a very special class of functions emerges as the solution for the optimization problem, despite the generality of the approach taken. This class of functions are the so-called *natural cubic splines*, a special subset of cubic polynomial splines.

8.5 Natural Cubic Splines

The function $f(z)$ is a natural cubic spline based on the knots, $a \le \kappa_1 < \ldots < \kappa_m \le b$, if:

(i) $f(z)$ is a cubic polynomial spline for the given knots.
(ii) $f(z)$ satisfies the boundary conditions $f''(a) = f''(b) = 0$, i.e., $f(z)$ is linear in the intervals $[a, \kappa_2]$ and $[\kappa_{m-1}, b]$.

Clearly, every natural cubic spline is also a cubic polynomial spline. However, due to the boundary conditions, we have additional restrictions such that m basis functions are sufficient to represent a natural cubic spline (in comparison to the $m+2$ basis functions for standard cubic polynomial splines). The boundary conditions are also the motivation for the name "natural cubic spline." Originally, the word "spline" designated a curve template, which means a flexible ruler that allows one to draw a smooth function that interpolates a given set of points. In our example, these points would be the function values evaluated at the knots. Outside of the knot set, the ruler then again takes its "natural," linear form.

The optimal solution for Eq. (8.8) is now given by a natural cubic spline with knots at the d ordered and unique covariate values $z_{(1)} < \ldots < z_{(d)}$. Such a spline is then also called a *smoothing spline*.

Since natural cubic splines are special polynomial splines, we can represent them using the basis functions already discussed in previous sections. Note, however, that d basis functions are sufficient for describing a natural cubic spline, and therefore the representation through common cubic polynomial spline functions contains two redundant basis functions. We obtain a "real" basis for natural cubic splines by introducing suitable modifications of the polynomial spline basis functions at the boundary of the domain. In the representation

$$f(z) = \sum_{j=1}^{d} \gamma_j B_j(z),$$

the basis functions $B_3, \ldots, B_{d-2}$ remain unchanged, while the basis functions B_1, B_2, B_{d-1}, and B_d are adjusted such that the natural boundary conditions are

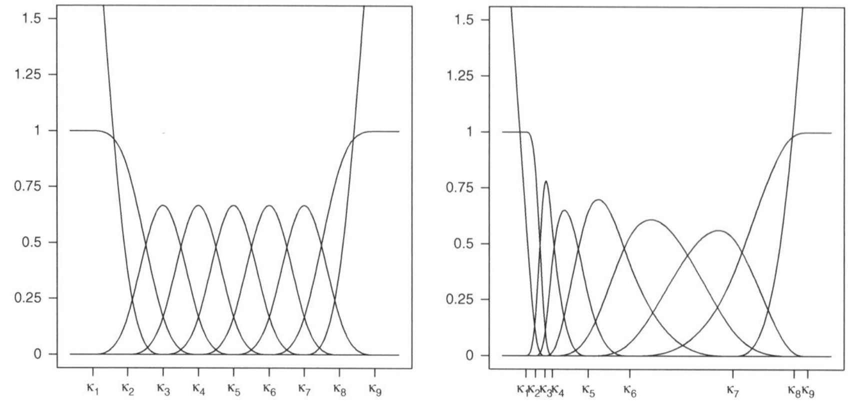

Fig. 8.19 Basis of a natural cubic spline with equidistant knots (*left panel*) and with unevenly distributed knots (*right panel*)

fulfilled. We restrain from an explicit derivation of the resulting basis functions, and rather provide a visualization in Fig. 8.19, using the same knot configuration that was used for constructing the B-spline basis in Fig. 8.12 (p. 428). We find that the basis functions are indeed the same in the interior, whereas the outer basis functions are modified to fulfill the boundary conditions of the natural cubic spline.

For minimizing the PLS criterion (8.8), we need an expression for the penalty based on the integrated squared second derivative. With natural cubic splines using the basis functions presented in Fig. 8.19, we again have

$$\int (f''(z))^2 dz = \sum_{i=1}^{d}\sum_{j=1}^{d} \gamma_i \gamma_j \int B_i''(z) B_j''(z) dz = \boldsymbol{\gamma}'\boldsymbol{K}\boldsymbol{\gamma}.$$

Thus the PLS criterion to be minimized is

$$(\boldsymbol{y} - \boldsymbol{Z}\boldsymbol{\gamma})'(\boldsymbol{y} - \boldsymbol{Z}\boldsymbol{\gamma}) + \lambda \boldsymbol{\gamma}'\boldsymbol{K}\boldsymbol{\gamma},$$

where $\boldsymbol{Z}$ contains the basis functions evaluated at the observed z values. We find that the form of the expression to be minimized corresponds exactly to the PLS criterion discussed in the previous sections of this chapter. Therefore, the PLS estimate is again given by

$$\hat{\boldsymbol{\gamma}} = (\boldsymbol{Z}'\boldsymbol{Z} + \lambda \boldsymbol{K})^{-1}\boldsymbol{Z}'\boldsymbol{y}.$$

Alternatively, we can directly write the smoothing spline minimization problem in terms of the function evaluations $f(z_{(j)})$. If we collect the function evaluations in the vector $\boldsymbol{f} = (f(z_{(1)}), \ldots, f(z_{(d)}))'$, and define the respective design matrix $\boldsymbol{Z}_f$ as an incidence matrix with entries

8.6 Smoothing Splines

We seek the function f, which minimizes the least squares criterion

$$\sum_{i=1}^{n}(y_i - f(z_i))^2 + \lambda \int (f''(z))^2 dz,$$

among all twice continuously differentiable functions. The solution for this optimality criterion is a natural cubic spline with knots located at the observed covariate values. We represent natural cubic splines using a modified B-spline basis, so that the PLS estimate can be calculated in the same way as for P-splines, i.e.,

$$\hat{\boldsymbol{y}} = (\boldsymbol{Z}'\boldsymbol{Z} + \lambda \boldsymbol{K})^{-1}\boldsymbol{Z}'\boldsymbol{y}.$$

$$\boldsymbol{Z}_f[i,j] = \begin{cases} 1 & \text{if } z_i = z_{(j)}, \\ 0 & \text{otherwise,} \end{cases}$$

we obtain the PLS criterion

$$(\boldsymbol{y} - \boldsymbol{Z}_f \boldsymbol{f})'(\boldsymbol{y} - \boldsymbol{Z}_f \boldsymbol{f}) + \lambda \boldsymbol{f}'\boldsymbol{K}_f \boldsymbol{f},$$

where the penalty matrix can be constructed based on the distances $\delta_j = z_{(j+1)} - z_{(j)}$ (see, e.g., Fahrmeir and Tutz (2001), p. 179–180). Minimizing the PLS criterion for $\boldsymbol{f}$ results in

$$\hat{\boldsymbol{f}} = (\boldsymbol{Z}'_f \boldsymbol{Z}_f + \lambda \boldsymbol{K}_f)^{-1}\boldsymbol{Z}'_f \boldsymbol{y},$$

which corresponds to the PLS estimate $\hat{\boldsymbol{y}}$.

Smoothing splines are typically based on a high-dimensional basis so that a large set of parameters needs to be estimated. In particular, the basis usually grows with the sample size so that for large data sets the practical calculations become demanding. Consequently, a reduced basis is used in many implementations of smoothing splines. A naive approach, implemented, for example, in PROC GAM in SAS, is to simply use a subset of the distinct covariate values to determine the set of knots. However, for this reduced basis, the optimality property that led us to the consideration of smoothing splines is no longer fulfilled. An alternative, corresponding to the default smoother in the mgcv package in R, is to determine a reduced basis in terms of an optimal approximation of the original basis utilizing a spectral decomposition of the design matrix. We will discuss this possibility in somewhat more detail in Sect. 8.2.2 on thin plate splines in the context of bivariate smoothing and utilizing a different basis since smoothing splines are a special case of thin plate splines.

8.1.5 Random Walks

Random walks are not only used for the construction of prior distributions in Bayesian regression models but can also be employed to directly model nonlinear effects. In particular, random walks provide a univariate counterpart to Markov random fields that we will discuss in the context of spatial smoothing in Sect. 8.2.4.

Random walk models are especially useful when analyzing time series data, i.e., we are assuming a temporal trend function $f(t)$ that is to be modeled nonparametrically, similarly as with function $f(z)$:

$$y_t = f(t) + \varepsilon_t, \quad t = 1, \ldots, T.$$

In a random walk approach, we identify every function evaluation $f(t)$ with the parameter γ_t and then assume a random walk prior for these parameters, i.e.,

$$f(t) = \gamma_t = \gamma_{t-1} + u_t, \quad u_t \sim \mathrm{N}(0, \tau^2),$$

or

$$f(t) = \gamma_t = 2\gamma_{t-1} - \gamma_{t-2} + u_t, \quad u_t \sim \mathrm{N}(0, \tau^2),$$

with random walks of first and second order, respectively. Random walks can also be considered as special P-splines if we use a B-spline basis of degree zero ($l = 0$) and if we further identify the knots with the observed time points. As with Bayesian P-splines, a PLS criterion results from this setting, which we can explicitly write for the RW2 as

$$\mathrm{PLS}(\lambda) = \sum_{t=1}^{T} (y_t - \gamma_t)^2 + \lambda \sum_{t=3}^{T} (\gamma_t - 2\gamma_{t-1} + \gamma_{t-2})^2.$$

In econometrics, the solution of the respective minimization problem is also known as the Hodrick–Prescott filter (see Hodrick & Prescott, 1997), but actually dates back to Whittaker (1923). In Hodrick and Prescott (1997), rules of thumb for the choice of the smoothing parameter are motivated by theoretical considerations regarding the signal-to-noise ratio. In contrast to this, embedding random walks into the general context of nonparametric regression offers the opportunity to use the approaches for an optimal choice of smoothing parameters as discussed in Sect. 8.1.9.

While in time series the assumption of equally spaced time intervals is often satisfied, this is usually not the case when transferring the random walk approach to nonparametric smoothing of general covariate effects $f(z)$. In this case, a weighted random walk definition is often considered with the weights derived from the distances between covariate values. We will not discuss this case further here. A detailed discussion can be found in Rue and Held (2005) within the scope of more general Gauß–Markov random fields.

8.1.6 Kriging

In this section, we discuss a smoothing approach, which was not directly developed for univariate smoothing, but which has its origin in spatial statistics. The method is referred to as *kriging*, named after the South African mining engineer D. G. Krige, who used the approach when modeling the course of stone cores based on depth drillings. Even though kriging is rarely used for univariate smoothing, we introduce it at this point to discuss the general idea and its close relationship to basis function smoothing.

Classical Kriging

The main idea of kriging is to express (originally spatial) correlations among the data using parametric correlation functions and to employ this information when fitting a regression model. This can be achieved through a stationary and multivariate normally distributed stochastic process, which induces the desired correlations analogous to random effects models. Therefore kriging is often thought of as synonymous to the inclusion of a stationary Gaussian process (or stationary Gaussian field in spatial statistics) in the regression model. Since we focus for the moment on univariate smoothing methods, we consider time series data instead of spatially aligned data, which will be discussed in more detail in Sect. 8.2.3.

We assume that a time series in continuous time $t \in \mathbb{R}$,

$$y_t = \boldsymbol{x}_t'\boldsymbol{\beta} + \epsilon_t$$

is given. In contrast to the classical linear model, the assumption of uncorrelated errors is usually violated in this situation, and it seems plausible to rather consider temporally correlated errors. Therefore we assume that the error $\boldsymbol{\epsilon}$ is multivariate normal with expectation $\boldsymbol{0}$ and covariance matrix $\sigma^2\boldsymbol{I} + \tau^2\boldsymbol{R}$. The matrix $\boldsymbol{R}$ is a correlation matrix, described in terms of a parametric correlation function ρ, leading to the elements

$$\boldsymbol{R}[t,s] = \text{Corr}(\epsilon_t, \epsilon_s) = \rho(t,s).$$

Hence, the covariance matrix is composed of an uncorrelated part $\sigma^2\boldsymbol{I}$ and a correlated part $\tau^2\boldsymbol{R}$. The former part corresponds to the usual covariance matrix in the classical linear model, while the later part models temporal dependencies. If the parameters σ^2 and τ^2, as well as the correlation function ρ, are fully known, we obtain a general linear model with covariance matrix $\text{Cov}(\boldsymbol{y}) = \boldsymbol{V} = \sigma^2\boldsymbol{I} + \tau^2\boldsymbol{R}$, which can be estimated as outlined in Sect. 4.1. In particular, the weighted least squares estimate for $\boldsymbol{\beta}$ results as

$$\hat{\boldsymbol{\beta}} = (\boldsymbol{X}'\boldsymbol{V}^{-1}\boldsymbol{X})^{-1}\boldsymbol{X}'\boldsymbol{V}^{-1}\boldsymbol{y}.$$

Similar as in Chap. 7, we can estimate the variance parameters σ^2 and τ^2 (and also possibly additional parameters for the correlation function), by maximizing the profile likelihood or the corresponding restricted likelihood function.

Most commonly, *stationary* error distributions are assumed, i.e., $\rho(t, s)$ only depends on the time difference $t - s$, and not on the exact locations of the time points t and s. This yields correlation functions $\rho(h)$ with scalar argument $h = t - s$.

Clearly, we have $\rho(0) = 1$ and $\rho(h) = \rho(-h)$ for a valid correlation function. Moreover, some further properties are often assumed:

- As $h \to \infty$, we have $\rho(h) \to 0$, i.e., for observations having large temporal distance, the correlation becomes negligible.
- The correlation function $\rho(h)$ decreases monotonically. This corresponds to the intuitive idea that observations that are temporally close to each other should always show a higher correlation than observations that are more distant apart in time.

Some frequently used parametric correlation functions belong to one of the following classes:

The family of *spherical correlation functions* is defined by

$$\rho(h; \phi) = \begin{cases} 1 - \frac{3}{2}|h/\phi| + \frac{1}{2}|h/\phi|^3 & 0 \leq h \leq \phi, \\ 0 & h > \phi. \end{cases}$$

The only parameter of the spherical correlation function is given by $\phi > 0$, which controls the distance at which the correlation decreases to 0. See Fig. 8.20a which shows spherical correlation functions for different values of ϕ. Due to this interpretation, ϕ is also called the *range* of the correlation function. The non-differentiability of the spherical correlation function in $h = \phi$ can be problematic when considering maximum likelihood estimation of the parameter ϕ. Obviously, the change from positive correlations to non-existent correlations occurs in a continuous but not differentiable form.

The *power exponential family* is defined by

$$\rho(h; \phi, \kappa) = \exp\left(-|h/\phi|^{\kappa}\right)$$

with parameters $\phi > 0$ and $0 < \kappa \leq 2$. The basic exponential correlation function results for $\kappa = 1$. The Gaussian correlation function, which resembles the normal density, follows for $\kappa = 2$. Figure 8.20b shows the power exponential family for various values of κ (and fixed $\phi = 1$). In contrast to the spherical correlation function, the power exponential correlation between two points remains positive regardless of their distance, i.e., the range is unlimited. Therefore, one often considers the *effective range* instead, i.e., the distance where the correlation function falls below a predetermined small value (e.g., 0.01). In applications, correlations between points that are farther apart than the effective range are then usually considered negligible and are therefore set to zero to achieve a sparser correlation matrix $\boldsymbol{R}$. The effective range of the power exponential family is mostly determined

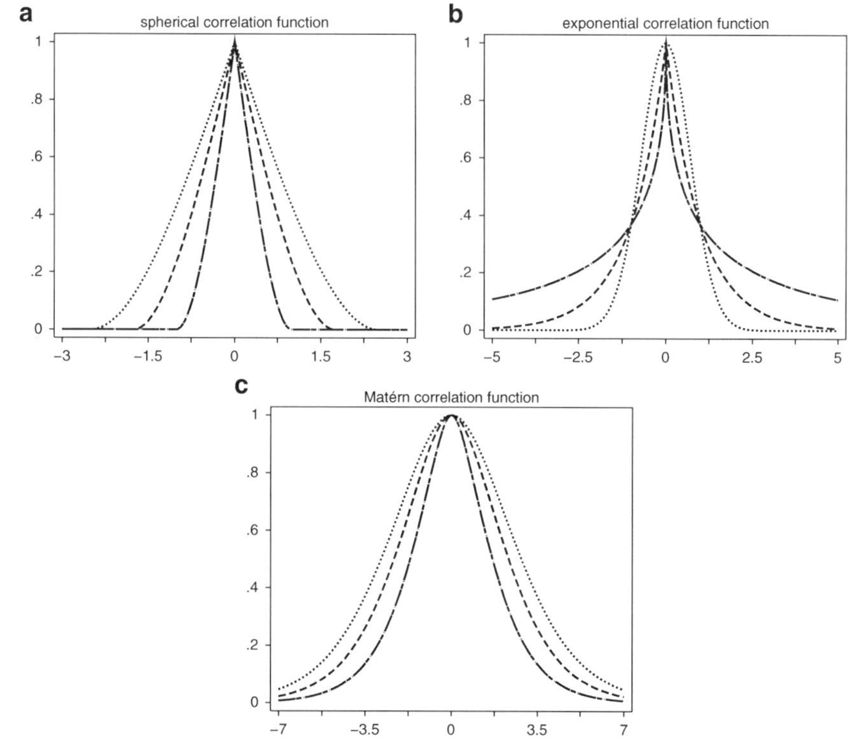

Fig. 8.20 Examples of parametric correlation functions: Panel (**a**) shows spherical correlation functions for $\phi = 1$ (—), $\phi = 1.75$ (- - -), and $\phi = 2.5$ (···); panel (**b**) visualizes exponential correlation functions (with fixed $\phi = 1$) for $\kappa = 0.5$ (—), $\kappa = 1$ (- - -) and $\kappa = 2$ (···); and panel (**c**) shows Matérn correlation functions (with fixed $\phi = 1$) for $\kappa = 1.5$ (—), $\kappa = 2.5$ (- - -), and $\kappa = 3.5$ (···). Note that the range of the correlation functions varies over the different panels, i.e., the range of the *horizontal axis* is adapted to facilitate the comparison of correlation functions within one panel

by the parameter ϕ, which controls the scaling of the time axis, but also varies depending on the specific value of κ.

The basic exponential correlation function also has an interesting connection with linear models having autocorrelated errors; see also Sect. 4.1.4. In this case, the time index t and also the observed time difference h are assumed to be discrete. In this setting, the simplest and the most frequently used model to account for temporal correlations is an AR(1)-process, which assumes

$$\epsilon_t = \rho\epsilon_{t-1} + u_t, \quad u_t \sim \mathrm{N}(0, \tau^2), \tag{8.9}$$

for the errors. The parameter ρ controls the influence of the lagged values ϵ_{t-1} on ϵ_t, so that positively correlated errors result for $\rho > 0$, with correlation function

$$\mathrm{Corr}(\epsilon_t, \epsilon_{t+h}) = \rho^h, \quad h = 1, 2, \ldots.$$

Hence, the correlation decays exponentially in the time distance h similar to the time-continuous exponential correlation function. If we define the parameter $\rho = \exp(-1/\phi)$, we can rewrite the exponential correlation function as

$$\rho(h) = \exp(-h/\phi) = \exp(-1/\phi)^h = \rho^h.$$

Therefore, the autoregressive error process Eq. (8.9) can also be interpreted as a discretized version of a continuous-time process with exponential correlation function.

One of the most flexible families of correlation functions, which in practice became the standard of spatial statistics, is the *Matérn* family $\rho(h;\phi,\kappa)$ with $\phi > 0$, $\kappa > 0$. The general representation of correlation functions from this family is, however, only possible with the help of modified Bessel functions of the order κ, which can only be evaluated numerically, but cannot be derived explicitly. Consequently, one often restricts the attention to a subset of the Matérn correlation functions with $\kappa = 0.5, 1.5, 2.5, \ldots$. For these cases, the correlation function can be derived analytically and, for example, takes the form

$$\begin{aligned}
\rho(h;\phi,\kappa = 0.5) &= \exp(-|h/\phi|),\\
\rho(h;\phi,\kappa = 1.5) &= \exp(-|h/\phi|)(1 + |h/\phi|),\\
\rho(h;\phi,\kappa = 2.5) &= \exp(-|h/\phi|)(1 + |h/\phi| + \tfrac{1}{3}|h/\phi|^2),\\
\rho(h;\phi,\kappa = 3.5) &= \exp(-|h/\phi|)(1 + |h/\phi| + \tfrac{2}{5}|h/\phi|^2 + \tfrac{1}{15}|h/\phi|^3).
\end{aligned}$$

For $\kappa = 0.5$, the Matérn correlation function again simplifies to the basic exponential correlation function. Figure 8.20c shows the remaining three examples. Larger values of κ evidently lead to correlation functions with a larger effective range, and thus stronger correlations for points in time with large temporal distance. If we let κ go to infinity, we obtain the Gaussian correlation function as the limiting case, i.e., yet another special case of the power exponential correlation function.

Kriging as Time Series Smoother

In order to motivate the use of correlation functions and stationary Gaussian processes in nonparametric function estimation, we decompose the error ϵ_t into a temporally correlated component γ_t and the remaining independent and identically distributed component ε_t:

$$y_t = \boldsymbol{x}_t'\boldsymbol{\beta} + \gamma_t + \varepsilon_t, \quad t \in \{t_{(1)}, \ldots, t_{(d)}\}.$$

With adequate distributional specifications for γ_t and ε_t, we obtain exactly the covariance structure for the response variable $\boldsymbol{y}$ considered above in the previous paragraph.

In matrix notation, the model can be represented as

$$\boldsymbol{y} = \boldsymbol{X}\boldsymbol{\beta} + \boldsymbol{Z}\boldsymbol{\gamma} + \boldsymbol{\varepsilon},$$

where $\boldsymbol{\gamma} = (\gamma_1, \ldots, \gamma_d)'$ defines the vector of the d unique and temporally ordered errors γ_t and $\boldsymbol{Z} = \boldsymbol{I}_d$ is the d-dimensional identity matrix. In a second step, we further reparameterize the model to

$$\boldsymbol{y} = \boldsymbol{X\beta} + \boldsymbol{Z R} \cdot \boldsymbol{R}^{-1}\boldsymbol{\gamma} + \boldsymbol{\varepsilon} = \boldsymbol{X\beta} + \tilde{\boldsymbol{Z}}\tilde{\boldsymbol{\gamma}} + \boldsymbol{\varepsilon}$$

with $\tilde{\boldsymbol{Z}} = \boldsymbol{Z R} = \boldsymbol{R}$ and $\tilde{\boldsymbol{\gamma}} = \boldsymbol{R}^{-1}\boldsymbol{\gamma}$.

This reparameterization evidently leads to the same distribution of $\boldsymbol{y}$ and, thus, to an equivalent model formulation. However, the interpretation of the design matrix $\tilde{\boldsymbol{Z}}$ changes. Due to the special structure of $\boldsymbol{Z}$, we obtain

$$\tilde{\boldsymbol{Z}}[i, j] = \rho(t_{(i)}, t_{(j)})$$

for the individual entries. If we compare this definition with the construction of the design matrix for B-splines or TP-splines, we find that the correlation function ρ has the role of a basis function, and that the observed time points $t_{(j)}$ adopt the role of knots. Thus, for the temporal trend, we have

$$f(t) = \gamma_t = \sum_{j=1}^{d} \tilde{\gamma}_j \rho(t, t_{(j)}).$$

The joint distribution of the temporally correlated effects $\tilde{\boldsymbol{\gamma}}$ is

$$\tilde{\boldsymbol{\gamma}} \sim \mathrm{N}(0, \tau^2 \boldsymbol{R}^{-1})$$

and therefore has a density of the form

$$p(\tilde{\boldsymbol{\gamma}} \mid \tau^2) \propto \exp\left(-\frac{1}{2\tau^2}\tilde{\boldsymbol{\gamma}}'\boldsymbol{R}\tilde{\boldsymbol{\gamma}}\right).$$

This form exactly corresponds to the smoothing prior that we discussed in connection with the Bayesian formulation of penalty approaches. It also yields an equivalent PLS criterion defined as

$$(\boldsymbol{y} - \boldsymbol{X\beta} - \tilde{\boldsymbol{Z}}\tilde{\boldsymbol{\gamma}})'(\boldsymbol{y} - \boldsymbol{X\beta} - \tilde{\boldsymbol{Z}}\tilde{\boldsymbol{\gamma}}) + \frac{\sigma^2}{\tau^2}\tilde{\boldsymbol{\gamma}}'\boldsymbol{R}\tilde{\boldsymbol{\gamma}}.$$

In this interpretation, we assume that all parameters and in particular the range parameter ϕ in the correlation function are either given or determined before estimating the regression coefficients $\tilde{\boldsymbol{\gamma}}$ so that the basis functions (and also the penalty matrix) do no longer contain any unknown parameters. In Example 8.3 below, we will provide a simple data-driven rule to determine the range parameter when applying kriging as a smoothing approach.

In summary, even though the kriging method was derived from a purely stochastic model to describe correlated data, the method is formally equivalent to a basis function approach with basis functions ρ and penalty matrix $\boldsymbol{R}$. This explains why the simultaneous modeling of a temporal trend through a nonparametric approach (or a simple polynomial), together with temporally correlated errors, can lead to identification problems. Due to the temporally correlated errors, a nonparametric trend function is also implicitly estimated, so that the additional assumption of a further trend leads to a nearly non-identifiable model. The representation of kriging as a basis function approach also illustrates that the estimate of the temporal effects inherits the smoothness properties of the correlation function ρ. When the correlation function is differentiable or continuous, the same holds for the estimated function.

Kriging as a Nonparametric Smoothing Procedure

In order to use kriging to estimate nonlinear covariate effects, we return to the model

$$y_i = f(z_i) + \varepsilon_i.$$

If $z_{(1)} < \ldots < z_{(d)}$ are the ordered covariate values, we define the parameters $\gamma_j = f(z_{(j)})$ and assume a stationary Gaussian process prior, with expectation 0, variance τ^2, and correlation function

$$\rho(\gamma_j, \gamma_r) = \rho(|z_{(j)} - z_{(r)}|).$$

Accordingly, we can also view the correlation function as a basis function, and, as with smoothing splines, we can identify the knot positions as the covariate locations $z_{(j)}$. We thus obtain the representation

$$\boldsymbol{y} = \tilde{\boldsymbol{Z}}\tilde{\boldsymbol{\gamma}} + \boldsymbol{\varepsilon}$$

with $\boldsymbol{Z}[i,j] = \rho(|z_i - z_{(j)}|)$ and $\boldsymbol{\gamma} = (\gamma_1, \ldots, \gamma_d)'$. We then have to optimize the penalized residual sum of squares

$$\text{PLS}(\lambda) = (\boldsymbol{y} - \tilde{\boldsymbol{Z}}\tilde{\boldsymbol{\gamma}})'(\boldsymbol{y} - \tilde{\boldsymbol{Z}}\tilde{\boldsymbol{\gamma}}) + \lambda\tilde{\boldsymbol{\gamma}}'\boldsymbol{K}\tilde{\boldsymbol{\gamma}}$$

with $\boldsymbol{K}[j,r] = \rho(|z_{(j)} - z_{(r)}|)$ and smoothing parameter $\lambda = \sigma^2/\tau^2$. Hence, the usual results for PLS estimation carry over to kriging and, in particular, we obtain the known form of the PLS estimate.

Example 8.3 Malnutrition in Tanzania: Kriging

Figure 8.21 shows several kriging estimates for the effect of age on the malnutrition score. These estimates are obtained with the four previously presented special Matérn correlation functions, with scale parameter determined via

$$\hat{\phi} = \max_{j,r} |z_{(j)} - z_{(r)}|/c. \tag{8.10}$$

8.7 Kriging as a Nonparametric Smoothing Procedure

The function f is expanded in terms of basis functions resulting from a parametric correlation function $\rho(h)$. This yields the PLS criterion

$$\text{PLS}(\lambda) = (\boldsymbol{y} - \tilde{\boldsymbol{Z}}\tilde{\boldsymbol{\gamma}})'(\boldsymbol{y} - \tilde{\boldsymbol{Z}}\tilde{\boldsymbol{\gamma}}) + \lambda\tilde{\boldsymbol{\gamma}}'\boldsymbol{K}\tilde{\boldsymbol{\gamma}},$$

with $\tilde{\boldsymbol{Z}}[i,j] = \rho(|z_i - z_{(j)}|)$ and $\boldsymbol{K}[j,r] = \rho(|z_{(j)} - z_{(r)}|)$. Smoothing properties of the chosen correlation function carry over to the estimate $\hat{f}(z)$.

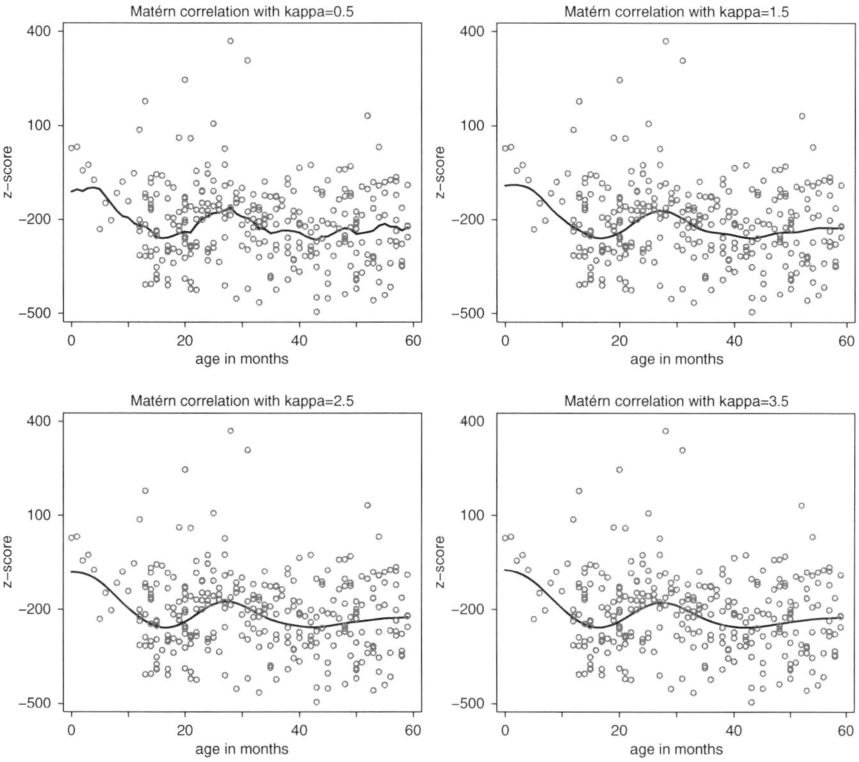

Fig. 8.21 Malnutrition in Tanzania: kriging estimates

The value of the constant c is chosen such that $\rho(c; \phi = 1)$ is small, i.e., in essence, c determines the effective range of the correlation function. Choosing c in this manner ensures that the correlation functions are well distributed across the covariate domain. For polynomial splines, the choice of knots automatically guarantees such a distribution while in the case of kriging an effective range that is too small can result in non-overlapping correlation functions. This would in turn induce unstable estimation of the nonparametric effects in the corresponding regions. Moreover, Eq. (8.10) guarantees scale invariance of the resulting estimates, i.e., when scaling the covariates z_i with a constant factor, this scaling is compensated by rescaling the range parameter ϕ. The variance τ^2 (and, thus, also the smoothing parameter λ) is determined using REML estimation as already introduced in the context of linear mixed models in Sect. 7.3.2. Section 8.1.9 will show that indeed REML estimation can also be used to determine smoothing parameters in penalized basis function approaches or more generally in nonparametric regression models.

The estimated age effects $\hat{f}(z)$ also show the impact of specific properties of the chosen correlation function: For $\kappa = 0.5$ we obtain a rather rough, non-differentiable function estimate, while for larger values of κ, the function is estimated much smoother. Moreover, the differences between the estimates decrease as the correlation function varies beyond $\kappa = 0.5$.

△

8.1.7 Local Smoothing Procedures

We next examine another class of smoothers, which, in general, cannot be derived from a global regression formulation but are defined *locally*. Many of these estimates are extremely intuitive and easy to understand, and thus are often used in explorative analyses.

Nearest Neighbor Estimates

A widely used method for the (descriptive) smoothing of time series is *running means*. For a time series y_t, $t = 1, \dots, T$, running means of order 3 are, for example, defined by

$$\hat{y}_t = \frac{1}{3}(y_{t-1} + y_t + y_{t+1}),$$

with appropriate modifications at the boundaries. Obviously, taking averages of observed values in time generally smooths the random fluctuations of a time series and gives a first impression of the underlying trend.

Nearest neighbor estimates extend the concept of running means into a more general framework, and also enable the application in nonparametric regression models. In general, a nearest neighbor estimate is defined by

$$\hat{f}(z) = \underset{j \in N(z)}{\text{Ave}} \; y_j,$$

where Ave defines some averaging operator and $N(z)$ is an appropriate neighborhood of z. In the example of running means of order 3, the averaging operator is the arithmetic mean and the two (temporally) adjacent values $t - 1$ and $t + 1$ (as well as t itself) form the neighborhood of time point t. In general, observations are considered to be their own neighbors, so that $z_i \in N(z_i)$ holds.

The following averaging operators are often used for the determination of nearest neighbor estimates:

8.8 Nearest Neighbor Estimates

The general form of a nearest neighbor estimate is

$$\hat{f}(z) = \underset{j \in N(z)}{\text{Ave}}\; y_j,$$

with a user-defined averaging operator Ave and a local neighborhood $N(z)$. As averaging operator, we often use the arithmetic mean, the median, or the linear regression. Neighborhoods can either be defined symmetrically around z or can be based on the k-nearest neighbors.

1. Arithmetic mean (running mean): Determine the arithmetic mean of the response variable in the neighborhood of z, i.e.,

$$\hat{f}(z) = \frac{1}{|N(z)|} \sum_{j \in N(z)} y_j,$$

   where $|N(z)|$ is the number of neighbors of z.
2. Median (running median): Determine the median of the response variables in the neighborhood of z, i.e.,

$$\hat{f}(z) = \text{Median}\{y_j, j \in N(z)\}.$$

3. Linear regression (running line): Estimate a linear regression based on the observations in the neighborhood of z and use the prediction from this model as the estimate, i.e.,

$$\hat{f}(z) = \hat{\gamma}_{0,z} + \hat{\gamma}_{1,z} z,$$

   where $\hat{\gamma}_{0,z}$ and $\hat{\gamma}_{1,z}$ are the least squares estimates using the data $\{(y_j, z_j), j \in N(z)\}$.

To completely determine a nearest neighbor estimate, we also need an adequate neighborhood definition. Commonly used variants are the following:

1. Symmetric neighborhoods of order k
2. Neighborhoods that consist of k-nearest neighbors

Running means of order 3 provide a simple example for a symmetric neighborhood. When defining symmetric neighborhoods (with an uneven order k), we generally proceed as follows: When estimating $f(z_i)$, we use the nearest $(k-1)/2$ observations to the right and to the left from z_i (in addition to z_i itself) as the neighborhood. Hence, the order is divided symmetrically to the area left and right of z_i, yielding a justification for the term *symmetric neighborhood*. If, for the sake of simplicity, we assume ordered observations with $z_1 \leq \ldots \leq z_n$, the symmetric neighborhood is given by

$$N(z_i) = \{\max(1, i - (k-1)/2), \ldots, i-1, i, i+1, \ldots, \min(n, i + (k-1)/2)\}.$$

This definition already includes the necessary corrections at the boundaries of the defined domain, as there are not enough observations available to apply the standard definition. Note that the neighborhood at the boundaries is no longer symmetric and also does not contain k observations.

Instead of using the order k, symmetric neighborhoods can also be defined based on the fraction $\omega \in (0, 1)$ of observations contained in the neighborhood (also referred to as the *bandwidth*). The corresponding order k is then given by ωn, where some rounding scheme has to be applied for non-integer values of ωn. The order k, as well as the fraction ω, can be interpreted as smoothing parameters of the nearest neighbor estimate. Figure 8.22 demonstrates the effect of varying ω for the Tanzania data set using local arithmetic means. The closer the bandwidth ω is to 1, the more neighbors are included in the estimate, leading to smoother functions. As ω gets smaller and finally approaches 0, the estimate $\hat{f}(z_i)$ is based on only very few observations, resulting in a very rough estimate.

The definition of a neighborhood based on k-nearest neighbors generally leads to asymmetric neighborhoods. In this case, the neighborhood $N(z)$ is given by

$$N(z) = \{i : d_i \in \{d_{(1)}, \ldots, d_{(k)}\}\},$$

where $d_{(1)}, \ldots, d_{(n)}$ represents the ordered distances, $d_i = |z_i - z|$. In this setting, the same number of neighbors is used, even at the boundaries, i.e., modifications at the boundaries are not necessary. Similar to the symmetric neighborhoods, we can interpret the number of neighbors k as a smoothing parameters. If k is small, a rough estimate results while, on the other hand, a very smooth estimate is obtained with k chosen close to the sample size n. In the latter case, we end up with the value of the averaging operator applied to the complete data set.

Local Polynomial Regression and the Nadaraya–Watson Estimator

Although, at the beginning of this chapter, we already discussed the difficulties arising from the global approximation of nonlinear functions $f(z)$, it is often possible to *locally* approximate $f(z)$ with polynomials (as already illustrated in Fig. 2.12 on p. 46 for a local linear approximation). To motivate such an approximation, we consider the (local) approximation of an l-times continuously differentiable function $f(z_i)$ using a *Taylor series expansion* around z, yielding

$$f(z_i) \approx f(z) + (z_i - z) f'(z) + (z_i - z)^2 \frac{f''(z)}{2!} + \ldots + (z_i - z)^l \frac{f^{(l)}(z)}{l!}.$$

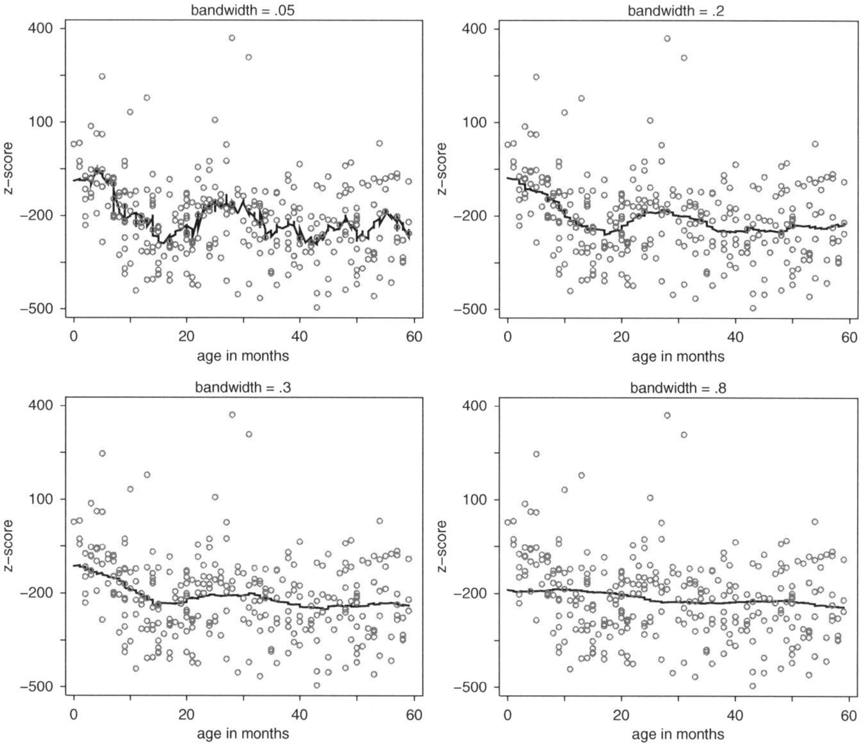

Fig. 8.22 Malnutrition in Tanzania: running mean with different bandwidths

Hence, we approximate the function $f(z_i)$ with polynomials of the form $(z_i - z)^j$ in a neighborhood of z_i. The polynomials are weighted by the derivatives $f^{(j)}(z)/j!$ evaluated at the expansion point z. Note that $f(z)$ is considered as the zero derivative of f. In practice, $l = 1$, i.e., local linear regression is used most frequently.

If we apply the Taylor series approximation to the problem of nonparametric function estimation for $f(z)$ at a given point z, we obtain

$$\begin{aligned} y_i &= f(z_i) + \varepsilon_i \\ &\approx f(z) + (z_i - z) f'(z) + (z_i - z)^2 \frac{f''(z)}{2!} + \ldots + (z_i - z)^l \frac{f^{(l)}(z)}{l!} + \varepsilon_i \\ &= \gamma_0 + (z_i - z)\gamma_1 + (z_i - z)^2 \gamma_2 + \ldots + (z_i - z)^l \gamma_l + \varepsilon_i, \end{aligned}$$

for each observation (y_i, z_i) with expansion point z. Therefore, a polynomial regression model for y_i results, which is based on polynomials of the form $(z_i - z)^j$ and regression coefficients $\gamma_j = f^{(j)}(z)/j!$. If we now estimate the regression parameters from this model, we obtain an implicit estimate for the function value

$f(z)$ through $\hat{\gamma}_0 = \hat{f}(z)$, and more generally, we even obtain estimates for the derivatives through $j!\hat{\gamma}_j = \hat{f}^{(j)}(z)$.

Hence, an estimate for $f(z)$ can in principle be determined by fitting a specific linear model. However, since the Taylor series approximation is only valid locally, i.e., close to the expansion point z, estimation is usually based on a weighted version of the residual sum of squares. This results in a weighted least squares criterion of the form

$$\sum_{i=1}^{n} \left(y_i - \sum_{j=0}^{l} \gamma_j (z_i - z)^j \right)^2 w_\lambda(z, z_i)$$

with weights $w_\lambda(z, z_i)$. These are typically constructed based on the distances $|z_i - z|$ such that larger weights result for observations with a small distance. A general class of such weights results with the use of *kernel functions* K in

$$w_\lambda(z, z_i) = K\left(\frac{z_i - z}{\lambda}\right). \tag{8.11}$$

Typical examples of kernel functions include

$$K(u) = \begin{cases} \frac{1}{2} & -1 \leq u \leq 1 \\ 0 & \text{otherwise} \end{cases} \qquad \text{Uniform kernel,}$$

$$K(u) = \begin{cases} \frac{3}{4}(1 - u^2) & -1 \leq u \leq 1 \\ 0 & \text{otherwise} \end{cases} \qquad \text{Epanechnikov kernel,}$$

$$K(u) = \frac{1}{\sqrt{2\pi}} \exp\left(-\frac{1}{2}u^2\right) \qquad \text{Gaussian kernel;}$$

see Fig. 8.23. The Epanechnikov kernel, as well as the uniform kernel, set all weights outside the interval $[-1, 1]$ to zero, whereas the Gaussian kernel maintains positive weights for observations of arbitrary distance. The additional parameter λ (the bandwidth of the kernel) controls how quickly the weights approach zero and can be viewed as the smoothing parameter of local polynomial regression. Kernel functions are also commonly applied in nonparametric density estimation (see Fahrmeir, Künstler, Pigeot, & Tutz, 2007 or Härdle, 1990) and are therefore also densities themselves. Most commonly, kernel functions are assumed to be symmetric about zero.

In order to determine the weighted least squares estimate in local polynomial regression modeling, we first write the weighted residual sum of squares in matrix notation as

$$(\boldsymbol{y} - \boldsymbol{Z}\boldsymbol{\gamma})'\boldsymbol{W}(\boldsymbol{y} - \boldsymbol{Z}\boldsymbol{\gamma})$$

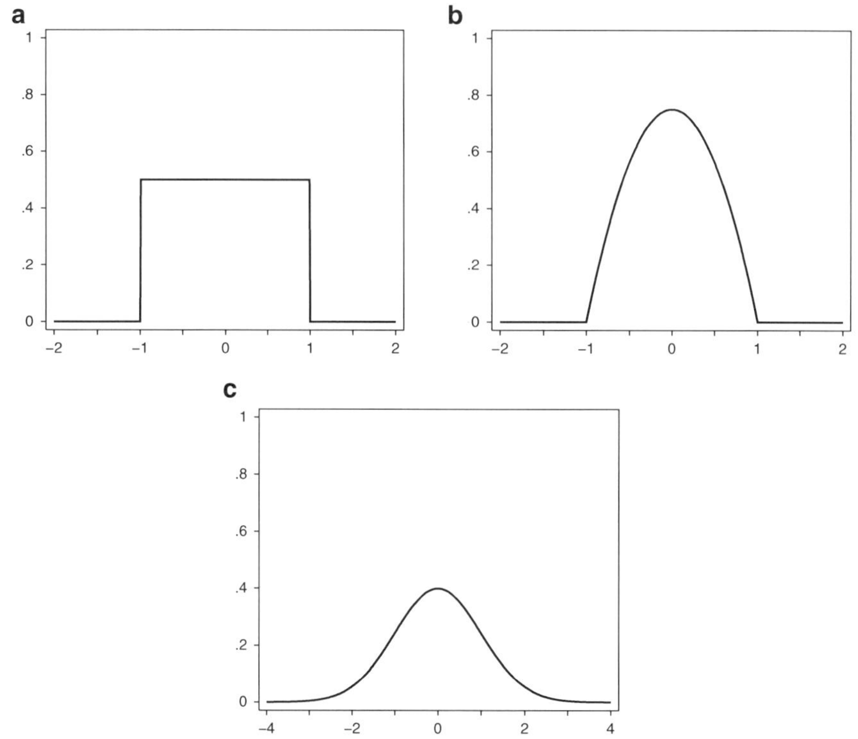

Fig. 8.23 Examples of kernel functions. (**a**) Rectangular kernel, (**b**) Epanechnikov kernel, (**c**) Gaussian kernel

with design matrix

$$\boldsymbol{Z} = \begin{pmatrix} 1 & (z_1 - z) & \dots & (z_1 - z)^l \\ \vdots & \vdots & & \vdots \\ 1 & (z_n - z) & \dots & (z_n - z)^l \end{pmatrix},$$

vector of the regression coefficients $\boldsymbol{\gamma} = (\gamma_0, \dots, \gamma_l)'$, and weight matrix

$$\boldsymbol{W} = \text{diag}(w_\lambda(z, z_1), \dots, w_\lambda(z, z_n)).$$

Minimizing this residual sum of squares corresponds to the estimation of a general linear model as discussed in Sect. 4.1. Therefore, we obtain the weighted least squares estimate

$$\hat{\boldsymbol{\gamma}} = (\boldsymbol{Z}'\boldsymbol{W}\boldsymbol{Z})^{-1}\boldsymbol{Z}'\boldsymbol{W}\boldsymbol{y} \quad \text{and} \quad \hat{f}(z) = \hat{\gamma}_0. \tag{8.12}$$

8.9 Local Polynomial Regression

Based on the Taylor series expansion around the point z, we obtain a local representation

$$y_i \approx \gamma_0 + (z_i - z)\gamma_1 + (z_i - z)^2\gamma_2 + \ldots + (z_i - z)^l \gamma_l + \varepsilon_i$$

with regression coefficients $\gamma_j = f^{(j)}(z)/j!$. To determine the corresponding estimate, we minimize the weighted least squares criterion,

$$\sum_{i=1}^{n} \left(y_i - \sum_{j=0}^{l} \gamma_j (z_i - z)^j \right)^2 w_\lambda(z, z_i),$$

where the weights $w_\lambda(z, z_i)$ are defined based on a kernel function. From the resulting weighted least squares estimate, we obtain $\hat{f}(z) = \hat{\gamma}_0$. The Nadaraya–Watson estimate results as a special case with $l = 0$.

Since both the design matrix $\boldsymbol{Z}$ and the weight matrix $\boldsymbol{W}$ depend on the location of z where we aim to estimate f, each point requires a separate model fit. In practice, one usually considers estimates at an equidistant grid of covariate values.

An interesting special case of local polynomial regression results for the local constant polynomial model. In this case, we obtain the weighted residual sum of squares

$$\sum_{i=1}^{n} (y_i - \gamma_0)^2 w_\lambda(z, z_i),$$

and the estimate $\hat{f}(z)$ has the explicit representation

$$\hat{f}(z) = \frac{\sum_i w_\lambda(z, z_i) y_i}{\sum_i w_\lambda(z, z_i)}. \tag{8.13}$$

Alternatively, we can also derive the same estimate from nonparametric density estimation ideas. The estimate is then called the Nadaraya–Watson estimator; see, for example, Fahrmeir and Tutz (2001, Chap. 5) or Härdle (1990).

Loess

Nearest neighbor estimates often result in relatively rough function estimates, even if the chosen smoothing parameter is based on a suitable optimality criterion. This is due to the definition of local estimates based on neighborhoods: While the observations within the neighborhood have relatively large impact on the estimates,

8.10 Loess Estimation

1. Determine $N(z)$, the set of the k-nearest neighbors of z.
2. Determine the largest distance between any two data points within the neighborhood
$$\Delta(z) = \max_{i,j \in N(z)} |z_i - z_j|.$$
3. Define weights
$$w_{\Delta(z)}(z, z_i) = K\left(\frac{|z - z_i|}{\Delta(z)}\right)$$
with the tri-cubic kernel function
$$K(u) = \begin{cases} (1 - |u|^3)^3 & -1 \leq u \leq 1, \\ 0 & \text{otherwise.} \end{cases}$$
4. Determine $\hat{f}(z)$ using weighted linear regression based on the data points in the neighborhood $N(z)$.

all of the other observations are completely neglected. If we combine the use of weights as introduced for local polynomial regression with nearest neighbor estimates, we may expect smoother estimates due to the smoothly decaying weights instead of weights dropping to zero outside the neighborhood. This combination leads to *locally weighted regression (loess)*, originally proposed by Cleveland (1979) and summarized in Box 8.10. The term loess is derived both from an abbreviation of LOcal regrESSion as well as from the German word *Löß*. In geology, the term Löß defines sediment that resulted from the erosion of rocks and that was accumulated by the wind. Levels of Löß form a smooth surface, a fact that has led to the corresponding term for the smooth function estimates resulting from the loess procedure. The term *lowess* is also often used instead of loess, being an abbreviation of LOcally WEighted Scatter plot Smoothing. This highlights that loess is a weighted estimate for nonparametric regression. In the original loess proposal, asymptotic considerations motivated the use of the tri-cubic kernel function but of course other kernel functions can be used as well. In statistical software packages, different loess variants are available either under the name loess or lowess. Even though they differ from each other in detail, their main approach corresponds to the algorithm provided in Box 8.10.

Sometimes, a robustified version of loess is also applied where the definition of weights is iterated to reduce the influence of observations with large residuals. More precisely, in each iteration, new weights are computed as

$$w_{\Delta(z)}(z, z_i) = \delta_i K\left(\frac{|z - z_i|}{\Delta(z)}\right)$$

where δ_i is derived from the residuals of the previous iterations such that δ_i is small for large residuals. For example, $\delta_i = \exp(-|\hat{\varepsilon}_i|)$ is a possible choice.

8.1.8 General Scatter Plot Smoothing

To enable a thorough comparison of different smoothing procedures, this section introduces a general formulation of scatter plot smoothers, followed by the presentation of properties derived from this formulation.

Linear Smoothers

When reviewing the various smoothing procedures discussed so far, we find, with very few exceptions, that most of the approaches can be represented as

$$\hat{f}(z) = \sum_{i=1}^{n} s(z, z_i) y_i, \tag{8.14}$$

with weights $s(z, z_i)$. Accordingly, we obtain the estimate $\hat{f}(z)$ as a weighted sum of the observations y_i, where the weights depend on the location z, as well as on the observed values of the covariates $z_1, \ldots, z_n$. Since $\hat{f}(z)$ can be expressed as a linear combination of the observed response values, we also refer to smoothing procedures with property (8.14) as *linear smoothing procedures*.

It is particularly straightforward to verify property (8.14) for nearest neighbor estimates (see p. 460ff). The estimated function $\hat{f}(z)$ was, in this case, defined as a local average of all observed dependent variables within the neighborhood of z. If we use a linear averaging operator, we also obtain a linear smoother. This implies that both the local arithmetic mean and local linear regression are linear smoothers. For the local arithmetic mean, we have

$$s(z, z_i) = \begin{cases} \frac{1}{|N(z)|} & \text{if } i \in N(z), \\ 0 & \text{otherwise.} \end{cases}$$

Since the median is a nonlinear averaging operator, it is not possible to determine the weights required in sum (8.14) and, thus the local median is an example of a nonlinear smoother.

Linear smoothers can also be written in matrix notation as

$$\hat{f}(z) = \boldsymbol{s}(z)'\boldsymbol{y},$$

with vector $\boldsymbol{s}(z) = (s(z, z_1), \ldots, s(z, z_n))'$. In particular, we have $\hat{f}(z_i) = \boldsymbol{s}(z_i)'\boldsymbol{y}$ for the observed covariate values. If we apply this equation to the n covariate values, we obtain the linear smoother

$$\hat{\boldsymbol{f}} = \boldsymbol{S}\boldsymbol{y},$$

with $\hat{\boldsymbol{f}} = (\hat{f}(z_1), \ldots, \hat{f}(z_n))'$ and the $n \times n$- *smoother matrix* $\boldsymbol{S}$ consisting of the row vectors $\boldsymbol{s}(z_i)'$. The smoother matrix will form the basis for many of the tools to characterize nonparametric smoothers; see below.

The linear model discussed in Chap. 3, as well as the general linear model introduced in Sect. 4.1, can actually be viewed as linear smoothers. In these cases, the smoother matrix coincides with the prediction or hat matrix $\boldsymbol{H}$ (see p. 107) and is therefore given by

$$\boldsymbol{S} = \boldsymbol{H} = \boldsymbol{Z}(\boldsymbol{Z}'\boldsymbol{Z})^{-1}\boldsymbol{Z}'$$

for the classical linear model and by

$$\boldsymbol{S} = \boldsymbol{Z}(\boldsymbol{Z}'\boldsymbol{W}\boldsymbol{Z})^{-1}\boldsymbol{Z}'\boldsymbol{W}$$

for the general linear model. As a consequence, polynomial splines and any other basis function approach (without penalization) are linear smoothers. In order to estimate the function f at a specific (possibly unobserved) covariate value z, we obtain

$$\boldsymbol{s}(z)' = \boldsymbol{z}'(\boldsymbol{Z}'\boldsymbol{Z})^{-1}\boldsymbol{Z}',$$

where $\boldsymbol{z}$ is the vector of basis functions evaluated at z (see p. 423).

For any roughness penalty approach derived from a model of the form

$$\boldsymbol{y} = \boldsymbol{Z}\boldsymbol{\gamma} + \boldsymbol{\varepsilon}$$

with penalty matrix $\boldsymbol{K}$ and smoothing parameter λ, we only have to replace the least squares estimate with the PLS estimate $\hat{\boldsymbol{\gamma}}$; see formula (8.3) on p. 439. Thus, we obtain

$$\boldsymbol{S} = \boldsymbol{Z}(\boldsymbol{Z}'\boldsymbol{Z} + \lambda\boldsymbol{K})^{-1}\boldsymbol{Z}'$$

and

$$\boldsymbol{s}(z)' = \boldsymbol{z}'(\boldsymbol{Z}'\boldsymbol{Z} + \lambda\boldsymbol{K})^{-1}\boldsymbol{Z}'.$$

It follows that all penalization approaches derived from polynomial splines, but also smoothing splines, random walks, and kriging methods are linear smoothers.

Moreover, the two remaining methods (local polynomial regression and loess) can also be cast in the framework of linear smoothing methods. For local polynomial regression, the function estimate $\hat{f}(z)$ was derived as the weighted least squares estimate (8.12). Thus, we obtain the corresponding weight vector $\boldsymbol{s}(z)$ as

$$\hat{f}(z) = \hat{\gamma}_0 = \boldsymbol{e}_1'(\boldsymbol{Z}'\boldsymbol{W}\boldsymbol{Z})^{-1}\boldsymbol{Z}'\boldsymbol{W}\boldsymbol{y}, = \boldsymbol{s}(z)'\boldsymbol{y}$$

with $\boldsymbol{e}_1 = (1, 0, 0, \ldots)'$. The vector $\boldsymbol{e}_1$ extracts the first component of the weighted least squares estimate that corresponds to the function estimate and drops all values corresponding to derivatives. The linearity of the loess estimate directly follows from its construction as a weighted nearest neighbor estimate.

8.11 Confidence Intervals and Confidence Bands for Linear Smoothers

1. Pointwise confidence intervals with level α:

$$\hat{f}(z) \pm z_{1-\alpha/2}\sigma\sqrt{s(z)'s(z)},$$

where $z_{1-\alpha/2}$ is the $(1-\alpha/2)$-quantile of the standard normal distribution.

2. Simultaneous confidence bands at $\{z_1, \dots, z_j, \dots, z_r\}$ with Bonferroni correction:

$$\hat{f}(z_j) \pm z_{1-\alpha/(2r)}\sigma\sqrt{s(z_j)'s(z_j)}.$$

3. Simultaneous confidence bands at $\{z_1, \dots, z_j, \dots, z_r\}$ based on the joint distribution of $(f(z_1), \dots, f(z_r))'$:

$$\hat{f}(z_j) \pm m_{1-\alpha}\sigma\sqrt{s(z_j)'s(z_j)},$$

where the quantile $m_{1-\alpha}$ of Eq. (8.18) is determined via simulation.

The same formulae can be used when replacing σ^2 with a consistent estimate $\hat{\sigma}^2$.

We can summarize as follows: With the exception of the local median, all smoothing methods discussed thus far are in fact linear smoothers. In the following, we will therefore focus on linear smoothers and derive properties for this general class.

Confidence Intervals and Confidence Bands

In addition to point estimates for $f(z)$, information on the variability of the estimate or corresponding confidence intervals is typically required in practice. Considering the case of the function estimate at a fixed covariate value z, it is rather straightforward to derive a variance formula for linear smoothers. Since $\hat{f}(z) = s(z)'y$ and $\text{Var}(\varepsilon) = \sigma^2 I$, it directly follows

$$\text{Var}(\hat{f}(z)) = \sigma^2 s(z)'s(z).$$

If we additionally assume normally distributed errors, a $(1-\alpha)$-confidence interval is given by

$$\hat{f}(z) \pm z_{1-\alpha/2}\sigma\sqrt{s(z)'s(z)}, \qquad (8.15)$$

where $z_{1-\alpha/2}$ is the $(1-\alpha/2)$-quantile of the standard normal distribution. This construction is based on the assumption

$$\hat{f}(z) - f(z) \overset{a}{\sim} \mathrm{N}(0, \sigma^2 s(z)' s(z)).$$

While both the variance of the estimate $\hat{f}(z)$ and the normal distribution follow from the distributional assumptions for the error terms, we also have to assume that $\hat{f}(z)$ is (approximately) unbiased. In general, this property will only be fulfilled asymptotically and therefore the coverage probability of the constructed confidence intervals is also asymptotic in nature. If the error terms are not assumed to be normally distributed, the normal distribution for the estimates will also hold only asymptotically.

One often uses the confidence intervals (8.15) to provide an impression about the variability of the estimated function in graphical representations. In addition to the estimated function, we also plot the lower and upper limit of the confidence interval, as functions of the covariate value z; see Fig. 8.24. However, when interpreting such figures, we need to keep in mind that the confidence intervals have been constructed for a single covariate value z and therefore the coverage probability is only valid pointwise and not simultaneously for the complete function. More specifically, for a pointwise level $(1-\alpha)$-confidence interval $[L(z), U(z)]$, we have

$$\mathrm{P}\Big(L(z) \leq f(z) \leq U(z)\Big) \geq 1 - \alpha,$$

i.e., the probability that the (random) interval covers the (fixed) true function value at the prespecified covariate value z is at least $1-\alpha$. It is critical that the expression only applies for a fixed value z, and not for several covariate values $z_1, \ldots, z_r$ at the same time. However, often such simultaneous expressions are of interest, and thus the construction of (simultaneous) *confidence bands* is necessary.

Such a simultaneous confidence band $[L(z), U(z)]$ of level $1-\alpha$ should fulfill

$$\mathrm{P}\Big(L(z) \leq f(z) \leq U(z) \text{ for all } z \in \{z_1, \ldots, z_r\}\Big) \geq 1 - \alpha. \qquad (8.16)$$

In this case, the coverage probability does not apply pointwise but rather simultaneously for all covariate values in $\{z_1, \ldots, z_r\}$. A simple option to derive such a simultaneous confidence band is given by the *Bonferroni correction* of the level α. The Bonferroni correction is motivated by the fact that, for the simultaneous occurrence of events $A_1, \ldots, A_r$, we have

$$\mathrm{P}(A_1 \cap \ldots \cap A_r) \leq \sum_{j=1}^{r} \mathrm{P}(A_j).$$

In our case, A_j corresponds to the event "the function evaluation $f(z_j)$ is not contained in the interval $[L(z_j), U(z_j)]$." Now the aim is to limit the probability that (at least) one of these events occurs by α. This can be achieved by combining pointwise confidence intervals with levels $\alpha_1, \ldots, \alpha_r$, such that $\sum_j \alpha_j \leq \alpha$. The most common approach for determining such levels α_j is to uniformly distribute the confidence level as $\alpha_j = \alpha/r$.

In summary, the Bonferroni correction provides a simple way to define simultaneous confidence bands. However, despite maintaining the desired confidence level, these confidence bands are often rather wide, and therefore the information provided by the confidence bands may be imprecise. We therefore explore a second option for constructing confidence bands, which is numerically more complex, but provides more exact results translating into narrower confidence bands. First, we determine the covariance matrix of the vector $\hat{\boldsymbol{f}}_r = (\hat{f}(z_1), \ldots, \hat{f}(z_r))'$, which can be written as a linear smoother $\hat{\boldsymbol{f}}_r = \boldsymbol{S}_r \boldsymbol{y}$, where $\boldsymbol{S}_r$ consists of the vectors $\boldsymbol{s}(z_1), \ldots, \boldsymbol{s}(z_r)$. The covariance matrix can then be expressed as

$$\mathrm{Cov}(\hat{\boldsymbol{f}}_r) = \sigma^2 \boldsymbol{S}_r \boldsymbol{S}_r'$$

and therefore

$$\hat{\boldsymbol{f}}_r - \boldsymbol{f}_r \overset{a}{\sim} \mathrm{N}(\boldsymbol{0}, \sigma^2 \boldsymbol{S}_r \boldsymbol{S}_r'). \tag{8.17}$$

From this joint distribution we obtain a simultaneous confidence band with level $1-\alpha$ as

$$\hat{f}(z_j) \pm m_{1-\alpha}\sigma\sqrt{\boldsymbol{s}(z_j)'\boldsymbol{s}(z_j)},$$

where $m_{1-\alpha}$ defines the $(1-\alpha)$-quantile of the distribution of the random variable

$$\max_{1 \le j \le r} \left| \frac{\hat{f}(z_j) - f(z_j)}{\sigma\sqrt{\boldsymbol{s}(z_j)'\boldsymbol{s}(z_j)}} \right|. \tag{8.18}$$

In Eq. (8.18), we consider the maximum absolute standardized deviation between the true function value and the estimate. Since the distribution of this random variable and corresponding quantiles are difficult to obtain analytically, we approximate the distribution using simulation. To do so, we first draw N random vectors from the asymptotic distribution (8.17) and then calculate the respective N realizations from Eq. (8.18). If N is chosen large enough, we obtain an estimate for $m_{1-\alpha}$ based on the corresponding empirical $(1-\alpha)$-quantile.

In order to determine any of the three different confidence intervals or bands, we need the error variance σ^2, which in practice is generally unknown and thus an estimate is needed. Since the coverage probability is only valid asymptotically, we can, however, replace σ^2 with a consistent estimate $\hat{\sigma}^2$. In the next section, we will present a possible estimate for σ^2.

Example 8.4 Malnutrition in Tanzania—Confidence Bands for the Age Effect

To clarify the differences between pointwise confidence intervals and simultaneous confidence bands, we consider the age effect in the Tanzania data set. For estimation, we use a cubic P-spline with 20 interior knots and a second-order difference penalty, while fixing the smoothing parameter at $\lambda = 20$ and choosing a level of $\alpha = 0.05$. The goal is to construct a confidence band for the entire function, and thus we base our calculations of the confidence bands on all 60 observed distinct covariate values.

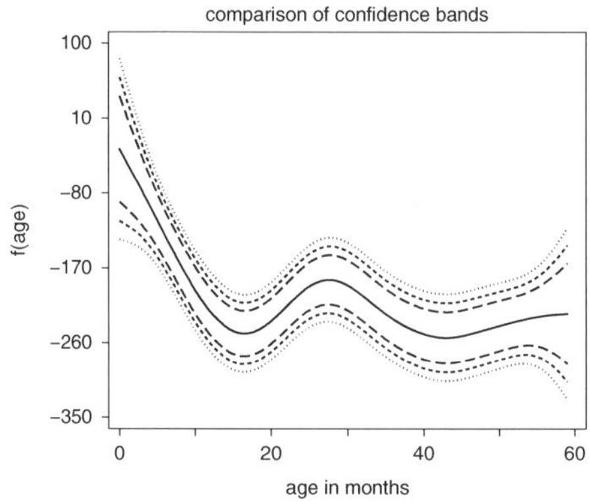

Fig. 8.24 Malnutrition in Tanzania: pointwise confidence intervals (– – –), simultaneous confidence band based on the joint distribution (- - -), and simultaneous confidence band based on the Bonferroni correction (· · ·) for the age effect

Figure 8.24 shows the resulting estimated function, together with the three confidence areas. The pointwise confidence intervals correspond to the narrowest confidence region, since they only consider uncertainty in estimation at one given point. The other extreme results from the use of the Bonferroni correction, which leads to the widest confidence region. In this case, we distributed the level equally on all 60 observed covariate settings, yielding a level of $\alpha_j = 0.0008333$. Hence, the Bonferroni confidence bands are based on the 0.9995833-quantile of the standard normal distribution, which results in a value of 3.34. In comparison the pointwise confidence intervals are based on the 0.975-quantile of the standard normal distribution given by $z_{0.975} \approx 1.96$. The result is nearly a 1.7 times wider confidence area for the Bonferroni correction.

For the construction of a confidence region based on the joint distribution of the 60 estimated function values, we need to derive the corresponding quantile of the distribution in Eq. (8.18). A simulation with $N = 100,000$ repetitions provides $m_{0.95} = 2.66$ so that (as expected) the result is a confidence band in between the pointwise confidence intervals and the confidence band obtained with the Bonferroni correction.

△

Equivalent Degrees of Freedom (Effective Number of Parameters)

Thus far, all smoothing methods that we have discussed have in common that the smoothness of the estimated function is controlled by (at least) one smoothing parameter. For basis function methods without penalty, smoothness is mostly determined by the number of basis functions. For penalization procedures, an additional parameter was included to explicitly determine the impact of the penalty on the model fit. For nearest neighbor estimates, the number of the nearest neighbors or the order of the symmetric neighborhood determined the smoothness of the estimated function. Clearly, the smoothing parameters of different approaches are not directly comparable despite their common nature. Consequently, it would be desirable to have a general measure to evaluate the approximate dimension of the smoothness for an estimated function.

The derivation of such a measure is easiest by considering analogies to the linear model. The complexity of a linear model is provided by the number of regression coefficients in the model. The number of parameters can, for example, be recovered by the trace of the prediction matrix $\boldsymbol{H} = \boldsymbol{X}(\boldsymbol{X}'\boldsymbol{X})^{-1}\boldsymbol{X}'$ since $\text{trace}(\boldsymbol{H}) = p$. Since the prediction matrix of the linear model corresponds to the smoothing matrix $\boldsymbol{S}$ of a linear smoother, the *equivalent degrees of freedom* are analogously defined as

$$\text{df}(\boldsymbol{S}) = \text{trace}(\boldsymbol{S}). \tag{8.19}$$

The equivalent degrees of freedom are also often interpreted as the *effective number of parameters* of a smoother. In fact, we reproduce the number of parameters in the model for basis function approaches without penalization, while in the case of penalization, the penalty effectively reduces the number of parameters.

Figure 8.25 shows the connection between the original smoothing parameter and the resulting effective degrees of freedom for a selection of smoothing approaches and the simulated data example. For the running mean, a larger neighborhood leads to a smoother estimate and, hence, to a smaller effective number of parameters. As a limiting case, we obtain only one parameter, when the neighborhood contains all observations. For polynomial splines, the effective number of parameters corresponds exactly to the number of basis functions, so that a linear relationship results. When considering penalized splines, the effective number of parameters decreases with an increasing smoothing parameter (note that a logarithmic scale for the smoothing parameter is used to facilitate presentation). For $\lambda = 0$, we default back to the unpenalized case, so that the number of parameters corresponds to the number of basis functions used. For a very large smoothing parameter, the effective number of parameters apparently approaches a limiting value, which depends on the order of differences used when constructing the penalty. We saw in Sect. 8.1.2 that a polynomial of degree $k - 1$ results as $\lambda \to \infty$ when using a kth order difference penalty. Such a polynomial is described by k parameters so that we obtain one or two parameters as limiting cases in Fig. 8.25.

If we want to determine the smoothing parameter corresponding to given effective degrees of freedom, we can in principle simply read them from Fig. 8.25. However, for polynomial splines and running means, the smoothing parameter is discrete and can only adopt a limited number of values. Thus, not every value of the effective degrees of freedom is actually achievable. For P-splines, however, we are able to choose every value for the effective degrees of freedom that lies between the difference order k and the number of basis functions used.

The equivalent degrees of freedom in Eq. (8.19) are of course not the only possibility to quantify the complexity of nonparametric function estimates. At least two further propositions exist, which can also be derived from analogies to the linear model but are less commonly used in practice. They are given by

$$\text{df}_{var}(\boldsymbol{S}) = \frac{1}{\sigma^2}\sum_{i=1}^{n} \text{Var}(\hat{f}(z_i)) = \text{trace}(\boldsymbol{S}\boldsymbol{S}')$$

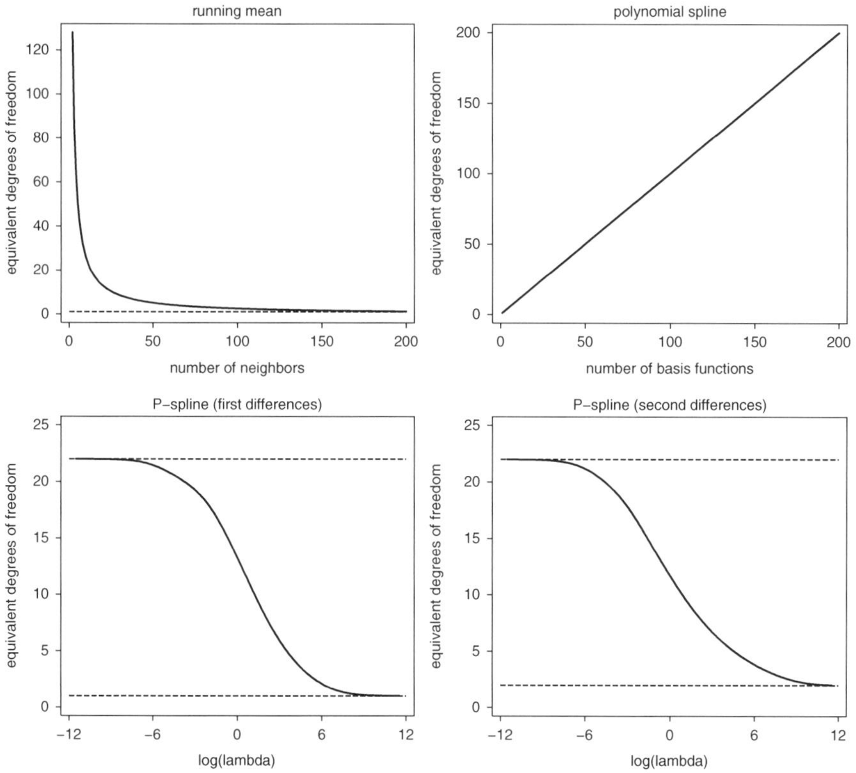

Fig. 8.25 Equivalent degrees of freedom as a function of the smoothing parameter. For P-splines, the log-transformed smoothing parameter is presented. The *dashed lines* indicate limiting values of the equivalent degrees of freedom

and

$$\mathrm{df}_{res}(\boldsymbol{S}) = \mathrm{trace}(2\boldsymbol{S} - \boldsymbol{S}\boldsymbol{S}'),$$

and result from considerations regarding the sum of variances of predicted values and the expected value of the mean squared error, respectively; see Hastie and Tibshirani (1990), Sect. 3.5 for a detailed derivation. If the smoothing matrix $\boldsymbol{S}$ is symmetric and idempotent (i.e., $\boldsymbol{S} = \boldsymbol{S}'$ and $\boldsymbol{S}^2 = \boldsymbol{S}$), the definitions of df, df_{var}, and df_{res} coincide. This is, for example, the case for simple basis function approaches, but does not apply for procedures involving penalization.

The simplicity of the calculation for the first definition of effective degrees of freedom usually makes it favored in practice. In particular, we can avoid the calculation of the complete $n \times n$-matrix $\boldsymbol{S}$, since the trace operator is invariant to cyclical permutations of matrix products (see Theorem A.18 in Appendix A.4). For example, we obtain

$$\mathrm{df}(\boldsymbol{S}) = \mathrm{trace}(\boldsymbol{Z}(\boldsymbol{Z}'\boldsymbol{Z} + \lambda\boldsymbol{K})^{-1}\boldsymbol{Z}') = \mathrm{trace}(\boldsymbol{Z}'\boldsymbol{Z}(\boldsymbol{Z}'\boldsymbol{Z} + \lambda\boldsymbol{K})^{-1})$$

in case of P-splines so that the trace can be computed from the product of two matrices with dimension given by the number of basis functions.

Estimates for the Error Variance

In the linear model, an unbiased estimate for the error variance σ^2 was given by

$$\hat{\sigma}^2 = \frac{1}{n-p} \sum_{i=1}^{n} (y_i - \hat{y}_i)^2.$$

In nonparametric regression, it seems plausible to replace the number of parameters p with the equivalent degrees of freedom, as outlined in the previous section. In fact, when assuming that at least the two first moments of the model are correctly specified, one can show that the expected value of the residual sum of squares can be written as

$$\mathrm{E}\left(\sum_{i=1}^{n} (y_i - \hat{f}(z_i))^2\right) = (n - \mathrm{tr}(2\boldsymbol{S} - \boldsymbol{S}\,\boldsymbol{S}'))\sigma^2 + \sum_{i=1}^{n} b_i^2,$$

where b_i defines the bias of the function estimate at the point z_i. If we can neglect the bias, i.e., if the smoother is approximately unbiased, we obtain an unbiased error variance estimate as

$$\hat{\sigma}^2 = \frac{1}{n - \mathrm{df}_{res}} \sum_{i=1}^{n} (y_i - \hat{f}(z_i))^2.$$

Even though unbiasedness for $\hat{\sigma}^2$ is only assured when using df_{res}, the simpler definition df is typically used in practice for the estimation of the error variance.

Bias–Variance Trade-Off

In this section, we will more closely examine how varying specific parameters of a smoother affects the mean squared error of the function estimate. For the sake of simplicity, we first discuss the running mean estimate

$$\hat{f}(z_i) = \frac{1}{k} \sum_{j \in N(z_i)} y_j$$

that is based on the k-nearest neighbors. For this estimate, the expected value and the variance can be easily calculated as

$$\mathrm{E}(\hat{f}(z_i)) = \frac{1}{k} \sum_{j \in N(z_i)} f(z_j)$$

and

$$\mathrm{Var}(\hat{f}(z_i)) = \frac{\sigma^2}{k}$$

with $k = |N(z)|$. Since MSE is additively composed of both the squared bias and the variance, we could try to choose the smoothing parameter k in such a way that bias and variance are simultaneously minimized. The formulae above, however, show that this approach is not feasible, since the variation of k has an opposite effect on the bias relative to the variance. If we increase the neighborhood, the variance clearly decreases. On the other hand, a larger bias results as more and more terms are contained in the expected value of $\hat{f}(z_i)$, which are different from $f(z_i)$. A neighborhood of $k = 1$ with $N(z_i) = \{i\}$ results in an unbiased estimate which, however, has the highest possible variance. We found a similar dilemma in Sect. 3.4 within the context of model choice in linear models and identified it as a classical example of the bias–variance trade-off.

In general, smoother function estimates resulting from the choice of a smaller number of effective degrees of freedom usually lead to reduced variability, but increasing bias. On the other hand, a more complex model with a larger number of effective degrees of freedom typically reduces the bias, while at the same time resulting in larger variability. We can also empirically identify this effect when revisiting the plots (e.g., Figs. 8.10 on p. 425, 8.16 on p. 440 or 8.22 on p. 463) that illustrate the impact of the smoothing parameter. In contrast to the running mean estimate, the explicit derivation of the expected value and the variance of most scatter plot smoothers are actually more complicated. Hence, we will only examine local polynomial models as a further example for the bias–variance trade-off, where at least asymptotic results are available; see Fahrmeir and Tutz (2001), Sect. 5.1.3, or Fan and Gijbels (1996).

The smoothness in local polynomial regression is mainly controlled by the bandwidth λ of the kernel function. As a consequence, we examine the influence of λ on bias and variance of the resulting estimate. We limit ourselves to asymptotic results, i.e., we will examine the behavior of the estimate for a large sample size, $n \to \infty$. Note that in order to obtain meaningful asymptotic expressions, we have to impose conditions not only on the sample size itself but also on the smoothing parameter. The underlying reason is that an increasing sample size also yields more precise estimates and therefore requires a smaller bandwidth. More specifically, we examine the limiting case $\lambda \to 0$ and $n\lambda \to \infty$, i.e. even though we let the bandwidth approach zero, the sample size increases at a rate such that $n\lambda$ still diverges to infinity. Under these conditions, we obtain

$$\mathrm{E}(\hat{f}(z) - f(z)) \approx \frac{\lambda^{l+1} f^{(l+1)}(z)}{(l+1)!} \mu_{l+1}(K)$$

as an approximation for the asymptotic bias with odd polynomial degree l and

$$\mathrm{E}(\hat{f}(z) - f(z)) \approx \left(\frac{\lambda^{l+2} f^{(l+1)}(z) d_z'(z)}{(l+1)! d_z(z)} + \frac{\lambda^{l+2} f^{(l+2)}(z)}{(l+2)!} \right) \mu_{l+2}(K)$$

in case of an even polynomial degree. In the latter case, $d_z(z)$ and $d_z'(z)$ define the density of the distribution of z and its first derivative, respectively, and $\mu_q(K) = \int u^q K(u)du$ is the q-th moment for the chosen kernel. For the variance, we obtain the asymptotic approximation

$$\text{Var}(\hat{f}(z)) \approx \frac{\sigma^2}{n\lambda d_z(z)} \int K^2(u)du.$$

These two results, provide us with very interesting and useful information, which again illustrate the mentioned trade-off between bias and variance associated with nonparametric regression estimates. Regarding the bias, we can identify the following main points:

- The bias of $\hat{f}(z)$ decreases with decreasing bandwidth λ.
- For even l, the bias depends on the distribution of z (expressed through the density $d_z(z)$). The bias decreases when $d_z(z)$ is large, i.e., when a large number of observations are expected in the neighborhood of z. On the contrary, the bias increases in regions with a lower expected number of observations (i.e., a lower density $d_z(z)$).
- For odd l, the bias is independent of the distribution of the covariates.
- For $l = 0$, the bias depends on the first and second derivative of the function $f(z)$. A larger bias results in regions with a steep gradient $f'(z)$. Since the second derivative is a measure for the curvature of the function, we additionally obtain more bias in regions having large $f''(z)$. More precisely, we find underestimation (a negative bias) in local maxima and overestimation (a positive bias) in local minima.
- For even l, we get an additional term in the bias formula. This generally yields a preference for an odd polynomial degree.

For the variance, we have the following main results:

- An increase of the smoothing parameter leads to reduced variability in estimation.
- For both even or odd l, the variance depends on the distribution of the covariates. Analogous to the bias, we observe a decrease in the variance for large $d_z(z)$, and an increase for small $d_z(z)$. However, the precise form of the effect differs for bias and variance.

In summary, it is again not possible to decrease both variance and bias simultaneously. The use of the mean squared error is an appropriate compromise, which especially could be used to choose an optimal smoothing parameter. However, since the MSE generally depends on the unknown true function $f(z)$, this approach is typically not feasible. Nevertheless it provides the basis for strategies that rely on an approximation of the mean squared error.

8.1.9 Choosing the Smoothing Parameter

Our considerations of nonparametric regression models thus far yield one important remaining question: How do we "optimally" choose the smoothing parameter to properly describe the data? Based on our findings related to the bias–variance trade-off presented in the previous section, we can equivalently formulate this question as the problem of how to obtain an appropriate compromise between bias and variance of the function estimate. The mean squared error offers an immediate option, and we will first focus on this criterion in choosing smoothing parameters. Moreover, we will briefly discuss the use of AIC to determine optimal smoothing parameters. The second and third part of this section focus on penalization approaches and derive general possibilities to estimate the smoothing parameter in this context. Specifically, on the one hand, we will rely on the close connection between penalization approaches and mixed models, and on the other hand, we will derive Bayesian Markov chain Monte Carlo simulation procedures.

Choosing the Smoothing Parameter Based on Optimality Criteria

Our first option for choosing the smoothing parameter results from our considerations on the bias–variance trade-off. We discovered that both the bias and variance associated with a fitted smooth function depend on the smoothing parameter, and that both cannot be simultaneously decreased. Thus, an appropriate compromise has to be found, for example, based on the mean squared error

$$\mathrm{MSE}(\hat{f}(z)) = \mathrm{E}\left((\hat{f}(z) - f(z))^2\right) = \left(\mathrm{E}(\hat{f}(z) - f(z))\right)^2 + \mathrm{Var}(\hat{f}(z)),$$

which consists of both the squared bias and variance. From the pointwise MSE, we obtain a measure for the quality of the entire estimated function by averaging over the observed covariate values, i.e.,

$$\frac{1}{n}\mathrm{E}\left(\sum_{i=1}^{n}(\hat{f}(z_i) - f(z_i))^2\right).$$

A naive approximation is the residual sum of squares

$$\frac{1}{n}\sum_{i=1}^{n}(y_i - \hat{f}(z_i))^2.$$

However, this choice is not useful to determine an optimal smoothing parameter, as it can be minimized by $\hat{f}(z_i) = y_i$. The optimization of the residual sum of squares would therefore lead to nothing more than a trivially interpolating estimate; see also Sect. 3.4.

Consequently, the squared prediction error for new observations y^* is preferred when choosing the smoothing parameter. Typically, such new data are not available, and we therefore resort to a squared error approximation that results from cross validation. The procedure is as follows: We first fix the smoothing parameter, remove one of the observations from the data, and estimate the smooth function using the remaining $n - 1$ observations. Using this estimated function, we next predict the function value $f(z_i)$ at the location of the eliminated observation. Let $\hat{f}^{(-i)}(z)$ denote the estimated function obtained when removing observation (y_i, z_i). Cycling over all the observation, we then obtain the cross validation criterion

$$\text{CV} = \frac{1}{n}\sum_{i=1}^{n}(y_i - \hat{f}^{(-i)}(z_i))^2.$$

Note that the cross validation criterion depends on the chosen smoothing parameter and therefore choosing the smoothing parameter to minimize the cross validation criterion provides a way to determine an optimal smoothing parameter minimizing the prediction error. Furthermore, the use of the cross validation criterion is also theoretically justified due to the fact that

$$\text{E(CV)} \approx \frac{1}{n}\sum_{i=1}^{n}(y_{n+i} - \hat{f}(z_i))^2,$$

where y_{n+i} define new observations at the points z_i. Hence, the mean squared prediction error results as the expectation of the CV criterion.

At first glance, it appears that for computation of the cross validation criterion, we have to perform n separate nonparametric regression model fits (for each value of the smoothing parameter). However, we can actually obtain the CV score by only performing one fit using all the data (similar as in the linear model in Sect. 3.4). Utilizing the diagonal elements s_{ii} of the smoother matrix $\boldsymbol{S}$, CV can in fact be calculated via

$$\text{CV} = \frac{1}{n}\sum_{i=1}^{n}\left(\frac{y_i - \hat{f}(z_i)}{1 - s_{ii}}\right)^2.$$

Although this statement does not hold in full generality for all types of smoothers, it is still routinely used to determine an approximation to CV. The calculation of the smoothing matrix and its diagonal elements can, however, still be numerically complex (especially for large data sets). For this reason one often replaces the diagonal elements with their average, yielding the generalized cross validation criterion (GCV):

$$\text{GCV} = \frac{1}{n}\sum_{i=1}^{n}\left(\frac{y_i - \hat{f}(z_i)}{1 - \text{tr}(\boldsymbol{S})/n}\right)^2.$$

The sum of the diagonal elements exactly corresponds to our first definition of the equivalent degrees of freedom, namely the trace of the smoothing matrix. As we saw for P-splines, it is straightforward to calculate this trace, since products of matrices can be cyclically permuted without affecting the trace (see p. 475). In addition to a simpler calculation, the GCV criterion, in comparison to the CV criterion, also has the theoretical advantage to be invariant under orthogonal transformations of the data; compare Wood (2006), Sect. 4.5.2.

As an alternative to the generalized cross validation criterion, we can also use other criteria developed in the context of model choice for the determination of smoothing parameters. A typical example is Akaike's information criterion AIC, which was discussed in Sect. 3.4. The AIC includes the number of parameters as a form of penalization to correct the likelihood for the model complexity; see also Sect. B.4.5. In nonparametric regression, the effective number of parameters replaces the actual number of parameters leading to the criterion

$$\text{AIC} = n\log(\hat{\sigma}^2) + 2(\text{df} + 1),$$

which needs to be minimized with respect to the smoothing parameter. As in the linear model, $\hat{\sigma}^2 = \sum(y_i - \hat{f}(z_i))^2/n$ refers to the ML estimate for the error variance.

Figure 8.26 shows GCV and AIC as functions of the smoothing parameter for a cubic P-spline with 20 inner knots and a second-order difference penalty, using the Tanzania data together with the resulting optimal estimated functions. To improve readability of the GCV and AIC plots, the minimal value of AIC or GCV was set to zero so that the curves reflect differences relative to the optimal model. It appears that GCV and AIC show a very similar relationship with the smoothing parameter λ. In fact, both criteria are actually asymptotically equivalent although in specific applications clear differences may be observed. For the Tanzania data, both criteria provide similar, although not identical, optimal smoothing parameters and therefore also yield very similar function estimates.

How can we concretely determine the optimal smoothing parameter for a given criterion? So far, we only examined the simplest case, i.e., univariate nonparametric regression. Thus it is possible to directly optimize the criterion of interest, for example, using a grid search over λ (or using more efficient methods for numerical minimization). To do so, we generate a grid of candidate smoothing parameters, calculate the criterion for each of these candidates, and then choose the value corresponding to the minimal criterion. It can also be useful to iteratively refine the grid based on a first rough discretization to assure that the optimal value is obtained precisely.

In more complex problems, as the ones that we will examine in Chap. 9, direct optimization is no longer possible or at least very time consuming, since several smoothing parameters have to be determined simultaneously. However, even in these settings, algorithms exist for an efficient minimization of the GCV or the AIC. Such approaches are based on the Newton method, i.e., rely on the derivative of the criterion with respect to the smoothing parameters. Since these methods are mathematically and algorithmically complex, we do not discuss them here in detail but rather refer to the corresponding literature, especially Wood (2000, 2004, 2006, 2008). The R package `mgcv` provides implementations.

Mixed Model Representation of Penalization Approaches

The methods based on model choice criteria that we discussed in the previous section can be generally used for the choice of smoothing parameters for any scatter plot smoother. In the following two paragraphs, we rather focus on penalization approaches, with penalties

$$\lambda \boldsymbol{\gamma}' \boldsymbol{K} \boldsymbol{\gamma}.$$

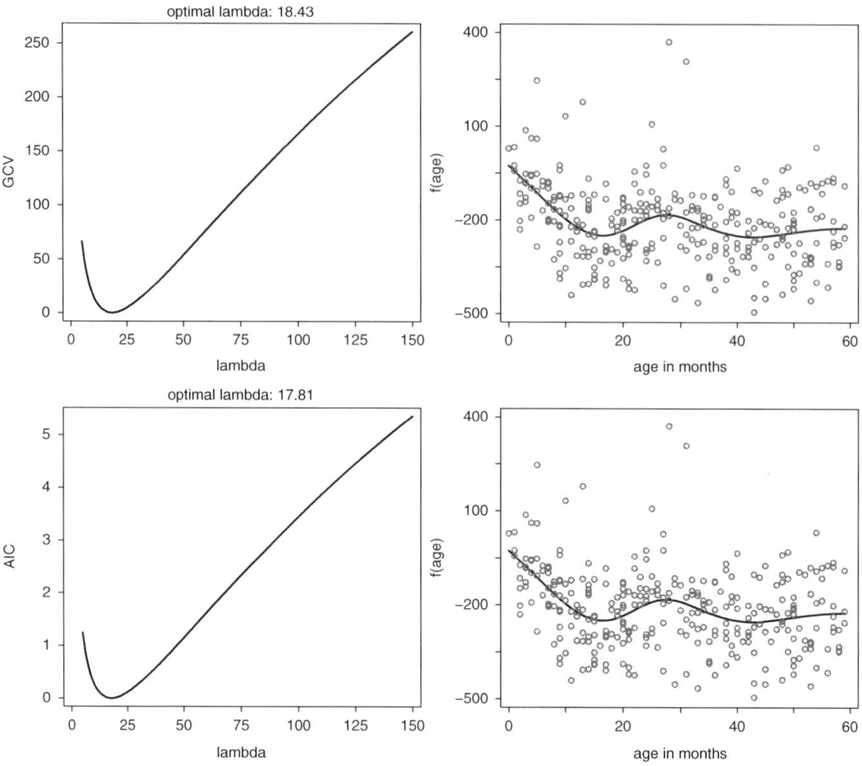

Fig. 8.26 Malnutrition in Tanzania: GCV and AIC (*left panel*) and cubic P-splines fits resulting with the corresponding optimal smoothing parameters (*right panel*)

As seen, the penalty is a quadratic form in the parameters and depends on a (symmetric, positive semidefinite) penalty matrix $\boldsymbol{K}$. This limitation is, however, not very restrictive, because it includes all methods examined in Sects. 8.1.2–8.1.6, especially penalized splines, smoothing splines, random walks, and kriging.

We start our considerations with P-splines in truncated power series representation. In this case, the PLS criterion is given by

$$\text{PLS}(\lambda) = (\boldsymbol{y} - \boldsymbol{Z}\boldsymbol{\gamma})'(\boldsymbol{y} - \boldsymbol{Z}\boldsymbol{\gamma}) + \lambda \sum_{j=l+2}^{d} \gamma_j^2$$

(see p. 432). To establish a connection to the mixed models of Chap. 7, we first partition the parameter vector $\boldsymbol{\gamma}$ into $\boldsymbol{\beta} = (\gamma_1, \ldots, \gamma_{l+1})'$ and $\tilde{\boldsymbol{\gamma}} = (\gamma_{l+2}, \ldots, \gamma_d)'$. The vector $\boldsymbol{\beta}$, thus, consists of the non-penalized parameters corresponding to the global polynomial, whereas $\tilde{\boldsymbol{\gamma}}$ contains the parameters of the truncated power functions that are included as squared values in the penalty. If we split the design matrix accordingly into $\boldsymbol{X}$ and $\boldsymbol{U}$ so that $\boldsymbol{Z} = [\boldsymbol{X}\ \boldsymbol{U}]$, the PLS criterion can be reexpressed as

$$(y - X\beta - U\tilde{\gamma})'(y - X\beta - U\tilde{\gamma}) + \lambda\tilde{\gamma}'\tilde{\gamma}. \tag{8.20}$$

In Sect. 7.3.1, we found that in the mixed model formulation,

$$y = X\beta + U\tilde{\gamma} + \varepsilon, \quad \varepsilon \sim \mathrm{N}(\mathbf{0}, \sigma^2 I), \quad \tilde{\gamma} \sim \mathrm{N}(\mathbf{0}, \tau^2 I),$$

estimates for the regression coefficients can also be obtained by minimizing the PLS criterion

$$(y - X\beta - U\tilde{\gamma})'(y - X\beta - U\tilde{\gamma}) + \frac{\sigma^2}{\tau^2}\tilde{\gamma}'\tilde{\gamma}.$$

If we now compare the penalized least squares criterion arising from mixed models with Eq. (8.20), we find that β can be interpreted as a vector of fixed effects while $\tilde{\gamma}$ corresponds to a vector of random effects. The variance of the random effects is then defined by $\tau^2 = \sigma^2/\lambda$ and, as a consequence, we can interpret the smoothing parameter λ as the ratio of error variance and random effects variance similar to our considerations regarding Bayesian P-splines (p. 441ff).

Embedding P-splines in the scope of mixed models has the advantage that we can use the estimation procedures discussed in Chap. 7 to determine estimates of σ^2 and τ^2 and therefore also the smoothing parameter λ. Specifically, we can use either maximum likelihood or restricted maximum likelihood estimation as introduced in Sect. 7.3.2. Based on the estimates $\hat{\sigma}^2$ and $\hat{\tau}^2$, we obtain the optimal smoothing parameter $\hat{\lambda} = \hat{\sigma}^2/\hat{\tau}^2$. Although smoothing parameter selection via restricted maximum likelihood is usually done in the context of mixed models a direct approach without resorting to the connection with mixed models is possible; see Wood (2011).

In summary, a close connection exists between P-splines in TP-representation and mixed models yielding an alternative perspective on P-splines. We next examine this connection more generally for arbitrary penalization approaches. Starting with the penalized sum of squared residuals

$$\mathrm{PLS}(\lambda) = (y - Z\gamma)'(y - Z\gamma) + \lambda\gamma' K\gamma,$$

it may be tempting to define the corresponding mixed model as

$$y = Z\gamma + \varepsilon, \quad \varepsilon \sim \mathrm{N}(\mathbf{0}, \sigma^2 I), \quad \gamma \sim \mathrm{N}(\mathbf{0}, \tau^2 K^{-1}) \tag{8.21}$$

with $\lambda = \sigma^2/\tau^2$. However, in this direct approach, we encounter the problem that the inverse K^{-1} generally does not exist, since the penalty matrix K often does not have full rank (e.g., in case of P-splines based on B-splines). For kriging, the formulation in Eq. (8.21) is immediately applicable since the penalty matrix has full rank.

To relate arbitrary penalization approaches to mixed models, we therefore have to proceed differently than we did for P-splines with a TP basis. The basic problem is that the density of the random effects resulting from representation (8.21),

$$p(\boldsymbol{\gamma}) \propto \exp\left(-\frac{1}{2\tau^2}\boldsymbol{\gamma}'\boldsymbol{K}\boldsymbol{\gamma}\right)$$

cannot be normalized. Such a density is also called *improper*. More precisely, we obtain a partially improper density, since $\text{rk}(\boldsymbol{K}) > 0$, yet $\text{rk}(\boldsymbol{K}) \neq \dim(\boldsymbol{\gamma})$. In a mixed model, this partial impropriety is avoided by separating the density into a noninformative distribution for fixed effects and a nonsingular normal distribution with a normalized density for the random effects. It is our goal to derive such a separation for any penalization approach (see also Sect. B.3.2 in Appendix B).

To do so, we need to split $\boldsymbol{\gamma}$ into two subvectors $\boldsymbol{\beta}$ and $\tilde{\boldsymbol{\gamma}}$ with dimensionalities depending on the rank deficiency of $\boldsymbol{K}$. Let $r = \text{rk}(\boldsymbol{K})$ denote the rank of the penalty matrix and $d = \dim(\boldsymbol{\gamma})$ the dimension of the coefficient vectors. We then consider reparameterizations

$$\boldsymbol{\gamma} = \tilde{\boldsymbol{X}}\boldsymbol{\beta} + \tilde{\boldsymbol{U}}\tilde{\boldsymbol{\gamma}}$$

with $d\times(d-r)$- and $d\times r$-dimensional design matrices $\tilde{\boldsymbol{X}}$ and $\tilde{\boldsymbol{U}}$ and corresponding $(d-r)$ and r-dimensional parameter vectors $\boldsymbol{\beta}$ and $\tilde{\boldsymbol{\gamma}}$. Our aim is to choose the design matrices $\tilde{\boldsymbol{X}}$ and $\tilde{\boldsymbol{U}}$ so that we can rewrite the penalty $\lambda\boldsymbol{\gamma}'\boldsymbol{K}\boldsymbol{\gamma}$ as $\lambda\tilde{\boldsymbol{\gamma}}'\tilde{\boldsymbol{\gamma}}$. As such $\boldsymbol{\beta}$ can be considered a vector of fixed effects while $\tilde{\boldsymbol{\gamma}}$ corresponds to a vector of independent and identically distributed (i.i.d.) random effects. More precisely, this requires the following properties of the design matrices:

1. The composite matrix $[\tilde{\boldsymbol{X}}, \tilde{\boldsymbol{U}}]$ should have full rank to yield a one-to-one transformation.
2. $\tilde{\boldsymbol{X}}'\boldsymbol{K} = \boldsymbol{0}$, so that $\boldsymbol{K}$ does not penalize $\boldsymbol{\beta}$.
3. $\tilde{\boldsymbol{U}}'\boldsymbol{K}\tilde{\boldsymbol{U}} = \boldsymbol{I}$, so that $\tilde{\boldsymbol{\gamma}}$ consists of i.i.d. random effects.

For the penalty, we then have

$$\begin{aligned}\boldsymbol{\gamma}'\boldsymbol{K}\boldsymbol{\gamma} &= (\tilde{\boldsymbol{X}}\boldsymbol{\beta} + \tilde{\boldsymbol{U}}\tilde{\boldsymbol{\gamma}})'\boldsymbol{K}(\tilde{\boldsymbol{X}}\boldsymbol{\beta} + \tilde{\boldsymbol{U}}\tilde{\boldsymbol{\gamma}})\\ &= \boldsymbol{\beta}'\underbrace{\tilde{\boldsymbol{X}}'\boldsymbol{K}}_{=\boldsymbol{0}}\tilde{\boldsymbol{X}}\boldsymbol{\beta} + 2\boldsymbol{\beta}'\underbrace{\tilde{\boldsymbol{X}}'\boldsymbol{K}}_{=\boldsymbol{0}}\tilde{\boldsymbol{U}}\tilde{\boldsymbol{\gamma}} + \tilde{\boldsymbol{\gamma}}'\underbrace{\tilde{\boldsymbol{U}}'\boldsymbol{K}\tilde{\boldsymbol{U}}}_{=\boldsymbol{I}}\tilde{\boldsymbol{\gamma}}\\ &= \tilde{\boldsymbol{\gamma}}'\tilde{\boldsymbol{\gamma}}.\end{aligned}$$

The model can therefore be rewritten as a mixed model, yielding

$$\boldsymbol{y} = \boldsymbol{Z}\boldsymbol{\gamma} + \boldsymbol{\varepsilon} = \boldsymbol{Z}(\tilde{\boldsymbol{X}}\boldsymbol{\beta} + \tilde{\boldsymbol{U}}\tilde{\boldsymbol{\gamma}}) + \boldsymbol{\varepsilon} = \boldsymbol{X}\boldsymbol{\beta} + \boldsymbol{U}\tilde{\boldsymbol{\gamma}} + \boldsymbol{\varepsilon}$$

with design matrices $\boldsymbol{X} = \boldsymbol{Z}\tilde{\boldsymbol{X}}$ and $\boldsymbol{U} = \boldsymbol{Z}\tilde{\boldsymbol{U}}$, as well as fixed effects $\boldsymbol{\beta}$ and random effects $\tilde{\boldsymbol{\gamma}} \sim \text{N}(\boldsymbol{0}, \tau^2\boldsymbol{I}_r)$. As for P-splines in a TP-representation, we are now able to use mixed models methodology to determine the variance parameters, and thus the smoothing parameter.

The remaining question is the construction of design matrices $\tilde{\boldsymbol{X}}$ and $\tilde{\boldsymbol{U}}$ that actually fulfill conditions (1) and (2) listed above. For the design matrix of the fixed effects, we can easily obtain the desired orthogonality to the penalty matrix by using a basis of the null space of $\boldsymbol{K}$ for the columns of $\tilde{\boldsymbol{X}}$; see Definition A.16 in Appendix A.2). For P-splines with a B-spline basis, this null space can also be easily characterized, since polynomials of degree $k-1$ remain unpenalized by the difference penalty. Thus, we can define $\tilde{\boldsymbol{X}}$ as

$$\tilde{\boldsymbol{X}} = \begin{pmatrix} 1 & \kappa_1 & \dots & \kappa_1^{k-1} \\ \vdots & \vdots & & \vdots \\ 1 & \kappa_d & \dots & \kappa_d^{k-1} \end{pmatrix},$$

where $\kappa_1, \dots, \kappa_d$ are the knots of the spline basis.

The design matrix of the random effects can be derived from the spectral decomposition of the penalty matrix (see Theorem A.25 in Appendix A). Using the decomposition $\boldsymbol{K} = \boldsymbol{\Gamma}\boldsymbol{\Omega}_+\boldsymbol{\Gamma}'$, where $\boldsymbol{\Omega}_+$ is the matrix of the positive eigenvalues and $\boldsymbol{\Gamma}$ is the corresponding orthonormal matrix of eigenvectors, we can define an appropriate $\tilde{\boldsymbol{U}}$ by $\tilde{\boldsymbol{U}} = \boldsymbol{L}(\boldsymbol{L}'\boldsymbol{L})^{-1}$ with $\boldsymbol{L} = \boldsymbol{\Gamma}\boldsymbol{\Omega}_+^{1/2}$. It follows then that

$$\tilde{\boldsymbol{U}}'\boldsymbol{K}\tilde{\boldsymbol{U}} = (\boldsymbol{L}'\boldsymbol{L})^{-1}\boldsymbol{L}'\boldsymbol{L}\boldsymbol{L}'\boldsymbol{L}(\boldsymbol{L}'\boldsymbol{L})^{-1} = \boldsymbol{I}.$$

In some specific situations, we can avoid the use of the spectral decomposition. For P-splines based on a B-spline basis, for example, we can also choose $\boldsymbol{L} = \boldsymbol{D}'$ with difference matrix $\boldsymbol{D}$. This also indicates that the mixed model decomposition of $\boldsymbol{\gamma}$ is not unique.

In summary, we have found a possibility to transfer basically any penalization approach into a mixed model and, in doing so, to obtain an optimal smoothing parameter using ML or REML estimation. In the following Sect. 8.2, we will further apply this knowledge to more penalization methods in the context of bivariate smoothing. A further advantage of the representation as a mixed model that is often claimed is that standard mixed model software can be used. We must, however, be aware of the fact that mixed models derived from penalization approaches differ significantly in structure from most of the mixed models discussed in Chap. 7, even though they can formally be written in the same way. The mixed models arising in the context of smoothing do not have any grouping structure, which is one reason why numerical problems may occur when using standard software. Thus, generally speaking, estimation using specialized software that is based on the mixed model representation should be preferred. This is especially the case with the more complex models that will be discussed in Chap. 9.

Finally, we will now consider a methodological problem that arises when estimating the smoothing parameter using mixed models. From a frequentist perspective, the parameters $\boldsymbol{\gamma}$ to be estimated in a penalization approach are fixed, unknown coefficients. In the reformulation as a mixed model, one part of the vector $\boldsymbol{\gamma}$ will be converted into random effects and does, consequently, no longer formally represent a fixed parameter, but rather a random variable. According to the frequentist interpretation, however, $\tilde{\boldsymbol{\gamma}}$ should not be interpreted as random. Strictly speaking, any representation as a mixed model is then to be understood as an algorithmic trick and not as an actual reformulation of the model. From a Bayesian perspective, this problem does not occur, since all parameters are viewed as random anyway and either representation defines an equivalent formulation of the same prior assumption. According to the distinction made in Sect. 7.4, the estimates resulting from the Bayesian interpretation of mixed model-based smoothing have to be interpreted as empirical Bayes estimates since $\boldsymbol{\gamma}$ is viewed as random while the variance parameters are estimated from a likelihood and therefore in a frequentist setting.

The difference between the Bayesian and the frequentist understanding of penalization methods is also evident in the covariance matrix of the estimates; see also the different covariance matrices that were obtained in Sect. 7.3.3 (p. 378) for general mixed models. Inserting the specific quantities

$$\boldsymbol{R} = \sigma^2 \boldsymbol{I}, \quad \boldsymbol{G} = \tau^2 \boldsymbol{I}, \quad \boldsymbol{C} = (\boldsymbol{X}, \boldsymbol{U}), \quad \boldsymbol{B} = \begin{pmatrix} \boldsymbol{0} & \boldsymbol{0} \\ \boldsymbol{0} & 1/\tau^2 \boldsymbol{I} \end{pmatrix},$$

in Eq. (7.30) of Sect. 7.3.3, we obtain the Bayesian covariance matrix

$$\text{Cov}\begin{pmatrix} \hat{\boldsymbol{\beta}} \\ \hat{\hat{\boldsymbol{\gamma}}} \end{pmatrix} = \sigma^2 \boldsymbol{H}^{-1},$$

with

$$\boldsymbol{H} = \begin{pmatrix} \boldsymbol{X}'\boldsymbol{X} & \boldsymbol{X}'\boldsymbol{U} \\ \boldsymbol{U}'\boldsymbol{X} & \boldsymbol{U}'\boldsymbol{U} + \lambda \boldsymbol{I} \end{pmatrix}.$$

Note that $\boldsymbol{H}$ is required for the estimation of the regression coefficients anyway. The frequentist version (7.32) of the covariance is given by

$$\text{Cov}\begin{pmatrix} \hat{\boldsymbol{\beta}} \\ \hat{\hat{\boldsymbol{\gamma}}} \end{pmatrix} = \boldsymbol{H}^{-1} \begin{pmatrix} \boldsymbol{X}' \\ \boldsymbol{U}' \end{pmatrix} \text{Cov}(\boldsymbol{y})(\boldsymbol{X}\ \boldsymbol{U})\boldsymbol{H}^{-1} = \sigma^2 \boldsymbol{H}^{-1} \boldsymbol{H}_1 \boldsymbol{H}^{-1},$$

with $\text{Cov}(\boldsymbol{y}) = \sigma^2 \boldsymbol{I}$ and the cross product matrix

$$\boldsymbol{H}_1 = \begin{pmatrix} \boldsymbol{X}'\boldsymbol{X} & \boldsymbol{X}'\boldsymbol{U} \\ \boldsymbol{U}'\boldsymbol{X} & \boldsymbol{U}'\boldsymbol{U} \end{pmatrix}.$$

In the frequentist approach, we obtain the "sandwich" form for the covariance estimate that was already discussed in Sect. 7.3.3. In contrast, a simpler covariance matrix results from the Bayesian approach, which leads to somewhat wider confidence intervals.

Bayesian Smoothing Parameter Choice Based on MCMC

After defining empirical Bayes estimates for the smoothing parameter based on mixed models, we now discuss a fully Bayesian alternative relying on MCMC simulations. We again restrict ourselves to penalization approaches with quadratic penalties. For Bayesian P-splines, we already found that such a penalty can be derived equivalently by specifying an appropriate smoothness prior. More generally, the following equivalence holds: Frequentist regression models of the form $\boldsymbol{y} = \boldsymbol{Z}\boldsymbol{\gamma} + \boldsymbol{\varepsilon}$ with quadratic penalty

$$\frac{\sigma^2}{\tau^2}\boldsymbol{\gamma}'\boldsymbol{K}\boldsymbol{\gamma}$$

correspond to Bayesian regression models of the same structure with multivariate normal prior distribution

$$p(\boldsymbol{\gamma} \mid \tau^2) \propto \left(\frac{1}{\tau^2}\right)^{\mathrm{rk}(\boldsymbol{K})/2} \exp\left(-\frac{1}{2\tau^2}\boldsymbol{\gamma}'\boldsymbol{K}\boldsymbol{\gamma}\right)$$

for the regression coefficients. Note that, in general, this density cannot be normalized, since $\boldsymbol{K}$ is often not of full rank and therefore $p(\boldsymbol{\gamma} \mid \tau^2)$ is partially improper. Taking the logarithm of the density shows that it is equivalent (up to a change in the sign and an additive constant) to the penalty in the PLS criterion. Therefore, the posterior mode $\hat{\boldsymbol{\gamma}}$ is equivalent to the PLS estimate for given smoothing parameter (as was the case with Bayesian P-splines). In analogy to the penalization of certain function types, we can interpret the prior distribution as a smoothness prior, which enforces prior beliefs regarding the smoothness of the function to be estimated.

Rather than examining the posterior mode estimate, we will now use MCMC simulation techniques to determine the posterior mean. Therefore we require a fully Bayesian formulation of the nonparametric regression problem where adequate prior distributions are provided for all unknown parameters. This includes the variance parameters, which were estimated with ML or REML in the empirical Bayes approach. Since the inverse gamma distribution $\mathrm{IG}(a,b)$ is the conjugate prior distribution in the normal setting, the choice of IG priors for σ^2 and τ^2 yields feasible models; see also Sect. 4.4 on Bayesian linear models. More specifically, we assume

$$\sigma^2 \sim \mathrm{IG}(a,b) \quad \text{and} \quad \tau^2 \sim \mathrm{IG}(a_1,b_1).$$

Assuming (conditional) independence a priori, the joint posterior distribution is given by

$$\begin{aligned}
p(\boldsymbol{\gamma},\sigma^2,\tau^2 \mid \boldsymbol{y}) &\propto p(\boldsymbol{y} \mid \boldsymbol{\gamma},\sigma^2)p(\boldsymbol{\gamma} \mid \tau^2)p(\sigma^2)p(\tau^2)\\
&\propto (\sigma^2)^{-\frac{n}{2}} \exp\left(-\frac{1}{2\sigma^2}(\boldsymbol{y}-\boldsymbol{Z}\boldsymbol{\gamma})'(\boldsymbol{y}-\boldsymbol{Z}\boldsymbol{\gamma})\right)\\
&\quad \cdot(\tau^2)^{-\frac{\mathrm{rk}(\boldsymbol{K})}{2}} \exp\left(-\frac{1}{2\tau^2}\boldsymbol{\gamma}'\boldsymbol{K}\boldsymbol{\gamma}\right)\\
&\quad \cdot\frac{1}{(\sigma^2)^{a+1}} \exp\left(-\frac{b}{\sigma^2}\right) \frac{1}{(\tau^2)^{a_1+1}} \exp\left(-\frac{b_1}{\tau^2}\right).
\end{aligned}$$

Due to the conjugate IG priors, we obtain known distributions for all full conditional distributions so that a Gibbs sampler can be used for estimation. More precisely, we obtain a normal distribution

$$\boldsymbol{\gamma} \mid \boldsymbol{y}, \sigma^2, \tau^2 \sim \mathrm{N}(\boldsymbol{\mu}_\gamma, \boldsymbol{\Sigma}_\gamma),$$

for $\boldsymbol{\gamma}$ with expected value and covariance matrix

$$\boldsymbol{\mu}_\gamma = \mathrm{E}(\boldsymbol{\gamma} \mid \cdot) = \left(\frac{1}{\sigma^2}\boldsymbol{Z}'\boldsymbol{Z} + \frac{1}{\tau^2}\boldsymbol{K}\right)^{-1} \frac{1}{\sigma^2}\boldsymbol{Z}'\boldsymbol{y}$$

$$\boldsymbol{\Sigma}_\gamma = \mathrm{Cov}(\boldsymbol{\gamma} \mid \cdot) = \left(\frac{1}{\sigma^2}\boldsymbol{Z}'\boldsymbol{Z} + \frac{1}{\tau^2}\boldsymbol{K}\right)^{-1}.$$

For the variance parameters, the full conditional distributions are of the inverse gamma type, and we find

$$\tau^2 \mid \boldsymbol{y}, \boldsymbol{\gamma}, \sigma^2 \sim \mathrm{IG}(a_1 + 0.5\mathrm{rk}(\boldsymbol{K}), b_1 + 0.5\boldsymbol{\gamma}'\boldsymbol{K}\boldsymbol{\gamma}),$$

$$\sigma^2 \mid \boldsymbol{y}, \boldsymbol{\gamma}, \tau^2 \sim \mathrm{IG}(a + 0.5n, b + 0.5(\boldsymbol{y} - \boldsymbol{Z}\boldsymbol{\gamma})'(\boldsymbol{y} - \boldsymbol{Z}\boldsymbol{\gamma})).$$

Based on these full conditionals, we can now derive an MCMC sampler by iteratively drawing random numbers with current parameters inserted. However, $\boldsymbol{\gamma}$ may be of relatively high dimension in nonparametric regression models; a naive implementation can be rather time-consuming as we have to solve high-dimensional systems of equations in every iteration to sample from $\boldsymbol{\gamma} \mid \boldsymbol{y}, \sigma^2, \tau^2$. To considerably shorten the computing times, efficient algorithms for drawing multivariate normal random numbers can be used. These are based on the fact that both the penalty matrix $\boldsymbol{K}$ and the cross product of the design matrix $\boldsymbol{Z}'\boldsymbol{Z}$ have a very specific and typically sparse structure that can be exploited in the calculations; see Lang, Umlauf, Wechselberger, Harttgen, and Kneib (2012) for details.

Besides point estimates $\hat{f}(z)$ obtained as the median or mean of the samples

$$f^{(t)}(z) = \sum_{j=1}^{d} \gamma_j^{(t)} B_j(z),$$

the MCMC samples also form the basis for defining appropriate credible regions, i.e., the Bayesian analogues to confidence intervals. Pointwise credible intervals are typically computed by taking empirical quantiles of the samples, e.g., for obtaining a 95 % pointwise credible interval for $f(z)$, the 2.5 % and the 97.5 % quantile of the samples $f^{(t)}(z)$, $t = 1, \ldots, T$, define the boundaries of the interval. Simultaneous credible bands as discussed on p. 471 in the frequentist context are more difficult to obtain since there is no common ordering of the samples for a vector of function evaluations $\boldsymbol{f}_r = (f(z_1), \ldots, f(z_r))'$, where $z_1, \ldots, z_r$ are a representative selection of covariate values (e.g., the subset of all distinct covariate values or a grid of equidistant values).

An early suggestion for constructing simultaneous credible bands has been made by Besag, Green, Higdon, and Mengersen (1995) credible bands based on the order statistics of the samples. An alternative approach based on assuming posterior normality has been proposed by Crainiceanu, Ruppert, Carroll, Adarsh, and Goodner (2007). Let $\hat{\boldsymbol{f}}_r$ denote the posterior mean and $\hat{s}_i$, $i = 1, \ldots, r$, the posterior standard deviation for each element of $\boldsymbol{f}_r$ computed from the samples. Then after assuming approximate posterior normality and deriving the $(1 - \alpha)$ sample quantile $m_{1-\alpha}$ of

$$\max_{i=1,\ldots,r} \left| \frac{f(z_i)^{(t)} - \hat{f}(z_i)}{\hat{s}_i} \right|, \quad t = 1, \ldots, T, \tag{8.22}$$

a simultaneous credible band for $\boldsymbol{f}_r$ is given by the hyperrectangular

$$\left[\hat{f}(z_i) - \hat{s}_i m_{1-\alpha},\ \hat{f}(z_i) + \hat{s}_i m_{1-\alpha}\right], \quad i = 1, \ldots, r.$$

In fact, this credible band is the sampling-based analogue to the simultaneous confidence band derived from Eq. (8.18) on p. 472. It implicitly depends on the assumption of posterior normality and uses the posterior standard deviation as a measure of uncertainty (also assuming symmetry of the posterior distribution) and the posterior mean as a point estimate. As a consequence, the posterior information contained in the samples is not fully utilized but reduced to measures of location and scale.

Krivobokova, Kneib, and Claeskens (2010) propose another alternative definition of simultaneous Bayesian credible bands that form a kind of compromise between the approaches by Besag et al. (1995) and Crainiceanu et al. (2007) and thereby avoids their disadvantages. As a starting point, the pointwise credible intervals derived from the $\alpha/2$ and $1 - \alpha/2$ quantiles of the samples $f(z_1)^{(t)}, \ldots, f(z_r)^{(t)}$, $t = 1, \ldots, T$, are used to measure pointwise uncertainty in the estimated function instead of using the pointwise standard deviation as in Crainiceanu et al. (2007). Then these pointwise credible intervals are scaled with a constant factor until a fraction of $(1 - \alpha)$ of all sampled curves is contained in the credible band. This can be achieved by monitoring for each of the sampled curves whether it is already contained in the credible band and by modifying the scaling factor until the desired confidence level $1 - \alpha$ is achieved. By construction, this new credible band is still based on pointwise measures of uncertainty (reflected by the pointwise credible intervals) but avoids the assumption of posterior normality and in particular the assumption of symmetry for the posterior distribution.

Krivobokova et al. (2010) also perform a simulation study comparing the three suggested credible bands. It turns out that the approach of Besag et al. (1995) often does not yield the desired coverage properties but understates uncertainty and therefore yields coverages below the nominal level. In contrast, both the approaches by Crainiceanu et al. (2007) and Krivobokova et al. (2010) work well in case of normally distributed responses. Since this is the ideal situation for the credible intervals based on posterior normality, the similar performance is not too surprising. However, in general, the approach by Krivobokova et al. (2010) requires less assumptions and is therefore probably preferable in practice. It also has the advantage that no additional computations for deriving the quantiles in Eq. (8.22) are required.

Example 8.5 Malnutrition in Tanzania—Credible Bands for the Age Effect

We demonstrate the difference between pointwise credible intervals and simultaneous credible bands by analyzing the age effect in the Tanzania example, providing a Bayesian supplement to the frequentist results achieved in Example 8.4 (p. 472). Figure 8.27 shows the posterior mean together with a pointwise 95 % credible interval and a simultaneous 95 % credible band computed according to the approach of Krivobokova et al. (2010). The scaling factor required to move from the pointwise to the simultaneous credible band is about 1.6, which seems to be a common value in many analyses. As can be seen, the simultaneous band picks up local variability in the pointwise bands but scales them to a larger magnitude to cover the overall variability of the curve samples.

△

8.1.10 Adaptive Smoothing Approaches

In Sect. 8.1.1, we first introduced polynomial splines without penalization and then motivated the need of appropriate regularization implemented via different penalization approaches. Alternatively, algorithms can be constructed for estimating polynomial splines without penalty but with optimally selected number and position of the knots. In this setting, regularization is achieved directly through model choice strategies, rather than indirectly through the addition of a penalty term. Such approaches are then often referred to as *adaptive procedures*, since the construction of the polynomial spline is adaptively controlled by the observed response values. Note that with penalization approaches, the number and position of the knots are chosen based on either the covariate values alone or some partition of the covariate domain.

Various different adaptive procedures exist in both Bayesian and frequentist formulations, and we will not be able to explain all of them in detail here. Instead, we restrict the presentation to a broad overview and develop the main ideas while providing additional references in Sect. 8.4.

Multivariate Adaptive Regression Splines

Many frequentist approaches for adaptive nonparametric regression can be developed analogously to the variable selection ideas that have been discussed in Sect. 3.4. As a representative example, we describe a variant of the popular Multivariate Adaptive Regression Splines (MARS) algorithm. The basis function

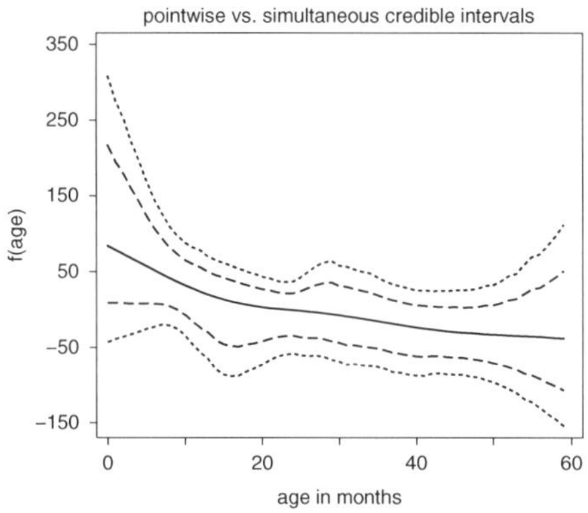

Fig. 8.27 Malnutrition in Tanzania: pointwise credible intervals (– – –) and simultaneous confidence band (- - -) for the age effect

representation

$$y_i = f(z_i) + \varepsilon_i = \sum_{j=1}^{d} \gamma_j B_j(z_i) + \varepsilon_i,$$

serves as the foundation for the MARS algorithm. However, the basis functions are no longer fixed but are rather selected from a large set of possible candidates. The model fit is then evaluated using one of the model choice criteria already discussed, such as GCV or AIC. A possible search strategy for determining the optimal model is presented in the following algorithm:

1. Start with the minimal model, i.e., the smallest basis to be considered.
2. Iteratively extend the model until the full model is reached:
   - For all basis functions B_j that are not yet included in the model, calculate the score statistic for the test on $\gamma_j = 0$ and add the basis function with the highest score statistic.
   - Estimate the new model and save the respective model fit criterion.
3. Based on the full model, iteratively eliminate basis functions until the minimal model is reached:
   - For each basis function B_j still included in the model, calculate the Wald statistic for the test on $\gamma_j = 0$ and remove the basis function with the smallest Wald statistic from the model.
   - Estimate the new model and save the respective model fit criterion.
4. From the resulting sequence of models choose the one which optimizes the model fit criterion as the best model.

The algorithm mainly consists of two steps in which a forward selection and a backward elimination are conducted through the space of possible models. In doing so, different criteria are used for the inclusion of new terms and for the elimination of terms already included in the model so that two different sequences of models result. The test criteria are chosen to ease calculation: While the score statistic for the inclusion of a new basis function can be determined without actually estimating the extended model, the Wald statistic can be calculated without having to estimate the reduced model. Thus, the execution of the algorithm is feasible even for a large number of basis functions.

For demonstration purposes, we analyze the simulated data example using a MARS implementation available in function `polymars` from the R package `polspline`. The implementation uses a linear TP basis and starts with the intercept as the minimal model. All observed covariate values are candidate knot positions resulting in a maximum number of 17 basis functions. To evaluate the model fit, we chose the GCV criterion. Figure 8.28 (left) shows the resulting GCV for the sequence of selected models. Large values appear at both the beginning and the end of the model sequence corresponding to simple models with a small number of basis functions that clearly fit the data poorly. A similar statement holds for very complex models that overfit the data and therefore also yield larger values for the GCV. In addition, the GCV sequence shows that the forward and backward search actually result in different model sequences and, as a consequence, the GCV curve is not completely symmetric. In more complex models with a larger number of regressor variables, even stronger deviations from symmetry are to be expected. Figure 8.28 (right) shows the optimal function estimate with eight basis functions obtained from the MARS algorithm. In the example, a satisfactory model fit results even with linear splines. In general, a basis of higher degree will be recommended to ensure smoother fits, but such basis functions are not available in `polymars`.

Regression Trees

As a second example for adaptive smoothing, we examine a procedure that approximates $f(z)$ using piecewise constant functions. In this approach, the domain of the covariate z is partitioned in such a way that the observations within the resulting groups are as homogeneous as possible (with respect to the response variable) while the groups themselves are as heterogeneous as possible. Within the scope of nonparametric regression, this problem can be formulated as follows: The function $f(z)$ is expressed as

$$f(z) = \sum_{j=1}^{d} \gamma_j I(\kappa_{j-1} < z \leq \kappa_j),$$

i.e., the domain of z will be divided into intervals $R_j = (\kappa_{j-1}, \kappa_j]$ based on the cut points $\kappa_0 < \kappa_1 < \ldots < \kappa_d$, and the function f takes the constant value γ_j on such an interval. The cut points $\kappa_0, \ldots, \kappa_d$, as well as the function values γ_j, are then to be chosen to minimize the residual sum of squares $\sum_i (y_i - f(z_i))^2$. Without further restrictions, this will lead to a perfect fit, in which every observation is represented by one individual group. As a consequence, additional constraints are usually imposed on $f(z)$, as we will discuss in more detail in the rest of this section.

For a piecewise constant function, the residual sum of squares can be written as

$$\sum_{j=1}^{d} \sum_{i: z_i \in R_j} (y_i - \gamma_j)^2, \qquad (8.23)$$

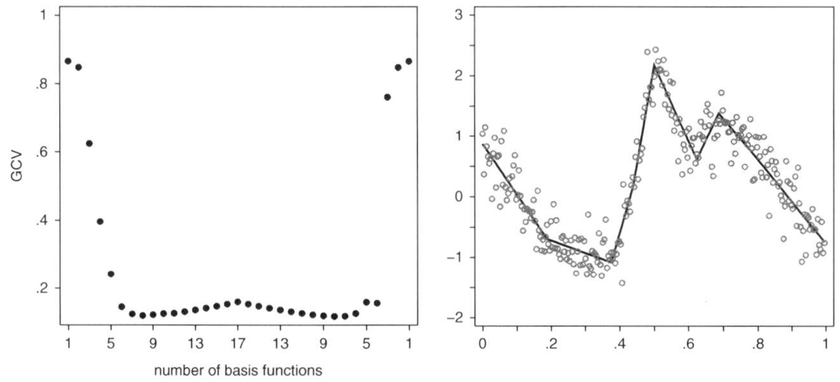

Fig. 8.28 MARS estimates for the simulated data: GCV criterion as a function of the number of basis functions (*left*) and corresponding optimal estimate with eight basis functions (*right*)

and therefore estimates for the function values γ_j are simply given by the local means of the response variable for $z_i \in R_j$, i.e., $\hat{\gamma}_j = \bar{y}(R_j)$. The optimal determination of the cut points is, however, more difficult and not directly attainable. Thus, one often resorts to recursive algorithms, which successively split the covariate domain into intervals. Figure 8.29 illustrates this approach using the malnutrition example (and also motivates the notion *regression tree*). In a first step, a split of the observations is performed based on a single cut point κ so that the variability within the resulting two groups is minimized. In our example, we obtain an optimal cut point of $\kappa = 7.5$, as denoted in the top branch of Fig. 8.29. The same splitting principle is then iteratively applied to the resulting subgroups and finally yields a decomposition of the covariate domain. The resulting structure can then be represented as a binary tree, in which every branch corresponds to a specific cut point. In Fig. 8.29, the cut points are provided on top of each branch. In addition, the lower part of the tree provides information about the number of observations in the final leafs of the tree and the estimated function value $\hat{\gamma}_j = \bar{y}(R_j)$ assigned to the group.

We now define regression trees in a more formal way: The tree is initialized based on a decomposition in the intervals

$$R_1 = \{z : z \leq \kappa\}, \qquad R_2 = \{z : z > \kappa\},$$

where κ is chosen to minimize the criterion

$$\sum_{\{i:z_i \in R_1\}} (y_i - \bar{y}(R_1))^2 + \sum_{\{i:z_i \in R_2\}} (y_i - \bar{y}(R_2))^2.$$

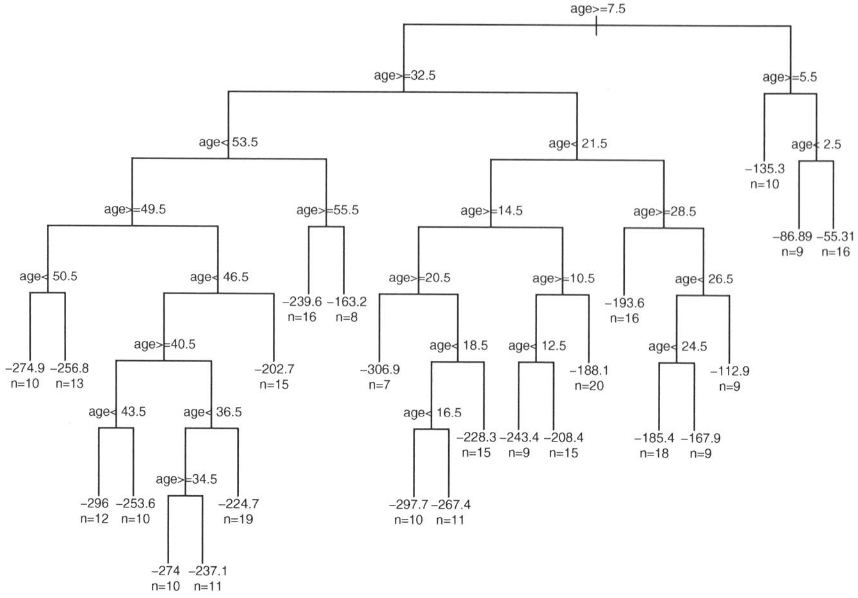

Fig. 8.29 Malnutrition in Tanzania: regression tree. No further splits were performed once a leaf of the tree contained less than 20 observations

Since we only need to determine one single split, the numerical determination of κ is rather simple. We can choose from all unique observed covariate levels z_i so that a direct search is possible. In the next step, the same principle is applied to the two resulting intervals R_1 and R_2 and proceeding iteratively through the intervals finally yields a decomposition of the form (8.23).

The specification of a regression tree is completed by defining a stopping criterion that determines whether or not additional branches are to be included. Two simple strategies to do so are as follows:

1. The given interval is only eligible for a further split if it contains at least a pre-specified number of observations.
2. The given interval is only eligible for a further split if a certain (absolute or relative) reduction of the residual sum of squares can be achieved.

We used the first criterion with a minimal number of 20 observations for the example in Fig. 8.29.

In general, regression trees that result from these simple breakdown criteria have too many branches, so that it is necessary to use an additional strategy to reduce model complexity. To do so, we first define a complex tree with a large number of branches, and then, with the help of a complexity measure, we eliminate superfluous branches (pruning). A complexity-adjusted model fit criterion is given by

$$\text{PLS}_\lambda(T) = \text{LS}(T) + \lambda|T|,$$

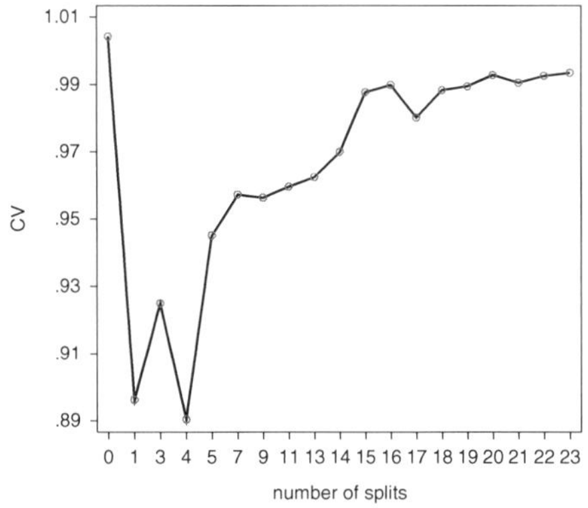

Fig. 8.30 Malnutrition in Tanzania: cross validation criterion as a function of the number of splits

where $\mathrm{LS}(T)$ defines the residual sum of squares of tree T, and $|T|$ is the number of branches in the tree. This construction corresponds to our previous considerations regarding penalization methods and, in particular, λ can be interpreted as a smoothing or complexity parameter. In varying λ, we obtain trees of different complexity with increasing complexity resulting from smaller values of λ. The optimal value for λ can then be determined using cross validation criteria.

Figure 8.30 plots the cross validation criterion for the Tanzania data set against the number of branches in the tree. Note that the value of the smoothing parameter λ implicitly defines the number of branches. In fact, not all numbers of branches are actually observed since a change in λ may lead to a model with more than just one additional branch. It appears that four splits are optimal while one split also leads to relatively small value of the cross validation criterion, with a slight preference towards the more complex model. Figure 8.31 provides a visualization of this model with four splits. We again recognize the main features of the data, namely a higher Z-score, i.e., a lower malnutrition risk for younger children, and a pronounced increase in f after the $24th$ month when the reference population is changing.

One of the main advantages of regression trees is their rather simple interpretation arising from the piecewise constant approximation. In more complex models, regression trees also allow to detect interactions fairly easily. The assumption of a piecewise constant function, however, can be problematic in many applications. Moreover, regression trees are unstable, i.e., a slight change of the data set can lead to an entirely different optimal tree. In addition, such an instability can also strongly influence the choice of an optimal smoothing parameter determined through cross validation when considering small variations in the cross validation folds. Modern extensions of tree-based models rectify this problem by using resampling methods in combination with weighted averaging (random forests). Further comments on such approaches are given in Sect. 8.4.

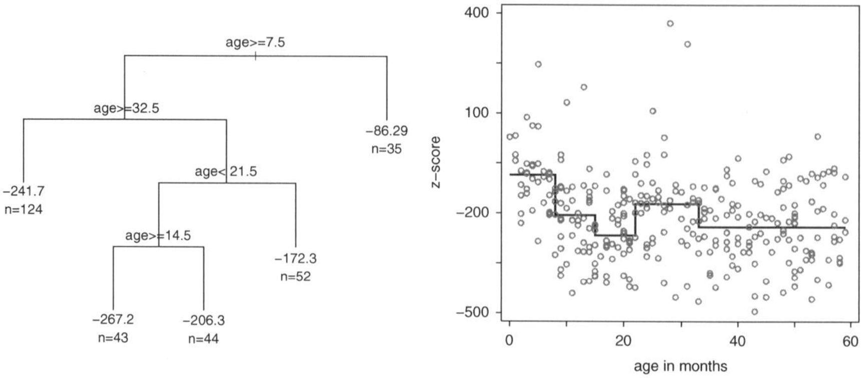

Fig. 8.31 Malnutrition in Tanzania: pruned regression tree and corresponding piecewise constant function estimate

Bayesian Adaptive Smoothing I: Model Averaging

To motivate Bayesian adaptive smoothing approaches, we start with the basis function formulation

$$y_i = f(z_i) + \varepsilon_i = \sum_{j=1}^{d} \gamma_j B_j(z_i) + \varepsilon_i,$$

using a large number of candidate basis functions, from which an adequate subset shall be chosen. In other words, we are interested in determining which of the coefficients γ_j differ from zero and are therefore associated with basis functions that have non-negligible impact on the response. Exactly the same question has already been discussed in Sect. 4.4.3 where we developed Bayesian linear models with built-in variable selection. In fact, the same models can be used for adaptive knot selection (with minor modifications). In the following, we briefly review some of the main ideas of Sect. 4.4.3 in the context of basis function selection.

In a Bayesian approach, the question of basis function selection can be approached by inserting indicator variables $\boldsymbol{\delta} = (\delta_1, \ldots, \delta_d)'$, such that

$$\delta_j = \begin{cases} 1 & \text{if } \gamma_j \neq 0, \\ 0 & \text{if } \gamma_j = 0. \end{cases}$$

When no prior information regarding the plausibility of the models exists, a uniform prior is a straightforward choice. In our example, every possible indicator vector $\boldsymbol{\delta}$ would then obtain the prior probability $1/2^d$. Compare, however, the remarks in Sect. 4.4.3 regarding the prior expected model size and the difficulties arising from this (naive) prior specification.

The original model can be rewritten as

$$f(z) = \sum_{j=1}^{d} \delta_j \gamma_j B_j(z_i).$$

For a given vector of indicators, define $\boldsymbol{\gamma}_{\boldsymbol{\delta}}$ as the regression coefficient vector, containing only the components that remain after eliminating coefficients associated with zero indicators. Similarly, let $\boldsymbol{Z}_{\boldsymbol{\delta}}$ denote the corresponding design matrix resulting from the basis functions with $\delta_j = 1$. In matrix notation, we now obtain the vector of function evaluations $\boldsymbol{f}_{\boldsymbol{\delta}} = \boldsymbol{Z}_{\boldsymbol{\delta}}\boldsymbol{\gamma}_{\boldsymbol{\delta}}$ that is based on the model defined by $\boldsymbol{\delta}$.

If we neglect the error variance σ^2 for the moment, the posterior distribution for the vector $\boldsymbol{f}$ (without conditioning on a specific model) can be written as

$$p(\boldsymbol{f} \mid \boldsymbol{y}) = \sum_{\boldsymbol{\delta} \in \{0,1\}^d} p(\boldsymbol{f}_{\boldsymbol{\delta}} \mid \boldsymbol{\delta}, \boldsymbol{y})\, p(\boldsymbol{\delta} \mid \boldsymbol{y}), \tag{8.24}$$

i.e., we obtain a mixture of all possible configurations of the indicator vector $\boldsymbol{\delta}$, where the posterior distributions $p(\boldsymbol{f}_{\boldsymbol{\delta}} \mid \boldsymbol{\delta}, \boldsymbol{y})$ are weighted by the posterior probabilities $p(\boldsymbol{\delta} \mid \boldsymbol{y})$. In particular, the marginal posterior expectation of $\boldsymbol{f}$ is also a mixture of the conditional posterior expected values:

$$\mathrm{E}(\boldsymbol{f} \mid \boldsymbol{y}) = \sum_{\boldsymbol{\delta} \in \{0,1\}^d} \mathrm{E}(\boldsymbol{f}_{\boldsymbol{\delta}} \mid \boldsymbol{\delta}, \boldsymbol{y})\, p(\boldsymbol{\delta} \mid \boldsymbol{y}). \tag{8.25}$$

This is a special example of the concept of *model averaging*; see Appendix B.5.5 for an introduction and Sect. 4.4.3 for details in the context of variable selection in linear models.

In our model choice problem, we obtain normal posterior distributions for $\boldsymbol{\gamma}_{\boldsymbol{\delta}}$ and the function evaluations $\boldsymbol{f}_{\boldsymbol{\delta}}$ when using normal prior distributions. Direct access through Eq. (8.24) is, however, limited by another difficulty. For a given number of basis functions, we have to consider 2^d models so that, in practice, the determination of all posterior probabilities cannot be realized, even for a modest number of basis functions. This is the reason why we generally do not use the explicit formulae for model averaging, but an approximation based on MCMC simulations. To do so, first, the indicators $\boldsymbol{\delta}$ are updated to obtain a current model, at each iteration. Based on this current model, we are able to simulate $\boldsymbol{f}_{\boldsymbol{\delta}}$. More precisely, we obtain the following algorithm, which also contains the estimation of the error variance:

1. Choose adequate starting values $\boldsymbol{\delta}^{(0)}$, $\boldsymbol{\gamma}^{(0)}$, and $\sigma^{2(0)}$ as well as a maximal number of iterations T and set $t = 1$.
2. For $j = 1, \dots, d$ generate $\delta_j^{(t)}$ from the marginal full conditional density $p(\delta_j \mid \boldsymbol{\delta}_{-j}^{(t-1)}, \boldsymbol{y})$, where

$$\delta_{-j}^{(t-1)} = (\delta_1^{(t)}, \ldots, \delta_{j-1}(t), \delta_{j+1}^{(t-1)}, \ldots, \delta_d^{(t-1)})$$

is the current index vector without the jth entry.

3. Simulate $\boldsymbol{f}_{\delta}^{(t)}$ according to the full conditional density $p(\boldsymbol{f}_{\delta} \mid \boldsymbol{\delta}^{(t)}, \sigma^{2(t-1)}, \boldsymbol{y})$.
4. Simulate $\sigma^{2(t)}$ from the full conditional density $p(\sigma^2 \mid \boldsymbol{\delta}^{(t)}, \boldsymbol{f}_{\delta}^{(t)}, \boldsymbol{y})$.
5. If $t < T$, set $t = t + 1$ and go back to 2.

The use of the marginal full conditional densities for the simulation of δ_j (and therefore a distribution that is independent from $\boldsymbol{\gamma}_{\delta}$ and σ^2) generally leads to a more favorable behavior of the generated Markov chain. If we use conjugate prior distributions, i.e., a normal prior for the regression parameters and an inverse gamma prior for σ^2, steps 3 and 4 can be realized with the help of Gibbs sampling. A more detailed discussion of adequate prior distributions and a description of the resulting MCMC algorithm can be found in Smith and Kohn (1996). Note also that the algorithm differs from the MC3 algorithm of Box 4.8 (p. 247). In contrast to the work by Smith and Kohn (1996) an accessible implementation of the MC3 algorithm in form of the R package `BMS` is available. The implementation is, however, not specialized to nonparametric regression. Hence the use in the context of basis function selection is tedious.

Based on realizations of the MCMC algorithm, the posterior expected value (8.25) can be approximated through

$$\frac{1}{T} \sum_{t=1}^{T} \boldsymbol{f}_{\delta}^{(t)}.$$

Since models with a high posterior probability are visited more often within the run of the MCMC algorithm, the weighting of the mean value corresponds (at least approximately) to the theoretical weighting in Eq. (8.25). The empirical frequencies

$$\frac{1}{T} \sum_{t=1}^{T} \boldsymbol{\delta}^{(t)}$$

also provide estimates for the posterior probabilities $\mathrm{P}(\delta_j = 1 \mid \boldsymbol{y})$, i.e., for the posterior inclusion probabilities for the individual basis functions. If we apply a threshold to these probabilities, we also gain the possibility to choose an adequate individual model based on the MCMC output, which does only contain basis function important in terms of their posterior probabilities.

Bayesian Adaptive Smoothing II: Reversible Jump MCMC

A second Bayesian approach yielding adaptive smoothing in nonparametric regression results from including the number and the location of the knots as additional parameters to be estimated. Therefore, we first need to find adequate prior distributions for these new parameters. The number of knots m can, for example, be assigned to a Poisson distribution with parameter λ, which represents prior information on the model size. For a large parameter λ, more complex models are preferred while a small parameter favors simpler models with a smaller number of knots. In practice, the Poisson distribution is often truncated at an upper limit $m_{\max}$, i.e., the maximal number of basis functions is bounded upwards so that $m \leq m_{\max}$. A uniform distribution on $\{0, \ldots, m_{\max}\}$ is an alternative to the Poisson distribution. The prior distribution for the position of knots is often defined on a large number of candidate knots, from which a selection is made during the model definition. All knot positions can be assigned equal prior probability. An alternative prior distribution is obtained from the distribution of the order statistic of an i.i.d. sample from the uniform distribution on the domain of z.

The resulting model formulation, enhanced by the new parameters corresponding to the number and location of the knots, can now again be transformed into an MCMC algorithm. However, the construction of such an algorithm is complicated by the fact that the dimension of the parameter vector to be estimated varies from iteration to iteration. Therefore standard MCMC methods have to be extended to so-called *reversible jump MCMC* (RJMCMC) approaches that include new steps for simulating the number and the position of knots in addition to the steps already known to be necessary when simulating the regression coefficients and the error variance at a given knot configuration. In an RJMCMC algorithm, this simulation is achieved indirectly, by adequately modifying the current values instead of directly proposing new values. There are three possibilities to do so:

1. *Creation of a new knot (birth step):* Randomly choose a knot that so far has not been included in the model and add this knot to the knot set.
2. *Elimination of an existing knot (death step):* Randomly choose a knot currently included in the model and eliminate it from the knot set.
3. *Relocation of a present knot:* Randomly choose a knot currently in the model and relocate it within the range defined by the two adjacent knots.

While the first two approaches modify the number of knots, the third approach allows for variation in the knot positions without actually changing their number. Within the RJMCMC algorithm, we have to randomly decide, at each iteration, whether we simulate new coefficients from the current model specification or whether to vary the knot configuration using one of the three available options. When modifying the knot configuration, the probability of acceptance must also be adjusted. We will not discuss this in further detail, but refer to the appropriate literature (Green, 1995) for the basic RJMCMC algorithm, as well as Biller (2000) and Denison, Mallick, and Smith (1998) for applications within the scope of Bayesian adaptive smoothing.

Similar to Bayesian model averaging, a sample of possible models results from the RJMCMC simulation, weighted according to their posterior probabilities. In addition, we obtain the posterior distribution for the number and position of knots so that these can be analyzed using standard approaches for MCMC algorithms.

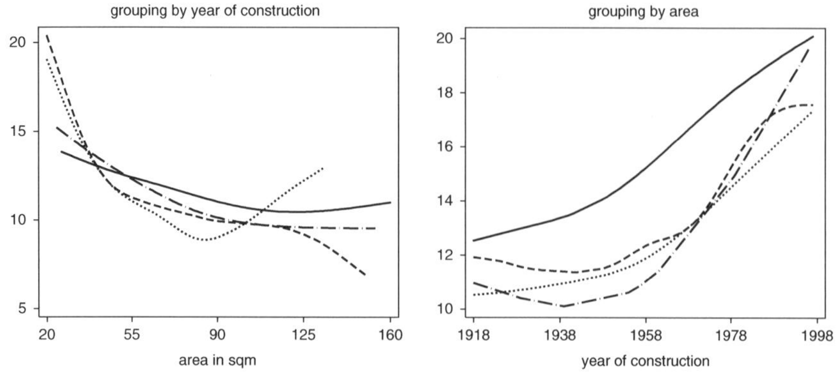

Fig. 8.32 Munich rent index: separate estimates for the effect of living area grouped according to year of construction (*left*) and separate estimates for the effect of year of construction grouped according to living area (*right*). The grouping structure has been determined as follows: $yearc \leq 1938$ (—), $1938 < yearc \leq 1958$ (- - -), $1958 < yearc \leq 1978$ (···), $yearc > 1978$ (- · - ·); $area \leq 55$ (—), $55 < area \leq 90$ (- - -), $90 < area \leq 135$ (···), $area > 135$ (- · - ·)

8.2 Bivariate and Spatial Smoothing

Thus far, we investigated the effect of *one single* continuous regressor in nonparametric regression. In this section, we examine approaches for bivariate smoothing (i.e., models with two continuous regressors) and for modeling spatial effects. To get an impression of the different possible problems and data situations, we start with various representative examples.

Example 8.6 Munich Rent Index—Interaction Between Living Area and Year of Construction

In Chap. 3 (Example 3.5), we saw that the year of construction and the living area have possibly nonlinear effects on the net rent per square meter. With the methods discussed in Sect. 8.1, we would only be able to examine each nonparametric effect of the two variables separately. However, when doing so, we assume that no interaction exists between the two effects. To evaluate this assumption, we grouped the data according to year of construction and living area, respectively, and estimated the effects in the corresponding groups using nonparametric methods. The results shown in Fig. 8.32 suggest that interaction effects occur between the two covariates, since the effects differ clearly for the individual groups. A flexible model for the description of such interaction effects has the form

$$rentsqm_i = f(area_i, yearc_i) + \varepsilon_i,$$

where f is a smooth function that, as in Sect. 8.1, is to be estimated flexibly from the given data, but now depending on two covariates. Since the interaction between living area and year of construction is modeled by a surface, we also refer to such a model as an interaction surface.

△

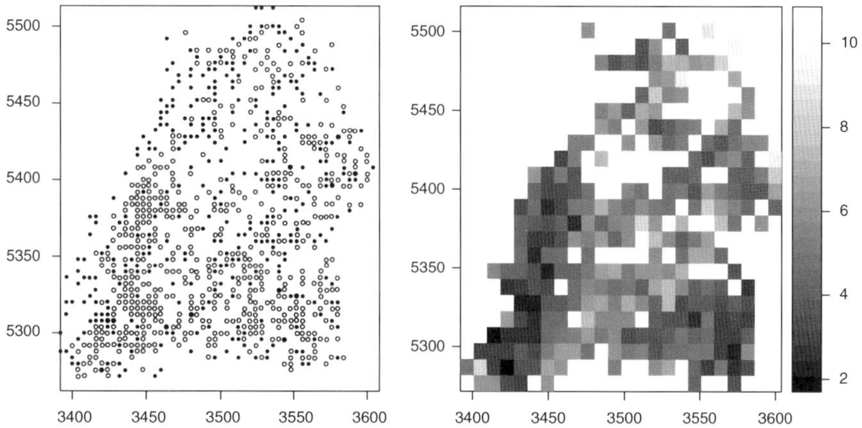

Fig. 8.33 Forest health status: *Left panel*, *open circles* indicate locations with known calcium concentration while predictions for the calcium concentration are required at locations represented as *solid circles*. *Right panel*, local averages based on a division in *regular rectangles*

Example 8.7 Forest Health Status—Spatial Smoothing of Chemical Concentration

This example shows an analysis of forest health in Baden–Württemberg, one of the federal states of Germany. The health status of trees is to be related to the concentration of different chemicals in the soil. However, the chemical concentrations have not been determined at all locations where the forest condition was measured. Figure 8.33 illustrates this situation with the concentration of calcium. In the left panel, every point represents a location where the forest condition was evaluated. Open circles indicate locations where calcium concentrations were also determined, whereas solid circles represent locations where calcium concentrations were not measured. To consider all observations in the analysis, we use the model

$$Ca_i = f(x_i, y_i) + \varepsilon_i$$

to predict the calcium concentration even at locations where it was not determined. In this model, Ca denotes the measured calcium concentration, whereas x and y correspond to the locations of the measurements in terms of longitude and latitude. In fact, we are then in the same situation as in Example 8.6, i.e., a continuous response variable is flexibly modeled in relation to two continuous explanatory variables. However, the problem differs from Example 8.6 for two reasons: First, the covariates now represent spatial information so that special methodology of spatial statistics can be used. Secondly, we need a representation of f, which can be evaluated at arbitrary locations. The latter is difficult to achieve, e.g., through data grouping in a regular grid and local averaging, because the data show relatively large gaps (again see Fig. 8.33). However, the figure illustrates that the calcium concentration changes continuously, and thus a smooth surface seems to be an appropriate model.

△

Example 8.8 Human Brain Mapping

Human brain mapping aims at identifying areas within the brain that are associated with specific tasks such as processing a visual stimulus. Therefore, a number of subjects were examined in an experiment using functional magnetic resonance imaging (fMRI), in which periods of visual stimulation are alternated with rest periods. Figure 8.34 shows a single layer of the brain at two different time points of visual stimulation (and therefore a very similar spatial activation structure). Regions with an increased activation are represented by the lighter colored pixels. We note that these light pixels are primarily located towards the back portion of the brain, i.e., the location of the visual cortex, which is responsible for processing visual stimuli. With the help of spatial smoothing methods, we aim at filtering the noise out of the data in an effort to more clearly differentiate between activated and deactivated areas. Hence, we again find a spatial smoothing problem where a bivariate (or spatial) function is to be estimated. Although, in a strict sense, spatial information in the human brain mapping example is only available discretely, the large number of pixels allows us to disregard the discreteness and to identify every pixel with the coordinates at its center.

△

In the examples presented thus far, the covariates or spatial coordinates were always measured on a continuous scale or could at least be interpreted on a quasi-continuous scale. Within the scope of spatial statistics, we will also refer to these coordinates as *continuous location variables*. With other typical spatial statistics data, only discrete information is actually provided, and in such cases we refer to the covariates as *discrete location variables*. An example of such discrete spatial information could be the location of a household or residence, e.g., at a county level. It is then our goal to use the spatial allocation of the counties in order to estimate a spatially smooth function. We already encountered similar situations in Sect. 2.8. We next consider another spatial example using the rent index in Munich.

Example 8.9 Munich Rent Index—Examining Spatial Dependence

The rent index data also contain information regarding the subquarters in Munich in which the apartments are located. Figure 8.35 shows the average net rent per square meter for each of these subquarters separately. For example, the graphical presentation reveals somewhat reduced rents in the northern part of Munich. The goal of a spatial analysis in this context is to more clearly show such spatially structured effects by taking into account the spatial proximity of the subquarters in a smoothing approach.

△

Generally speaking, we can distinguish two different problems:

- Estimation of two-dimensional surfaces to model interactions or spatial effects of location variables measured on a continuous scale.
- Estimation of spatial effects based on discrete spatial information, represented, for example, in form of regions or spatial locations on a discrete grid.

Sections 8.2.1–8.2.3 focus on the former problem. We first extend the univariate penalization approaches discussed in Sect. 8.1 to the bivariate setting, especially penalized splines and the kriging method. Approaches for modeling discrete spatial information, which actually can be viewed as extensions of the random walk models discussed in Sect. 8.1.5, will be further investigated in Sect. 8.2.4. In Sect. 8.2.6, we briefly discuss some bivariate extensions of local and adaptive methods.

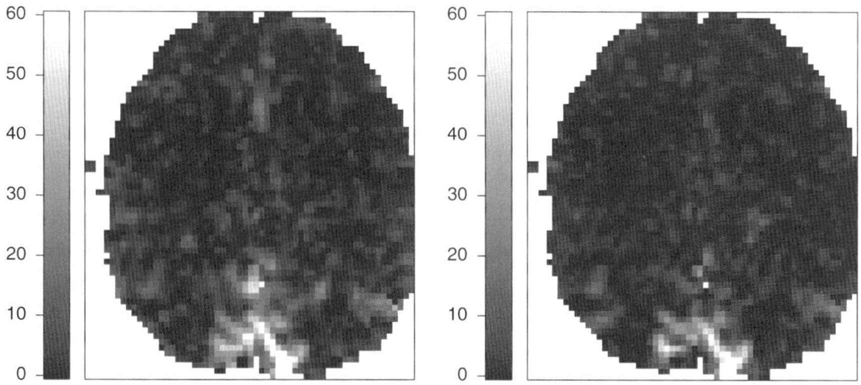

Fig. 8.34 Human brain mapping: pixelwise activity as a response to visual stimulation. The *left panel* shows the activity at time point $t = 18$ while the *right panel* refers to activity at time point $t = 38$

Fig. 8.35 Munich rent index: average net rent per square meter per subquarter. *Striped regions* correspond to subquarters without any observations (such as parks or industrial areas)

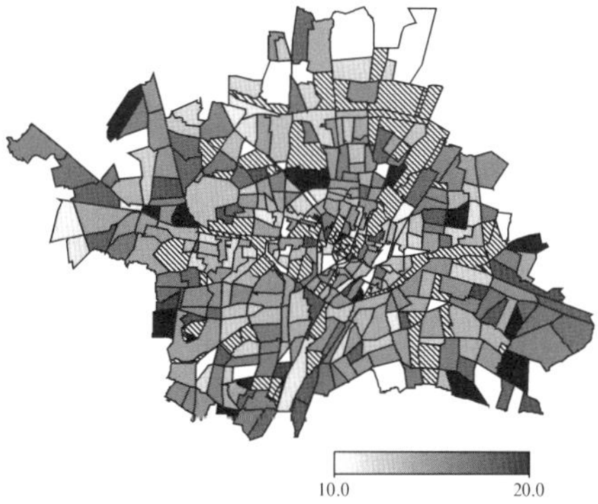

8.2.1 Tensor Product P-Splines

Tensor Product Bases

In Sect. 3.1.3, we have been modeling interactions between covariates as products of the corresponding design vectors. When using basis functions in nonparametric regression, this idea can be extended to bivariate interaction surfaces through the use of *tensor product bases*. Such a bivariate basis is obtained by considering all pairwise products of two univariate bases constructed for univariate smooths.

More specifically, we consider the following situation: The response variable y is to be described in terms of a two-dimensional surface $f(z_1, z_2)$, where z_1 and z_2 can be continuous covariates, as well as coordinates in the case of a spatial model. We then first construct the univariate bases for z_1 and z_2, yielding the basis functions $B_j^{(1)}(z_1)$, $j = 1, \ldots, d_1$, and $B_r^{(2)}(z_2)$, $r = 1, \ldots, d_2$. The tensor product basis then consists of all basis functions of the form

$$B_{jr}(z_1, z_2) = B_j^{(1)}(z_1) \cdot B_r^{(2)}(z_2), \quad j = 1, \ldots, d_1, \quad r = 1, \ldots, d_2,$$

so that we obtain the following representation for $f(z_1, z_2)$:

$$f(z_1, z_2) = \sum_{j=1}^{d_1} \sum_{r=1}^{d_2} \gamma_{jr} B_{jr}(z_1, z_2).$$

For the case of polynomial splines, we refer to the tensor product basis as *tensor product splines* or bivariate polynomial splines. To illustrate the construction of a tensor product basis, Fig. 8.36 shows linear tensor product splines based on the univariate TP basis functions

$$B_1^{(1)}(z_1) = 1, \quad B_2^{(1)}(z_1) = z_1, \quad B_3^{(1)}(z_1) = (z_1 - \kappa_1)_+$$

and

$$B_1^{(2)}(z_2) = 1, \quad B_2^{(2)}(z_2) = z_2, \quad B_3^{(2)}(z_2) = (z_2 - \kappa_2)_+.$$

The constant function on the upper left results from the product of the two univariate constant basis functions $B_1^{(1)}$ and $B_1^{(2)}$. The first row and the first column are then obtained by multiplying the constant basis function in z_1-direction with the basis functions in z_2-direction or vice versa. The remaining four basis functions correspond to the products of the remaining univariate basis functions.

For regularizing tensor product TP-splines, we again use penalties derived from squared coefficients, similar as in the univariate setting discussed in Sect. 8.1.3. However, the numerical difficulties discussed there are even more pronounced in the bivariate setting, and we will therefore resort to tensor products of the numerically more stable B-spline basis. Figure 8.37 shows individual tensor product B-splines for the degrees $l = 0, 1, 2,$ and 3. We again notice that a higher spline degree leads to more smoothness. In particular, tensor product splines of degree $l = 0$ are not continuous, while tensor product splines of degree $l = 1$ are continuous but not differentiable (see the definition of bivariate polynomial splines in Dierckx (1993) for a more detailed description of continuity and differentiability properties of tensor product splines). Figure 8.38 shows a larger number of cubic B-spline basis functions. In order to get a clearer picture, not all basis functions of a complete basis are mapped, as a considerably strong overlapping would occur, similar to what we saw for univariate B-splines in Sect. 8.1.1.

If we look at the contour plots (i.e., a graphical representation of the contour lines) of tensor product B-splines (Fig. 8.39), we find that the contour lines clearly deviate from circles, especially for the lower spline degrees. Consequently, tensor product B-splines are not radial. Radial basis functions will be discussed in Sects. 8.2.2 and 8.2.3.

Although at first sight, tensor product approaches appear to be much more complex than univariate basis function approaches, it is nevertheless straightforward to represent them in the form of large linear models. To do so, we define the design

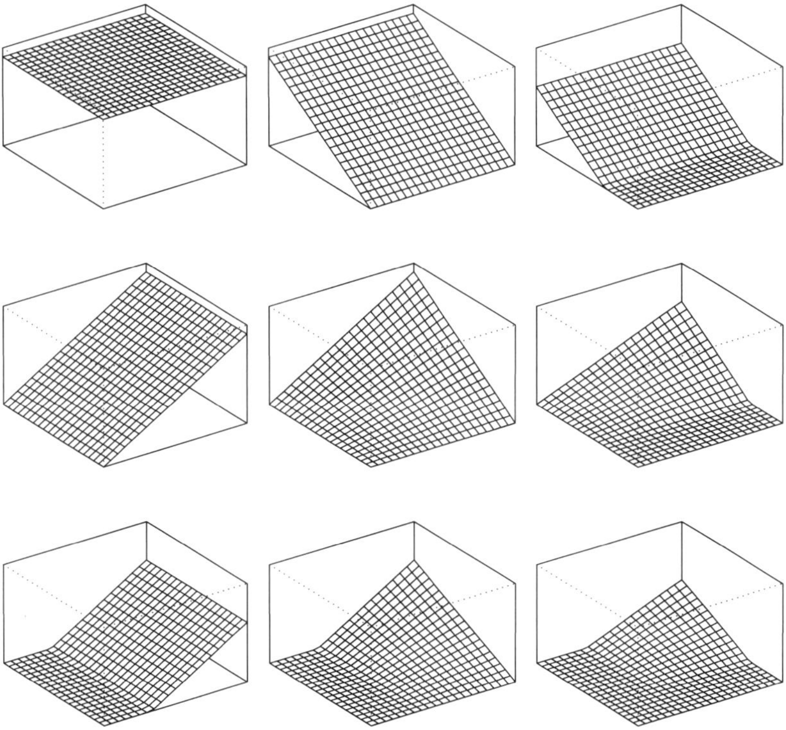

Fig. 8.36 Tensor product basis obtained from univariate linear TP bases

matrix $\boldsymbol{Z}$ with rows

$$z_i' = (B_{11}(z_{i1}, z_{i2}), \ldots, B_{d_1 1}(z_{i1}, z_{i2}), \ldots, B_{1 d_2}(z_{i1}, z_{i2}), \ldots, B_{d_1 d_2}(z_{i1}, z_{i2}))$$

and express the vector of the corresponding regression coefficients as

$$\boldsymbol{\gamma} = (\gamma_{11}, \ldots, \gamma_{d_1 1}, \ldots, \gamma_{1 d_2}, \ldots, \gamma_{d_1 d_2})'.$$

We finally obtain the standard regression equation

$$\boldsymbol{y} = \boldsymbol{Z}\boldsymbol{\gamma} + \boldsymbol{\varepsilon}. \tag{8.26}$$

In principle, bivariate smoothing approaches can therefore also be estimated within the scope of linear models. However, in comparison to the univariate case, the number of the parameters to be estimated is typically much larger. Numerically efficient computation of $\hat{\boldsymbol{\gamma}}$ is thus even more important, which can be implemented utilizing the sparse structure of the design matrix for tensor product B-splines (see also the comments on p. 430 and p. 488).

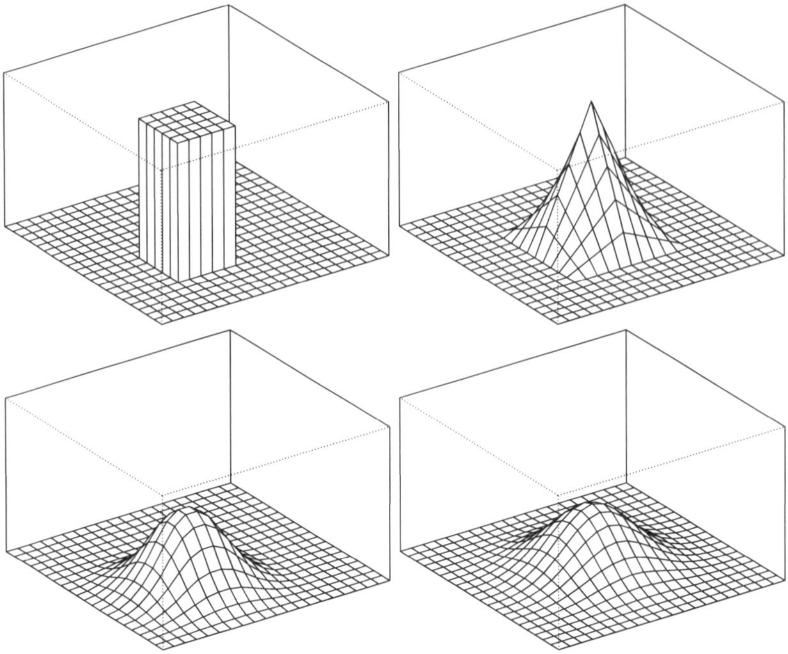

Fig. 8.37 Tensor product basis functions obtained from univariate B-splines of degrees $l = 0, 1, 2$, and 3

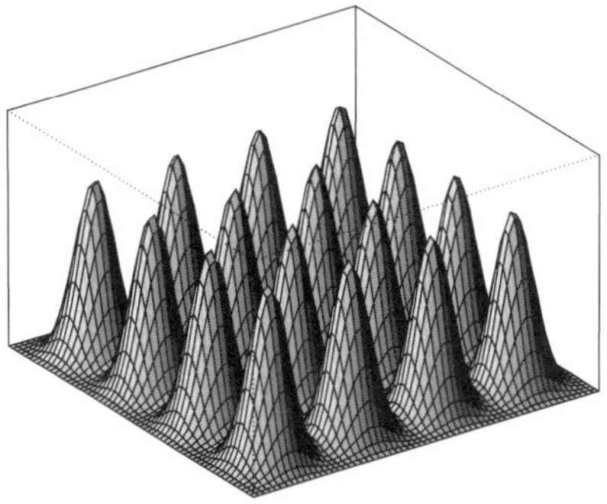

Fig. 8.38 A partial tensor product basis obtained from cubic univariate B-splines

As with univariate polynomial splines, we need to determine the optimal number and position of knots to construct tensor product splines. Moreover, we often encounter the problem that certain data regions may not have any observations. In such cases, it is impossible to estimate the coefficients of the basis functions associated with these regions. This is the case in the human brain mapping example. With tensor product splines, we construct basis functions over the entire area, $[\min(z_1), \max(z_1)] \times [\min(z_2), \max(z_2)]$. However, due to the shape of the brain,

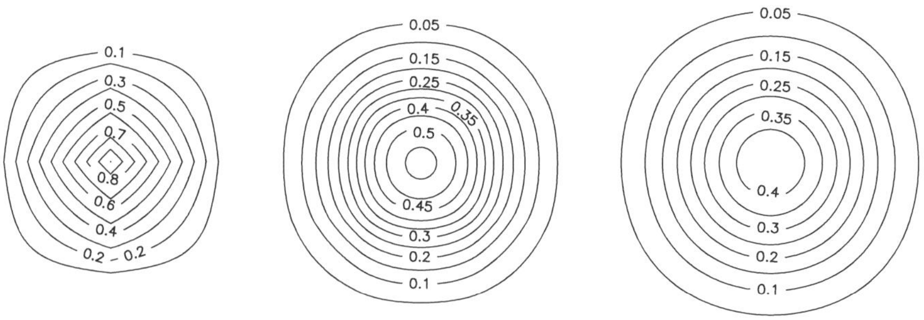

Fig. 8.39 Contour plots of tensor product B-splines of degree $l = 1, 2$, and 3

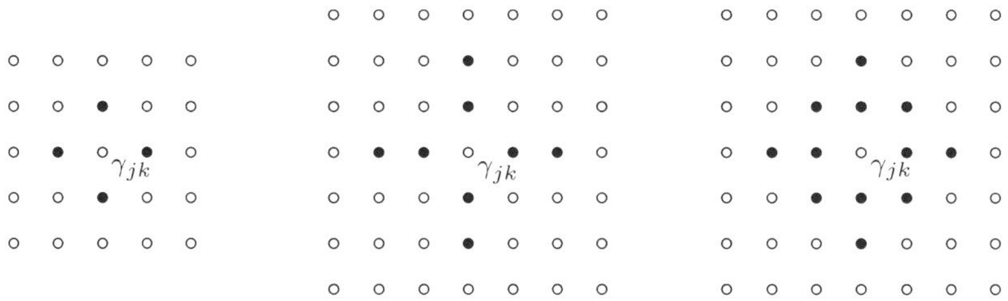

Fig. 8.40 Spatial neighborhoods on a regular lattice: The neighbors of γ_{jk} are marked as *filled dots*

no observations can be made in the corners of the resulting rectangle. Consequently, we are not able to estimate the corresponding regression coefficients. The same problems can occur, in principle, for univariate B-splines, e.g., when the covariate data has large gaps. Such problems are not quite as common in the univariate setting compared to the two-dimensional case. Both the problem of determining the number and position of the knots and the non-identifiability resulting from gaps in the data can be eliminated by adding a penalty for regularizing the estimation problem.

2D Penalties

Even though we could choose a ridge-type penalty for the tensor product TP basis, the spatial alignment of the basis functions and the associated regression coefficients may need to be incorporated appropriately when using tensor product B-splines. For univariate B-splines, penalties have been constructed based on squared differences of coefficients associated with neighboring basis functions. To transfer this concept to the two-dimensional case, we first define appropriate spatial neighborhoods. Figure 8.40 shows possible neighborhood definitions for four, eight, and twelve neighbors, respectively. We will now introduce several penalties that are based on these neighborhoods.

We start with the simplest case of a neighborhood consisting of four neighbors. A reasonable choice of the penalty is based on the squared differences between γ_{jk} and these four neighbors. Therefore, let $\boldsymbol{D}_1$ and $\boldsymbol{D}_2$ denote the univariate difference matrices of first order in z_1- and z_2-direction, respectively. Row-wise first-order differences are then obtained by applying the expanded difference matrix $\boldsymbol{I}_{d_2} \otimes \boldsymbol{D}_1$ to the vector $\boldsymbol{\gamma}$, where $\boldsymbol{I}_d$ is the d-dimensional identity matrix and $\otimes$ denotes the Kronecker product (see Definition A.10 in Appendix A.1). In fact, applying this difference matrix to the vector of regression coefficients yields

$$\boldsymbol{\gamma}'(\boldsymbol{I}_{d_2} \otimes \boldsymbol{D}_1)'(\boldsymbol{I}_{d_2} \otimes \boldsymbol{D}_1)\boldsymbol{\gamma} = \sum_{r=1}^{d_2} \sum_{j=2}^{d_1} (\gamma_{jr} - \gamma_{j-1,r})^2,$$

which is the sum of all squared row-wise differences. Analogously, we obtain the squared column-wise differences

$$\boldsymbol{\gamma}'(\boldsymbol{D}_2 \otimes \boldsymbol{I}_{d_1})'(\boldsymbol{D}_2 \otimes \boldsymbol{I}_{d_1})\boldsymbol{\gamma} = \sum_{j=1}^{d_1} \sum_{r=2}^{d_2} (\gamma_{jr} - \gamma_{j,r-1})^2.$$

Summing up all squared row-wise and column-wise differences then finally yields the penalty

$$\lambda \boldsymbol{\gamma}'\boldsymbol{K}\boldsymbol{\gamma} = \lambda \boldsymbol{\gamma}' \left[(\boldsymbol{I}_{d_2} \otimes \boldsymbol{D}_1)'(\boldsymbol{I}_{d_2} \otimes \boldsymbol{D}_1) + (\boldsymbol{D}_2 \otimes \boldsymbol{I}_{d_1})'(\boldsymbol{D}_2 \otimes \boldsymbol{I}_{d_1}) \right] \boldsymbol{\gamma}. \quad (8.27)$$

Based on properties of Kronecker products (see Theorem A.4 in Appendix A.1), one can show that the penalty can equivalently be defined as

$$\lambda \boldsymbol{\gamma}'\boldsymbol{K}\boldsymbol{\gamma} = \lambda \boldsymbol{\gamma}' \left[\boldsymbol{I}_{d_2} \otimes \boldsymbol{K}_1 + \boldsymbol{K}_2 \otimes \boldsymbol{I}_{d_1} \right] \boldsymbol{\gamma},$$

with the univariate penalty matrices $\boldsymbol{K}_1 = \boldsymbol{D}_1'\boldsymbol{D}_1$ and $\boldsymbol{K}_2 = \boldsymbol{D}_2'\boldsymbol{D}_2$.

A Bayesian derivation of this penalty can be given as follows: As in the univariate case, we can define the difference penalties that were constructed row by row or column by column by using first-order random walks. Consequently, we can interpret $\boldsymbol{K}$ as the precision matrix of the entire distribution of the vector $\boldsymbol{\gamma}$ when assuming a two-dimensional random walk of first order. More precisely, we obtain the prior distribution for $\boldsymbol{\gamma}$ as

$$p(\boldsymbol{\gamma} \mid \tau^2) \propto \left(\frac{1}{\tau^2} \right)^{\mathrm{rk}(\boldsymbol{K})/2} \exp\left(-\frac{1}{2\tau^2} \boldsymbol{\gamma}'\boldsymbol{K}\boldsymbol{\gamma} \right). \quad (8.28)$$

Based on this joint prior, we can calculate the conditional distribution of γ_{jr} given all other coefficients as

$$\gamma_{jr} \,|\, \cdot \sim \mathrm{N}\left(\frac{1}{4}(\gamma_{j-1,r} + \gamma_{j+1,r} + \gamma_{j,r-1} + \gamma_{j,r+1}), \frac{\tau^2}{4}\right)$$

(apart from points at the boundary of the coefficient space). Thus, a spatial form of the Markov property holds for the vector $\boldsymbol{\gamma}$, since the conditional distribution of γ_{jr} only depends on the four nearest (spatial) neighbors. For the expected value of the conditional distribution, we obtain the local mean computed from the four nearest neighbors. Hence, the penalty for two-dimensional surfaces generalizes the properties that we discussed in Sect. 8.1.2, p. 441ff for univariate P-splines.

The principle for building two-dimensional penalties, based on row-wise or column-wise differences, can also be applied to difference matrices of a higher order. We then obtain penalties of the form

$$\lambda \boldsymbol{\gamma}' \boldsymbol{K} \boldsymbol{\gamma} = \lambda \boldsymbol{\gamma}' \left[\boldsymbol{I}_{d_2} \otimes \boldsymbol{K}_1^{(k_1)} + \boldsymbol{K}_2^{(k_2)} \otimes \boldsymbol{I}_{d_1} \right] \boldsymbol{\gamma}$$

with univariate penalty matrices $\boldsymbol{K}_1^{(k_1)}$ and $\boldsymbol{K}_2^{(k_2)}$ of orders k_1 and k_2. For example, with $k_1 = k_2 = 2$, we obtain a penalty based on squared second-order differences whose neighborhood structure consists of the eight nearest neighbors along the coordinate axis (see Fig. 8.40).

In conclusion, we find that the approaches discussed in this section for bivariate penalties (and also some extensions) in combination with tensor product B-splines can be expressed in the general form seen in Sect. 8.1: The vector of function evaluations can be represented as a large linear model $\boldsymbol{Z}\boldsymbol{\gamma}$ with a quadratic penalty $\lambda \boldsymbol{\gamma}' \boldsymbol{K} \boldsymbol{\gamma}$ or an equivalent Gaussian prior for $\boldsymbol{\gamma}$ with density (8.28). It follows that the estimation procedures discussed in Sect. 8.1.9 are again applicable. Since the number of parameters in the two-dimensional model is, however, much larger than in the univariate case, numerically efficient implementations are even more crucial, especially when considering MCMC algorithms.

For mixed model-based inference, we also need a partition of the regression coefficients as

$$\boldsymbol{\gamma} = \tilde{\boldsymbol{X}} \boldsymbol{\beta} + \tilde{\boldsymbol{U}} \tilde{\boldsymbol{\gamma}}.$$

Recall from Sect. 8.1.9 that $\tilde{\boldsymbol{X}}$ results from a basis of the null space of the matrix $\boldsymbol{K}$. The null space is determined by the space of functions that remains unpenalized by the penalty matrix $\boldsymbol{K}$. When using a Kronecker product penalty based on first-order differences, the addition of a constant term does not affect either the row-wise or the column-wise penalty. One can actually show that indeed $\mathrm{rk}(\boldsymbol{K}) = d_1 d_2 - 1$ still holds, so that it is only the constant term that remains unpenalized, and the null space of the matrix $\boldsymbol{K}$ is given by a $d_1 d_2$-dimensional vector of ones. In general, the null space of a Kronecker product penalty matrix $\boldsymbol{K}$ is given by the tensor product

8.12 Tensor Product P-Splines

The goal is to estimate the bivariate function f in the model

$$y = f(z_1, z_2) + \varepsilon$$

based on continuous covariates z_1 and z_2. Tensor product splines are constructed through forming all pairwise products of univariate polynomial splines for z_1 and z_2, i.e.,

$$B_{jr}(z_1, z_2) = B_j^{(1)}(z_1) \cdot B_r^{(2)}(z_2), \quad j = 1, \ldots, d_1, \quad r = 1, \ldots, d_2.$$

This yields the representation

$$f(z_1, z_2) = \sum_{j=1}^{d_1} \sum_{r=1}^{d_2} \gamma_{jr} B_{jr}(z_1, z_2)$$

and therefore the model

$$\boldsymbol{y} = \boldsymbol{Z}\boldsymbol{\gamma} + \boldsymbol{\varepsilon},$$

where $\boldsymbol{Z}$ consists of the evaluated basis functions and $\boldsymbol{\gamma}$ are the corresponding regression coefficients.
Penalties can be constructed from Kronecker products of univariate penalty matrices:

$$\boldsymbol{K} = \boldsymbol{I}_{d_2} \otimes \boldsymbol{K}_1 + \boldsymbol{K}_2 \otimes \boldsymbol{I}_{d_1}.$$

This results in the quadratic penalty

$$\lambda \boldsymbol{\gamma}' \boldsymbol{K} \boldsymbol{\gamma},$$

and therefore the methods discussed in Sect. 8.1.9 can be used for estimating λ.

of the null spaces of the univariate penalty matrices. Thus, second-order differences lead to a four-dimensional null space, containing the constant, linear effects in z_1- and z_2-direction, and the interaction $z_1 \cdot z_2$. This null space basis can be represented based on the univariate knots $\kappa_1^{(1)}, \ldots, \kappa_{d_1}^{(1)}$ and $\kappa_1^{(2)}, \ldots, \kappa_{d_2}^{(2)}$, leading to

$$\begin{pmatrix} 1 & \kappa_1^{(1)} & \kappa_1^{(2)} & \kappa_1^{(1)}\kappa_1^{(2)} \\ 1 & \kappa_2^{(1)} & \kappa_1^{(2)} & \kappa_2^{(1)}\kappa_1^{(2)} \\ \vdots & \vdots & \vdots & \vdots \\ 1 & \kappa_{d_1}^{(1)} & \kappa_1^{(2)} & \kappa_{d_1}^{(1)}\kappa_1^{(2)} \\ 1 & \kappa_1^{(1)} & \kappa_2^{(2)} & \kappa_1^{(1)}\kappa_2^{(2)} \\ \vdots & \vdots & \vdots & \vdots \\ 1 & \kappa_{d_1}^{(1)} & \kappa_2^{(2)} & \kappa_{d_1}^{(1)}\kappa_2^{(2)} \\ \vdots & \vdots & \vdots & \vdots \\ 1 & \kappa_1^{(1)} & \kappa_{d_2}^{(2)} & \kappa_1^{(1)}\kappa_{d_2}^{(2)} \\ \vdots & \vdots & \vdots & \vdots \\ 1 & \kappa_{d_1}^{(1)} & \kappa_{d_2}^{(2)} & \kappa_{d_1}^{(1)}\kappa_{d_2}^{(2)} \end{pmatrix}.$$

For the construction of $\tilde{\boldsymbol{U}}$, we can proceed as in Sect. 8.1.9, i.e., we define $\tilde{\boldsymbol{U}}$ based on the spectral decomposition of $\boldsymbol{K}$. Note that we actually need to use the spectral decomposition in this case, since it is not possible to represent $\boldsymbol{K}$ as $\boldsymbol{K} = \boldsymbol{D}'\boldsymbol{D}$.

Example 8.10 Munich Rent Index—Interaction Between Year of Construction and Living Area

We use tensor product splines with different penalties to examine the interaction effect between year of construction and living area (see Example 8.6, p. 500) while neglecting all other covariate information and in particular the spatial information on the location of the flats in specific subquarters of Munich. Figure 8.41 shows such estimates based on cubic B-splines with 20 inner knots, using first- and second-order differences.

Both estimates basically reflect the same structure, i.e., higher estimated net rents per square meter for new and small apartments. It is rather clear that an interaction effect exists, and thus estimation based on univariate functions of year of construction and living area would be insufficient. If we compare the smoothness of the estimated surfaces, the use of higher differences in the penalty, as already seen in the univariate case, tends to lead to smoother functions.

△

Example 8.11 Forest Health Status—Spatial Smoothing of Chemicals

As a second example, we analyze the spatial distribution of calcium concentrations in Baden–Württemberg; see Example 8.7. Figure 8.42 shows the estimated surface using cubic tensor product B-splines with 15 inner knots for each of the longitudinal and latitudinal coordinates, and a penalty based on first-order differences. The estimated surface does reproduce the regional trend found in the descriptive analysis very well (Fig. 8.33 on p. 501) but does have the particular advantage that it can be visualized in a very high resolution, and it can also be extrapolated into regions without observations. Augustin, Lang, Musio, and von Wilpert (2007) use these extrapolated values in their analysis of forest health in Baden–Württemberg. Note, however, that with tensor product splines, we always estimate the unknown surface over a rectangular domain which in case of the calcium concentrations means that we are sometimes extrapolating very far from the observed data (e.g., in the

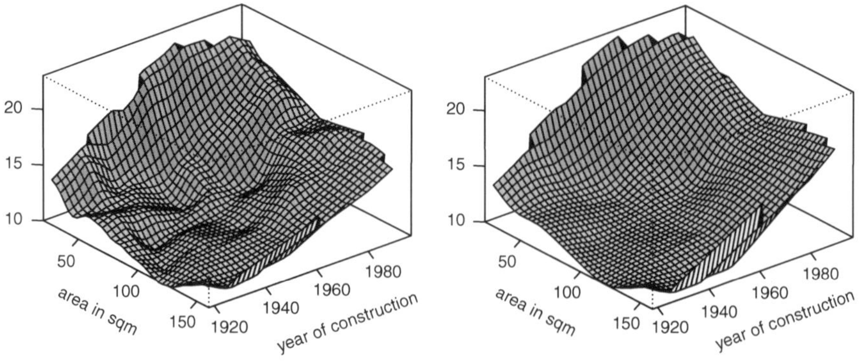

Fig. 8.41 Munich rent index: estimated interaction surface between living area in square meters and year of construction based on cubic B-splines with 20 inner knots and first (*left*) and second (*right*) order difference penalty

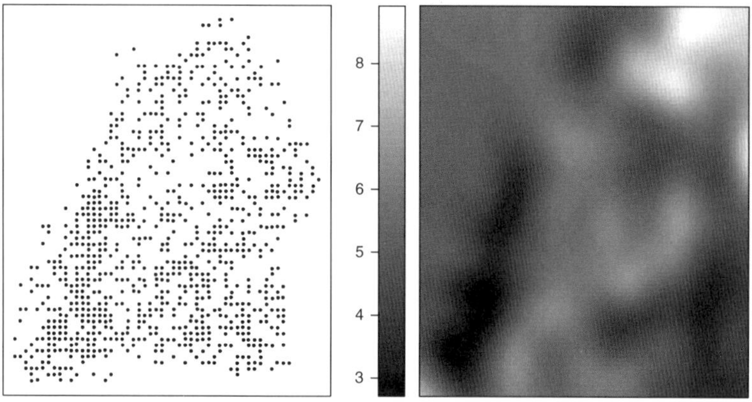

Fig. 8.42 Forest health status: estimated spatial calcium concentration based on cubic tensor product B-splines with 15 inner knots and first-order difference penalty

upper left corner of the right panel in Fig. 8.42). In these cases, information on the estimated surface is very scarce, which will typically be reflected by very wide confidence intervals. △

8.2.2 Radial Basis Functions and Thin Plate Splines

An alternative way of constructing bivariate basis functions is obtained with *radial bases*. Generally speaking, a radial basis function is defined as a function of the Euclidian distance between a knot $\boldsymbol{\kappa} = (\kappa_1, \kappa_2)$ and an observation point $\boldsymbol{z} = (z_1, z_2)$, i.e.,

$$B_{\boldsymbol{\kappa}}(\boldsymbol{z}) = B(||\boldsymbol{z} - \boldsymbol{\kappa}||) = B(r),$$

with a suitably chosen scalar function B and Euclidean distance $r = ||z - \kappa|| = ((z_1 - \kappa_1)^2 + (z_2 - \kappa_2)^2)^{0.5}$. The term radial basis function results from the fact that, due to their construction, the contour plots of radial basis functions consist of circular contour lines.

All radial basis functions have the same functional form and, in contrast to tensor product B-splines, are attributed to one specific knot. Typically, the knots of a radial basis are a subset of the observation points, i.e., $\{\kappa_1, \ldots, \kappa_d\} \subset \{z_1, \ldots, z_n\}$. Thus, the distribution of the radial basis functions adapts automatically to the distribution of the data, whereas for tensor product bases, a large number of basis functions can be placed in regions in which no observations were made. For example, in case of the Baden–Württemberg forest health data, a large number of knots were actually placed outside the region of interest when using tensor products, since the knots are equally distributed over $[\min(z_1), \max(z_1)] \times [\min(z_2), \max(z_2)]$. However, this extended knot set also provides us with the possibility to extrapolate into these regions, whereas the extrapolation region for radial basis functions is more limited.

The most well-known example of radial basis functions can be obtained through optimizing the criterion

$$\sum_{i=1}^{n} (y_i - f(z_i))^2 + \lambda \int\int \left[\left(\frac{\partial^2}{\partial^2 z_1} + 2\frac{\partial^2}{\partial z_1 \partial z_2} + \frac{\partial^2}{\partial^2 z_2} \right) f(z_1, z_2) \right]^2 dz_1 dz_2 \to \min_f \tag{8.29}$$

over the class of all twice continuously differentiable functions $f(z)$. In this case,

$$\int\int \left[\left(\frac{\partial^2}{\partial^2 z_1} + 2\frac{\partial^2}{\partial z_1 \partial z_2} + \frac{\partial^2}{\partial^2 z_2} \right) f(z_1, z_2) \right]^2 dz_1 dz_2 \tag{8.30}$$

is the bivariate analogue of the integrated squared second derivative, which we discussed in the context of smoothing splines. Hence, this approach aims at generalizing smoothing splines to bivariate surface smoothing. As a solution, we obtain the *thin plate splines*, a generalization of natural cubic splines that also fulfills the natural boundary conditions, i.e., it behaves linearly outside of the observation domain. A thin plate spline can be represented as

$$f(z_1, z_2) = \beta_0 + \beta_1 z_1 + \beta_2 z_2 + \sum_{j=1}^{n} \gamma_j B_j(z_1, z_2)$$

with

$$B_j(z_1, z_2) = B(||z - z_j||) = ||z - z_j||^2 \log(||z - z_j||)$$

and subject to an identifiability restriction on the coefficients γ_j that we will work out below. Thus, the thin plate spline relies on linear effects in the directions z_1 and z_2 and radial basis functions

$$B(r) = r^2 \log(r)$$

centered at the n covariate values; see Green and Silverman (1993) for more details on thin plate splines and a proof of their optimality property. In summary, the thin plate spline is a bivariate generalization of the smoothing spline which would result as a special case when restricting the surface to a univariate regression function.

In matrix notation, the above representation of the thin plate spline induces the model equation

$$\boldsymbol{y} = \boldsymbol{X}\boldsymbol{\beta} + \boldsymbol{Z}\boldsymbol{\gamma} + \boldsymbol{\varepsilon},$$

where $\boldsymbol{X}$ is the design matrix containing a constant as well as the linear effects of z_1 and z_2, $\boldsymbol{\beta} = (\beta_0, \beta_1, \beta_2)'$ collects the corresponding regression coefficients,

$$\boldsymbol{Z}[i,j] = B_j(z_{i1}, z_{i2})$$

contains the radial basis functions evaluated at the observed covariate values, and $\boldsymbol{\gamma}$ is the vector of the basis coefficients. When counting the number of regression coefficients in this model, we obtain $n+3$ parameters and therefore the model is overspecified. To obtain an identifiable version, the restriction $\boldsymbol{X}'\boldsymbol{\gamma} = \boldsymbol{0}$ has to be imposed, basically ensuring that the linear part of the model is orthogonal to the part represented by the radial basis functions. In addition, it can be shown that the integral penalty (8.30) can equivalently be represented as

$$\boldsymbol{\gamma}'\boldsymbol{Z}\boldsymbol{\gamma},$$

where the penalty matrix coincides with the design matrix. Hence, the overall estimation problem (8.29) can be rewritten as

$$(\boldsymbol{y} - \boldsymbol{X}\boldsymbol{\beta} - \boldsymbol{Z}\boldsymbol{\gamma})'(\boldsymbol{y} - \boldsymbol{X}\boldsymbol{\beta} - \boldsymbol{Z}\boldsymbol{\gamma}) + \lambda\boldsymbol{\gamma}'\boldsymbol{Z}\boldsymbol{\gamma} \to \min_{\boldsymbol{\beta},\boldsymbol{\gamma}}$$

subject to the constraint $\boldsymbol{X}'\boldsymbol{\gamma} = \boldsymbol{0}$.

Similar as in the context of smoothing splines, the number of regression coefficients associated with thin plate splines is often too large in practice since $(n+3) \times (n+3)$ systems of equations have to be solved. It is therefore of interest to obtain low rank approximations to the thin plate spline while still remaining as close to the optimal solution as possible. While simple rules for choosing a subset of observations as knots may often also produce sensible solutions, Wood (2003) proposed an optimal approximation based on a spectral decomposition of the design matrix $\boldsymbol{Z}$ that is also implemented in the R package `mgcv`. Basically, we first compute the spectral decomposition

$$\boldsymbol{Z} = \boldsymbol{\Gamma}\boldsymbol{\Omega}\boldsymbol{\Gamma}',$$

where $\boldsymbol{\Gamma}$ is an orthonormal matrix of eigenvectors and $\boldsymbol{\Omega}$ contains the corresponding (nonnegative) eigenvalues in descending order. It can then be shown that

$$\boldsymbol{Z}_d = \boldsymbol{\Gamma}_d\boldsymbol{\Omega}_d\boldsymbol{\Gamma}_d',$$

where $\boldsymbol{\Gamma}_d$ and $\boldsymbol{\Omega}_d$ are the submatrices of $\boldsymbol{\Gamma}$ and $\boldsymbol{\Omega}$ associated with the d largest eigenvalues, is the best rank d approximation to $\boldsymbol{Z}$ in the sense of the spectral norm $||\boldsymbol{Z} - \boldsymbol{Z}_d||$ (with $||\boldsymbol{A}||$ corresponding to the square root of the largest eigenvalue of the positive semidefinite matrix $\boldsymbol{A}$). The idea is now to replace $\boldsymbol{Z}$ with $\boldsymbol{Z}_d$ which basically projects the original estimation problem into the optimal d-dimensional subspace where d can be chosen such that the approximation error is small. In fact, fast numerical algorithms for truncated spectral decompositions exist that avoid the need to compute the full spectral decomposition.

Other commonly used radial basis functions are, for example,

$$B(r) = r^l, \quad l \text{ odd},$$

or

$$B(r) = \sqrt{r^2 + c^2} \quad \text{for a constant } c > 0.$$

In this case, the penalty matrix results from the integrated squared second derivatives of the basis functions, similar to those in the univariate case. In the following section, we will see that the stationary Gaussian fields that we previously discussed in Sect. 8.1.6 can also be interpreted as radial basis function approaches.

8.2.3 Kriging: Spatial Smoothing with Continuous Location Variables

In Sect. 8.1.6 we already discovered that temporal correlations can be described using stationary Gaussian processes and parametric correlation functions. We will now transfer this approach to spatial correlations using Gaussian random fields. In contrast to the previously discussed approaches for modeling interaction surfaces based on basis functions, we now consider a probabilistic modeling framework, i.e., spatial effects are described in terms of stochastic processes. However, as in the univariate case, we will see that the probabilistic model formulation is in fact equivalent to a special basis function approach.

In general, a Gaussian field $\{\gamma(s), s \in \mathbb{R}^2\}$ is characterized by the expectation function $\mu(s) = \mathrm{E}(\gamma(s))$, the variance function $\tau^2(s) = \mathrm{Var}(\gamma(s))$, and the correlation function $\rho(s,t) = \mathrm{Corr}(\gamma(s), \gamma(t))$. For *stationary* Gaussian fields, the expected value and variance are spatially constant ($\mu(s) \equiv \mu$ and $\tau^2(s) \equiv \tau^2$), and the correlation function only depends on the difference $s - t$. Thus, we obtain $\rho(s,t) = \rho(s-t) = \rho(h)$ with $h = s - t$. When analyzing spatial correlations, we often limit ourselves to the special case of *isotropic* correlation functions, where

$$\rho(s,t) = \rho(||s - t||) = \rho(r)$$

with $r = ||s - t||$. It follows that the correlation between two points s and t only depends on their Euclidean distance but not on the exact position of the points or on the direction of the vector between the two points. When assuming isotropy, we can use all parametric classes discussed in Sect. 8.1.6 (p. 453) for the correlation function ρ, since the actual bivariate spatial correlation function only depends on the scalar r.

An easy way to incorporate anisotropy is to replace the Euclidean distance

$$||s - t|| = \sqrt{(s-t)'(s-t)}$$

with

$$\sqrt{(s-t)' R(\psi)' D(\delta) R(\psi)(s-t)}, \tag{8.31}$$

where $R(\psi)$ defines a rotation matrix

$$R(\psi) = \begin{pmatrix} \cos(\psi) & \sin(\psi) \\ -\sin(\psi) & \cos(\psi) \end{pmatrix},$$

with anisotropy angle $\psi \in [0, 2\pi]$ and $D(\delta)$ defines a prolongation matrix

$$D(\delta) = \begin{pmatrix} \delta^{-1} & 0 \\ 0 & 1 \end{pmatrix}$$

with anisotropy ratio $\delta \geq 1$. Figure 8.43 shows the resulting contour lines for various combinations of ψ and δ. Note that Eq. (8.31) can only be used to produce correlation functions with elliptic contour lines. More general approaches to the construction of anisotropic correlation functions also allow other deviations from radial contour lines.

Classical Geostatistics

In spatial statistics, kriging was developed for smoothing or interpolating spatial phenomena. The classical geostatistical model is defined as

$$y(s) = \boldsymbol{x}(s)'\boldsymbol{\beta} + \gamma(s) + \varepsilon(s),$$

where

- $\boldsymbol{x}(s)'\boldsymbol{\beta}$ is the spatial trend parameterized by covariates $\boldsymbol{x}$,
- $\gamma(s)$ is a stationary Gaussian process with expected value 0, variance τ^2, and correlation function $\rho(h)$, and
- $\varepsilon(s)$ is the usual i.i.d. error $\varepsilon(s) \sim \mathrm{N}(0, \sigma^2)$ (independent of $\gamma(s)$).

If the spatial trend is constant, i.e., $\boldsymbol{x}(s)'\boldsymbol{\beta} \equiv \beta_0$, the model simplifies to the case of ordinary kriging, whereas the general case is referred to as universal kriging.

In matrix notation, the kriging model can be written as

$$\boldsymbol{y} = \boldsymbol{X\beta} + \boldsymbol{Z\gamma} + \boldsymbol{\varepsilon},$$

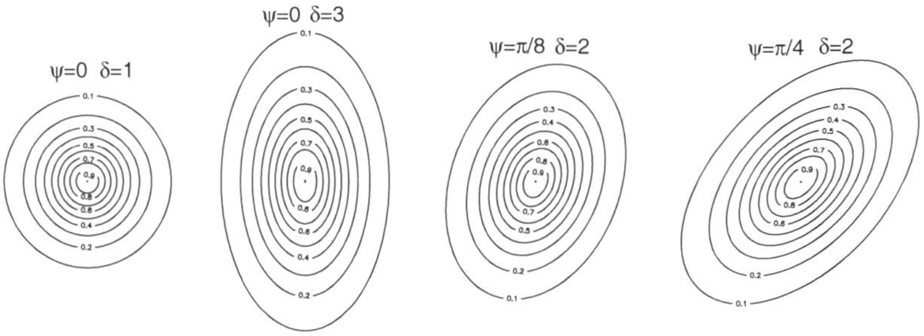

Fig. 8.43 Contour lines of anisotropic correlation functions constructed based on the distance measure (8.31)

where $\boldsymbol{\gamma} = (\boldsymbol{\gamma}(s_{(1)}), \ldots, \boldsymbol{\gamma}(s_{(d)}))'$ defines the values of the stationary Gaussian process at the d unique observed spatial locations $s_{(1)}, \ldots, s_{(d)}$. The matrix $\boldsymbol{Z}$ corresponds to an incidence matrix with

$$\boldsymbol{Z}[i,j] = \begin{cases} 1 & \text{if } y_i \text{ is observed at point } s_{(j)} \text{ (i.e., } s_i = s_{(j)}), \\ 0 & \text{otherwise.} \end{cases}$$

Based on the model specification, the covariance matrix of the response vector $\boldsymbol{y}$ becomes

$$\text{Cov}(\boldsymbol{y}) = \tau^2 \boldsymbol{Z}\boldsymbol{R}\boldsymbol{Z}' + \sigma^2 \boldsymbol{I}_n,$$

where the correlation matrix of the spatial effects $\boldsymbol{\gamma}$,

$$\boldsymbol{R} = (\text{Corr}(\gamma(s_{(i)}), \gamma(s_{(j)})) = (\rho(s_{(i)} - s_{(j)})),$$

introduces spatial correlation and is embedded in $\text{Cov}(\boldsymbol{y})$.

The goal of classical geostatistics is to obtain optimal predictions for $\gamma(s_0)$ or $\eta(s_0) = \boldsymbol{x}(s_0)'\boldsymbol{\beta} + \gamma(s_0)$ at new locations s_0. Using the definitions $\boldsymbol{r} = (\rho(s_1 - s_0), \ldots, \rho(s_n - s_0))'$, $\eta_0 = \eta(s_0)$ and $\mu_0 = \boldsymbol{x}(s_0)'\boldsymbol{\beta}$, the optimal prediction can be derived from the joint normal distribution

$$\begin{pmatrix} \boldsymbol{y} \\ \eta_0 \end{pmatrix} \sim \text{N}\left(\begin{bmatrix} \boldsymbol{X}\boldsymbol{\beta} \\ \mu_0 \end{bmatrix}, \begin{bmatrix} \tau^2 \boldsymbol{Z}\boldsymbol{R}\boldsymbol{Z}' + \sigma^2 \boldsymbol{I}_n & \tau^2 \boldsymbol{r}' \\ \tau^2 \boldsymbol{r} & \tau^2 \end{bmatrix} \right)$$

(see Theorem B.6 in Appendix B.3). The mean squared error optimal prediction is simply the conditional expectation of η_0 given $\boldsymbol{y}$, i.e.,

$$\hat{\eta}_0 = \text{E}(\eta_0 \mid \boldsymbol{y}) = \mu_0 + \tau^2 \boldsymbol{r}' (\tau^2 \boldsymbol{Z}\boldsymbol{R}\boldsymbol{Z}' + \sigma^2 \boldsymbol{I}_n)^{-1} (\boldsymbol{y} - \boldsymbol{X}\boldsymbol{\beta}).$$

The conditional variance of $\hat{\eta}_0$ can also be derived from the joint normal distribution, yielding

$$\text{Var}(\eta_0 \mid \boldsymbol{y}) = \tau^2 - \tau^2 \boldsymbol{r}' (\tau^2 \boldsymbol{Z}\boldsymbol{R}\boldsymbol{Z}' + \sigma^2 \boldsymbol{I}_n)^{-1} \boldsymbol{r} \tau^2.$$

Even without the explicit assumption of normality, one can show that $\hat{\eta}_0$ still fulfills certain optimality properties. Similar to the estimates of random effects presented in Sect. 7.3.1 (p. 371), $\hat{\eta}_0$ is the best linear unbiased predictor.

Kriging as a Basis Function Approach

With the use of classical geostatistics, we obtain optimal predictions of the spatial effects at arbitrary locations s_0. However, in contrast to the basis function approaches discussed thus far, we do not obtain a compact representation for such predictions. A possible scenario arises from the developments presented in Sect. 8.1.6 (p. 453), where we reexpressed the stochastic kriging approach using basis functions derived from the correlation function $\rho(\cdot)$. This reparameterization can also be used for the spatial kriging model, so that the classical geostatistics model can be expressed as

$$\boldsymbol{y} = \boldsymbol{X}\boldsymbol{\beta} + \tilde{\boldsymbol{Z}}\tilde{\boldsymbol{\gamma}} + \boldsymbol{\varepsilon},$$

with design matrix

$$\tilde{\boldsymbol{Z}}[i, j] = \rho(s_i, s_{(j)}).$$

For a single observation, this yields the model specification

$$y(s) = \boldsymbol{x}(s)'\boldsymbol{\beta} + f_{geo}(s) + \varepsilon(s)$$

with spatial effect

$$f_{geo}(s) = \sum_{j=1}^{d} \tilde{\gamma}_j B_j(s)$$

and basis functions

$$B_j(s) = \rho(s, s_{(j)}),$$

obtained from the correlation function.

For isotropic correlation functions, we obtain the special case of radial basis functions $B_j(s) = \rho(||s - s_{(j)}||)$. Similar to Sect. 8.1.6, the penalty for $\tilde{\boldsymbol{\gamma}}$ results from the correlation matrix $\boldsymbol{R}$ as

$$\lambda \tilde{\boldsymbol{\gamma}}'\boldsymbol{K}\tilde{\boldsymbol{\gamma}} = \frac{\sigma^2}{\tau^2}\tilde{\boldsymbol{\gamma}}'\boldsymbol{R}\tilde{\boldsymbol{\gamma}}.$$

If we choose Matérn correlation functions, we obtain the class of *Matérn splines*. Figure 8.44 shows two isotropic Matérn spline basis functions, which result from the hyperparameters $\kappa = 0.5$ and $\kappa = 1.5$. Contour plots, which illustrate the radiality of the basis functions, are provided as well.

From a Bayesian point of view, the kriging approach in the basis function representation is equivalent to a certain smoothing prior, which is given by

$$\tilde{\boldsymbol{\gamma}} \sim \mathrm{N}(\boldsymbol{0}, \tau^2 \boldsymbol{R}^{-1}). \tag{8.32}$$

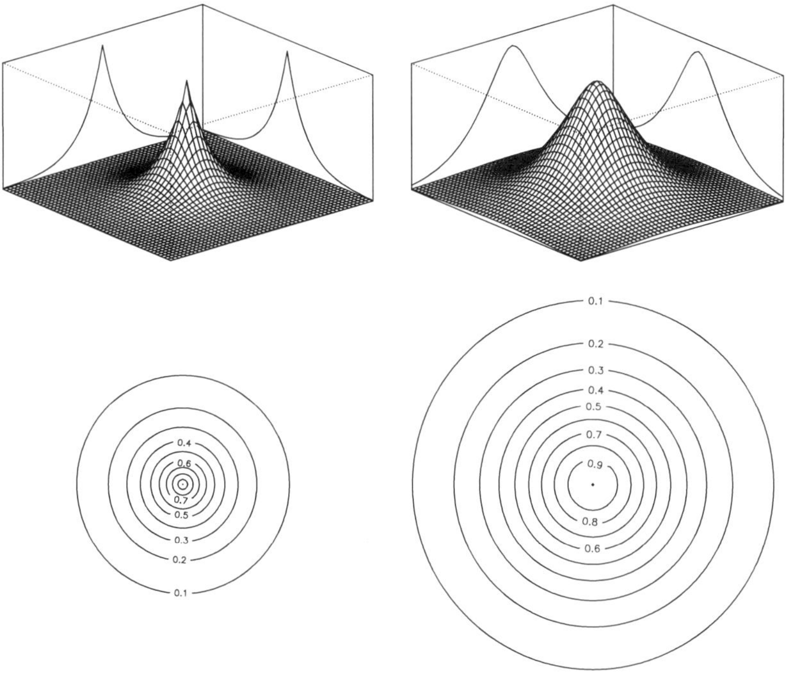

Fig. 8.44 Perspective plots and contour lines of Matérn spline basis functions with $\kappa = 0.5$ (*left*) and $\kappa = 1.5$ (*right*)

In contrast to the tensor product methods that we discussed for polynomial splines, the knots for the kriging approach are determined from the observed data locations. As the number of knots is usually close to the sample size, there are typically a large number of coefficients to estimate. Therefore, to alleviate computing demands, one therefore often considers only a subset of the observed locations as knots, i.e.,

$$\mathcal{D} = \{\kappa_1, \ldots, \kappa_m\} \subset \mathcal{C} = \{s_{(1)}, \ldots, s_{(d)}\}.$$

Based on adequate criteria, this subset can be determined as representative for the original observations; see Johnson, Moore, and Ylvisaker (1990) and Nychka and Saltzman (1998) for details. Through $\tilde{\boldsymbol{Z}}\tilde{\boldsymbol{\gamma}}$, we have an approximate kriging surface, where the design matrix is defined by

$$\tilde{\boldsymbol{Z}}[i,j] = \rho(s_i - \kappa_j), \tag{8.33}$$

and the m-dimensional coefficient vector has the distribution

$$\tilde{\boldsymbol{\gamma}} \sim \mathrm{N}(\boldsymbol{0}, \tau^2 \boldsymbol{R}^{-1}) \quad \text{with} \quad \boldsymbol{R}[i,j] = \rho(\kappa_i, \kappa_j). \tag{8.34}$$

8.13 Kriging

Classical Geostatistics Model

$$y(s) = \mu(s) + \gamma(s) + \varepsilon(s), \quad s \in \mathbb{R}^2,$$

where $\mu(s) = \boldsymbol{x}(s)'\boldsymbol{\beta}$ defines the spatial trend and $\gamma(s)$ is a (stationary) Gaussian field with expectation 0, variance τ^2, and parametric correlation function $\rho(h)$.

Estimation

For given variances, the best linear unbiased predictions for $\eta_0 = \boldsymbol{x}(s_0)'\boldsymbol{\beta} + \gamma(s_0)$ result from the joint normal distribution of $\boldsymbol{y}$ and η_0. Since $\gamma(s)$ corresponds to a spatially correlated random effect, estimation can be approached more generally using the mixed model methods discussed in Chap. 7.

Basis Function Representation

As with the univariate kriging method, we also obtain a bivariate basis function representation with basis functions

$$B_j(s) = \rho(s, s_{(j)}) \quad \text{or} \quad B_j(s) = \rho(||s - s_{(j)}||).$$

This yields the model specification

$$\boldsymbol{y} = \boldsymbol{X}\boldsymbol{\beta} + \tilde{\boldsymbol{Z}}\tilde{\boldsymbol{\gamma}} + \boldsymbol{\varepsilon},$$

with $\tilde{\boldsymbol{Z}}[i,j] = \rho(s_i, s_{(j)})$ or $\tilde{\boldsymbol{Z}}[i,j] = \rho(||s_i - s_{(j)}||)$. The penalty for $\tilde{\boldsymbol{\gamma}}$ is determined by the correlation matrix $\boldsymbol{R}$ with elements $\boldsymbol{R}[i,j] = \rho(s_{(i)}, s_{(j)})$ yielding

$$\lambda \tilde{\boldsymbol{\gamma}}'\boldsymbol{K}\tilde{\boldsymbol{\gamma}} = \frac{\sigma^2}{\tau^2}\tilde{\boldsymbol{\gamma}}'\boldsymbol{R}\tilde{\boldsymbol{\gamma}}.$$

An alternative would be to use a low rank approximation based on the spectral decomposition of $\boldsymbol{Z}$ similar as discussed for thin plate splines.

Estimation of Kriging Models

Estimation of kriging models can be achieved in one of two ways resulting from the different perspectives on the geostatistical model. If we consider the geostatistical model in the original stochastic interpretation, the assumption of a Gaussian field for the spatial effect basically corresponds to the assumption of a spatially correlated random effect. The methods discussed in Chap. 7 can therefore be used to determine ML or REML estimates of all parameters of the correlation function (including for example the range parameter).

When interpreting kriging as a basis function method, all parameters of the correlation function are typically fixed at prespecified values or chosen based on simple rules (as done in Example 8.3). The variance τ^2 then remains as the only unknown parameter, which acts as a smoothing parameter for the basis function method, while the penalty matrix $\boldsymbol{R}$ is completely specified. To estimate τ^2, we can use the methods discussed in Sect. 8.1.9, in particular the mixed model representation or MCMC-based algorithms.

Example 8.12 Human Brain Mapping—Kriging

We next apply the spatial kriging method in the human brain mapping example with a visual stimulus (see Example 8.8 on p. 501). Figure 8.45 shows two estimates for this activation profile based on Matérn splines with parameter $\kappa = 1.5$ using either 100 or 200 knots. The range parameter of the correlation function was chosen according to the rule of thumb, which we already used in Example 8.8.

For both 100 and 200 knots, strong activation is clearly visible within the visual cortex. The basis function approach does not only allow for pixelwise identification of activation, but in fact provides a continuous activation surface. Comparing the two knot sets (100 or 200 knots), we find that the higher knot density leads to a somewhat rougher estimate. However, in general, the estimated function is not highly sensitive to the number of knots. △

8.2.4 Markov Random Fields: Spatial Smoothing with Discrete Location Variables

To this point, we discussed spatial effects based on spatial coordinates, i.e., cases where continuous spatial information is available. As we saw in Example 8.9 (p. 502), it is also possible to have discrete spatial information, e.g., spatial regions $s \in \{1, \dots, d\}$. Discrete spatial information also arises naturally in case of data arranged on a regular grid. Note that the human brain mapping example actually consists of such gridded data. However, with a large number of grid points, it is also possible to analyze such data using a continuous spatial model, identifying each pixel with its respective coordinates.

Spatial Neighborhoods

Whereas in the case of continuous spatial information we can easily determine the distance between two locations using, for example, the Euclidean distance, this is no longer the case with discrete spatial data. Consequently, we use a different concept to describe the spatial arrangement of the data, which relies on the definition of adequate forms of neighborhoods. Such neighborhoods can actually be constructed in different ways:

- For spatial data where the spatial covariate s defines the membership of an observation to a particular region s, neighborhoods are usually defined by

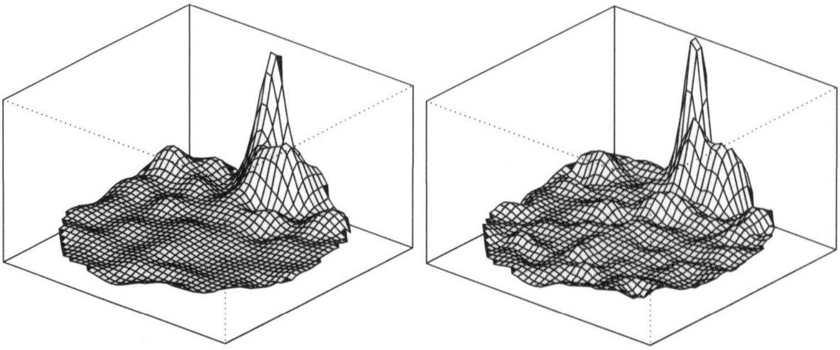

Fig. 8.45 Human brain mapping: kriging estimates for the activation at time point $t = 38$ obtained from Matérn splines with $\kappa = 1.5$ and 100 (*left*) or 200 (*right*) knots

common boundaries (see Fig. 8.46). Modifications of this definition may be necessary when some regions are islands, or when the observation domain is divided into separate subdomains.

- On regular grids, one often uses nearest neighbors on the grid, e.g., the four or eight nearest neighbors. In this case, it is also possible to define neighborhoods of a higher order, as we have already seen when discussing bivariate polynomial splines. However, in the following, we limit ourselves to simple, direct neighborhoods.

In the following we will use the notation $s \sim r$ to denote that regions s and r are neighbors.

Similarly, as we proceeded in the case of random walks in Sect. 8.1.5, every region s is assigned its own regression coefficient $f_{geo}(s) = \gamma_s$, $s = 1, \ldots, d$. Due to the large number of coefficients that usually results from this approach, we need an appropriate structure to regain smooth spatial effects and thus to reduce the effective number of parameters. In obtaining smooth spatial effects, the coefficients of nearby regions should not divert too strongly from each other. Thus, we create a penalty that is based on squared differences between the parameters of nearby regions. Consider the PLS criterion

$$\text{PLS}(\lambda) = \sum_{i=1}^{n} \left(y_i - f_{geo}(s_i)\right)^2 + \lambda \sum_{s=2}^{d} \sum_{r \in N(s), r<s} (\gamma_r - \gamma_s)^2, \tag{8.35}$$

where $N(s)$ defines the set of neighbors for region s. The penalty consists of squared differences of all possible combinations of neighboring regions, where each combination is considered only once. This in fact yields a penalty that discourages large deviations of effects associated with neighboring regions.

To include this approach within the scope of general penalization methods, we first rewrite the PLS criterion in matrix notation. To do so, we define the design matrix $\boldsymbol{Z}$ with the entries

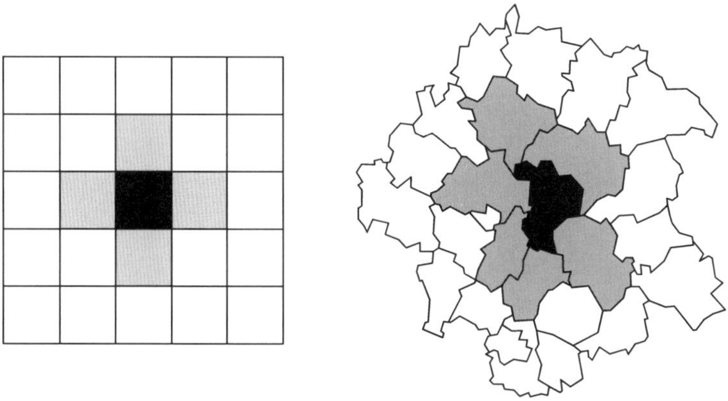

Fig. 8.46 First-order neighborhoods on a regular grid (*left*) and for irregular regional data (*right*). The neighbors of the *black* region are shaded in *grey*

$$\boldsymbol{Z}[i,s] = \begin{cases} 1 & \text{if } y_i \text{ was observed in region } s \text{ (i.e., } s_i = s\text{) and} \\ 0 & \text{otherwise.} \end{cases}$$

With this definition, we can express the vector of the function evaluations $\boldsymbol{f}_{geo} = (f_{geo}(s_1), \ldots, f_{geo}(s_n))'$ as a linear model $\boldsymbol{Z}\boldsymbol{\gamma}$. In turn, the penalty consists of squared differences and can be compactly written as the quadratic form $\lambda\boldsymbol{\gamma}'\boldsymbol{K}\boldsymbol{\gamma}$ with

$$\boldsymbol{K}[s,r] = \begin{cases} -1 & s \neq r, s \sim r, \\ 0 & s \neq r, s \nsim r, \\ |N(s)| & s = r. \end{cases}$$

By minimizing the PLS criterion, we then again obtain the PLS estimate $\hat{\boldsymbol{\gamma}} = (\boldsymbol{Z}'\boldsymbol{Z} + \lambda\boldsymbol{K})^{-1}\boldsymbol{Z}'\boldsymbol{y}$.

According to its definition, the penalty matrix $\boldsymbol{K}$ has the structure of an adjacency matrix, since the entries $\boldsymbol{K}[s,r]$ only differ from zero when s and r are neighbors. Efficient numerical methods for storing and processing $\boldsymbol{K}$ can be used since $\boldsymbol{K}$ is a sparse matrix. However, we should be aware that the order of the regions within the vector $\boldsymbol{\gamma}$ plays a crucial role for the structure of $\boldsymbol{K}$. Figure 8.47 illustrates this phenomenon for a large number of regions in two different orderings. In the left panel, the regions are simply arranged according to their number, whereas in the right panel, the arrangement results from the attempt to reorder the regions such that the nonzero elements are as close as possible to the diagonal, yielding a matrix that is close to a band matrix with small bandwidth. This ordering was achieved with the Cuthill–McKee algorithm; see, for example, George and Liu (1981), who also offer further details regarding numerically efficient algorithms for sparse matrices. Rue and Held (2005) also discuss some of these algorithms from a statistical point of view.

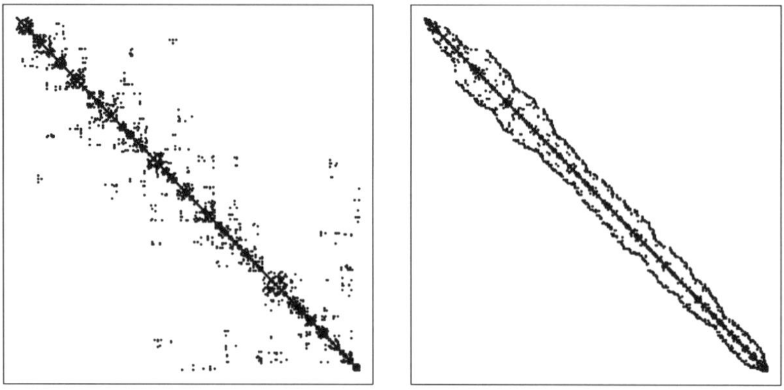

Fig. 8.47 Adjacency matrix for a large set of regions in two different orderings

Bayesian Model Formulation

As with the other penalization approaches discussed thus far, the above model formulation can also be interpreted in a Bayesian context. This yields so-called Markov random fields, which we will now introduce in a somewhat more general context. In the literature, penalization methods for discrete spatial locations are commonly motivated in this Bayesian framework.

We start with a general definition of Markov random fields (MRF) that generalize the temporal Markov property to the spatial situation that we discussed for random walks in Sect. 8.1.5 (p. 452). Let $D = \{1, \dots, s, \dots, d\}$ denote the set of all regions. Then, $\boldsymbol{\gamma} = \{\gamma_s, s \in D\}$ follows an MRF if the conditional distribution of γ_s given all other effects γ_r, $r \neq s$, depends only on its neighbors. The corresponding (conditional) density can then be written as

$$p(\gamma_s \mid \gamma_r, r \neq s) = p(\gamma_s \mid \gamma_r, r \in N(s)).$$

If we consider the model specification

$$y_i = f_{geo}(s_i) + \varepsilon_i, \quad \varepsilon_i \sim \mathrm{N}(0, \sigma^2),$$

the aim is to assign an MRF to the distribution of $f_{geo}(s) = \gamma_s$. Gauß–Markov random fields (GMRF), i.e., MRF where the conditional distributions (and also the joint distribution) are normal, provide an adequate class of latent MRF to describe such spatial effects. One then often proceeds as follows: Start by specifying the conditional distributions $\gamma_s \mid \gamma_r$, $r \in N(s)$ and then derive the joint distribution of the entire vector $\boldsymbol{\gamma} = (\gamma_1, \dots, \gamma_d)'$ from these conditional distributions. However, we must be aware that not all specifications of conditional distributions are actually allowed to obtain a valid joint distribution. At this point, however, we do not intend to discuss the theoretical considerations that are necessary for the formulation of adequate classes of conditional distributions, since we are able to derive a valid specification directly from the PLS criterion.

This particular class has the form

$$\gamma_s \mid \gamma_r, r \in N(s) \sim \mathrm{N}\left(\frac{1}{|N(s)|} \sum_{r:r\sim s} \gamma_r, \frac{\tau^2}{|N(s)|}\right), \tag{8.36}$$

where $|N(s)|$ is the number of neighbors for region s. According to this specification, the conditional expectation of the spatial effect in region s is given by the arithmetic mean of the neighboring regional effects. The variance τ^2 controls how much the spatial effect γ_s can deviate from this expectation. Thus, we obtain a stochastic formulation of the desired property that neighboring regions show a similar spatial effect. The joint distribution of all spatial effects can now be derived from the conditional distributions (which in this case actually define a system of consistent distributions) yielding the density

$$p(\boldsymbol{\gamma} \mid \tau^2) \propto \left(\frac{1}{\tau^2}\right)^{(d-1)/2} \exp\left(-\frac{1}{2\tau^2}\boldsymbol{\gamma}'\boldsymbol{K}\boldsymbol{\gamma}\right). \tag{8.37}$$

This density is only defined upon proportionality since, in fact, the joint distribution is partially improper. The precision matrix $\boldsymbol{K}$ exactly corresponds to the penalty matrix from the previously given PLS criterion (8.35), and thus the Bayesian formulation (8.36) in fact leads to an equivalent model. In the literature, GMRFs with a partially improper distribution are also referred to as *intrinsic* GMRF (IGMRF). It is, however, also possible to define MRF with a proper joint distribution; see, for example, Banerjee, Carlin, and Gelfand (2003, Chap. 3).

Thus far, we have only used the spatial neighborhood information when constructing an MRF. It can also be meaningful to generalize this definition to some extent and appropriately weigh the influence of the neighbors. Therefore we extend Eq. (8.36) to

$$\gamma_s \mid \gamma_r, r \in N(s) \sim \mathrm{N}\left(\sum_{r:r\sim s} \frac{w_{sr}}{w_{s+}}\gamma_r, \frac{\tau^2}{w_{s+}}\right),$$

with symmetric weights $w_{sr} = w_{rs}$ and $w_{s+} = \sum_{r:r\sim s} w_{sr}$. Now, the conditional expectation of γ_s is given by the weighted average of the neighboring coefficients. The following strategies are commonly used to define weights w_{sr}:

- Use the same weight for all neighbors, i.e., $w_{sr} = 1$. This specification yields our original MRF definition.
- Use weights that are inversely proportional to the distance of the centroids, e.g., $w_{sr} \propto \exp(-d(s,r))$, where $d(s,r)$ defines the Euclidean distance between the centroids of the regions s and r.
- Use weights proportional to the length of the common boundary of regions s and r.

If weights are to be considered, then the penalty or precision matrix $\boldsymbol{K}$ has to be modified, taking the form

$$\boldsymbol{K}[s,r] = \begin{cases} -w_{sr} & s \neq r, s \sim r, \\ 0 & s \neq r, s \nsim r, \\ w_{s+} & s = r. \end{cases}$$

Example 8.13 Munich Rent Index—Analysis of Spatial Dependence

Figure 8.48 shows the MCMC-based estimate of an unweighted IGMRF for the Munich rent index data. If we compare the estimated effect with the separate means obtained in Fig. 8.35 (p. 503), we clearly recognize the smoothing effect that is introduced by the IGMRF assumption. This approach has the advantage that, for districts where only a few apartments were sampled, the estimated net rent is stabilized and less variable due to the integration of neighborhood information. Moreover, the classification of subquarters into areas with rent above or below average is more sensible based on smoothed effects.

Surprisingly, the results shown in Fig. 8.48 also indicate that the rent per square meter for apartments in the city center is smaller than for those in the suburbs. This demonstrates that the sole use of a regional effect can lead to results that are difficult to interpret due to the fact that other important explanatory variables are not taken into consideration. In our example, the reason for the surprising result is that the apartments in the city center are typically older than those in the suburbs. If we take this effect into consideration, as we will in some examples of Chap. 9, the effect changes and the apartments in the city center will actually turn out to be more expensive than those in the suburbs after adjusting for year of construction. This shows that the *simultaneous* consideration of covariate effects and regional effects is necessary to obtain improved estimates.

△

Spatially Autoregressive Processes

In econometrics, MRFs are widely applied in a variety of models other than the one we have considered thus far. In applications, autoregressive processes in time are generalized such that the MRF is assumed for the response variable y instead of a latent spatial process. This leads to spatially autoregressive processes (SAR), with the simplest special case given by

$$y_s = \alpha \sum_{r \in N(s)} w_{sr} y_r + \varepsilon_s, \qquad \varepsilon_s \sim \mathrm{N}(0, \sigma^2),$$

with autoregressive parameter $\alpha \in [0,1)$ and symmetric weights w_{sr} fulfilling $w_{ss} = 0$. We then obtain the conditional distributions

$$y_s \mid y_r, r \neq s \sim \mathrm{N}\left(\alpha \sum_{r \in N(s)} w_{sr} y_r, \sigma^2\right),$$

such that the vector of the response variable $\boldsymbol{y} = \{y_s, s = 1, \ldots, d\}$ appears to follow an MRF. If covariate effects are to be included in a SAR model, this is

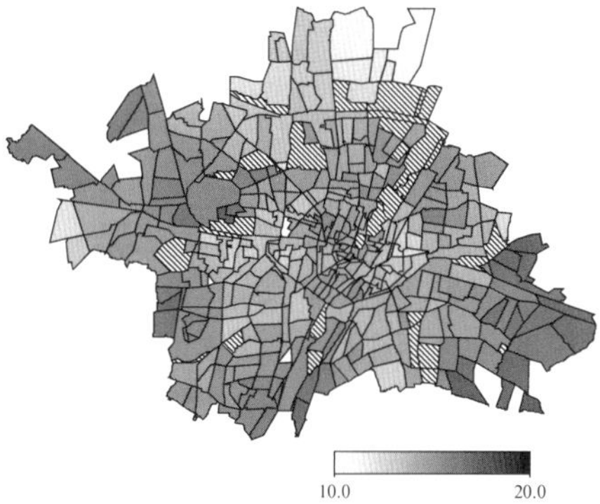

Fig. 8.48 Munich rent index: spatial effect obtained with an IGMRF without weights estimated with MCMC in a fully Bayesian framework

typically achieved by applying the SAR specification to the residuals $y_s - \boldsymbol{x}'_s\boldsymbol{\beta}$. Alternatively, it is also possible to add covariate effects directly to the SAR model, yielding

$$y_s = \boldsymbol{x}'_s\boldsymbol{\beta} + \alpha \sum_{r \in N(s)} w_{sr} y_r + \varepsilon_s, \quad \varepsilon_s \sim \mathrm{N}(0, \sigma^2).$$

However, the interpretation of covariate effects differs greatly between the two model specifications: The covariates either effect the response adjusted by the regional effects or the response variable directly.

Since SAR models do not fit into the general scope of penalization methods that we are considering, we will not discuss them in further detail (see, e.g., Anselin (1988) or Banerjee et al. (2003), Chap. 3, for more information).

8.2.5 Summary of Roughness Penalty Approaches

We now provide a brief summary of the bivariate smoothing methods that have been discussed thus far. As we have seen, bivariate basis function methods, stationary Gaussian fields, and GMRFs can all be subsumed in the general framework introduced in Sect. 8.1.3. In every case, we obtained a linear model of the form

$$\boldsymbol{y} = \boldsymbol{Z}\boldsymbol{\gamma} + \boldsymbol{\varepsilon},$$

with a large number of regression coefficients, representing either the coefficients of the basis functions or the spatial effects. To regularize estimation of $\boldsymbol{\gamma}$, we introduced quadratic penalties

$$\lambda \boldsymbol{\gamma}' \boldsymbol{K} \boldsymbol{\gamma}.$$

Depending on the method, the penalties were constructed based on either differences of coefficients, derivative operators, correlations, or neighborhood structures. Due to the general formulation we can use the methods discussed in Sect. 8.1.9 to determine a data-driven choice for the smoothing parameter.

8.14 Gauß–Markov Random Fields

Model

$$y = Z\gamma + \varepsilon$$

with

$$Z[i,s] = \begin{cases} 1 & \text{if } y_i \text{ was observed in region } s \text{ (i.e., } s_i = s\text{) and} \\ 0 & \text{otherwise} \end{cases}$$

and $\gamma = (f_{geo}(s_1), \dots, f_{geo}(s_d))'$.

IGMRF for the Spatial Effect

$$\gamma_s \mid \gamma_r, r \in N(s) \sim \mathrm{N}\left(\sum_{r:r\sim s} \frac{w_{sr}}{w_{s+}}\gamma_r, \frac{\tau^2}{w_{s+}}\right)$$

with $\gamma_s = f_{geo}(s)$ and $w_{sr} = w_{rs}$. The (partially improper) joint distribution is

$$p(\gamma \mid \tau^2) \propto \left(\frac{1}{\tau^2}\right)^{(d-1)/2} \exp\left(-\frac{1}{2\tau^2}\gamma' K \gamma\right)$$

with

$$K[s,r] = \begin{cases} -w_{sr} & s \neq r, s \sim r, \\ 0 & s \neq r, s \nsim r, \\ w_{s+} & s = r. \end{cases}$$

In the bivariate case, differences between the approaches are more apparent than in the univariate case. First of all, we generally distinguish between continuous and discrete regional information, where specific approaches were developed in either context. In our case, these are basis function approaches and stationary Gaussian fields on the one hand, and Markov random fields on the other hand. However, the distinction is not as obvious as it might seem. Methods which require continuous spatial information can be used, e.g., for regional data, where the centroids of the regions are used as the locations in the analysis. Models that require discrete spatial information can also be used in the continuous case, for example, by discretizing the observation domain on a regular grid.

Secondly, if we compare the different approaches regarding their numerical properties, the results vary depending on which inferential concept is used. If we use MCMC methods, it is necessary to solve a high-dimensional system of equations for the simulation of $\boldsymbol{\gamma}$ in every iteration. With the help of sophisticated methods for sparse matrices, these calculations can be performed efficiently. Sparse cross product matrices $\boldsymbol{Z}'\boldsymbol{Z}$ and penalty matrices $\boldsymbol{K}$ result, in particular, for Gauß–Markov random fields and bivariate P-splines. However, if stationary Gaussian fields or radial bases are used, then $\boldsymbol{Z}'\boldsymbol{Z}$ and $\boldsymbol{K}$ are typically dense. Recently, Lang et al. (2012) proposed a highly efficient Gibbs sampler even in the case of dense matrices. The alternative sampling scheme works with a transformed parameterization such that the cross product of the design matrix and the penalty matrix are *diagonal* resulting in a diagonal posterior precision matrix.

If inference is based on a mixed model representation, then the structure of the penalty matrix $\boldsymbol{K}$ is less important since the penalty matrix only needs to be evaluated once during the reparameterization. Rather, the dimension of the parameter vector $\boldsymbol{\gamma}$ is more important. This is the reason why radial basis function approaches and stationary Gaussian fields, without an adequate strategy for choosing a small set of knots, are less useful for this method than bivariate P-splines and Gauß–Markov random fields. If we, however, use a strategy for reducing the number of knots, it is often possible to reduce the dimension of $\boldsymbol{\gamma}$ such that every estimation method shows a similar level of complexity.

8.2.6 Local and Adaptive Smoothing

Local smoothers are also often used for the analysis of interactions and spatial effects, relying on extensions of the methods discussed in Sect. 8.1.7. This extension is particularly straightforward for nearest neighbor estimates, since the k-nearest neighbors can also be determined in bivariate problems using, for example, the Euclidean distance. Consequently, it is also possible to generalize the loess procedure, as well as the Nadaraya–Watson estimate that is based on kernel density estimates, to bivariate smoothing. The choice of an optimal smoothing parameter can be chosen either via a grid search or with the help of adequate numerical procedures; see, for example, Kauermann and Opsomer (2004). However, the use of local smoothing procedures as building blocks in more complex models is difficult when combining, for example, nonparametric effects and random effects. This is the reason why we will not discuss local methods in detail, but instead refer to the literature listed in Sect. 8.4.

Adaptive procedures can also be efficiently used for bivariate smoothing and more complex models. These procedures have the particular advantage that not only are they able to determine how to model the joint effect of z_1 and z_2 but they also are able to detect whether or not an interaction effect is in fact present. This can be especially useful for the extensions presented in Chap. 9, where potentially even more than two effects need to be determined simultaneously.

For the MARS algorithm, we obtain a bivariate extension by examining, in every iteration, the univariate basis functions for z_1 and z_2, as well as their corresponding interactions for potential improvements of the model fit. This results in a sequence of models, which can comprise complex interaction models, as well as simple models containing only univariate covariate effects. With the help of the model choice criterion, we adaptively decide which of the models is the most adequate for modeling the effects of the covariates.

For regression trees, extensions for more complex models result when considering, in every iteration, partitions in all variables. Hence, in each iteration step, we optimally choose both the variable to be split and the split point for this variable. Thus, we obtain adaptive splits of the data into homogeneous sets of similar response values, which correspond to a piecewise constant function estimate for $f(z_1, z_2)$.

In a similar way, we can also generalize the Bayesian adaptive procedures. Since we will concentrate on penalization methods in Chap. 9, we will not discuss adaptive methods in more detail at this point. Further information about adaptive procedures will be provided at the end of this chapter in Sect. 8.4.

8.3 Higher-Dimensional Smoothing

In principle, the ideas developed in Sect. 8.2 can also be applied for modeling higher-dimensional surfaces, i.e., in models of the form

$$y = f(z_1, \dots, z_q) + \varepsilon.$$

For example, higher-dimensional tensor product splines result from considering all possible interactions of univariate splines for $z_1, \dots, z_q$. The respective penalty matrices can then be constructed analogous to those in Sect. 8.2.1 based on Kronecker products leading to

$$\boldsymbol{I}_{d_3} \otimes \boldsymbol{I}_{d_2} \otimes \boldsymbol{K}_1 + \boldsymbol{I}_{d_3} \otimes \boldsymbol{K}_2 \otimes \boldsymbol{I}_{d_1} + \boldsymbol{K}_3 \otimes \boldsymbol{I}_{d_2} \otimes \boldsymbol{I}_{d_1},$$

in the special case of $q = 3$, where $\boldsymbol{K}_1$, $\boldsymbol{K}_2$, and $\boldsymbol{K}_3$ denote univariate penalty matrices and $\boldsymbol{I}_d$ are identity matrices of appropriate dimension. For radial bases and kriging terms that are based on isotropic correlation functions, the construction of higher-dimensional surfaces is even simpler. In such cases, the basis functions rely only on the Euclidean distance between two points, which is defined in $\mathbb{R}^q$, as well as in the special case of $\mathbb{R}^2$. By extending the concept of neighborhoods, the construction of Markov random fields can also immediately be applied to problems in higher dimensions.

Nevertheless, regardless of the chosen approach, problems occur when estimating higher-dimensional functions nonparametrically. In general, such functions rely on a large number of parameters and the respective estimation procedures are computationally expensive. For example, a tensor product spline described by a 20-dimensional basis in each dimension has already $20^3 = 8,000$ parameters in $\mathbb{R}^3$. Even though a penalty reduces the effective number of parameters, we are still required to solve an 8,000-dimensional system of equations (often multiple times) for obtaining the estimates.

Another difficulty arising in higher-dimensional approaches, is the "curse of dimensionality." Consider the following problem: Suppose we are given a q-dimensional unit cube (i.e., a cube with edges of length 1), where all data are uniformly distributed within this cube. What will be the edge length l of a partial cube that contains a fraction r of the data? For $q = 1$, we obviously have $l = r$, whereas for $q = 2$, we have $l = \sqrt{r}$. In general, it follows that

$$l = r^{\frac{1}{q}},$$

and therefore for $r = 0.1$

$$\begin{array}{ll} q = 1 & l = 0.1 \\ q = 2 & l \approx 0.3 \\ q = 3 & l \approx 0.47 \\ \vdots & \vdots \\ q = 10 & l \approx 0.8. \end{array}$$

For a three-dimensional estimation problem, this implies that the partial cube must already have an edge length of about 0.5 to contain 10 % of the data. Obviously, our intuition of "localness" no longer works in higher dimensions. Moreover, the curse of dimensionality suggests that in higher-dimensional smoothing problems, we only obtain a good coverage of the observation domain when the number of parameters for each dimension is actually greater than what we used in univariate smoothing problems. Consequently, we again face the problem that a large number of parameters need to be estimated.

This is the reason why, in practice, we rarely consider surfaces that have a dimension higher than $q = 2$. Rather, we assume an additive structure for the function $f(z_1, \ldots, z_q)$, leading to

$$f(z_1, \ldots, z_q) = f_1(z_1) + \ldots + f_q(z_q).$$

This yields the class of additive models and their extensions, which we will examine in detail in Chap. 9.

8.4 Bibliographic Notes

In the following, we give an overview of selected literature, which will provide either more detailed or more general information regarding some aspects of Chap. 8. Many of the references listed will also be of interest for the extensions to be discussed in Chap. 9.

Hastie and Tibshirani (1990) introduce and systematically examine univariate smoothing procedures within the scope of additive models and, in particular, discuss general questions regarding linear smoothing procedures in detail. Fahrmeir and Tutz (2001, Chap. 5) provide a good overview of modern extensions. The overview by Wand (2000) compares some univariate smoothing procedures via simulation, and also includes adaptive procedures. Wood (2006) describes nonparametric regression models based on penalization approaches and splines, along with an introduction to the R package `mgcv`. Mathematical properties of splines are discussed in De Boor (2001) and Dierckx (1993); see also Eilers and Marx (2010) for a nice (and for most practical purposes sufficient) introduction to splines. Detailed descriptions of local regression models are given in Härdle (1990), Fan and Gijbels (1996), Loader (1999), and Härdle, Müller, Sperlich, and Werwatz (2004).

Extensive discussions of the connection between penalized splines and mixed models used in Sect. 8.1.9 can be found in Ruppert et al. (2003), Wand (2003), or Kauermann (2006). Fahrmeir, Kneib, and Lang (2004) describe this connection for general penalization approaches and in particular for smoothing procedures developed in spatial statistics. Extensions for survival models are discussed in Kauermann and Khomski (2006) or Kneib and Fahrmeir (2007). Bayesian approaches for the selection of smoothing parameters that are based on MCMC simulations are discussed in detail in Lang and Brezger (2004), Brezger and Lang (2006), and Lang et al. (2012).

Friedman (1991), Stone, Hansen, Kooperberg, and Truong (1997), and Hansen and Kooperberg (2002) discuss adaptive approaches that are based on frequentist model choice strategies. The articles by Biller and Fahrmeir (2001), Biller (2000), Yau, Kohn, and Wood (2003), Denison et al. (1998), or DiMatteo, Genovese, and Kass (2001) describe corresponding Bayesian approaches. An introduction to regression trees is provided in Tutz (2011, Chap. 9); the classical reference in the field of regression and classification trees is Breiman, Friedman, Stone, and Olshen (1984). A conceptually powerful extension of regression trees is proposed in Hothorn, Hornik, and Zeileis (2006). Here the leaves of a tree are not constant but full linear or generalized linear models. As already mentioned in Sect. 8.1.10, regression trees are unstable due to the fact that even small changes within the data structure can result in completely different trees. Resampling methods, which build not only one but an entire ensemble of regression trees, mitigate this instability. Examples are bagging (Breiman, 1996) or random forests (Breiman, 2001). We can actually employ similar methods in regression and smoothing procedures based on extensions of boosting algorithms; see, for example, Bühlmann and Yu (2003, 2006), Tutz and Binder (2006), or Bühlmann and Hothorn (2007).

Finally, we want to refer to some approaches which we were unable to discuss in detail in this chapter. Wavelets are a smoothing procedure similar in spirit to Fourier analysis, which can be used especially for the representation of functions with pronounced (local) variability. In principle, we can also understand wavelets as a basis function approach, but estimation relies on efficient algorithms, which employ special properties of wavelet basis functions, and therefore differ greatly from the methods that we discussed. Wavelets are especially interesting from a numerical point of view but have the disadvantage that complex mathematical tools are required. Hastie et al. (2009) offer a brief introduction to wavelets, while Ogden (1997) or Gençay, Selçuk, and Whitcher (2002) provide more challenging descriptions.

Another class of nonparametric regression models is considered in the context of functional data analysis. In such a setting, the parameters and/or the response variable are viewed as functions, and methods are used that are especially designed for this situation. Ramsay and Silverman (2005) provide a detailed introduction while Ramsay and Silverman (2002) describe functional data analysis within the scope of some case studies.

Banerjee et al. (2003) and Schabenberger and Gotway (2005) are introductory textbooks on spatial statistics, i.e., the methods discussed in Sect. 8.2. Rue and Held (2005), as well as Chiles and Delfiner (1999) and Stein (1999), elaborate on aspects of Gauß–Markov random fields and kriging, respectively. Ecker and Gelfand (2003) and Zimmermann (1993) discuss possibilities as how to consider anisotropy in geostatistical models. Nychka (2000) offers a rigorous mathematical access to the relationship between kriging and smoothing procedures that are based on radial basis functions, while Eilers and Marx (2003) and Dierckx (1993) discuss bivariate polynomial splines in more detail.

Structured Additive Regression

9

In the previous chapter, we illustrated how to flexibly model and estimate the effect of a continuous covariate z on the response variable y without specifying a restrictive functional form of the effect $f(z)$. We also showed how to extend the concepts for two continuous covariates z_1 and z_2 or for a location variable s. As shown in Chap. 3 through various examples, a linear effect was reasonable for a considerable number of the available covariates. However, there are often one or more continuous covariates $z_1, \dots, z_q$, whose effects cannot, at least not a priori, be described with a simple functional form. Hence, we are generally interested in flexibly modeling the effect of these covariates in form of a function $f(z_1, \dots, z_q)$. As seen in Sect. 8.3, estimation of these high-dimensional functions is problematic and requires a very large sample size (to circumvent the curse of dimensionality). Therefore, we often assume a more restrictive additive structure for the effect of the covariates, i.e.,

$$f(z_1, \dots, z_q) = f_1(z_1) + \dots + f_q(z_q).$$

Moreover, models with nonlinear interaction effects or geoadditive models, which consider spatial information in form of a location variable, or models with random effects are also of interest. Sections 9.1–9.4 discuss these models for standard regression problems with continuous responses and approximately normally distributed errors. The models can be formulated in a unified framework as structured additive regression models and can be generalized for non-normal and discrete response variables; compare Sects. 9.5 and 9.6. We will primarily concentrate on the penalization methods and the corresponding Bayesian approaches that were outlined in Sects. 8.1 and 8.2, as they allow for unified estimation of the model parameters, but will also consider boosting in Sect. 9.7. Section 9.8 presents an extensive case study to summarize various aspects of regression modeling in general and structured additive regression in particular.

L. Fahrmeir et al., *Regression*, DOI 10.1007/978-3-642-34333-9_9,

9.1 Additive Models

We assume a data setting which is similar to that of linear regression. Consider observations $(y_i, x_{i1}, \ldots, x_{ik})$, $i = 1, \ldots, n$, of a continuous response variable y and covariates $x_1, \ldots, x_k$, whose effect on y can be modeled through a linear predictor. Additionally, we have observations $(z_{i1}, \ldots, z_{iq})$, $i = 1, \ldots, n$, of *continuous* covariates $z_1, \ldots, z_q$ whose effects are to be modeled and analyzed nonparametrically. For instance in the rent index application, y represents the net rent per square meter, z_1 the living area, z_2 the year of construction, and the covariates $x_1, \ldots, x_k$ the binary coded specific characteristics of the apartment, as well as the experts' assessment of the location in three categories: average, good, and top location.

Additive models are defined by

$$\begin{aligned} y_i &= f_1(z_{i1}) + \ldots + f_q\left(z_{iq}\right) + \beta_0 + \beta_1 x_{i1} + \ldots + \beta_k x_{ik} + \varepsilon_i \\ &= f_1(z_{i1}) + \ldots + f_q\left(z_{iq}\right) + \eta_i^{lin} + \varepsilon_i \\ &= \eta_i^{add} + \varepsilon_i \end{aligned} \tag{9.1}$$

with

$$\eta_i^{lin} = \beta_0 + \beta_1 x_{i1} + \ldots + \beta_k x_{ik}, \qquad \eta_i^{add} = f_1(z_{i1}) + \ldots + f_q\left(z_{iq}\right) + \eta_i^{lin}.$$

The functions $f_1(z_1), \ldots, f_q(z_q)$ are nonlinear smooth effects of the covariates $z_1, \ldots, z_q$, and are modeled and estimated in a nonparametric way (see Sect. 9.6). In principle, all scatter plot smoothers from Sect. 8.1 can be used as building blocks. At this point, however, we limit ourselves to penalization methods and their corresponding Bayesian approaches.

As already mentioned in Sect. 2.6, additive models have an *identification problem*. More specifically, if we add a constant $a \neq 0$ to the function $f_1(z_1)$ and subtract at the same time a from a second function $f_2(z_2)$, the sum

$$f_1(z_1) + f_2(z_2) = f_1(z_1) + a + f_2(z_2) - a$$

remains the same, i.e., the predictor η does not change if $f_1(z_1)$ changes to $\tilde{f}_1(z_1) = f_1(z_1) + a$ and $f_2(z_2)$ changes to $\tilde{f}_2(z_2) = f_2(z_2) - a$. Although the functional form of $f_1(z_1)$ and $\tilde{f}_1(z_1)$ or $f_2(z_2)$ and $\tilde{f}_2(z_2)$ is unchanged, the overall level of the functions is arbitrary without imposing further restrictions. Hence it is necessary to fix the level of the functions. This is usually obtained by "centering the functions around zero," such that

$$\sum_{i=1}^{n} f_1(z_{i1}) = \ldots = \sum_{i=1}^{n} f_q(z_{iq}) = 0$$

holds.

9.1 Standard Additive Regression Model

Data

$(y_i, z_{i1}, \ldots, z_{iq}, x_{i1}, \ldots, x_{ik})$, $i = 1, \ldots, n$, with y and $x_1, \ldots, x_k$ as in the linear regression model and additional continuous covariates $z_1, \ldots, z_q$.

Model

$$y_i = f_1(z_{i1}) + \ldots + f_q(z_{iq}) + \beta_0 + \beta_1 x_{i1} + \ldots + \beta_k x_{ik} + \varepsilon_i.$$

The functions $f_1(z_1), \ldots, f_q(z_q)$ are assumed to be smooth nonlinear effects of the continuous covariates $z_1, \ldots, z_q$. The same assumptions are made for the errors ε_i as with the classical linear model.

For the standard additive model, the same error assumptions are made as in the classical linear model, i.e., the errors ε_i are i.i.d. and normally distributed with $\mathrm{E}(\varepsilon_i) = 0$, $\mathrm{Var}(\varepsilon_i) = \sigma^2$. As with the linear model, the assumptions regarding the error variables ε_i carry over to the response variables y_i, i.e., the y_i are (conditionally) independent with

$$\begin{aligned} \mathrm{E}(y_i) &= \mu_i = f_1(z_{i1}) + \ldots + f_q(z_{iq}) + \beta_0 + \beta_1 x_{i1} + \ldots + \beta_k x_{ik}, \\ \mathrm{Var}(y_i) &= \sigma^2 \end{aligned}$$

and, if applicable, normally distributed with

$$y_i \sim \mathrm{N}(\mu_i, \sigma^2).$$

Analogous to linear models, more general assumptions, e.g., heteroscedastic errors, are possible. However, we do not pursue these extensions here.

The literature often refers the special case

$$y_i = f_1(z_{i1}) + \ldots + f_q(z_{iq}) + \varepsilon_i, \tag{9.2}$$

without the additional linear predictor η_i^{lin} as the additive model. The model (9.1) is then called *partial linear model* or *semiparametric model*.

In either case, additive models (9.1) or (9.2) do not contain interactions between covariates. Models with interactions will be discussed in Sect. 9.3.

The nonlinear and linear effects can be interpreted as described already in Sect. 2.6. Further illustration is provided with the rent index data, as we will see in Example 9.1 below.

For modeling and estimation of nonlinear functions, this chapter primarily focuses on penalization and Bayesian methods with basis functions as summarized in Sect. 8.1.3. Hence, the functions f_j, $j = 1, \ldots, q$, will be approximated by

$$f_j(z_j) = \sum_{l=1}^{d_j} \gamma_{jl} B_l(z_j),$$

where the basis functions B_l represent TP- or B-spline bases of polynomial splines, smoothing splines, or even kriging methods. In principle, it is possible to choose different types of basis functions for the different functions f_j, e.g., polynomial splines of different order. For simplicity in presentation, such variations are notationally suppressed. The vector $\boldsymbol{f}_j = (f_j(z_{1j}), \ldots, f_j(z_{nj}))'$ of function values evaluated at the observed covariate values $z_{1j}, \ldots, z_{nj}$ can then be expressed as

$$\boldsymbol{f}_j = \boldsymbol{Z}_j \boldsymbol{\gamma}_j,$$

where $\boldsymbol{\gamma}_j = (\gamma_{j1}, \ldots, \gamma_{jd_j})'$ is the vector of regression coefficients. The design matrix $\boldsymbol{Z}_j$ consists of the basis functions evaluated at the observed covariate values, i.e., $\boldsymbol{Z}_j[i, l] = B_l(z_{ij})$.

As in the linear model, $\boldsymbol{y} = (y_1, \ldots, y_n)'$ defines the vector of the response values, $\boldsymbol{\varepsilon} = (\varepsilon_1, \ldots, \varepsilon_n)'$ the vector of the error values, $\boldsymbol{\beta} = (\beta_0, \ldots, \beta_k)'$ the vector of regression coefficients for the linear part of the predictor, and $\boldsymbol{X}$ the corresponding design matrix. The additive model can then be written in matrix notation in the form

$$\boldsymbol{y} = \boldsymbol{Z}_1 \boldsymbol{\gamma}_1 + \ldots + \boldsymbol{Z}_q \boldsymbol{\gamma}_q + \boldsymbol{X}\boldsymbol{\beta} + \boldsymbol{\varepsilon}. \tag{9.3}$$

At first glance, Eq. (9.3) appears to be a large linear model. In fact, this would be the case if the coefficient vectors $\boldsymbol{\gamma}_1, \ldots, \boldsymbol{\gamma}_q$ of basis functions are estimated, unrestrictedly, using (ordinary) least squares. In general, however, estimation is regularized by penalties of the form $\lambda_j \boldsymbol{\gamma}_j' \boldsymbol{K}_j \boldsymbol{\gamma}_j$ (or corresponding smoothness priors), as described for a single function f in Sect. 8.1.3, to enforce specific smoothness properties of the estimates and to achieve regularization. The form of the penalty matrix depends on the specific choice of basis functions. Details on how to estimate the model will be provided in Sect. 9.6.

As in the example to follow, we will use penalized splines (Sect. 8.1.2) or the corresponding Bayesian version, in which the Gaussian prior $p(\boldsymbol{\gamma}_j \mid \tau_j^2) \propto \exp(-0.5 \boldsymbol{\gamma}_j' \boldsymbol{K}_j \boldsymbol{\gamma}_j / \tau_j^2)$ replaces the penalty function.

Example 9.1 Munich Rent Index—Additive Model

We choose the net rent per square meter as the response variable. As covariates we include the living area (*area*), the year of construction (*yearc*), and the location (*location*). This leads to the additive model

$$rentsqm = f_1(area) + f_2(yearc) + \beta_0 + \beta_1 glocation + \beta_2 tlocation + \varepsilon.$$

9.2 Basis Function Approach to Additive Regression

The functions f_j are approximated using basis functions of the form

$$f_j(z_j) = \sum_{l=1}^{d_j} \gamma_{jl} B_l(z_j).$$

Regardless of the chosen basis, we obtain the additive model

$$\boldsymbol{y} = \boldsymbol{Z}_1\boldsymbol{\gamma}_1 + \ldots + \boldsymbol{Z}_q\boldsymbol{\gamma}_q + \boldsymbol{X}\boldsymbol{\beta} + \boldsymbol{\varepsilon},$$

where the design matrices $\boldsymbol{Z}_1, \ldots, \boldsymbol{Z}_q$ consist of the basis functions evaluated at the given covariate values. The design matrix $\boldsymbol{X}$ is constructed as in the linear model.

The functions f_1 and f_2 are modeled by P-splines with 20 interior knots and second-order difference penalties. For the location, the model includes the dummy variables *glocation* for good locations and *tlocation* for top location. The average location serves as the reference category. Note that the model is a pure main effects model, i.e., we do not consider possible interactions, e.g., between living area and location. As mentioned, such interaction models are discussed in Sect. 9.3.

Figure 9.1 shows the estimated nonlinear effects of living area and year of construction. The results are obtained using `remlreg objects` of the software `BayesX`. In the left panels the partial residuals are additionally included. Their calculation is completely analogous to partial residuals in linear models described in section "Statistical Properties of Residuals" of Sect. 3.2.3. For example, the partial residuals with regard to the apartment's living area are defined by

$$\hat{\varepsilon}_{area,i} = \hat{\varepsilon}_i + \hat{f}_1(area_i).$$

Also see Sect. 9.5 for further details regarding partial residuals in additive models.

As expected, the estimated function of the living area shows a nonlinear decreasing effect on rent with increasing area. For living area greater than about 110 sqm, the width of pointwise confidence intervals increases due to fewer larger apartments. Consequently, we should not over-interpret the slightly increasing effect at the right margin: A horizontal trend, from about 110 sqm onward, easily fits within the confidence limits. Initially, the estimated effect for the year of construction is almost constant. However, from 1945 onward, the effect is almost linearly increasing. All in all, the estimated nonlinear effects are very similar to those found in the parametric model of Example 3.5 (p. 90) in Chap. 3. Nevertheless, the fit to the data is better with the additive model used here. If we rather use the transformation $1/area$ for the living area and a cubic polynomial for modeling the effect of the year of construction, as presented in Example 3.5, we obtain an AIC of 7,380.3. With the nonparametric model, we obtain a lower AIC of 7,362.6 such that the nonparametric model is preferable. The calculation of the AIC and other goodness-of-fit measures for additive models are detailed in Sect. 9.5. In contrast to the parametric model of Example 3.5, the nonparametric approach has yet another distinct advantage: Nonlinear effects are obtained in an automatic way, i.e., it is not necessary to reflect upon, e.g., the type of variable transformation or the order of the polynomial.

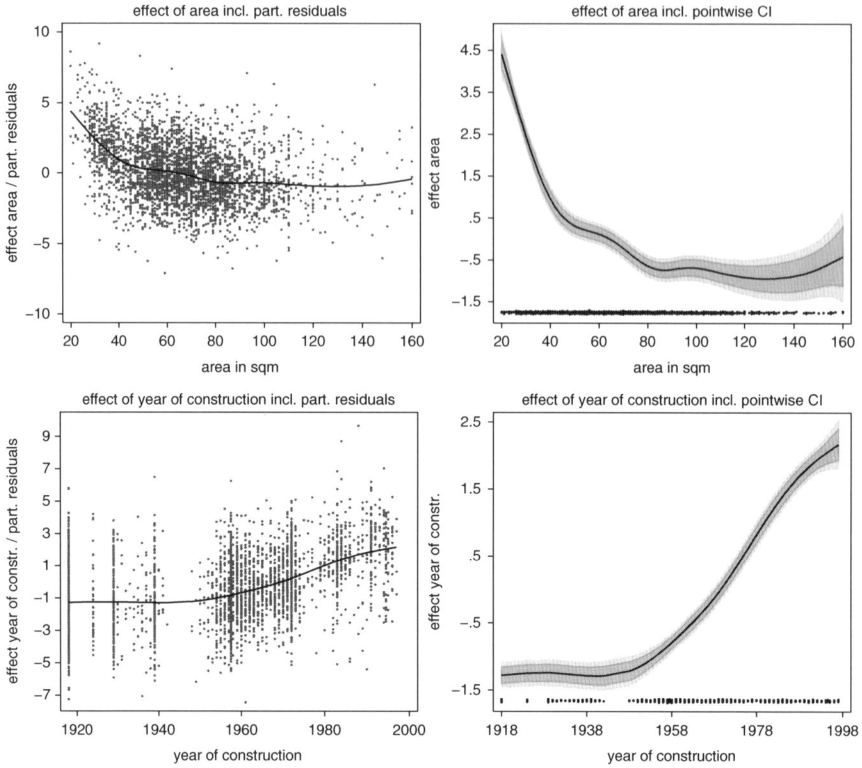

Fig. 9.1 Munich rent index: estimated nonlinear effects of area and year of construction

Table 9.1 Munich rent index: estimated effects of location

Variable	Coefficient	Standard error	p-value	95 % confidence interval	
intercept	7.083	0.120	<0.001	6.848	7.317
glocation	0.633	0.076	<0.001	0.484	0.783
tlocation	1.503	0.233	<0.001	1.046	1.961

Finally, we observe comparably strong and highly significant additional charge of approximately 0.63 Euro for a good location and of approximately 1.5 Euro for top location, compared to average location (see Table 9.1).

△

9.2 Geoadditive Regression

In addition to the three-categorical expert assessment of location, the rent index data also contain information in which of the 336 districts in Munich the apartment is located. If we want to use the more precise location variable *district* rather than the expert assessment, we arrive at the geoadditive model

$$rentsqm = \beta_0 + f_1(area) + f_2(yearc) + f_{geo}(district) + \varepsilon.$$

It is also possible to additionally include the expert assessment in the model. In this case, we obtain the modified geoadditive model

$$\begin{aligned} rentsqm = \; & f_1(area) + f_2(yearc) + f_{geo}(district) + \beta_0 + \beta_1\, glocation \\ & + \beta_2\, tlocation + \varepsilon. \end{aligned}$$

In general, the data for geoadditive models are given through observations $(y_i, \boldsymbol{x}_i, z_i)$, $i = 1, \ldots, n$, for the response variable y and the covariates $\boldsymbol{x}, z$ (as in the additive model), plus the additional values s_i, $i = 1, \ldots, n$, associated with a geographic location index s. As already outlined in Sect. 8.2, the variable s is either discrete, taking the values $s \in \{1, \ldots, d\}$ (see Sect. 8.2.4), or varies continuously in a subset of $\mathbb{R}^2$ (see Sects. 8.2.1–8.2.3). In the discrete case, s_i defines a specific region, in which the individual or the unit i was observed, e.g., the district in Munich, districts in Zambia, counties, and area code districts in Germany. In the continuous case, s is usually given by the coordinates of $\mathbb{R}^2$, i.e., the location s_i of the unit i is (almost) exactly known. In the original sample of the Munich rent index, the exact address of the apartment is provided. However, due to data confidentiality, we are only given the district in which the apartment is located. Data derived from insurance policies are handled in a similar way. The insurance companies know the exact address of the apartment, but due to data confidentiality, this data is aggregated to a discrete location variable, for example, for counties or licensing districts.

In geoadditive models, the predictor is extended by the spatial effect $f_{geo}(s)$ of the location variable s, and we obtain

$$\begin{aligned} y_i &= \eta_i^{add} + f_{geo}(s_i) + \varepsilon_i \\ &= f_1(z_{i1}) + \ldots + f_q(z_{iq}) + f_{geo}(s_i) + \boldsymbol{x}_i'\boldsymbol{\beta} + \varepsilon_i, \end{aligned} \tag{9.4}$$

with $\boldsymbol{x}_i'\boldsymbol{\beta} = \beta_0 + \beta_1 x_{i1} + \ldots + \beta_k x_{ik}$. For the covariates $\boldsymbol{x}_i, z_i$ and the error variables ε_i, the same assumptions apply as for the additive model.

The spatial effect $f_{geo}(s)$ can be understood as a surrogate for unobserved spatial variables not included in the data. The approaches presented in Sect. 8.2 can be used as building blocks for the modeling and estimation of $f_{geo}(s)$. Markov random fields (see Sect. 8.2.4) are especially useful for discrete $s \in \{1, \ldots, d\}$. It is also possible to apply smoothers for continuous s, in particular kriging (see Sect. 8.2.3), e.g., by using the region centroids as coordinates.

For discrete locations, $s_i = s$ denotes that the ith observation belongs to region s, with $s \in \{1, \ldots, d\}$, then $f_{geo}(s_i)$ is the spatial effect of region $s = s_i$. Similar to the basis function approaches discussed in the previous section, we can represent the vector $\boldsymbol{f}_{geo} = (f_{geo}(s_1), \ldots, f_{geo}(s_n))'$ of the spatial effect for the observed units $i = 1, \ldots, n$, as

$$\boldsymbol{f}_{geo} = \boldsymbol{Z}_{geo}\boldsymbol{\gamma}_{geo}. \tag{9.5}$$

Here, $\boldsymbol{\gamma}_{geo} = (\gamma_{geo,1}, \ldots, \gamma_{geo,d})'$ is the vector of regression coefficients and the $n \times d$ design matrix $\boldsymbol{Z}_{geo}$ is an incidence matrix, i.e., with $\boldsymbol{Z}_{geo}[i,s] = 1$ for the sth element of the ith row if $s_i = s$, and 0 otherwise. For a continuous s, $\boldsymbol{f}_{geo}$ can also be written in the form (9.5). The specific forms of $\boldsymbol{Z}_{geo}$ and $\boldsymbol{\gamma}_{geo}$ can be found in Sects. 8.2.1–8.2.3. When using kriging, we usually work with a representative number of knots (see p. 518).

We can now write geoadditive models in matrix notation as

$$\boldsymbol{y} = \boldsymbol{Z}_1\boldsymbol{\gamma}_1 + \ldots + \boldsymbol{Z}_q\boldsymbol{\gamma}_q + \boldsymbol{Z}_{geo}\boldsymbol{\gamma}_{geo} + \boldsymbol{X}\boldsymbol{\beta} + \boldsymbol{\varepsilon}, \tag{9.6}$$

extending the additive model (9.3) by the spatial effect $\boldsymbol{Z}_{geo}\boldsymbol{\gamma}_{geo}$.

Similar as for the coefficient vectors $\boldsymbol{\gamma}_1, \ldots, \boldsymbol{\gamma}_q$, estimation of $\boldsymbol{\gamma}_{geo}$ is again regularized using quadratic penalties of the form $\lambda_{geo}\boldsymbol{\gamma}'_{geo}\boldsymbol{K}_{geo}\boldsymbol{\gamma}_{geo}$ or with the Gaussian smoothness priors as described in Sect. 8.2. See Sect. 9.6 for further details.

Example 9.2 Munich Rent Index—Geoadditive Model

We consider the two geoadditive models

$$rentsqm = f_1(area) + f_2(yearc) + f_{geo}(district) + \beta_0 + \varepsilon \tag{9.7}$$

and

$$rentsqm = f_1(area) + f_2(yearc) + f_{geo}(district) + \beta_0 + \beta_1\, glocation + \beta_2\, tlocation + \varepsilon \tag{9.8}$$

with the spatial effect $f_{geo}(district)$ of the districts in Munich.

Whereas model (9.7) does not consider the dummies for the expert assessment level of location, they are included within the predictor of model (9.8). In both cases, we choose to use kriging with the district centroids as coordinates for modeling the spatial effect $f_{geo}(district)$. Comparing the estimates $\hat{f}_{geo}$ of the two models allows us to evaluate the expert assessment of location. If the experts assessment regarding the apartment's location is valid the location dummies *glocation* and *tlocation* should absorb most of the spatial heterogeneity and the estimated spatial effect $\hat{f}_{geo}$ should be much weaker than in model (9.7).

Figure 9.2 displays the estimated spatial effects. The results are again obtained using `remlreg objects` of `BayesX`. The left panel very well reflects the spatial variation of the rent market in the city of Munich, adjusted to the effects of living area and year of construction. Districts in the southern part of Munich, e.g., Grünwald and Menterschwaige, or those that are close to the Nymphenburg Park and the English Garden, show a rent index that is much higher than the average. Northern districts close to or north of the Frankfurter Ring show a rent index that is much lower than the average. Overall the effect varies between -1 and 1. An effect of, e.g., 0.5 implies that apartments in this particular subquarter are on average 50 cents more expensive than apartments in subquarters with zero effect (keeping all other covariates constant).

If the location dummies, according to the expert assessment, are additionally included in the predictor, the spatially smooth district effect clearly diminishes. The kernel density estimators of the smooth district effects, illustrated in Fig. 9.3, provide a good impression of the decreased variability. Considering the expert assessment, the kernel density estimator of the district effect shows a higher concentration around zero, but a visible variability still remains. Although we find that the expert assessment absorbs a considerable portion of the spatial heterogeneity, there appears to be opportunity for improvement in expert assessment.

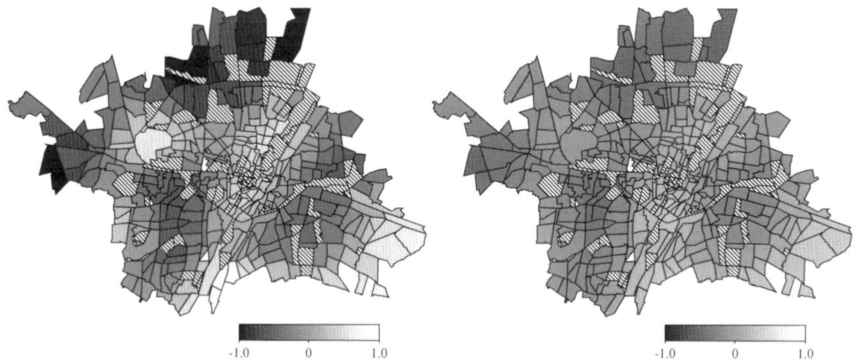

Fig. 9.2 Munich rent index: spatial effect based on kriging. *Left*: expert assessment of location omitted. *Right*: expert assessment of location included

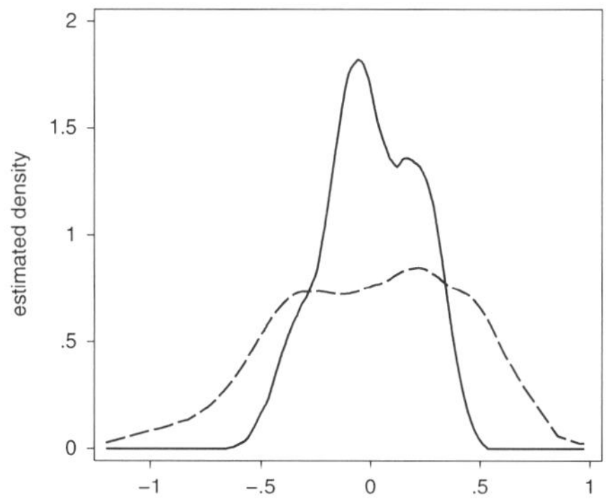

Fig. 9.3 Munich rent index: kernel density estimators for the spatial effect. *Solid line*: model with expert assessment of location included. *Dashed line*: model with expert assessment of location excluded

The quality of the expert assessment of the location can also be seen from the goodness-of-fit measured by the AIC. For model (9.7) we obtain AIC $= 7,359$, while the second model (9.8), yields AIC $= 7,336$. Compared to the model in Example 9.1, where the spatial effect was modeled parametrically with two location dummy variables, the value of the AIC in model (9.7) is only smaller by approximately three units. Only by combining the experts assessment and the nonparametric spatial effect the AIC value significantly decreases by almost 30 units.

△

9.3 Models with Interactions

Section 3.1.3 (p. 98) already illustrated how (pairwise) covariate interactions, along with main effects, can be considered in a linear model. The additive and geoadditive models of the previous sections are purely main effect models with respect to the nonparametric components $f_1, \ldots, f_q$ and the spatial effect f_{geo}. In this section, we will show how these models can be further extended using (pairwise) nonparametric interaction terms. At first, we will consider the case of a continuous covariate z and a binary or categorical covariate x. Due to the specific type of interaction, we also speak of models with varying coefficients. Models of this type can also be modified in a way that an interaction between a location variable s and x is possible. Finally, we will address pairwise interactions between two continuous covariates z_1 and z_2. We use the rent index data for illustration.

9.3.1 Models with Varying Coefficient Terms

In additive models, the interaction between a continuous variable z_1 and a binary variable x_1 can be considered by including the interaction term $f_{z_1|x_1}(z_1) \cdot x_1$. Here, $f_{z_1|x_1}(z_1)$ is a nonparametric smooth function, which similar to the main effect $f_1(z_1)$ needs to be estimated from the data. This results in the model

$$y_1 = f_1(z_{i1}) + \ldots + f_q(z_{iq}) + f_{z_1|x_1}(z_{i1})\, x_{i1} + \boldsymbol{x}_i'\boldsymbol{\beta} + \varepsilon_i, \qquad (9.9)$$

where the same assumptions apply as for the additive model (9.1). Since the model (9.9) also contains the main effects $f_1(z_1)$ and $\beta_1 x_1$, the function $f_{z_1|x_1}$ must be centered around zero in the same way as the other functions $f_1, \ldots, f_q$. Otherwise, the transformations $\tilde{f}_{z_1|x_1} = f_{z_1|x_1}(z_1) + a$ and $\tilde{\beta}_1 = \beta_1 - a$ would result in the same predictor, i.e., the model would not be identifiable. If the main effect $\beta_1 x_1$ is excluded from the model, then centering of $f_{z_1|x_1}$ around zero is not necessary. The varying coefficients model can be interpreted as follows:

- $f_1(z_1)$ is the nonlinear effect of z_1 if $x_1 = 0$.
- $f_1(z_1) + f_{z_1|x_1}(z_1) + \beta_1$ is the nonlinear effect of z_1 if $x_1 = 1$.
- $f_{z_1|x_1}(z_1) + \beta_1$ is a *varying effect* for $x_1 = 1$ (relative to $x_1 = 0$) depending on z_1. The variable z_1 is also called an *effect modifier* of x_1, and x_1 is called the *interaction variable*.

Instead of the conventional fixed coefficient β_1 for x_1, we rather obtain a varying coefficient $f_{z_1|x_1}(z_1) + \beta_1$ depending on z_1. To get an intuitive understanding about the effects, it is useful to visualize both effects $f_1(z_1)$ and $f_1(z_1) + f_{z_1|x_1}(z_1) + \beta_1$ in a combined graph. Additionally, the varying effect $f_{z_1|x_1}(z_1) + \beta_1$ of x_1, depending on z_1, can be plotted in a separate graph.

Similarly, we can define an interaction between z_1 and a three-level categorical variable $x \in \{1, 2, 3\}$, where x is coded by the dummy variables x_1 and x_2, with the last value 3 as the reference category. This leads to the model

$$y_i = f_1(z_{i1}) + \ldots + f_{z_1|x_1}(z_{i1})\, x_{i1} + f_{z_1|x_2}(z_{i1})\, x_{i2} + \beta_0 + \beta_1 x_{i1} + \beta_2 x_{i2} + \ldots + \varepsilon_i.$$

Interpretation and visualization extends similarly: $f_1(z_1) + f_{z_1|x_1}(z_1) + \beta_1$ is the effect of z_1 if $x = 1$, and $f_1(z_1) + f_{z_1|x_2}(z_1) + \beta_2$ is the effect of z_1 if $x = 2$, whereas $f_1(z)$ is the nonlinear effect of z_1 if $x = 3$. For interpretation, it is again advisable to visualize the effects.

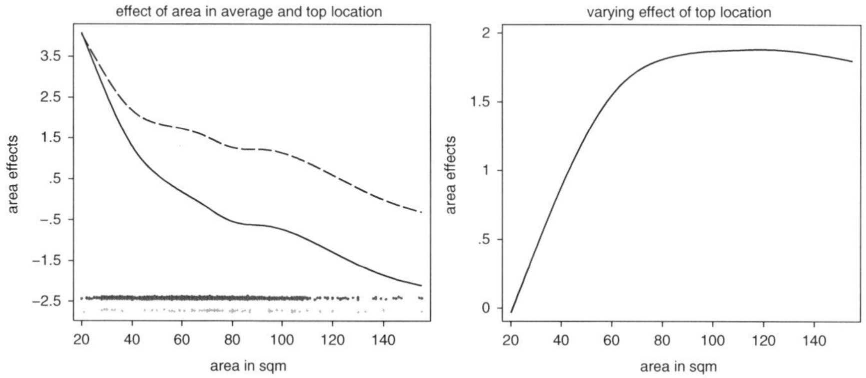

Fig. 9.4 Munich rent index: *left*, effect of area in average (*solid line*) and top location (*dashed line*). The *dots* show the distribution of the observations in average (*dark dots*) and top location (*bright dots*). *Right*, varying effect of top location

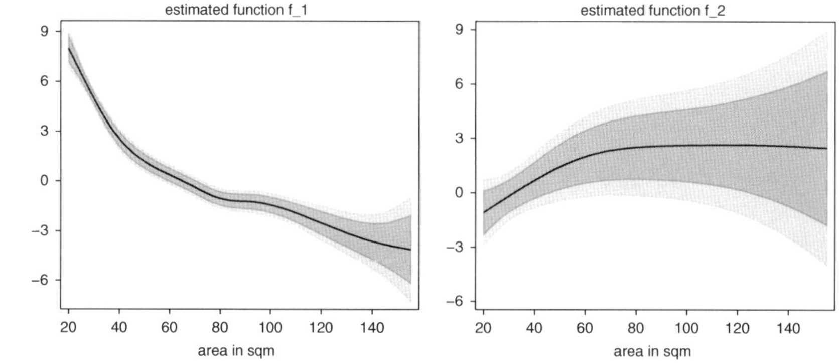

Fig. 9.5 Munich rent index: estimated functions f_1 and f_2 including pointwise confidence intervals in the model (9.10)

Example 9.3 Munich Rent Index—Interaction Between Living Area and Location

This example can be viewed as the nonparametric extension to the strictly parametric Example 3.8 (p. 102). As in Example 3.8, we use only the data for apartments in average location (reference category) and top location and estimate the model

$$rentsqm = f_1(area) + f_2(area)\, tlocation + \beta_0 + \beta_1\, tlocation + \varepsilon. \qquad (9.10)$$

The function $f_1(area)$ can be interpreted as the nonlinear effect of the living area in average location ($tlocation = 0$), and the term $f_1(area) + f_2(area) + \beta_1$ can be interpreted as the nonlinear effect of the living area in top location ($tlocation = 1$). The term $f_2(area) + \beta_1$ represents the varying effect of living area in top location. Figures 9.4 and 9.5 show the estimated effects, as well as the functions $\hat{f}_1$, $\hat{f}_2$. The results are based on `remlreg objects` of `BayesX`.

In the left panel of Fig. 9.4, the estimated effects of the living area for average and top location show the extent of interaction. If the functions are parallel, no interaction is present, and the function $f_2(area)$ (Fig. 9.5, right) should then be (almost) zero. In our case, we do find a relatively strong interaction. For apartments with a small living area, the effect of average and top location is nearly the same. The larger the living area, the more the effects drift apart, i.e., the greater the difference in rent per square meter in average and top locations.

The right panel of Fig. 9.4 shows that the effect of top location is at first small, but then strongly increases as living area increases. From 70 sqm living area onward, the effect remains relatively constant, with an increase of rent of approximately 2 Euro. Hence, the effect of living area clearly depends on location, which can be interpreted as an interaction effect between living area and location.

As mentioned, the model was already estimated with strictly parametric terms in Example 3.8 (p. 102). Recall that the interaction effect was modeled in a linear way. In comparison to this parametric model, we find differences for apartments larger than 70 sqm regarding the varying effect of top location. The AIC is 4,656 in the parametric model and 4,658 in the nonparametric model. Thus the parametric model performs equally well or even slightly better than the nonparametric model.

△

In the case that a basis function approach is used for modeling the functions $f_1, \ldots, f_q$ and $f_{z_1|x_1}$ (and $f_{z_2|x_2}$), we obtain

$$f_{z_1|x_1}(z_1) = \sum_{l=1}^{d} \gamma_{int,l} B_l(z_1) \quad \text{and} \quad f_{z_1|x_1}(z_1) \cdot x_1 = \sum_{l=1}^{d} \gamma_{int,l} B_l(z_1)\, x_1.$$

The vector of the interaction term $\boldsymbol{f}_{int} = (f_{z_1|x_1}(z_{11})\, x_{11}, \ldots, f_{z_1|x_1}(z_{n1})\, x_{n1})'$ can be represented with the vector $\boldsymbol{\gamma}_{int} = (\gamma_{int,1}, \ldots, \gamma_{int,d})'$ of the regression coefficients using

$$\boldsymbol{f}_{int} = \boldsymbol{Z}_{int} \boldsymbol{\gamma}_{int}.$$

The design matrix now contains the values of the basis functions multiplied with the respective x_1 values, i.e., $\boldsymbol{Z}_{int}[i, l] = B_l(z_{i1})\, x_{i1}$. Thus, we can write model (9.9) in matrix notation in the form of a large linear model

$$\boldsymbol{y} = \boldsymbol{Z}_1 \boldsymbol{\gamma}_1 + \ldots + \boldsymbol{Z}_q \boldsymbol{\gamma}_q + \boldsymbol{Z}_{int} \boldsymbol{\gamma}_{int} + \boldsymbol{X}\boldsymbol{\beta} + \boldsymbol{\varepsilon}.$$

This also applies if several interaction or varying coefficient terms are included in the model.

The concept also remains the same, if, instead of z_1, an interaction $f_{geo|x}(s) \cdot x$ between the location variable s and a binary (or multicategorical) variable x is included in geoadditive models. In this case, we speak of *models with spatially varying coefficients* or *geographically weighted regression*.

9.3.2 Interactions Between Two Continuous Covariates

For two continuous covariates z_1 and z_2, nonparametric interactions can be considered by adding a smooth two-dimensional function $f_{1|2}(z_1, z_2)$ to the predictor. There are principally two alternatives to do so:

- *Two-dimensional function without main effects*: The effect of z_1 and z_2 is modeled by the two-dimensional surface $f_{1|2}(z_1, z_2)$. One-dimensional main effects $f_1(z_1)$ and $f_2(z_2)$ are not included in the model. This results in the model

$$y_i = f_{1|2}(z_1, z_2) + f_3(z_{i3}) + \ldots + f_q\left(z_{iq}\right) + \eta_i^{lin} + \varepsilon_i,$$

  where once again the level of the function $f_{1|2}(z_1, z_2)$ is not identified. Thus, the surface needs to be centered around zero, i.e.,

$$\sum_{i=1}^{n} f_{1|2}(z_{i1}, z_{i2}) = 0.$$

  The interpretation of the results is more involved than in models with pure main effects. In the following Example 9.4 we will show how an appropriate presentation of results can help with interpretation.

  Prior to estimating a two-dimensional interaction, we should be sure that we have enough data combinations of z_1 and z_2; a simple scatter plot between z_1 and z_2 can be useful to help determine if this is the case. If there are data gaps in subareas, then the results from these areas should be interpreted carefully. It may be possible that the existing data set is not sufficient to estimate a two-dimensional surface between z_1 and z_2. Then the specification of a pure main effects model may be preferable.
- *Two-dimensional function with main effects:* In this case, we include the two-dimensional function $f_{1|2}(z_1, z_2)$ *in addition* to the one-dimensional main effects $f_1(z_1)$ and $f_2(z_2)$ into the model. We then obtain

$$y_i = f_1(z_{i1}) + f_2(z_{i2}) + f_{1|2}(z_1, z_2) + f_3(z_{i3}) + \ldots + f_q\left(z_{iq}\right) + \eta_i^{lin} + \varepsilon_i.$$

  With the inclusion of the main effects, the problem of identifiability becomes more complicated. The type of constraints for $f_{1|2}$ depends on the choice of basis functions used for the main effects and the interaction and also depends on penalization. In general, identifiability is guaranteed if (in addition to centering all functions around zero) also all slices of the interaction $f_{1|2}(z_1, z_2)$, i.e., all one-dimensional smooths with fixed value of z_1 or z_2, are centered around zero. This approach also facilitates interpretation. The interaction $f_{1|2}(z_1, z_2)$ can then be interpreted as a deviation from the main effects. This modeling variant requires an even larger sample size in order to obtain reliable results. Moreover, a meaningful interpretation of the interaction surface requires some experience.

For modeling $f_{1|2}$, we generally use tensor product-P-splines, radial basis functions, or kriging approaches as described in Sects. 8.2.1–8.2.3. As stated there, the vector

$$\boldsymbol{f}_{1|2} = (f_{1|2}(z_{11}, z_{12}), \ldots, f_{1|2}(z_{n1}, z_{n2}))'$$

of function evaluations can be written as

$$\boldsymbol{f}_{1|2} = \boldsymbol{Z}_{1|2}\boldsymbol{\gamma}_{1|2},$$

using an appropriately defined design matrix $\boldsymbol{Z}_{1|2}$ and with corresponding coefficient vector $\boldsymbol{\gamma}_{1|2}$. We again obtain a large linear model

$$\boldsymbol{y} = \boldsymbol{Z}_1\boldsymbol{\gamma}_1 + \ldots + \boldsymbol{Z}_q\boldsymbol{\gamma}_q + \boldsymbol{Z}_{1|2}\boldsymbol{\gamma}_{1|2} + \boldsymbol{X}\boldsymbol{\beta} + \boldsymbol{\varepsilon},$$

where the coefficients $\boldsymbol{\gamma}_1, \ldots, \boldsymbol{\gamma}_q, \boldsymbol{\gamma}_{1|2}$ need to be estimated while considering the penalty restrictions; refer to Sect. 9.6.

Example 9.4 Munich Rent Index—Interaction Between Living Area and Year of Construction

In the geoadditive model of Example 9.2, we replace the two main effects of the living area and the year of construction with a two-dimensional surface $f_{1|2}(area, yearc)$ and obtain the model

$$rentsqm = f_{1|2}(area, yearc) + f_{geo}(district) + \beta_0 + \beta_1\, glocation + \beta_2\, tlocation + \varepsilon.$$

As in Example 8.6 (p. 500), the two-dimensional function $f_{1|2}$ is modeled using tensor product P-splines. First- and second-order differences were considered for penalization. Estimates were carried out using `remlreg objects` of `BayesX`. With first differences, we obtain an AIC of 7,317, whereas second-order differences yield an AIC of 7,304. In either case, we find that the AIC is lower than what was found in the geoadditive model of Example 9.2 (AIC $= 7,336$). In the following, we limit our comparison to the models based on second-order difference penalties.

Before we interpret the results, we first have a look at the distribution of the observations across the living area and year of construction, as seen in the left panel of Fig. 9.6. The plot illustrates the potential difficulties when estimating two-dimensional covariate effects. It appears that we do not have enough observations for apartments with a living area smaller than 30 or greater than 110 sqm to make reliable statements. In these regions, poorly interpretable effects would not be surprising. For apartments with a living space between 30 and 110 sqm, the amount of data appears to be sufficient (Fig. 9.6, right). Consequently, we limit the interpretation to this interval. In practice we mainly have the choice between four alternatives: The collection of additional data, the consideration of additional data from previous studies, limiting the validity of the rent index to apartments with a living area between 30 and 110 sqm, or the use of models with less complexity (in particular models with only main effects). Perhaps the last choice is one of our best alternatives.

We limit the presentation of results to the interaction between the living area and year of construction. In comparison to Example 9.2, the spatial effects remained almost unchanged. Figure 9.7 (left) shows the estimated two-dimensional surface of the living area and the year of construction. The interpretation of two-dimensional surfaces can be quite involved. A helpful device is often to plot *slices* of the two-dimensional function. The right panel of Fig. 9.7 shows various slices of the living area for the year of construction in 1918, 1957, 1972, and 1991, respectively. The slices show whether or not an interaction is necessary.

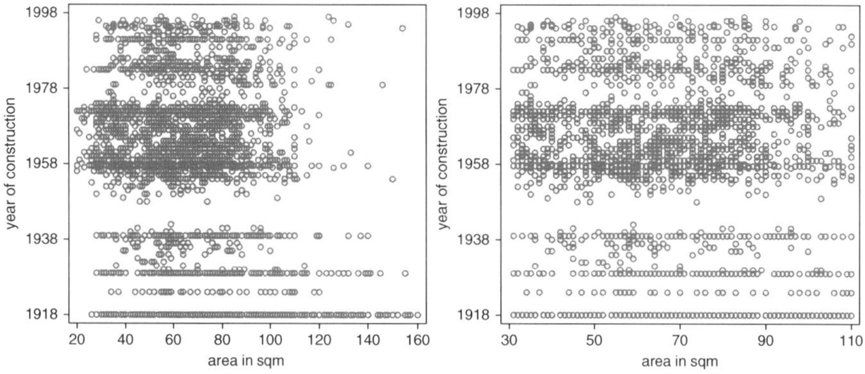

Fig. 9.6 Munich rent index: distribution of area and year of construction. *Left*, complete dataset. *Right*, only observations with area between 30 and 110 square meters

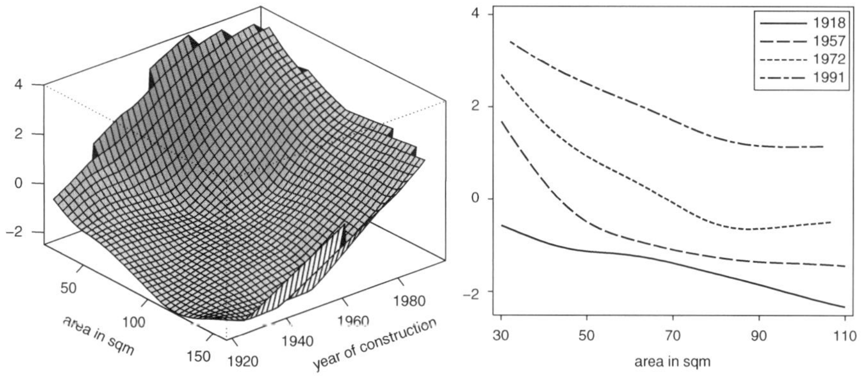

Fig. 9.7 Munich rent index: estimated two-dimensional function $\hat{f}_{1|2}(area, yearc)$ (*left*) and slices for various years of construction (*right*)

If the slices run nearly parallel to each other, a noteworthy interaction does not exist, and a model with the two main effects may be sufficient. In our case, the slices do not run parallel to each other and the inclusion of an interaction seems justified. The lower AIC values for the interaction models compared to that of the main effects model also support our conclusion. The slices nicely show that smaller apartments have a larger interaction effect. The decrease of the average rent per sqm is much weaker for older and more modern apartments than for apartments intermediate in age. It is interesting to note that for very old apartments the effect of the living area is almost linear.

△

9.4 Models with Random Effects

In Chap. 7, we discussed random intercept and slope models for longitudinal and cluster data. In this section we show how the semiparametric models considered thus far can be extended in a straightforward way by random coefficient terms. We will show that such terms have the same structure as nonlinear functions of continuous covariates or spatial effects in a geoadditive model.

Suppose we have longitudinal data for individuals $i = 1, \dots, m$ at observed times $t_{i1} < \dots < t_{ij} < \dots < t_{in_i}$ or cluster data for subjects $j = 1, \dots, n_i$ in clusters $i = 1, \dots, m$ as in Chap. 7.

Consider the simple random coefficient model

$$y_{ij} = \gamma_{0i} + \gamma_{1i} x_{ij} + \beta_0 + \beta_1 x_{ij} + \varepsilon_{ij}, \quad i = 1, \dots, m, \quad j = 1, \dots, n_i,$$

with fixed intercept β_0 and slope β_1 of covariate x, cluster-specific random intercept γ_{0i}, and random slope γ_{1i}. We restrict ourselves to a priori independent random intercepts $\gamma_{0i} \sim \mathrm{N}(0, \tau_0^2)$ and slopes $\gamma_{1i} \sim \mathrm{N}(0, \tau_1^2)$. Similarly as in Chap. 7 we express the model in matrix notation in the form

$$\boldsymbol{y}_i = \mathbf{1}_i \gamma_{0i} + \boldsymbol{x}_i \gamma_{1i} + \mathbf{1}_i \beta_0 + \boldsymbol{x}_i \beta_1 + \boldsymbol{\varepsilon}_i,$$

where $\boldsymbol{y}_i = (y_{i1}, \dots, y_{in_i})'$, $\mathbf{1}_i = (1, \dots, 1)'$ is a n_i-dimensional vector of ones, $\boldsymbol{x}_i = (x_{i1}, \dots, x_{in_i})'$, $\boldsymbol{\varepsilon}_i = (\varepsilon_{i1}, \dots, \varepsilon_{in_i})'$. Defining $\boldsymbol{\gamma}_0 = (\gamma_{01}, \dots, \gamma_{0i}, \dots, \gamma_{0m})'$ and $\boldsymbol{\gamma}_1 = (\gamma_{11}, \dots, \gamma_{1i}, \dots, \gamma_{1m})'$ as the vectors of random intercepts and random slopes, respectively, $\boldsymbol{\beta} = (\beta_0, \beta_1)'$, and the design matrices

$$\boldsymbol{Z}_0 = \begin{pmatrix} \mathbf{1}_1 & & & & \mathbf{0} \\ & \ddots & & & \\ & & \mathbf{1}_i & & \\ & & & \ddots & \\ \mathbf{0} & & & & \mathbf{1}_m \end{pmatrix} \qquad \boldsymbol{Z}_1 = \begin{pmatrix} \boldsymbol{x}_1 & & & & \mathbf{0} \\ & \ddots & & & \\ & & \boldsymbol{x}_i & & \\ & & & \ddots & \\ \mathbf{0} & & & & \boldsymbol{x}_m \end{pmatrix}$$

as well as

$$\boldsymbol{X} = \begin{pmatrix} \mathbf{1}_1 & \boldsymbol{x}_1 \\ \vdots & \vdots \\ \mathbf{1}_i & \boldsymbol{x}_i \\ \vdots & \vdots \\ \mathbf{1}_m & \boldsymbol{x}_m \end{pmatrix} \qquad \boldsymbol{y} = \begin{pmatrix} \boldsymbol{y}_1 \\ \vdots \\ \boldsymbol{y}_i \\ \vdots \\ \boldsymbol{y}_m \end{pmatrix} \qquad \boldsymbol{\varepsilon} = \begin{pmatrix} \boldsymbol{\varepsilon}_1 \\ \vdots \\ \boldsymbol{\varepsilon}_i \\ \vdots \\ \boldsymbol{\varepsilon}_m \end{pmatrix},$$

we obtain

$$\boldsymbol{y} = \boldsymbol{Z}_0 \boldsymbol{\gamma}_0 + \boldsymbol{Z}_1 \boldsymbol{\gamma}_1 + \boldsymbol{X}\boldsymbol{\beta} + \boldsymbol{\varepsilon}.$$

Note the close relationship of random slopes with varying coefficient terms. The effect of x varies with respect to the clusters $i = 1, \dots, m$ as the effect modifier.

As pointed out in Sect. 7.1.1 on p. 354, random coefficient models have an alternative penalized least squares view. Specifically, for fixed variance parameters τ_0^2 and τ_1^2, the parameters of the model can be estimated by minimizing the penalized least squares criterion

$$PLS(\boldsymbol{\gamma}_0, \boldsymbol{\gamma}_1, \boldsymbol{\beta}) = (\boldsymbol{y} - \boldsymbol{Z}_0\boldsymbol{\gamma}_0 - \boldsymbol{Z}_1\boldsymbol{\gamma}_1 - \boldsymbol{X}\boldsymbol{\beta})'(\boldsymbol{y} - \boldsymbol{Z}_0\boldsymbol{\gamma}_0 - \boldsymbol{Z}_1\boldsymbol{\gamma}_1 - \boldsymbol{X}\boldsymbol{\beta})$$
$$+\lambda_0\boldsymbol{\gamma}_0'\boldsymbol{K}_0\boldsymbol{\gamma}_0 + \lambda_1\boldsymbol{\gamma}_1'\boldsymbol{K}_1\boldsymbol{\gamma}_1,$$

where $\boldsymbol{K}_0 = \boldsymbol{K}_1 = \boldsymbol{I}$.

Summarizing, we can express random intercepts and slopes in exactly the same form as nonlinear functions of continuous covariates or the spatial effects in geoadditive models. They can be expressed in the form $\boldsymbol{Z}\boldsymbol{\gamma}$ with specific penalty $\boldsymbol{\gamma}'\boldsymbol{K}\boldsymbol{\gamma} = \boldsymbol{\gamma}'\boldsymbol{I}\boldsymbol{\gamma}$ or in a Bayesian approach with prior $p(\boldsymbol{\gamma} \mid \tau^2) \propto \exp(-0.5\boldsymbol{\gamma}'\boldsymbol{K}\boldsymbol{\gamma}/\tau^2) = \exp(-0.5\boldsymbol{\gamma}'\boldsymbol{I}\boldsymbol{\gamma}/\tau^2)$. In contrast to most other smoothing priors, the random effects priors are not rank deficient.

The next example shows how nonlinear covariate effects and random coefficients can be combined quite naturally in a single regression model with an additive predictor. Such models are often called *additive mixed models.*

Example 9.5 Sales of Orange Juice—Additive Price Effects

This example continues the case study on linear mixed models in Sect. 7.7 (p. 401). There, we estimated the linear mixed model

$$\begin{aligned} y_{ij} &= \beta_0 + \beta_1 \cdot logp_{ij} + \beta_2 \cdot logpc1_{ij} + \beta_3 \cdot logpc2_{ij} + \beta_4 \cdot logpc3_{ij} \\ &\quad +\gamma_{0i} + \gamma_{1i} \cdot logp_{ij} + \gamma_{2i} \cdot logpc1_{ij} + \gamma_{3i} \cdot logpc2_{ij} + \gamma_{4i} \cdot logpc3_{ij} + \varepsilon_{ij} \end{aligned} \tag{9.11}$$

with fixed effects for the (centered) log own-item price and the log-cross prices of competitors, a store-specific random intercept γ_{1i} and random slopes $\gamma_{2i} - \gamma_{4i}$ of the log prices for the $m = 81$ stores, $i = 1, \ldots, 81$.

A somewhat unsatisfactory aspect of this model is that the type of nonlinearity of the price effects is determined in advance using a log-transformation of prices. To investigate whether the log-transformation is really appropriate, we can replace the linear fixed effects in Eq. (9.11) by P-splines leading to the model

$$\begin{aligned} y_{ij} &= f_1(logp_{ij}) + f_2(logpc1_{ij}) + f_3(logpc2_{ij}) + f_4(logpc3_{ij}) \\ &\quad +\gamma_{0i} + \gamma_{1i} logp_{ij} + \gamma_{2i} logpc1_{ij} + \gamma_{3i} logpc2_{ij} + \gamma_{4i} logpc3_{ij} + \beta_0 + \varepsilon_{ij}. \end{aligned} \tag{9.12}$$

This is a model that combines nonlinear covariate effects with random coefficients. Since models with nonlinear covariate effects and additional random effects are well suited for a Bayesian approach using MCMC for inference, we use a fully Bayesian approach and `bayesreg objects` of `BayesX` for estimation. See Sects. 8.1.9 (p. 486) and 9.6.3 below for details. We assume a priori a monotonically decreasing effect of the own price and monotonically increasing effects of cross prices. Imposing monotonicity constraints is relatively straightforward in a Bayesian approach; see Brezger and Steiner (2008) for

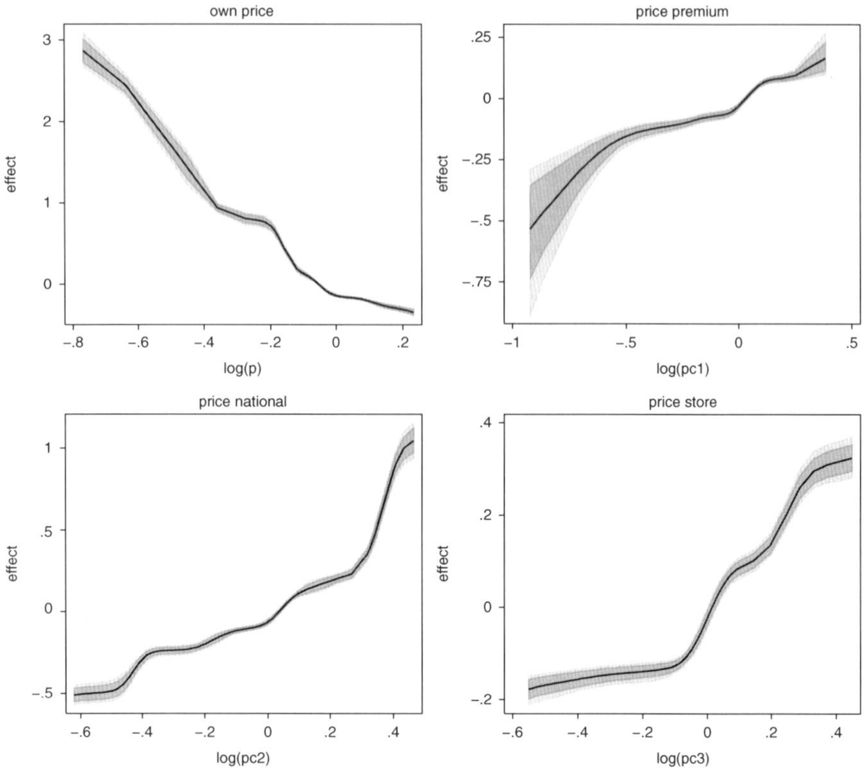

Fig. 9.8 Sales of orange juice: estimated price effects in the model (9.12). Shown is the posterior mean together with 80 % and 95 % pointwise credible intervals

details. If the log-transformation is sufficient to capture the nonlinearity, the estimated curves $\hat{f}_1, \ldots, \hat{f}_4$ should be approximately linear.

In matrix notation model (9.12) can be expressed in additive form as

$$y = Z_1\gamma_1 + \ldots + Z_9\gamma_9 + X\beta + \varepsilon,$$

where $Z_1\gamma_1$ to $Z_4\gamma_4$ correspond to the P-spline smooth terms of the own-item price and cross prices, $Z_5\gamma_5$ corresponds to the random intercept, $Z_6\gamma_6$ to $Z_9\gamma_9$ correspond to the random slopes of the price effects, and $X\beta$ corresponds to the fixed effects. The matrices Z_1 to Z_4 are the standard P-spline design matrices, as outlined in Sect. 8.1.2.

Figure 9.8 shows the estimated price effects. Although considerable additional nonlinearity remains, the plots reveal that the log-transformation is a reasonable approximation to the nonlinear effects. The size of the random effects measured through the random effects variances is in the same range as in the parametric model of Sect. 7.7.

To demonstrate the outlet-specific heterogeneity, Fig. 9.9 shows some outlet-specific log-sales own-price curves which are now additively composed of the random intercept, the nonlinear log-price effect f_1, and the linear log-price random effect $\gamma_{1i} logp_{ij}$.

△

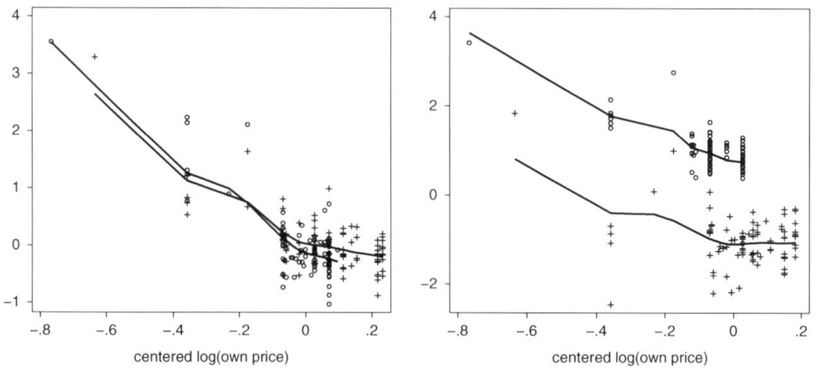

Fig. 9.9 Sales of orange juice: log-sales versus (*centered*) log-price curves in model (9.12) for four different outlets, together with corresponding partial residuals

9.5 Structured Additive Regression

The various terms within the model classes considered thus far can be combined arbitrarily. For example, a particular model could include all main effects, a spatial effect, and one or more terms with varying coefficients for modeling interactions. As shown in the previous section, it is also possible that additional random effects can be included as in mixed models. The resulting models can be described in a unified and general form as *structured additive regression (STAR) models*,

$$y = f_1(v_1) + \ldots + f_q(v_q) + \beta_0 + \beta_1 x_1 + \ldots + \beta_k x_k + \varepsilon, \tag{9.13}$$

where $v_1, \ldots, v_q$ are one- or multidimensional covariates of different types constructed from the original variables. The functions can also be of different structure. Specific types of covariates and functions yield the following special cases:

$f_1(v_1) = f_1(z_1)$,	$v_1 = z_1$,	nonlinear effect of z_1,
$f_2(v_2) = f_{geo}(s)$,	$v_2 = s$,	spatial effect of location variable s,
$f_3(v_3) = f(z)\,x$,	$v_3 = (z, x)$,	varying effect of x with z,
$f_4(v_4) = f_{1\|2}(z_1, z_2)$,	$v_4 = (z_1, z_2)$,	nonlinear interaction between z_1 and z_2,
$f_5(v_5) = \gamma_i$,	$v_5 = i$	individual-specific random intercept,
$f_6(v_6) = \gamma_i u$,	$v_6 = (u, i)$,	individual-specific random slope of u.

Recall that it is also possible to express the linear part $\beta_0 + \beta_1 x_1 + \ldots + \beta_k x_k = \boldsymbol{x}'\boldsymbol{\beta}$ of the predictor as the function $f_0(\boldsymbol{v}_0) = \boldsymbol{x}'\boldsymbol{\beta}$, where $\boldsymbol{v}_0 = (1, x_1, \ldots, x_k)'$. Since

Table 9.2 Overview of commonly used model terms in STAR models

Term type	Design matrix $\boldsymbol{V}$	Penalty matrix $\boldsymbol{K}$	Section
P-spline	Basis functions evaluated at the observations	$\boldsymbol{K} = \boldsymbol{D}_r'\boldsymbol{D}_r$, with $\boldsymbol{D}_r$ a rth order difference matrix	8.1.2
General basis functions approach	Basis functions evaluated at the observations	Based on the integral of squared second derivatives	8.1.3
Varying coefficient term	Basis functions multiplied with the interaction variable	$\boldsymbol{K} = \boldsymbol{D}_r'\boldsymbol{D}_r$ as for P-splines	9.3.1
2D-P-spline	2D-basis functions evaluated at the observations	$\boldsymbol{K} = \boldsymbol{I} \otimes \boldsymbol{K}_1 + \boldsymbol{K}_2 \otimes \boldsymbol{I}$ with identity matrix $\boldsymbol{I}$ and penalty matrices $\boldsymbol{K}_1$ and $\boldsymbol{K}_2$ as for univariate P-splines	8.2.1
Kriging	Basis functions based on the correlation function	$\boldsymbol{K} = \boldsymbol{R}$, with correlation matrix $\boldsymbol{R}$	8.1.6, 8.2.3
Markov random field	0/1-incidence matrix, that links observations and regions	$\boldsymbol{K}$ = neighborhood matrix	8.2.4
Random intercept	0/1-incidence matrix, that links observations and clusters	$\boldsymbol{K} = \boldsymbol{I}$, with identity matrix $\boldsymbol{I}$	Chap. 7, 9.4

the linear part, or at least the constant β_0, is always part of the model, it is explicitly listed in Eq. (9.13).

If the functions $f_1, \ldots, f_q$ are modeled and estimated with basis functions, we obtain a large linear model, which is in the spirit of Eilers and Marx (2002). Using the vectors $\boldsymbol{f}_1, \ldots, \boldsymbol{f}_q$ of functions evaluated at the observations of the covariates $v_1, \ldots, v_q$, we obtain

$$\boldsymbol{y} = \boldsymbol{f}_1 + \ldots + \boldsymbol{f}_q + \boldsymbol{X}\boldsymbol{\beta} + \boldsymbol{\varepsilon} = \boldsymbol{V}_1\boldsymbol{\gamma}_1 + \ldots + \boldsymbol{V}_q\boldsymbol{\gamma}_q + \boldsymbol{X}\boldsymbol{\beta} + \boldsymbol{\varepsilon},$$

with appropriately defined design matrixes $\boldsymbol{V}_j$ and coefficient vectors $\boldsymbol{\gamma}_j$. The smoothness of the functions will be regularized with penalties of the form $\lambda_j \boldsymbol{\gamma}_j' \boldsymbol{K}_j \boldsymbol{\gamma}_j$, with smoothing parameter λ_j and penalty matrix $\boldsymbol{K}_j$. In a fully Bayesian approach, each penalty corresponds to a smoothness prior of the form

$$p(\boldsymbol{\gamma}_j|\tau_j^2) \propto \left(\frac{1}{\tau^2}\right)^{\mathrm{rk}(\boldsymbol{K}_j)/2} \exp\left(-\frac{1}{2\tau_j^2}\boldsymbol{\gamma}_j'\boldsymbol{K}_j\boldsymbol{\gamma}_j\right).$$

Thus every model term in STAR models is characterized by the form of the design matrix $\boldsymbol{V}_j$ and the penalty matrix $\boldsymbol{K}_j$. Table 9.2 gives an overview of commonly used terms. Details of parameter estimation will be provided in the next section.

The unified concept of STAR models can also be extended to non-normal regression settings, in particular binary, discrete, and categorical responses. As with generalized linear models, we assume that the response variables y_i are (conditionally) independent given the predictor η_i. The structured additive predictor

$$\eta_i^{struct} = f(v_{i1}) + \ldots + f(v_{iq}) + \boldsymbol{x}_i'\boldsymbol{\beta}$$

is linked to the (conditional) mean $\mathrm{E}(y_i) = \mu_i$ through

$$\mathrm{E}(y_i) = \mu_i = h(\eta_i^{struct}),$$

with the response function h. We refer to the resulting model class as *generalized STAR models*.

This generalized model class comprises all model classes discussed so far as special cases. For example, choosing

$$\eta_i^{struct} = \eta_i^{add} = f_1(z_{i1}) + \cdots + f_q(z_{iq}) + \boldsymbol{x}_i'\boldsymbol{\beta}$$

results in a *generalized additive model* (GAM). With

$$\eta_i^{struct} = \eta_i^{add} + f_{geo}(s_i),$$

we obtain *generalized geoadditive models*, and so on.

Example 9.6 Vehicle Insurance—Claim Frequency

We have already discussed examples of generalized STAR models in Chap. 2, as seen in the Examples 2.12, 2.13, and 2.15. In Example 2.15 (p. 57), we modeled the claim frequency of a vehicle insurance policy using a *structured additive Poisson model* with the predictor

$$\eta^{struct} = f_1(age) + f_2(age\_v) + f_3(hp) + f_4(bm) + f_5(district) + \beta_0 + \beta_1 gender + \ldots.$$

In a case study, Denuit and Lang (2005) modeled an additional interaction between gender and policyholder's age in form of an age-varying gender effect. This yields the extended structured additive predictor

$$\begin{aligned}\eta^{struct} = {} & f_1(age) + f_2(age\_v) + f_3(hp) + f_4(bm) + f_5(district) \\ & + f_6(age, gender) + \beta_0 + \beta_1 gender + \ldots,\end{aligned}$$

where the interaction f_6 has the form $f_6(age, gender) = f(age) \cdot gender$ of a varying coefficient term with a nonparametric function f of age. All nonlinear functions are modeled as P-splines with 20 interior knots and second-order difference penalty. The district effect is based on a Markov random field. All results are based on mixed model technology as will be outlined in Sect. 9.6.2. We used `relmreg objects` of `BayesX` for estimation.

Figure 9.10 (left panel) shows the estimated age effect $\hat{f}_1$ for women and the age effect $\hat{f}_1 + \hat{f}_6 + \hat{\beta}_1$ for men ($gender = 1$). The right panel of Fig. 9.10 displays the effect $\hat{f}_6 + \hat{\beta}_1$ of men compared to women varying with respect to age.

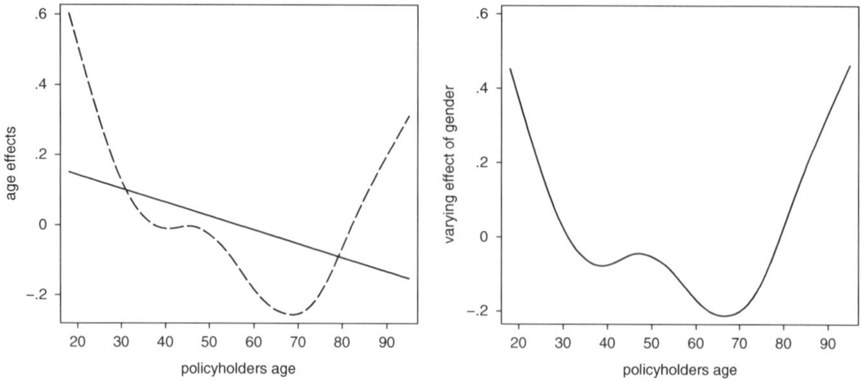

Fig. 9.10 Vehicle insurance: The *left panel* shows the effect of the policyholders age for females (*solid line*) and males (*dashed line*). The *right panel* shows the effect of males in comparison to females varying with respect to age

As women get older, the average damage frequency decreases linearly. For men we obtain the nonlinear form that was already discussed in the examples of Chap. 2. Younger and very old men show a strongly increased tendency towards having accidents. Men between the ages of 30 and 80 years cause less damage than women of the same age group. Overall, the interaction effect is quite pronounced. Further discussion of the other results do not differ much from those outlined in Chap. 2.

△

In the next example we will show how to extend the STAR approach to structured additive categorical regression models. The example also shows that care has to be taken when modeling spatial effects, especially if some covariates (e.g., in ecological applications) are *location-specific*. It is then possible that the spatial heterogeneity effect and the influence of these variables cannot be separated. Problems of this kind have only been recently resolved in the literature. See Lawson and Liu (2007), Hodges and Reich (2010), and Paciorek (2010) for various investigations and possible solutions.

Example 9.7 Forest Health Status—Ordinal Logit Model

We revisit the Examples 1.4 (p. 9) and 6.5 (p. 337) of Chaps. 1 and 6 and extend the linear predictors $\eta_{itr} = \theta_{0r} + \boldsymbol{x}_{it}'\boldsymbol{\beta}$, $r = 1, 2$ of the three-categorical ordinal logit model $P(Y_{it} = r)$ for tree i in year t to the additive predictors

$$\eta_{itr} = \theta_{0r} + f_1(age_{it}) + f_2(gradient_i) + f_3(canopyd_{it}) + f_4(t) + f_5(alt_i) + \boldsymbol{x}_{it}'\boldsymbol{\beta}.$$

In comparison to Example 6.5, the parametric effects of most of the continuous covariates are replaced by possibly nonlinear functions. Only the effect of soil depth (*depth*) and pH-value (*ph*) are still modeled linearly, since previous nonparametric analyses resulted in linear effects for these variables. The nonlinear effects f_1 through f_5 are based on cubic P-splines with 20 interior knots and second-order difference penalties. The covariate vector $\boldsymbol{x}_{it}$ contains the remaining regressors with linear effects $\boldsymbol{\beta}$, as listed in Table 9.3. Figure 9.11 contains the estimated nonlinear effects f_1 through f_5 (estimated using a mixed

9.3 Generalized STAR Models

Model

The response variables are (conditionally) independent and the (conditional) density of y_i belongs to an exponential family. The (conditional) mean μ_i is connected to the structured additive predictor

$$\eta_i^{struct} = f(v_{i1}) + \ldots + f(v_{iq}) + \boldsymbol{x}_i'\boldsymbol{\beta}$$

through

$$\mu_i = h(\eta_i^{struct}) \quad \text{or} \quad \eta_i^{struct} = g(\mu_i).$$

The $v_1, \ldots, v_q$ are one- or multidimensional covariates of differing types constructed from the original variables. See Table 9.2 for an overview of the various model terms.

Software

- R package `mgcv`: See Wood (2006) for details. The package provides various uni- and multidimensional smoothers to estimate univariate STAR models. Smoothing parameter estimation is based on minimizing GCV or REML estimation. A similar class of models is available in the package `amer`.
- Software package `BayesX` and R package `R2BayesX`: Estimation is carried out either using mixed model methodology (`remlreg` objects), fully Bayesian inference based on MCMC (`bayesreg` and `mcmcreg` objects), or penalized likelihood in combination with variable selection techniques for smoothing parameter selection and model choice (`stepwisereg` objects). The software also provides methodology for categorical responses.
- R package `gamlss`: The package even allows the specification of additive predictors for other parameters associated with the response distribution (e.g., the mean and the variance for Gaussian responses). The software is limited to pure additive predictors.
- R packages: `mboost` and `gamboostLSS` for estimating STAR models with boosting.

model representation, see Sect. 9.6.2). All results are based on `relmreg objects` of the software `BayesX`.

We obtain an AIC of 1,682.8, which indicates a clear improvement in goodness of fit compared to the parametric model with an AIC of 2,029.16 (see Example 6.5 on p. 337).

The covariate effects do, however, show some surprising results:

- As in the parametric model, younger trees appear to be less damaged than older trees. However, we observe a decreasing effect for very old trees which contradicts our expectations on a monotonically increasing relationship.

Table 9.3 Forest health status: estimated linear effects in the model without spatial effects (left) and in the model with spatial effects (right)

	Coefficient	p-value	Coefficient	p-value
θ_1	−4.000	0.009	−1.515	0.453
θ_2	1.107	0.470	4.846	0.019
depth	−0.048	<0.001	−0.006	0.838
ph	−0.718	0.018	−0.249	0.491
type	−0.644	<0.001	−0.202	0.373
fert	−0.571	0.004	−0.570	0.253
humus0	−0.419	0.008	−0.453	0.014
humus2	0.413	0.002	0.279	0.057
humus3	0.087	0.573	0.218	0.223
humus4	−0.032	0.879	0.192	0.443
watermoisture1	−0.801	<0.001	−0.762	0.167
watermoisture3	0.483	0.001	0.812	0.031
alkali1	1.165	<0.001	0.920	0.144
alkali3	−0.966	<0.001	−0.359	0.523
alkali4	0.195	0.507	−0.201	0.798

- The effect of the gradient of slope is extremely rough and not interpretable.
- Trees in a lower and higher altitude appear to be less damaged than trees in moderate altitude. This effect is somewhat unexpected, as we might expect a monotonically increasing effect.
- Compared to the results obtained from Example 6.5, the effects of the linear covariates are relatively unstable; see, for example, the effect of the alkalinity.

How can we explain these somewhat peculiar results? As a first step, we extend the regression model by an additional spatial effect using a two-dimensional P-spline. Table 9.3 again shows the resulting linear effects. Figure 9.12 shows the resulting nonlinear effects, which dramatically change with the inclusion of the spatial effect. Figure 9.13 displays the spatial effect $f_{geo}(s)$. We observe the following:

- The decrease of the age effect for very old trees disappears.
- The effects of soil depth (*depth*), pH-value (*ph*), type of forest (*type*), alkalinity (*alkali*), gradient (*gradient*), and altitude (*alt*) either disappear, or are at least considerably reduced. All these variables are location-specific covariates whose value either does not change at all or very little over time.

In particular, the last observation provides essential information towards explaining the surprising results found with our initial estimates. In the first model without spatial effects, spatial heterogeneity was entirely captured by location-specific covariates. However, the location-specific effects appear to be rather complicated in this application. Apparently, the gradient and the altitude do absorb the effect of further, unobserved or unrecognized, covariates. As a result, their effects are difficult to interpret as they are confounded with other effects. In the second model, the flexibly modeled spatial heterogeneity effect covers the effects of the location-specific covariates and to a large extent absorbs their influences. However, this is not desirable, since the spatial effect is included to absorb the effects of unobserved covariates, but not the effects of observed covariates.

Which conclusion can we draw from this? It seems that in this application it is not possible or at least very difficult to separate the different effects. A possible solution

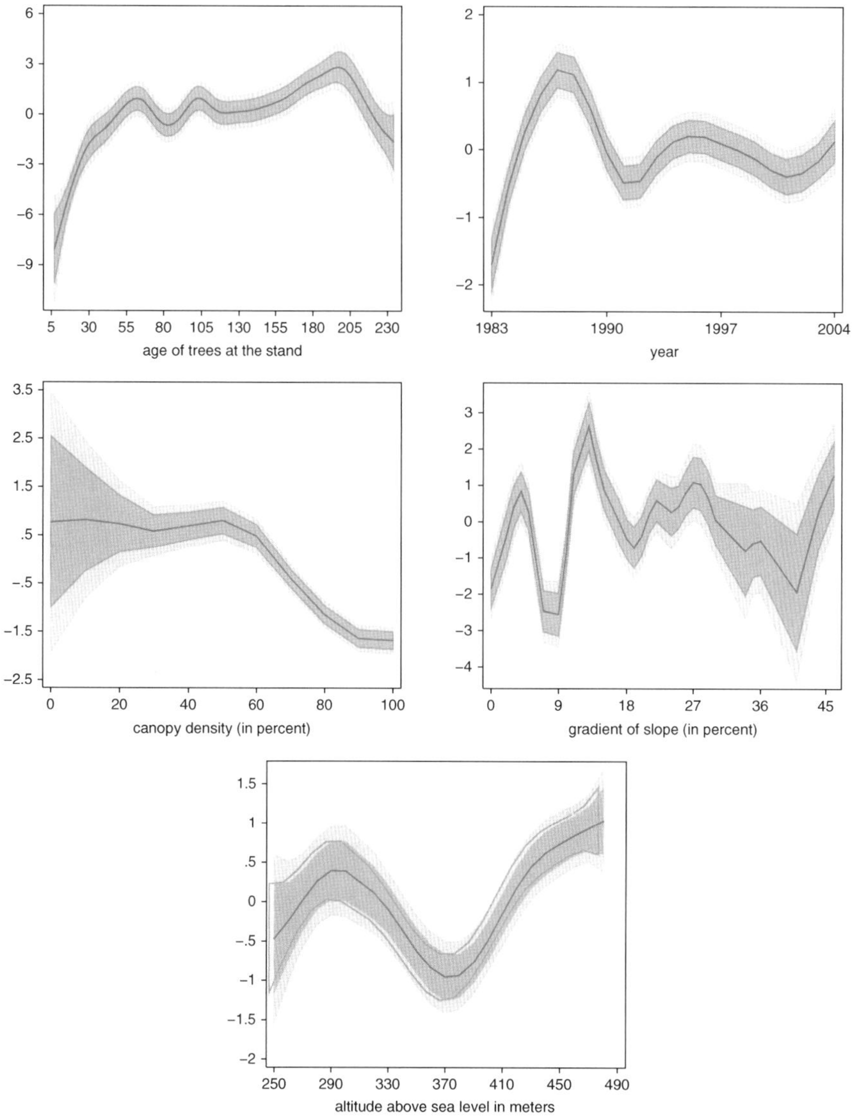

Fig. 9.11 Forest health status: estimated nonlinear effects in the model excluding spatial effects. Additionally included are pointwise 80 % and 95 % confidence intervals

would be the estimation of a model with spatial effects, but without location-specific covariates. By carefully examining the properties of the locations with especially large (or small) spatial effects, we could then try to extract the relevant location-specific covariates. The definition of new covariates, e.g., well-chosen linear combinations, could be useful. Again we defer to the review of recent literature on spatial confounding provided above.

△

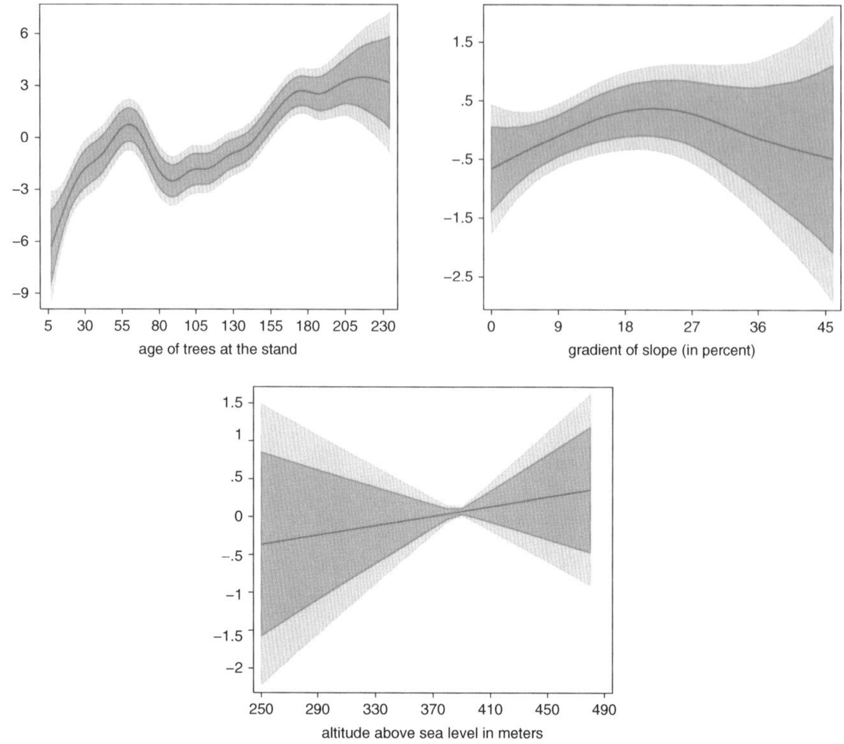

Fig. 9.12 Forest health status: selected nonlinear effects in the model with spatial effect included, along with pointwise 80 % and 95 % confidence intervals

Fig. 9.13 Forest health status: estimated spatial effect

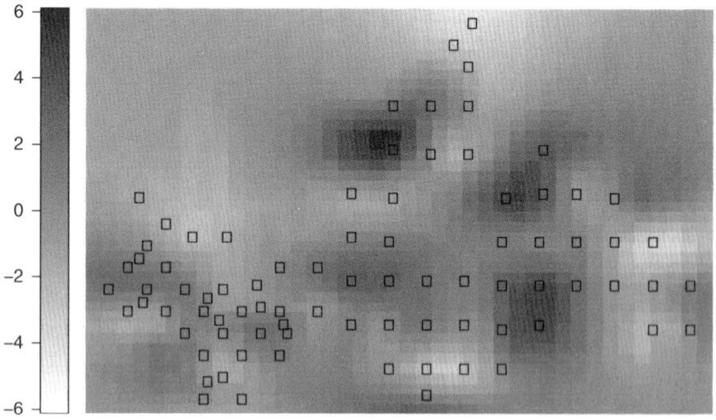

9.6 Inference

We restrict ourselves to the estimation of STAR models that are based on basis functions approaches. We will consider four different inferential concepts:

- Penalized least squares or likelihood estimates in combination with the minimization of a model choice criterion for the choice of smoothing parameters (Sect. 9.6.1)
- Inference based on a mixed model representation (Sect. 9.6.2)
- Fully Bayesian inference based on MCMC simulation techniques (Sect. 9.6.3)
- Boosting for structured additive regression models (Sect. 9.7)

9.6.1 Penalized Least Squares or Likelihood Estimation

We first consider STAR models

$$\boldsymbol{y} = \boldsymbol{f}_1 + \ldots + \boldsymbol{f}_q + \boldsymbol{X}\boldsymbol{\beta} + \boldsymbol{\varepsilon} = \boldsymbol{V}_1\boldsymbol{\gamma}_1 + \ldots + \boldsymbol{V}_q\boldsymbol{\gamma}_q + \boldsymbol{X}\boldsymbol{\beta} + \boldsymbol{\varepsilon} \qquad (9.14)$$

with a continuous response variable y. Extending the penalized least squares criterion of Chap. 8, we minimize the criterion

$$\mathrm{LS}_{pen}(\boldsymbol{\gamma}_1, \ldots, \boldsymbol{\gamma}_q, \boldsymbol{\beta}) = \mathrm{LS}(\boldsymbol{\gamma}_1, \ldots, \boldsymbol{\gamma}_q, \boldsymbol{\beta}) + \sum_{j=1}^{q} \lambda_j \boldsymbol{\gamma}_j' \boldsymbol{K}_j \boldsymbol{\gamma}_j \qquad (9.15)$$

with respect to $\boldsymbol{\gamma}_1, \ldots, \boldsymbol{\gamma}_q, \boldsymbol{\beta}$. In this setting, the least squares criterion is defined as

$$\mathrm{LS}(\boldsymbol{\gamma}_1, \ldots, \boldsymbol{\gamma}_q, \boldsymbol{\beta}) = (\boldsymbol{y} - \boldsymbol{V}_1\boldsymbol{\gamma}_1 - \ldots - \boldsymbol{V}_q\boldsymbol{\gamma}_q - \boldsymbol{X}\boldsymbol{\beta})'(\boldsymbol{y} - \boldsymbol{V}_1\boldsymbol{\gamma}_1 - \ldots - \boldsymbol{V}_q\boldsymbol{\gamma}_q - \boldsymbol{X}\boldsymbol{\beta}).$$

As in Chap. 8, the form of the penalty matrix $\boldsymbol{K}_j$ depends on the specific type of f_j and also on the chosen penalty. An overview is provided in Table 9.2. The smoothing parameter λ_j governs the trade-off between fidelity to the data and smoothness controlled by the penalty $\boldsymbol{\gamma}_j' \boldsymbol{K}_j \boldsymbol{\gamma}_j$. We first assume that λ_j is known.

To minimize Eq. (9.15), we can choose between two alternatives: iterative minimization using the backfitting algorithm or direct minimization of the penalized least squares criterion. The backfitting algorithm has the advantage that estimation of complex models can be traced back to one- or two-dimensional smoothing as described in Chap. 8. In addition, the backfitting algorithm allows any combination of smoothers as building blocks for the estimation of STAR models. This implies that basis function approaches as well as local smoothers can be used as building blocks or combinations thereof.

9.4 Backfitting Algorithm

1. Initialization: set, for example, $\hat{\boldsymbol{f}}_1 = \ldots = \hat{\boldsymbol{f}}_q = \mathbf{0}, \hat{\boldsymbol{\beta}} = \mathbf{0}$.
2. Obtain updated estimates $\hat{\boldsymbol{f}}_j, j = 1, \ldots, q$, through

$$\hat{\boldsymbol{f}}_j = \boldsymbol{S}_j(\boldsymbol{y} - \sum_{l \neq j} \hat{\boldsymbol{f}}_l - \boldsymbol{X}\hat{\boldsymbol{\beta}}).$$

3. Obtain updated estimates $\hat{\boldsymbol{\beta}}$, through

$$\hat{\boldsymbol{\beta}} = (\boldsymbol{X}'\boldsymbol{X})^{-1}\boldsymbol{X}'(\boldsymbol{y} - \hat{\boldsymbol{f}}_1 - \ldots - \hat{\boldsymbol{f}}_q).$$

4. Iterate the steps 2–3 until the estimated functions do not deviate more than a small given increment in two subsequent iterations.

Backfitting

The idea of backfitting is to obtain estimates $\hat{\boldsymbol{f}}_1, \ldots, \hat{\boldsymbol{f}}_q, \hat{\boldsymbol{\beta}}$ iteratively by the smoothing of partial residuals. Neglecting the errors $\boldsymbol{\varepsilon}$ in Eq. (9.14), the following approximately holds:

$$\boldsymbol{f}_j \approx \boldsymbol{y} - \boldsymbol{f}_1 - \ldots - \boldsymbol{f}_{j-1} - \boldsymbol{f}_{j+1} - \ldots - \boldsymbol{f}_q - \boldsymbol{X}\boldsymbol{\beta}.$$

In the case that we already have estimates for $\hat{\boldsymbol{\beta}}$ and $\hat{\boldsymbol{f}}_l$, $l \neq j$, we can view $\boldsymbol{y} - \sum_{l \neq j} \hat{\boldsymbol{f}}_l - \boldsymbol{X}\hat{\boldsymbol{\beta}}$ as a (partial) residual vector (without $\boldsymbol{f}_j$). In order to estimate $\boldsymbol{f}_j$, we apply the scatter plot or bivariate smoother $\boldsymbol{S}_j$ associated with $\boldsymbol{f}_j$ to this residual vector. We obtain

$$\hat{\boldsymbol{f}}_j = \boldsymbol{S}_j(\boldsymbol{y} - \sum_{l \neq j} \hat{\boldsymbol{f}}_l - \boldsymbol{X}\hat{\boldsymbol{\beta}}).$$

All estimators that are based on basis function approaches have the form

$$\hat{\boldsymbol{f}}_j = \boldsymbol{V}_j(\boldsymbol{V}_j'\boldsymbol{V}_j + \lambda_j \boldsymbol{K}_j)^{-1}\boldsymbol{V}_j'(\boldsymbol{y} - \sum_{l \neq j} \hat{\boldsymbol{f}}_l - \boldsymbol{X}\hat{\boldsymbol{\beta}}).$$

The particular form of $\boldsymbol{V}_j$ and $\boldsymbol{K}_j$ depends on the type of smoother used, as outlined in Table 9.2. Given starting values for $\hat{\boldsymbol{\beta}}, \hat{\boldsymbol{f}}_1, \ldots, \hat{\boldsymbol{f}}_q$, this concept is applied iteratively and results in the *backfitting algorithm* of Box 9.4.

Note that we have deliberately ignored the dependence of the smoothers $\boldsymbol{S}_j$ on the unknown smoothing parameters. The choice of the smoothing parameters will be further discussed below.

Direct Minimization of the Penalized Least Squares Criterion

In many cases, it is possible to compute the penalized least squares estimator through direct minimization of Eq. (9.15) with a non-iterative procedure. For instance the `mgcv` package of R works with direct optimization. Minimization can be achieved by computing the first derivatives of the criterion with respect to the unknown parameters, setting the derivatives to zero and finally solving the resulting system of equations.

Setting the first derivatives to zero results in the system of equations

$$\begin{pmatrix} \boldsymbol{V}_1'\boldsymbol{V}_1 + \lambda_1 \boldsymbol{K}_1 & \cdots & \boldsymbol{V}_1'\boldsymbol{V}_q & \boldsymbol{V}_1'\boldsymbol{X} \\ \vdots & \ddots & & \\ \boldsymbol{V}_q'\boldsymbol{V}_1 & & \boldsymbol{V}_q'\boldsymbol{V}_q + \lambda_q \boldsymbol{K}_q & \boldsymbol{V}_q'\boldsymbol{X} \\ \boldsymbol{X}'\boldsymbol{V}_1 & \cdots & \boldsymbol{X}'\boldsymbol{V}_q & \boldsymbol{X}'\boldsymbol{X} \end{pmatrix} \begin{pmatrix} \boldsymbol{\gamma}_1 \\ \vdots \\ \boldsymbol{\gamma}_q \\ \boldsymbol{\beta} \end{pmatrix} = \begin{pmatrix} \boldsymbol{V}_1'\boldsymbol{y} \\ \vdots \\ \boldsymbol{V}_q'\boldsymbol{y} \\ \boldsymbol{X}'\boldsymbol{y} \end{pmatrix}. \tag{9.16}$$

The direct (non-iterative) solution of the system of equations works well with a comparably small number of regression coefficients, e.g., if q is small and the functions $f_1, \ldots, f_q$ are specified by P-splines having a relatively modest number of knots. For high-dimensional $\boldsymbol{\gamma}_1, \ldots, \boldsymbol{\gamma}_q$, the numerical solution of the system of Eq. (9.16) can be obtained through the iterative *Gauß–Seidel algorithm*, which in turn is identical to the backfitting algorithm described above; see Buja, Hastie, and Tibshirani (1989).

Similar to linear models, we can estimate the (conditional) mean $\mathrm{E}(\boldsymbol{y})$ of the response by $\hat{\boldsymbol{y}} = \boldsymbol{H}\boldsymbol{y}$ using the estimated regression coefficients. The hat (prediction) matrix or smoother matrix $\boldsymbol{H}$ is given by

$$\boldsymbol{H} = (\boldsymbol{V}_1\, \boldsymbol{V}_2 \ldots \boldsymbol{V}_q\, \boldsymbol{X}) \begin{pmatrix} \boldsymbol{V}_1'\boldsymbol{V}_1 + \lambda_1 \boldsymbol{K}_1 & \cdots & \boldsymbol{V}_1'\boldsymbol{V}_q & \boldsymbol{V}_1'\boldsymbol{X} \\ \vdots & \ddots & & \\ \boldsymbol{V}_q'\boldsymbol{V}_1 & & \boldsymbol{V}_q'\boldsymbol{V}_q + \lambda_q \boldsymbol{K}_q & \boldsymbol{V}_q'\boldsymbol{X} \\ \boldsymbol{X}'\boldsymbol{V}_1 & \cdots & \boldsymbol{X}'\boldsymbol{V}_q & \boldsymbol{X}'\boldsymbol{X} \end{pmatrix}^{-1} \begin{pmatrix} \boldsymbol{V}_1' \\ \vdots \\ \boldsymbol{V}_q' \\ \boldsymbol{X}' \end{pmatrix}.$$

Generalized STAR Models

For generalized structured additive models with non-normally distributed response and the predictor $\boldsymbol{\eta}^{struct} = \boldsymbol{V}_1\boldsymbol{\gamma}_1 + \ldots + \boldsymbol{V}_q\boldsymbol{\gamma}_q + \boldsymbol{X}\boldsymbol{\beta}$, estimation is achieved by maximizing the penalized log-likelihood criterion

$$l_{pen}(\boldsymbol{\gamma}_1, \ldots, \boldsymbol{\gamma}_q, \boldsymbol{\beta}) = l(\boldsymbol{\gamma}_1, \ldots, \boldsymbol{\gamma}_q, \boldsymbol{\beta}) - \frac{1}{2}\sum_{j=1}^{q} \lambda_j \boldsymbol{\gamma}_j' \boldsymbol{K}_j \boldsymbol{\gamma}_j$$

with respect to $\boldsymbol{\gamma}_1, \ldots, \boldsymbol{\gamma}_q, \boldsymbol{\beta}$. Thereby $l(\boldsymbol{\gamma}_1, \ldots, \boldsymbol{\gamma}_q, \boldsymbol{\beta})$ denotes the usual log-likelihood of generalized linear models but with the linear predictor $\boldsymbol{\eta}^{lin}$ replaced by $\boldsymbol{\eta}^{struct}$. Setting the first derivatives to zero leads to the high-dimensional system of equations

$$s_{pen,0}(\boldsymbol{\gamma}_1, \ldots, \boldsymbol{\gamma}_q, \boldsymbol{\beta}) = s_0(\boldsymbol{\gamma}_1, \ldots, \boldsymbol{\gamma}_q, \boldsymbol{\beta}) = \mathbf{0}$$

$$s_{pen,j}(\boldsymbol{\gamma}_1, \ldots, \boldsymbol{\gamma}_q, \boldsymbol{\beta}) = s_j(\boldsymbol{\gamma}_1, \ldots, \boldsymbol{\gamma}_q, \boldsymbol{\beta}) - \lambda_j \boldsymbol{K}_j \boldsymbol{\gamma}_j = \mathbf{0}, \qquad j = 1, \ldots, q.$$

The functions $s_0(\boldsymbol{\gamma}_1, \ldots, \boldsymbol{\gamma}_q, \boldsymbol{\beta}) = \partial l / \partial \boldsymbol{\beta}$, $s_j(\boldsymbol{\gamma}_1, \ldots, \boldsymbol{\gamma}_q, \boldsymbol{\beta}) = \partial l / \partial \boldsymbol{\gamma}_j$ have the form of the score function in generalized linear models. The structure of the system of equations is similar to that of Eq. (9.16). However, similar to generalized linear models, the equations depend on working weights and working observations. As presented in Chap. 5, the system of equations can be solved with the Fisher scoring algorithm or in combination with backfitting; see, for example, Fahrmeir and Tutz (2001, Chap. 5), Tutz (2011, Chap. 8), Eilers and Marx (2002), and Marx and Eilers (1998).

Choice of the Smoothing Parameters

The smoothing parameters are estimated by optimizing an appropriate model choice criterion, e.g., AIC or GCV.

To minimize AIC, we have to take the effective dimension of the model or effective number of parameters into consideration. The effective number of parameters can be defined in a generalized way, as outlined in Sect. 8.1.8 (p. 473), with the trace of the full STAR model smoother matrix $\boldsymbol{H}$. For Gaussian responses, the smoother matrix projects the vector $\boldsymbol{y}$ into $\hat{\boldsymbol{y}} = \boldsymbol{H}\boldsymbol{y}$. For non-Gaussian response, we use the corresponding matrix from the working model. For STAR models with Gaussian errors, the AIC criterion is then given by

$$\text{AIC} = n \log(\hat{\sigma}^2) + 2(\text{tr}(\boldsymbol{H}) + 1),$$

where $\hat{\sigma}^2 = \sum (y_i - \hat{\eta}_i^{struct})^2 / n$ is the ML estimator of the error variance.

For the generalized cross validation score GCV, we obtain

$$\text{GCV} = \frac{1}{n} \sum_{i=1}^{n} \left(\frac{y_i - \hat{\eta}_i^{struct}}{1 - \text{tr}(\boldsymbol{H})/n} \right)^2.$$

If we use the iterative backfitting algorithm to estimate the regression coefficients, we do not obtain the full smoother matrix and (exact) computation of the criteria is, thus, not possible. In practice, the trace of the full smoother matrix is often approximated via the sum of the traces of individual smoother matrices. Provided that the correlation among individual terms is not too strong, this approximation works well in practice.

Until approximately the year 2000, a simple but not very effective grid search was widely used to minimize AIC or GCV, and thus problems with more than, say four or five smoothing parameters, were barely manageable. Currently, much more efficient (albeit technically complex) algorithms for the optimization of model choice criteria are available. The monograph by Wood (2006) offers a detailed description of the

9.5 Residuals in STAR Models with Continuous Response

Ordinary Residuals

$$\hat{\varepsilon}_i = y_i - \hat{\eta}_i^{struct} = y_i - \hat{f}_1(v_{i1}) - \ldots - \hat{f}_q(v_{iq}) - \boldsymbol{x}_i'\hat{\boldsymbol{\beta}}.$$

Standardized Residuals

The ith standardized residual is given by

$$r_i = \frac{y_i - \hat{\eta}_i^{struct}}{\hat{\sigma}\sqrt{1 - h_{ii}}},$$

where h_{ii} is the ith diagonal element of the full smoother matrix $\boldsymbol{H}$.

Partial Residuals

Partial residuals regarding v_j are defined by

$$\begin{aligned}\hat{\varepsilon}_{v_j,i} &= y_i - \hat{f}_1(v_{i1}) - \ldots - \hat{f}_{j-1}(v_{i,j-1}) - \hat{f}_{j+1}(v_{i,j+1}) - \ldots - \hat{f}_q(v_{iq}) - \boldsymbol{x}_i'\hat{\boldsymbol{\beta}} \\ &= y_i - \hat{\eta}_i^{struct} + \hat{f}_j(v_{ij}).\end{aligned}$$

methodological background; see also Wood (2000, 2004, 2008). An implementation can be found in the R package mgcv.

Model Choice and Diagnosis

Relatively little literature exists about model choice and diagnosis for nonparametric and semiparametric models, and user-friendly software is even rarer. However, some of the concepts regarding model choice in linear models, discussed in Sect. 3.4, easily carry over to STAR models. For instance, competing models can be compared via standard model choice criteria such as AIC or GCV, even though there are some restrictions to be considered when the model has been estimated using the mixed model representation of penalized regression; see the next section.

Model diagnostics centrally relies on the various types of residuals. Ordinary, standardized, and partial residuals can be defined analogously to those of the linear model. Refer to Box 9.5 for continuous responses. However, an analogue to studentized residuals does not exist. In the case of generalized models, see the books of McCullagh and Nelder (1989), Tutz (2011), and Collett (1991) for the definition of appropriate residuals. In principle, influence measures and outlier analysis, as well as Cook's distance, are also available. However, to date, they have not been thoroughly examined and are not part of common software packages.

There is a contemporary approach for automatic variable selection and model choice in generalized STAR models using goodness-of-fit criteria such as AIC or cross validation, as developed in Belitz and Lang (2008). The proposed algorithms are able to:

- Determine whether or not a particular covariate is to be included in the model
- Determine whether a continuous covariate enters the model linearly or nonlinearly
- Determine whether or not a spatial effect enters the model
- Determine whether or not a unit- or cluster-specific heterogeneity effect is included in the model
- select complex interaction effects (two-dimensional surfaces, varying coefficient terms)
- Select the degree of smoothness of nonlinear covariate, spatial, or cluster-specific heterogeneity effects

Particular emphasis is devoted to modeling and the selection of interaction terms. The approach is included in the software package BayesX (`stepwisereg objects`); also see the R interface `R2BayesX` described in Umlauf, Kneib, Lang, and Zeileis (2012). For automatic estimation and model selection in STAR models, also see Sect. 9.7 on boosting.

9.6.2 Inference Based on Mixed Model Representation

A second approach for estimating STAR models is based on their representation as mixed models. In Sect. 8.1.9, we already discussed the close relationship between penalization approaches and mixed models. We now briefly discuss the generalization to STAR models.

For every model component $\boldsymbol{f}_j = \boldsymbol{V}_j \boldsymbol{\gamma}_j$ the vector $\boldsymbol{\gamma}_j$ can be decomposed as

$$\boldsymbol{\gamma}_j = \tilde{\boldsymbol{X}}_j \boldsymbol{\beta}_j + \tilde{\boldsymbol{U}}_j \tilde{\boldsymbol{\gamma}}_j.$$

Through appropriate choices of the design matrices $\tilde{\boldsymbol{X}}$ and $\tilde{\boldsymbol{U}}$, we obtain a vector $\boldsymbol{\beta}_j$ of fixed effects and a vector $\tilde{\boldsymbol{\gamma}}_j \sim \mathrm{N}(\boldsymbol{0}, \tau_j^2 \boldsymbol{I})$ of independent and identically distributed random effects. Analogously, the vector of function evaluations can be expressed as

$$\boldsymbol{f}_j = \boldsymbol{V}_j(\tilde{\boldsymbol{X}}_j \boldsymbol{\beta}_j + \tilde{\boldsymbol{U}}_j \tilde{\boldsymbol{\gamma}}_j) = \boldsymbol{X}_j \boldsymbol{\beta}_j + \boldsymbol{U} \tilde{\boldsymbol{\gamma}}_j,$$

with $\boldsymbol{X}_j = \boldsymbol{V}_j \tilde{\boldsymbol{X}}_j$ and $\boldsymbol{U}_j = \boldsymbol{V}_j \tilde{\boldsymbol{U}}_j$. Applying this decomposition to all model components, we obtain a large mixed model whose fixed effects consist of the original parametric effects $\boldsymbol{\beta}$ and the fixed effects $\boldsymbol{\beta}_j$, $j = 1, \dots, q$, resulting from the reparameterization. The random effects are composed of the vectors $\tilde{\boldsymbol{\gamma}}_j$, $j = 1, \dots, q$, with joint distribution $\mathrm{N}(\boldsymbol{0}, \mathrm{diag}(\tau_1^2 \boldsymbol{I}, \dots, \tau_q^2 \boldsymbol{I}))$. Thus, the resulting mixed model is a variance components model. To estimate the model, in particular the variance or smoothing parameters, we can use the methodology presented in Sect. 8.1.9. Note that the reparameterization generally causes redundant columns in the design matrix of the fixed effects. The reason for the redundancy is the identification problem of additive models. Every nonparametric model term has its own level corresponding to a constant in the fixed effects part of the reparameterization. After the reparameterization, the model then has these further constants, in addition to the intercept in $\boldsymbol{\beta}$, so that the joint design matrix is no longer of full rank. In order to obtain an identifiable model, redundant columns need to be eliminated from the design matrix. For the constants, this mainly corresponds to the previously discussed centering conditions on f_j.

For additive and geoadditive models, inference based on mixed models is extensively discussed in Ruppert, Wand, and Carroll (2003). Mixed model inference in generalized STAR models is proposed in Fahrmeir, Kneib, and Lang (2004). Corresponding extensions to categorical responses are given in Kneib and Fahrmeir (2006). A detailed presentation can also be found in Kneib (2005).

Model Choice and Diagnosis

The remarks mentioned above regarding model choice and diagnosis mostly remain valid with inference based on mixed models.

To verify whether or not the flexible nonparametric modeling of the individual covariates effects is necessary, we can in principle apply statistical tests regarding the variance parameters. If we want to test, e.g., a P-spline with a second-order difference penalty against a linear effect, then this corresponds to the hypothesis

$$H_0 : \tau^2 = 0 \quad \text{versus} \quad H_1 : \tau^2 > 0.$$

At first thought, the likelihood ratio statistic that compares the likelihoods under H_0 and H_1 may be used to perform the test. However, the parameter to be tested is on the boundary of the parameter space under the null hypotheses. This contradicts the usual regularization conditions which are used when deriving the asymptotic χ^2-distribution of the likelihood ratio test statistic. Thus, the likelihood ratio test statistic does not follow a χ^2-distribution in this case. Currently, the exact distribution of the likelihood ratio statistic can only be derived in relatively simple special cases with Gaussian responses. Further information can be found, for example, in Crainiceanu and Ruppert (2004) or Crainiceanu et al. (2005). A simulation-based approach for obtaining p-values that works also for models with more than one variance parameter has been proposed in Scheipl et al. (2008) and is implemented in the R package `RLRsim`; see also Sect. 7.3.4, p. 381.

Similar problems with violated regularity conditions occur when considering the AIC for comparing models with estimated smoothing variance and the simpler model obtained when setting the variance equal to zero. This includes the important case of comparing a simple linear model for a specific covariate with a model comprising a P-spline with second-order difference penalty with smoothing variance estimated from the mixed model representation. More specifically, Greven and Kneib (2010) considered two variants of the AIC derived from the marginal and the conditional perspective on mixed models (compare Sect. 7.1.4) and investigated their properties in this situation. The marginal AIC is obtained when integrating out the random effects and relying on the marginal normal distribution with correlations between the responses for deriving the likelihood involved in the AIC. It turns out that basic regularity conditions of the AIC are violated in this situation and therefore the marginal AIC cannot be used to compare models with estimated variance components and models where the variances are set to zero. Nevertheless, the marginal AIC is routinely returned by several mixed model implementations in statistical software such as the `nlme` package in R.

The AIC we have used so far in this book is the conditional AIC that is specified as

$$-2l(\cdot) + 2\mathrm{tr}(\boldsymbol{H})$$

where $l(\cdot)$ denotes the log-likelihood of the estimated model (conditional on the random effects) and $\boldsymbol{H}$ is the corresponding hat matrix. While this standard definition is valid in case of given smoothing variances, it has a bias towards the more complex model in case of estimated smoothing variances. In fact, the conditional AIC always decides for the model with nonparametric effect when the smoothing variance is estimated to be non-zero. This defect results from the fact that uncertainty due to the estimation of smoothing variances is not taken into account in the conditional AIC. As a consequence, the conditional AIC in its standard form also does not enable proper model choice in situations where the decision problem corresponds to differentiating between estimated smoothing variances and smoothing variances set to zero. Greven and Kneib (2010) therefore introduce a corrected conditional AIC that circumvents this problem, but this corrected AIC is not yet available in standard software packages and is also restricted to models with Gaussian responses. We therefore give the following recommendations for practical work with the conditional AIC:

- If there is strong evidence for one model in the model comparison (i.e., if one model has significantly lower AIC), then the standard conditional AIC should be reliable enough since the problems described above mostly occur when the smoothing variance is close to zero.
- The standard conditional AIC can still be used to compare models that only differ in the specification for the parametric effect or that compare different structures for the random effect (e.g., an i.i.d. random effect versus a Markov random field for a spatial effect based on geographical regions).

In that sense, the model comparisons made so far in this chapter should be valid. The only borderline case appeared in Example 9.3. However, since we found a preference for the simpler model in this example anyway, the decision should still be fine.

9.6.3 Bayesian Inference Based on MCMC

Gaussian Response

We first assume Gaussian responses, i.e.,

$$\boldsymbol{y} \mid \boldsymbol{\gamma}_1, \ldots, \boldsymbol{\gamma}_q, \boldsymbol{\beta}, \sigma^2 \sim \mathrm{N}(\boldsymbol{\eta}^{struct}, \sigma^2 \boldsymbol{I}).$$

The density of the observation model is then proportional to the likelihood and is given by

$$p(\boldsymbol{y} \mid \boldsymbol{\gamma}_1, \ldots, \boldsymbol{\gamma}_q, \boldsymbol{\beta}, \sigma^2) \propto \frac{1}{(\sigma^2)^{\frac{n}{2}}} \exp\left(-\frac{1}{2\sigma^2}\left(\boldsymbol{y} - \boldsymbol{\eta}^{struct}\right)'\left(\boldsymbol{y} - \boldsymbol{\eta}^{struct}\right)\right).$$

Bayesian inference is based on the smoothness prior

$$p(\boldsymbol{\gamma}_j \mid \tau_j^2) \propto \left(\frac{1}{\tau_j^2}\right)^{\mathrm{rk}(\boldsymbol{K}_j)/2} \exp\left(-\frac{1}{2\tau_j^2}\boldsymbol{\gamma}_j'\boldsymbol{K}_j\boldsymbol{\gamma}_j\right), \tag{9.17}$$

introduced in the Box given in Sect. 8.1.9 on p. 486 for the regression coefficients $\boldsymbol{\gamma}_j$ of the jth term of the structured additive predictor. For the variance parameters τ_j^2, we specify an inverse gamma distribution with hyperparameters a_j and b_j, i.e.,

$$\tau_j^2 \sim \mathrm{IG}(a_j, b_j),$$

as the prior distribution. For the variance σ^2 of the error term, we similarly define an inverse gamma distribution with hyperparameters a and b. In practice, small values are often chosen for a_j and b_j, e.g., $a_j = b_j = 0.001$ or $a_j = b_j = 0.0001$. This recommendation is only meaningful if the response values are in a specific range that is obtained by dividing the original response observations by their empirical standard deviation.

For the linear effects regression coefficients $\boldsymbol{\beta}$, in this chapter, we restrict ourselves to noninformative priors, i.e., $p(\boldsymbol{\beta}) \propto$ const. A multivariate Gaussian prior $\boldsymbol{\beta} \sim \mathrm{N}(\boldsymbol{m}, \boldsymbol{M})$ could alternatively be used, as seen in Sect. 5.6.

The density of the prior distribution of all parameters can now be written as the product of the densities in Eq. (9.17) and the densities of the inverse gamma distributions for the variance parameters. The posterior distribution is proportional to the product resulting from the likelihood with the prior distribution. We obtain

$$\begin{aligned}
p(\boldsymbol{\theta} \mid \boldsymbol{y}) \propto\ & \frac{1}{(\sigma^2)^{\frac{n}{2}}} \exp\left(-\frac{1}{2\sigma^2}\left(\boldsymbol{y} - \boldsymbol{\eta}^{struct}\right)'\left(\boldsymbol{y} - \boldsymbol{\eta}^{struct}\right)\right) \\
& \prod_{j=1}^{q} \frac{1}{(\tau_j^2)^{\mathrm{rk}(\boldsymbol{K}_j)/2}} \exp\left(-\frac{1}{2\tau_j^2}\boldsymbol{\gamma}_j'\boldsymbol{K}_j\boldsymbol{\gamma}_j\right) \prod_{j=1}^{q} (\tau_j^2)^{-a_j-1} \exp\left(-\frac{b_j}{\tau_j^2}\right) \\
& (\sigma^2)^{-a-1} \exp\left(-\frac{b}{\sigma^2}\right),
\end{aligned}$$

where $\boldsymbol{\theta}$ is the vector of all model parameters (including the variance parameters). We will next describe a Gibbs sampler (see Sect. B.5.3 in Appendix B) for drawing random numbers from the posterior. To do so, the entire parameter vector is decomposed into the blocks $\boldsymbol{\gamma}_1, \ldots, \boldsymbol{\gamma}_q, \boldsymbol{\beta}, \tau_1^2, \ldots, \tau_q^2, \sigma^2$. A requirement of the Gibbs-sampler is that the full conditionals of the blocks must represent known distributions, from which sampling of random numbers is fast and easily available. Similar to and as already described in Sect. 8.1.9, the conditional densities of the regression parameters are multivariate normal distributions, and the conditional densities of the variance parameters are inverse gamma distributions.

More precisely, we have

$$\boldsymbol{\gamma}_j \mid \cdot \sim \mathrm{N}(\boldsymbol{m}_j, \boldsymbol{\Sigma}_j)$$

with expectation and covariance matrix

$$\boldsymbol{m}_j = \mathrm{E}(\boldsymbol{\gamma}_j \mid \cdot) = \left(\frac{1}{\sigma^2}\boldsymbol{V}'_j\boldsymbol{V}_j + \frac{1}{\tau_j^2}\boldsymbol{K}_j\right)^{-1} \frac{1}{\sigma^2}\boldsymbol{V}'_j(\boldsymbol{y} - \boldsymbol{\eta}_{-j}^{struct})$$

$$\boldsymbol{\Sigma}_j = \mathrm{Cov}(\boldsymbol{\gamma}_j \mid \cdot) = \left(\frac{1}{\sigma^2}\boldsymbol{V}'_j\boldsymbol{V}_j + \frac{1}{\tau_j^2}\boldsymbol{K}_j\right)^{-1}.$$

The vector $\boldsymbol{\eta}_{-j}^{struct} = \boldsymbol{\eta}^{struct} - \boldsymbol{V}_j\boldsymbol{\gamma}_j$ is the current predictor $\boldsymbol{\eta}^{struct}$ without the jth term. Similarly, for the vector $\boldsymbol{\beta}$, we obtain a multivariate normal distribution with expectation and covariance matrix given by

$$\boldsymbol{m}_\beta = \mathrm{E}(\boldsymbol{\beta} \mid \cdot) = \left(\boldsymbol{X}'\boldsymbol{X}\right)^{-1} \boldsymbol{X}'(\boldsymbol{y} - \tilde{\boldsymbol{\eta}}^{struct})$$

$$\boldsymbol{\Sigma}_\beta = \mathrm{Cov}(\boldsymbol{\beta} \mid \cdot) = \frac{1}{\sigma^2}(\boldsymbol{X}'\boldsymbol{X})^{-1},$$

where $\tilde{\boldsymbol{\eta}}^{struct} = \boldsymbol{\eta}^{struct} - \boldsymbol{X}\boldsymbol{\beta}$ is the predictor $\boldsymbol{\eta}^{struct}$ without the linear effects. For the variance parameters, we obtain

$$\tau_j^2 \mid \cdot \sim \mathrm{IG}(a_j + 0.5\mathrm{rk}(\boldsymbol{K}_j), b_j + 0.5\boldsymbol{\gamma}'_j\boldsymbol{K}_j\boldsymbol{\gamma}_j),$$

$$\sigma^2 \mid \cdot \sim \mathrm{IG}(a + 0.5n, b + 0.5(\boldsymbol{y} - \boldsymbol{\eta}^{struct})'(\boldsymbol{y} - \boldsymbol{\eta}^{struct})).$$

The derivation of all full conditionals is similar to the classical linear model in Sect. 4.4. It is now straightforward to implement a Gibbs sampler that successively draws samples from the full conditionals. Numerically efficient sampling of random numbers from the high-dimensional Gaussian distributions is guaranteed with methodology for banded or sparse matrices. More details can be found in Rue and Held (2005), Lang and Brezger (2004), and Lang et al. (2012).

MCMC-Based Inference Using Data Augmentation

As already discussed in Sect. 5.6.3 (p. 316), the Gibbs sampler for Gaussian responses can be often modified for MCMC inference with non-Gaussian responses. Applying the data augmentation schemes of Sect. 5.6.3 to STAR models is straightforward; its description would be identical to Chap. 5 and is therefore omitted.

Non-Gaussian Responses

For some distributions, e.g., gamma-distributed responses, a Gibbs sampler can no longer be derived. In this case, MCMC inference can be based on MH algorithms with IWLS proposals as described for generalized linear models in Sect. 5.6. Details for generalized STAR models can be found in Brezger and Lang (2006) and Lang et al. (2012).

Model Choice and Diagnosis

For comparing competing models, we can use the DIC (see Sect. B.5.4 in Appendix B) as a global goodness-of-fit measure. The DIC can be computed as a by-product of the MCMC sampler. However, the DIC is subject to sampling variation and therefore shows (slightly) different values in each MCMC run. In the case that two competing models have very different DIC values, problems due to the sampling variability of the DIC should not be an issue. Some care has to be taken if DIC values are close to each other. In such a case, comparison is only possible by means of several (possibly time intensive) MCMC runs (at least 20) and then averaging the DIC values.

For model diagnosis, we can again use the various forms of residuals. With Bayesian inference based on MCMC methods, some additional diagnostic checks regarding convergence of the samplers are necessary. The literature provides a huge variety of diagnostic tools; refer to Cowles and Carlin (1996) and Mengersen, Robert, and Guihenneuc-Jouyaux (1999) for an overview. Their use is, however, restricted to relatively simple models with only a few parameters. In any case, including complex models with many parameters, it is extremely important to examine the sampling paths and autocorrelation functions of *all* samples. Striking features, especially non convergence, low acceptance rates, and high autocorrelations, become apparent with these tools; see Example B.10 in Appendix B. When using the MCMC algorithms described in this chapter, problems usually only occur when the models that are being estimated are too complex for the data at hand.

Another important diagnostic step for Bayesian models is to perform a *sensitivity analysis* regarding the effect of slight changes in the priors (particularly the hyperparameters) on the posterior. In the case of STAR models, this requires examining the sensitivity of results on the choice of the hyperparameters a_j and b_j of the inverse gamma prior distribution for the variance parameters τ_j^2. Depending on the choice of hyperparameters, it is possible that the estimated results differ considerably. This is especially the case if we do not have sufficient information, i.e., a limited number of observations in subdomains of the covariate range. In most cases, however, the results are relatively stable. The possible dependence on the choice of hyperparameters is not a disadvantage of Bayesian inference for STAR models. On the contrary, the Bayesian approach allows a careful sensitivity analysis when examining the stability of the chosen model.

Recently, a Bayesian approach based on spike and slab priors for STAR models, with built-in model and variable selection options, has been proposed; see Scheipl,

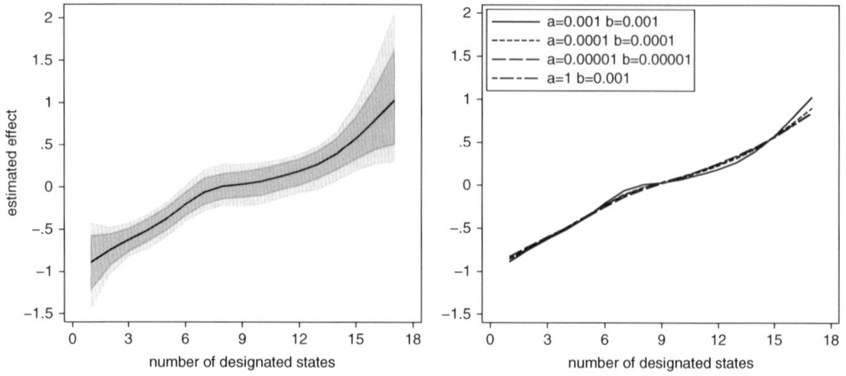

Fig. 9.14 Patent opposition: The *left panel* shows the estimated nonlinear effect of the number of designated states for the patent. The *right panel* shows the estimated effect of *ncountry* based on four different combinations of the hyperparameters a and b for the prior of the variance parameter

Fahrmeir, and Kneib (2012) for details. Software is provided in the R package `spikeSlabGAM`, outlined in Scheipl (2011).

Lastly, we want to mention that there are some extensions for the Bayesian adaptive procedures described in Sect. 8.1.10, which allow for simultaneous estimation and model choice. Since generally usable software for this methods is not available we restrain from presenting detailed algorithms but rather point to the references given in Sect. 8.1.10 for details.

Example 9.8 Patent Opposition—Additive Model

In the Examples 2.8 (p. 35), 5.1 (p. 275) and 5.4 (p. 288), we modeled the effect of the covariate *ncountry* on the binary response *opp* using a cubic polynomial. For comparison, we now estimate the effect $f(ncountry)$ nonparametrically. This leads to the additive logit model

$$P(opp_i = 1) = \frac{\exp(\eta_i^{add})}{1 + \exp(\eta_i^{add})}$$

with the predictor

$$\eta_i^{add} = f(ncountry_i) + \beta_0 + \beta_1 \, yearc_i + \beta_2 \, ncit_i + \beta_3 \, nclaims_i + \beta_4 \, ustwin_i$$
$$+ \beta_5 \, patus_i + \beta_6 patgsgr_i .$$

The (possibly) nonlinear function $f(ncountry)$ is modeled by cubic P-splines and estimated using MCMC. We included the centered year (variable *yearc*) rather than the original variable *year* because otherwise the intercept and the effect of year would be highly correlated. We used `bayesreg objects` of the software `BayesX`. Figure 9.14 (left) shows the posterior mean together with 80 % and 95 % (pointwise) confidence bands. Clearly, the function does not greatly differ from a straight line. We obtain DIC = 2, 993. Modeling the effect of *ncountry* linearly or with a polynomial, we obtain DIC = 2, 993 and DIC = 2, 995, respectively. Thus, a linear effect is apparently sufficient. This also confirms the result of the statistical test conducted in Example 5.4.

The right panel of Fig. 9.14 shows the result of a sensitivity analysis, in which different hyperparameters a and b of the prior for the variance parameter τ^2 were tested. We used the combinations $(a = 0.001, b = 0.001)$, $(a = 0.0001, b = 0.0001)$, $(a = 0.00001, b = 0.00001)$, and $(a = 1, b = 0.001)$. We find that estimated results do not differ much from each other. Moreover, the figure once again confirms the linearity assumption for the effect of *ncountry*.

△

9.7 Boosting STAR Models

In Sect. 4.3, we introduced boosting as a regularized estimation technique that had the particular advantage to combine model estimation with automatic model choice and variable selection. Such a combination would be very desirable in the context of STAR models, especially since several competing modeling possibilities and model formulations are available. We will therefore extend boosting approaches to STAR models.

In the context of linear regression models, boosting estimates were obtained by repeatedly applying simple ordinary least squares fits to iteratively updated residuals. In *componentwise* boosting, only the best-fitting covariate was updated in each iteration of the boosting process. To generalize this componentwise approach to STAR models, we basically have to replace the least squares base-learners with penalized least squares base-learners.

Therefore, we assume a structured additive regression model with predictor

$$\eta_i = f_1(v_{i1}) + \ldots + f_q(v_{iq}) + \beta_0 + \beta_1 x_{i1} + \ldots + \beta_k x_{ik},$$

where as before the functions f_j may represent different types of nonlinear regression effects. In matrix notation, the model can be reexpressed as

$$\boldsymbol{\eta} = \boldsymbol{V}_1\boldsymbol{\gamma}_1 + \ldots + \boldsymbol{V}_q\boldsymbol{\gamma}_q + \boldsymbol{X}_1\boldsymbol{\beta}_1 + \ldots + \boldsymbol{X}_L\boldsymbol{\beta}_L,$$

where we have grouped covariates with parametric effects in $L \leq k + 1$ suitable blocks of coefficients. This grouping may result, for example, from considering dummy or effect coding of categorical covariates where all dummy variables should be assigned to a common block, such that when performing model choice, either the complete block is selected or the complete block is deselected from the model. A similar situation arises when considering a polynomial expansion of a continuous covariate. In this case, it may be useful to combine all polynomials in one coefficient block. However, in the simplest situation, each block will only comprise exactly one covariate and therefore $L = k + 1$.

We further assume that the estimation problem at hand is described in terms of a suitable loss function ρ. Typically, this will be the least squares criterion in case of Gaussian responses or the negative log-likelihood for exponential family regression. More general situations are also conceivable, for example, to cast median or robust regression in the context of boosting.

Similar to the presentation in Sect. 4.3, the base-learning procedure for estimating parametric linear effects for the blocks $\boldsymbol{X}_l \boldsymbol{\beta}_l$ is given by ordinary least squares fits and can therefore be characterized by the corresponding hat matrices

$$\boldsymbol{H}_l = \boldsymbol{X}_l(\boldsymbol{X}_l'\boldsymbol{X}_l)^{-1}\boldsymbol{X}_l', \quad l = 1, \ldots, L.$$

Analogously, all base-learning procedures for nonlinear effects $\boldsymbol{V}_j \boldsymbol{\gamma}_j$ can be described in terms of penalized least squares fits as characterized by the hat matrices

$$\boldsymbol{H}_j = \boldsymbol{V}_j(\boldsymbol{V}_j'\boldsymbol{V}_j + \lambda_j \boldsymbol{K}_j)^{-1}\boldsymbol{V}_j', \quad j = 1, \ldots, q,$$

with smoothing parameters $\lambda_j \geq 0$. Note that the smoothing parameters are now no longer part of the estimation process but will be specified a priori to assign a certain flexibility to the base-learners.

Based on these definitions, a generic boosting algorithm for structured additive regression models is given in Box 9.6.

In each iteration of the algorithm, all base-learning procedures are fitted to the current negative gradient vector $\boldsymbol{u}$, but only the best-fitting base-learner is actually updated. Note that this decision is always based on the L_2 loss (as indicated in Box 9.6) regardless of the specific loss function describing the estimation problem since the negative gradient vectors is treated as a continuous response in the base-learner. Updating only the best fitting base-learner allows us to implement model choice by setting an optimal, data-driven stopping iteration m_{stop}. Typically, the important and influential covariates will be selected first, while those covariates that are only weakly associated with the response will be selected only in very late boosting iterations. Hence, when stopping early enough, such covariates will effectively drop out of the estimated model. An optimal stopping iteration will usually be determined based on cross validation techniques.

Note that the model choice possibilities of boosting can be further improved when considering competing modeling alternatives for the same covariate. For example, a continuous covariate can be included linearly and nonlinearly in the candidate model such that in each iteration the algorithm compares the performance of the linear and the nonlinear modeling alternative and decides for the better fitting alternative. However, care has to be taken when actually implementing this approach since the nonlinear effect naturally contains the linear effect as a special case and therefore has a greater chance to be selected simply because the corresponding base-learner is more flexible. It is therefore important to make the base-learners comparable in terms of their complexity. This can be achieved by a suitable choice of the smoothing parameter in the penalized least squares base-learners. Note that these are no longer hyperparameters to be estimated but rather constitute the flexibility of the base-learner. Typically, user-specified degrees of freedom will be assigned to each base-learner and the corresponding smoothing parameter will be numerically determined accordingly. Assigning the same degrees of freedom to each base-learner yields a comparable complexity for all model terms. This may also require an assignment of ridge-type base-learners to the parametric

9.6 Boosting in STAR Models

(i) Initialize all parameter blocks $\boldsymbol{\beta}_l$ and vectors of function evaluations $\boldsymbol{f}_j$ with suitable starting values $\hat{\boldsymbol{\beta}}_l^{(0)}$ and $\hat{\boldsymbol{f}}_j^{(0)}$ and set the predictor $\boldsymbol{\eta}^{(0)} =$ offset with a suitable offset (such as the population minimizer of the loss function). Choose a maximum number of iterations m_{stop} and set the iteration index to $m = 1$.

(ii) Compute the negative gradients (i.e., partial derivatives of the loss function)

$$u_i = -\frac{\partial}{\partial \eta} \rho(y_i, \eta)\Big|_{\eta=\hat{\eta}_i^{(m-1)}}, \quad i = 1, \ldots, n,$$

that will serve as working responses for the base-learning procedures.

(iii) Fit all base-learning procedures to the negative gradients to obtain estimates $\hat{\boldsymbol{b}}_l^{(m)}$ and $\hat{\boldsymbol{g}}_j^{(m)}$ and find the best-fitting base-learning procedure, i.e., the one that minimizes the L_2 loss

$$(\boldsymbol{u} - \hat{\boldsymbol{u}})'(\boldsymbol{u} - \hat{\boldsymbol{u}}),$$

inserting either $\boldsymbol{X}_l \hat{\boldsymbol{b}}_l^{(m)}$ or $\hat{\boldsymbol{g}}_j^{(m)}$ for $\hat{\boldsymbol{u}}$.

(iv) If the best-fitting base-learner is the linear effect with index l^*, update the corresponding coefficient vector as

$$\hat{\boldsymbol{\beta}}_{l^*}^{(m)} - \hat{\boldsymbol{\beta}}_{l^*}^{(m-1)} + \nu \hat{\boldsymbol{b}}_{l^*}^{(m)},$$

where $\nu \in (0, 1]$ is a given step size, and keep all other effects constant, i.e.,

$$\hat{\boldsymbol{\beta}}_l^{(m)} = \hat{\boldsymbol{\beta}}_l^{(m-1)}, \; l \neq l^* \quad \text{and} \quad \hat{\boldsymbol{f}}_j^{(m)} = \hat{\boldsymbol{f}}_j^{(m-1)}, \; j = 1, \ldots, q.$$

Correspondingly, if the best-fitting base-learner is the nonlinear effect with index j^*, update the vector of function evaluations as

$$\hat{\boldsymbol{f}}_{j^*}^{(m)} = \hat{\boldsymbol{f}}_{j^*}^{(m-1)} + \nu \hat{\boldsymbol{g}}_{j^*}^{(m)}$$

and keep all other effects constant, i.e.,

$$\hat{\boldsymbol{\beta}}_l^{(m)} = \hat{\boldsymbol{\beta}}_l^{(m-1)}, \; l = 1, \ldots, L, \quad \text{and} \quad \hat{\boldsymbol{f}}_j^{(m)} = \hat{\boldsymbol{f}}_j^{(m-1)}, \; j \neq j^*.$$

(v) Unless $m = m_{\text{stop}}$, increase m by one and go back to (ii).

effects of categorical covariates with a large number of levels to prevent a preference for coefficient blocks having a larger number of coefficients and therefore more flexibility.

A further improvement of the model choice capabilities of boosting can be achieved when decomposing certain types of effects. For example, a nonlinear effect $f(z)$ of a continuous covariate z may be represented as

$$z\beta + \tilde{f}(z),$$

i.e., as the sum of its linear effect $z\beta$ and the deviation from the linear effect $\tilde{f}(z) = f(z) - \beta z$. This has the advantage that the boosting algorithm can automatically differentiate between no effect of z (neither the linear nor the nonlinear effect are selected), linear effect of z (only the linear part is selected), and nonlinear effect of z (the nonlinear part is also selected). Technically, such a decomposition can be accomplished based on the mixed model representation of penalized least squares estimates introduced in Sect. 8.1.9.

We will illustrate these extended model choice abilities of boosting, as well as the general principles of the algorithm for structured additive regression models in the following case study on malnutrition in Zambia.

9.8 Case Study: Malnutrition in Zambia

Using the data on malnutrition in Zambia this section illustrates how the methods presented in this book can be applied in practice. In the next section, we first provide some general guidelines on how to carry out a regression analysis. We then present the case study on malnutrition in Zambia.

9.8.1 General Guidelines

Regression problems can be generally divided into the following main steps:

Descriptive Analysis of Raw Data

The first important step in *every* statistical application is the careful analysis of the raw data. We first need to get an overview of the (univariate) distribution of all variables in the data set. Useful tools are the graphical devices and summary statistics outlined in Sect. 1.2.1. The following goals are pursued in this first analysis:

- *Description of the distribution of all variables in the data set.*
- *Discovery of data anomalies.* This includes incorrect coding, implausible and extreme data values. In regression problems, the inspection of the covariate domain is especially important. For some continuous covariates the observed data are often sparse, particularly near the boundaries. With such limited information in some regions, it is often difficult to make valid statements regarding the effect

of a covariate on the response. Sometimes it is even appropriate to exclude extreme observations from further analysis. A typical example was the analysis of the data on patent opposition in Example 2.8 (p. 35), where we could observe an extremely positively skewed distribution for the covariate *nclaims*. The majority of observations lie in the region 0–60. However, some observations were much larger, with the maximum value of 355. Due to the fact that very little information is available in the interval 61–355, all observations with *ncountry* > 60 were excluded from further analysis.

The problem of unequally distributed values can also exist with categorical covariates. For some categories, absolute or relative frequencies are often very low. In the extreme case, a covariate may be useless for regression analysis. This is, for example, the case with binary covariates when one of the two categories is not populated at all, or perhaps only very sparsely. The missing or limited variability of the covariate then precludes examination of the effect on the response variable. In many cases, the problem of sparsely populated categories can be solved by combining individual categories (when meaningful).

Data Preparation

After the initial analysis of the raw data, the data must be prepared for the regression analysis. This means that some extreme observations, possibly even some covariates, may be excluded from further analysis. Outlier analysis and plausibility checks are useful tools to assist in this regard. Furthermore, existing data errors need to be corrected, and possibly new covariates derived from existing covariates need to be created (e.g., orthogonal polynomials or other transformations).

Finally, we must ensure that the raw data does not get lost during data preparation. This may sound rather trivial, but lost raw data are a frequent phenomenon in practice. It is strongly recommended to create a new data set out of the raw data containing the prepared observations and variables ready for regression analysis. Also, the necessary steps for data preparation should be recorded and documented.

Graphical Two-Dimensional Correlation Analysis

In the case of continuous responses, two-dimensional graphical correlation analyses between the response values and each covariate can provide important information about the type and strength of the relationship. However, any interpretation should be taken with care, since at this stage we are only investigating two-way correlations, and we are generally not controlling for the effect of other covariates. Hence, it is absolutely possible that the further regression analysis provides different results. Useful devices for the correlation analysis are the graphical tools outlined in Sect. 1.2.2 (scatter plots, box plots). We should also take care to find possible outliers (see Sect. 3.4.4). Often—but not always—such observations can be identified with the help of scatter plots.

In the case of discrete responses, in particular categorical variables, graphical exploratory tools can be of rather limited use.

Estimation of Preliminary Working Models

The semiparametric regression models discussed above and in the previous chapter are particularly useful for exploratory data analysis and for getting a first working model. With these models, nonlinear covariate effects, as well as spatial and cluster-specific heterogeneity, can be detected quickly and automatically. Usually, the starting point is a pure main effects model. Interactions can be considered in subsequent steps. In addition to the results from the univariate and bivariate exploratory analyses, expertise knowledge and knowledge from former examinations should be considered in the model building process. It is generally not advisable to include all available variables in an unreflected way into the regression model.

Generally, a number of competing model specifications will be available. For example, it is possible to have models with and without spatial components or perhaps the effect of a variable is doubtful, so we may estimate models with and without the covariate of concern. We can then compare the competing models with a goodness-of-fit or model choice criterion. Based on this criterion, model specifications with little impact on the fit can then be excluded from further analysis.

In the end, there is generally not the *one perfect model* which dominates the other models in all aspects. Rather there are several competing models which fit the data (almost) equally well.

Model Diagnostics, Model Evaluation, and Improvement of the Working Models

The main goal of this step is to improve or to refine the models of the previous step. The following steps can be taken:

Reduce to a Simple Parametric Model

In a number of cases, the covariate effects, which were discovered through the semiparametric analysis, can be nearly linear or rather simple nonlinear functions. In such a case, parts of the model can be replaced by simple parametric terms, even using piecewise linear or constant terms.

Exploratory Analysis of the Spatial or Cluster-Specific Effects

The spatial or cluster effects can be interpreted as surrogates for unobserved or unavailable covariate effects. In some cases, a careful analysis of the heterogeneity effects provides information about new covariates or aspects, which thus far had not been taken into consideration. For example, when analyzing the claim frequency of the vehicle insurance data (see Example 2.12, p. 52), we discovered that the claim frequency is higher in urban than in rural areas. In such a case, we might be able to replace the nonparametric spatial effect with a binary dummy variable, which identifies the urban areas. The ideal outcome is, in any case, the replacement or at least the decrease of the spatial or cluster-specific effects. This can be achieved by including new and reasonably interpretable covariates derived by investigating the heterogeneity effects.

As Example 9.7 (p. 556) shows, the inclusion of spatial or cluster-specific effects can lead to identifiability problems. In particular, the effects of other covariates can be concealed by the spatial or cluster effects. Hence, it is advisable to carefully examine the differences in results of models with and without spatial or cluster-specific components. As mentioned, this problem is subject to growing research; see the references given on p. 556.

Model Diagnostics

Model diagnostic tools aim at checking the adequacy of a working model and whether certain assumptions, such as homoscedasticity of the error terms, are fulfilled. This may ultimately lead to the consideration of alternative model formulations and appropriate refinements of the working model. For parametric linear models, there are many useful model diagnostic tools; see Sect. 3.4.4. For semiparametric models, many of the tools, in particular residual analysis, can be easily adapted. Some diagnostics for outlier analysis are, however, either not at all or methodologically less developed or not included in software packages. In models with discrete responses, the use of residual plots is also limited.

In the case of continuous responses, residual plots provide information about omitted or misspecified covariate effects, outliers, and heteroscedastic or correlated errors. Depending on the findings, different alternatives are conceivable:

- *Misspecified covariate effect.* In case of misspecified covariate effects, we have the following options:
  - *Nonlinear effect.* The effect of a continuous covariate, initially modeled as linear, proves to be nonlinear. In this case, more flexible modeling of the effect, e.g., using spline-based methods is in order.
  - *Missing covariates.* Available covariates that were initially not incorporated can be included in the model at any time. The situation is more difficult with unobserved covariates. In this case, spatial or cluster-specific effects can be used for modeling unobserved heterogeneity.
  - *Missing interactions.* Complex interactions can be estimated either with the parametric approaches discussed in Sect. 3.4, or with the nonparametric alternatives discussed in Sect. 9.3. However, interactions should not be unnecessarily used, as they can greatly increase the complexity of the model.
- *Heteroscedastic errors.* In the presence of heteroscedastic errors, we can generally use the alternatives discussed in Sect. 4.1.3 (p. 186), i.e., variance stabilizing transformation of the response, two-step estimation, or simultaneous estimation using, e.g., the R package `gamlss`. The latter is restricted to purely additive models. In parametric linear models, we can also use the corrected covariance matrix of White presented in Sect. 4.1.3 on p. 190.
- *Correlated errors.* As discussed in Sects. 3.1.2 and 4.1.4, the existence of autocorrelation in the residuals indicates model misspecification. The primary goal should be to eliminate the misspecification as much as possible. The use of semiparametric models with correlated error terms is still in the development phase. For instance, identifiability problems are to be expected, e.g., when a

nonparametric time trend and correlated errors are simultaneously modeled; see Fahrmeir and Kneib (2008).

- *Outliers.* In case of outliers, we refer to the guidelines discussed in the context of linear models, see Sect. 3.4.4, p. 163. However, the use of robust procedures is limited with semiparametric models as the methodology is much less developed for this model class. A notable exception is median regression as a special case of quantile regression which in turn has undergone rapid development in recent years. Apart from the search for input data errors and/or possible explanations, probably the most important measure in the context of outliers is to refit the model without the outlying observations and to describe the differences in model fit.

Check Stability of Estimates Under Different Estimation Concepts

In many cases the models under consideration can be estimated using alternative estimation concepts. For instance, variable selection in the linear model can be performed by minimizing a model choice criterion (Sect. 3.4), with the LASSO (Sect. 4.2.3), using boosting (Sect. 4.3) or based on Bayesian model averaging (Sect. 4.4). (Generalized) linear mixed models can be estimated using maximum likelihood (Sect. 7.3), or fully Bayesian inference (Sect. 7.4). In structured additive regression models estimation may be based on either penalized likelihood in combination with the optimization of a model choice criterion (Sect. 9.6.1), a mixed model representation (Sect. 9.6.2), a fully Bayesian approach (Sect. 9.6.3), or boosting (Sect. 9.7). We generally recommend to estimate the models under consideration with as many estimation concepts as possible to check the stability of estimation results.

Presentation of Results

As already mentioned, the result of a regression analysis is generally not the one "best" model which dominates the other models in all aspects. Rather there are a few models which fit the data (almost) equally well. Usually, these models only differ in some minor aspects, which often are not relevant to the problem. When outlining and describing the results, the common core of all models should be pointed out. On the other hand, the differences between the competing models should also be presented. Regarding these differences, we face uncertainty about our final conclusions.

Finally we want to point out that any analysis should be comprehensible and reproducible for those who did not participate in the investigation. It is also essential that the software, in particular the code implemented for data analysis, is archived and well documented. Compare also Koenker and Zeileis (2009) who propose software tools that may be helpful for reproducible research.

9.8.2 Descriptive Analysis

When presenting the malnutrition data in Chap. 1, we already provided summary statistics and graphical visualization for the variables in the data set; see Table 1.3 (p. 7) and Fig. 1.4 (p. 14). The results can be interpreted as follows:

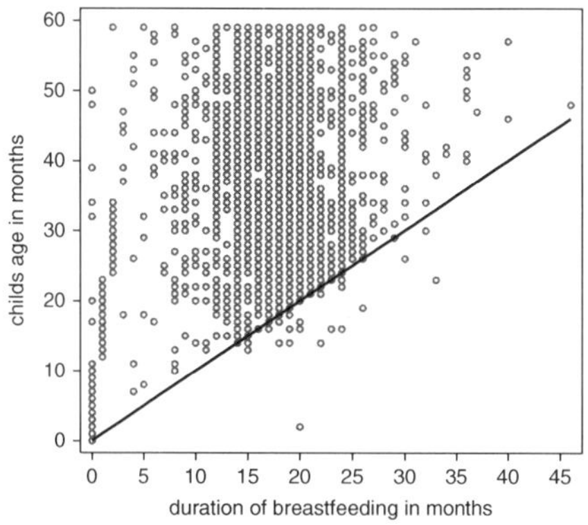

Fig. 9.15 Malnutrition in Zambia: *scatter plot* between the age of the child and the duration of breast-feeding. Observation below the main diagonal are implausible

- *Z-score:* The Z-score has a nearly symmetric distribution with values between −600 and 503, with the majority of the observations between −600 and 300. The average Z-score is −171.19, i.e., on average the children in Zambia are much smaller compared to children of the same age in the reference population. Implausible or extreme values cannot be found.
- *Child's age and duration of breast-feeding:* Since the duration of breast-feeding cannot last longer than the age of the child, these two variables are related and should be considered together.

  A preliminary univariate analysis reveals that age is nearly uniformly distributed. There are only slightly fewer older children than younger ones in the data set. The unusually rough form of the histogram appears to be more of an artifact, as is confirmed by a closer look at the frequency table or through refitted histograms based on different bandwidths.

  Duration of breast-feeding varies in the data set from 0 to 46 months. The distribution is clearly bimodal with modes at 0 and 18 months. Almost 25 % of the children were never breastfed. It is very rare that the duration of breast-feeding lasts longer than 30 months (only 24 observations).

  As a plausibility check, Fig. 9.15 includes a scatter plot between the age of the child and the duration of breast-feeding. As the duration of breast-feeding cannot last longer than the child's age, all observations below the bisecting line are implausible and thus are excluded from further analysis. The observations with $c\_breastf > 30$ are also excluded for the following two reasons: First, breast-feeding rarely lasts longer than 30 months; secondly, two-dimensional smoothers are used for modeling the effect of age and duration of breast-feeding. Plausible results with two-dimensional smoothers are obtained only with a solid data base.
- *Mother's age at birth:* The mother's age varies from 13 to 48 years, with a mean value of 26.40 years. The distribution is clearly skewed to the right. Some very small and large values are unusual, but not impossible.

- *Mother's height:* The mother's height varies between 134 and 185 cm, with a mean value of 158.06 cm. The distribution is almost symmetric, without any apparent data anomalies.
- *Mother's BMI :* The body mass index is a known (but somewhat controversial) measure for the nutritional status of adults. A BMI between 18.5 and 25 is considered to be normal. A BMI less than 18.5 is an indication of underweight. A BMI between 25 and 30 is an indication of (some) overweight. A BMI of 30 is an indication of extreme overweight or obesity. In our data set, the BMI values vary from 13.15 to 39.29, with a mean value of 21.99. The majority of mothers are in the range of normal weight to minor overweight. A percentage that should not be neglected is that of underweighted mothers (nearly 10 %). A smaller percentage of mothers are extremely overweight (about 2.6 %). No particularly implausible or striking values are found in these data.
- *Mother's education and work status:* The predominant majority of the mothers (about 80 %) have not completed their education. About 17 % have completed primary education, and only 1.75 % have a higher school degree. Slightly more than half of the mothers work, about 45 % of the mothers do not work. In both variables, all categories have sufficient data, so that it is neither necessary to combine, nor exclude, categories.

In Chap. 1, we also performed graphical correlation analyses, as seen in Figs. 1.8 (p. 17) and 1.9 (p. 18). The plots can be interpreted as follows:

- *Child's age:* Initially the average Z-score decreases linearly and then remains almost constant for children of age higher than about 15–20 months. The observed relationship is postulated similarly in the malnutrition literature. It is assumed that the nutritional status is best during the first 6 months after birth since the child is well nourished first in the womb, and then after birth through the mother's milk. From this point a deterioration in the nutritional status is expected, as the child is more and more exposed to insufficient, qualitatively inferior, and more unsanitary food. Some stabilization is expected during the child's second year, but on a very low level.
- *Duration of breast-feeding:* The WHO recommends to breast-feed newborns for at least 12 months, since the mother's milk is the best form of nutrition for the baby during the first year. The literature about malnutrition, thus, assumes a generally positive effect of breast-feeding. If children are breast-fed longer than for 1 year, it is assumed that the effect decreases or may even become negative. Breast-feeding for a long time is then understood as an indicator for insufficient food supply.

  In the scatter plot a relation between the Z-score and the duration of breast-feeding is barely detectable. There is a slight tendency towards a linearly decreasing relationship, i.e., with an increase in duration of breast-feeding, the average Z-score decreases. This would, however, contradict the literature. In any case, we should be careful when evaluating the effect of the duration of breast-feeding as we have to expect an interaction with the child's age. The effect of 1 month of breast-feeding for a 1-month-old child should differ from the same effect on a child of 3 years. In the first case, the duration of breast-feeding for

1 month is absolutely normal. In the second case, breast-feeding was stopped very early. We will therefore model the effect of child's age and duration of breast-feeding with a two-dimensional surface.

- *Mother's BMI:* The literature suggests an inverse U-form for the effect of the nutritional status of the mother, i.e., children of mothers with underweight or extreme overweight show a worse nutritional status than children of mothers who are of normal weight. The scatter plot shows largely a linear relationship with the Z-score, i.e., on average the Z-score increases linearly with BMI. From a BMI of 30–35 upward, i.e., in the case of obese mothers, the average effect seems to decrease, so that the postulated inverse U-form is reflected to some extent.
- *Mother's age at birth:* It seems as if the age of the mother has a weakly linear effect, i.e., the older the mother at the time of birth, the better the average nutritional status. In fact, the literature assumes that older mothers may take better care of a child than younger ones. These women have a greater chance of having completed vocational education, and they have in general more work experience leading to a higher income.
- *Mother's height:* It is well known that the height of the parents has an effect on the height of the children. Since the age-standardized height is used as a measure for the nutritional status, it is not surprising that we find a linearly increasing effect for the height of the mother in the scatter plot.
- *Education and work status:* We find a better nutritional status for children of better educated mothers. A relationship with the employment status cannot be observed.
- *Sex:* Gender-specific differences are not detectable (plot is not provided).

9.8.3 Modeling Variants

Based on the formulated hypotheses for the effect of the explanatory variables and the results of the descriptive analysis, the following modeling variants were examined. In order to be able to present the results in a compact way, we differentiate between different aspects of the models using the corresponding symbols:

- *Spatial effect:* Possible spatial heterogeneity was modeled either by a Markov random field (modeling variant M), by i.i.d. random effects (R), or parametrically using dummy variables for variable *region* (P).
- *Child's age and duration of breast-feeding:* As stated, the effect of the duration of breastfeeding interacts with the child's age. Therefore, the effect of both variables was estimated by a two-dimensional surface based on kriging (modeling variant I). We preferred kriging over other two-dimensional smoothers, for example, two-dimensional P-splines, as kriging appears to be most appropriate for the non-rectangular range of values associated with these two variables; see Fig. 9.15. Besides the two-dimensional modeling approach, also a pure nonparametric effect of age excluding the effect of the duration of breast-feeding was considered (modeling variant N). We used a P-spline with 20 interior knots based on a second-order difference penalty.

- *Mother's age at birth, height, and BMI:* The effects of the remaining continuous variables were modeled in four different ways:
  - All effects in a nonparametric way with P-splines (model variant N)
  - Nonparametric effect for the mother's body mass index and linear effects for the two other variables (B)
  - Assuming a quadratic polynomial for the body mass index and a linear effect for the two other variables (Q)
  - Assuming all effects to be of simple linear, parametric form (P)
- *Categorical covariates:* Every model includes the categorical variables *c_gender*, *m_education*, and *m_work* using appropriate dummy variables

The examined models can be uniquely characterized by a combination of three letters. For instance, the combination (MIN) denotes the model with a Markov random field for the spatial effect, a kriging approach for the two-dimensional interaction effect of age with the duration of breast-feeding, and nonparametrically modeled effects of the remaining continuous covariates. The model (RNQ) consists of i.i.d. random effects for the spatial effect, a nonparametric effect of the child's age, a quadratic effect for the mother's body mass index, and a linear effect for the remaining covariates.

9.8.4 Estimation Results and Model Evaluation

All models under consideration were estimated using mixed model technology and with `remlreg objects` of the software `BayesX`. Refer to Sects. 8.1.9 and 9.6.2. Also, we use the (conditional) AIC to assess the goodness of fit. For comparison, the models were also estimated with a fully Bayesian approach using MCMC techniques and `bayesreg objects` of the software `BayesX`. The differences in the resulting estimates are relatively small, so that the following presentation of results is restricted mostly to mixed model based inference. We will point out cases with noteworthy differences.

Of all examined models, the model (MIB), based on a Markov random field for the spatial effect, two-dimensional kriging for the effect of the child's age and the duration of breast-feeding, a nonlinear BMI effect, and otherwise linear effects, has minimal AIC. For all models, Fig. 9.16 shows AIC differences relative to the AIC of this "best" model. The models are sorted in ascending order according to their AIC values, so that models performing better in terms of AIC appear on the left. For the sake of comparison, a strictly parametric model was fitted as well. In this model, the age effect was modeled through a linear spline with one interior knot at $k\_age = 23.5$. For this model, we obtain an AIC difference of 308, which is by far the worst fitting model.

Figure 9.16 reveals the following conclusions:

- The models which include a two-dimensional kriging effect of age and the duration of breast-feeding are superior to the models with only a nonparametric effect of the child's age. The differences in AIC are large between these two types of models so that we can safely decide for the model comprising an interaction term.

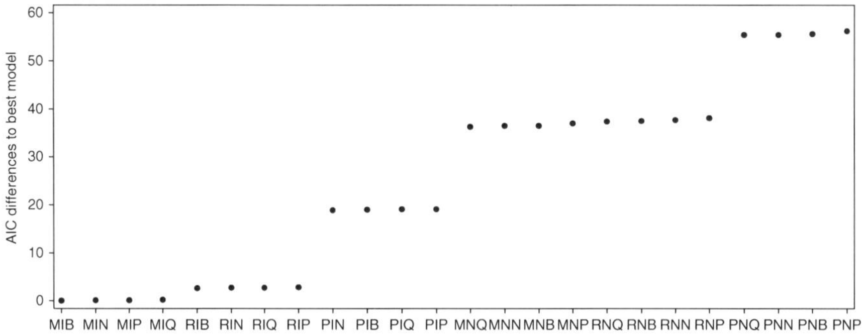

Fig. 9.16 Malnutrition in Zambia: AIC-differences to the best model for various model specifications

- Regarding the spatial effect, we obtain the best results when using a Markov random field. The use of i.i.d. random effects provides slightly higher AIC values of approximately 2.5 points. The parametric approach based on region dummies is not competitive. Therefore we can conclude that a more detailed spatial effect than the one based on dummy variables is indeed needed, but that the correlation structure specified for the spatial effect does not make too much of a difference.
- As Fig. 2.15 (p. 51) shows, the effects of the remaining three continuous variables are very smooth and do not differ much from a straight line. Therefore the four modeling variants for these variables have approximately the same AIC values. In all cases, the estimated effects are more or less indistinguishable. Only for the mother's BMI there is uncertainty whether a linear or a weakly nonlinear effect is more appropriate. In this situation, the restrictions of the conditional AIC in Sect. 9.6.2 (p. 567) have to be taken into account when interpreting the results. There is a bias in the conditional AIC towards more complex models and therefore we would probably decide for linear effects in case of age and height of the mother due to the very small difference in the AICs between linear and nonlinear models. For the effect of the BMI of the mother, the situation is less clear and the detected minor nonlinearity should only be interpreted with care. It can then be useful to consider alternative estimation approaches or model choice techniques to validate the results from the AIC (and we will do so in the following section).

Figures 9.17 and 9.18 show the estimated effects of the continuous covariates and the spatial effect for the model (MIN), where the effects of all continuous variables were modeled in a nonparametric way. This is the model with the second best AIC value. Table 9.4 shows the results for the categorical covariates. We finally point out that Fig. 2.15 (p. 51) illustrates the results for the model (MNN), which differs from model (MIN) in that it does not consider the duration of breast-feeding.

The results can be interpreted as follows:

- *Child's age and duration of breast-feeding:* Initially the average Z-score decreases linearly as age increases. For children older than 18 months, the

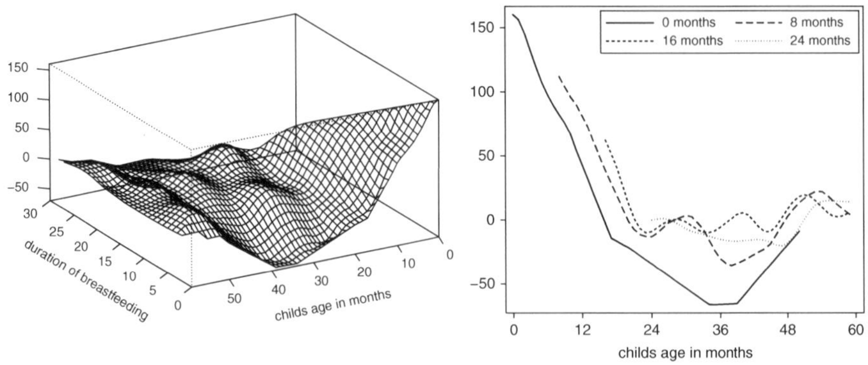

Fig. 9.17 Malnutrition in Zambia: interaction of the child's age and duration of breast-feeding in the model (MIN) (*left*). The *right panel* shows the age effect for various durations of breast-feeding

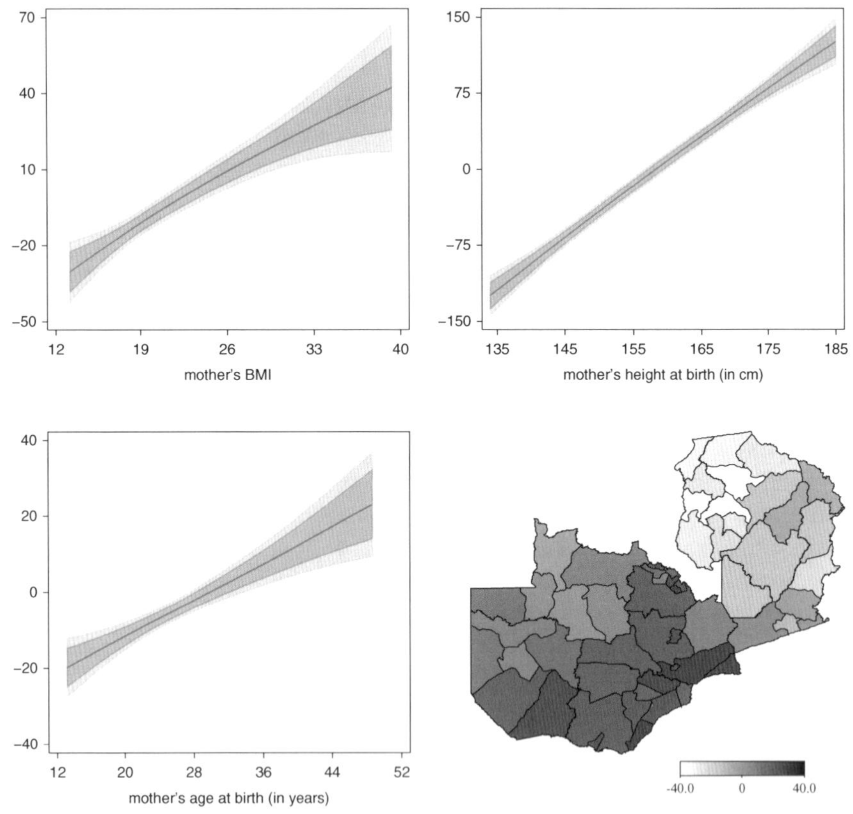

Fig. 9.18 Malnutrition in Zambia: nonparametric effects and spatial effect for the model (MIN)

Table 9.4 Estimated parametric effects in the model (MIN)

Variable	Coefficient	Standard-error	95 % confidence-interval	
intercept	−186.926	22.685	−231.398	−142.454
c_gender	−15.144	3.606	−22.213	−8.075
m_education1	−11.043	5.042	−20.927	−1.158
m_education3	21.774	5.055	11.863	31.684
m_education4	80.814	14.184	53.008	108.620
m_work	−4.420	3.807	−11.883	3.044

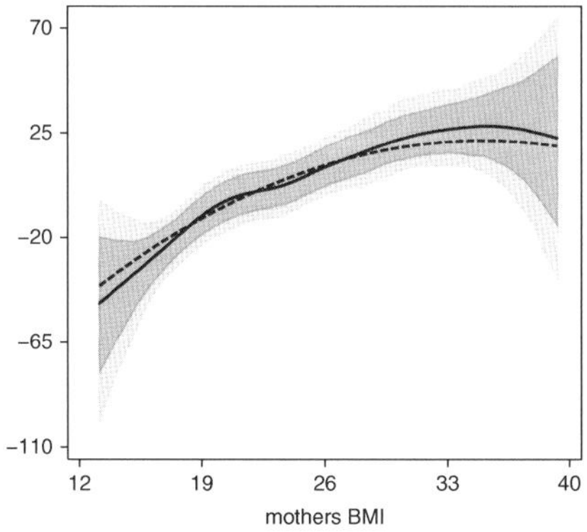

Fig. 9.19 Malnutrition in Zambia: estimated effect of the mother's BMI for some alternative model specifications. The graph displays a nonlinear estimator (including 80 % and 95 % credible intervals) based on a fully Bayesian approach estimated with MCMC techniques. Additionally included is a quadratic polynomial (*dashed line*) in the model (MIB)

nutritional status stabilizes at a low level. For children older than 3 years, a slight improvement can be observed. The curve for children who were never breast-fed is clearly below that of children who were breast-fed for 8, 16, and 24 months. We, thus, observe a positive effect of breast-feeding. However, differences regarding the duration of breast-feeding are not visible. Overall, the hypotheses formulated above can be mostly confirmed.

- *Mother's BMI:* The estimated effect demonstrates that the nutritional status of children with underweight mothers is worse when compared to children of mothers who are of normal weight or mothers who are slightly overweight. For children of obese mothers, the strong linear effect diminishes. Based on these results, the average nutritional status is not decreasing with obese mothers as was initially postulated with the inverse U-form of the effect.

  However, within the range of obese women, we find much wider confidence intervals caused by a small sample size. In this situation, it can be useful to take a look at the results of alternative models. In the model (MIQ), with AIC value almost equal to the best model, the BMI effect is modeled by a quadratic polynomial. Here, a tendency towards the postulated inverse U-form is visible and the average Z-score slightly decreases for extremely overweight women, as provided in Fig. 9.19. For the effect of mother's BMI, we also find the most visible differences compared to fully Bayesian estimation based on MCMC methods (Fig. 9.19): We obtain an even more pronounced U-form compared to the quadratic polynomial.

What conclusions can be drawn from these contradictory results? It is common practice to present the results that are closest to the initially formulated hypotheses (in our case the results from the Bayesian approach) and to dismiss all other results. This is of course not correct. We recommend to point out the common core of the various results and then to describe the differences among them. In our case study, we observe an almost linear effect for mothers with underweight, normal weight, and slight overweight. For obese mothers, the BMI effect is less clear. The linear effect diminishes or even reverses and slightly decreases. For an effect of inverse U-form effect postulated in the literature, we have some evidence but no compelling proof. To obtain more precise results, we could, for example, overrepresent extremely overweight mothers in subsequent studies. If this is not possible, we could further examine data from other countries. We will reiterate the question on the precise form of the effect of mother's BMI in the next section when discussing approaches for automatic function selection.

- *Mother's height:* This effect is clearly linear. The taller the mother, the taller the child is, on average, and so is the Z-score. This effect, however, does not provide any conclusions about the nutritional status. On the contrary, this relatively large effect, that is mostly onset through heredity, questions the Z-score as (the only) measure for chronic malnutrition.
- *Mother's age at birth:* This effect is essentially linear as well. On average, older mothers have better nourished children. For this variable, we also found slight differences in results for the alternative Bayesian approach, which finds a steep increase of the average Z-score for mothers older than about 40 years of age. Thus, the positive effect would increase for very old mothers. Hence, our results are consistent with the hypotheses formulated in the previous section.
- *Spatial effect:* We notice a pronounced north–south gradient, i.e., the nutritional situation is much better in the north relative to the south. One reason for these pronounced differences could be that the climatic conditions are worse in the south, as these regions show a lower altitude than those in the north.
- *Categorical covariates:* The estimated parameters associated with the education dummies demonstrate a strong education effect. The more educated the mother, the better is the nutritional status of the child. The difference in Z-score between children whose mothers have completed primary education, compared to those whose mothers have not completed education, is on average only 11 points. The differences for children, whose mothers have been further educated are much more pronounced. Children whose mothers have higher education show on average a Z-score that is 90 points higher compared to children whose mother has not completed primary education.

  The employment status of the mother does not seem to be of great importance. The effect is clearly nonsignificant.

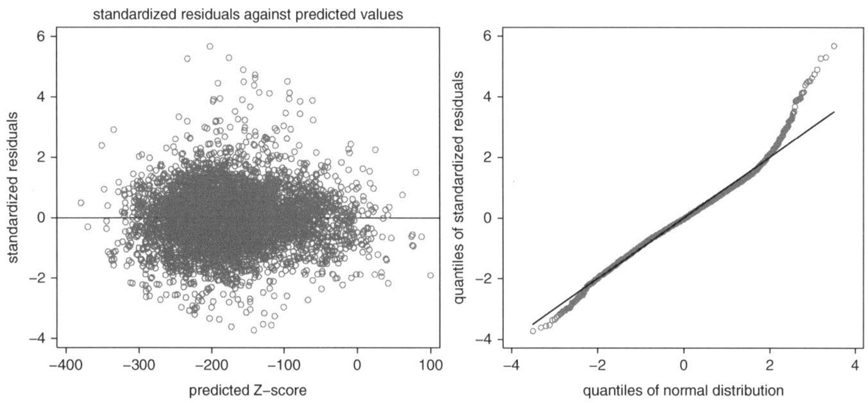

Fig. 9.20 Malnutrition in Zambia: standardized residuals versus predicted values (*left*) and a QQ-plot of standardized residuals (*right*) in the base model

As often stated in the literature, girls seem to be more robust than boys. Boys have a Z-score which on average is 15.14 points lower than girls.

Taking a look at the standardized residuals against the predicted Z-scores provides evidence about existing heteroscedasticity (left panel of Fig. 9.20). A two-step estimation approach that considers heteroscedasticity does not provide any noteworthy differences in results. The QQ-plot of the residuals provides clear evidence about departures from the normality assumption, especially for very large residuals with values greater than 2. This finding mainly concerns children with a positive Z-score, i.e., children with a very good nutritional status. Generally, minor violations of the normality assumption are not problematic since estimators are often relatively robust against such departures from normality. Possible alternatives for further analysis could be approaches that allow for simultaneous modeling of the mean and the skewness of the distributions; see our sketch of additive models for location, scale, and shape (GAMLSS) in Sect. 2.9. More details can be found in Rigby and Stasinopoulos (2005), Rigby and Stasinopoulos (2009), and the GAMLSS homepage http://gamlss.org/.

9.8.5 Automatic Function Selection

In the previous section, we have manually implemented function selection based on the comparison of different model specifications using AIC. However, recent developments in function selection also allow for the simultaneous determination of the unknown effects in a structured additive regression model combined with automatic function selection. In the following, we will illustrate such an approach relying on boosting utilizing function `gamboost` from the R package `mboost`. The approach proposed in Belitz and Lang (2008) yields comparable results.

For the boosting approach, we make use of the penalized least squares base-learners approach described in Sect. 9.7. Therefore, we define a model of maximum complexity that comprises all candidate terms that should be considered in the component-wise fits during each boosting iterations. More specifically, we include the following effects:

- All categorical covariates (gender of the child, education, and work status of the mother) are represented as dummy-coded variables, where the complete block of dummies for each variable is treated as one term that can be selected or deselected in each boosting iteration.
- The interaction between age of the child and duration of breast-feeding is modeled in a kriging specification exactly as specified in the previous section.
- All remaining continuous covariates (mother's age at birth, height, and BMI) are modeled as potentially nonlinear effects based on penalized splines. To allow for a more detailed investigation of the nonlinear effects, we apply a decomposition into a linear model component and the nonlinear deviation from the linear model for each effect based on the mixed model representation as discussed in Sect. 8.1.9. This has the advantage that the boosting algorithms compares linear and nonlinear effects of a continuous covariate in each iteration and only selects the nonlinear effect if it is really required. Note that the nonlinear effect should be defined as the deviation from the linear effect such that it does not comprise the linear effect as a special case. This allows for a proper discrimination, while using a naive penalized spline specification would always select the nonlinear effect since it contains additional flexibility as compared to the linear effect.
- For the spatial effect, we simultaneously included the dummy-coded regional variable, a Markov random field for the districts, and i.i.d. random effects for the districts as competitors within the same model. Since boosting will include function selection along with estimation, this allows us to determine automatically whether only one of the three types of spatial effects is required or whether a combination of two or even all three types of spatial effects is preferable. This is in contrast to the analyses in the previous section, where we exclusively considered one spatial modeling possibility at a time.

To further improve the selection properties, we chose the smoothing parameters in the penalized least squares base-learners such that each base-learner is assigned exactly one degree of freedom. This effectively makes each base-learner comparable to a linear effect of one single covariate. Note that we also have to use a ridge-penalized base-learner for categorical covariates with more than two categories. See Hofner, Hothorn, Schmid, and Kneib (2012) for details.

Figure 9.21 shows the cross-validated prediction error for this model specification for 10-fold cross validation. The minimum is reached within 7,563 iterations based on a total of 10,000 initial iterations. Note, however, that the risk is already very flat much earlier, and therefore the prediction performance seems to be rather insensitive to the exact values of the boosting iterations. In the following, we will present all results based on the model restricted to the optimal 7,563 iterations.

The main advantage of the boosting approach is its automated variable and function selection ability. Table 9.5 therefore shows the selection frequencies over the boosting iterations (second column) obtained for the different model terms. Any term not included in the table has never been selected during the first 7,563 iterations

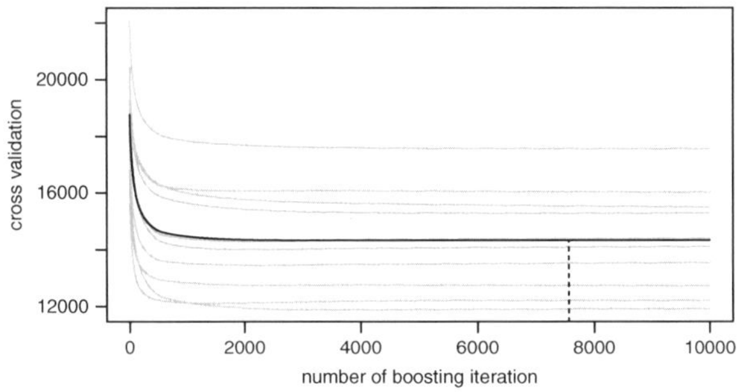

Fig. 9.21 Malnutrition in Zambia: cross validation criterion (*black line*) as a function of the boosting iterations. The *grey lines* indicate the prediction error obtained within a single cross validation fold. The *dashed vertical line* shows the optimal boosting iteration, i.e., the minimal cross validation criterion

Table 9.5 Selection frequencies in the boosting iterations and in 50-fold cross validation for the model comprising Markov random field and random effects

Variable / Term	Iterations	CV folds
$f(c\_age, c\_breastf)$	0.452	1.00
random effects	0.333	1.00
m_education	0.073	1.00
Markov random field	0.038	0.74
region	0.033	1.00
$f(m\_bmi)$	0.028	1.00
$f(m\_agebirth)$	0.014	0.06
c_gender	0.008	1.00
$f(m\_height)$	0.007	0.54
m_agebirth	0.007	1.00
m_work	0.002	0.74
m_height	0.002	1.00
m_bmi	0.001	1.00

and is therefore dropped from the model. We can summarize the main findings from this table as follows:

- All three types of spatial effects are selected by the boosting algorithm. Hence, none of the three effects seems to be able to represent the spatial correlations present in the data to a satisfactory extent when being considered as the sole representative for the spatial effect.
- All three continuous covariates are included with both linear and nonlinear effects. This seems to indicate that nonlinear modeling is required in all three cases, which is in contrast to our findings from the previous paragraph, where at least mother's age at birth and height of the mother were identified to be adequately represented by linear effects. However, the selection frequencies of both these covariates are rather small and the estimated effects are still very close to linear (not shown).

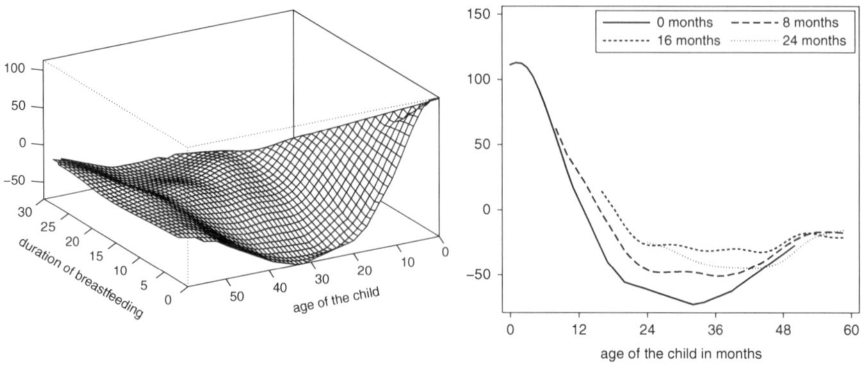

Fig. 9.22 Malnutrition in Zambia: interaction of the child's age and duration of breast-feeding obtained with boosting (*left*). The *right panel* shows the age effect for various durations of breast-feeding

- The interaction between age of the child and duration of breast-feeding (visualized in Fig. 9.22) is clearly identified with rather high selection frequency. The estimated function is much smoother than in the previous section, but still recovers the same general structure.
- Both categorical covariates are included in the model.

We close the discussion of Table 9.5 with a word of caution: though it is tempting to interpret the selection frequencies as a measure of importance for the corresponding term, this is not formally justifiable. The reason is that early inclusions typically contribute more strongly to the reduction of the risk, while later inclusions often only marginally improve the risk. Correspondingly, a variable may be selected very often in late boosting iterations and may still contribute only a negligible part to the overall estimated predictor. On the other hand, an important covariate may be selected in a relatively small number of early iterations and may still be very important for the complete model fit. As a consequence, a combination of the selection frequencies and the estimated effects should be interpreted to get an impression of the importance of the various estimated effects. As a further indication of a variable's importance, we computed 50-fold cross validation results with a small number of only 2,000 boosting iterations and included the frequencies for inclusion of an effect in the 50 folds in column three of Table 9.5. While we do not try to optimize the number of boosting iterations but rather use a small number of iterations, we will be able to identify those variables that are selected early in the boosting algorithm and are therefore important for predicting the nutritional status. In fact, most of the effects selected by the boosting iteration are also selected in all of the 50 folds but some variables seem to be of somewhat minor importance. In particular, the nonlinear effects of age at birth and height of the mother are ranked lower, which is in accordance with the selection results from the previous section. In addition, the Markov random field has smaller selection frequency then the random

Table 9.6 Malnutrition in Zambia: selection frequencies in the boosting iterations and in 50-fold cross validation for the model comprising only i.i.d. random effects for the spatial effect

Variable / Term	Iterations	CV folds
Random effects	0.504	1.00
f(c_age, c_breastf)	0.270	1.00
m_education	0.123	1.00
region	0.068	1.00
c_gender	0.015	1.00
m_agebirth	0.007	1.00
m_height	0.007	1.00
m_bmi	0.003	1.00
m_work	0.003	0.74
f(m_bmi)	0.001	1.00
f(m_height)	–	0.54
f(m_agebirth)	–	0.06

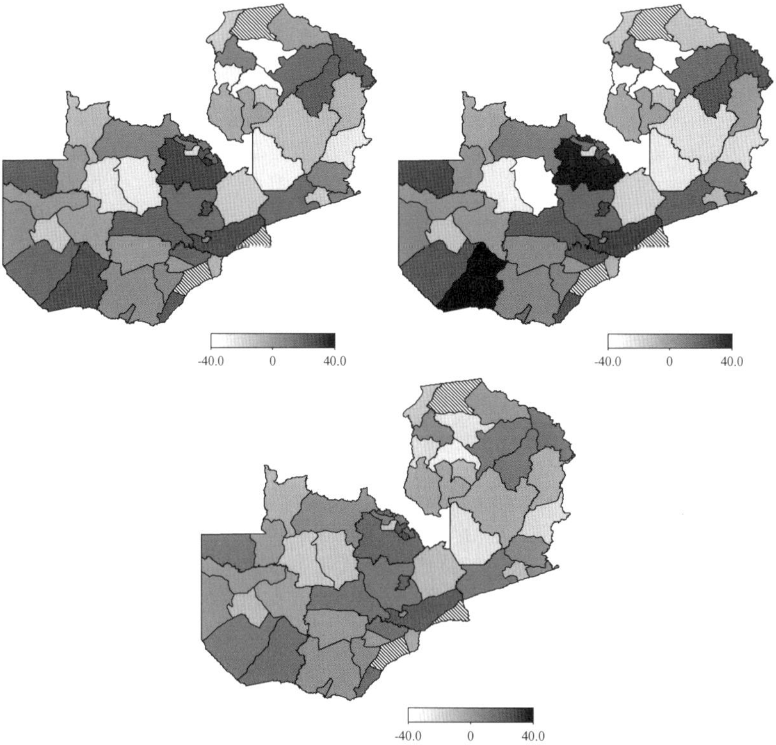

Fig. 9.23 Malnutrition in Zambia: estimated spatial effects obtained with boosting for the three different model specifications comprising Markov random field and random effect (*top left*), Markov random field only (*top right*), and random effects only (*bottom*)

effects term which may give some indication that an unstructured spatial effect is more important than a structured effect.

As a variation of our initial model we re-estimated two more models with different specifications for the spatial effect. More precisely, we include only the Markov and random field or the i.i.d. district-specific random effects for the district-level spatial information and not the combination of the two. For the two resulting models, the optimal boosting iterations were determined as 21,823 (Markov random fields) and 2,723 (i.i.d. random effects) via 10-fold cross validation. Especially in the latter case, a much simpler model specification arises as shown by the selection frequencies in Table 9.6. In this case, the chosen model no longer comprises nonlinear effects of mother's height and age at birth, while a slightly nonlinear effect remains for the body mass index of the mother. This model is therefore even closer to the one identified in the previous section. When considering selection frequencies from a 50-fold cross validation again, we find a similar picture. While most effects identified by the boosting algorithm are included in all 50 folds, the nonlinear effects of age at birth and height of the mother are only selected in only a relatively small number of folds.

Figure 9.23 shows the estimated spatial effects on district level for the three different model variants. In case of the first model comprising the Markov random field and i.i.d. random effects, the sum of both effects is shown to facilitate the comparison between the models. Obviously, all models identify the same spatial structure especially when comparing the combined model with the model consisting of only the Markov random field. The model containing only i.i.d. random effects also shows a similar spatial distribution of the estimated effects, but the magnitude of the effects is somewhat lower (which most probably will also be related to the much smaller number of boosting iterations identified for this model).

9.9 Bibliographic Notes

The literature references about nonparametric regression provided in Sect. 8.4 of Chap. 8 remain valid for this chapter. We will only briefly summarize the key references which specifically treat additive models and their extensions.

The classical textbook about GAMs (which is still worth reading) is that of Hastie and Tibshirani (1990). Marx and Eilers (1998) developed P-spline strategies to directly fit GAMs (without backfitting), while Eilers and Marx (2002) presented P-spline variants of STAR models, referred to as generalized linear additive smooth structures (GLASS). Wood (2006) describes GAMs based on P-splines and their application using the R package package `mgcv`. Ruppert et al. (2003) describe GAMs and their extensions in a unified approach as mixed models. Estimation is based on methods and software for this model framework. Fahrmeir and Kneib (2011) provide a recent overview on Bayesian smoothing approaches for longitudinal, spatial, and event history data comprising both fully Bayes and empirical Bayes methods.

Fan and Gijbels (1996), Loader (1999), and Härdle et al. (2004) propose approaches based on local smoothers. The monograph by Fotheringham, Brunsdon, and Charlton (2002) describes models with spatially varying coefficients where estimation is also based on local smoothers. The book by Yatchew (2003) describes the use of additive models in econometrics. The mathematically demanding monograph by Gu (2002) imbeds additive models into the theory of reproducing Hilbert spaces. Finally, Denison, Holmes, Mallick, and Smith (2002) describe Bayesian adaptive regression splines as building blocks for GAMs and their extensions. The book by Hastie et al. (2009) on "The Elements of Statistical Learning" provides a lucid presentation of contemporary statistical techniques, including boosting.

Quantile Regression 10

Essentially all regression models that we have dealt with thus far have been mean regression models since they relate the predictor η of a regression model to only one specific quantity of the response y, namely the expected value. For example, in case of a generalized linear model (or its extensions) with predictor η, we have

$$E(y) = h(\eta),$$

where h is a known response function. The distribution of the response was then, depending on this mean parameter, completely characterized (sometimes up to a scale parameter common to all observations and potentially with some prespecified weights) by the regression model. In case of Gaussian responses $y \sim \mathrm{N}(\mu, \sigma^2)$, we have already discussed in Sects. 2.9 and 4.1.3 that it can be useful to not only model the expectation μ, but also the variance σ^2 in terms of covariates to deal with heteroscedastic errors. We next develop a regression approach that is completely distribution-free to estimate the effect of a regressor on the *quantiles* of the response distribution. As such, we will also relax a number of assumptions of usual normal regression.

As an example to illustrate the utility of such regression models, consider the effects of living area and year of construction on the monthly net rent in the Munich rent index data. Figure 10.1 shows the corresponding scatter plots together with parametric least squares fits (for a linear model in case of living area and a quadratic fit in case of year of construction). Note that we are not dealing with the rent per square meter that we have previously used for most analyses in this book, but rather with the rent itself since this variable is better suited to illustrate the usefulness of quantile regression. Figure 10.1 clearly indicates that a simple mean regression line obtained from least squares is not sufficient to completely describe the dependency between rent and the covariates. In particular, the variance of the response is changing with the covariate values, at least in case of the living area where we find increased variability for larger apartments. This could be taken into account in a location-scale model

L. Fahrmeir et al., *Regression*, DOI 10.1007/978-3-642-34333-9_10,

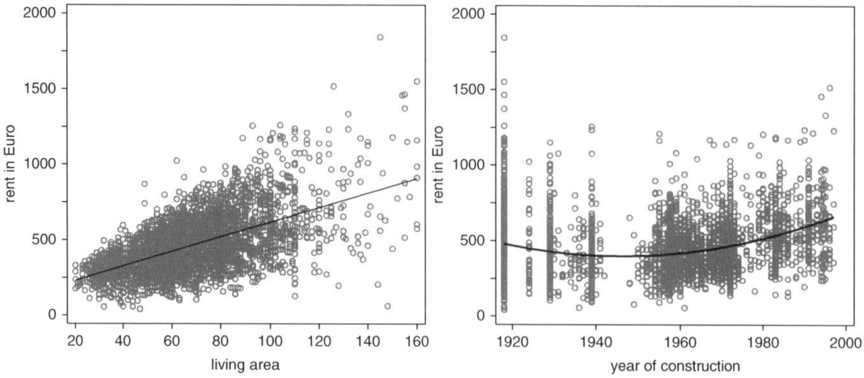

Fig. 10.1 Munich rent index: scatter plots of rents in Euro versus living area (*left panel*) and year of construction (*right panel*) together with a linear (*left panel*) and a quadratic (*right panel*) least squares fit

$$y = \boldsymbol{x}'\boldsymbol{\beta} + \exp(\boldsymbol{z}'\boldsymbol{\alpha})\,\varepsilon, \quad \varepsilon \sim \mathrm{N}(0,1), \tag{10.1}$$

where not only the mean $\mathrm{E}(y) = \boldsymbol{x}'\boldsymbol{\beta}$, but also the variance $\mathrm{Var}(y) = \exp(\boldsymbol{z}'\boldsymbol{\alpha})^2$ depends on covariates. We have already discussed similar approaches in Sect. 4.1.3 (including estimation procedures) although we have modeled the variance there instead of the standard deviation. In any case, this model still may not be flexible enough, e.g., if in addition to the variance, the skewness of the response also depends on covariates. Instead of explicitly modeling the variance of the response in addition to the mean, our aim in this chapter is to introduce a regression tool to portray essentially all of the characteristics of the response in terms of covariates at once.

We therefore focus on the quantiles of the response distribution and relate these quantiles to covariate effects. The basic idea is that a dense set of quantiles completely describes any given distribution. Hence, when computing enough quantile regression results, we will be able to analyze virtually any property of the response distribution based on these results. Prior to introducing the details corresponding to estimation, we outline some of the advantages of quantile regression:

- Quantile regression allows investigation of covariate effects, not only on the mean of a response variable but on the complete conditional distribution of the response given covariates.
- Quantile regression avoids some of the restrictive assumptions of the linear model (or more generally mean regression models). More specifically, we will not require homoscedasticity or a specific type of distribution for the responses (or equivalently the error terms).
- In applications, there often is a genuine interest in regression quantiles that describe "extreme" observations in terms of covariates. For example, in case of the Munich rent index, one is often interested in interval estimates that cover a range of "usual" rents for a given set of covariates. While such a prediction interval can also be computed from mean regression results (under certain

assumptions on the error variance), it may be more strategic to directly aim at the estimation of the interval boundaries based on regression quantiles.

To illustrate the last point, it is important to note that, in fact, any regression model can also determine estimates for the quantiles of the response. For example, in case of a normally distributed response as in the location-scale model (10.1), the τ-quantile, $\tau \in (0, 1)$, is given by

$$\boldsymbol{x}'\boldsymbol{\beta} + z_\tau \exp(\boldsymbol{z}'\boldsymbol{\alpha}), \tag{10.2}$$

where z_τ denotes the τ-quantile of the standard normal distribution. However, the normal distribution for the error terms implies that the quantiles are symmetric around $\boldsymbol{x}'\boldsymbol{\beta}$, i.e., $\boldsymbol{x}'\boldsymbol{\beta} + z_\tau \exp(\boldsymbol{z}'\boldsymbol{\alpha}) = \boldsymbol{x}'\boldsymbol{\beta} - z_{1-\tau} \exp(\boldsymbol{z}'\boldsymbol{\alpha})$. If, in addition, the variance is in fact constant and does not depend on covariates such that $\exp(\boldsymbol{z}'\boldsymbol{\alpha}) = \sigma^2 = \text{const}$, the quantile curves for varying covariates are parallel to each other and only shifted according to the standard normal quantiles. Consequently, more flexible approaches that aim at the direct estimation of quantiles appear to be more appropriate than first starting with a parametric regression model.

10.1 Quantiles

Before actually introducing quantile regression models, we recall the definition of quantiles, both theoretically for distributions and for observed random samples, and discuss some properties that are useful to understand quantile regression. The theoretical quantiles q_τ, $\tau \in (0, 1)$, of a random variable y are commonly and implicitly defined by the equations

$$\text{P}(y \leq q_\tau) \geq \tau \quad \text{and} \quad \text{P}(y \geq q_\tau) \geq 1 - \tau,$$

i.e., the probability of observing a value below (or equal to) q_τ should be (at least) τ while the probability of observing a value above (or equal to) q_τ should be (at least) $1 - \tau$. For quantile regression, it is useful to reformulate this implicit definition as the optimization problem

$$q_\tau = \underset{q}{\arg\min}\, \text{E}\left(w_\tau(y, q)|y - q|\right) \tag{10.3}$$

with weights

$$w_\tau(y, q) = \begin{cases} 1 - \tau & y < q, \\ 0 & y = q, \\ \tau & y > q. \end{cases}$$

Such weights define q_τ as the minimizer of an asymmetrically weighted absolute deviations criterion; see the appendix for a proof of the equivalence. Note that the weights $w_\tau(y, q)$ are defined differently for values above and below q and thus effectively shift the solution upwards or downwards depending on the choice of τ.

If y is continuous with strictly increasing cumulative distribution function $F(y)$ and density $f(y)$, the theoretical quantile is unique and is simply given by the inverse of the cumulative distribution function evaluated at τ, i.e.,

$$q_\tau = F^{-1}(\tau) \qquad \text{and} \qquad F(q_\tau) = \tau.$$

The function $Q(\tau) = F^{-1}(\tau) = q_\tau$ is also called the quantile function of the distribution of y. Note that the assumption of a strictly increasing cumulative distribution function is not very restrictive since it basically implies that the distribution is continuous and that there are no gaps in the domain of the response where the density is zero.

Empirical quantiles correspond to the estimated quantiles $\hat{q}_\tau$ determined from an i.i.d. sample $y_1, \ldots, y_n$ of observations from the corresponding distribution. An implicit definition can again be given as follows: At least a fraction of τ observations should be smaller or equal than $\hat{q}_\tau$ and at least a fraction of $1-\tau$ observations should be larger or equal than $\hat{q}_\tau$, i.e.,

$$\frac{1}{n}\sum_{i=1}^{n} I(y_i \le \hat{q}_\tau) \ge \tau \quad \text{and} \quad \frac{1}{n}\sum_{i=1}^{n} I(y_i \ge \hat{q}_\tau) \ge 1 - \tau,$$

where $I(\cdot)$ denotes the indicator function. For estimation purposes and in particular the generalization to the regression context, it is again better to use the equivalent definition as the solution of an optimization criterion, where

$$\hat{q}_\tau = \arg\min_q \sum_{i=1}^{n} w_\tau(y_i, q)|y_i - q| \tag{10.4}$$

is the empirical analogue to Eq. (10.3).

An interesting quantity that is associated with the estimation of quantiles is the *influence function* $I(y)$, given by the weights associated with a specific value of y in the estimating equations for determining a quantile:

$$I(y) = \begin{cases} 1-\tau & y < q_\tau, \\ 0 & y = q_\tau, \\ \tau & y > q_\tau. \end{cases}$$

The influence function is studied in detail in the area of robust statistics, since it measures the impact of an observation y on the determination of a quantity of interest, the τ-quantile in our case. The influence function related to the quantile function can be interpreted as follows: The distance between y and q_τ does not affect the influence function, which only depends on the sign of the deviation between y and q_τ. This gives a theoretical justification for the robustness of quantiles with respect to outliers since, regardless of the size of an observation, the impact is limited to $\max(\tau, 1 - \tau)$. For comparison, the influence function for the arithmetic mean is given by

$$I(y) = |y - \mu|,$$

where $\mu = \mathrm{E}(y)$ and therefore the impact of values further away from the expectation is much larger than the impact of values that are close to μ.

10.2 Linear Quantile Regression

10.2.1 Classical Quantile Regression

Estimation

We will now transfer regression techniques from mean regression to regression for quantiles. Recall that the standard linear regression model takes the form

$$y = \boldsymbol{x}'\boldsymbol{\beta} + \varepsilon$$

and (in addition to further properties of the error terms) assumes that $\mathrm{E}(\varepsilon) = 0$. This implies that the regression coefficients impact the expectation of the response since

$$\mathrm{E}(y) = \boldsymbol{x}'\boldsymbol{\beta} + \mathrm{E}(\varepsilon) = \boldsymbol{x}'\boldsymbol{\beta}.$$

Hence, the assumption on the expectation of the error term implies a specific interpretation of the regression coefficients. For quantile regression, we also assume the model

$$y = \boldsymbol{x}'\boldsymbol{\beta}_\tau + \varepsilon_\tau$$

but will now make specific assumptions on the error term for the quantile τ of interest to determine quantile-specific regression coefficients $\boldsymbol{\beta}_\tau$ and therefore also explicitly add a τ subscript to both quantities. If we assume that the cumulative distribution function F_{ε_τ} of the error term ε_τ fulfills $F_{\varepsilon_\tau}(0) = \tau$ or equivalently that the τ-quantile of the error term is zero, then this implies

$$\tau = F_{\varepsilon_\tau}(0) = \mathrm{P}(\varepsilon_\tau \leq 0) = \mathrm{P}(\boldsymbol{x}'\boldsymbol{\beta}_\tau + \varepsilon_\tau \leq \boldsymbol{x}'\boldsymbol{\beta}_\tau) = \mathrm{P}(y \leq \boldsymbol{x}'\boldsymbol{\beta}_\tau) = F_y(\boldsymbol{x}'\boldsymbol{\beta}_\tau)$$

and therefore the τ-quantile of the response is given by the predictor $\boldsymbol{x}'\boldsymbol{\beta}_\tau$. Estimation of the quantile-specific regression coefficients $\boldsymbol{\beta}_\tau$ is then achieved by generalizing the asymmetrically weighted absolute error criterion (10.4) to

$$\hat{\boldsymbol{\beta}}_\tau = \arg\min_{\boldsymbol{\beta}} \sum_{i=1}^{n} w_\tau(y_i, \eta_{i\tau})|y_i - \eta_{i\tau}|, \tag{10.5}$$

10.1 Quantile Regression

Model

$$y_i = \boldsymbol{x}_i'\boldsymbol{\beta}_\tau + \varepsilon_{i\tau}, \quad i = 1, \dots, n,$$

with assumptions

1. $F_{\varepsilon_{i\tau}}(0) = \tau$.
2. $\varepsilon_{1\tau}, \dots, \varepsilon_{n\tau}$ are independent.

Estimation of Regression Coefficients

The regression coefficients $\boldsymbol{\beta}_\tau$ are determined by minimizing

$$\sum_{i=1}^{n} w_\tau(y_i, \eta_{i\tau})|y_i - \eta_{i\tau}|,$$

where $\eta_{i\tau} = \boldsymbol{x}_i'\boldsymbol{\beta}_\tau$.

Software

- R package `quantreg`: Implements linear programming for determining $\hat{\boldsymbol{\beta}}_\tau$. See Koenker (2005) for details.
- R package `mboost`: Implements functional gradient descent boosting for determining $\hat{\boldsymbol{\beta}}_\tau$. See Fenske, Kneib, and Hothorn (2011) for details.

where $\eta_{i\tau} = \boldsymbol{x}_i'\boldsymbol{\beta}_\tau$ and

$$w_\tau(y_i, \eta_{i\tau}) = \begin{cases} 1-\tau & y_i < \eta_{i\tau}, \\ 0 & y_i = \eta_{i\tau}, \\ \tau & y_i > \eta_{i\tau}. \end{cases}$$

Basically, instead of assuming that the average error in the regression model is zero, we assume that the weighted "average" of the errors with weights defined as above is zero. A summary of the quantile regression setup is given in Box 10.1.

The most important feature of quantile regression to be noted here is its generality with respect to the error distribution. Apart from independence of the error terms between individual observations and the quantile restriction, there are no further assumptions on the corresponding distribution. In particular, the error terms may even follow different types of distributions and further, are not assumed to be homoscedastic. While usual least squares estimation in mean regression can also be accomplished without assuming specific types of distributions for the error terms, it always requires constant variances across the errors (unless generalized to weighted least squares with weights specified in advance). Such generality of quantile regression is one of its specific strengths and essentially leads us to a distribution-free regression approach.

To minimize the asymmetrically weighted absolute error criterion (10.5), we would like to proceed as in the case of mean regression, i.e., taking partial derivatives with respect to the regression coefficients and setting the derivatives to zero. Unfortunately, this is no longer possible since Eq. (10.5) is not differentiable in the origin. We therefore expand the estimation problem by introducing $2n$ auxiliary variables

$$u_{i\tau} = (y_i - \boldsymbol{x}_i'\boldsymbol{\beta}_\tau)_+,$$
$$v_{i\tau} = (\boldsymbol{x}_i'\boldsymbol{\beta}_\tau - y_i)_+,$$

where $(x)_+ = \min(x, 0)$ and therefore $\varepsilon_{i\tau} = u_{i\tau} - v_{i\tau}$. The minimization problem (10.5) can then be rewritten as

$$\sum_{i=1}^{n} w_\tau(y_i, \eta_{i\tau})|y_i - \eta_{i\tau}| = \sum_{i=1}^{n} \tau u_{i\tau} + \sum_{i=1}^{n}(1-\tau)v_{i\tau} = \tau \boldsymbol{1}'\boldsymbol{u}_\tau + (1-\tau)\boldsymbol{1}'\boldsymbol{v}_\tau,$$

given the constraint that

$$y_i = \eta_{i\tau} + \varepsilon_{i\tau} = \boldsymbol{x}_i'\boldsymbol{\beta}_\tau + u_{i\tau} - v_{i\tau},$$

or in matrix notation

$$\boldsymbol{y} - \boldsymbol{X}\boldsymbol{\beta}_\iota + \boldsymbol{u}_\iota + \boldsymbol{v}_\iota,$$

with $\boldsymbol{u}_\tau = (u_{1\tau}, \dots, u_{n\tau})'$, $\boldsymbol{v}_\tau = (v_{1\tau}, \dots, v_{n\tau})'$, and $\boldsymbol{1} = (1, \dots, 1)'$. In summary, the optimization problem after introducing the auxiliary variables is now given by

$$\min_{\boldsymbol{\beta}_\tau, \boldsymbol{u}_\tau, \boldsymbol{v}_\tau} \{\tau \boldsymbol{1}'\boldsymbol{u}_\tau + (1-\tau)\boldsymbol{1}'\boldsymbol{v}_\tau \mid \boldsymbol{X}\boldsymbol{\beta}_\tau + \boldsymbol{u}_\tau - \boldsymbol{v}_\tau = \boldsymbol{y}\}.$$

This is a constrained minimization problem with polyhedric constraints, i.e., the constraint defines a geometric object with flat faces and straight edges, such that the constrained problem corresponds to the original quantile restriction, but written in terms of the auxiliary variables. After augmenting the auxiliary variables and given the constraints, the minimization problem is now linear in the parameters and can therefore be tackled with linear programming techniques that allow to incorporate the polyhedric constraints. Refer to Lange (2000) for a brief introduction to linear programming with a focus on statistics-related applications and Lange (2004) for a detailed exposition.

An alternative estimation approach for minimizing the asymmetrically weighted absolute deviations criterion (10.5), which utilizes the boosting approach developed in Sect. 4.3, has been proposed in Fenske et al. (2011). This approach only requires the following modification: substitute the residuals that have served as working responses in (linear model) boosting with the corresponding negative gradients of the loss function

$$\rho(y, \eta) = w_\tau(y, \eta)|y - \eta|,$$

which define the individual contributions to Eq. (10.5). These specific negative gradients are given by

$$u = \frac{\partial}{\partial \eta}\rho(y, \eta) = \begin{cases} \tau - 1 & y < \eta, \\ 0 & y = \eta, \\ \tau & y > \eta. \end{cases} \tag{10.6}$$

The case $y = \eta$ occurs with probability zero, and therefore the non-differentiability of Eq. (10.5) does not cause a problem. Apart from that minor change, the rest of the boosting approach, and also the extensions discussed in Sect. 9.7, can immediately be used.

It is interesting to compare quantile regression estimation and least squares estimation in somewhat more detail. For the latter, we are considering the optimization problem

$$\min_{\boldsymbol{\beta}} (\boldsymbol{y} - \boldsymbol{X}\boldsymbol{\beta})'(\boldsymbol{y} - \boldsymbol{X}\boldsymbol{\beta})$$

and the solution $\hat{\boldsymbol{\beta}}$ is characterized by the normal equations

$$\boldsymbol{X}'(\boldsymbol{y} - \boldsymbol{X}\boldsymbol{\beta}) \stackrel{!}{=} \boldsymbol{0}.$$

For quantile regression, the derivative of the optimization criterion

$$R(\boldsymbol{\beta}_\tau) = \sum_{i=1}^{n} w_\tau(y_i, \eta_{i\tau})|y_i - \eta_{i\tau}|$$

only exists at points with $y_i - \eta_{i\tau} \neq 0$, $i = 1 \dots, n$, since the absolute value function is differentiable everywhere except in the origin. However, we can always define direction-specific derivatives in direction $\boldsymbol{w} \in \mathbb{R}^p$

$$\left.\frac{\partial}{\partial t} R(\boldsymbol{\beta}_\tau + t\boldsymbol{w})\right|_{t=0} = -\sum_{i=1}^{n} \psi_\tau(y_i - \boldsymbol{x}_i'\boldsymbol{\beta}, -\boldsymbol{x}_i'\boldsymbol{w})\boldsymbol{x}_i'\boldsymbol{w},$$

where

$$\psi_\tau(u, v) = \begin{cases} \tau - I(u < 0) & \text{if } u \neq 0 \\ \tau - I(v < 0) & \text{if } u = 0. \end{cases}$$

The optimal solution then is the value $\hat{\boldsymbol{\beta}}_\tau$ where all direction-specific derivatives are positive, indicating that whenever we move from $\hat{\boldsymbol{\beta}}_\tau$ towards any alternative candidate solution, the optimization criterion will increase.

Properties

One important property of the quantile regression estimate, inherited from estimated quantiles (determined from an i.i.d. sample), is its invariance under monotonic transformations. More specifically, if $\boldsymbol{x}'\hat{\boldsymbol{\beta}}_\tau$ is an estimate for the τ-quantile of the distribution of the response y, given covariates $\boldsymbol{x}$, then for any monotonically increasing transformation h the transformed estimate $h(\boldsymbol{x}'\hat{\boldsymbol{\beta}}_\tau)$ is an estimate for the τ-quantile of the distribution of $h(y)$. This simply follows from the fact that quantiles are invariant under monotonically increasing transformations, as the order of the data are preserved.

It is also possible to derive asymptotic results for the distribution of quantile regression coefficients under different assumptions for the error terms that may then form the basis for statistical inference. For example, in case of i.i.d. errors, the asymptotic distribution of $\hat{\boldsymbol{\beta}}_\tau$ is given by

$$\hat{\boldsymbol{\beta}}_\tau \overset{a}{\sim} \mathrm{N}\left(\boldsymbol{\beta}_\tau, \frac{\tau(1-\tau)}{f_{\varepsilon_\tau}(0)^2}(\boldsymbol{X}'\boldsymbol{X})^{-1}\right).$$

Although the asymptotic covariance matrix seems to suggest that estimation of regression quantiles becomes increasingly more precise when very small ($\tau \to 0$) or large ($\tau \to 1$) quantiles are considered (due to the product $\tau(1-\tau)$ in the numerator), in fact this effect is typically dominated by the error $f_{\varepsilon_\tau}(0)$ which will usually be close to zero for such values of τ. Hence, the estimation of interior quantiles close to the median will be more precise for most of the standard distributions, while estimation of outer quantiles is associated with larger uncertainty. The asymptotic normality result can also be generalized to non i.i.d. situations (which typically fit better to situations where quantile regression is of interest). For more information, see Chap. 3 in Koenker (2005). While the asymptotic normal distribution provides one possibility for performing inferences on the estimated regression coefficients $\hat{\boldsymbol{\beta}}_\tau$, there are also alternatives based on the relationship between quantiles and ranks. These rank-based inferential procedures are typically preferred in practice and also allow the construction of confidence intervals by inverting rank-based tests. Again, see Chap. 3 in Koenker (2005).

In theory, the quantiles of the distribution of a response should be ordered such that

$$\boldsymbol{x}'\boldsymbol{\beta}_{\tau_1} \leq \boldsymbol{x}'\boldsymbol{\beta}_{\tau_2} \quad \text{for} \quad \tau_1 \leq \tau_2$$

holds for any covariate vector $\boldsymbol{x}$. It can be shown that the ordering is preserved for the average covariate vector

$$\bar{\boldsymbol{x}} = \frac{1}{n}\sum_{i=1}^{n} \boldsymbol{x}_i,$$

when replacing the theoretical quantiles with estimated quantiles, i.e.,

$$\bar{\boldsymbol{x}}'\hat{\boldsymbol{\beta}}_{\tau_1} \leq \bar{\boldsymbol{x}}'\hat{\boldsymbol{\beta}}_{\tau_2} \quad \text{for} \quad \tau_1 \leq \tau_2.$$

However, in general these results will not transfer to arbitrary vectors $\boldsymbol{x}$ and will not even hold for all observed covariate vectors. This is due to the fact that each estimate is obtained separately for each specific choice of the quantile τ. In fact, the phenomenon of crossing quantiles will always occur when the regression lines for two quantiles are not exactly parallel. Note that parallel quantile curves can only occur when the error terms are independent and identically distributed, and as such the quantile of interest merely induces a shift in the intercept parameter. While in this case no quantile-crossing will be observed, the results of quantile regressions are then only mildly interesting (since the covariate is independent of the quantile). Yet, the result for the average covariate vector is reassuring and indicates that quantile-crossing will most likely occur at the boundaries of the covariate space, where the data may be sparse.

Example 10.1 Munich Rent Index—Quantile Regression

To illustrate some simple quantile regression models, we use the data from the Munich rent index shown in Fig. 10.1. We fit three different linear models for the effect of living area and three different quadratic models for the effect of the year of construction, in each case to determine the 11 quantiles $\tau = 0.01, 0.1, 0.2, \ldots, 0.8, 0.9, 0.99$ of the net rent distribution. The three different approaches are summarized as:

- Quantile regression: $y = \boldsymbol{x}'\boldsymbol{\beta}_\tau + \varepsilon_\tau$ as introduced in this chapter
- Homoscedastic linear model with i.i.d. Gaussian error terms: $y = \boldsymbol{x}'\boldsymbol{\beta} + \varepsilon, \varepsilon \sim \mathrm{N}(0, \sigma^2)$
- Heteroscedastic linear model with independent Gaussian error terms: $y = \boldsymbol{x}'\boldsymbol{\beta} + \exp(\boldsymbol{x}'\boldsymbol{\alpha})\varepsilon, \varepsilon \sim \mathrm{N}(0, 1)$

For the quadratic effect of the year of construction, we use an orthonormal polynomial. For both linear models, the regression quantiles are determined according to Eq. (10.2). The heteroscedastic linear model is fitted within the GAMLSS framework of Rigby and Stasinopoulos (2005).

The resulting estimates are shown in Fig. 10.2. For the quantile regression approach, we find a clear dependency of the specified quantile on the covariate effects, and that the quantiles correspond very well to the structure in the scatter plots. In particular, the effect of the living area almost disappears for apartments in the lower price segment (corresponding to a small value for the quantile τ), while the slope steadily increases when moving toward higher price segments (corresponding to large quantiles τ). For the year of construction, the effect is very close to linear for the lower quantiles, while it gets more and more U-shaped when moving towards higher quantiles. In this analysis, quantile-crossing does not occur, since we do not consider quantiles that are very close to each other and since the data set is relatively large (with approximately 3000 observations).

When comparing these results with those from a homoscedastic linear model, we find distinct differences. In particular, since we are assuming i.i.d. errors here, all quantile curves are exactly parallel to each other and only shifted according to the quantiles of a normal distribution. In contrast, results obtained in the heteroscedastic case are closer to the ones from the quantile regression. Note that the formula for the variance of the errors involves the exponential function and therefore induces nonlinear quantile curves, even in case of a linear model as for the living area. While the fit for living area seems to be sufficient for both the quantile regression model and the heteroscedastic linear model, the latter seems to be inadequate for describing the effect of the year of construction (at least for extreme quantiles). While large quantiles are underestimated, the curve for the lowest quantile seems to be much too low. This may be seen as an indication that, in addition to heteroscedasticity, there is also a dependence of skewness and/or kurtosis on the covariates. For these data, it appears to pays off to use a completely distribution-free approach that does not assume any specific form for the error distribution.

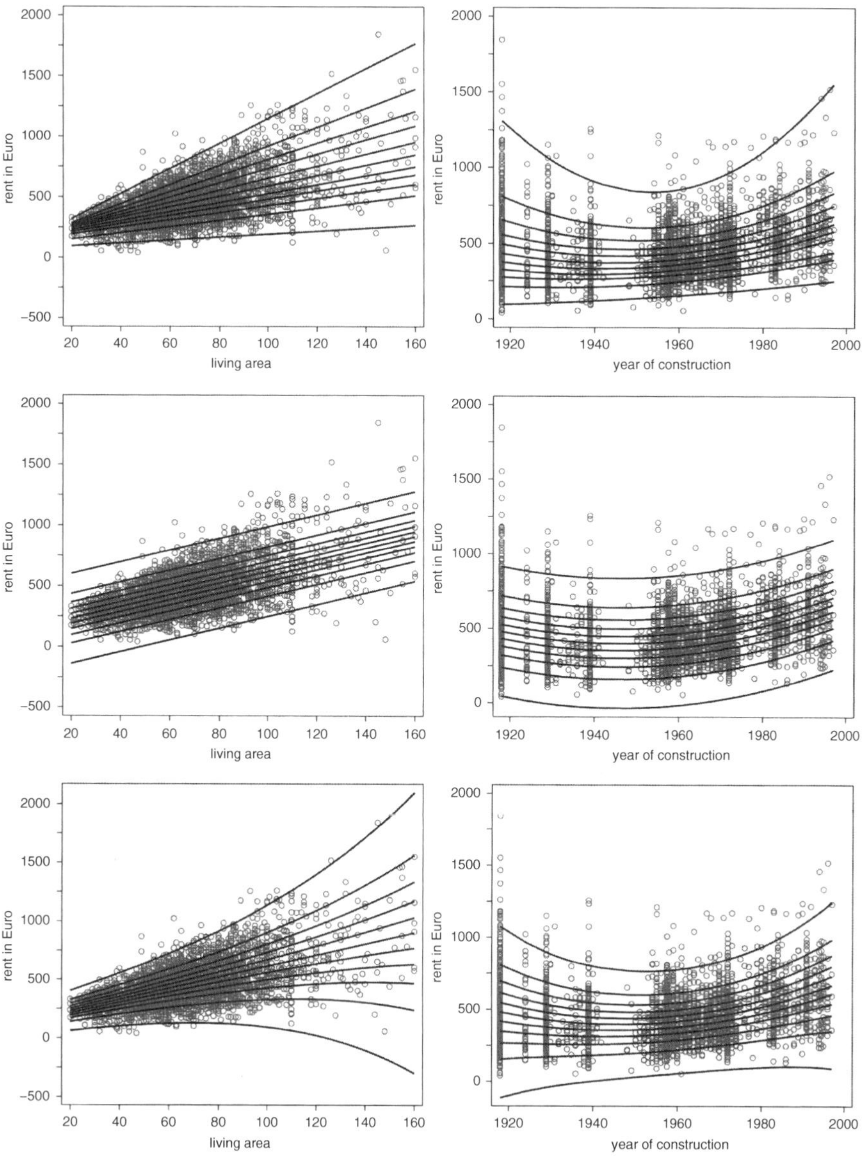

Fig. 10.2 Munich rent index: scatter plots of the rents in Euro versus living area (*left column*) and year of construction (*right panel*) together with linear/quadratic quantile regression fits for 11 quantiles (*top row*), quantiles determined from a homoscedastic linear model (*middle row*), and quantiles determined from a heteroscedastic linear model (*bottom row*)

While thus far we have only considered models consisting of one covariate with either linear or quadratic impact on the quantiles of the response, we will now consider a multiple regression model containing both effects simultaneously. This does not cause any difficulties with respect to the estimation of the regression coefficients, but the presentation

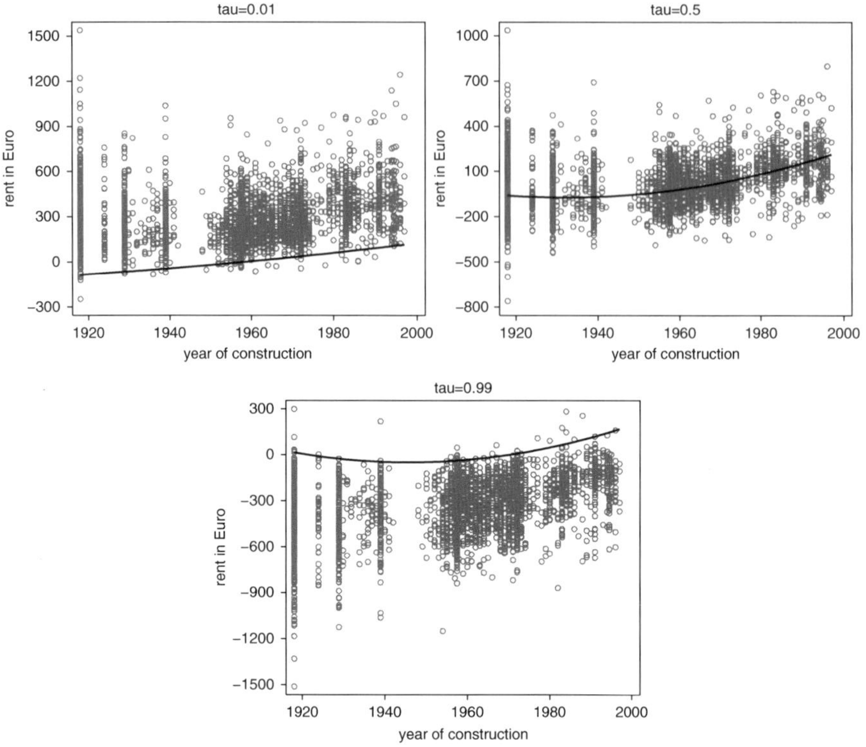

Fig. 10.3 Munich rent index: estimated effects of year of construction together with partial residuals for different quantiles

of the results is more involved. One problem is that we can no longer simply draw a scatter plot of the data along with the fits for the different quantiles, since we now must adjust for the effects of the other covariates. In the linear model, we used partial residuals to achieve this (see Sect. 3.2.3, p. 126), where all effects apart from those associated with the covariate of interest are subtracted from the response. This can also be done in quantile regression but yields different partial residuals for each value of τ. Therefore, we can no longer visualize all estimated effects simultaneously in one scatter plot. Figure 10.3 shows the estimated effect of the year of construction for three different quantiles along with the corresponding partial residuals. In all three cases, the fit appears to be quite satisfactory but with somewhat reduced curvature as compared to the results from the univariate model.

Figure 10.4 illustrates an alternative way to graphically present the results of quantile regressions, where the estimated coefficients and confidence intervals are plotted against the quantiles. The variation over the quantiles then gives indication for the necessity of quantile regression. If some or all of the estimated coefficients are almost constant, then they are independent of the quantile, and therefore only the mean but not the quantiles seem to depend on the covariate of interest. In addition, the figure shows the estimates from mean regression as a reference. Note that the magnitude of the estimated coefficients for living area and year of construction is very different since living area enters the model in untransformed linear fashion while year of construction is represented as an orthonormal polynomial.

△

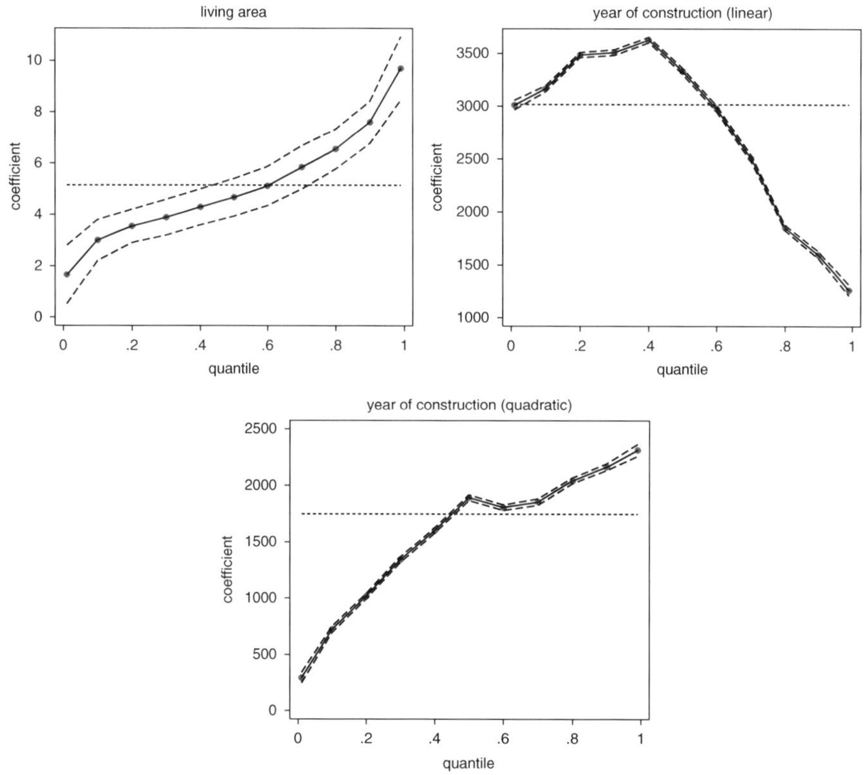

Fig. 10.4 Munich rent index: paths of estimated coefficients (*solid line*) together with 95 % confidence intervals (*dashed lines*) obtained from inverting a rank-based test for various quantiles τ. The horizontal dotted line corresponds to the least squares estimate

10.2.2 Bayesian Quantile Regression

Bayesian quantile regression has been developed utilizing the equivalence between posterior mode and maximum likelihood estimation under noninformative priors $\boldsymbol{\beta}_\tau \propto \text{const}$ (Yu & Moyeed, 2001; Yue & Rue, 2011). Therefore we have to define a specific distributional assumption for the error terms (or equivalently the responses) to make the Bayesian standard machinery work. If we start with the model

$$y_i = \boldsymbol{x}_i' \boldsymbol{\beta}_\tau + \varepsilon_{i\tau}, \quad i = 1, \ldots, n,$$

we will assume independent and identically distributed errors following an asymmetric Laplace distribution, i.e., $\varepsilon_{i\tau} \,|\, \sigma^2$ i.i.d. $\text{ALD}(0, \sigma^2, \tau)$ with density

$$p(\varepsilon_{i\tau} \,|\, \sigma^2) = \frac{\tau(1-\tau)}{\sigma^2} \exp\left(-w_\tau(\varepsilon_{i\tau}, 0) \frac{|\varepsilon_{i\tau}|}{\sigma^2} \right).$$

For the responses, the error distribution induces $y_i \mid \boldsymbol{\beta}_\tau, \sigma^2 \sim \text{ALD}(\boldsymbol{x}_i'\boldsymbol{\beta}_\tau, \sigma^2, \tau)$, such that the density of the responses is given by

$$p(y_i \mid \boldsymbol{\beta}_\tau, \sigma^2) = \frac{\tau(1-\tau)}{\sigma^2} \exp\left(-w_\tau(y_i, \boldsymbol{x}_i'\boldsymbol{\beta}_\tau)\frac{|y_i - \boldsymbol{x}_i'\boldsymbol{\beta}_\tau|}{\sigma^2}\right).$$

It then turns out that maximizing the corresponding posterior (for fixed σ^2)

$$\begin{aligned} p(\boldsymbol{\beta}_\tau \mid \boldsymbol{y}, \sigma^2) &\propto \prod_{i=1}^{n} p(y_i \mid \boldsymbol{\beta}_\tau, \sigma^2) \\ &\propto \exp\left(-\sum_{i=1}^{n} w_\tau(y_i, \boldsymbol{x}_i'\boldsymbol{\beta}_\tau)\frac{|y_i - \boldsymbol{x}_i'\boldsymbol{\beta}_\tau|}{\sigma^2}\right), \end{aligned}$$

with respect to $\boldsymbol{\beta}_\tau$, is equivalent to minimizing the optimization criterion (10.5).

While the asymmetric Laplace distribution allows to conveniently express quantile regression in a Bayesian framework, it complicates inference based on Markov chain Monte Carlo simulations due to the absolute value contained in its definition. It is therefore advantageous to represent the asymmetric Laplace distribution as a scale mixture of normal distributions as suggested in Yue and Rue (2011): Let $z_i \mid \sigma^2 \sim \text{Expo}(1/\sigma^2)$, $i = 1, \ldots, n$, be i.i.d. exponentially distributed with rate parameter σ^2 and

$$y_i \mid z_i, \boldsymbol{\beta}_\tau, \sigma^2 \sim N(\boldsymbol{x}_i'\boldsymbol{\beta}_\tau + \xi z_i, \sigma^2/w_i)$$

with

$$\xi = \frac{1-2\tau}{\tau(1-\tau)}, \qquad w_i = \frac{1}{\delta^2 z_i}, \qquad \delta^2 = \frac{2}{\tau(1-\tau)}.$$

Then the marginal distribution $y_i \mid \boldsymbol{\beta}_\tau, \sigma^2$ is obtained by integrating out z_i and is indeed an asymmetric Laplace distribution, i.e.,

$$y_i \mid \boldsymbol{\beta}_\tau, \sigma^2 \sim \text{ALD}(\boldsymbol{x}_i'\boldsymbol{\beta}_\tau, \sigma^2, \tau).$$

Bayesian inference can now efficiently be implemented after imputing the scale variables z_i as additional unknowns, similar to the approach that was outlined for binary regression models in Sect. 5.6.3. Basically, the resulting model is a conditionally Gaussian regression model with offsets ξz_i and weights w_i. Box 10.2 summarizes the Bayesian model and provides the full conditionals required for the Gibbs sampler. We thereby assume noninformative priors $\boldsymbol{\beta}_\tau \propto \text{const}$ and the usual inverse gamma priori for σ^2, i.e., $\sigma^2 \sim \text{IG}(a, b)$ with hyperparameters a and b. A derivation of the full conditionals can be found at the end of this chapter on p. 618.

When comparing these results with the ones from the basic Bayesian linear model in Sect. 4.4, the only change in the full conditional for the regression coefficients is that the classical linear model is replaced by a general linear model

10.2 Bayesian Linear Quantile Regression Based on the ALD Distribution

Observation Model

The observations $y_i, i = 1, \ldots, n$, are conditionally independent following an asymmetric Laplace distribution, i.e., y_i i.i.d. $\text{ALD}(\boldsymbol{x}_i'\boldsymbol{\beta}_\tau, \sigma^2, \tau)$. The scale mixture representation of the asymmetric Laplace distribution yields

$$y_i \mid z_i, \boldsymbol{\beta}_\tau, \sigma^2 \sim N(\boldsymbol{x}_i'\boldsymbol{\beta}_\tau + \xi z_i, \sigma^2/w_i),$$

where

$$\xi = \frac{1-2\tau}{\tau(1-\tau)}, \qquad w_i = \frac{1}{\delta^2 z_i}, \qquad \delta^2 = \frac{2}{\tau(1-\tau)}.$$

Priors

$$\begin{aligned} \boldsymbol{\beta}_\tau &\propto \text{const} \\ z_i \mid \sigma^2 &\sim \text{Expo}(1/\sigma^2) \\ \sigma^2 &\sim \text{IG}(a, b) \end{aligned}$$

All priors are mutually independent.

Full Conditionals Required for the Gibbs Sampler

- *Full conditional for the regression coefficients:* $\boldsymbol{\beta}_\tau \mid \cdot\ \text{N}(\boldsymbol{\mu}_{\boldsymbol{\beta}_\tau}, \boldsymbol{\Sigma}_{\boldsymbol{\beta}_\tau})$ with

$$\boldsymbol{\Sigma}_{\boldsymbol{\beta}_\tau} = \sigma^2(\boldsymbol{X}'\boldsymbol{W}\boldsymbol{X})^{-1}, \qquad \boldsymbol{\mu}_{\boldsymbol{\beta}_\tau} = (\boldsymbol{X}'\boldsymbol{W}\boldsymbol{X})^{-1}\boldsymbol{X}'\boldsymbol{W}(\boldsymbol{y} - \xi\boldsymbol{z}),$$

  where $\boldsymbol{W} = \text{diag}(w_1, \ldots, w_n)$ and $\boldsymbol{z} = (z_1, \ldots, z_n)'$.
- *Full conditional for the scale parameters:*

$$z_i^{-1} \mid \cdot \sim \text{InvGauss}\left(\sqrt{\frac{\xi^2 + 2\delta^2}{(y_i - \boldsymbol{x}_i'\boldsymbol{\beta}_\tau)^2}}, \frac{\xi^2 + 2\delta^2}{\sigma^2\delta^2}\right),$$

  refer to Definition B.7 in Appendix B.1 for the inverse Gaussian distribution.
- Full conditional for the error variance:

$$\sigma^2 \mid \cdot \sim \text{IG}\left(a + \frac{3n}{2}, b + \frac{1}{2}\sum_{i=1}^{n} w_i(y_i - \boldsymbol{x}_i'\boldsymbol{\beta}_\tau - \xi z_i)^2 + \sum_{i=1}^{n} z_i\right).$$

Software

Software package `BayesX` (see also the R interface `R2BayesX`) with weights w_i and offset ξz_i. This also indicates that it is easy to extend the Bayesian quantile regression model beyond linear regression specifications, as outlined in the following section by including, for example, Bayesian P-splines or spatial effects. The full conditional for the error variance also changes since the exponential prior of the weights depends on σ^2.

The presented Bayesian quantile regression approach relies on the asymmetric Laplace distribution as an auxiliary error distribution that yields a formal equivalence between posterior modes and usual quantile regression estimates. However, in a strict sense, this auxiliary error distribution can never be simultaneously true for all quantiles since only one quantile of the error distribution can in fact be zero. Consequently, the resulting point estimates are often close to those from frequentist analyses, while the corresponding credible intervals may be misleading when the error distribution is (greatly) misspecified. However, the Bayesian quantile regression approach still has considerable advantages with respect to generalized model specifications as will be discussed in the next section. Other Bayesian quantile regression approaches try to circumvent the auxiliary error distribution problem by including the error distribution as a part of the estimation problem and to determine quantiles from the estimated error distribution. See, for example, Reich, Bondell, and Wang (2010) and Kottas and Krnjajic (2009) for approaches based on mixtures of Gaussians and Dirichlet process mixtures, respectively.

10.3 Additive Quantile Regression

Additive quantile regression models result when replacing the assumed linear effects of covariates thus far, with potentially nonlinear effects, similar to those found in Chaps. 8 and 9. An approach for estimating nonlinear quantile functions $f_\tau(z_i)$ of continuous covariates z_i in the scatter plot smoothing model

$$y_i = f_\tau(z_i) + \varepsilon_{i\tau}$$

that still fits into the framework of linear programming (which is usually employed for estimating linear quantile regression models) has been suggested by Koenker, Ng, and Portnoy (1994). Such an approach relies on the fitting criterion

$$\arg\min_{f_\tau} \sum_{i=1}^{n} w_\tau(y_i, f_\tau(z_i))|y_i - f_\tau(z_i)| + \lambda V(f'_\tau). \qquad (10.7)$$

This criterion is inspired by the penalized least squares criterion employed in the derivation of smoothing splines (Sect. 8.1.4, p. 448) but relies on asymmetrically weighted absolute errors instead of least squares. Additionally it considers a somewhat different penalty that fits well with the absolute value criterion. More specifically, $V(f'_\tau)$ denotes the total variation of the derivative f'_τ defined as

$$V(f'_\tau) = \sup \sum_{i=1}^{n} |f'_\tau(z_{i+1}) - f'_\tau(z_i)|,$$

where the sup is taken over all partitions $a \leq z_1 < \ldots < z_n < b$. For twice continuously differentiable functions f_τ, the total variation penalty can be written as

$$V(f_\tau') = \int |f_\tau''(z)| dz$$

indicating that the penalty results from replacing the squared second derivative (applied in the context of smoothing splines) with the absolute second derivative. Utilizing the squared second derivative would inhibit the use of linear programming techniques since it would lead to a combination of an L_1 norm for the error terms and an L_2 norm for the penalty. In contrast, Koenker et al. (1994) showed that the solution to Eq. (10.7) can still be obtained by linear programming when considering a somewhat larger function space that also comprises functions with derivatives existing almost everywhere. Within this function space, the minimizer of Eq. (10.7) is a piecewise linear spline function with knots at the observations z_i, similar to those found with the natural cubic spline resulting in the context of smoothing splines.

This approach can also be extended to additive models that consist of more than one nonparametric function or the estimation of surfaces based on triograms Koenker and Mizera (2004). However, it is typically difficult to determine the smoothing parameters along with the estimated functions in an automatic and data-driven way. As a consequence, alternative approaches based on boosting or the Bayesian approach to quantile regression have been suggested, for example, in Fenske et al. (2011), Yue and Rue (2011), or Waldmann, Kneib, Lang, and Yue (2012).

For the boosting approach, we basically have to combine the penalized least squares base-learners that were introduced for structured additive regression in Sect. 9.7, with the loss function defining quantile regression estimates. Hence, the only change relates to different negative gradients employed in the boosting algorithm. More specifically, in the algorithm summarized in Box 9.6, the negative gradients are replaced by Eq. (10.6) where η can now be any structured additive predictor. All the remaining parts of the algorithm can be reused without any modifications.

For Bayesian additive quantile regression, we can extend the Gibbs-sampler outlined in the previous section in a similar spirit as in Sect. 9.6.3 for structured additive regression, relying on Markov chain Monte Carlo simulation techniques. Most importantly, the full conditionals for nonparametric effects represented as $\boldsymbol{V}_j \boldsymbol{\gamma}_j$ are now given by

$$\boldsymbol{\gamma}_j \,|\, \cdot \sim \mathrm{N}(\boldsymbol{m}_j, \boldsymbol{\Sigma}_j),$$

with expectation and covariance matrix

$$\boldsymbol{m}_j = \mathrm{E}(\boldsymbol{\gamma}_j \,|\, \cdot) = \left(\boldsymbol{V}_j' \boldsymbol{W} \boldsymbol{V}_j + \frac{\sigma^2}{\tau_j^2} \boldsymbol{K}_j \right)^{-1} \boldsymbol{V}_j \boldsymbol{W}' (\boldsymbol{y} - \boldsymbol{\eta}_{-j}^{struct} - \xi \boldsymbol{z})$$

$$\boldsymbol{\Sigma}_j = \mathrm{Cov}(\boldsymbol{\gamma}_j \,|\, \cdot) = \sigma^2 \left(\boldsymbol{V}_j' \boldsymbol{W} \boldsymbol{V}_j + \frac{\sigma^2}{\tau_j^2} \boldsymbol{K}_j \right)^{-1},$$

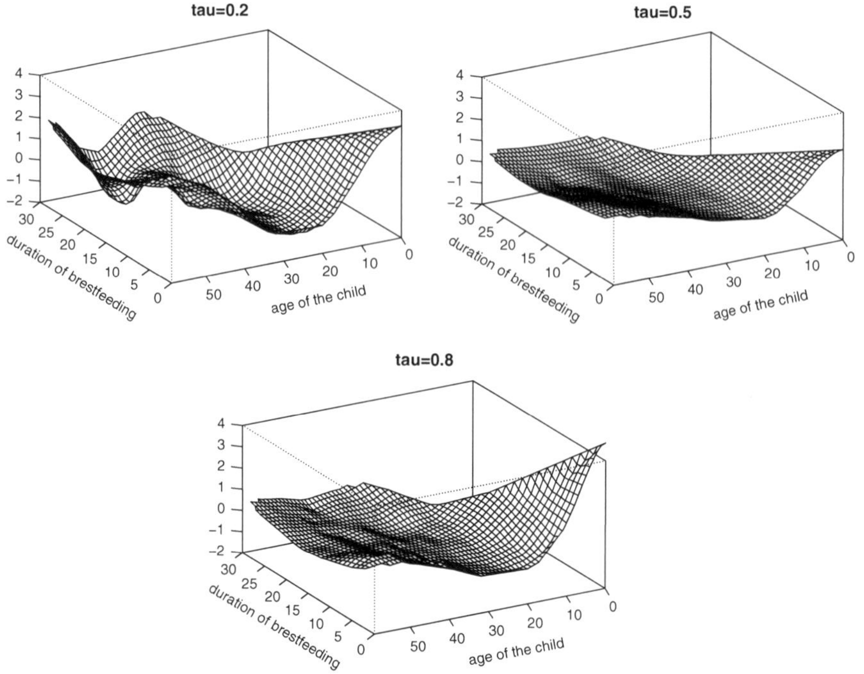

Fig. 10.5 Malnutrition in Zambia: interaction of the child's age and duration of breast-feeding obtained in a geoadditive Bayesian quantile regression model for $\tau = 0.2$ (*top left*), $\tau = 0.5$ (*top right*), and $\tau = 0.8$ (*bottom*)

where $\boldsymbol{W} = \text{diag}(w_1, \ldots, w_n)$ and $\boldsymbol{z} = (z_1, \ldots, z_n)'$ as in Box 10.2. Similarly, the full conditional for the error variance has to be adjusted while the full conditionals for the smoothing variances remain unchanged.

Example 10.2 Childhood Malnutrition in Zambia—Geoadditive Bayesian Quantile Regression

We illustrate the results acquired from a Bayesian geoadditive quantile regression model based on the asymmetric Laplace distribution as auxiliary error distribution by extending the case study presented in Sect. 9.8. Therefore, we estimate a quantile regression model

$$\begin{aligned} zscore_i &= f_{1\tau}(m\_agebirth_i) + f_{2\tau}(m\_height_i) + f_{3\tau}(m\_bmi_i) \\ &\quad + f_{4\tau}(c\_breastf_i, c\_age_i) + f_{\text{spat}\tau}(s_i) + \boldsymbol{x}_i'\boldsymbol{\beta}_\tau + \varepsilon_{i\tau} \end{aligned}$$

comprising a kriging term $f_{4\tau}(c\_breastf, c\_age)$ for the interaction of duration of breast-feeding and age of the child, penalized splines $f_{1\tau}(m\_agebirth)$, $f_{2\tau}(m\_height)$, and $f_{3\tau}(m\_bmi)$ for the nonlinear effects of the body mass index, height, and age of the mother, a Markov random field $f_{\text{spat}\tau}(s)$ for the spatial effects of the districts s, and parametric effects $\boldsymbol{x}_i'\boldsymbol{\beta}_\tau$ for the remaining (categorical) covariates. All hyperparameter settings (such as number of knots and degrees of the spline) have been chosen analogously to those in the case study in the previous chapter.

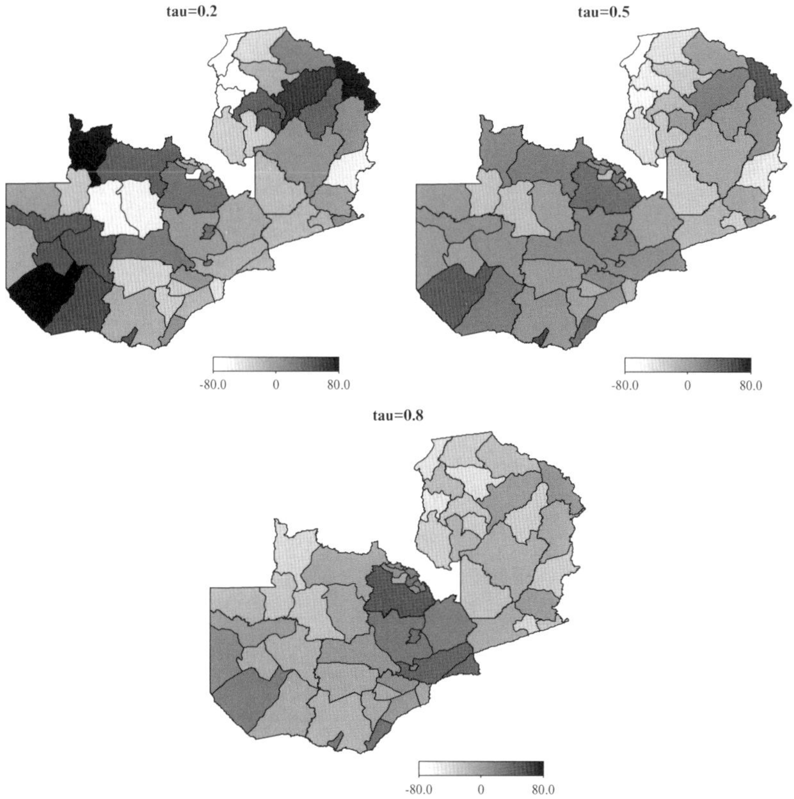

Fig. 10.6 Malnutrition in Zambia: spatial effect obtained in a geoadditive Bayesian quantile regression model for $\tau = 0.2$ (*top left*), $\tau = 0.5$ (*top right*), and $\tau = 0.8$ (*bottom*)

To implement the Bayesian approach, we make use of the latent Gaussian representation for the asymmetric Laplace distribution as introduced in Sect. 10.2.2. Refer to Waldmann et al. (2012) for details. In the following, we discuss only some exemplary results for the 20 %, 50 %, and 80 % quantiles.

Figure 10.5 shows estimates for the interaction effect between the age of the child and the duration of breast-feeding. While the principal form of the effect is not too different between the quantiles, and it also matches closely with the results from the mean regression model presented in the last chapter, the magnitude of the effects is much larger for the 20 % and the 80 % quantile as compared to the median regression results. Similarly, the 20 % quantile estimates for the spatial effect (Fig. 10.6) closely resemble the form of the spatial effect for the median but again show a larger magnitude. In contrast, the 80 % quantile also shows a different spatial pattern, indicating that the spatial distribution in case of well-nourished children is different from the one for malnourished children and children with average nutritional status. Finally, Fig. 10.7 shows estimates for the effect of the body mass index of the mother. There seems to be a slight indication for a nonlinear effect but with some uncertainty, especially for large and small values of the body mass index. Note, however, that the estimated credible intervals have to be treated with caution since they rely

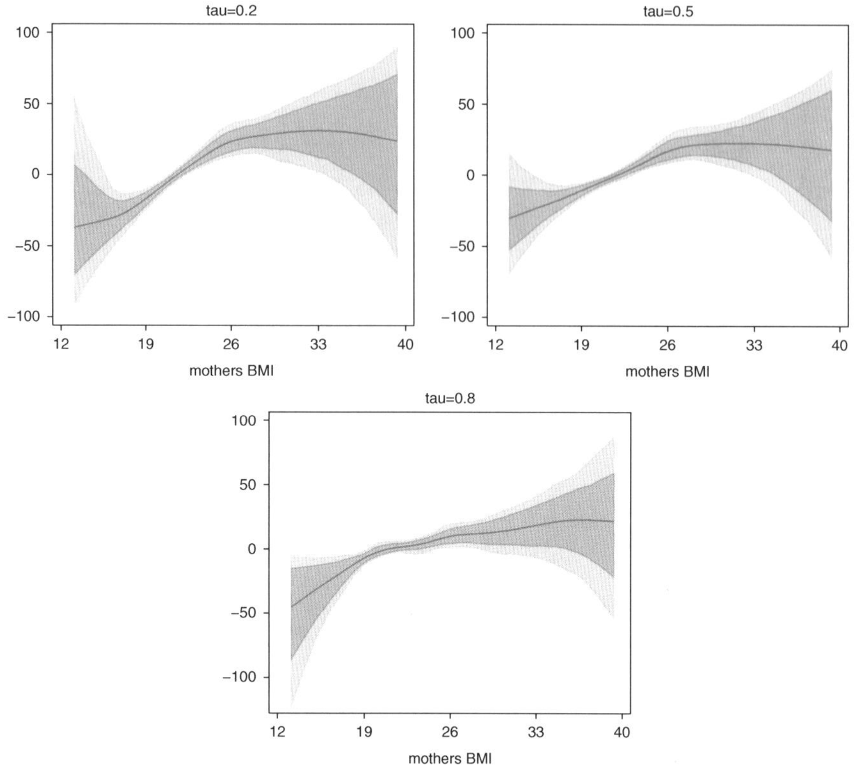

Fig. 10.7 Malnutrition in Zambia: nonparametric effect of the body mass index of the mother obtained in a geoadditive Bayesian quantile regression model for $\tau = 0.2$ (*top left*), $\tau = 0.5$ (*top right*), and $\tau = 0.8$ (*bottom*)

on the assumption of an asymmetric Laplace distribution for the errors which will usually not be fulfilled in practice.

△

10.4 Bibliographic Notes and Proofs

10.4.1 Bibliographic Notes

Non-crossing Quantiles

Approaches that circumvent the problem of crossing quantiles are typically based on joint estimation for several quantiles. The simplest example is the location-scale model

$$y_i = \boldsymbol{x}_i'\boldsymbol{\beta} + \exp(\boldsymbol{z}_i'\boldsymbol{\alpha})\varepsilon_i, \quad \text{Var}(\varepsilon_i) = 1,$$

where not only the mean but also the standard deviation of the response variable depend on covariates. This model basically reflects the situation of heteroscedasticity, but is less suited when more general effects of the covariates, e.g., on the skewness or kurtosis, are present as well. Estimation of the location-scale model can either be based on a distributional assumption for ε_i (such as a normal distribution) or include an additional estimation step for fitting the quantile curves to the data as suggested in different variants in He (1997) and Schnabel and Eilers (2012). A more sophisticated approach is quantile regression sheets where the regression coefficient β_τ corresponding to a specific covariate is treated as a function $\beta(\tau)$ that is estimated as a penalized spline with penalization along the quantile-domain; see Schnabel and Eilers (2012).

Expectile Regression

Some of the difficulties associated with extending linear quantile regression to more complex settings and developing completely data-driven estimation approaches for these settings can be attributed to the fact that the absolute deviation criterion (10.5) is not differentiable and therefore requires more specialized optimization approaches, such as linear programming or boosting. As a computationally attractive alternative, Newey and Powell (1987) have proposed expectiles that replace asymmetrically weighted absolute deviations with asymmetrically weighted squared deviations yielding the optimality criterion

$$\hat{\boldsymbol{\beta}}_\tau = \arg\min_{\boldsymbol{\beta}_\tau} \sum_{i=1}^{n} w_\tau(y_i, \eta_{i\tau})(y_i - \eta_{i\tau})^2.$$

This approach has recently gained considerable interest since the solutions can be computed by simple iteratively weighted least squares updates

$$\hat{\boldsymbol{\beta}}_\tau^{(t+1)} = (\boldsymbol{X}'\boldsymbol{W}_\tau^{(t)}\boldsymbol{X})^{-1}\boldsymbol{X}'\boldsymbol{W}_\tau^{(t)}\boldsymbol{y},$$

with iteratively recomputed weights

$$\boldsymbol{W}_\tau^{(t)} = \operatorname{diag}(w_\tau(y_1, \eta_{1\tau})^{(t)}, \ldots, w_\tau(y_n, \eta_{n\tau})^{(t)}).$$

This has particular advantages when considering complex models defined, for example, in terms of several nonlinear functions and quadratic penalties, as in the context of structured additive regression. In particular, the smoothing parameters can then still be automatically computed, for example, utilizing the mixed model representation of structured additive regression models; see Sobotka and Kneib (2012) for details. Although expectiles do not enjoy the intuitive interpretation of quantiles, they still provide a possibility to characterize the conditional distribution of a response distribution and therefore contain similar information as a set of quantile regressions. An implementation of structured additive expectile regression models is provided in the R package `expectreg`.

10.4.2 Proofs

Optimality Criterion for Quantiles (p. 599)

We show the equivalence of the implicit definition of quantiles and the optimality criterion (10.5) for a continuous random variable y with strictly increasing cumulative distribution function $F(y)$ and density $f(y)$ such that the τ-quantile q_τ is unique. Therefore, we first rewrite the expectation of the minimization criterion in terms of the density $f(y)$ as

$$\begin{aligned} \mathrm{E}\left(w_\tau(y)|y-q|\right) &= \int_{-\infty}^{\infty} w_\tau(y)|y-q| f(y)dy \\ &= \int_{-\infty}^{q} (1-\tau)(y-q)f(y)dy + \int_{q}^{\infty} \tau(y-q)f(y)dy. \end{aligned}$$

The minimum can then be found by differentiating with respect to q and setting the derivative to zero. As the limits of integration depend on q, we have to apply Leibniz integral rule, yielding

$$\begin{aligned} \frac{\partial}{\partial q}\mathrm{E}\left(w_\tau(y)|y-q|\right) &= -(1-\tau)\int_{-\infty}^{q_\tau} f(y)dy + \tau\int_{q_\tau}^{\infty} f(y)dy \\ &= -\int_{-\infty}^{q_\tau} f(y)dy + \tau\int_{-\infty}^{\infty} f(y)dy \\ &= -\int_{-\infty}^{q_\tau} f(y)dy + \tau. \end{aligned}$$

Setting this expression equal to zero, we obtain

$$\tau = \int_{-\infty}^{q_\tau} f(y)dy = F(q_\tau) = \mathrm{P}(y \leq \tau).$$

Since we assumed that the τ-quantile is unique, this also implies $\mathrm{P}(y \geq q_\tau) = 1-\tau$ and therefore proofs the equivalence between both definitions.

Derivation of the Full Conditionals in the Bayesian Linear Quantile Regression Model of Box 10.2 on p. 611

We derive the full conditionals for the model

$$\boldsymbol{y} \,|\, \boldsymbol{z}, \boldsymbol{\beta}_\tau, \sigma^2 \sim \mathrm{N}(\boldsymbol{X}\boldsymbol{\beta}_\tau + \xi\boldsymbol{z}, \sigma^2\boldsymbol{W}^{-1}), \qquad \boldsymbol{W} = \mathrm{diag}(w_1, \ldots, w_n), \qquad \boldsymbol{z} = (z1, \ldots, z_n)'$$

with

$$\xi = \frac{1-2\tau}{\tau(1-\tau)}, \qquad w_i = \frac{1}{\delta^2 z_i}, \qquad \delta^2 = \frac{2}{\tau(1-\tau)},$$

and the priors

$$\boldsymbol{\beta}_\tau \propto \text{const}, \qquad z_i \,|\, \sigma^2 \sim \text{Expo}(1/\sigma^2), \qquad \sigma^2 \sim \text{IG}(a,b).$$

Using $\boldsymbol{\Sigma}_{\boldsymbol{\beta}_\tau} = \sigma^2(\boldsymbol{X}'\boldsymbol{W}\boldsymbol{X})^{-1}$ and $\boldsymbol{\mu}_{\boldsymbol{\beta}_\tau} = (\boldsymbol{X}'\boldsymbol{W}\boldsymbol{X})^{-1}\boldsymbol{X}'\boldsymbol{W}(\boldsymbol{y}-\xi\boldsymbol{z})$, we obtain

$$\begin{aligned}
p(\boldsymbol{\beta}_\tau \,|\, \cdot) &\propto p(\boldsymbol{y}\,|\,\boldsymbol{z},\boldsymbol{\beta}_\tau,\sigma^2) \\
&\propto \exp\{-\frac{1}{2\sigma^2}(\boldsymbol{X}\boldsymbol{\beta}_\tau + \xi\boldsymbol{z} - \boldsymbol{y})'\boldsymbol{W}(\boldsymbol{X}\boldsymbol{\beta}_\tau + \xi\boldsymbol{z} - \boldsymbol{y})\} \\
&\propto \exp\{-\frac{1}{2}\boldsymbol{\beta}_\tau'\frac{1}{\sigma^2}\boldsymbol{X}'\boldsymbol{W}\boldsymbol{X}\boldsymbol{\beta}_\tau + \frac{1}{\sigma^2}\boldsymbol{\beta}_\tau'\boldsymbol{X}'\boldsymbol{W}(\boldsymbol{y}-\xi\boldsymbol{z})\} \\
&= \exp\{-\frac{1}{2}\boldsymbol{\beta}_\tau'\boldsymbol{\Sigma}_{\boldsymbol{\beta}_\tau}^{-1}\boldsymbol{\beta}_\tau + \boldsymbol{\beta}_\tau'\boldsymbol{\Sigma}_{\boldsymbol{\beta}_\tau}^{-1}(\boldsymbol{X}'\boldsymbol{W}\boldsymbol{X})^{-1}\boldsymbol{X}'\boldsymbol{W}(\boldsymbol{y}-\xi\boldsymbol{z})\} \\
&= \exp\{-\frac{1}{2}\boldsymbol{\beta}_\tau'\boldsymbol{\Sigma}_{\boldsymbol{\beta}_\tau}^{-1}\boldsymbol{\beta}_\tau + \boldsymbol{\beta}_\tau'\boldsymbol{\Sigma}_{\boldsymbol{\beta}_\tau}^{-1}\boldsymbol{\mu}_{\boldsymbol{\beta}_\tau}\}
\end{aligned}$$

for the full conditional of $\boldsymbol{\beta}_\tau$. According to Definition B.20 in Appendix B.3.1 this is a multivariate normal distribution with mean $\boldsymbol{\mu}_{\boldsymbol{\beta}_\tau}$ and covariance matrix $\boldsymbol{\Sigma}_{\boldsymbol{\beta}_\tau}$.

The full conditional for σ^2 is given by

$$\begin{aligned}
p(\sigma^2 \,|\, \cdot) &\propto p(\boldsymbol{y}\,|\,\boldsymbol{z},\boldsymbol{\beta}_\tau,\sigma^2)\, p(\boldsymbol{z}\,|\,\sigma^2)\, p(\sigma^2) \\
&\propto \left(\frac{1}{\sigma^2}\right)^{\frac{n}{2}} \exp\left\{-\frac{1}{\sigma^2}\frac{1}{2}(\boldsymbol{X}\boldsymbol{\beta}_\tau + \xi\boldsymbol{z} - \boldsymbol{y})'\boldsymbol{W}(\boldsymbol{X}\boldsymbol{\beta}_\tau + \xi\boldsymbol{z} - \boldsymbol{y})\right\} \\
&\quad \prod_{i=1}^{n}\frac{1}{\sigma^2}\exp\left(-\frac{1}{\sigma^2}z_i\right) \\
&\quad \left(\frac{1}{\sigma^2}\right)^{a-1}\exp\left(-\frac{1}{\sigma^2}b\right) \\
&= \left(\frac{1}{\sigma^2}\right)^{n+\frac{n}{2}+a-1} \exp\{-\frac{1}{\sigma^2}(b + \frac{1}{2}(\boldsymbol{X}\boldsymbol{\beta}_\tau + \xi\boldsymbol{z} - \boldsymbol{y})'\boldsymbol{W}(\boldsymbol{X}\boldsymbol{\beta}_\tau + \xi\boldsymbol{z} - \boldsymbol{y}) + \sum_{i=1}^{n} z_i)\}.
\end{aligned}$$

This is the density of an inverse gamma distribution with parameters

$$a' = a + \frac{3n}{2} \text{ and } b' = b + \frac{1}{2}(\boldsymbol{X}\boldsymbol{\beta}_\tau + \xi\boldsymbol{z} - \boldsymbol{y})'\boldsymbol{W}(\boldsymbol{X}\boldsymbol{\beta}_\tau + \xi\boldsymbol{z} - \boldsymbol{y}).$$

Finally the full conditionals for the latent variables z_i are obtained as

$$
\begin{aligned}
p(z_i \mid \cdot) &\propto p(y_i \mid z_i, \boldsymbol{\beta}_\tau, \sigma^2)\, p(z_i \mid \sigma^2) \\
&\propto \frac{1}{\sqrt{z_i}} \exp\left\{-\frac{1}{2}\frac{(y_i - \boldsymbol{x}_i'\boldsymbol{\beta}_\tau - \xi z_i)^2}{\delta^2 z_i \sigma^2}\right\} \exp\left(-\frac{1}{\sigma^2} z_i\right) \\
&= \frac{1}{\sqrt{z_i}} \exp\left\{-\frac{1}{2}\frac{(y_i - \boldsymbol{x}_i'\boldsymbol{\beta}_\tau)^2 - 2(y_i - \boldsymbol{x}_i'\boldsymbol{\beta}_\tau)\xi z_i + z_i^2(\xi^2 + 2\delta^2)}{\delta^2 z_i \sigma^2}\right\} \\
&= \frac{1}{\sqrt{z_i}} \exp\left\{-\frac{(y_i - \boldsymbol{x}_i'\boldsymbol{\beta}_\tau)^2}{2\sigma^2\delta^2} \cdot \frac{1 - \frac{2\xi z_i}{y_i - \boldsymbol{x}_i'\boldsymbol{\beta}_\tau} + z_i^2 \frac{\xi^2 + 2\delta^2}{(y_i - \boldsymbol{x}_i'\boldsymbol{\beta}_\tau)^2}}{z_i}\right\} \\
&= \frac{1}{\sqrt{z_i}} \exp\left\{-\frac{(y_i - \boldsymbol{x}_i'\boldsymbol{\beta}_\tau)^2}{2\sigma^2\delta^2} \cdot \frac{(z_i^{-1})^2 - \frac{2\xi}{(y_i - \boldsymbol{x}_i'\boldsymbol{\beta}_\tau)} z_i^{-1} + \mu_i^2}{z_i^{-1}}\right\} \\
&= \frac{1}{\sqrt{z_i}} \exp\left\{-\frac{(y_i - \boldsymbol{x}_i'\boldsymbol{\beta}_\tau)^2(\xi^2 + 2\delta^2)}{2\sigma^2\delta^2(\xi^2 + 2\delta^2)} \cdot \frac{(z_i^{-1})^2 - \frac{2\xi}{(y_i - \boldsymbol{x}_i'\boldsymbol{\beta}_\tau)} z_i^{-1} + \mu_i^2}{z_i^{-1}}\right\} \\
&= \frac{1}{\sqrt{z_i}} \exp\left\{-\frac{\lambda}{2\mu_i^2} z_i \cdot \left(\frac{1}{z_i^2} - \frac{2\xi}{(y_i - \boldsymbol{x}_i'\boldsymbol{\beta}_\tau)}\frac{1}{z_i} + \mu_i^2\right)\right\} \\
&\propto \frac{1}{\sqrt{z_i}} \exp\left\{-\frac{\lambda}{2\mu_i^2} z_i \cdot \left(\frac{1}{z_i^2} - 2\mu_i \frac{1}{z_i} + \mu_i^2\right)\right\} \\
&= \frac{1}{\sqrt{z_i}} \exp\left\{-\frac{\lambda}{2\mu_i^2} z_i \cdot \left(\frac{1}{z_i} - \mu_i\right)^2\right\}
\end{aligned}
$$

where

$$
\mu_i = \left(\frac{\xi^2 + 2\delta^2}{(y_i - \boldsymbol{x}_i'\boldsymbol{\beta}_\tau)^2}\right)^{1/2} \quad \text{and} \quad \lambda = \frac{\xi^2 + 2\delta^2}{\sigma^2\delta^2}.
$$

This is not a standard distribution. However, it turns out that the distribution of $1/z_i$ is inverse Gaussian with location parameter μ_i and scale parameter λ. Updating of z_i is then obtained by first sampling $1/z_i$ from the inverse Gaussian distribution and then inverting the result to obtain z_i.

To derive the distribution of $1/z_i$ we apply the change of variables Theorem B.1 of Appendix B.1. With $g(z_i) = 1/z_i$, $g^{-1}(z_i) = z_i$ and $g'(z_i) = -1/z_i^2$, we obtain

$$
p(1/z_i \mid \cdot) \propto \left(\frac{1}{(1/z_i)^3}\right)^{1/2} \exp\left(-\frac{\lambda}{2\mu_i(1/z_i)}(1/z_i - \mu_i)^2\right)
$$

which has the form of the proposed inverse Gaussian distribution.

Matrix Algebra

A

This appendix gives a summary of basic definitions and results in matrix algebra which are used in this book. The presentation is restricted to important definitions and theorems, without examples or proofs. There are many books that have an appendix on matrix algebra; see, for instance, Mardia, Kent, and Bibby (1999) or Rao, Toutenburg, Shalabh, and Heumann (2008). More detailed expositions, including proofs, are in Graybill (1961), Magnus and Neudecker (2002), Schott (2005), and Searle (2006).

A.1 Definition and Elementary Matrix Operations

Definition A.1 Matrix

A real matrix of order $n \times p$ (short: $n \times p$-matrix) is a rectangular array

$$\boldsymbol{A} = \begin{pmatrix} a_{11} & a_{12} & \cdots & a_{1p} \\ a_{21} & \ddots & & \vdots \\ \vdots & & \ddots & \vdots \\ a_{n1} & a_{n2} & \cdots & a_{np} \end{pmatrix}$$

of $n \cdot p$ real numbers a_{ij}, arranged in n rows and p columns. We often write $\boldsymbol{A} = (a_{ij}), i = 1, \ldots, n, j = 1, \ldots, p$.

Definition A.2 The Transpose of a Matrix

The transpose $\boldsymbol{A}'$ of a matrix $\boldsymbol{A}$ is formed by interchanging rows and columns:

L. Fahrmeir et al., *Regression*, DOI 10.1007/978-3-642-34333-9,

$$\boldsymbol{A}' = \begin{pmatrix} a_{11} & a_{21} & \cdots & a_{n1} \\ a_{12} & \ddots & & \vdots \\ \vdots & & \ddots & \vdots \\ a_{1p} & a_{2p} & \cdots & a_{np} \end{pmatrix}.$$

By definition, we have $(\boldsymbol{A}')' = \boldsymbol{A}$.

Definition A.3 Column Vector, Row Vector

A matrix with column order $p = 1$ is called a column vector. Thus,

$$\boldsymbol{a} = \begin{pmatrix} a_1 \\ a_2 \\ \vdots \\ a_n \end{pmatrix}$$

is a column vector with n components $a_1, a_2, \ldots, a_n$.

Row vectors are written as column vectors transposed, i.e.,

$$\boldsymbol{a}' = (a_1, a_2, \ldots, a_n).$$

Thus, an $n \times p$-matrix has n row vectors and p column vectors.

Definition A.4 Square Matrix

A matrix $\boldsymbol{A}$ is a square matrix if it is of order $n \times n$. Its diagonal, with elements $a_{11}, \ldots, a_{nn}$, is sometimes called main diagonal.

Definition A.5 Diagonal Matrix

A square matrix $\boldsymbol{D}$ is a diagonal matrix if the elements below and above the main diagonal are zero. Thus,

$$\boldsymbol{D} = \begin{pmatrix} d_1 & 0 & \ldots & 0 \\ \vdots & \ddots & & \vdots \\ \vdots & & \ddots & \vdots \\ 0 & \ldots & \ldots & d_n \end{pmatrix}$$

or $\boldsymbol{D} = \text{diag}(d_1, \ldots, d_n)$ in short notation.

Definition A.6 Identity Matrix

The $n \times n$-diagonal matrix

$$I_n = \text{diag}(1, \ldots, 1) = \begin{pmatrix} 1 & 0 & \ldots & 0 \\ \vdots & \ddots & & \vdots \\ \vdots & & \ddots & \vdots \\ 0 & \ldots & \ldots & 1 \end{pmatrix}$$

is called the identity matrix of order $n \times n$. In situations where no confusion can arise, we also drop the index n and simply write $\boldsymbol{I}$.

Definition A.7 Symmetric Matrix

A square matrix $\boldsymbol{A}$ is called symmetric if $\boldsymbol{A} = \boldsymbol{A}'$.

Definition A.8 Sums and Scalar Multiplication of Matrices

The sum $\boldsymbol{A} + \boldsymbol{B}$ of two $n \times p$-matrices $\boldsymbol{A} = (a_{ij})$ and $\boldsymbol{B} = (b_{ij})$ is defined as

$$\boldsymbol{A} + \boldsymbol{B} = (a_{ij} + b_{ij}).$$

The multiplication of $\boldsymbol{A}$ with a scalar $\lambda \in \mathbb{R}$ is defined as

$$\lambda \boldsymbol{A} = (\lambda\, a_{ij}).$$

Theorem A.1 Rules for Sums and Scalar Multiplication

Let $\boldsymbol{A}, \boldsymbol{B}, \boldsymbol{C}$ be $n \times p$-matrices, and let $r, k \in \mathbb{R}$ be scalars. We then have the following rules:

1. $\boldsymbol{A} + (\boldsymbol{B} + \boldsymbol{C}) = (\boldsymbol{A} + \boldsymbol{B}) + \boldsymbol{C}$
2. $\boldsymbol{A} + \boldsymbol{B} = \boldsymbol{B} + \boldsymbol{A}$
3. $(k + r)\boldsymbol{A} = k\boldsymbol{A} + r\boldsymbol{A}$ and $k(\boldsymbol{A} + \boldsymbol{B}) = k\boldsymbol{A} + k\boldsymbol{B}$
4. $(kr)\boldsymbol{A} = k(r\boldsymbol{A})$
5. $(k\boldsymbol{A})' = k\boldsymbol{A}'$
6. $(\boldsymbol{A} + \boldsymbol{B})' = \boldsymbol{A}' + \boldsymbol{B}'$

Definition A.9 Multiplication of Matrices

The product of the $n \times p$-matrix $\boldsymbol{A} = (a_{ij})$ with the $p \times m$-matrix $\boldsymbol{B} = (b_{ij})$ is the $n \times m$-matrix

$$\boldsymbol{A}\boldsymbol{B} = \boldsymbol{C} = (c_{ik}) \qquad \text{where} \qquad c_{ik} = \sum_{j=1}^{p} a_{ij} b_{jk}.$$

Thus,

$$A \cdot B = \begin{pmatrix} \sum_{j=1}^{p} a_{1j} b_{j1} & \cdots & \sum_{j=1}^{p} a_{1j} b_{jm} \\ \vdots & \ddots & \vdots \\ \sum_{j=1}^{p} a_{nj} b_{j1} & \cdots & \sum_{j=1}^{p} a_{nj} b_{jm} \end{pmatrix}.$$

Note that two matrices $\boldsymbol{A}$ and $\boldsymbol{B}$ can only be multiplied if the number of columns of $\boldsymbol{A}$ equals the number of rows of $\boldsymbol{B}$. In general, multiplication of two matrices is not commutative, i.e., in general $\boldsymbol{B} \cdot \boldsymbol{A} \neq \boldsymbol{A} \cdot \boldsymbol{B}$.

Theorem A.2 Representing Sums as Products of Vectors

Let $\boldsymbol{x}, \boldsymbol{y} \in \mathbb{R}^n$ be column vectors and let $\mathbf{1}$ be the $n \times 1$-(column) vector $(1, \ldots, 1)'$. It follows:

1. $\sum_{i=1}^{n} x_i = \mathbf{1}'\boldsymbol{x} = \boldsymbol{x}'\mathbf{1}$
2. Scalar product of $\boldsymbol{x}$ and $\boldsymbol{y}$: $\sum_{i=1}^{n} x_i y_i = \boldsymbol{x}'\boldsymbol{y} = \boldsymbol{y}'\boldsymbol{x}$
3. Squared Euclidean length of $\boldsymbol{x}$: $\sum_{i=1}^{n} x_i^2 = \boldsymbol{x}'\boldsymbol{x}$

Theorem A.3 Rules for Multiplication of Matrices

Let $\boldsymbol{A}$, $\boldsymbol{B}$, and $\boldsymbol{C}$ be matrices of appropriate dimensions. We then have the following rules:

1. $\boldsymbol{A}(\boldsymbol{B} + \boldsymbol{C}) = \boldsymbol{A}\boldsymbol{B} + \boldsymbol{A}\boldsymbol{C}$
2. $(\boldsymbol{A}\boldsymbol{B})\boldsymbol{C} = \boldsymbol{A}(\boldsymbol{B}\boldsymbol{C})$
3. $(\boldsymbol{A}\boldsymbol{B})' = \boldsymbol{B}'\boldsymbol{A}'$
4. $\boldsymbol{A}\boldsymbol{I} = \boldsymbol{A}$ and $\boldsymbol{I}\boldsymbol{A} = \boldsymbol{A}$

Definition A.10 Kronecker Product

The Kronecker product $\boldsymbol{A} \otimes \boldsymbol{B}$ of the $n \times p$-matrix $\boldsymbol{A}$ and the $r \times q$-matrix $\boldsymbol{B}$ is defined as the $nr \times pq$-matrix

$$\boldsymbol{C} = \boldsymbol{A} \otimes \boldsymbol{B} = \begin{pmatrix} a_{11}\boldsymbol{B} & a_{12}\boldsymbol{B} & \cdots & a_{1p}\boldsymbol{B} \\ \vdots & \vdots & & \vdots \\ a_{n1}\boldsymbol{B} & a_{n2}\boldsymbol{B} & \cdots & a_{np}\boldsymbol{B} \end{pmatrix}.$$

Theorem A.4 Rules for the Kronecker Product

Let $\boldsymbol{A}$, $\boldsymbol{B}$, $\boldsymbol{C}$, and $\boldsymbol{D}$ be matrices of appropriate order and let k be a scalar. We then have the following rules:

1. $k(\boldsymbol{A} \otimes \boldsymbol{B}) = (k\boldsymbol{A}) \otimes \boldsymbol{B} = \boldsymbol{A} \otimes (k\boldsymbol{B})$
2. $\boldsymbol{A} \otimes (\boldsymbol{B} \otimes \boldsymbol{C}) = (\boldsymbol{A} \otimes \boldsymbol{B}) \otimes \boldsymbol{C}$
3. $\boldsymbol{A} \otimes (\boldsymbol{B} + \boldsymbol{C}) = (\boldsymbol{A} \otimes \boldsymbol{B}) + (\boldsymbol{A} \otimes \boldsymbol{C})$
4. $(\boldsymbol{A} \otimes \boldsymbol{B})' = \boldsymbol{A}' \otimes \boldsymbol{B}'$
5. $(\boldsymbol{A}\boldsymbol{B}) \otimes (\boldsymbol{C}\boldsymbol{D}) = (\boldsymbol{A} \otimes \boldsymbol{C})(\boldsymbol{B} \otimes \boldsymbol{D})$

Definition A.11 Orthogonal Matrices

A square matrix $\boldsymbol{A}$ is called orthogonal if $\boldsymbol{A}\boldsymbol{A}' = \boldsymbol{A}'\boldsymbol{A} = \boldsymbol{I}$.

Theorem A.5 Properties of Orthogonal Matrices

Let $\boldsymbol{A}$ be orthogonal. We then have:

1. Row and column vectors span an orthonormal basis, i.e., the vectors have (Euclidean) length 1 and are pairwise orthogonal.
2. If $\boldsymbol{A}$ and $\boldsymbol{B}$ are orthogonal, then $\boldsymbol{A}\boldsymbol{B}$ is orthogonal.

Definition A.12 Idempotent Matrices

A square matrix $\boldsymbol{A}$ is called idempotent if $\boldsymbol{A}\boldsymbol{A} = \boldsymbol{A}^2 = \boldsymbol{A}$.

The $n \times n$-matrix

$$\boldsymbol{C} := \boldsymbol{I}_n - \frac{1}{n}\mathbf{1}\mathbf{1}',$$

a special idempotent matrix, is often used in statistics. The following holds:

1. Multiplication of $\boldsymbol{C}$ with an $n \times 1$-vector $\boldsymbol{a}$ gives

$$\boldsymbol{C}\boldsymbol{a} = \begin{pmatrix} a_1 - \bar{a} \\ \vdots \\ a_n - \bar{a} \end{pmatrix},$$

   i.e., one obtains the vector centered about the mean $\bar{a}$ of its elements.
2. Multiplication of $\boldsymbol{C}$ with an $n \times p$-matrix $\boldsymbol{A}$ gives

$$\boldsymbol{C}\boldsymbol{A} = \begin{pmatrix} a_{11} - \bar{a}_1 & \cdots & a_{1p} - \bar{a}_p \\ \vdots & & \vdots \\ a_{n1} - \bar{a}_1 & \cdots & a_{np} - \bar{a}_p \end{pmatrix},$$

   where $\bar{a}_1, \ldots, \bar{a}_p$ are the means of the columns of A.
3. $\boldsymbol{C}\mathbf{1} = \mathbf{0}$.
4. $\mathbf{1}'\boldsymbol{C} = \mathbf{0}'$.
5. $\mathbf{1}\mathbf{1}'\boldsymbol{C} = \boldsymbol{C}\mathbf{1}\mathbf{1}' = \mathbf{0}$.
6. $\sum_{i=1}^{n}(x_i - \bar{x})^2 = \boldsymbol{x}'\boldsymbol{C}\boldsymbol{x}$ where $\boldsymbol{x} = (x_1, \ldots, x_n)'$.

Theorem A.6 Properties of Idempotent Matrices

If $\boldsymbol{A}$ and $\boldsymbol{B}$ are idempotent, then:
1. $\boldsymbol{AB} = \boldsymbol{BA}$, and $\boldsymbol{AB}$ is also idempotent.
2. $\boldsymbol{I} - \boldsymbol{A}$ is idempotent.
3. $\boldsymbol{A}(\boldsymbol{I} - \boldsymbol{A}) = (\boldsymbol{I} - \boldsymbol{A})\boldsymbol{A} = \boldsymbol{0}$.

A.2 Rank of a Matrix

Definition A.13 Row Rank, Column Rank

Let $\boldsymbol{A}$ be of order $n \times p$. The column rank $\text{rkc}(\boldsymbol{A})$ of $\boldsymbol{A}$ is defined as the maximum number of linearly independent columns of $\boldsymbol{A}$. Correspondingly, the row rank $\text{rkr}(\boldsymbol{A})$ is the maximum number of linearly independent rows of $\boldsymbol{A}$.

Theorem A.7

The column rank equals the row rank, i.e.,

$$\text{rkc}(\boldsymbol{A}) = \text{rkr}(\boldsymbol{A}).$$

Definition A.14 Rank of a Matrix

The rank of an $n \times p$-matrix is defined as

$$\text{rk}(\boldsymbol{A}) := \text{rkc}(\boldsymbol{A}) = \text{rkr}(\boldsymbol{A}) \leq \min\{n, p\}.$$

We denote $\text{rk}(\boldsymbol{A})$ also by $\text{r}(\boldsymbol{A})$.

If $\text{rk}(\boldsymbol{A}) = \min\{n, p\}$, then $\boldsymbol{A}$ has full rank and is called regular.

Theorem A.8 Properties of Ranks

Let $\boldsymbol{A}$, $\boldsymbol{B}$, and $\boldsymbol{C}$ be matrices of appropriate order. It follows:
1. $\text{rk}(\boldsymbol{A}) = \text{rk}(-\boldsymbol{A})$
2. $\text{rk}(\boldsymbol{A}') = \text{rk}(\boldsymbol{A})$
3. $\text{rk}(\boldsymbol{A} + \boldsymbol{B}) \leq \text{rk}(\boldsymbol{A}) + \text{rk}(\boldsymbol{B})$
4. $\text{rk}(\boldsymbol{AB}) \leq \min\{\text{rk}(\boldsymbol{A}), \text{rk}(\boldsymbol{B})\}$
5. $\text{rk}(\boldsymbol{I}_n) = n$

Definition A.15 Row Space, Column Space

The row space $R(\boldsymbol{A})$ of an $n \times p$-matrix is the subspace of $\mathbb{R}^n$ spanned by the rows of $\boldsymbol{A}$, i.e.,

$$R(\boldsymbol{A}) := \{\boldsymbol{x} \in \mathbb{R}^n : \boldsymbol{x} = \boldsymbol{A}\boldsymbol{y} \text{ for some } \boldsymbol{y} \in \mathbb{R}^p\}.$$

The column space is defined correspondingly.

Definition A.16 Null Space

The null space $N(\boldsymbol{A})$ of an $n \times p$-matrix $\boldsymbol{A}$ is defined as the set

$$N(\boldsymbol{A}) := \{\boldsymbol{x} \in \mathbb{R}^p : \boldsymbol{A}\boldsymbol{x} = \boldsymbol{0}\}.$$

Theorem A.9 Properties of the Null Space

Let $\boldsymbol{A}$ be an $n \times p$-matrix. We then have:

1. The null space is a subspace of $\mathbb{R}^p$.
2. $\text{rk}(\boldsymbol{A}) + \dim(N(\boldsymbol{A})) = p$.
3. $N(\boldsymbol{A}'\boldsymbol{A}) = N(\boldsymbol{A})$.

Definition A.17 Inverse of a Matrix

Let $\boldsymbol{A}$ be a square matrix. A matrix $\boldsymbol{A}^{-1}$ is called the inverse of $\boldsymbol{A}$ if

$$\boldsymbol{A}\boldsymbol{A}^{-1} = \boldsymbol{A}^{-1}\boldsymbol{A} = \boldsymbol{I}.$$

Theorem A.10 Existence and Uniqueness of the Inverse

The inverse $\boldsymbol{A}^{-1}$ of a square $n \times n$-matrix $\boldsymbol{A}$ exists if and only if $\text{rk}(\boldsymbol{A}) = n$, i.e., if $\boldsymbol{A}$ is regular. The inverse is unique, and $\boldsymbol{A}$ is called invertible, regular, or nonsingular. If no inverse of $\boldsymbol{A}$ exists, it is called singular.

Theorem A.11 Rules for Inverses

Let $\boldsymbol{A}$, $\boldsymbol{B}$, and $\boldsymbol{C}$ be invertible matrices of the same order, and let $k \neq 0$ be a scalar. The following rules then hold:

1. $(\boldsymbol{A}^{-1})^{-1} = \boldsymbol{A}$
2. $(k\boldsymbol{A})^{-1} = k^{-1}\boldsymbol{A}^{-1} = \frac{1}{k}\boldsymbol{A}^{-1}$
3. $(\boldsymbol{A}')^{-1} = (\boldsymbol{A}^{-1})'$
4. $(\boldsymbol{A}\boldsymbol{B})^{-1} = \boldsymbol{B}^{-1}\boldsymbol{A}^{-1}$
5. $(\boldsymbol{A}\boldsymbol{B}\boldsymbol{C})^{-1} = \boldsymbol{C}^{-1}\boldsymbol{B}^{-1}\boldsymbol{A}^{-1}$
6. If $\boldsymbol{A}$ is symmetric, $\boldsymbol{A}^{-1}$ is also symmetric.
7. The inverse of a diagonal matrix $\boldsymbol{A} = \text{diag}(a_1, \ldots, a_n)$ is

$$\boldsymbol{A}^{-1} = \text{diag}(a_1^{-1}, \ldots, a_n^{-1}).$$

8. If $\boldsymbol{A}$ is orthogonal, then $\boldsymbol{A}^{-1} = \boldsymbol{A}'$.

A.3 Block Matrices and the Matrix Inversion Lemma

Definition A.18 Block Matrix, Block-Diagonal Matrix

A matrix $\boldsymbol{A}$ is called a block (or partitioned) matrix if it is partitioned into

$$\boldsymbol{A} = \begin{bmatrix} \boldsymbol{A}_1 & \boldsymbol{A}_{12} & \dots & \boldsymbol{A}_{1p} \\ \boldsymbol{A}_{21} & \boldsymbol{A}_2 & \dots & \boldsymbol{A}_{2p} \\ \vdots & \vdots & \ddots & \vdots \\ \boldsymbol{A}_{p1} & \dots & \dots & \boldsymbol{A}_p \end{bmatrix},$$

where $\boldsymbol{A}_1, \dots, \boldsymbol{A}_p, \boldsymbol{A}_{12}, \dots, \boldsymbol{A}_{p1}, \dots$ are (non-overlapping) matrix sub-blocks of suitable dimensions. A special case, appearing several times in this book, is block-diagonal matrices

$$\boldsymbol{A} = \begin{bmatrix} \boldsymbol{A}_1 & \boldsymbol{0} & \dots & \boldsymbol{0} \\ \boldsymbol{0} & \boldsymbol{A}_2 & \dots & \boldsymbol{0} \\ \vdots & \vdots & \ddots & \vdots \\ \boldsymbol{0} & \dots & \dots & \boldsymbol{A}_p \end{bmatrix},$$

where all off-diagonal matrices are $\boldsymbol{0}$. We often denote such block-diagonal matrices by

$$\boldsymbol{A} = \text{blockdiag}(\boldsymbol{A}_1, \boldsymbol{A}_2, \dots, \boldsymbol{A}_p).$$

Theorem A.12 Sums and Products of Block Matrices

Rules for sums and products of matrices can be generalized to block matrices. For example,

$$\begin{bmatrix} A & B \\ C & D \end{bmatrix} + \begin{bmatrix} E & F \\ G & H \end{bmatrix} = \begin{bmatrix} A + E & B + F \\ C + G & D + H \end{bmatrix}$$

and

$$\begin{bmatrix} A & B \\ C & D \end{bmatrix} \cdot \begin{bmatrix} E & F \\ G & H \end{bmatrix} = \begin{bmatrix} AE + BG & AF + BH \\ CE + DG & CF + DH \end{bmatrix},$$

assuming that all dimensions of submatrices fit together appropriately.

Theorem A.13 Inversion of Block-Diagonal Matrices

Matrices can also be inverted blockwise. If all sub-blocks $\boldsymbol{A}_1, \boldsymbol{A}_2, \dots, \boldsymbol{A}_p$ in a block-diagonal matrix $\boldsymbol{A} = \text{blockdiag}(\boldsymbol{A}_1, \boldsymbol{A}_2, \dots, \boldsymbol{A}_p)$ are quadratic and invertible, then

$$A^{-1} = \begin{bmatrix} A_1^{-1} & \mathbf{0} & \dots & \mathbf{0} \\ \mathbf{0} & A_2^{-1} & \dots & \mathbf{0} \\ \vdots & \vdots & \ddots & \vdots \\ \mathbf{0} & \dots & \dots & A_p^{-1} \end{bmatrix} = \text{blockdiag}(A_1^{-1}, A_2^{-1}, \dots, A_p^{-1}).$$

Theorem A.14 Matrix Inversion Lemma

In general, blockwise inversion can be based on the following analytic inversion formula:

$$\begin{bmatrix} A & B \\ C & D \end{bmatrix}^{-1} = \begin{bmatrix} A^{-1} + A^{-1}B(D - CA^{-1}B)^{-1}CA^{-1} & -A^{-1}B(D - CA^{-1}B)^{-1} \\ -(D - CA^{-1}B)^{-1}CA^{-1} & (D - CA^{-1}B)^{-1} \end{bmatrix},$$

where A and D are square matrices and A and $(D - CA^{-1}B)^{-1}$ are nonsingular. Exchanging the roles of A, B and C, D, an analogous formula is

$$\begin{bmatrix} A & B \\ C & D \end{bmatrix}^{-1} = \begin{bmatrix} (A - BD^{-1}C)^{-1} & -(A - BD^{-1}C)^{-1}BD^{-1} \\ -D^{-1}C(A - BD^{-1}C)^{-1} & D^{-1} + D^{-1}C(A - BD^{-1}C)^{-1}BD^{-1} \end{bmatrix}.$$

Equating the left upper blocks gives

$$(A - BD^{-1}C)^{-1} = A^{-1} + A^{-1}B(D - CA^{-1}B)^{-1}CA^{-1}$$

which is the *matrix inversion lemma*. It appears in several variants in the literature. One of them is the expression

$$(A + B)^{-1} = A^{-1} - A^{-1}B(B + BA^{-1}B)^{-1}BA^{-1}$$

for the inverse of the sum of two matrices, where B is not necessarily nonsingular.

A.4 Determinant and Trace of a Matrix

Definition A.19 Determinant

The determinant of a square matrix A of order $n \times n$ is defined as

$$|A| = \sum_{i=1}^{n} (-1)^{i+j} a_{ij} |A_{-ij}|,$$

where $\boldsymbol{A}_{-ij}$ is the $((n-1)\times(n-1))$-matrix obtained by deleting row i and column j from $\boldsymbol{A}$. For scalar matrices $\boldsymbol{A} = (a_{11})$ we get $|\boldsymbol{A}| = a_{11}$. For a 2×2-matrix, $|\boldsymbol{A}| = a_{11}a_{22} - a_{12}a_{21}$.

Theorem A.15 Determinant of the Transpose

Let $\boldsymbol{A}$ be a square matrix, then $|\boldsymbol{A}'| = |\boldsymbol{A}|$.

Theorem A.16 Determinants of Some Matrices

Let $\boldsymbol{A}$ be a square matrix. It follows:

1. If a row (or column) of $\boldsymbol{A}$ consists of zeros only, then $|\boldsymbol{A}| = 0$.
2. If $\boldsymbol{A}$ has two identical rows (columns), then $|\boldsymbol{A}| = 0$.
3. Let $\boldsymbol{A}$ be a triangular matrix, i.e., all elements above or all elements below the main diagonal are zero. It follows that $|\boldsymbol{A}|$ is then the product of all diagonal elements.
4. $|\boldsymbol{I}| = 1$.

Theorem A.17 Properties of Determinants

1. $|k\boldsymbol{A}| = k^n|\boldsymbol{A}|$
2. $|\boldsymbol{A}| \neq 0 \iff \text{rk}(\boldsymbol{A}) = n$
3. $|\boldsymbol{A}\boldsymbol{B}| = |\boldsymbol{A}| \cdot |\boldsymbol{B}|$
4. $|\boldsymbol{A}^{-1}| = \dfrac{1}{|\boldsymbol{A}|}$
5. If $\boldsymbol{A}$ is orthogonal, then $|\boldsymbol{A}| = \pm 1$
6. Sylvester's theorem: $|\boldsymbol{A} + \boldsymbol{B}\boldsymbol{C}| = |\boldsymbol{A}||\boldsymbol{I} + \boldsymbol{C}\boldsymbol{A}^{-1}\boldsymbol{B}|$

Definition A.20 Trace of a Matrix

Let $\boldsymbol{A} = (a_{ij})$ be an $n \times n$-matrix. The sum of the diagonal elements is then called trace of $\boldsymbol{A}$, i.e.,

$$\text{tr}(\boldsymbol{A}) = \sum_{i=1}^{n} a_{ii}.$$

Theorem A.18 Properties of the Trace

The trace has the following properties:

1. $\text{tr}(\boldsymbol{A} + \boldsymbol{B}) = \text{tr}(\boldsymbol{A}) + \text{tr}(\boldsymbol{B})$.
2. $\text{tr}(\boldsymbol{A}) = \text{tr}(\boldsymbol{A}')$.
3. $\text{tr}(k\boldsymbol{A}) = k \cdot \text{tr}(\boldsymbol{A})$.
4. $\text{tr}(\boldsymbol{A}\boldsymbol{B}) = \text{tr}(\boldsymbol{B}\boldsymbol{A})$. This remains valid if $\boldsymbol{A}$ is of order $n \times p$ and $\boldsymbol{B}$ is of order $p \times n$.
5. Let $\boldsymbol{x}, \boldsymbol{y}$ be vectors $\in \mathbb{R}^n$. It follows that $\text{tr}(\boldsymbol{x}\boldsymbol{y}') = \text{tr}(\boldsymbol{y}\boldsymbol{x}') = \text{tr}(\boldsymbol{x}'\boldsymbol{y}) = \boldsymbol{x}'\boldsymbol{y}$.

A.5 Generalized Inverse

Definition A.21 Generalized Inverse

Let A be an $n \times p$-matrix with $n \leq p$. The $p \times n$-matrix A^- is then called generalized inverse (or g-inverse) of A if

$$A A^- A = A .$$

Theorem A.19 Existence of a Generalized Inverse

A generalized inverse always exists, but it is not unique in general.

Theorem A.20 Properties of the Generalized Inverse

For any $n \times p$-matrix A and any $p \times n$-generalized inverse, we have
1. $\text{rk}(A) = \text{rk}(A A^-) = \text{rk}(A^- A)$.
2. $\text{rk}(A) \leq \text{rk}(A^-)$.
3. If A is regular, then $A^- = A^{-1}$, and A^{-1} is unique.
4. $A^- A$ and $A A^-$ are idempotent.

A.6 Eigenvalues and Eigenvectors

Definition A.22 Eigenvalue and Eigenvector

Let A be an $n \times n$-square matrix. The (possibly complex) number $\lambda \in \mathbb{C}$ is called an eigenvalue of A, if there exists a nonzero vector $x \in \mathbb{C}^n$ (with possibly complex elements) such that

$$A x = \lambda x$$

or equivalently

$$(A - \lambda I)x = \mathbf{0}.$$

The vector x is then called a (right) eigenvector for the eigenvalues λ.

Definition A.23 Characteristic Polynomials

Let A be an $n \times n$-square matrix; then

$$q(\lambda) := |A - \lambda I|$$

is called the characteristic polynomial of A.

Definition A.19 of the determinant of a matrix shows that $q(\lambda)$ is indeed a polynomial of order n. Therefore, $q(\lambda)$ has the form

$$q(\lambda) = (-\lambda)^n + \alpha_{m-1}(-\lambda)^{m-1} + \cdots + \alpha_1(-\lambda) + \alpha_0, \tag{A.1}$$

where the scalar coefficients are unspecified. The polynomial can always be written as

$$q(\lambda) = |A - \lambda I| = \prod_{i=1}^{n}(\lambda_i - \lambda), \tag{A.2}$$

where $\lambda_1, \ldots, \lambda_n$ are the n roots of the characteristic equation $q(\lambda) = |A - \lambda I| = 0$. The fundamental theorem of algebra states that $q(\lambda) = 0$ has exactly n roots, which are not necessarily different and may be complex numbers.

Theorem A.21 Computation of Eigenvalues via the Characteristic Polynomial

The eigenvalues $\lambda_1, \ldots, \lambda_n$ of a square matrix A are the roots of the characteristic equation

$$|A - \lambda I| = 0.$$

Theorem A.22 Properties of the Eigenvalues

The eigenvalues of an $n \times n$-matrix have the following properties:

1. $|A| = \prod_{i=1}^{n} \lambda_i$.
2. $\mathrm{tr}(A) = \sum_{i=1}^{n} \lambda_i$.
3. A is regular if and only if all eigenvalues are nonzero.
4. The matrices A and A' have the same eigenvalues.
5. If λ is an eigenvalue of a regular matrix A, then $\frac{1}{\lambda}$ is an eigenvalue of A^{-1}.
6. The eigenvalues of a diagonal matrix are the elements of the diagonal.
7. The eigenvalues of an orthogonal matrix A are either 1 or -1.
8. The eigenvalues of an idempotent matrix A are either 1 or 0.

Definition A.24 Eigenspace

Let λ be an eigenvalue of a square matrix A. The set

$$A_\lambda := \{x \in \mathbb{C}^n | x \text{ eigenvector for } \lambda\} \cup \{0\}$$

is called eigenspace for λ.

Definition A.25 Similar Matrices

The matrices A and B are called similar (denoted as $A \sim B$), if there exists a regular matrix C such that $B = CAC^{-1}$.

Theorem A.23 Eigenvalues of Similar Matrices

Let $\boldsymbol{A}$ and $\boldsymbol{B}$ be similar, then it follows:

1. $\boldsymbol{A}$ and $\boldsymbol{B}$ have the same characteristic polynomial and the same eigenvalues.
2. If $\boldsymbol{x}$ is an eigenvector of $\boldsymbol{A}$ for the eigenvalue λ, then $\boldsymbol{C}\boldsymbol{x}$ is an eigenvector of $\boldsymbol{B} = \boldsymbol{C}\boldsymbol{A}\boldsymbol{C}^{-1}$.

Theorem A.24 Eigenvalues and Eigenvectors of Symmetric Matrices

Let $\boldsymbol{A}$ be a symmetric $n \times n$-matrix, then it follows:

1. All eigenvalues are real numbers.
2. Eigenvectors for different eigenvalues are pairwise orthogonal.

Theorem A.25 Spectral Decomposition Theorem

Let $\boldsymbol{A}$ be a symmetric $n \times n$-matrix with rank $\mathrm{rk}(\boldsymbol{A}) = r$; then $\boldsymbol{A}$ can be written as

$$\boldsymbol{A} = \boldsymbol{P}\mathrm{diag}(\lambda_1, \dots, \lambda_r)\boldsymbol{P}',$$

where $\lambda_1, \dots, \lambda_r$ are the nonzero eigenvalues of $\boldsymbol{A}$ and $\boldsymbol{P}$ is an orthogonal $n \times r$-matrix whose columns are the corresponding orthonormal eigenvectors. Equivalently,

$$\boldsymbol{P}'\boldsymbol{A}\boldsymbol{P} = \mathrm{diag}(\lambda_1, \dots, \lambda_r).$$

Theorem A.26 Spectral Decomposition of an Idempotent Matrix

Let the $n \times n$-matrix $\boldsymbol{A}$ be symmetric and idempotent with $\mathrm{rk}(\boldsymbol{A}) = r$; then

$$\boldsymbol{P}'\boldsymbol{A}\boldsymbol{P} = \boldsymbol{I}_r,$$

and $\mathrm{rk}(\boldsymbol{A}) = \mathrm{tr}(\boldsymbol{A})$.

A.7 Quadratic Forms

Definition A.26 Quadratic Form

Let $\boldsymbol{A}$ be a symmetric $n \times n$-matrix. A quadratic form in the vector $\boldsymbol{x}$ is a function of the form

$$Q(\boldsymbol{x}) = \boldsymbol{x}'\boldsymbol{A}\boldsymbol{x} = \sum_{i=1}^{n}\sum_{j=1}^{n} a_{ij}x_i x_j = \sum_{i=1}^{n} a_{ii}x_i^2 + 2\sum_{i=1}^{n}\sum_{j>i} a_{ij}x_i x_j .$$

Definition A.27 Definite Matrices

The quadratic form $\boldsymbol{x}'\boldsymbol{A}\boldsymbol{x}$ and the matrix $\boldsymbol{A}$ are called:

1. Positive definite (p.d.), if $\boldsymbol{x}'\boldsymbol{A}\boldsymbol{x} > 0$ for all $\boldsymbol{x} \neq 0$, notation: $\boldsymbol{A} > 0$
2. Positive semidefinite (p.s.d.), if $\boldsymbol{x}'\boldsymbol{A}\boldsymbol{x} \geq 0$ and $\boldsymbol{x}'\boldsymbol{A}\boldsymbol{x} = 0$ for at least one $\boldsymbol{x} \neq 0$

3. Nonnegative definite, if $\boldsymbol{x}'\boldsymbol{A}\boldsymbol{x}$ and $\boldsymbol{A}$ are either p.d. or p.s.d., notation: $\boldsymbol{A} \geq 0$
4. Negative definite (n.d.), if $-\boldsymbol{A}$ is positive definite
5. Negative semidefinite (n.s.d.), if $-\boldsymbol{A}$ is p.s.d.
6. Indefinite in all other cases

Theorem A.27 Criteria for Definite Matrices

Let $\boldsymbol{A}$ be a symmetric matrix with real eigenvalues $\lambda_1, \ldots, \lambda_n$. It then follows that $\boldsymbol{A}$ is:

1. Positive definite, if and only if $\lambda_i > 0$ for $i = 1, \ldots, n$
2. Positive semidefinite, if and only if $\lambda_i \geq 0$ for $i = 1, \ldots, n$ and $\lambda_i = 0$ for at least one eigenvalue
3. Negative definite, if and only if $\lambda_i < 0$ for $i = 1 \ldots, n$
4. Negative semidefinite, if and only if $\lambda_i \leq 0$ for $i = 1, \ldots, n$ and at least one $\lambda_i = 0$
5. Indefinite, if and only if $\boldsymbol{A}$ has at least one positive and one negative eigenvalue

Theorem A.28 Properties of Positive Definite Matrices

For any positive definite matrix $\boldsymbol{A}$ the following properties hold:

1. $\boldsymbol{A}$ is regular and, thus, invertible.
2. $\boldsymbol{A}^{-1}$ is positive definite.
3. The diagonal elements $a_{ii}, i = 1, \ldots, n$, are positive, i.e. $a_{ii} > 0$.
4. $\text{tr}(\boldsymbol{A}) > 0$.
5. If $\boldsymbol{B}$ is positive semidefinite, then $\boldsymbol{A} + \boldsymbol{B}$ is positive definite.

Theorem A.29

Let $\boldsymbol{A}$ be an $n \times n$-matrix and $\boldsymbol{Q}$ an $n \times m$-matrix. The following then holds:

1. If $\boldsymbol{A}$ is nonnegative definite, then $\boldsymbol{Q}'\boldsymbol{A}\,\boldsymbol{Q}$ is nonnegative definite.
2. If $\boldsymbol{A}$ is positive definite and if $\text{rk}(\boldsymbol{Q}) = m$, then $\boldsymbol{Q}'\boldsymbol{A}\,\boldsymbol{Q}$ is positive definite.

Theorem A.30

Let $\boldsymbol{B}$ be an $n \times p$-matrix. The matrix $\boldsymbol{B}'\boldsymbol{B}$ is then symmetric and nonnegative definite. If $\text{rk}(\boldsymbol{B}) = p$, then $\boldsymbol{B}'\boldsymbol{B}$ is positive definite.

Theorem A.31 Eigenvalues of $\boldsymbol{B}'\boldsymbol{B}$ and $\boldsymbol{B}\boldsymbol{B}'$

Let $\boldsymbol{B}$ be an $n \times p$-matrix with $\text{rk}(\boldsymbol{B}) = r$. It follows that:

1. $\boldsymbol{B}\boldsymbol{B}'$ and $\boldsymbol{B}'\boldsymbol{B}$ have the same r positive eigenvalues λ_j, $j = 1, \ldots, r$.
2. If $\boldsymbol{v}$ is an eigenvector of $\boldsymbol{B}'\boldsymbol{B}$ for the eigenvalue λ, then

$$\boldsymbol{u} := \frac{1}{\sqrt{\lambda}}\boldsymbol{B}\boldsymbol{v}$$

is an eigenvector of $\boldsymbol{B}\boldsymbol{B}'$ for λ.

Theorem A.32 Cholesky Decomposition

Any symmetric and positive definite $n \times n$-matrix $\boldsymbol{A}$ can be uniquely decomposed into

$$\boldsymbol{A} = \boldsymbol{L}\boldsymbol{L}',$$

where $\boldsymbol{L}$ is a lower triangular matrix with positive diagonal elements. $\boldsymbol{L}$ is called Cholesky factor of $\boldsymbol{A}$.

A.8 Differentiation of Matrix Functions

Definition A.28 Differentiation with Respect to a Vector

Let $\boldsymbol{x} = (x_1, \ldots, x_n)'$ be a $n \times 1$-vector and $f(\boldsymbol{x})$ a real function differentiable with respect to the elements x_i of $\boldsymbol{x}$.

The $n \times 1$-vector

$$\frac{\partial f}{\partial \boldsymbol{x}} = \begin{pmatrix} \frac{\partial f}{\partial x_1} \\ \frac{\partial f}{\partial x_2} \\ \vdots \\ \frac{\partial f}{\partial x_n} \end{pmatrix}$$

is then called differential of f with respect to $\boldsymbol{x}$. We denote by

$$\frac{\partial f}{\partial \boldsymbol{x}'} = \left(\frac{\partial f}{\partial x_1}, \ldots, \frac{\partial f}{\partial x_n} \right)$$

the transpose of $\frac{\partial f}{\partial \boldsymbol{x}}$.

To give an example, suppose that

$$f(\boldsymbol{x}) = \boldsymbol{y}'\boldsymbol{x} = \sum_{i=1}^{n} y_i x_i,$$

where $\boldsymbol{y} = (y_1, \ldots, y_n)'$ is constant. Then we have

$$\frac{\partial f}{\partial \boldsymbol{x}} = \frac{\partial \boldsymbol{y}'\boldsymbol{x}}{\partial \boldsymbol{x}} = \begin{pmatrix} \frac{\partial \left(\sum y_i x_i\right)}{\partial x_1} \\ \frac{\partial \left(\sum y_i x_i\right)}{\partial x_2} \\ \vdots \\ \frac{\partial \left(\sum y_i x_i\right)}{\partial x_n} \end{pmatrix} = \begin{pmatrix} y_1 \\ y_2 \\ \vdots \\ y_n \end{pmatrix} = \boldsymbol{y}.$$

Definition A.29 Differentiation of a Vector Function with Respect to a Vector

Let $\boldsymbol{x} = (x_1, \ldots, x_n)'$ be a $n \times 1$-vector and $\boldsymbol{f}(\boldsymbol{x}) = (f_1(\boldsymbol{x}), \ldots, f_m(\boldsymbol{x}))'$ a $m \times 1$ vector function differentiable with respect to the elements x_i of $\boldsymbol{x}$.

The $n \times m$-matrix

$$\frac{\partial \boldsymbol{f}}{\partial \boldsymbol{x}} = \left(\frac{\partial f_j}{\partial x_i}\right) = \begin{pmatrix} \frac{\partial f_1}{\partial x_1} & \cdots & \frac{\partial f_m}{\partial x_1} \\ \vdots & & \vdots \\ \frac{\partial f_1}{\partial x_n} & \cdots & \frac{\partial f_m}{\partial x_n} \end{pmatrix}$$

is then called differential of $\boldsymbol{f}$ with respect to $\boldsymbol{x}$. We denote by

$$\frac{\partial \boldsymbol{f}}{\partial \boldsymbol{x}'} = \left(\frac{\partial \boldsymbol{f}}{\partial \boldsymbol{x}}\right)' = \begin{pmatrix} \frac{\partial f_1}{\partial x_1} & \cdots & \frac{\partial f_1}{\partial x_n} \\ \vdots & & \vdots \\ \frac{\partial f_m}{\partial x_1} & \cdots & \frac{\partial f_m}{\partial x_n} \end{pmatrix}$$

the transpose of $\frac{\partial f}{\partial x}$.

Theorem A.33 Differentiation Rules

Assume that $\boldsymbol{A}$ is a matrix and $\boldsymbol{a}$, $\boldsymbol{x}$, and $\boldsymbol{y}$ are vectors. Furthermore, we assume that the following expressions exist and all matrices and vectors are of appropriate order. The following rules then hold:

1. $\frac{\partial \boldsymbol{y}'\boldsymbol{x}}{\partial \boldsymbol{x}} = \boldsymbol{y}$.
2. $\frac{\partial \boldsymbol{x}'\boldsymbol{A}\boldsymbol{x}}{\partial \boldsymbol{x}} = (\boldsymbol{A} + \boldsymbol{A}')\boldsymbol{x}$.
3. If $\boldsymbol{A}$ is symmetric, then

$$\frac{\partial \boldsymbol{x}'\boldsymbol{A}\boldsymbol{x}}{\partial \boldsymbol{x}} = 2\boldsymbol{A}\boldsymbol{x} = 2\boldsymbol{A}'\boldsymbol{x}.$$

4. $\frac{\partial \boldsymbol{A}\boldsymbol{x}}{\partial \boldsymbol{x}} = \boldsymbol{A}'$.
5. $\frac{\partial \boldsymbol{A}\boldsymbol{x}}{\partial \boldsymbol{x}'} = \boldsymbol{A}$.

Theorem A.34 Local Extremes

Let $\boldsymbol{x} = (x_1, \ldots, x_n)'$ be a $(n \times 1)$-vector and $f(\boldsymbol{x})$ a real function differentiable with respect to the elements x_i of $\boldsymbol{x}$. Define the vector

$$s(x) = \frac{\partial f(x)}{\partial x}$$

of first derivatives and the matrix

$$H(x) = \frac{\partial s(x)}{\partial x'} = \begin{pmatrix} \frac{\partial s_1(x)}{\partial x_1} & \cdots & \frac{\partial s_1(x)}{\partial x_n} \\ \vdots & & \vdots \\ \frac{\partial s_n(x)}{\partial x_1} & \cdots & \frac{\partial s_n(x)}{\partial x_n} \end{pmatrix} = \begin{pmatrix} \frac{\partial^2 f(x)}{\partial x_1 \partial x_1} & \cdots & \frac{\partial^2 f(x)}{\partial x_1 \partial x_n} \\ \vdots & & \vdots \\ \frac{\partial^2 f(x)}{\partial x_n \partial x_1} & \cdots & \frac{\partial^2 f(x)}{\partial x_n \partial x_n} \end{pmatrix}$$

of second derivatives. $H(x)$ is also called *Hessian matrix* or simply the *Hessian*.

A necessary condition for $x = x_0$ being a local extreme of f is

$$s(x_0) = 0. \tag{A.3}$$

If Eq. (A.3) is true, the following sufficient condition holds:

- If $H(x_0)$ is positive definite x_0 is a local minimum.
- If $H(x_0)$ is negative definite x_0 is a local maximum.

Probability Calculus and Statistical Inference B

This appendix contains (in concise form) parts of probability calculus and statistical inference that are used in this book but may not be sufficiently covered in introductory courses or textbooks. Apart from some univariate distributions, this appendix mainly considers multivariate random variables, as well as likelihood and Bayesian inference for multidimensional parameters.

B.1 Some Univariate Distributions

Definition B.1 Binomial Distribution

A discrete random variable X is said to have a binomial distribution with parameters $n \in \{1, 2, \ldots\}$ and $\pi \in [0, 1]$ if it has (discrete) probability function

$$f(x) = \binom{n}{x} \pi^x (1 - \pi)^x, \qquad x = 0, 1, \ldots, n.$$

The mean and variance are given by

$$\mathrm{E}(X) = n\,\pi,$$

$$\mathrm{Var}(X) = n\,\pi\,(1 - \pi).$$

We write $X \sim \mathrm{B}(n, \pi)$. For $n = 1$ we also speak of a Bernoulli distribution.

A binomial distributed random variable results as the sum of independent and identically distributed binary random variables $X_1, \ldots, X_n$ with $\mathrm{P}(X_i = 1) = \pi$ and $\mathrm{P}(X_i = 0) = 1 - \pi$. Then

$$X = X_1 + \ldots + X_n \sim \mathrm{B}(n, \pi).$$

Note that the independence of the X_i is crucial for this result.

L. Fahrmeir et al., *Regression*, DOI 10.1007/978-3-642-34333-9,

Definition B.2 Beta Distribution

A continuous random variable X is said to have a beta distribution with parameters $a > 0$ and $b > 0$ if it has probability function

$$f(x) = \frac{\Gamma(a+b)}{\Gamma(a)\Gamma(b)} x^{a-1}(1-x)^{b-1}, \qquad x \in (0,1),$$

where $\Gamma(\cdot)$ is the gamma function. The mean and the variance are given by

$$\mathrm{E}(X) = \frac{a}{a+b},$$
$$\mathrm{Var}(X) = \frac{ab}{(a+b)^2(a+b+1)}.$$

We write $X \sim \mathrm{Beta}(a,b)$. For $a = b = 1$, we obtain a uniform distribution in the interval $(0,1)$.

Definition B.3 Beta-Binomial Distribution

A discrete random variable X is said to have a beta-binomial distribution with parameters $n \in \{1,2,\dots\}$, $a > 0$, $b > 0$ if it has probability function

$$f(x) = \frac{\Gamma(a+b)}{\Gamma(a)\Gamma(b)\Gamma(a+b+n)} \binom{n}{x} \Gamma(a+x)\Gamma(a+n-x) \qquad x = 0,1,2,\dots,n.$$

The mean and the variance are given by

$$\mathrm{E}(X) = n\frac{a}{a+b},$$
$$\mathrm{Var}(X) = n\frac{ab}{(a+b)^2}\frac{a+b+n}{a+b+1}.$$

We write $X \sim \mathrm{BetaB}(n,a,b)$. For $a = b = 1$, the beta-binomial distribution corresponds to a discrete uniform distribution on $0,1,\dots,n$, i.e., $f(x) = 1/(n+1)$.

The beta-binomial distribution arises as a mixture distribution. Suppose $X \mid \pi \sim \mathrm{B}(n,\pi)$ and $\pi \sim \mathrm{Beta}(a,b)$, then $X \sim \mathrm{BetaB}(n,a,b)$.

Definition B.4 Poisson Distribution

A discrete random variable X is said to have a Poisson distribution with parameter $\lambda > 0$ if it has probability function

$$f(x) = \frac{\lambda^x}{x!}\exp(-\lambda), \qquad x = 0,1,2\dots.$$

The mean and the variance are given by

$$\mathrm{E}(X) = \lambda,$$

$$\mathrm{Var}(X) = \lambda.$$

We write $X \sim \mathrm{Po}(\lambda)$.

Definition B.5 Normal Distribution and Truncated Normal Distribution

A continuous random variable X is said to have a normal (or Gaussian) distribution if it has probability density function (p.d.f)

$$f(x) = \frac{1}{\sigma\sqrt{2\pi}} \exp\left(-\frac{(x-\mu)^2}{2\sigma^2}\right).$$

The mean and variance are given by

$$\mathrm{E}(X) = \mu,$$

$$\mathrm{Var}(X) = \sigma^2.$$

We write $X \sim \mathrm{N}(\mu, \sigma^2)$. If $\mu = 0, \sigma^2 = 1$, then X is said to have a standard normal distribution.

If X is restricted to $a \le X \le b$, then it is said to have a truncated normal distribution. We write $X \sim \mathrm{TN}_{a,b}(\mu, \sigma^2)$. Its p.d.f. is given by

$$g(x) = \begin{cases} \dfrac{f(x)}{\mathrm{P}(a \le X \le b)}, & a \le x \le b, \\ 0, & \text{else.} \end{cases}$$

The support of X is restricted to the interval $[a, b]$ and the density has to be renormalized. For $a = -\infty$ or $b = \infty$, X is said to be left or right truncated.

Definition B.6 Lognormal Distribution

A continuous nonnegative random variable X is said to have a lognormal distribution if $Y = \log(X)$ follows a $\mathrm{N}(\mu, \sigma^2)$ distribution. Its p.d.f. is

$$f(x) = \frac{1}{\sqrt{2\pi}\sigma} \frac{1}{x} \exp\left(-(\log(x) - \mu)^2/2\sigma^2\right), \qquad x > 0.$$

The mean and variance are given by

$$\mathrm{E}(X) = \exp(\mu + \sigma^2/2),$$

$$\mathrm{Var}(X) = \exp(2\mu + \sigma^2) \cdot (\exp(\sigma^2) - 1).$$

We write $X \sim \mathrm{LN}(\mu, \sigma^2)$.

Definition B.7 Inverse Gaussian Distribution

A continuous nonnegative random variable X is said to have an inverse Gaussian distribution with parameters $\mu > 0$ and $\lambda > 0$ if it has p.d.f.

$$f(x) = \left(\frac{\lambda}{2\pi x^3}\right)^{\frac{1}{2}} \exp\left(-\frac{\lambda(x-\mu)^2}{2\mu^2 x}\right) \qquad x > 0.$$

The mean and variance are given by

$$\begin{aligned} \mathrm{E}(X) &= \mu, \\ \mathrm{Var}(X) &= \mu^3/\lambda. \end{aligned}$$

We write $X \sim \mathrm{invGauss}(\mu, \lambda)$.

Definition B.8 Laplace Distribution

A continuous random variable X is said to have a Laplace distribution (or double exponential distribution) with location parameter μ and scale parameter $s > 0$ if it has p.d.f.

$$f(x) = \frac{1}{2s} \exp\left(-\frac{|x-\mu|}{s}\right).$$

The mean and variance are given by

$$\begin{aligned} \mathrm{E}(X) &= \mu, \\ \mathrm{Var}(X) &= 2s^2. \end{aligned}$$

We write $X \sim \mathrm{La}(\mu, s)$.

Definition B.9 Gamma Distribution

A continuous nonnegative random variable X is said to have a gamma distribution with parameters $a > 0$ and $b > 0$ if it has p.d.f.

$$f(x) = \frac{b^a}{\Gamma(a)} x^{a-1} \exp(-bx), \qquad x > 0.$$

The mean and variance are given by

$$\begin{aligned} \mathrm{E}(X) &= a/b, \\ \mathrm{Var}(X) &= a/b^2. \end{aligned}$$

The mode is $(a-1)/b$ (for $a > 1$). We write $X \sim \mathrm{G}(a, b)$.

An alternative parameterization of the p.d.f., depending on $\mu = \mathrm{E}(X)$ and a scale parameter $\nu > 0$, is

$$f(x) = \frac{1}{\Gamma(\nu)} \left(\frac{\nu}{\mu}\right)^{\nu} x^{\nu-1} \exp\left(-\frac{\nu}{\mu} x\right), \qquad x > 0.$$

This alternative p.d.f. is used, for example, in Chap. 5 for gamma regression models.

Definition B.10 Exponential Distribution

A continuous nonnegative random variable X is said to have an exponential distribution with parameter $\lambda > 0$ if it has p.d.f.

$$f(x) = \lambda \exp(-\lambda x), \qquad x > 0.$$

The mean and variance are given by

$$\mathrm{E}(X) = 1/\lambda,$$
$$\mathrm{Var}(X) = 1/\lambda^2.$$

We write $X \sim \mathrm{Expo}(\lambda)$. The exponential distribution is a special gamma distribution with $a = 1, b = \lambda$.

Definition B.11 Chi-Squared Distribution

A continuous nonnegative random variable X is said to have a chi-squared distribution with n degrees of freedom if it has p.d.f.

$$f(x) = \frac{1}{2^{\frac{n}{2}} \Gamma(\frac{n}{2})} x^{\frac{n}{2}-1} \exp\left(-\frac{1}{2} x\right), \qquad x > 0.$$

The mean and variance are given by

$$\mathrm{E}(X) = n,$$
$$\mathrm{Var}(X) = 2n.$$

We write $X \sim \chi^2_n$. The chi-squared distribution is a special gamma distribution with $a = n/2$ and $b = 1/2$.

If $X_1, \ldots, X_n$ are i.i.d. $\mathrm{N}(0, 1)$ variables, then

$$Y_n = \sum_{i=1}^{n} X_i^2$$

is χ^2_n-distributed. By the (strong) law of large numbers, this representation implies $Y_n/n \to 1$ for $n \to \infty$ (almost surely).

Definition B.12 Inverse Gamma Distribution

If $Y \sim \mathrm{G}(a, b)$, then $X = 1/Y$ has an inverse gamma distribution with p.d.f.

$$f(x) = \frac{b^a}{\Gamma(a)} x^{-(a+1)} \exp(-b/x), \qquad x > 0.$$

The mean and variance are given by

$$\mathrm{E}(X) = b/(a-1), \quad a > 1,$$

$$\mathrm{Var}(X) = b^2/((a-1)^2(a-2)), \quad a > 2.$$

We write $X \sim \mathrm{IG}(a, b)$.

Definition B.13 t-Distribution

A continuous random variable X is said to have a t-distribution with n degrees of freedom if it has p.d.f.

$$f(x) = \frac{\Gamma(n+1)/2}{\sqrt{n\pi}\,\Gamma(n/2)(1+x^2/n)^{(n+1)/2}}.$$

The mean and variance are given by

$$\mathrm{E}(X) = 0, \quad n > 1,$$

$$\mathrm{Var}(X) = n/(n-2), \quad n > 2.$$

We write $X \sim \mathrm{t}_n$. The t_1-distribution is also called Cauchy distribution.

If $X \sim \mathrm{N}(0, 1)$ and $Y \sim \chi^2_n$ are independent, then

$$T = \frac{X}{\sqrt{\frac{Y}{n}}} \sim \mathrm{t}_n. \tag{B.1}$$

If $X_1, \dots, X_n$ are i.i.d. $\mathrm{N}(\mu, \sigma^2)$ random variables, then

$$\frac{\bar{X} - \mu}{S}\sqrt{n} \sim \mathrm{t}_{n-1}$$

with

$$S = \frac{1}{n-1}\sum_{i=1}^{n}(X_i - \bar{X})^2 \quad \text{and} \quad \bar{X} = \frac{1}{n}\sum_{i=1}^{n} X_i.$$

Definition B.14 F-Distribution

A continuous random variable X is said to have a F-distribution with n and m degrees of freedom if it has p.d.f.

$$f(x) = n^{n/2} m^{m/2} \frac{\Gamma(n/2 + m/2)}{\Gamma(n/2)\Gamma(m/2)} \frac{x^{n/2-1}}{(nx+m)^{(n+m)/2}}, \qquad x \geq 0.$$

We write $F \sim \mathrm{F}_{n,m}$.

If $X_1 \sim \chi^2_n$ and $X_2 \sim \chi^2_m$ are independent, then

$$X = \frac{X_1/n}{X_2/m}$$

has an F-distribution with n and m degrees of freedom.

If Y is t-distributed with m degrees of freedom then $X = Y^2 \sim \mathrm{F}_{1,m}$.

Theorem B.1 Change of Variables

Let X be a continuous random variable with p.d.f. $f_X(x)$ and $Y = g(X)$ a monotonic transformation of X. It follows that Y has p.d.f.

$$f_Y(y) = \frac{f_X(g^{-1}(y))}{|g'(g^{-1}(y))|}.$$

B.2 Random Vectors

Definition B.15 Random Vector

The p-dimensional vector $\boldsymbol{X} = (X_1, \ldots, X_p)'$ is called random vector or p-dimensional random variable if the components $X_1, \ldots, X_p$ are univariate (scalar) random variables. The random vector $\boldsymbol{X}$ is called continuous if there is a function $f(\boldsymbol{x}) = f(x_1, \ldots, x_p) \geq 0$ such that

$$\mathrm{P}(a_1 \leq X_1 \leq b_1, \ldots, a_p \leq X_p \leq b_p) = \int_{a_p}^{b_p} \cdots \int_{a_1}^{b_1} f(x_1, \ldots, x_p)\, dx_1 \ldots dx_p.$$

The function f is called (joint) probability density function (p.d.f.) of $\boldsymbol{X}$.

The random vector $\boldsymbol{X}$ is called discrete, if $\boldsymbol{X}$ has only values in a finite or countable set $\{\boldsymbol{x}_1, \boldsymbol{x}_2, \ldots\} \subset \mathbb{R}^p$. The function f with

$$f(\boldsymbol{x}) = \begin{cases} \mathrm{P}(\boldsymbol{X} = \boldsymbol{x}) & \boldsymbol{x} \in \{\boldsymbol{x}_1, \boldsymbol{x}_2, \ldots\} \\ 0 & \text{else} \end{cases}$$

is called probability function or discrete p.d.f. of $\boldsymbol{X}$.

Definition B.16 Marginal and Conditional Distributions

Let the p-dimensional random vector $\boldsymbol{X} = (X_1, \ldots, X_p)'$ be partitioned into the p_1-dimensional vector $\boldsymbol{X}_1$ and the p_2-dimensional vector $\boldsymbol{X}_2$, i.e., $\boldsymbol{X} = (\boldsymbol{X}_1', \boldsymbol{X}_2')'$. The p_1-dimensional p.d.f. or probability function $f_{X_1}(\boldsymbol{x}_1)$ of $\boldsymbol{X}_1$ is then called marginal p.d.f. or marginal probability function of $\boldsymbol{X}$. It is given by

$$f_{X_1}(\boldsymbol{x}_1) = \int_{-\infty}^{\infty} \cdots \int_{-\infty}^{\infty} f(\boldsymbol{x}_1, \boldsymbol{x}_2)\, dx_{p_1+1} \ldots dx_p$$

for continuous random vectors and

$$f_{X_1}(\boldsymbol{x}_1) = \textstyle\sum_{x_2} f(\boldsymbol{x}_1, \boldsymbol{x}_2)$$

for discrete random vectors. The conditional p.d.f. or probability function of $\boldsymbol{X}_1$ given $\boldsymbol{X}_2 = \boldsymbol{x}_2$ is defined as

$$f(\boldsymbol{x}_1|\boldsymbol{x}_2) = \begin{cases} \dfrac{f(\boldsymbol{x}_1, \boldsymbol{x}_2)}{f_{X_2}(\boldsymbol{x}_2)} & \text{for } f_{X_2}(\boldsymbol{x}_2) > 0 \\ 0 & \text{else.} \end{cases}$$

The marginal and conditional p.d.f.'s or probability functions for $\boldsymbol{X}_2$ are defined in complete analogy.

Definition B.17 Mean Vector

Let $\boldsymbol{X} = (X_1, \ldots, X_p)'$ be a p-dimensional random vector. Then

$$\mathrm{E}(\boldsymbol{X}) = \boldsymbol{\mu} = (\mu_1, \ldots, \mu_p)' = (\mathrm{E}(X_1), \ldots, \mathrm{E}(X_p))'$$

is called mean vector of $\boldsymbol{X}$.

Definition B.18 Covariance Matrix, Correlation Matrix, and Precision Matrix

The covariance matrix $\mathrm{Cov}(\boldsymbol{X}) = \boldsymbol{\Sigma}$ of a p-dimensional random vector $\boldsymbol{X}$ is defined as

$$\mathrm{Cov}(\boldsymbol{X}) = \boldsymbol{\Sigma} = \mathrm{E}(\boldsymbol{X} - \boldsymbol{\mu})(\boldsymbol{X} - \boldsymbol{\mu})' = \begin{pmatrix} \sigma_{11} & \ldots & \sigma_{1p} \\ \vdots & & \vdots \\ \sigma_{p1} & \ldots & \sigma_{pp} \end{pmatrix},$$

where $\sigma_{ij} = \mathrm{Cov}(X_i, X_j)$, $i \neq j$, is the covariance between X_i and X_j, and $\sigma_{ii} = \sigma_i^2 = \mathrm{Var}(X_i)$ is the variance of X_i.

The correlation matrix $\boldsymbol{R}$ of $\boldsymbol{X}$ is defined as

$$\boldsymbol{R} = \begin{pmatrix} 1 & \rho_{12} & \dots & \rho_{1p} \\ \vdots & & & \vdots \\ \rho_{p1} & \rho_{p2} & \dots & 1 \end{pmatrix},$$

where

$$\rho_{ij} = \frac{\mathrm{Cov}(X_i, X_j)}{\sqrt{\mathrm{Var}(X_i) \cdot \mathrm{Var}(X_j)}}.$$

The covariance matrix $\boldsymbol{\Sigma}$ as well as the correlation matrix $\boldsymbol{R}$ are symmetric and positive semidefinite. They are positive definite if the components $\boldsymbol{X}_1, \dots, \boldsymbol{X}_p$ are linearly independent, i.e., no component can be expressed as a linear combination of remaining components. If $\boldsymbol{\Sigma}$ is positive definite, then the inverse $\boldsymbol{P} = \boldsymbol{\Sigma}^{-1}$ is called precision matrix. In case that $\boldsymbol{\Sigma}$ is only positive semidefinite and therefore not invertible, the precision matrix is a generalized inverse, i.e., $\boldsymbol{P} = \boldsymbol{\Sigma}^{-}$.

Theorem B.2 Rules for Mean Vectors and Covariance Matrices

Let $\boldsymbol{X}$, $\boldsymbol{Y}$, and $\boldsymbol{Z}$ be random vectors; $\boldsymbol{A}, \boldsymbol{B}, \boldsymbol{a}, \boldsymbol{b}$ matrices or vectors of appropriate order; and $\mathrm{E}(\boldsymbol{X}) = \boldsymbol{\mu}$, $\mathrm{Cov}(\boldsymbol{X}) = \boldsymbol{\Sigma}$. The following rules then hold:

1. $\mathrm{E}(\boldsymbol{X} + \boldsymbol{Y}) = \mathrm{E}(\boldsymbol{X}) + \mathrm{E}(\boldsymbol{Y})$
2. $\mathrm{E}(\boldsymbol{A}\boldsymbol{X} + \boldsymbol{b}) = \boldsymbol{A} \cdot \mathrm{E}(\boldsymbol{X}) + \boldsymbol{b}$
3. $\mathrm{Cov}(\boldsymbol{X}) = \mathrm{E}(\boldsymbol{X}\boldsymbol{X}') - \mathrm{E}(\boldsymbol{X})\mathrm{E}(\boldsymbol{X})'$
4. $\mathrm{Var}(\boldsymbol{a}'\boldsymbol{X}) = \boldsymbol{a}'\mathrm{Cov}(\boldsymbol{X})\boldsymbol{a} = \sum_{i=1}^{p}\sum_{j=1}^{p} a_i a_j \sigma_{ij}$
5. $\mathrm{Cov}(\boldsymbol{A}\boldsymbol{X} + \boldsymbol{b}) = \boldsymbol{A}\mathrm{Cov}(\boldsymbol{X})\boldsymbol{A}'$
6. $\mathrm{Cov}(\boldsymbol{X} + \boldsymbol{Y}) = \mathrm{Cov}(\boldsymbol{X}) + \mathrm{Cov}(\boldsymbol{Y})$ provided that $\boldsymbol{X}$ and $\boldsymbol{Y}$ are uncorrelated
7. $\mathrm{Cov}(\boldsymbol{X} + \boldsymbol{Y}, \boldsymbol{Z}) = \mathrm{Cov}(\boldsymbol{X}, \boldsymbol{Z}) + \mathrm{Cov}(\boldsymbol{Y}, \boldsymbol{Z})$
8. $\mathrm{E}(\boldsymbol{X}'\boldsymbol{A}\boldsymbol{X}) = \mathrm{tr}(\boldsymbol{A}\boldsymbol{\Sigma}) + \boldsymbol{\mu}'\boldsymbol{A}\boldsymbol{\mu}$
9. Law of total (or iterated) expectation: $\mathrm{E}(\boldsymbol{X}) = \mathrm{E}(\mathrm{E}(\boldsymbol{X} \mid \boldsymbol{Y}))$
10. Law of total covariance: $\mathrm{Cov}(\boldsymbol{X}) = \mathrm{E}(\mathrm{Cov}(\boldsymbol{X} \mid \boldsymbol{Y})) + \mathrm{Cov}(\mathrm{E}(\boldsymbol{X} \mid \boldsymbol{Y}))$
11. Law of total variance for scalar random variables X: $\mathrm{Var}(X) = \mathrm{E}(\mathrm{Var}(X \mid \boldsymbol{Y})) + \mathrm{Var}(\mathrm{E}(X \mid \boldsymbol{Y}))$

Definition B.19 Empirical Means, Covariance, and Correlation Matrices

Let $\boldsymbol{x}_1, \dots \boldsymbol{x}_n$ be realizations of i.i.d. random vectors $\boldsymbol{X}_1, \dots, \boldsymbol{X}_n$ from the distribution of $\boldsymbol{X}$. The mean vector then can be estimated by the empirical mean

$$\hat{\boldsymbol{\mu}} = (\hat{\mu}_1, \dots, \hat{\mu}_p)' = (\bar{x}_1, \dots, \bar{x}_p)' = \bar{\boldsymbol{x}},$$

where

$$\hat{\mu}_j = \frac{1}{n}\sum_{i=1}^{n} x_{ij} = \bar{x}_j.$$

Correspondingly, the empirical covariance matrix $\hat{\boldsymbol{\Sigma}}$ is defined by estimating variances and covariances in $\boldsymbol{\Sigma}$ through the empirical variances

$$\hat{\sigma}_j^2 = \frac{1}{n-1}\sum_{i=1}^{n}(x_{ij} - \bar{x}_j)^2$$

and the empirical covariances

$$\hat{\sigma}_{jk} = \frac{1}{n-1}\sum_{i=1}^{n}(x_{ij} - \bar{x}_j)(x_{ik} - \bar{x}_k).$$

The empirical correlation matrix $\hat{\boldsymbol{R}}$ is defined analogously through the empirical correlation coefficients.

B.3 Multivariate Normal Distribution

B.3.1 Definition and Properties

Definition B.20 Multivariate Normal Distribution

A continuous p-dimensional random vector $\boldsymbol{X} = (X_1, X_2, \ldots, X_p)'$ is said to have a multivariate normal (or Gaussian) distribution if it has p.d.f.

$$f(\boldsymbol{x}) = (2\pi)^{-\frac{p}{2}}|\boldsymbol{\Sigma}|^{-\frac{1}{2}} \exp\left[-\tfrac{1}{2}(\boldsymbol{x} - \boldsymbol{\mu})'\boldsymbol{\Sigma}^{-1}(\boldsymbol{x} - \boldsymbol{\mu})\right] \tag{B.2}$$

with $\boldsymbol{\mu} \in \mathbb{R}^p$ and positive definite $(p \times p)$-matrix $\boldsymbol{\Sigma}$.

For deriving posterior distributions in Bayesian models it is often convenient to rewrite the p.d.f. in alternative form. Omitting all factors in Eq. (B.2) that do not depend on $\boldsymbol{x}$, we obtain

$$\begin{aligned} f(\boldsymbol{x}) &\propto \exp\left(-\tfrac{1}{2}(\boldsymbol{x} - \boldsymbol{\mu})'\boldsymbol{\Sigma}^{-1}(\boldsymbol{x} - \boldsymbol{\mu})\right) \\ &= \exp\left(-\tfrac{1}{2}\boldsymbol{x}'\boldsymbol{\Sigma}^{-1}\boldsymbol{x} + \boldsymbol{x}'\boldsymbol{\Sigma}^{-1}\boldsymbol{\mu} - \tfrac{1}{2}\boldsymbol{\mu}'\boldsymbol{\Sigma}^{-1}\boldsymbol{\mu}\right) \\ &\propto \exp\left(-\tfrac{1}{2}\boldsymbol{x}'\boldsymbol{\Sigma}^{-1}\boldsymbol{x} + \boldsymbol{x}'\boldsymbol{\Sigma}^{-1}\boldsymbol{\mu}\right). \end{aligned} \tag{B.3}$$

If $\boldsymbol{X}$ is multivariate normal, then its p.d.f. is always proportional to Eq. (B.3).

Theorem B.3 Mean and Covariance Matrix

It can be shown that $\mathrm{E}(\boldsymbol{X}) = \boldsymbol{\mu}$ and $\mathrm{Cov}(\boldsymbol{X}) = \boldsymbol{\Sigma}$. We write

$$\boldsymbol{X} \sim \mathrm{N}_p(\boldsymbol{\mu}, \boldsymbol{\Sigma}),$$

analogous to the univariate normal distribution. The index p is often suppressed if the dimension p is obvious from the context. The special case $\boldsymbol{\mu} = \mathbf{0}$ and $\boldsymbol{\Sigma} = \boldsymbol{I}$ is called the (multivariate) standard normal distribution.

Theorem B.4 Product of Two Normal Densities

Suppose that the density of a random vector $\boldsymbol{X}$ is given by

$$f(\boldsymbol{x}) \propto \exp(-\frac{1}{2}(\boldsymbol{x} - \boldsymbol{a})'\boldsymbol{A}(\boldsymbol{x} - \boldsymbol{a}))\exp(-\frac{1}{2}(\boldsymbol{x} - \boldsymbol{b})'\boldsymbol{B}(\boldsymbol{x} - \boldsymbol{b})).$$

$\boldsymbol{X}$ is then multivariate normal with covariance matrix and mean

$$\boldsymbol{\Sigma} = (\boldsymbol{A} + \boldsymbol{B})^{-1} \qquad \boldsymbol{\mu} = \boldsymbol{\Sigma}(\boldsymbol{A}\boldsymbol{a} + \boldsymbol{B}\boldsymbol{b}).$$

Theorem B.5 Linear Transformation

Let $\boldsymbol{X} \sim \mathrm{N}_p(\boldsymbol{\mu}, \boldsymbol{\Sigma})$, $\boldsymbol{d} \in \mathbb{R}^q$, and $\boldsymbol{D}$ a $(q \times p)$-matrix with $\mathrm{rk}(\boldsymbol{D}) = q \leq p$; then

$$\boldsymbol{Y} = \boldsymbol{d} + \boldsymbol{D}\boldsymbol{X} \sim \mathrm{N}_q(\boldsymbol{d} + \boldsymbol{D}\boldsymbol{\mu}, \boldsymbol{D}\boldsymbol{\Sigma}\boldsymbol{D}').$$

Theorem B.6 Marginal and Conditional Distributions

Let the multivariate normal random variable $\boldsymbol{X} \sim \mathrm{N}(\boldsymbol{\mu}, \boldsymbol{\Sigma})$ be partitioned into the subvectors $\boldsymbol{Y} = (X_1, \ldots, X_r)'$ and $\boldsymbol{Z} = (X_{r+1}, \ldots, X_p)'$, i.e.,

$$\boldsymbol{X} = \begin{pmatrix} \boldsymbol{Y} \\ \boldsymbol{Z} \end{pmatrix}, \quad \boldsymbol{\mu} = \begin{pmatrix} \boldsymbol{\mu}_Y \\ \boldsymbol{\mu}_Z \end{pmatrix}, \quad \boldsymbol{\Sigma} = \begin{pmatrix} \boldsymbol{\Sigma}_Y & \boldsymbol{\Sigma}_{YZ} \\ \boldsymbol{\Sigma}_{ZY} & \boldsymbol{\Sigma}_Z \end{pmatrix}.$$

Then $\boldsymbol{Y}$ has an r-dimensional normal distribution $\boldsymbol{Y} \sim \mathrm{N}(\boldsymbol{\mu}_Y, \boldsymbol{\Sigma}_Y)$.

The conditional distribution of $\boldsymbol{Y}$ given $\boldsymbol{Z}$ is again multivariate normal with mean

$$\boldsymbol{\mu}_{Y|Z} = \boldsymbol{\mu}_Y + \boldsymbol{\Sigma}_{YZ} \cdot \boldsymbol{\Sigma}_Z^{-1}(\boldsymbol{Z} - \boldsymbol{\mu}_Z)$$

and covariance matrix

$$\boldsymbol{\Sigma}_{Y|Z} = \boldsymbol{\Sigma}_Y - \boldsymbol{\Sigma}_{YZ}\boldsymbol{\Sigma}_Z^{-1}\boldsymbol{\Sigma}_{ZY}.$$

Furthermore, $\boldsymbol{Y}$ and $\boldsymbol{Z}$ are independent if and only if $\boldsymbol{Y}$ and $\boldsymbol{Z}$ are uncorrelated, i.e., if $\boldsymbol{\Sigma}_{YZ} = \boldsymbol{\Sigma}_{ZY} = \mathbf{0}$. This equivalence is generally not true for non-normal random vectors: If $\boldsymbol{Y}$ and $\boldsymbol{Z}$ are independent they are also uncorrelated, but in general $\boldsymbol{\Sigma}_{ZY} = \mathbf{0}$ does not imply independence.

B.3.2 The Singular Multivariate Normal Distribution

Up to now, we have assumed that $\text{rk}(\boldsymbol{\Sigma}) = p$ so that $\boldsymbol{\Sigma}$ is positive definite and $\boldsymbol{\Sigma}^{-1}$ exists. In the following, we consider the case $\text{rk}(\boldsymbol{\Sigma}) < p$.

Definition B.21 Singular Multivariate Normal Distribution

Let $\boldsymbol{X} \sim \text{N}_p(\boldsymbol{\mu}, \boldsymbol{\Sigma})$. Then $\boldsymbol{X}$ follows a singular multivariate normal distribution if $\text{rk}(\boldsymbol{\Sigma}) = r < p$. The distribution of a singular normal distribution is often expressed in terms of the precision matrix $\boldsymbol{P}$ with $\text{rk}(\boldsymbol{P}) = r < p$. The density is then given by

$$f(\boldsymbol{x}) \propto \exp\left[-\frac{1}{2}(\boldsymbol{x}-\boldsymbol{\mu})'\boldsymbol{P}(\boldsymbol{x}-\boldsymbol{\mu})\right].$$

The precision matrix $\boldsymbol{P}$ may be chosen as a generalized inverse $\boldsymbol{\Sigma}^{-}$ of $\boldsymbol{\Sigma}$. This form of the singular normal distribution will be used frequently in the book, in particular in Chaps. 8 and 9.

The virtue of specifying in statistical models the precision matrix rather than the covariance matrix is that the elements of the precision matrix can be interpreted as conditional correlations. More specifically, for multivariate normal distributions we have

$$\text{Corr}(X_i, X_j \mid \boldsymbol{X}_{-ij}) = -\frac{p_{ij}}{\sqrt{p_{ii}\,p_{jj}}},$$

where $\boldsymbol{X}_{-ij}$ is the vector $\boldsymbol{X}$ with the ith and jth element excluded and p_{ij} is the element of $\boldsymbol{P}$ in the ith row and jth column. Thus, $p_{ij} = 0$ in the precision matrix implies that X_i and X_j are conditionally uncorrelated and (in case of the normal distribution) also conditionally independent. Note that conditional independence does not imply (unconditional) independence.

Theorem B.7 Characterization of the Singular Normal Distribution

Let $\boldsymbol{X}$ be a singular normal random vector, i.e., $\boldsymbol{X} \sim \text{N}_p(\boldsymbol{\mu}, \boldsymbol{\Sigma})$, with $\text{rk}(\boldsymbol{\Sigma}) = \text{rk}(\boldsymbol{P}) = r < p$. Assume that $(\boldsymbol{G}\ \boldsymbol{H})$ is an orthogonal matrix, where the columns of the $(p \times r)$-matrix $\boldsymbol{G}$ are a basis for the column space of $\boldsymbol{\Sigma}$ and the columns of $\boldsymbol{H}$ are a basis of the null space of $\boldsymbol{\Sigma}$. Consider the transformation

$$\begin{pmatrix} \boldsymbol{Y}_1 \\ \boldsymbol{Y}_2 \end{pmatrix} = (\boldsymbol{G}\ \boldsymbol{H})'\boldsymbol{X} = \begin{pmatrix} \boldsymbol{G}'\boldsymbol{X} \\ \boldsymbol{H}'\boldsymbol{X} \end{pmatrix}.$$

It follows that $\boldsymbol{Y}_1$ is the stochastic part of $\boldsymbol{X}$ and is nonsingular normal with

$$\boldsymbol{Y}_1 \sim \text{N}(\boldsymbol{G}'\boldsymbol{\mu}, \boldsymbol{G}'\boldsymbol{\Sigma}\boldsymbol{G}).$$

$\boldsymbol{Y}_2$ is the deterministic part of $\boldsymbol{X}$ with

$$\text{E}(\boldsymbol{Y}_2) = \boldsymbol{H}'\boldsymbol{\mu}, \qquad \text{Var}(\boldsymbol{Y}_2) = \boldsymbol{0}.$$

Here, $\mathrm{Var}(\boldsymbol{Y}_2)$ is the vector of variances of $\boldsymbol{Y}_2$. The p.d.f. of the stochastic part $\boldsymbol{Y}_1 = \boldsymbol{G}'\boldsymbol{X}$ has the form

$$f(\boldsymbol{y}_1) = \frac{1}{(2\pi)^{\frac{r}{2}}(\prod_{i=1}^{r}\lambda_i)^{\frac{1}{2}}} \exp\left[-\frac{1}{2}(\boldsymbol{y}_1 - \boldsymbol{G}'\boldsymbol{\mu})'(\boldsymbol{G}'\boldsymbol{\Sigma}\boldsymbol{G})^{-1}(\boldsymbol{y}_1 - \boldsymbol{G}'\boldsymbol{\mu})\right], \tag{B.4}$$

where $\lambda_1, \ldots, \lambda_r$ are the r nonzero eigenvalues of $\boldsymbol{\Sigma}$.

B.3.3 Distributions of Quadratic Forms

Quadratic forms of normal random vectors appear in tests of linear hypotheses. Refer to Sect. 3.3 in Chap. 3.

Theorem B.8 Distributions of Quadratic Forms

1. Let $\boldsymbol{X} \sim \mathrm{N}_p(\boldsymbol{\mu}, \boldsymbol{\Sigma})$ with $\boldsymbol{\Sigma} > 0$. It follows that

$$Y = (\boldsymbol{X} - \boldsymbol{\mu})'\boldsymbol{\Sigma}^{-1}(\boldsymbol{X} - \boldsymbol{\mu}) \sim \chi^2_p.$$

2. Let $\boldsymbol{X} \sim \mathrm{N}_p(\boldsymbol{0}, \boldsymbol{I})$, $\boldsymbol{B}$ an $(n \times p)$-matrix $(n \leq p)$ and, $\boldsymbol{R}$ a symmetric idempotent $(p \times p)$-matrix with $\mathrm{rk}(\boldsymbol{R}) = r$; then
   - $\boldsymbol{X}'\boldsymbol{R}\boldsymbol{X} \sim \chi^2_r$.
   - $\boldsymbol{B}\boldsymbol{R} = \boldsymbol{0}$ implies independence of $\boldsymbol{X}'\boldsymbol{R}\boldsymbol{X}$ and $\boldsymbol{B}\boldsymbol{X}$.
3. Assume that $X_1, \ldots, X_n$ are i.i.d. $\sim \mathrm{N}(\mu, \sigma^2)$ variables and

$$S^2 = \frac{1}{n-1}\sum_{i=1}^{n}(X_i - \bar{X})^2.$$

   It then follows that:
   - $\dfrac{n-1}{\sigma^2}S^2 \sim \chi^2_{n-1}$.
   - S^2 and $\bar{X}$ are independent.
4. Assume that $\boldsymbol{X} \sim \mathrm{N}_n(\boldsymbol{0}, \boldsymbol{I})$, and let $\boldsymbol{R}$ and $\boldsymbol{S}$ be symmetric and idempotent $n \times n$ matrices with $\mathrm{rk}(\boldsymbol{R}) = r$, $\mathrm{rk}(\boldsymbol{S}) = s$, and $\boldsymbol{R}\boldsymbol{S} = \boldsymbol{0}$. It follows that
   - $\boldsymbol{X}'\boldsymbol{R}\boldsymbol{X}$ and $\boldsymbol{X}'\boldsymbol{S}\boldsymbol{X}$ are independent.
   - $\dfrac{s}{r}\dfrac{\boldsymbol{X}'\boldsymbol{R}\boldsymbol{X}}{\boldsymbol{X}'\boldsymbol{S}\boldsymbol{X}} \sim \mathrm{F}_{r,s}$.

B.3.4 Multivariate t-Distribution

A continuous p-dimensional random vector $\boldsymbol{X} = (X_1, \ldots, X_p)'$ is said to have a multivariate t-distribution with ν degrees of freedom, location parameter $\boldsymbol{\mu}$, and (positive definite) dispersion matrix $\boldsymbol{\Sigma}$, if it has p.d.f.

$$f(\boldsymbol{x}) = |\boldsymbol{\Sigma}|^{-\frac{1}{2}}(\nu\pi)^{-\frac{p}{2}} \frac{\Gamma((\nu+p)/2)}{\Gamma(\nu/2)} \left(1 + \frac{(\boldsymbol{x}-\boldsymbol{\mu})'\boldsymbol{\Sigma}^{-1}(\boldsymbol{x}-\boldsymbol{\mu})}{\nu}\right)^{-(\nu+p)/2}.$$

We write $\boldsymbol{X} \sim \mathrm{t}(\nu, \boldsymbol{\mu}, \boldsymbol{\Sigma})$. The expectation is $\boldsymbol{\mu}$ (provided that $\nu > 1$) and the covariance matrix is $\nu/(\nu-2)\boldsymbol{\Sigma}$ (provided that $\nu > 2$). Note that a diagonal dispersion matrix $\boldsymbol{\Sigma}$ corresponds to uncorrelated components of the random vector $\boldsymbol{X}$. In contrast to the multivariate normal distribution the components are, however, not stochastically independent.

Any subvector of $\boldsymbol{X}$ has a (multivariate) t-distribution with ν degrees of freedom and the corresponding subvector of $\boldsymbol{\mu}$ and the submatrix of $\boldsymbol{\Sigma}$ as location parameter and dispersion matrix, respectively.

In analogy to the constructive definition (B.1) of the univariate t-distribution, the multivariate t-distribution can be defined constructively, based on a multivariate normal random vector and a chi-squared distributed random variable.

B.3.5 Normal-Inverse Gamma Distribution

Let $\boldsymbol{Y}$ be a $p \times 1$ dimensional random vector and S be a random variable. The random vector $\boldsymbol{X} = (\boldsymbol{Y}, S)'$ is said to have a normal-inverse gamma distribution with parameters $\boldsymbol{\mu}$, $\boldsymbol{\Sigma}$, a, and b if

$$\boldsymbol{Y} \mid S \sim \mathrm{N}(\boldsymbol{\mu}, S\boldsymbol{\Sigma}),$$
$$S \sim \mathrm{IG}(a, b).$$

We write $\boldsymbol{X} = (\boldsymbol{Y}, S)' \sim \mathrm{NIG}(\boldsymbol{\mu}, \boldsymbol{\Sigma}, a, b)$. The density of the distribution is given by

$$f(\boldsymbol{y}, s) = \frac{1}{(2\pi)^{\frac{p}{2}}|\boldsymbol{\Sigma}|^{\frac{1}{2}}} \exp\left(-\frac{1}{2}(\boldsymbol{y}-\boldsymbol{\mu})'\boldsymbol{\Sigma}^{-1}(\boldsymbol{y}-\boldsymbol{\mu})\right)$$
$$\frac{b^a}{\Gamma(a)} \frac{1}{(\sigma^2)^{a+1}} \exp\left(-\frac{b}{\sigma^2}\right).$$

The $\mathrm{NIG}(\boldsymbol{\mu}, \boldsymbol{\Sigma}, a, b)$-distribution has the following properties:

1. $\mathrm{E}(\boldsymbol{Y}) = \boldsymbol{\mu}$
2. $\mathrm{Cov}(\boldsymbol{Y}) = b/(a-1)\boldsymbol{\Sigma}$
3. $\mathrm{E}(S) = b/(a-1)$ provided that $a > 1$
4. $\mathrm{Var}(S) = b^2/[(a-1)^2(a-2)]$ provided that $a > 2$
5. $\boldsymbol{Y} \sim \mathrm{t}(2a, \boldsymbol{\mu}, b/a\boldsymbol{\Sigma})$

B.4 Likelihood Inference

This section describes the method of maximum likelihood (ML) for estimation of unknown parameters in statistical models and likelihood-based tests for linear hypotheses about these parameters. More detailed introductions can be found, for example, in Migon and Gamerman (1999) and Held and Sabanés Bové (2012).

B.4.1 Maximum Likelihood Estimation

Let $Y_1, \dots, Y_n$ be a random sample with realizations $y_1, \dots, y_n$. Let the joint probability (for discrete $Y_1, \dots, Y_n$)

$$\mathrm{P}(Y_1 = y_1, \dots, Y_n = y_n \,|\, \boldsymbol{\theta})$$

or the joint p.d.f. (for continuous $Y_1, \dots, Y_n$)

$$f(Y_1 = y_1, \dots, Y_n = y_n \,|\, \boldsymbol{\theta})$$

depend on an unknown vector $\boldsymbol{\theta} = (\theta_1, \dots, \theta_p)' \in \boldsymbol{\Theta}$ that has to be estimated. For given realizations $y_1, \dots, y_n$ this joint probability or the value of the p.d.f. is considered as a function of $\boldsymbol{\theta}$ and is called likelihood, denoted by $L(\boldsymbol{\theta})$:

$$L(\boldsymbol{\theta}) = \mathrm{P}(Y_1 = y_1, \dots, Y_n = y_n \,|\, \boldsymbol{\theta}),$$

$$L(\boldsymbol{\theta}) = f(Y_1 = y_1, \dots, Y_n = y_n \,|\, \boldsymbol{\theta}).$$

The *principle of maximum likelihood* postulates the *maximum likelihood estimator* (MLE) $\hat{\boldsymbol{\theta}}$ for $\boldsymbol{\theta}$ as the value $\hat{\boldsymbol{\theta}}$ which maximizes the likelihood $L(\boldsymbol{\theta})$.

In the discrete case, this principle says: Determine the MLE $\hat{\boldsymbol{\theta}}$ such that the probability of observing the realized sample $y_1, \dots, y_n$ assumes its maximum for $\boldsymbol{\theta} = \hat{\boldsymbol{\theta}}$ and, therefore, this sample becomes as likely as possible. In analogy, the p.d.f. assumes its maximum for $\boldsymbol{\theta} = \hat{\boldsymbol{\theta}}$, making the sample as plausible as possible. In most cases, in particular for almost all models in this book (excluding quantile regression), the likelihood $L(\boldsymbol{\theta})$ is differentiable with respect to $\boldsymbol{\theta}$, and the maximum can be determined by setting the first derivatives to zero and solving the resulting system of equations for $\boldsymbol{\theta} = \hat{\boldsymbol{\theta}}$. For technical reasons, maximization is usually not carried out for the likelihood but for the log-likelihood $l(\boldsymbol{\theta}) = \log(L(\boldsymbol{\theta}))$, obtained by taking the logarithm of the likelihood. Since the logarithm is a strictly increasing function, $l(\boldsymbol{\theta})$ attains its maximum at the same value $\boldsymbol{\theta} = \hat{\boldsymbol{\theta}}$ as $L(\boldsymbol{\theta})$.

In the most simple situation, $Y_1, \dots, Y_n$ are an i.i.d. sample from $f(y \,|\, \theta)$, i.e., they are independent and identically distributed as the typical random variable $Y \sim f(y|\theta)$. To unify notation, we denote probability functions (Y discrete) and p.d.f.'s (Y continuous) with $f(y|\theta)$. Because $Y_1, \dots, Y_n$ are independent, their joint density is the product of the densities $f(y_i \mid \theta)$. Therefore the likelihood is

$$L(\boldsymbol{\theta}) = f(y_1 \mid \boldsymbol{\theta}) \cdot \ldots \cdot f(y_n \mid \boldsymbol{\theta})$$

and the log-likelihood is the sum

$$l(\boldsymbol{\theta}) = \log f(y_1 \mid \boldsymbol{\theta}) + \ldots + \log f(y_n \mid \boldsymbol{\theta}) = \sum_{i=1}^{n} l_i(\boldsymbol{\theta})$$

with the log-likelihood contributions $l_i(\boldsymbol{\theta}) = \log f(y_i \mid \boldsymbol{\theta})$. The following example illustrates this situation.

Example B.1 Poisson Distribution: MLE

Let $Y \sim \text{Po}(\lambda)$ follow a Poisson distribution with unknown parameter λ. To estimate λ, we assume an i.i.d. sample $Y_1, \ldots, Y_n$ with $Y_i \sim \text{Po}(\lambda)$ and a realized sample $y_1, \ldots, y_n$. The MLE for λ is obtained in the following steps:

1. *Likelihood*

   Because $Y_1, \ldots, Y_n$ are independent, the joint probability for the realized sample factorizes into the product of the separate marginal probabilities, and we obtain the likelihood

$$\begin{aligned} L(\lambda) &= \text{P}(Y_1 = y_1, \ldots, Y_n = y_n \mid \lambda) \\ &= \text{P}(Y_1 = y_1 \mid \lambda) \cdot \ldots \cdot \text{P}(Y_n = y_n \mid \lambda) \\ &= \frac{\lambda^{y_1}}{y_1!} \exp(-\lambda) \cdot \ldots \cdot \frac{\lambda^{y_n}}{y_n!} \exp(-\lambda). \end{aligned}$$

   The factors $1/y_1!, \ldots, 1/y_n!$ do not depend on λ and can therefore be omitted when maximizing the likelihood. Therefore

$$L(\lambda) \propto \exp(-n\lambda) \cdot \lambda^{y_1} \cdot \ldots \cdot \lambda^{y_n}.$$

   Often the right-hand side, after omitting constant factors, is also called likelihood, and we write

$$L(\lambda) = \exp(-n\lambda) \cdot \lambda^{y_1} \cdot \ldots \cdot \lambda^{y_n}.$$

2. *Log-likelihood*

   Taking logarithms, we get

$$l(\lambda) = -n\lambda + \sum_{i=1}^{n} y_i \log(\lambda).$$

3. *First derivative and setting it to zero*

$$\frac{\partial l(\lambda)}{\partial \lambda} = -n + \sum_{i=1}^{n} y_i \frac{1}{\lambda} \stackrel{!}{=} 0.$$

   The first derivative $s(\lambda) = \partial l(\lambda)/\partial \lambda$ is called *score function*.

   Solving for the unknown parameter gives the MLE:

$$\hat{\lambda} = \frac{1}{n}\sum_{i=1}^{n} y_i = \bar{y}.$$

4. *Second derivative to check for a maximum*
   The second derivative of the log-likelihood is

$$\frac{\partial^2 l(\lambda)}{\partial \lambda^2} = -\sum_{i=1}^{n} y_i \frac{1}{\lambda^2} < 0.$$

Therefore the estimator of step 3 is indeed a maximizer of the log-likelihood (and the likelihood). The negative second derivative is called *observed Fisher information*. △

The score function and the Fisher information appearing in this example play an important role in likelihood theory. Generally, the score function is defined as the vector

$$\boldsymbol{s}(\boldsymbol{\theta}) = \left(s_1(\boldsymbol{\theta}), \dots, s_p(\boldsymbol{\theta})\right)' = \left(\frac{\partial l(\boldsymbol{\theta})}{\partial \theta_1}, \dots, \frac{\partial l(\boldsymbol{\theta})}{\partial \theta_p}\right)' = \frac{\partial l(\boldsymbol{\theta})}{\partial \boldsymbol{\theta}}$$

of partial first derivatives of the log-likelihood. The observed Fisher information (matrix) is defined as

$$\boldsymbol{H}(\boldsymbol{\theta}) = -\begin{pmatrix} \frac{\partial^2 l(\boldsymbol{\theta})}{\partial \theta_1 \partial \theta_1} & \cdots & \frac{\partial^2 l(\boldsymbol{\theta})}{\partial \theta_1 \partial \theta_p} \\ \vdots & & \vdots \\ \frac{\partial^2 l(\boldsymbol{\theta})}{\partial \theta_p \partial \theta_1} & \cdots & \frac{\partial^2 l(\boldsymbol{\theta})}{\partial \theta_p \partial \theta_p} \end{pmatrix} = -\begin{pmatrix} \frac{\partial s_1(\boldsymbol{\theta})}{\partial \theta_1} & \cdots & \frac{\partial s_1(\boldsymbol{\theta})}{\partial \theta_p} \\ \vdots & & \vdots \\ \frac{\partial s_p(\boldsymbol{\theta})}{\partial \theta_1} & \cdots & \frac{\partial s_p(\boldsymbol{\theta})}{\partial \theta_p} \end{pmatrix},$$

i.e., the negative matrix of second derivatives of the log-likelihood. In more compact notation we obtain

$$\boldsymbol{H}(\boldsymbol{\theta}) = -\left(\frac{\partial \boldsymbol{s}(\boldsymbol{\theta})}{\partial \boldsymbol{\theta}}\right)' = -\frac{\partial \boldsymbol{s}(\boldsymbol{\theta})}{\partial \boldsymbol{\theta}'} = -\frac{\partial^2 l(\boldsymbol{\theta})}{\partial \boldsymbol{\theta}\, \partial \boldsymbol{\theta}'},$$

thereby using Definition A.29 (p. 636) in Appendix A.8. As the name already indicates, the observed Fisher information matrix can be considered as a (local) measure of the information that the likelihood contains about the unknown parameter. The second derivative of a function is a measure of the curvature of the function at $\boldsymbol{\theta}$. The higher the curvature of the log-likelihood near its maximum, the more information is provided by the likelihood about the unknown parameter. Since the second derivative is negative at or near a maximum, the Fisher information is defined as the negative second derivative, i.e., it is positive at or near a maximum.

Since the likelihood (as well as the log-likelihood) depends on the realized values of the sample variables $Y_1, \dots, Y_n$, the likelihood function will have different values when samples are repeatedly drawn from the (joint) distribution of the sample.

Therefore, the likelihood as well as the log-likelihood and its derivatives can be interpreted as random variables, having mean, variance, etc. Under mild regularity conditions that are fulfilled for all examples and models in this book, the score function has mean $\mathbf{0}$, i.e.,

$$\mathrm{E}(\boldsymbol{s}(\boldsymbol{\theta})) = \mathbf{0}.$$

The mean is taken with respect to the sample variables $Y_1, \ldots, Y_n$. The mean of the observed Fisher information (matrix)

$$\boldsymbol{F}(\boldsymbol{\theta}) = -\begin{pmatrix} \mathrm{E}\left(\frac{\partial^2 l(\boldsymbol{\theta})}{\partial\theta_1\partial\theta_1}\right) & \cdots & \mathrm{E}\left(\frac{\partial^2 l(\boldsymbol{\theta})}{\partial\theta_1\partial\theta_p}\right) \\ \vdots & & \vdots \\ \mathrm{E}\left(\frac{\partial^2 l(\boldsymbol{\theta})}{\partial\theta_p\partial\theta_1}\right) & \cdots & \mathrm{E}\left(\frac{\partial^2 l(\boldsymbol{\theta})}{\partial\theta_p\partial\theta_p}\right) \end{pmatrix} = -\begin{pmatrix} \mathrm{E}\left(\frac{\partial s_1(\boldsymbol{\theta})}{\partial\theta_1}\right) & \cdots & \mathrm{E}\left(\frac{\partial s_1(\boldsymbol{\theta})}{\partial\theta_p}\right) \\ \vdots & & \vdots \\ \mathrm{E}\left(\frac{\partial s_p(\boldsymbol{\theta})}{\partial\theta_1}\right) & \cdots & \mathrm{E}\left(\frac{\partial s_p(\boldsymbol{\theta})}{\partial\theta_p}\right) \end{pmatrix}$$

is of particular interest and is called expected Fisher information. It can be considered as a global measure of information which can be determined prior to and independent from realized samples. Under mild regularity conditions fulfilled for all models throughout this book it can be shown that

$$\boldsymbol{F}(\boldsymbol{\theta}) = \mathrm{Cov}(\boldsymbol{s}(\boldsymbol{\theta})) = \mathrm{E}(\boldsymbol{s}(\boldsymbol{\theta})\boldsymbol{s}(\boldsymbol{\theta})').$$

The latter equality holds because $\mathrm{E}(\boldsymbol{s}(\boldsymbol{\theta})) = \mathbf{0}$.

For independent sample variables information is additive: Let $\boldsymbol{H}_{Y_i}(\boldsymbol{\theta})$ and $\boldsymbol{F}_{Y_i}(\boldsymbol{\theta})$ denote the observed and expected Fisher information with respect to the sample variable Y_i, i.e., the information contributed by Y_i alone; then the complete information provided by the sample is

$$\boldsymbol{H}_Y(\boldsymbol{\theta}) = \sum_{i=1}^{n} \boldsymbol{H}_{Y_i}(\boldsymbol{\theta}) \quad , \quad \boldsymbol{F}_Y(\boldsymbol{\theta}) = \sum_{i=1}^{n} \boldsymbol{F}_{Y_i}(\boldsymbol{\theta}).$$

Similarly, the score function is additive and composed of the individual score contributions:

$$\boldsymbol{s}(\boldsymbol{\theta}) = \sum_{i=1}^{n} \boldsymbol{s}_i(\boldsymbol{\theta}).$$

Example B.2 Poisson Distribution: Score Function and Fisher Information

Let $Y_1, \ldots, Y_n$ be an i.i.d. sample for a Poisson variable $Y \sim \mathrm{Po}(\lambda)$ with unknown parameter λ. The log-likelihood and score function contribution of y_i are (refer to Example B.1)

$$l_i(\lambda) = y_i \log(\lambda) - \log(y_i!) - \lambda \quad \text{and} \quad s_i(\lambda) = \frac{y_i}{\lambda} - 1.$$

It is easy to see that $\mathrm{E}(s(\lambda)) = 0$: Since

$$\mathrm{E}(s_i(\lambda)) = \mathrm{E}\left(\frac{Y_i}{\lambda} - 1\right) = \frac{1}{\lambda}\mathrm{E}(Y_i) - 1 = \frac{1}{\lambda}\lambda - 1 = 0,$$

we have

$$\mathrm{E}(s(\lambda)) = \mathrm{E}\left(\sum_{i=1}^{n} s_i(\lambda)\right) = \sum_{i=1}^{n} \mathrm{E}(s_i(\lambda)) = 0.$$

The observed Fisher information contributed by y_i is

$$H_{y_i}(\lambda) = -\frac{\partial^2 l_i(\lambda)}{\partial^2 \lambda} = -\frac{\partial s_i(\lambda)}{\partial \lambda} = -\left(-\frac{y_i}{\lambda^2}\right) = \frac{y_i}{\lambda^2}.$$

Due to additivity of information, we obtain the observed information

$$H_{\boldsymbol{Y}}(\lambda) = \frac{1}{\lambda^2}\sum_{i=1}^{n} y_i$$

for the entire sample. The expected Fisher information contributed by Y_i, with $\mathrm{E}(Y_i) = \lambda$, is

$$F_{Y_i}(\lambda) = \mathrm{E}\left(\frac{Y_i}{\lambda^2}\right) = \frac{\lambda}{\lambda^2} = \frac{1}{\lambda},$$

and we can verify additivity, i.e.,

$$F_{\boldsymbol{Y}}(\lambda) = \frac{n}{\lambda}$$

for the entire sample. This shows that expected information grows linearly with sample size n. Furthermore, it is inversely proportional to λ which is plausible because the variance of the MLE $\bar{Y}$ is $\mathrm{Var}(\bar{Y}) = \frac{1}{n}\lambda$, i.e., the inverse of the expected information increases linearly with λ.

△

In this book, we mostly consider regression situations with response variables $Y_1, \ldots, Y_n$ and with observed responses $y_1, \ldots, y_n$, as well as additional covariate values $\boldsymbol{x}_1, \ldots, \boldsymbol{x}_n$ as the realized sample. The vector $\boldsymbol{\theta}$ is then often the vector $\boldsymbol{\beta}$ of regression coefficients. Given $\boldsymbol{x}_1, \ldots, \boldsymbol{x}_n$, the response variables $Y_1, \ldots, Y_n$ are no longer identically distributed, but they are still assumed to be conditionally independent given the covariates. The densities of the Y_i depend on $\boldsymbol{x}_i$ and, therefore, on i, i.e., $Y_i \sim f_i(y_i \,|\, \boldsymbol{\theta}) = f(y_i \,|\, \boldsymbol{x}_i; \boldsymbol{\beta})$ with $\boldsymbol{\theta} = \boldsymbol{\beta}$. Due to independence of the sample variables, the likelihood is again the product

$$L(\boldsymbol{\theta}) = f_1(y_1 \,|\, \boldsymbol{\theta}) \cdot \cdots \cdot f_n(y_n \,|\, \boldsymbol{\theta})$$

of individual densities, and the log-likelihood is the sum

$$l(\boldsymbol{\theta}) = \sum_{i=1}^{n} l_i(\boldsymbol{\theta}) = \sum_{i=1}^{n} \log f_i(y_i \,|\, \boldsymbol{\theta})$$

of the log-likelihood contributions $l_i(\boldsymbol{\theta}) = \log f_i(y_i \,|\, \boldsymbol{\theta})$. This also implies that the score function and the Fisher information remain additive in terms of the observations as before.

Example B.3 Poisson Regression

We consider the following Poisson regression model:

$$\begin{aligned} Y_i &\sim \text{Po}(\lambda_i) \qquad i = 1, \ldots, n, \\ \lambda_i &= \exp(\beta_0 + \beta_1 x_i) = \exp(\eta_i), \\ \eta_i &= \beta_0 + \beta_1 x_i. \end{aligned}$$

Our aim is to compute the MLEs for β_0 and β_1. It will turn out that an analytical solution in closed form as in Example B.1 is no longer available. Therefore, computation has to be based on numerical optimization methods, as seen in the following section. The other steps for determining the MLE are conceptually the same as before:

1. *Likelihood*

   Omitting factors that do not depend on β_0 and β_1, the likelihood contribution of observation i is

   $$L_i(\beta_0, \beta_1) = \lambda_i^{y_i} \exp(-\lambda_i).$$

   The likelihood of the sample is the product

   $$L(\beta_0, \beta_1) = \prod_{i=1}^{n} L_i(\beta_0, \beta_1)$$

   of individual likelihood contributions.
2. *Log-likelihood*

   Taking logarithms, we obtain the individual log-likelihood contributions

   $$l_i(\beta_0, \beta_1) = y_i \log(\lambda_i) - \lambda_i = y_i(\beta_0 + \beta_1 x_i) - \exp(\beta_0 + \beta_1 x_i)$$

   and, summing up,

   $$l(\beta_0, \beta_1) = \sum_{i=1}^{n} (y_i(\beta_0 + \beta_1 x_i) - \exp(\beta_0 + \beta_1 x_i)).$$

3. *Score function*

   The partial first derivatives of individual log-likelihoods are

   $$\begin{aligned} \frac{\partial l_i(\beta_0, \beta_1)}{\partial \beta_0} &= y_i - \exp(\beta_0 + \beta_1 x_i) = y_i - \lambda_i, \\ \frac{\partial l_i(\beta_0, \beta_1)}{\partial \beta_1} &= y_i x_i - \exp(\beta_0 + \beta_1 x_i) \cdot x_i = x_i(y_i - \lambda_i). \end{aligned}$$

   Defining the vectors $\boldsymbol{y} = (y_1, \ldots, y_n)'$, $\boldsymbol{x} = (x_1, \ldots, x_n)'$, $\boldsymbol{\lambda} = (\lambda_1, \ldots, \lambda_n)'$, $\mathbf{1} = (1, \ldots, 1)'$ and the design matrix $\boldsymbol{X} = (\mathbf{1}\ \boldsymbol{x})$, we obtain the score function

   $$\boldsymbol{s}(\beta_0, \beta_1) = \begin{pmatrix} \sum_{i=1}^{n} (y_i - \lambda_i) \\ \sum_{i=1}^{n} x_i(y_i - \lambda_i) \end{pmatrix} = \begin{pmatrix} \mathbf{1}'(\boldsymbol{y} - \boldsymbol{\lambda}) \\ \boldsymbol{x}'(\boldsymbol{y} - \boldsymbol{\lambda}) \end{pmatrix} = \boldsymbol{X}'(\boldsymbol{y} - \boldsymbol{\lambda}).$$

Setting the score function to zero, we obtain a nonlinear system of equations that has to be solved numerically, as seen in the following section. The Fisher information matrix, determined in the next step, is an important component of the numerical algorithm.

4. *Fisher information matrix*

   The second partial derivatives of the individual log-likelihoods are

$$\frac{\partial^2 l_i(\beta_0, \beta_1)}{\partial \beta_0^2} = -\exp(\beta_0 + \beta_1 x_i) = -\lambda_i,$$

$$\frac{\partial^2 l_i(\beta_0, \beta_1)}{\partial \beta_1^2} = -\exp(\beta_0 + \beta_1 x_i) x_i^2 = -\lambda_i x_i^2,$$

$$\frac{\partial^2 l_i(\beta_0, \beta_1)}{\partial \beta_0 \partial \beta_1} = -\exp(\beta_0 + \beta_1 x_i) x_i = -\lambda_i x_i.$$

   Summing up and changing signs, we obtain the observed information matrix

$$\boldsymbol{H}(\boldsymbol{\beta}) = -\frac{\partial^2 l(\beta_0, \beta_1)}{\partial \boldsymbol{\beta} \partial \boldsymbol{\beta}'} = -\sum_{i=1}^{n} \frac{\partial^2 l_i(\beta_0, \beta_1)}{\partial \boldsymbol{\beta} \partial \boldsymbol{\beta}'} = \begin{pmatrix} \sum_{i=1}^{n} \lambda_i & \sum_{i=1}^{n} \lambda_i x_i \\ \sum_{i=1}^{n} \lambda_i x_i & \sum_{i=1}^{n} \lambda_i x_i^2 \end{pmatrix}.$$

   Defining the diagonal matrix $\boldsymbol{W} = \text{diag}(\lambda_1, \ldots, \lambda_n)$, we can write

$$\boldsymbol{H}(\boldsymbol{\beta}) = \boldsymbol{X}'\boldsymbol{W}\boldsymbol{X}$$

   in matrix notation. As $\boldsymbol{H}(\boldsymbol{\beta})$ does not depend on $\boldsymbol{y}$, the observed information matrix is equal to the expected information matrix in this example, i.e.,

$$\boldsymbol{F}(\boldsymbol{\beta}) = \text{E}_y(\boldsymbol{H}(\boldsymbol{\beta})) = \boldsymbol{H}(\boldsymbol{\beta}).$$

   In general, however, $\boldsymbol{F}(\boldsymbol{\beta}) \neq \boldsymbol{H}(\boldsymbol{\beta})$.

5. *Iterative numerical computation of the MLE* $\hat{\boldsymbol{\beta}}$

   $\hat{\boldsymbol{\beta}}$ is computed iteratively as the solution of the nonlinear system of equations

$$\boldsymbol{s}(\hat{\boldsymbol{\beta}}) \stackrel{!}{=} \boldsymbol{0}$$

   using the Newton or Fisher scoring algorithm sketched in the following section.

△

We finally remark that the likelihood principle can also be applied if the variables $Y_1, \ldots, Y_n$ are not independent. Such a situation appears, for example, in Sect. 7.3.2. There, the response vector $\boldsymbol{Y} = (Y_1, \ldots, Y_n)'$ follows a multivariate normal distribution

$$\boldsymbol{Y} \sim \text{N}(\boldsymbol{\mu}(\boldsymbol{\beta}), \boldsymbol{\Sigma}(\boldsymbol{\alpha})),$$

where the mean vector $\boldsymbol{\mu}(\boldsymbol{\beta})$ depends on $\boldsymbol{\beta}$ and the (non-diagonal) covariance matrix $\boldsymbol{\Sigma}(\boldsymbol{\alpha})$ depends on an unknown parameter vector $\boldsymbol{\alpha}$. The likelihood $L(\boldsymbol{\beta}, \boldsymbol{\alpha}) = L(\boldsymbol{\theta})$, $\boldsymbol{\theta} = (\boldsymbol{\alpha}, \boldsymbol{\beta})$ is then given by the p.d.f. of the multivariate normal distribution for $\boldsymbol{Y}$. It can no longer be factorized in the product of separate univariate p.d.f.'s for the variables $Y_1, \ldots, Y_n$.

B.4.2 Numerical Computation of the MLE

In most applications, the MLE cannot be determined analytically: the system of equations, obtained from setting the score function to zero, is usually nonlinear and cannot be solved for the unknown parameters in closed form; see Example B.3. Therefore, numerical methods for computing the roots of the score function are required. Various such methods have been developed; see, for example, Lange (2000) for a survey. Here we only sketch the two most popular algorithms: the *Newton (or Newton–Raphson) method* and the *Fisher scoring algorithm*.

We first illustrate the Newton–Raphson method for a scalar parameter. The aim is to numerically compute the root of the score function, i.e., the solution of the generally nonlinear score equation

$$s(\theta) = 0.$$

Beginning with an initial value $\theta^{(0)}$, an (approximate) solution is computed iteratively as follows (see Fig. B.1): At $\theta^{(0)}$ the score function is approximated through a straight line, the *tangent*. An improved approximate solution $\theta^{(1)}$ is then obtained as the root of the tangent. The tangent is given by

$$g(\theta) = s\left(\theta^{(0)}\right) + s'\left(\theta^{(0)}\right) \cdot \left(\theta - \theta^{(0)}\right),$$

obtained through a first-order Taylor expansion of $s(\theta)$ at $\theta^{(0)}$. The root of the tangent provides the improved iterate

$$\theta^{(1)} = \theta^{(0)} - \frac{1}{s'\left(\theta^{(0)}\right)} \cdot s\left(\theta^{(0)}\right).$$

Since $-s'(\theta)$ is the observed Fisher information $H(\theta)$, we can also write

$$\theta^{(1)} = \theta^{(0)} + H\left(\theta^{(0)}\right)^{-1} s\left(\theta^{(0)}\right).$$

Starting from $\theta^{(1)}$ we get the next improved approximate solution $\theta^{(2)}$ by constructing another tangent of $s(\theta)$ at $\theta^{(1)}$ and computing its root $\theta^{(2)}$. This algorithm continues iteratively until successive iterates do practically not change any more.

The extension of the algorithm to a multivariate parameter $\boldsymbol{\theta} = (\theta_1, \ldots, \theta_p)'$ is as follows: Let $\boldsymbol{\theta}^{(t)}$ be the current iterate of the approximate solution of $\boldsymbol{s}(\boldsymbol{\theta}) = \boldsymbol{0}$. An improved iterate is

$$\boldsymbol{\theta}^{(t+1)} = \boldsymbol{\theta}^{(t)} - \left(\frac{\partial \boldsymbol{s}\left(\boldsymbol{\theta}^{(t)}\right)}{\partial \boldsymbol{\theta}'}\right)^{-1} \boldsymbol{s}\left(\boldsymbol{\theta}^{(t)}\right) = \boldsymbol{\theta}^{(t)} + \boldsymbol{H}\left(\boldsymbol{\theta}^{(t)}\right)^{-1} \boldsymbol{s}\left(\boldsymbol{\theta}^{(t)}\right).$$

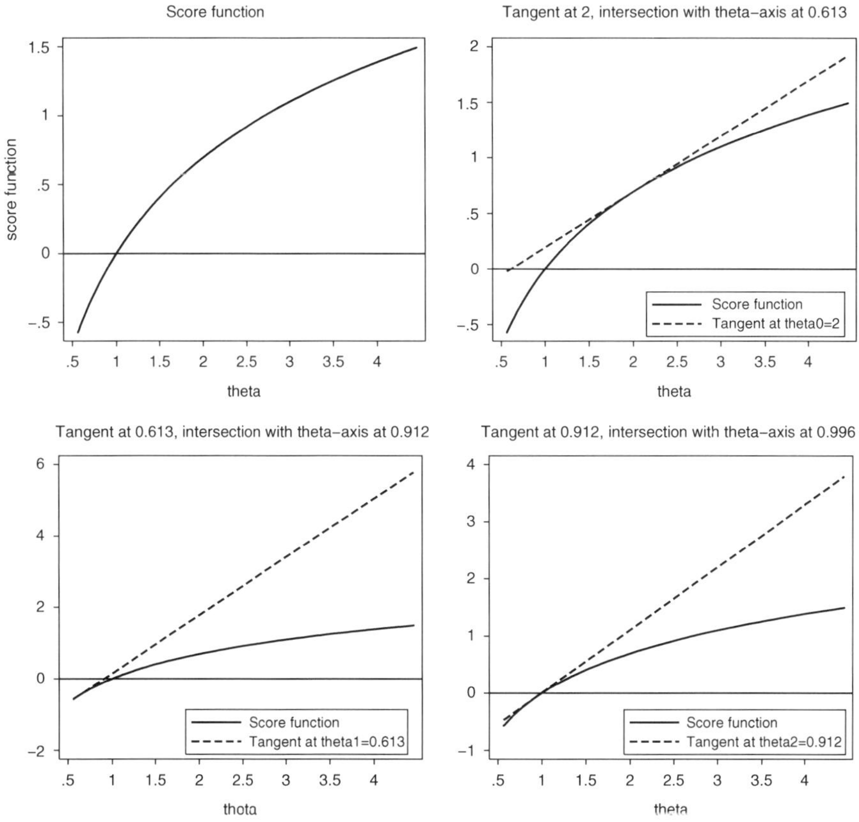

Fig. B.1 Illustration of the Newton method

The *Fisher scoring method* is obtained if the observed information matrix is replaced by the expected information matrix $\boldsymbol{F}(\boldsymbol{\theta}^{(t)})$. As an advantage, the expression for the expected information matrix is often simpler and faster to compute.

Example B.4 Poisson Regression

Continuing the Poisson regression example, the Newton–Raphson method for computing the MLE $\hat{\boldsymbol{\beta}} = (\hat{\beta}_0, \hat{\beta}_1)'$ works as follows:

1. Define the initial values $\hat{\boldsymbol{\beta}}^{(0)}$, e.g., $\hat{\boldsymbol{\beta}}^{(0)} = (0, 0)'$. Set $t = 0$.
2. Compute improved estimates $\hat{\boldsymbol{\beta}}^{(t)}$ through

$$\hat{\boldsymbol{\beta}}^{(t+1)} = \hat{\boldsymbol{\beta}}^{(t)} + \boldsymbol{H}\left(\hat{\boldsymbol{\beta}}^{(t)}\right)^{-1} \boldsymbol{s}\left(\hat{\boldsymbol{\beta}}^{(t)}\right) = \hat{\boldsymbol{\beta}}^{(t)} + (\boldsymbol{X}'\boldsymbol{W}\boldsymbol{X})^{-1}\boldsymbol{X}'(\boldsymbol{y} - \boldsymbol{\lambda}),$$

   where $\boldsymbol{W} = \boldsymbol{W}\left(\hat{\boldsymbol{\beta}}^{(t)}\right)$ and $\boldsymbol{\lambda} = \boldsymbol{\lambda}\left(\hat{\boldsymbol{\beta}}^{(t)}\right)$ depend on the current iterate $\hat{\boldsymbol{\beta}}^{(t)}$.
3. Stop if

$$\frac{||\hat{\boldsymbol{\beta}}^{(t+1)} - \hat{\boldsymbol{\beta}}^{(t)}||}{||\hat{\boldsymbol{\beta}}^{(t)}||} \leq \varepsilon$$

   for a (very) small value $\varepsilon > 0$. Otherwise set $t = t + 1$ and go to step 2.

△

B.4.3 Asymptotic Properties of the MLE

Under certain regularity assumptions, depending in detail on the specific model $f(y_1, \dots, y_n \mid \boldsymbol{\theta})$, the following asymptotic properties hold: For $n \to \infty$, or practically for large sample size, the MLE is consistent, asymptotically (or approximately) unbiased, and asymptotically (or approximately) multivariate normal:

$$\hat{\boldsymbol{\theta}} \overset{a}{\sim} \mathrm{N}(\boldsymbol{\theta}, \hat{\boldsymbol{V}}).$$

The (estimated) covariance matrix $\widehat{\mathrm{Cov}}(\hat{\boldsymbol{\theta}}) = \hat{\boldsymbol{V}}$ is obtained as the inverse of the observed or expected information matrix, i.e.,

$$\hat{\boldsymbol{V}} = \boldsymbol{H}^{-1}(\hat{\boldsymbol{\theta}}) \quad \text{or} \quad \hat{\boldsymbol{V}} = \boldsymbol{F}^{-1}(\hat{\boldsymbol{\theta}}),$$

evaluated at the MLE $\boldsymbol{\theta} = \hat{\boldsymbol{\theta}}$. These inverses are computed in the final iteration step of the Newton or Fisher scoring algorithm. The diagonal elements $\hat{v}_{jj}$ of $\hat{\boldsymbol{V}}$ are the estimated variances $\widehat{\mathrm{Var}}(\hat{\theta}_j)$ of the jth component of $\hat{\boldsymbol{\theta}}$, and $\sqrt{\hat{v}_{jj}}$ is the estimated standard deviation or standard error: $se_j = \sqrt{\hat{v}_{jj}}$.

Finally, MLEs are also asymptotically efficient, i.e., for large sample sizes their variance is smaller or at least not larger than the variance of other asymptotically unbiased estimators.

How large should the sample size n be so that these asymptotic properties hold sufficiently well? It is not possible to give a general definitive answer to this question. Basically, it depends on the quality of the approximation of the likelihood or the log-likelihood in a neighborhood of $\hat{\boldsymbol{\theta}}$ through a normal distribution or a quadratic function, respectively. As a rule of thumb, the sample size n has to be much larger (about a factor of 10–20) than the number p of unknown parameters. If in doubt, simulation studies are useful.

B.4.4 Likelihood-Based Tests of Linear Hypotheses

We consider testing linear hypotheses about $\boldsymbol{\theta} = (\theta_1, \dots, \theta_p)' \in \boldsymbol{\Theta}$ of the form

$$H_0 : \boldsymbol{C\theta} = \boldsymbol{d} \quad \text{versus} \quad H_1 : \boldsymbol{C\theta} \neq \boldsymbol{d},$$

where the $(r \times p)$-matrix $\boldsymbol{C}$ has full row rank $r \leq p$. In this book $\boldsymbol{\theta}$ is mostly the vector $\boldsymbol{\beta}$ of regression coefficients. An important special case are hypotheses about an r-dimensional subvector $\boldsymbol{\theta}_1$ of $\boldsymbol{\theta}$ of the form

$$H_0 : \boldsymbol{\theta}_1 = \boldsymbol{d} \quad \text{versus} \quad H_1 : \boldsymbol{\theta}_1 \neq \boldsymbol{d}.$$

If $\boldsymbol{\theta}_1 = \boldsymbol{\beta}_1$ is a subvector of regression coefficients and $\boldsymbol{d} = \boldsymbol{0}$, this is a test about significance of corresponding covariates. The three commonly used test statistics are the (log-)likelihood ratio statistic, the Wald statistic, and the score statistic.

The *(log-)likelihood ratio statistic*

$$lr = 2\{l(\hat{\boldsymbol{\theta}}) - l(\tilde{\boldsymbol{\theta}})\} = -2\{l(\tilde{\boldsymbol{\theta}}) - l(\hat{\boldsymbol{\theta}})\}$$

compares the unrestricted maximum $l(\hat{\boldsymbol{\theta}})$ of the log-likelihood with the maximum obtained for the restricted MLE $l(\tilde{\boldsymbol{\theta}})$, computed under the restriction $\boldsymbol{C}\boldsymbol{\theta} = \boldsymbol{d}$ of H_0. Since $l(\hat{\boldsymbol{\theta}}) \geq l(\tilde{\boldsymbol{\theta}})$, we have $lr \geq 0$. If the unrestricted maximum $l(\hat{\boldsymbol{\theta}})$ is significantly larger than $l(\tilde{\boldsymbol{\theta}})$, implying that lr is large, H_0 will be rejected in favor of H_1. The test statistic can also be motivated by considering the likelihood ratio

$$LR = \frac{L(\hat{\boldsymbol{\theta}})}{L(\tilde{\boldsymbol{\theta}})},$$

where $L(\hat{\boldsymbol{\theta}})$ and $L(\tilde{\boldsymbol{\theta}})$ are the maximum of the likelihood without and under H_0, respectively. Now $LR \geq 1$, and H_0 will be rejected if LR is significantly larger than 1. Taking logarithms and multiplying by the factor 2, one obtains the (log-)likelihood ratio test statistic.

To compute lr, the log-likelihood $l(\boldsymbol{\theta})$ has to be maximized under the restriction of H_0, which can require considerable effort for the general form $\boldsymbol{C}\boldsymbol{\theta} = \boldsymbol{d}$ of the null restriction. This can be avoided with the *Wald statistic*

$$w = (\boldsymbol{C}\hat{\boldsymbol{\theta}} - \boldsymbol{d})'(\boldsymbol{C}\hat{\boldsymbol{V}}\boldsymbol{C}')^{-1}(\boldsymbol{C}\hat{\boldsymbol{\theta}} - \boldsymbol{d}).$$

It measures the weighted distance between the unrestricted estimate $\boldsymbol{C}\hat{\boldsymbol{\theta}}$ of $\boldsymbol{C}\boldsymbol{\theta}$ and its hypothetical value $\boldsymbol{d}$ under H_0. The weight is the inverse of the (estimated) covariance matrix $\boldsymbol{C}\hat{\boldsymbol{V}}\boldsymbol{C}'$ of $\boldsymbol{C}\hat{\boldsymbol{\theta}} - \boldsymbol{d}$, where $\hat{\boldsymbol{V}}$ is the (estimated) covariance matrix of $\hat{\boldsymbol{\theta}}$. If H_0 is true, this weighted distance should be small, whereas H_0 should be rejected if w is large. For the special hypothesis $H_0 : \boldsymbol{\theta}_1 = \boldsymbol{d}$, the Wald statistic simplifies to

$$w = (\hat{\boldsymbol{\theta}}_1 - \boldsymbol{d})'\hat{\boldsymbol{V}}_1^{-1}(\hat{\boldsymbol{\theta}}_1 - \boldsymbol{d}),$$

where $\hat{\boldsymbol{V}}_1 = \widehat{\text{Cov}}(\hat{\boldsymbol{\theta}}_1)$ is (the estimated) covariance matrix of $\hat{\boldsymbol{\theta}}_1$.

The *score statistic*

$$u = \boldsymbol{s}'(\tilde{\boldsymbol{\theta}})\tilde{\boldsymbol{V}}\boldsymbol{s}(\tilde{\boldsymbol{\theta}})$$

measures the weighted distance between the value $\boldsymbol{0} = \boldsymbol{s}(\hat{\boldsymbol{\theta}})$ of the score function at the MLE $\hat{\boldsymbol{\theta}}$ and the value $\boldsymbol{s}(\tilde{\boldsymbol{\theta}})$ at the restricted MLE $\tilde{\boldsymbol{\theta}}$, where $\tilde{\boldsymbol{V}}$ is the estimated covariance matrix of $\tilde{\boldsymbol{\theta}}$.

Wald tests are computationally advantageous in cases when H_0 defines a submodel that has to be tested against a larger model (without the H_0 restriction) that has been estimated already. Additionally, score tests are advantageous if an already fitted model has to be tested against a larger model. Therefore, numerically efficient forward and backward variable selection may be based on Wald and score tests.

Under H_0, the three test statistics are asymptotically equivalent and asymptotically (or approximately) chi-squared distributed with r degrees of freedom:

$$lr, w, s \overset{a}{\sim} \chi^2_r.$$

Critical values or p-values are computed using the limiting χ^2_r-distribution. For medium and large sample sizes, this approximation is usually sufficient. For smaller sample sizes, the three test statistics may be comparably different from each other. Note that the Wald test is not restricted to genuine likelihood inference. It is applicable whenever an estimator $\hat{\boldsymbol{\theta}}$ is asymptotically normal, as, for example, in least squares inference without normality assumption or in quasi-likelihood inference.

B.4.5 Model Choice

For comparing and choosing between several competing statistical models with different predictors and parameters, we need a compromise between good fit to the data and model complexity, i.e., a (too) high effective number of parameters. In linear regression models, for example, the goodness-of-fit measure R^2 will be increased by including additional covariates, interactions, etc. However, this usually results in overfitting the (training) data and worsens predictive capabilities and generalization to new data. Therefore, criteria for goodness of fit, such as R^2 and, for likelihood inference, the deviance or the log-likelihood, have to be modified such that overfitting the data is avoided by penalizing high model complexity, i.e., a high effective number of parameters. For likelihood inference with p-dimensional parameter $\boldsymbol{\theta} = (\theta_1, \ldots, \theta_p)'$, Akaike's information criterion AIC is defined as

$$\text{AIC} = -2\, l(\hat{\boldsymbol{\theta}}) + 2\, p,$$

where the term $2\,p$ penalizes overly complex models. For model choice, models with small AIC value (and not with large log-likelihood $l(\hat{\boldsymbol{\theta}})$!) are favored. Instead of the AIC, also AIC$/n$, where n is the sample size, is used.

For non- and semiparametric models as in Chaps. 8 and 9, the fixed dimension p is replaced by the "effective" dimension

$$df = \text{tr}(\boldsymbol{S}),$$

where $\boldsymbol{S}$ is the smoother matrix (see Sect. 8.1.8, p. 473). Alternative penalty terms, for example, the adjusted AIC_{corr},

$$\text{AIC}_{corr} = -2\,l(\hat{\theta}) + \frac{2n(df + 1)}{n - df - 2},$$

are also suggested. Burnham and Anderson (2002) give a nice justification of the AIC.

B.5 Bayesian Inference

Due to the development of computer intensive and simulation-based Markov chain Monte Carlo (MCMC) methods, Bayesian inference has become an attractive approach for analyzing complex statistical models. We first introduce concepts of Bayesian inference and then describe MCMC methods. More detailed introductions can be found, e.g., in Migon and Gamerman (1999) and Held and Sabanés Bové (2012).

B.5.1 Basic Concepts of Bayesian Inference

The fundamental difference to likelihood-based inference is that the unknown parameters $\boldsymbol{\theta} = (\theta_1, \dots, \theta_p)'$ are not considered as fixed, deterministic quantities but as random variables with a *prior distribution*. A Bayesian model therefore consists of two parts:

- *Prior distribution:* Any (subjective) information about the unknown parameter $\boldsymbol{\theta}$ is expressed by specifying a probability distribution for $\boldsymbol{\theta}$. This distribution is called prior (distribution) of $\boldsymbol{\theta}$. The specification of a prior for $\boldsymbol{\theta}$ does not necessarily mean that the unknown parameters are actually stochastic quantities. Rather the prior describes the *degree of uncertainty* about the unknown parameters prior to the statistical analysis. The p.d.f. or probability function of the prior distribution will be denoted by $p(\boldsymbol{\theta})$.
- *Observation model:* The observation model specifies the conditional distribution of observable quantities, that is, the random sample variables $\boldsymbol{Y} = (Y_1, \dots, Y_n)'$, given the parameters. The p.d.f. or probability function of this conditional distribution is proportional to the likelihood $L(\boldsymbol{\theta})$ and will be denoted by $p(\boldsymbol{y} \,|\, \boldsymbol{\theta})$.

Based on the prior and the observation model, Bayes' theorem determines the distribution of $\boldsymbol{\theta}$ after the data are known through the statistical experiment, that is the conditional distribution of $\boldsymbol{\theta}$ given the observations $\boldsymbol{y} = (y_1, \dots, y_n)'$. We obtain

$$p(\boldsymbol{\theta} \,|\, \boldsymbol{y}) = \frac{p(\boldsymbol{y} \,|\, \boldsymbol{\theta})\, p(\boldsymbol{\theta})}{\int p(\boldsymbol{y} \,|\, \boldsymbol{\theta})\, p(\boldsymbol{\theta})\, d\boldsymbol{\theta}} = c \cdot p(\boldsymbol{y} \,|\, \boldsymbol{\theta})\, p(\boldsymbol{\theta}),$$

with the normalizing constant $c = [\int p(\boldsymbol{y} \,|\, \boldsymbol{\theta}) p(\boldsymbol{\theta})\, d\boldsymbol{\theta}]^{-1}$. This conditional distribution is called *posterior (distribution)*.

Example B.5 Poisson Distribution

Consider an i.i.d. sample $Y_1, \dots, Y_n$ from a Poisson distribution, i.e., $Y_i \sim \text{Po}(\lambda)$. The unknown parameter is to be estimated using Bayesian inference. As in Example B.1, the joint probability for the observed sample $\boldsymbol{y} = (y_1, \dots, y_n)'$ is

$$p(\boldsymbol{y} \mid \lambda) = \frac{1}{y_1! \cdots y_n!} \lambda^{\sum_{i=1}^n y_i} \exp(-n\lambda).$$

We specify a gamma distribution with parameters a and b for λ, i.e., $\lambda \sim \text{G}(a, b)$. A justification for the choice of this prior will be given later. It follows that λ has p.d.f.

$$p(\lambda) = k\, \lambda^{a-1} \exp(-b\lambda)$$

with $k = \frac{b^a}{\Gamma(a)}$. The posterior is obtained as

$$p(\lambda \mid \boldsymbol{y}) = \frac{p(\boldsymbol{y} \mid \lambda)\, p(\lambda)}{\int p(\boldsymbol{y} \mid \lambda)\, p(\lambda)\, d\lambda} = c\, \frac{1}{y_1! \cdots y_n!} \lambda^{\sum_{i=1}^n y_i} \exp(-n\lambda)\, k \lambda^{a-1} \exp(-b\lambda).$$

To determine the type of this distribution, we can ignore all factors that do not depend on λ. This gives

$$p(\lambda \mid \boldsymbol{y}) \propto \lambda^{\sum_{i=1}^n y_i} \exp(-n\lambda)\, \lambda^{a-1} \exp(-b\lambda) = \lambda^{a + \sum_{i=1}^n y_i - 1} \exp(-(b+n)\lambda).$$

This has the form of a gamma distribution with parameters $a' = a + \sum_{i=1}^n y_i$ and $b' = b + n$, i.e.,

$$\lambda \mid \boldsymbol{y} \sim \text{G}\left(a + \sum_{i=1}^n y_i, b + n\right),$$

and the posterior has the same type of distribution as the prior. This desirable case appears in many other simple Bayesian models. We call the prior as conjugate to the Poisson model because the posterior is of the same type as the prior.

△

The next example discusses a situation where we have rather limited prior knowledge about the unknown parameters. In this case we are confronted with the question how to specify a prior in the absence of prior knowledge. We call such priors *noninformative*.

Example B.6 Bayesian Logit Model—Diffuse Prior

We consider a logit model with a single covariate x:

$$Y_i = B(1, \pi_i), \quad \pi_i = \frac{\exp(\eta_i)}{1 + \exp(\eta_i)}, \quad \eta_i = \beta_0 + \beta_1 x_i, \quad i = 1, \dots, n.$$

Assuming, as usual, (conditionally) independent response variables, the observation model is given by

$$p(\boldsymbol{y} \mid \boldsymbol{\beta}) \propto L(\boldsymbol{\beta}) = \prod_{i=1}^n \pi_i^{y_i} (1 - \pi_i)^{1 - y_i},$$

where $\boldsymbol{\beta} = (\beta_0, \beta_1)'$ is the vector of regression coefficients. To fully specify the Bayesian model, we have to supplement the observation model by a prior for β_0 and β_1.

Since estimated regression coefficients are often approximately normally distributed, it is reasonable to assume a two-dimensional normal prior. Therefore, we specify the prior

$$p(\boldsymbol{\beta}) \sim \mathrm{N}(\boldsymbol{m}, \boldsymbol{M})$$

with prior mean $\boldsymbol{m}$ and prior covariance matrix $\boldsymbol{M}$. If, for example, results from a previous statistical analysis are available, we could choose the previous point estimate as $\boldsymbol{m}$ and its estimated covariance matrix as $\boldsymbol{M}$. If the previous analysis has been carried out some time ago, we may also multiply $\boldsymbol{M}$ with a factor $a > 1$ to express increased uncertainty.

In many situations, however, we may have no information about the regression coefficients. Increasing the variances in $\boldsymbol{M}$, the normal prior becomes very flat and approximates a uniform distribution. In the limiting case the prior becomes proportional to a constant, i.e.,

$$p(\boldsymbol{\beta}) \propto \text{const.}$$

Often we also write $p(\boldsymbol{\beta}) \propto 1$. The integral of this flat prior over $\mathbb{R}^2$ is not finite, so that $p(\boldsymbol{\beta})$ is not a density in the usual sense. Such a prior is called improper or diffuse. Nonetheless such diffuse priors are admissible as long as the posterior, resulting from Bayes' theorem, is a proper distribution, i.e., its integral over $\mathbb{R}^2$ is finite. In a Bayesian logit model this is the case if a finite MLE exists. With a flat, diffuse prior the posterior density is

$$p(\boldsymbol{\beta} \mid \boldsymbol{y}) = \frac{p(\boldsymbol{\beta}) p(\boldsymbol{y} \mid \boldsymbol{\beta})}{\int p(\boldsymbol{\beta}) p(\boldsymbol{y} \mid \boldsymbol{\beta}) d\, \boldsymbol{\beta}} \propto p(\boldsymbol{y} \mid \boldsymbol{\beta}) = \prod_{i=1}^{n} \pi_i^{y_i} (1 - \pi_i)^{1 - y_i}.$$

Although the posterior is proper, it has no known distributional type. This causes problems for statistical inference because characteristics of the posterior, such as the mean, are not available in closed analytical form. However inference based on MCMC simulation is possible. See Sect. B.5.3 and in particular, Example B.10.

△

The previous example suggests to assume a uniform prior distribution for the parameters if there is no prior knowledge. The next example, however, shows that the construction of noninformative priors is a bit more subtle.

Example B.7 Normal Distribution

Consider an i.i.d. sample $Y_1, \ldots, Y_n$ from a normal distribution, i.e., $Y_i \sim \mathrm{N}(\mu, \sigma^2)$. We first assume the expectation μ to be known. In the absence of any prior knowledge regarding the unknown parameter σ^2 one is tempted to assume a diffuse uniform prior over $\mathbb{R}^+$. It follows that $p(\sigma^2)$ is improper and proportional to a constant, i.e., $p(\sigma^2) \propto 1$. Suppose now, that we had parameterized the normal distribution in terms of the standard deviation σ rather than the variance. According to our recipe for a noninformative prior we would then assign a uniform distribution for σ, i.e., $p(\sigma) \propto 1$. From this assumption, we can derive the corresponding distribution of the variance σ^2 using the change of variables theorem of continuous random variables (Theorem B.1 in Appendix B.1). We now obtain $p(\sigma^2) \propto (\sigma^2)^{-1/2}$, which is no longer a uniform distribution over $\mathbb{R}^+$. Thus, if we parameterize the normal distribution with the variance our noninformative prior for σ^2 is a uniform distribution. If we parameterize in terms of the standard deviation we come out with a nonuniform distribution for σ^2. This means that the prior is not invariant with respect to the specific parameterization.

△

A class of priors that help to circumvent the invariance problem mentioned in the example are *Jeffreys' priors*. Jeffreys' prior for a scalar parameter θ is defined to be proportional to the square root of the expected Fisher information, i.e.,

$$p(\theta) \propto \sqrt{F(\theta)} = \sqrt{\mathrm{E}(-l''(\theta))}.$$

Jeffreys' prior solves the invariance problem as it is indeed invariant with respect to one-to-one transformations of the parameter. The remaining question is if Jeffreys' prior is also noninformative in some sense. The answer is yes, as Jeffreys' prior for a scalar parameter can be characterized as a *reference prior*. The concept of a reference prior has been introduced in Bernardo (1979). Informally, the reference prior can be characterized as the distribution that maximizes the influence of the data on the posterior. More specifically, the reference prior maximizes the expected Kullback–Leibler distance of the posterior relative to the prior. In this sense the reference prior is noninformative as the data get maximal weight and the influence of the prior is minimized.

In situations with vector-valued parameter $\boldsymbol{\theta}$, Jeffreys' prior generalizes to

$$p(\boldsymbol{\theta}) \propto \sqrt{|F(\boldsymbol{\theta})|}.$$

This prior is still translation invariant, but is in general not the reference prior. This is one of the reasons why Jeffreys' prior is usually not accepted as a valid noninformative prior for vector valued parameters. See, e.g., Held and Sabanés Bové (2012) for a discussion of the deficiencies of Jeffreys' prior in the multiparameter case. Instead, we can still base the choice of noninformative priors on the reference prior concept. However, it would be beyond the scope of this book to fully introduce this technically difficult concept; see, e.g., Held and Sabanés Bové (2012) for details.

Example B.8 Normal Distribution—Jeffreys' Prior—Reference Prior

We continue the previous Example B.7 and derive the Jeffreys' or reference prior. In the case of a normal distribution with known expectation μ, Jeffreys' prior or the reference prior for σ^2 is easily obtained as

$$p(\sigma^2) \propto \frac{1}{\sigma^2}.$$

Now the distribution of the standard deviation can be derived to be $p(\sigma) \propto 1/\sigma$ which is the same distribution as for σ^2.

If both parameters μ and σ^2 are unknown we obtain a noninformative prior in the form of the reference prior. This reference prior is given by

$$p(\mu, \sigma^2) \propto \frac{1}{\sigma^2}$$

implying a priori independence of μ and σ^2 with distributions $p(\mu) \propto 1$ and $p(\sigma^2) \propto 1/\sigma^2$.

△

B.5.2 Point and Interval Estimation

Point Estimation

The usual point estimators in Bayesian inference are the posterior mean, the posterior median, and the posterior mode. These three estimators can be discussed and justified from a decision theoretic perspective. We do not describe this here in detail; rather we only introduce the estimators.

The posterior mean is given by

$$\hat{\boldsymbol{\theta}} = \mathrm{E}(\boldsymbol{\theta} \mid \boldsymbol{y}) = \int \boldsymbol{\theta} \, p(\boldsymbol{\theta} \mid \boldsymbol{y}) \, d\boldsymbol{\theta} = c \cdot \int \boldsymbol{\theta} \, p(\boldsymbol{y} \mid \boldsymbol{\theta}) \, p(\boldsymbol{\theta}) \, d\boldsymbol{\theta}.$$

Thus, the analytical or numerical determination of the posterior mean requires the computation of (possibly high-dimensional) integrals. The resulting difficulties have been a main obstacle for applying Bayesian methods in practice. Simulation-based methods, in particular modern MCMC methods, have greatly reduced these problems. Refer to Sect. B.5.3.

The posterior mode is the value $\hat{\boldsymbol{\theta}}$ that (globally) maximizes the posterior density, i.e.,

$$\hat{\boldsymbol{\theta}} = \arg\max_{\boldsymbol{\theta}} \, p(\boldsymbol{\theta} \mid \boldsymbol{y}) = \arg\max_{\boldsymbol{\theta}} p(\boldsymbol{y} \mid \boldsymbol{\theta}) p(\boldsymbol{\theta}).$$

The second expression shows that no integration is necessary to compute the posterior mode, because the normalizing constant is not needed.

The posterior median, that is, the median of the posterior distribution, is sometimes preferred to the posterior mean because it is more robust against outliers.

Example B.9 Poisson Distribution

In Example B.5 we obtained a gamma distribution with parameters $a' = a + \sum_{i=1}^{n} y_i$ and $b' = b + n$ as the posterior for λ. The posterior mean is

$$\mathrm{E}(\lambda \mid \boldsymbol{y}) = \frac{a + \sum\limits_{i=1}^{n} y_i}{b + n}.$$

The smaller a (in relation to $\sum y_i$) and b (in relation to n), the closer the posterior mean is to the usual MLE $\hat{\lambda} = \bar{y}$. The larger the prior information, that is, the larger a and b are, the more the posterior mean and the MLE differ from each other.

△

Interval Estimation

Point estimators reduce the information of the posterior to a single number. In particular, it is not possible to see how precise these estimates are. Usual quantities for measuring variability of a random variable are also natural measures for assessing how precise, or imprecise, the estimators are. For the posterior mean, a natural measure is the posterior variance; for the posterior median, the interquartile distance seems to be appropriate to measure its variability. In case of the posterior mode, the curvature of the posterior at the mode, i.e., the observed Fisher information, is a natural choice.

Another popular way of assessing uncertainty is Bayesian confidence intervals or *credible intervals* or, more generally, *credible regions*:

A region $C \subset \boldsymbol{\Theta}$ of the parameter space is said to be a $(1-\alpha)$-credible region for $\boldsymbol{\theta}$ if

$$\mathrm{P}(\boldsymbol{\theta} \in C \mid \boldsymbol{y}) = 1 - \alpha.$$

If $C \subseteq \mathbb{R}$ is an interval it is called credible interval. In other words, a credible region contains (at least) a probability mass $1-\alpha$ of the posterior. Note that credible regions and classical confidence regions have rather different interpretations: The classical confidence interval does not provide a probability statement about $\boldsymbol{\theta}$, which is a deterministic quantity in the classical inferential concept. Rather, it is related with a probability statement about the random sample $\boldsymbol{y} = (y_1, \ldots, y_n)'$: The confidence region $C(\boldsymbol{y})$ contains the unknown true parameter $\boldsymbol{\theta}$ with probability $1-\alpha$. The frequentist interpretation is that confidence intervals cover the true parameter with a percentage of about $(1-\alpha) \cdot 100\,\%$ if the estimation procedure is repeated for many samples. In contrast, Bayesian credible intervals allow a probability statement about the random parameter $\boldsymbol{\theta}$: The credible region contains $(1-\boldsymbol{\alpha}) \cdot 100\,\%$ of the probability mass of the posterior, i.e., a random number $\boldsymbol{\theta}$ drawn from the posterior is within the credible region with probability $1-\alpha$.

Bayesian credible intervals are often difficult to determine analytically. However, they can be easily computed with MCMC algorithms for drawing random numbers from the posterior.

B.5.3 MCMC Methods

Historically, the main difficulty in applying Bayesian methods has been that in many situations the posterior is not available analytically or even numerically. Basically, MCMC is Monte Carlo integration using Markov chains. Although MCMC was known in physics very early (Metropolis, Rosenbluth, Rosenbluth, Teller, & Teller, 1953), it took nearly 40 years to enter mainstream statistical practice (Gelfand & Smith, 1990), in particular for Bayesian inference. Since then, MCMC has had an enormous effect on Bayesian statistics, and it allows to solve complex and high-dimensional inferential problems. Good introductions are, for example, Gilks, Richardson, and Spiegelhalter (1996) and Green (2001). Here we only describe the basic idea and the most important algorithms.

MCMC methods allow to draw samples from posterior distributions (and, in principle, from any distribution) that are usually not available analytically and to estimate characteristics of the posterior such as the mean, the variance or quantiles, or the posterior density itself. The most important advantage compared to more traditional methods of drawing a sample from a distribution, for example, importance or rejection sampling, is that samples can be drawn from high-dimensional densities, even for dimensions in the thousands. Another advantage is that the normalizing constant, often a high-dimensional integral that cannot be computed with traditional numerical methods, does not have to be known.

The basic idea of MCMC methods is comparably simple. Let $\boldsymbol{\theta}$ be the unknown vector of parameters in a Bayesian model and $p(\boldsymbol{\theta} \mid \boldsymbol{y})$ the posterior density (we assume here that $\boldsymbol{\theta}$ is continuous). Instead of directly drawing an i.i.d. sample from $p(\boldsymbol{\theta} \mid \boldsymbol{y})$, a Markov chain is generated such that the iterations of the transition kernel converge to the posterior of interest. In this way random numbers are generated that can be considered as a (correlated) sample from the posterior after some time of convergence, the *burn-in phase*. Before describing some algorithms in detail, we point out that MCMC may be used for drawing random numbers from any complex distribution, not only from posterior distributions. The posterior $p(\boldsymbol{\theta} \mid \boldsymbol{y})$ then has to be replaced by such a density in the following.

Metropolis–Hastings Algorithm

The basic algorithm, from which all other algorithms can be derived, works as follows: First, a starting value $\boldsymbol{\theta}^{(0)}$ is chosen. Instead of drawing directly from the posterior $p(\boldsymbol{\theta} \mid \boldsymbol{y})$, a new random number $\boldsymbol{\theta}^*$ is drawn from a *proposal density* q in each iteration. Usually, the proposal density depends on the current iterate (or state) $\boldsymbol{\theta}^{(t-1)}$, i.e., $q = q(\boldsymbol{\theta}^* \mid \boldsymbol{\theta}^{(t-1)})$. In principle, the choice of the proposal density is arbitrary. However, it should be comparably easy to draw random numbers from this density. Since the proposal density differs from the posterior, a proposed random number $\boldsymbol{\theta}^*$ cannot always be accepted as the new current state $\boldsymbol{\theta}^{(t)}$ but only with some acceptance probability α. Basically, this is the ratio of the posterior and the proposal density, evaluated at the current state $\boldsymbol{\theta}^{(t-1)}$ and the proposed value $\boldsymbol{\theta}^*$. More specifically, the acceptance probability is defined as

$$\alpha(\boldsymbol{\theta}^* \mid \boldsymbol{\theta}^{(t-1)}) = \min\left\{ \frac{p(\boldsymbol{\theta}^* \mid \boldsymbol{y})\, q(\boldsymbol{\theta}^{(t-1)} \mid \boldsymbol{\theta}^*)}{p(\boldsymbol{\theta}^{(t-1)} \mid \boldsymbol{y})\, q(\boldsymbol{\theta}^* \mid \boldsymbol{\theta}^{(t-1)})}, 1 \right\}.$$

If the proposed $\boldsymbol{\theta}^*$ is accepted, then the next state is $\boldsymbol{\theta}^{(t)} = \boldsymbol{\theta}^*$; otherwise $\boldsymbol{\theta}^{(t)} = \boldsymbol{\theta}^{(t-1)}$. The density $p(\boldsymbol{\theta} \mid \boldsymbol{y})$ does appear directly, not only in $\alpha(\boldsymbol{\theta}^* \mid \boldsymbol{\theta}^{(t-1)})$, but also in the ratio $p(\boldsymbol{\theta}^* \mid \boldsymbol{y})/p(\boldsymbol{\theta}^{(t-1)} \mid \boldsymbol{y})$. Therefore, all expressions in $p(\boldsymbol{\theta} \mid \boldsymbol{y})$ that are constant can be omitted. In particular, the normalizing constant of the posterior is not needed. This is one of the big advantages of MCMC methods compared to conventional methods for drawing random numbers. A summary of the MH algorithm is given in Box 2.1.

The Metropolis–Hastings algorithm simplifies for symmetric proposals with $q(\boldsymbol{\theta}^* \mid \boldsymbol{\theta}^{(t-1)}) = q(\boldsymbol{\theta}^{(t-1)} \mid \boldsymbol{\theta}^*)$. The acceptance probability then becomes

$$\alpha(\boldsymbol{\theta}^* \mid \boldsymbol{\theta}^{(t-1)}) = \min\left\{ \frac{p(\boldsymbol{\theta}^* \mid \boldsymbol{y})}{p(\boldsymbol{\theta}^{(t-1)} \mid \boldsymbol{y})}, 1 \right\}.$$

This is the original Metropolis algorithm, published in 1953, and generalized to nonsymmetric proposal densities by Hastings (1970).

2.1 Metropolis–Hastings Algorithm

To draw random numbers from the density $p(\boldsymbol{\theta} \mid \boldsymbol{y})$, the Metropolis–Hastings algorithm proceeds as follows:

1. Initialize $\boldsymbol{\theta}^{(0)}$ and the number T of iterations. Set $t = 1$.
2. Draw a random number $\boldsymbol{\theta}^*$ from the proposal density $q(\boldsymbol{\theta}^* \mid \boldsymbol{\theta}^{(t-1)})$ and accept it as the new state $\boldsymbol{\theta}^{(t)}$ with probability $\alpha(\boldsymbol{\theta}^* \mid \boldsymbol{\theta}^{(t-1)})$; otherwise set $\boldsymbol{\theta}^{(t)} = \boldsymbol{\theta}^{(t-1)}$.
3. Stop if $t = T$; otherwise set $t = t + 1$ and go to 2.

After a *burn-in phase* t_0, the random numbers $\boldsymbol{\theta}^{(t_0+1)}, \ldots, \boldsymbol{\theta}^{(T)}$ can be considered as a (correlated) sample from $p(\boldsymbol{\theta} \mid \boldsymbol{y})$.

Note that numerical computation of α is usually problematic in practice because the densities contain exponential functions. It is therefore more favorable to base acceptance/rejection on $\log(\alpha)$. Then in practice we generate a random number $u \sim U(0, 1)$ (uniform on $(0, 1)$) and accept $\boldsymbol{\theta}^*$ if $\log(u) \leq \log(\alpha)$ thereby avoiding the evaluation of exponential functions.

Although any proposal density will ultimately deliver samples from the posterior, it is important to choose appropriate proposal densities. In particular, acceptance probabilities should be large enough and correlation between successive iterates should be small. The smaller the correlation, the smaller is the required sample size to estimate characteristics of the posterior. We illustrate the construction of a Metropolis–Hastings algorithm for a Bayesian logit model.

Example B.10 Bayesian Logit Model

We consider the following simulated logit model with two covariates x_1 and x_2:

$$\begin{aligned} Y_i &= B(1, \pi_i) \qquad i = 1, \ldots, 500, \\ \pi_i &= \frac{\exp(\eta_i)}{1 + \exp(\eta_i)}, \\ \eta_i &= -0.5 + 0.6\,x_{i1} - 0.3\,x_{i2}. \end{aligned}$$

The covariates x_1 and x_2 are drawn independently from a standard normal distribution. We want to construct a Metropolis–Hastings algorithm to estimate the parameter $\boldsymbol{\beta} = (-0.5, 0.6, -0.3)'$ given this simulated data.

We specify independent diffuse priors $p(\beta_j) \propto$ const. The posterior is then proportional to the likelihood:

$$p(\boldsymbol{\beta} \mid \boldsymbol{y}) \propto \prod_{i=1}^{500} \pi_i^{y_i} (1 - \pi_i)^{1-y_i}.$$

As a proposal density for the Metropolis–Hastings algorithm we choose a three-dimensional normal distribution, with the current state $\boldsymbol{\beta}^{(t-1)}$ as its mean. For its covariance matrix,

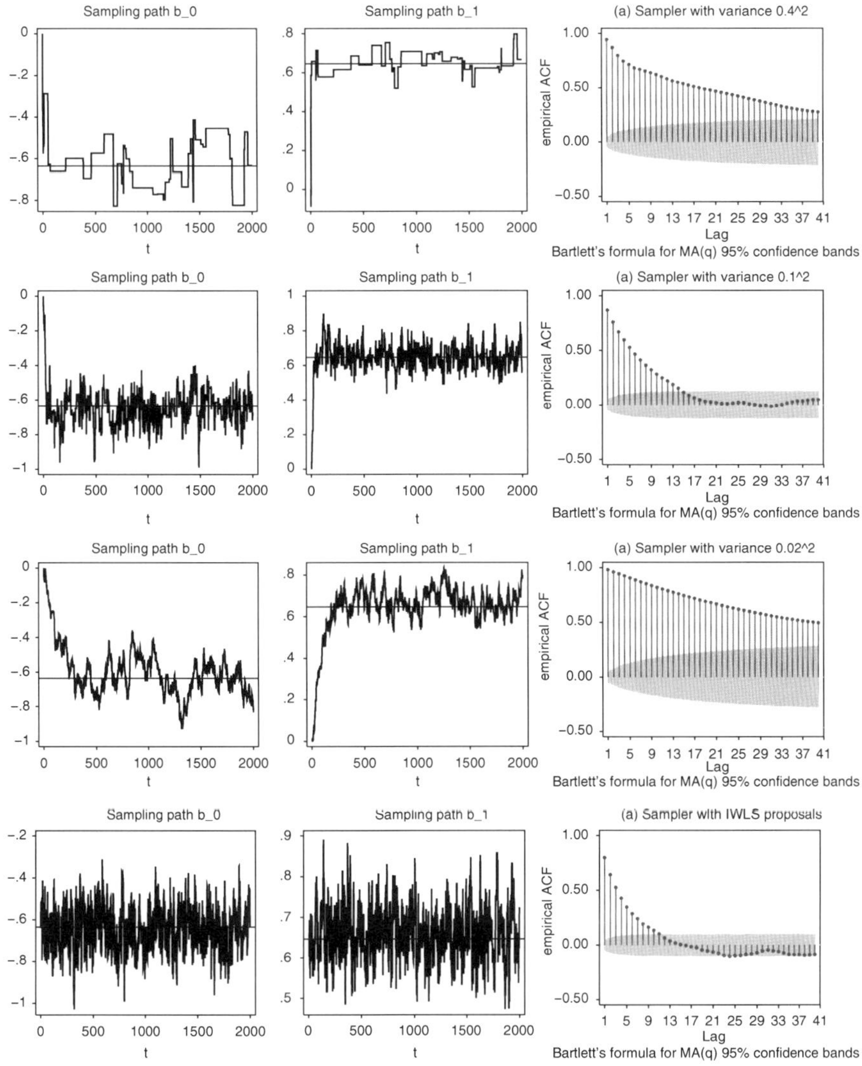

Fig. B.2 Sampling paths for β_0 and β_1 for different MH algorithms. The *third column* shows the respective autocorrelation functions for β_1

we start with the diagonal matrix $\boldsymbol{\Sigma} = \text{diag}(0.4^2, 0.4^2, 0.4^2)$. Figure B.2 (first row) shows the first 2,000 random numbers for β_0 and β_1 drawn with this proposal density. Since we have specified diffuse priors, Bayes estimates for the regression coefficients should not differ too much from the MLEs. Therefore, the MLEs are displayed as horizontal lines in the plots. Clearly, only a few of the proposed random numbers are accepted with this first algorithm; sometimes the state remains unchanged for more than 100 iterations. Thus, the acceptance probabilities are far too small. We obtain larger acceptance probabilities if the variances of the proposal density are decreased to $\boldsymbol{\Sigma} = \text{diag}(0.1^2, 0.1^2, 0.1^2)$. The second row in Fig. B.2 shows the first 2,000 random numbers for β_0 and β_1 resulting from this second MH algorithm. We recognize a short burn-in phase of about 50 iterations, followed by reasonable iterations with relatively large acceptance rates. If we further decrease the variance to $\boldsymbol{\Sigma} = \text{diag}(0.02^2, 0.02^2, 0.02^2)$, acceptance rates are further increased, but successive draws remain almost in the same state; see the third row in Fig. B.2.

A useful and important tool for assessing the quality of MCMC algorithms is the autocorrelation function of the sample [see section "First Order Autocorrelation" of Sect. 4.1.4 (p. 192), for the autocorrelation function]. Ideally, autocorrelations should rapidly converge to zero with increasing lags. The smaller the autocorrelation of successive parameters, the better the characteristics of the posterior can be estimated, based on the same length T of the sample. As an example, the right column of Fig. B.2 shows the autocorrelation functions for β_1, corresponding to the three MH algorithms. Obviously, correlations for the first and third algorithm are extremely large, even for a lag of 40. On the other hand, autocorrelation for lags larger than 20 are almost zero for the second algorithm. For practical work, "thinning" is carried out for the original sample, i.e., only every kth random number is kept in the sample, so that the remaining random numbers are almost uncorrelated. In this way, memory space can be saved without worsening estimation results. To generate an approximately uncorrelated sample of size 1,000 with our second MH algorithm, we would have to generate about 20,000 random numbers after a short burn-in phase and then keep only every 20th random number in the thinned sample.

We can conclude the following: Small variances of the proposal density lead to high acceptance rates. In contrast, acceptance rates become small for large variances. For very large or very small variances autocorrelations of successive random numbers are high. The art of designing good MH algorithms is therefore the choice of appropriate proposal densities that combine high acceptance rates with low autocorrelations. Furthermore, automated methods are desirable that do not require subjective tuning of parameters of the proposal density.

An algorithm with these desirable properties is the MH algorithm described in detail in Sect. 5.6 (p. 311). The last row of Fig. B.2 displays the sampling paths for β_0 and β_1 and the autocorrelation function for β_1. Obviously, this algorithm has the best properties among the four algorithms suggested so far. Autocorrelations are practically zero for lags about 13 and larger. Variances in the proposal density are chosen automatically and correlation of parameters is taken into account.

Using this algorithm a Markov chain was generated and, after the burn-in phase, $20,000$ random numbers were drawn. Saving every 20th random number led to a thinned sample of size 1,000. Based on this thinned sample all characteristics of interest of the posterior can be approximated. To approximate the posterior mean we compute the arithmetic means for the sample, resulting in $\hat{\boldsymbol{\beta}} = (-0.64, 0.65, -0.38)'$. This estimate is very close to the MLEs. Estimation of credible intervals can be based on the quantiles of the sampled random numbers. For example, we obtain 95 % credible intervals by choosing the 2.5 % quantiles as lower and 97.5 % quantiles as upper bounds. This results in the intervals $[-0.87, -0.42]$, $[0.52, 0.78]$, and $[-0.52, -0.26]$ for the sample generated in our example. They are also quite close to the confidence intervals of the regression coefficients obtained by likelihood inference (but have to be interpreted differently).

A further advantage of simulation-based inference is that inference is also easily possible for (complicated) nonlinear transformations of the original parameters. If, for example, we are interested in a credible interval for $\exp(\beta_0 + \beta_1 + \beta_2)$, then we simply compute the transformed quantity for the random numbers from the original sample. A 95 % credible interval for the transformed parameter is simply given by corresponding 2.5 % and 97.5 % quantiles.

△

2.2 Gibbs Sampler

Let $p(\boldsymbol{\theta} \mid \boldsymbol{y})$ be the posterior and assume that $\boldsymbol{\theta}$ is partitioned into S blocks $\boldsymbol{\theta}_1, \dots, \boldsymbol{\theta}_S$. The Gibbs sampler generates random numbers as follows:

1. Specify initial values $\boldsymbol{\theta}_1^{(0)}, \dots, \boldsymbol{\theta}_S^{(0)}$ and the number T of iterations. Set $t = 1$.
2. For $s = 1, \dots, S$, draw random numbers from the full conditionals

$$p(\boldsymbol{\theta}_s \mid \boldsymbol{\theta}_1^{(t)}, \dots, \boldsymbol{\theta}_{s-1}^{(t)}, \boldsymbol{\theta}_{s+1}^{(t-1)}, \dots, \boldsymbol{\theta}_S^{(t-1)}, \boldsymbol{y}).$$

   Note that the most actual states are used in the conditioning set of parameter blocks.
3. Stop if $t = T$; otherwise set $t = t + 1$ and go to 2.

Gibbs Sampler and Hybrid Algorithms

In many practical applications the parameter vector is high-dimensional. The acceptance rates then become rather small even for well-designed MH algorithms, because a high-dimensional random number has to be accepted or not. So-called hybrid algorithms provide a solution to this problem, using a "divide and conquer" strategy. The high-dimensional parameter vector $\boldsymbol{\theta}$ is partitioned into smaller blocks $\boldsymbol{\theta}_1, \boldsymbol{\theta}_2, \dots, \boldsymbol{\theta}_S$. Separate MH steps are then constructed for these subvectors.

The *Gibbs sampler* is the simplest special case of this strategy. In most cases no (simple) methods for directly drawing random numbers from the density $p(\boldsymbol{\theta} \mid \boldsymbol{y})$ of the entire parameter vector are available. Often, however, random numbers can be directly drawn from the conditional densities $p(\boldsymbol{\theta}_1 \mid \cdot), p(\boldsymbol{\theta}_2 \mid \cdot), \dots, p(\boldsymbol{\theta}_S \mid \cdot)$, where $p(\boldsymbol{\theta}_s \mid \cdot)$ denotes the conditional density of $\boldsymbol{\theta}_s$ given all other blocks $\boldsymbol{\theta}_1, \dots, \boldsymbol{\theta}_{s-1}, \boldsymbol{\theta}_{s+1}, \dots, \boldsymbol{\theta}_S$ and the data $\boldsymbol{y}$. These densities are called full conditionals. Gibbs sampling consists in successively drawing random numbers from the full conditionals in each iteration and accepting them (with probability one) as the next state of the Markov chain; see Box 2.2 for a summary. After a burn-in phase the sampled random numbers can be considered as realizations from the marginal posteriors $p(\boldsymbol{\theta}_1 \mid \boldsymbol{y}), p(\boldsymbol{\theta}_2 \mid \boldsymbol{y}), \dots, p(\boldsymbol{\theta}_S \mid \boldsymbol{y})$.

If it is not possible to directly draw random numbers from some of the full conditionals, then an MH step is included instead. For the corresponding block $\boldsymbol{\theta}_s$, a proposal density

$$q_s(\boldsymbol{\theta}_s^* \mid \boldsymbol{\theta}_1^{(t)}, \dots, \boldsymbol{\theta}_{s-1}^{(t)}, \boldsymbol{\theta}_s^{(t-1)}, \dots, \boldsymbol{\theta}_S^{(t-1)})$$

is chosen and random numbers $\boldsymbol{\theta}_s^*$ are drawn from it. They are accepted as new states of the Markov chain with probability

$$\alpha(\boldsymbol{\theta}_s^* \mid \boldsymbol{\theta}_s^{(t-1)}) = \min\left\{\frac{p(\boldsymbol{\theta}_s^* \mid \boldsymbol{\theta}_{-s}^{(t-1)})q_s(\boldsymbol{\theta}_s^{(t-1)} \mid \boldsymbol{\theta}_1^{(t)},\ldots,\boldsymbol{\theta}_{s-1}^{(t)},\boldsymbol{\theta}_s^*,\ldots,\boldsymbol{\theta}_S^{(t-1)})}{p(\boldsymbol{\theta}_s^{(t-1)} \mid \boldsymbol{\theta}_{-s}^{(t-1)})q_s(\boldsymbol{\theta}_s^* \mid \boldsymbol{\theta}_1^{(t)},\ldots,\boldsymbol{\theta}_{s-1}^{(t)},\boldsymbol{\theta}_s^{(t-1)},\ldots,\boldsymbol{\theta}_S^{(t-1)})}, 1\right\},$$

where $p(\boldsymbol{\theta}_s \mid \boldsymbol{\theta}_{-s}^{(t-1)}) = p(\boldsymbol{\theta}_s \mid \boldsymbol{\theta}_1^{(t)},\ldots,\boldsymbol{\theta}_{s-1}^{(t)},\boldsymbol{\theta}_{s+1}^{(t-1)},\ldots,\boldsymbol{\theta}_S^{(t-1)},\boldsymbol{y})$ denotes the full conditional of $\boldsymbol{\theta}_s$. Otherwise, $\boldsymbol{\theta}_s^{(t)} = \boldsymbol{\theta}_s^{(t-1)}$ as in the original MH algorithm. In summary, step 2 of the Gibbs sampler is replaced by:

2.* For $s = 1,\ldots,S$, draw random numbers $\boldsymbol{\theta}_s^*$ from the proposal densities $q_s(\boldsymbol{\theta}_s^* \mid \cdot)$ and accept them with probability $\alpha(\boldsymbol{\theta}_s^* \mid \boldsymbol{\theta}_s^{(t-1)})$; otherwise set $\boldsymbol{\theta}_s^{(t)} = \boldsymbol{\theta}_s^{(t-1)}$.

Note that MH and Gibbs steps in 2.* can be used in combination. Formally, the corresponding full conditional is directly chosen as proposal density. The acceptance probability then becomes $\alpha(\boldsymbol{\theta}_s^* \mid \boldsymbol{\theta}_s^{(t-1)}) = 1$, so that all proposed values are accepted.

B.5.4 Model Selection

The classical approach for Bayesian model choice is to compare competing models through the *posterior probabilities* of the models. Suppose we are given K competing models $M_1,\ldots,M_K$ with associated parameters $\boldsymbol{\theta}_1,\ldots,\boldsymbol{\theta}_K$. By a "model" we mean a set of probability distributions. For instance M_j, $j = 1,\ldots,K$, could denote the regression models $\boldsymbol{y} \mid \boldsymbol{\theta}_j, M_j \sim \mathrm{N}(\boldsymbol{X}_j\boldsymbol{\theta}_j, \sigma^2\boldsymbol{I})$ (with known variance σ^2 for simplicity). For each model M_j let $p(\boldsymbol{y} \mid \boldsymbol{\theta}_j, M_j)$ denote the observation model and $p(\boldsymbol{\theta}_j \mid M_j)$ be the prior for the model parameters $\boldsymbol{\theta}_j$. The posterior for $\boldsymbol{\theta}_j$ under model M_j is then given by

$$p(\boldsymbol{\theta}_j \mid \boldsymbol{y}, M_j) = \frac{p(\boldsymbol{y} \mid \boldsymbol{\theta}_j, M_j)\, p(\boldsymbol{\theta}_j \mid M_j)}{p(\boldsymbol{y} \mid M_j)},$$

where

$$p(\boldsymbol{y} \mid M_j) = \int p(\boldsymbol{y} \mid \boldsymbol{\theta}_j, M_j)\, p(\boldsymbol{\theta}_j \mid M_j)\, d\boldsymbol{\theta}_j \tag{B.5}$$

is the *marginal likelihood*. For model selection, we additionally have to assign prior probabilities $p(M_j)$ associated with each model M_j. Now the competing models can be compared through the posterior model probabilities given by

$$p(M_j \mid \boldsymbol{y}) = \frac{p(\boldsymbol{y} \mid M_j)\, p(M_j)}{p(\boldsymbol{y})} \propto p(\boldsymbol{y} \mid M_j)\, p(M_j)$$

with

$$p(\boldsymbol{y}) = \sum_{k=1}^{K} p(\boldsymbol{y} \mid M_k)\, p(M_k).$$

We prefer model M_j against model M_s if $p(M_j \mid \boldsymbol{y}) > p(M_s \mid \boldsymbol{y})$ or the posterior ratio

$$\frac{p(M_j \mid \boldsymbol{y})}{p(M_s \mid \boldsymbol{y})} = \frac{p(M_j)}{p(M_s)} \frac{p(\boldsymbol{y} \mid M_j)}{p(\boldsymbol{y} \mid M_s)}$$

is larger than 1. In cases of equal priors $p(M_1) = p(M_2) = \ldots = p(M_K) = 1/K$ the posterior ratio simplifies to the *Bayes factor*

$$BF_{js}(\boldsymbol{y}) = \frac{p(\boldsymbol{y} \mid M_j)}{p(\boldsymbol{y}|M_s)}.$$

If none of the competing models is favored prior to analysis of the data, model choice is based on Bayes factors.

Care has to be taken when using the Bayes factor in combination with improper priors. Suppose, we assume improper priors

$$p(\boldsymbol{\theta}_j \mid M_j) \propto c_j, \qquad p(\boldsymbol{\theta}_s \mid M_s) \propto c_s$$

for the model parameters $\boldsymbol{\theta}_j$ and $\boldsymbol{\theta}_s$ of models M_j and M_s. Here c_j and c_s are arbitrary constants. The Bayes factor then becomes

$$BF_{js}(\boldsymbol{y}) = \frac{c_j}{c_s} \frac{\int p(\boldsymbol{y} \mid \boldsymbol{\theta}_j, M_j)\, d\boldsymbol{\theta}_j}{\int p(\boldsymbol{y} \mid \boldsymbol{\theta}_s, M_s)\, d\boldsymbol{\theta}_s},$$

where c_j/c_s is an arbitrary constant. This in turn implies that the Bayes factor is not uniquely defined. Thus, improper priors cannot be used in the context of Bayesian model choice, at least if Bayes factors or in other words marginal likelihoods are involved. Improper priors may be appropriate only if the parameter $\boldsymbol{\theta}$ is the "same" under all models under consideration.

In many applications the exact computation of Bayes factors is difficult because computation of the marginal model likelihoods $p(\boldsymbol{y} \mid M_j)$ causes problems. An approximation (after multiplying with -2) is

$$-2p(\boldsymbol{y} \mid M_j) \approx -2 \cdot \log(p(\boldsymbol{y} \mid \hat{\boldsymbol{\theta}}_j, M_j)) + \log(n)\, p_j,$$

where p_j is the dimension of the parameter vector $\boldsymbol{\theta}_j$ and $\hat{\boldsymbol{\theta}}_j$ is the posterior mode. The approximation can be derived through a Laplace approximation of the integral in Eq. (B.5) and leads to the Bayesian information criterion (BIC). For a model with parameter $\boldsymbol{\theta}$, log-likelihood $l(\boldsymbol{\theta})$, and MLE $\hat{\boldsymbol{\theta}}$ the BIC is defined as

$$BIC = -2l(\hat{\boldsymbol{\theta}}) + \log(n)\, p.$$

Among a set of competing models, the model with the smallest BIC will be selected. Formally, the BIC is similar to the AIC: the factor 2 multiplying the number or parameters in the AIC is replaced by $\log(n)$. Derivation of the two criteria is quite different, however. In general, the BIC selects less complex models than the AIC because penalization of the number of parameters is stronger.

Although derived from a Bayesian perspective, the BIC is not very popular in Bayesian inference. The main reasons are: First, the assumptions underlying the derivation of the BIC as an approximation of marginal log-likelihoods are not sufficiently well fulfilled in complex high-dimensional models. Related to this is the problem of determining n in the factor $\log n$. It is not always the data sample size: For example, in mixed models n is the number of individuals. Second, when more complex Bayesian models are fitted with MCMC methods, the BIC is not directly available anyway.

A more recent criterion for model choice that has become quite popular in connection with MCMC inference is the deviance information criterion (DIC). Refer to Spiegelhalter, Best, Carlin, and van der Linde (2002). Its popularity is due to the fact that it can be easily computed from MCMC output. Let $\boldsymbol{\theta}^{(1)}, \dots, \boldsymbol{\theta}^{(T)}$ denote an MCMC sample from the posterior of the model. Computation of the DIC is based on two quantities: The first is the (unstandardized) deviance

$$D(\boldsymbol{\theta}) = -2\log(p(\boldsymbol{y} \mid \boldsymbol{\theta}))$$

of the model. The second is the *effective number* p_D of parameters in the model. It can be estimated through

$$p_D = \overline{D(\boldsymbol{\theta})} - D(\bar{\boldsymbol{\theta}}),$$

where

$$\overline{D(\boldsymbol{\theta})} = \frac{1}{T}\sum_{t=1}^{T} D(\boldsymbol{\theta}^{(t)})$$

is the average posterior deviance and $D(\bar{\boldsymbol{\theta}})$ is the deviance evaluated at $\bar{\boldsymbol{\theta}} = \frac{1}{T}\sum_{t=1}^{T}\boldsymbol{\theta}^{(t)}$. The DIC is then defined as

$$\mathrm{DIC} = \overline{D(\boldsymbol{\theta})} + p_D = 2\overline{D(\boldsymbol{\theta})} - D(\bar{\boldsymbol{\theta}}).$$

As a disadvantage, the DIC value changes for different MCMC random samples. Therefore it may happen that model choice by DIC can lead to selecting different models with different MCMC samples. This will be only the case, however, if the DIC values of the models are quite close. It then will be better to keep both (or several) competing models under consideration.

Example B.11 Bayesian Logit Model—DIC

We illustrate the use of DIC with the simulated data from Example B.10. If we mistakenly omit the covariate x_2 and fit a logit model with x_1 only, then we obtain the (estimated) values $p_D = 1.99$ and $\mathrm{DIC} = 571.6$. The effective number of parameters of about 2 is plausible, because we have estimated exactly two parameters β_0 and β_1. Fitting the correctly specified model, we obtain $p_D = 2.93$ and $\mathrm{DIC} = 540.3$. The effective number of parameters is now about 3, as had to be expected. The DIC is now considerably smaller than for the wrong model so that the more complex, true model is selected.

For illustration, we fit the correct model with five further MCMC runs. For p_D we obtain the values 3.05, 2.99, 3.15, 2.87, and 3.23 and for the DIC 540.56, 540.42, 540.73, 540.19, and 540.91, respectively. We see that the DIC varies between the different MCMC runs, but variability is usually quite low.

△

B.5.5 Model Averaging

If the focus of the analysis is not necessarily on selecting a single best model, model averaging could be a promising alternative. Suppose we are interested in inference regarding a quantity Δ. If we perform model selection, we first determine the model M_* with highest posterior probability. The inference is then based on the posterior $p(\Delta \mid \boldsymbol{y}, M_*)$ of Δ conditional on the "best" model M_*. Alternatively, inference could be based on the unconditional posterior $p(\Delta \mid \boldsymbol{y})$ which is given by

$$p(\Delta \mid \boldsymbol{y}) = \sum_{k=1}^{K} p(\Delta \mid \boldsymbol{y}, M_k)\, p(M_k \mid \boldsymbol{y}).$$

Hence, $p(\Delta \mid \boldsymbol{y})$ is obtained as a simple average of the posteriors $p(\Delta \mid \boldsymbol{y}, M_k)$ under model M_k weighted by the posterior model probabilities $p(M_k \mid \boldsymbol{y})$. Model averaging might be particularly favorable (in terms of predictive power) compared to selecting a single best model if the posterior probability mass is not concentrated on a particular model.

Bibliography

Agresti, A. (2002). *Categorical data analysis* (2nd ed.). New York: Wiley.

Agresti, A., & Finlay, B. (2008). *Statistical methods for the social sciences* (4th ed.). Upper Saddle River: Prentice Hall.

Albert, J., & Chib, S. (1993). Bayesian analysis of binary and polychotomous response data. *Journal of the American Statistical Association, 88*, 669–679.

Anderson, T. W. (2003). *An introduction to multivariate statistical analysis*. Dordrecht: Kluwer.

Anselin, L. (1988). *Spatial econometrics: Methods and models*. Dordrecht: Kluwer.

Augustin, N. H., Lang, S., Musio, M., & von Wilpert, K. (2007). A spatial model for the needle losses of pine-trees in the forests of Baden-Württemberg: An application of Bayesian structured additive regression. *Applied Statistics, 56*, 29–50.

Banerjee, S., Carlin, B. P., & Gelfand, A. E. (2003). *Hierarchical modelling and analysis for spatial data*. New York/Boca Raton: Chapman & Hall/CRC.

Barbieri, M., & Berger, J. (2004). Optimal predictive model selection. *The Annals of Statistics, 32*, 870–897.

Belitz, C., & Lang, S. (2008). Simultaneous selection of variables and smoothing parameters in structured additive regression models. *Computational Statistics and Data Analysis, 53*, 61–81.

Belsley, D. A., Kuh, E., & Welsch, R. E. (2003). *Regression diagnostics: Identifying influential data and sources of collinearity*. New York: Wiley.

Bernardo, J. (1979). Reference posterior distributions for Bayesian inference. *Journal of the Royal Statistical Society B, 41*, 113–147.

Besag, J., Green, P., Higdon, D., & Mengersen, K. (1995). Bayesian computation and stochastic systems. *Statistical Science, 10*, 3–66.

Biller, C. (2000). Adaptive Bayesian regression splines in semiparametric generalized linear models. *Journal of Computational and Graphical Statistics, 9*, 122–140.

Biller, C., & Fahrmeir, L. (2001). Bayesian varying-coefficient models using adaptive regression splines. *Statistical Modelling, 1*, 195–211.

Bondell, H. D., & Reich, B. J. (2008). Simultaneous regression shrinkage, variable selection, and supervised clustering of predictors with OSCAR. *Biometrics, 64*, 115–123.

Box, G., & Tiao, G. (1992). *Bayesian inference in statistical analysis*, New York: Wiley.

Breiman, L. (1996). Bagging predictors. *Machine Learning, 24*, 123–140.

Breiman, L. (2001). Random forests. *Machine Learning, 45*, 5–32.

Breiman, L., Friedman, J., Stone, C. J., & Olshen, R. A. (1984). *Classification and regression trees*. New York/Boca Raton: Chapman & Hall/CRC.

Breslow, N. E., & Clayton, D. G. (1993). Approximate inference in generalized linear mixed models. *Journal of the American Statistical Association, 88*, 9–25.

Breusch, T., & Pagan, A. (1979). A simple test for heteroscedasticity and random coefficient variation. *Econometrica, 47*, 1287–1294.

Brezger, A., & Lang, S. (2006). Generalized additive regression based on Bayesian P-splines. *Computational Statistics and Data Analysis, 50*, 967–991.

L. Fahrmeir et al., *Regression*, DOI 10.1007/978-3-642-34333-9,

Brezger, A., & Steiner, W. (2008). Monotonic regression based on Bayesian P-splines: An application to estimating price response functions from store-level scanner data. *Journal of Business and Economic Statistics, 26*, 90–104.

Brockwell, P. J., & Davis, R. A. (2002). *Introduction to time series and forecasting* (2nd ed.). New York: Springer.

Bühlmann, P. (2006). Boosting for high-dimensional linear models. *Annals of Statistics, 34*, 559–583.

Bühlmann, P., & Hothorn, T. (2007). Boosting algorithms: Regularization, prediction and model fitting (with discussion). *Statistical Science, 22*, 477–505.

Bühlmann, P., & Yu, B. (2003). Boosting with the L2 loss: Regression and classification. *Journal of the American Statistical Association, 98*, 324–339.

Bühlmann, P., & Yu, B. (2006). Sparse boosting. *Journal of Machine Learning Research, 7*, 1001–1024.

Buja, A., Hastie, T., & Tibshirani, R. (1989). Linear smoothers and additive models. *Annals of Statistics, 17*, 453–510.

Burnham, K. P., & Anderson, D. R. (2002). *Model selection and multimodal inference* (2nd ed.). New York: Springer.

Cameron, A. C., & Trivedi, P. K. (1998). *Regression analysis of count data*. Cambridge: Cambridge University Press.

Cameron, A. C., & Trivedi, P. K. (2005). *Microeconometrics: Methods and applications*. Cambridge: Cambridge University Press.

Carroll, R. J., Ruppert, D., Stefanski, L. A., & Crainiceanu, C. M. (2006). *Measurement error in nonlinear models* (2nd ed.). New York/Boca Raton: Chapman & Hall/CRC.

Chen, M. H., & Dey, D. K. (2000). Bayesian analysis for correlated ordinal data models. In D. K. Dey, S. K. Ghosh, & B. K. Mallick (Eds.), *Generalized linear models: A Bayesian perspective* (pp. 133–159). New York: Marcel Dekker.

Chen, Z., & Dunson, D. B. (2003). Random effects selection in linear mixed models. *Biometrics, 59*, 762–769.

Chiles, J.-P., & Delfiner, P. (1999). *Geostatistics: Modeling spatial uncertainty*. New York: Wiley.

Cleveland, W. S. (1979). Robust locally weighted regression and smoothing scatterplots. *Journal of the American Statistical Association, 74*, 829–836.

Collett, D. (1991). *Modelling binary data*. New York/Boca Raton: Chapman & Hall/CRC.

Collett, D. (2003). *Modelling survival data in medical research* (2nd ed.). New York/Boca Raton: Chapman & Hall/CRC.

Cowles, M., & Carlin, B. (1996). Markov chain Monte Carlo convergence diagnostics: A comparative review. *Journal of the American Statistical Association, 91*, 883–904.

Crainiceanu, C. M., & Ruppert, D. (2004). Likelihood ratio tests in linear mixed models with one variance component. *Journal of the Royal Statistical Society B, 66*, 165–185.

Crainiceanu, C. M., & Ruppert, D. (2004a). Restricted likelihood ratio tests in nonparametric longitudinal models. *Statistica Sinica, 14*, 713–729.

Crainiceanu, C., Ruppert, D., Claeskens, G., & Wand, M. (2005). Exact likelihood ratio tests for penalised splines. *Biometrika 92*, 91–103.

Crainiceanu, C. M., Ruppert, D., Carroll, R. J., Adarsh, J., & Goodner, B. (2007). Spatially adaptive penalized splines with heteroscedastic errors. *Journal of Computational and Graphical Statistics, 16*, 265–288.

De Boor, C. (2001). *A practical guide to splines*. New York: Springer.

Dellaportas, P., & Smith, A. F. M. (1993). Bayesian inference for generalized linear and proportional hazards models via Gibbs sampling. *Applied Statistics, 42*, 443–459.

Demidenko, E. (2004). *Mixed models*. New York: Wiley.

Denison, D., Holmes, C. C., Mallick, B. K., & Smith, A. F. M. (2002). *Bayesian methods for nonlinear classification and regression*. New York: Wiley.

Denison, D. G. T., Mallick, B. K., & Smith, A. F. M. (1998). Automatic Bayesian curve fitting. *Journal of the Royal Statistical Society B, 60*, 333–350.

Denuit, M., & Lang, S. (2005). Nonlife ratemaking with Bayesian GAM's. *Insurance: Mathematics and Economics, 35*, 627–647.
Dey, D., Gosh, S. K., & Mallick, B. K. (2000). *Generalized linear models: A Bayesian Perspective*. New York: Marcel Dekker.
Dierckx, P. (1993). *Curve and surface fitting with splines*. Oxford: Clarendon Press.
Diggle, P. J., Heagerty, P., Liang, K. L., & Zeger, S. L. (2002). *Analysis of longitudinal data* (2nd ed.). Oxford: Oxford University Press.
DiMatteo, I., Genovese, C. R., & Kass, R. E. (2001). Bayesian curve-fitting with free-knot splines. *Biometrika 88*, 1055–1071.
Drygas, H. (1976). Weak and strong consistency of the least squares estimators in regression models. *Zeitschrift fr Wahrscheinlichkeitstheorie und verwandte Gebiete, 34*, 119–127.
Durbin, J., & Watson, G. (1950). Testing for serial correlation in least squares regression–I. *Biometrika, 37*, 409–428.
Durbin, J., & Watson, G. (1951). Testing for serial correlation in least squares regression–II, *Biometrika, 38*, 159–178.
Durbin, J., & Watson, G. (1971). Testing for serial correlation in least squares regression–III, *Biometrika, 58*, 1–42.
Ecker, M. D., & Gelfand, A. E. (2003). Spatial modelling and prediction under stationary non-geometric range anisotropy. *Environmental and Ecological Statistics, 10*, 165–178.
Efron, B., Hastie, T., Johnstone, I., & Tibshirani, R. (2004). Least angle regression. *Annals of Statistics, 32*, 407–451.
Eilers, P. H. C., & Marx, B. D. (1996). Flexible smoothing using B-splines and penalized likelihood. *Statistical Science, 11*, 89–121.
Eilers, P. H. C., & Marx, B. D. (2002). Generalized linear additive smooth structures. *Journal of Computational and Graphical Statistics, 11*, 758–783.
Eilers, P. H. C., & Marx, B. D. (2003). Multidimensional calibration with temperature interaction using two-dimensional penalized signal regression. *Chemometrics and Intelligent Laboratory Systems, 66*, 159–174.
Eilers, P. H. C., & Marx, B.D. (2010). Splines, knots, and penalties. *Wiley interdisciplinary reviews: Computational statistics*. New York: Wiley.
Fahrmeir, L., Hamerle, A., & Tutz, G. (1996). *Multivariate Statistische Verfahren (second edition)*, De Gruyter.
Fahrmeir, L., & Kaufmann, H. (1985). Consistency and asymptotic normality of the maximum likelihood estimator in generalized linear models. *The Annals of Statistics, 13*, 342–368.
Fahrmeir, L., & Kneib, T. (2008). On the identification of trend and correlation in temporal and spatial regression. In Shalabh and C. Heumann (Eds.), *Recent advances in linear models and related areas* (pp. 1–28). Springer.
Fahrmeir, L., & Kneib, T. (2011). *Bayesian smoothing and regression for longitudinal, spatial and event history data*. Oxford: Oxford University Press.
Fahrmeir, L., Kneib, T., & Konrath, S. (2010). Bayesian regularization in structured additive regression: A unifying perspective on shrinkage, smoothing and predictor selection. *Statistics and Computing, 20*, 203–219.
Fahrmeir, L., Kneib, T., & Lang, S. (2004). Penalized structured additive regression for space-time data: A Bayesian perspective. *Statistica Sinica, 14*, 731–761.
Fahrmeir, L., Künstler, R., Pigeot, I., & Tutz, G. (2007). *Statistik: Der weg zur datenanalyse* (6th ed.). Berlin: Springer.
Fahrmeir, L., & Lang, S. (2001). Bayesian semiparametric regression analysis of multicategorical time-space data. *Annals of the Institute of Statistical Mathematics, 53*, 11–30.
Fahrmeir, L., & Tutz, G. (2001). *Multivariate statistical modelling based on generalized linear models* (2nd ed.). Berlin: Springer.
Fan, J., & Gijbels, I. (1996). *Local polynomial modelling and its applications*. Chapman & Hall/CRC.
Fan, J., & Li, R. (2001). Variable selection via nonconcave penalized likelihood and its oracle properties. *Journal of the American Statistical Association, 96*, 1348–1360.

Faraway, J. J. (2004). *Linear models with R*. New York/Boca Raton: Chapman & Hall/CRC.
Fenske, N., Kneib, T., & Hothorn, T. (2011). Identifying risk factors for severe childhood malnutrition by boosting additive quantile regression. *Journal of the American Statistical Association, 106*, 494–510.
Fernandez, C., Ley, E., & Steel, M. F. J. (2001). Benchmark priors for Bayesian model averaging. *Journal of Econometrics, 100*, 381–427.
Fisman, R., Iyengar, S. S., Kamenica, E., & Simonson, I. (2006). Gender differences in mate selection: Evidence from a speed dating experiment. *Quarterly Journal of Economics, 121*, 673–697.
Fitzmaurice, G., Davidian, M., Verbeke, G., & Molenberghs, G. (2003). *Longitudinal data analysis: A handbook of modern statistical methods*. New York/Boca Raton: Chapman & Hall/CRC.
Forthofer, R. N., & Lehnen, R. G. (1981). *Public program analysis: A new categorical data approach*. Belmont, CA: Lifetime Learning Publications.
Foster, D., & George, E. (1994). The risk inflation criterion for multiple regression. *The Annals of Statistics, 22*, 1947–1975.
Fotheringham, A., Brunsdon, C., & Charlton, M. (2002). *Geographically weighted regression: The analysis of spatially varying relationships*. New York: Wiley.
Friedman, J. (2001). Greedy function approximation: A gradient boosting machine. *The Annals of Statistics, 29*, 1189–1232.
Friedman, J. H. (1991). Multivariate adaptive regression splines. *The Annals of Statistics, 19*, 1–141.
Friedman, J., Hastie, T., & Tibshirani, R. (2000). Additive logistic regression: A statistical view of boosting. *The Annals of Statistics, 28*, 337–407.
Frühwirth-Schnatter, S., & Frühwirth, R. (2010). Data augmentation and MCMC for binary and multinomial logit models. In T. Kneib, & G. Tutz (Eds.), *Statistical modelling and regression structures: Festschrift in honour of Ludwig Fahrmeir* (pp. 111–132). Heidelberg: Springer.
Frühwirth-Schnatter, S., Frühwirth, R., Held, L., & Rue, H. (2009). Improved auxiliary mixture sampling for hierarchical models of non-Gaussian data. *Statistics and Computing, 19*, 479–492.
Frühwirth-Schnatter, S., & Tüchler, R. (2008). Bayesian parsimonious covariance estimation for hierarchical linear mixed models. *Statistics and Computing, 18*, 1–13.
Frühwirth-Schnatter, S., & Wagner, H. (2006). Gibbs sampling for parameter-driven models of time series of small counts with application to state space modelling. *Biometrika, 93*, 827–841.
Fu, W. J. (1998). Penalized regression: The bridge versus the LASSO. *Journal of Computational and Graphical Statistics, 7*, 397–416.
Furnival, G. M., & Wilson, R. W. (1974). Regression by leaps and bounds. *Technometrics, 16*, 499–511.
Galton, F. (1889). *Natural inheritance*. London: Macmillan.
Gamerman, D. (1997). Efficient sampling from the posterior distribution in generalized linear mixed models. *Statistics and Computing, 7*, 57–68.
Gelfand, A. E., & Smith, A. F. M. (1990). Sampling-based approaches to calculating marginal densities. *Journal of the American Statistical Association, 85*, 398–409.
Gelman, A. (2006). Prior distributions for variance parameters in hierarchical models. *Bayesian Analysis, 1*, 515–534.
Gelman, A., Carlin, J., Stern, H., & Rubin, D. (2003). *Bayesian data analysis*. New York/ Boca Raton: Chapman & Hall/CRC.
Gelman, A., & Hill, J. (2006). *Data analysis using regression and multilevel/hierarchical models*. Cambridge: Cambridge University Press.
Gençay, R., Selçuk, F., & Whitcher, B. (2002). *An introduction to wavelets and other filtering methods in finance and economics*. San Diego: Academic.
George, A., & Liu, J. W. (1981). *Computer solution of large sparse positive definite systems*. Upper Saddle River: Prentice Hall.
George, E., & Foster, D. (2000). Calibration and empirical Bayes variable selection. *Biometrica, 87*, 731–747.

George, E., & Mc Culloch, R. (1993). Variable selection via Gibbs sampling. *Journal of the American Statistical Association, 88*, 881–889.

George, E., & Mc Culloch, R. (1997). Approaches for Bayesian variable selection. *Statistica Sinica 7*, 339–373.

Geweke, J. (1991). Efficient simulation from the multivariate normal and student-t distribution subject to linear constraints. *Computer science and statistics: Proceedings of the twenty-third symposium on the interface* (pp. 571–578). Alexandria.

Giampaoli, V., & Singer, V. M. (2009). Likelihood ratio tests for variance components in linear mixed models. *Journal of Statistical Planning and Inference, 139*, 1435–1448.

Gilks, W. R., Richardson, S., & Spiegelhalter, D. J. (Eds.) (1996). *Markov chain Monte Carlo in practice*. New York/Boca Raton: Chapman & Hall/CRC.

Goeman, J. J. (2010). L1 penalized estimation in the Cox proportional hazards model. *Biometrical Journal, 52*, 70–84.

Graybill, F. A. (1961). *An introduction to linear statistical models*. New York: McGraw-Hill.

Green, P. J. (1995). Reversible jump Markov chain Monte Carlo computation and Bayesian model determination. *Biometrika, 82*, 711–732.

Green, P. J. (2001). A primer on Markov chain Monte Carlo. In O. Barndorff-Nielsen, D. Cox, & C. Klüppelberg (Eds.), *Complex stochastic systems* (pp. 1–51). New York/Boca Raton: Chapman & Hall/CRC.

Green, P. J., & Silverman, B. W. (1993). *Nonparametric regression and generalized linear models*. New York/Boca Raton: Chapman & Hall/CRC.

Greene, W. H. (2000). *Econometric analysis* (4th ed.). Upper Saddle River: Prentice Hall.

Greven, S., & Kneib, T. (2010). On the behavior of marginal and conditional Akaike information criteria in linear mixed models. *Biometrika, 97*, 773–789.

Gu, C. (2002). *Smoothing spline ANOVA models*. New York: Springer.

Hamilton, J. D. (1994). *Time series analysis*. Princeton: Princeton University Press.

Hansen, M. H., & Kooperberg, C. (2002). Spline adaptation in extended linear models. *Statistical Science, 17*, 2–51.

Härdle, W. (1990). *Smoothing techniques*. New York: Springer.

Härdle, W., Müller, M., Sperlich, S., & Werwatz, A. (2004). *Nonparametric and semiparametric models*. Heidelberg: Springer.

Harville, D. A. (1974). Bayesian inference for variance components using only error contrasts. *Biometrika, 61*, 383–385.

Harville, D. A. (1977). Maximum likelihood approaches to variance component estimation and to related problems. *Journal of the American Statistical Association, 72*, 320–338.

Hastie, T. J., & Tibshirani, R. J. (1990). *Generalized additive models*. New York/Boca Raton: Chapman & Hall/CRC.

Hastie, T. J., Tibshirani, R. J., & Friedman, J. (2009). *The elements of statistical learning*. Berlin: Springer.

Hastings, W. K. (1970). Monte Carlo sampling methods using Markov chains and their applications. *Biometrika, 57*, 97–109.

Hausmann, J. A. (1978). Specification tests in econometrics. *Econometrica, 46*, 1251–1271.

He, X. (1997). Quantile curves without crossing. *The American Statistician, 51*, 186–192.

Held, L., & Sabanés Bové, D. (2012). *Applied statistical inference*. Berlin: Springer.

Hodges, J. S., & Reich, B. J. (2010). Adding spatially-correlated errors can mess up the fixed effect you love. *The American Statistician, 64*, 325–334.

Hodrick, R. J., & Prescott, E. C. (1997). Postwar U.S. business cycles: An empirical investigation. *Journal of Money, Credit, and Banking, 29*, 1–16.

Hofner, B., Hothorn, T., Schmid, M., & Kneib, T. (2012). A framework for unbiased model selection based on boosting. *Journal of Computational and Graphical Statistics, 20*, 956–971.

Holmes, C. C., & Held, L. (2006). Bayesian auxiliary variable models for binary and multinomial regression. *Bayesian Analysis, 1*, 145–168.

Hosmer, D. W., Lemeshow, S., & May, S. (2008). *Applied survival analysis: Regression modeling of time to event data*. New York: Wiley.

Hothorn, T., Hornik, K., & Zeileis, A. (2006). Unbiased recursive partitioning: A conditional inference framework. *Journal of Computational and Graphical Statistics*, *15*, 651–674.

Hsiao, C. (2003). *Analysis of panel data*. Cambridge: Cambridge University Press.

Imai, K., & van Dyk, D. A. (2005). A Bayesian analysis of the multinomial probit model using marginal data augmentation. *Journal of Econometrics*, *124*, 311–334.

Ishwaran, H., & Rao, S. (2003). Detecting differentially expressed genes in microarrays using Bayesian model selection. *Journal of the American Statistical Association*, *98*, 438–455.

Ishwaran, H., & Rao, S. (2005). Spike and slab variable selection: Frequentist and Bayesian strategies. *Annals of Statistics*, *33*, 730–773.

Joe, H. (1997). *Multivariate models and dependence concepts*. New York/Boca Raton: Chapman & Hall/CRC.

Johnson, M. E., Moore, L. M., & Ylvisaker, D. (1990). Minimax and maximin distance designs. *Journal of Statistical Planning and Inference*, *26*, 131–148.

Judge, G., Griffith, W., Hill, R., Lütkepohl, H., & Lee, T. (1980). *The theory and practice of econometrics*. New York: Wiley.

Kauermann, G. (2006). Nonparametric models and their estimation. *Allgemeines Statistisches Archiv*, *90*, 135–150.

Kauermann, G., & Khomski, P. (2006). Additive two way hazards model with varying coefficients. *Computational Statistics and Data Analysis*, *51*, 1944–1956.

Kauermann, G., & Opsomer, J. (2004). Generalized cross-validation for bandwidth selection of backfitting estimates in generalized additive models. *Journal of Computational and Graphical Statistics*, *13*, 66–89.

Kleiber, C., & Zeileis, A. (2008). *Applied econometrics with R*. New York: Springer.

Klein, J. P., & Moeschberger, M. L. (2005). *Survival analysis* (2nd edn.). New York: Springer.

Kneib, T. (2005). *Mixed model based inference in structured additive regression*. Dr. Hut-Verlag München, available from http://edoc.ub.uni-muenchen.de/archive/00005011/.

Kneib, T., & Fahrmeir, L. (2006). Structured additive regression for multicategorical space-time data: A mixed model approach. *Biometrics*, *62*, 109–118.

Kneib, T., & Fahrmeir, L. (2007). A mixed model approach for geoadditive hazard regression. *Scandinavian Journal of Statistics*, *34*, 207–228.

Koenker, R. (2005). *Quantile regression*. Cambridge: Cambridge University Press.

Koenker, R., & Mizera, I. (2004). Penalized triograms: Total variation regularization for bivariate smoothing. *Journal of the Royal Statistical Society B*, *66*, 145–163.

Koenker, R., Ng, P., & Portnoy, S. (1994). Quantile smoothing splines. *Biometrika*, *81*, 673–680.

Koenker, R., & Zeileis, A. (2009). On reproducible econometric research. *Journal of Applied Econometrics*, *24*, 833–847.

Kottas, A., & Krnjajic, M. (2009). Bayesian semiparametric modelling in quantile regression. *Scandinavian Journal of Statistics*, *36*, 297–319.

Krivobokova, T., Kneib, T., & Claeskens, G. (2010). Simultaneous confidence bands for penalized spline estimators. *Journal of the American Statistical Association*, *105*, 852–863.

Lai, T., Robins, H., & Wei, C. (1979). Strong consistency of least squares estimates in multiple regression II. *Journal of Multivariate Analysis*, *9*, 343–361.

Lang, S., & Brezger, A. (2004). Bayesian P-splines. *Journal of Computational and Graphical Statistics*, *13*, 183–212.

Lang, S., Umlauf, N., Wechselberger, P., Harttgen, K., & Kneib, T. (2013). Multilevel structured additive regression. *Statistics and Computing*, to appear.

Lange, K. (2000). *Numerical analysis for statisticians*. New York: Springer.

Lange, K. (2004). *Optimization*. New York: Springer.

Lawson, A. B., & Liu, Y. (2007). Evaluation of Bayesian models for focused clustering in health data. *Environmetrics*, *18*, 871–887.

Leeflang, P. S. H., Wittink, D. R., Wedel, M., & Naert, P. A. (2000). *Building models for marketing decisions*. Dordrecht: Kluwer.

Lenk, P., & DeSarbo, W. (2000). Bayesian inference for finite mixtures of generalized linear models with random effects. *Psychometrika*, *65*, 93–119.

Ley, E., & Steel, F. (2009). On the effect of prior assumptions in Bayesian model averaging with applications to growth regression. *Journal of Applied Econometrics*, *24*, 651–674.

Liang, F., Paulo, R., Molina, G., Clyde, M., & Berger, J. (2008). Mixtures of g-priors for Bayesian variable selection. *Journal of the American Statistical Association*, *103*, 410–423.

Loader, C. (1999). *Local regression and likelihood*. New York: Springer.

Madigan, D., & York, J. (1995). Bayesian graphical models for discrete data. *International Statistical Review*, *63*, 215–232.

Magnus, J. R., & Neudecker, H. (2002). *Matrix differential calculus with applications in statistics and econometrics*. New York: Wiley.

Malsiner-Walli, G., & Wagner, H. (2011). Comparing spike and slab priors for Bayesian variable selection. *Austrian Journal of Statistics*, *40*, 241–264.

Mardia, K. V., Kent, J. T., & Bibby, J. M. (1999). *Multivariate analysis*. London: Academic.

Marx, B. D., & Eilers, P. H. C. (1998). Direct generalized additive modeling with penalized likelihood. *Computational Statistics and Data Analysis*, *28*, 193–209.

McCullagh, P., & Nelder, J. A. (1989). *Generalized linear models* (2nd ed.). New York/ Boca Raton: Chapman & Hall/CRC.

McCulloch, C. E., & Searle, S. R. (2001). *Generalized, linear, and mixed models*. New York: Wiley.

McFadden, D. (1973). Conditional logit analysis of qualitative choice behaviour. In P. Zarembka (Ed.), *Frontiers in Econometrics*. New York: Academic.

McFadden, D. (1984). Econometric analysis of qualitative response models. In Z. Griliches, & M. Intriligator (Eds.), *Handbook of econometrics* (pp. 1395–1457). Amsterdam: North Holland.

Mengersen, K. L., Robert, C., & Guihenneuc-Jouyaux, C. (1999). MCMC convergence diagnostics: A review. In J. M. Bernardo, J. O. Berger, A. P. Dawid, & A. F. M. Smith (Eds.), *Bayesian statistics 6*. Oxford: Oxford University Press.

Metropolis, N., Rosenbluth, A. W., Rosenbluth, M. N., Teller, A. H., & Teller, E. (1953). Equations of state calculations by fast computing machines. *Journal of Chemical Physics*, *21*, 1087–1091.

Migon, H. S., & Gamerman, D. (1999). *Statistical inference: An integrated approach*. London: Arnold.

Miller, A. (2002). *Subset selection in regression*. New York/Boca Raton: Chapman & Hall/CRC.

Nelder, J. A., & Wedderburn, R. W. M. (1972). Generalized linear models. *Journal of the Royal Statistical Society A*, *135*, 370–384.

Newey, W. K., & Powell, J. L. (1987). Asymmetric least squares estimation and testing. *Econometrica*, *55*, 819–847.

Nychka, D. (2000). Spatial-process estimates as smoothers. In M. Schimek (Ed.). *Smoothing and regression: Approaches, computation and application* (pp. 393–424). New York: Wiley.

Nychka, D., & Saltzman, N. (1998). Design of air quality monitoring networks. In D. Nychka, W. W. Piegorsch, & L. H. Cox (Eds.), *Case studies in environmental statistics* (pp. 51–76). New York: Springer.

Ogden, R. T. (1997). *Essential wavelets for statistical applications and data analysis*. Boston: Birkhäuser.

O'Hagan, A. (1994). *Kendall's advanced theory of statistics vol. 2b: Bayesian inference*. London: Arnold.

Paciorek, C. J. (2010). The importance of scale for spatial-confounding bias and precision of spatial regression estimators. *Statistical Science*, *25*, 107–125.

Park, T., & Casella, G. (2008). The Bayesian lasso. *Journal of the American Statistical Association*, *103*, 681–686.

Ramsay, J. O., & Silverman, B. W. (2002). *Applied functional data analysis: Methods and case studies*. New York: Springer.

Ramsay, J. O., & Silverman, B. W. (2005). *Functional data analysis* (2nd ed.). New York: Springer.

Rao, C. R., Toutenburg, H., Shalabh, & Heumann, C. (2008). *Linear models and generalizations - least squares and alternatives*. Berlin: Springer.

Rawlings, J. O., Pantula, S. G., & Dickey, A. D. (2001). *Applied regression analysis* (2nd ed.). New York: Springer.

Reich, B. J., Bondell, H. D., & Wang, H. J. (2010). Flexible Bayesian quantile regression for independent and clustered data. *Biostatistics, 11*, 337–352.

Rigby, R. A., & Stasinopoulos, D. M. (2005). Generalized additive models for location, scale and shape. *Applied Statistics, 54*, 507–554.

Rigby, R. A., & Stasinopoulos, D. M. (2009). *A flexible regression approach using GAMLSS in R*. Available at http://gamlss.org/.

Robert, C. P. (1995). Simulation of truncated normal variables. *Statistics and Computing, 5*, 121–125.

Rousseeuw, P. J., & Leroy, A. M. (2003). *Robust regression and outlier detection*. New York: Wiley.

Rue, H., & Held, L. (2005). *Gaussian Markov random fields*. New York/Boca Raton: Chapman & Hall/CRC.

Ruppert, D., Wand, M. P., & Carroll, R. J. (2003). *Semiparametric regression*. Cambridge: Cambridge University Press.

Schabenberger, O., & Gotway, C. A. (2005). *Statistical methods for spatial data analysis*. New York/Boca Raton: Chapman & Hall/CRC.

Scheipl, F. (2011). spikeslabgam: Bayesian variable selection, model choice and regularization for generalized additive mixed models in R. *Journal of Statistical Software, 43*(14), 1–24.

Scheipl, F., Fahrmeir, L., & Kneib, T. (2012). Spike-and-slab priors for function selection in structured additive regression models. *In revision for Journal of the American Statistical Association.*

Scheipl, S., Greven, S., & Küchenhoff, H. (2008). Size and power of tests for a zero random effect variance or polynomial regression in additive and linear mixed models. *Computational Statistics and Data Analysis, 52*, 3283–3299.

Schnabel, S., & Eilers, P. (2012). Simultaneous estimation of quantile curves using quantile sheets. Advances in Statistical Analysis, DOI: 10.1007/S10182-012-0198-1

Schott, J. R. (2005). *Matrix analysis for statistics*. New York: Wiley.

Scott, M. A., Simonoff, J. S., & Marx, B. D. (2012). *The SAGE handbook of multilevel modeling*. Thousand Oaks: Sage.

Searle, S. R. (2006). *Matrix algebra useful for statistics*. New York: Wiley.

Skrondal, A., & Rabe-Hesketh, S. (2004). *Generalized latent variable modelling*. New York/ Boca Raton: Chapman & Hall/CRC.

Skrondal, A., & Rabe-Hesketh, S. (2008). *Multilevel and longitudinal modeling using stata*. College Station: Stata Press.

Smith, M., & Kohn, R. (1996). Nonparametric regression using Bayesian variable selection. *Journal of Econometrics, 75*, 317–343.

Snijders, T. A. B., & Berkhof, J. (2004). Diagnostic checks for multilevel models. In J. de Leeuw & I. Kreft (Eds.), *Handbook of quantitative multilevel analysis*. Thousand Oaks: Sage.

Sobotka, F., & Kneib, T. (2012). Geoadditive expectile regression. *Computational Statistics & Data Analysis, 56*, 755–767.

Spiegelhalter, D. J., Best, N. G., Carlin, B. P., & van der Linde, A. (2002). Bayesian measures of model complexity and fit. *Journal of the Royal Statistical Society B, 65*, 583–639.

Stein, M. L. (1999). *Interpolation of spatial data: Some theory for Kriging*. New York: Springer.

Stone, C. J., Hansen, M. H., Kooperberg, C., & Truong, Y. K. (1997). Polynomial splines and their tensor products in extended linear modeling. *Annals of Statistics, 25*, 1371–1470.

Stram, D. O., & Lee, J. W. (1994). Variance components testing in the longitudinal mixed effects model. *Biometrics, 50*, 1171–1177.

Sydsaeter, K., Hammond, P., Seierstad, A., & Strom, A. (2005). *Further mathematics for economic analysis*. Upper Saddle River: Prentice Hall.

Therneau, T., & Grambsch, P. (2000). *Modeling survival data: Extending the Cox model*. New York: Springer.

Tibshirani, R. (1996). Regression shrinkage and selection via the lasso. *Journal of the Royal Statistical Society B, 58*, 267–288.

Train, K. E. (2003). *Discrete choice methods with simulation*. Cambridge: Cambridge University Press.

Tutz, G. (2011). *Regression for categorical data*. Cambridge: Cambridge University Press.

Tutz, G., & Binder, H. (2006). Generalized additive modelling with implicit variable selection by likelihood based boosting. *Biometrics*, *62*, 961–971.

Umlauf, N., Kneib, T., Lang, S., & Zeileis, A. (2012). Structured additive regression models: An R interface to BayesX. *Working papers in economics and statistics 2012–08*. Faculty of Economics and Statistics, University of Innsbruck.

Veaux, R. D., Velleman, P. F., & Bock, D. E. (2011). *Intro stats* (3rd ed.). Addison Wesley.

Verbeke, G., & Molenberghs, G. (2000). *Linear mixed models for longitudinal data*. New York: Springer.

Waldmann, E., Kneib, T., Lang, S., & Yue, Y. (2012). Bayesian semiparametric additive quantile regression. *Working papers in economics and statistics 2012–06*. Faculty of Economics and Statistics, University of Innsbruck.

Wand, M. P. (2000). A comparison of regression spline smoothing procedures. *Computational Statistics*, *15*, 443–462.

Wand, M. P. (2003). Smoothing and mixed models. *Computational Statistics*, *18*, 223–249.

Weisberg, S. (2005). *Applied linear regression* (3rd ed.). New York: Wiley.

White, H. (1980). A heteroscedasticity-consistent covariance matrix estimator and a direct test for heteroscedasticity. *Econometrica*, *48*, 817–838.

Whittaker, E. T. (1923). On a new method of graduation. *Proceedings of the Edinburgh Mathematical Society*, *41*, 63–75.

Winkelmann, R. (2010a). *Analysis of microdata* (2nd ed.). Heidelberg: Springer.

Winkelmann, R. (2010b). *Econometric analysis of count data* (5th ed.). Berlin: Springer.

Wood, S. N. (2000). Modelling and smoothing parameter estimation with multiple quadratic penalties. *Journal of the Royal Statistical Society B*, *62*, 413–428.

Wood, S. N. (2003). Thin-plate regression splines. *Journal of the Royal Statistical Society B*, *65*, 95–114.

Wood, S. N. (2004). Stable and efficient multiple smoothing parameter estimation for generalized additive models. *Journal of the American Statistical Association*, *99*, 673–686.

Wood, S. N. (2006). *Generalized additive models: An introduction with R*. New York/Boca Raton: Chapman & Hall/CRC.

Wood, S. N. (2008). Fast stable direct fitting and smoothness selection for generalized additive models. *Journal of the Royal Statistical Society B*, *70*, 495–518.

Wood, S. N. (2011). Fast stable restricted maximum likelihood and marginal likelihood estimation of semiparametric generalized linear models. *Journal of the Royal Statistical Society B*, *73*, 3–36.

Wooldridge, J. M. (2006). *Introductory econometrics* (3rd ed.). Mason, OH: Thomson.

Yatchew, A. (2003). *Semiparametric regression for the applied econometrician*. Cambridge: Cambridge University Press.

Yau, P., Kohn, R., & Wood, S. (2003). Bayesian variable selection and model averaging in high dimensional multinomial nonparametric regression. *Journal of Computational and Graphical Statistics*, *12*, 23–54.

Yu, K., & Moyeed, R. A. (2001). Bayesian quantile regression. *Statistics & Probability Letters*, *54*, 437–447.

Yue, Y., & Rue, H. (2011). Bayesian inference for additive mixed quantile regression models. *Computational Statistics and Data Analysis*, *55*, 84–96.

Zeileis, A., Kleiber, C., & Jackman, S. (2008). Regression models for count data in r. *Journal of Statistical Software*, *27*(8), 1–25.

Zellner, A. (1986). On assessing prior distributions and Bayesian regression analysis with g-prior distributions. In P. Goel, & A. Zellner (Eds.). *Bayesian inference and decision techniques: Essays in honour of Bruno de Finetti* (pp. 233–243). Amsterdam: North-Holland.

Zeugner, S. (2010). Bayesian model averaging with BMS. *Technical report*. Available at http://bms.zeugner.eu/tutorials/bms.pdf.

Zimmermann, D. L. (1993). Another look at anisotropy in geostatistics. *Mathematical Geology*, *25*, 453–470.

Zou, H., & Hastie, T. (2005). Regularization and variable selection via the elastic net. *Journal of the Royal Statistical Society Series B*, *67*, 301–320.

Index

Adaptive smoothing, 490, 529
Additive model, 49, 536
  basis functions approach, 538
  identification problem, 536
Additive models for location, scale and shape, 62
Adjacency matrix, 523
AIC, 664
  adjusted, 664
  linear model, 148
  nonparametric regression, 481
  structured additive regression, 564
Aitken-estimator, 180
AM. *See* Additive model
Anisotropy, 516
Asymmetric Laplace distribution, 609
Autocorrelated errors, 80, 191
  first order, 80, 192
  two-stage estimation, 197
Autocorrelation
  partial, 192
Autocorrelation function, 192
  partial, 192
Autoregressive models, 410

Backfitting, 562
Backward elimination, 151
Basic-splines, 426
Basis function, 421
Bayes factor, 677
Bayesian confidence interval, 669
Bayesian confidence region, 669
Bayesian covariance matrix, 378
Bayesian generalized linear model, 311
  MCMC, 314
  MCMC based on data augmentation, 316
  posterior mode estimation, 313
Bayesian GLM. *See* Bayesian generalized linear model
Bayesian inference, 665
  interval estimator, 669
  model choice, 676
  point estimator, 669
Baynesian information criterion (BIC), 149, 677
Bayesian LASSO, 239
Bayesian linear mixed models, 383
  MCMC, 385
Bayesian linear model, 225
  classical model choice, 243
  conjugate analysis, 227
  conjugate prior, 233
  Gibbs sampler, 234
  noninformative prior, 231
  posterior analysis, 232
  spike and slab priors, 253
Bayesian mixed models, 383
Bayesian model averaging, 243, 496, 679
  adaptive smoothing, 496
Bayesian P-splines, 441
Bayesian ridge regression, 238
Bernoulli distribution, 639
Best linear unbiased predictor, 372, 518
Beta-binomial distribution, 640
Beta distribution, 640
Between-cluster effect, 353
Bias-variance trade-off, 476
BIC, 149, 677
Binary regression model, 270
  latent linear model, 274
  maximum likelihood, 279
Binomial distribution, 639
Bivariate polynomial splines, 504
Block-diagonal matrix, 628
Block matrix, 628
  inverse, 628
  product, 628
  sum, 628
BLUP, 372, 518

L. Fahrmeir et al., *Regression*, DOI 10.1007/978-3-642-34333-9,

BMA. *See* Bayesian model averaging
Bonferroni correction, 471
Boosting, 217, 319, 603
  additive models, 573
  componentwise, 226
  GLM, 319
  linear models, 217
  structured additive regression models, 573
Boscovich, 105
Breusch-Pagan test, 184
B-splines, 426

Caesarean delivery, 278
Canonical link function, 304
Canonical parameter, 301
Case study
  Malnutrition in Zambia, 576
  prices of used cars, 152, 156, 158, 162, 166
  sales of orange juice, 403, 551
Categorical covariates, 26, 94
Categorical regression, 325
  estimation, 343
  latent utilities, 332
  ordered regression, 334
  ordinal models, 334
  testing linear hypotheses, 346
  unordered categories, 329
Centering, 92, 567
Change of variables theorem, 645
χ^2-distribution, 643
Characteristic polynomial, 631
Cholesky decomposition, 635
Classical linear model, 73
  Bayesian, 225
  confidence ellipsoids, 136
  confidence interval, 136
  definition, 73
  estimation of error variance, 108
  hypothesis testing, 125
  maximum likelihood, 107
  parameter estimation, 104
  prediction interval, 136
  prediction quality, 144
Cluster data, 38
Clustered scatter plots, 16
Cluster-specific effect, 360
Coding
  dummy, 26, 94
  effect, 97
  indicator, 26, 94
Coefficient of determination, 115
  corrected, 147
Collinearity, 157
Column rank, 626
Column regular, 626
Column space of a matrix, 626
Column vector, 621
Complementary log-log model, 271
Conditional distribution, 646
Confidence band, 470
Confidence ellipsoid, 136
Confidence interval, 470
Conjugate Bayesian linear model, 227
Conjugate priors, 666
Continuous random vector, 645
Corrected coefficient of determination, 147
Correlated errors, 80
Correlation function, 453
  exponential, 454
  Gaussian, 454
  isotropic, 515
  Matérn, 456
  power exponential, 454
  range, 454
  spherical, 454
  stationary, 454
Correlation matrix, 646
  empirical, 647
Count data, 293
Covariance matrix, 646
  empirical, 647
Covariate, 21
Credit scoring, 290
Cross validation, 149
  smoothing, 480
Cumulative extreme value model, 336
Cumulative model, 334
Curse of dimensionality, 531
CV. *See* Cross validation

Data augmentation, 316
Defect of a matrix, 627
Degrees of freedom, 473
Dependent variable, 21
Description of the distribution of the variables, 11
Design matrix, 75
Determinant, 629
Deviance, 287
Deviance information criterion, 678
Diagonal matrix, 622
DIC, 678
Difference matrix, 437
Differentiation of matrix functions, 635
Diffuse prior, 666
Discrete random vector, 645

Distribution of the variables, 11
Dummy coding, 26, 94
Dummy variables, 26, 94, 95
Durbin-Watson test, 195

Effect coding, 97
Effect modifier, 544
Eigenspace, 632
Eigenvalue, 631
  symmetric matrix, 633
Eigenvector, 631
Empirical Bayes estimate, 385, 395, 486
Empirical best linear predictors, 375
Empirical correlation matrix, 647
Empirical covariance matrix, 647
Epanechnikov kernel, 464
Equivalent degrees of freedom, 473
Error term, 74
  correlated, 80
  heteroscedastic, 78
  homoscedastic, 78
  multiplicative, 83
Expected information matrix, 655
Expected quadratic prediction error, 145
Expectiles, 617
Explanatory variables, 21
Exponential correlation function, 454
Exponential distribution, 643
Exponential family, 301
Exponential model, 83
Extreme value sequential model, 340

F-distribution, 645
First order autocorrelation, 192
Fisher information, 655
Fisher information matrix, 655
Fisher matrix, 655
Fisher scoring, 283, 346, 660
Forest health status, 9
Forward selection, 151
Francis Galton, 1
Frequentist covariance matrix, 379
F-test, 128

Galton, 1
GAM. *See* Generalized additive model
GAMLSS, 62
Gamma distribution, 642
Gamma regression, 300
Gaussian correlation function, 454
Gaussian kernel, 464
Gauß-Markov random fields, 524
Gauß-Markov theorem, 118
GCV, 480
Generalized additive model, 52
Generalized cross validation, 480
Generalized estimating equations, 410
Generalized inverse, 631
Generalized linear mixed model, 389
  Bayesian inference, 397
  likelihood inference, 395
  MCMC, 397
Generalized linear model, 301
  Bayesian, 311
  boosting, 319
  maximum likelihood, 306
  MCMC, 314
  MCMC based on data augmentation, 316
  variance function, 307
Generalized structured additive categorical regression, 556
Generalized structured additive regression, 553
  categorical, 556
  estimation, 561
General linear hypothesis, 128
General linear mixed model, 368
General linear model, 179
Geoadditive model, 55, 59, 540
Geometric properties of the least squares estimator, 112
Gibbs sampler, 675
G inverse, 631
GLM. *See* Generalized linear model
GLMM. *See also* Generalized linear mixed model
  Bayesian inference, 397
  MCMC, 397
GMRF, 524
g-prior, 230
Gram-Schmidt-orthogonalization, 112
Graphical association analysis, 13
  categorical covariates, 17
  continuous covariates, 13
Grouped Cox model, 336
Grouped data, 181, 276

Hat matrix, 108, 469, 474
Heteroscedastic errors, 78
  detection, 183
  maximum likelihood, 189
  testing, 183
  two-stage least squares, 187
  variable transformation, 186
  White-estimator, 190

Hodrick-Prescott filter, 452
Homoscedastic errors, 78
Hormone therapy with rats, 39

Idempotent matrix, 625
Identification problem additive model, 536
Identity matrix, 622
Indefinite, 633
Independent variable, 21
Indicator coding, 26, 94
Indicator variables, 26, 94
Individual-specific effect, 360
Influence analysis, 164
Influence function, 600
Information matrix, 655
  expected, 655
  observed, 655
Interactions between covariates, 30, 98, 543
Inverse gamma distribution, 644
Inverse Gaussian distribution, 642
Inverse matrix, 627
Inversion of block matrices, 628
Irreducible prediction error, 146
Irrelevant variables, 143
Isotropy, 515
Iteratively weighted least squares, 283, 306, 324, 346
IWLS proposals, 314

Jeffreys' prior, 667

Kernel function, 464
Knots, 418
Kriging, 453, 515
  ordinary, 516
  universal, 516
Kronecker product, 624

Laplace distribution, 642
LASSO, 208
  Bayesian, 239
  geometric properties, 211
Latent linear model, 274
Least Absolute Shrinkage and Selection Operator. *See* LASSO
Least squares estimator
  asymptotic properties, 120
  covariance matrix, 116
  distribution, 118
  expected value, 116
  geometric properties, 112
  irrelevant variables, 143
  missing variables, 143
  penalization, 435
  penalized, 371
  properties, 119
Leave one out cross validation, 149
Leverage, 165
Likelihood ratio test, 663
Linear estimator, 117
Linear hypothesis, 128
Linear mixed model. *See* Mixed model
Linear Mixed Model for Longitudinal and Clustered Data, 364
Linear model, 22, 73
  Bayesian, 225
  Bayesian model averaging, 243
  Bayesian model choice, 243
  case study, 152, 156, 158, 162, 166
  categorical covariates, 26
  confidence ellipsoids, 136
  confidence interval, 136
  definition, 73
  estimation of error variance, 108
  hypthesis testing, 125
  interactions between covariates, 30, 98
  maximum likelihood, 107
  model choice, 139
  multiple, 26
  nonlinear covariate effects, 28
  parameter estimation, 104
  polynomial regression, 89
  prediction interval, 136
  prediction quality, 144
  simple, 22
  variable selection, 139
  variable transformation, 28, 87
Linear Poisson model, 293
Linear predictor, 270
Linear probability model, 270
Linear programming, 603
Linear smoothing procedures, 468
Link function, 270, 301
  canonical, 304
  natural, 304
Locally weighted regression, 466
Local polynomial regression, 462
Local smoothing, 460, 529
Location-scale model, 597
Loess, 466, 467, 529
Logit model, 33, 271
Log-linear Poisson model, 293
Log-normal distribution, 641
Log-normal model, 299

Longitudinal data, 38
LS method. *See* Least squares estimator

Main diagonal, 622
Mallow's C_p, 148
Malnutrition in Zambia, 5
  case study, 576
Marginal distribution, 646
Marginal likelihood, 676
Markov-Chain-Monte-Carlo-methods. *See* MCMC
Markov property, 442, 509
MARS, 490, 530
Matérn correlation function, 456
Matérn splines, 518
Matrix, 621
  definiteness, 633
  diagonal, 622
  idempotent, 625
  inverse, 627
  orthogonal, 625
  regular, 626
  similar, 632
  square, 622
  symmetric, 623
  trace, 630
  transpose, 621
Matrix inversion lemma, 629
Matrix multiplication, 623
Maximum likelihood estimation, 653
  asymptotic properties, 662
  equations, 281, 343
  numerical computation, 660
  testing, 662
MC3 algorithm, 247
MCMC, 670
  data augmentation, 316
Mean vector, 646
Median regression, 105
Method of least squares. *See* Least squares estimator
Metropolis–Hastings algorithm, 671
Missing variables, 143
Mixed logit model, 390
Mixed model, 38, 349
  Bayesian, 383
  Bayesian covariance matrix, 378
  case study, 403, 551
  categorical responses, 392
  conditional formulation, 365, 392
  frequentist covariance matrix, 379
  hypothesis testing, 380
  likelihood inference, 371
  marginal formulation, 365, 392
  matrix notation, 361
  ML estimator, 372
  penalized least squares view, 355
  REML estimator, 373
  stochastic covariates, 366
  testing fixed effects, 380
  testing random effects, 381
Mixed model equations, 372
Mixed model for categorical responses, 392
Mixed model representation, 481, 566
Mixed Poisson model, 391
Mixed probit model, 390
Mixture distribution, 351
ML estimation. *See also* Maximum likelihood estimation
  numerical computation, 660
  testing, 662
Model choice, 139
  additive model, 565
  Bayesian, 676
  binary regression, 287
  generalized linear model, 308
  likelihood inference, 664
  linear model, 139
  Poisson regression, 297
  structured additive regression, 565
Model choice criteria, 664, 676
  generalized linear model, 308
  linear model, 146
Model diagnostics, 155
Models for location, scale and shape, 62
MRF, 521
Multi-categorical logit model, 330
Multilevel models, 369
Multinomial distribution, 327
Multinomial logit model, 330
Multiple linear regression model, 26
Multiplicative errors, 83
Multivariate adaptive regression splines, 490, 529
Multivariate normal distribution, 648
  conditional distributions, 649
  covariance matrix, 648
  linear transformations, 649
  marginal distributions, 649
  mean, 648
Multivariate t-distribution, 651
Munich rent index, 5

Nadaraya-Watson estimator, 462, 466, 529
Natural link function, 304
Natural parameter, 301

Nearest neighbor estimator, 45, 460, 529
Negative definite matrix, 633
Neighborhoods, 521
Newton method, 660
Newton-Raphson algorithm, 660
Nominal response, 325
Noninformative prior, 231, 667
Nonlinear covariate effects via polynomials, 95
Nonlinear covariate effects via variable transformation, 94
Nonparametric regression, 44, 413
Normal distribution
  multivariate, 648
  singular, 650
  truncated, 641
  univariate, 641
Normal-inverse Gamma distribution, 227
Normal-inverse Gamma prior, 227
Null space of a matrix, 627

Observation model, 665
Observed information matrix, 655
OLS. *See* Least squares estimator
Order of a matrix, 621
Ordinal model, 334
  cumulative, 334
  sequential, 337
Ordinal response, 325
Ordinary kriging, 516
Orthogonal design, 111
Orthogonal matrix, 625
Orthogonal polynomials, 92
Outlier, 160
Overdispersion, 279, 294
  estimation, 292, 307
  Poisson regression, 294

Partial autocorrelation function, 192
Partial linear model, 537
Partial residuals, 78, 126, 565
Partitioned matrix, 628
Patent opposition, 8
Patients suffering from leukemia, 57
Pearson statistic, 287
Penalized least squares estimator, 202, 371, 432, 435, 561
Penalized log-likelihood, 313, 395, 448
  criterion, 371
Penalized residual sum of squares, 432
Penalized splines. *See* P-splines
Poisson distribution, 640
Polynomial regression, 89
Polynomial splines, 415, 418
  bivariate, 504
Population effect, 42, 360
Positive definite matrix, 633
Posterior distribution, 665
Posterior mode estimator, 669
Posterior probabilities, 676
Power exponential family, 454
Precision matrix, 646
Prediction error
  expected quadratic, 145
  irreducible, 146
Prediction interval, 136
Prediction matrix, 469, 474
Prediction quality, 144
Prices of used cars, 152
Principal component regression, 159
Prior distribution, 665
Probit model, 271
Probit normal model, 393
Product of block matrices, 628
Profile likelihood, 373
Prolongation matrix, 516
Proportional odds model, 336
P-splines, 48, 431
  Bayesian, 441
Pulmonary function, 326

Quadratic form, 633
Quantile regression, 66, 105
Quasi-likelihood, 309, 410

Radial basis functions, 512, 530
Random coefficient, 357, 549
Random effects, 39, 349
  testing, 381
Random intercept, 350, 549
Random slope, 357, 549
Random variable, 639
  multidimensional, 645
Random vector, 645
  conditional distribution, 646
  continuous, 645
  discrete, 645
  marginal distribution, 646
Random walk, 452
  *k*th order, 441
Range, 454
Rank of a matrix, 626
Reference prior, 667
Regression splines, 415

Regression trees, 492, 530
Regressor, 21
Regular inverse, 627
Regularization priors, 237
  Gibbs sampler, 240, 242
Regular matrix, 626
REML, 109, 373
Residual plots, 183
Residuals, 77, 126
  matrix notation, 108
  partial, 78, 126, 565
  properties, 122
  standardized, 126, 565
  studentized, 126
Response function, 270, 301, 304
Response variable, 21
Restricted Maximum Likelihood. *See* REML
Reversible jump MCMC, 498
Ridge regression, 203
  Bayesian, 238
  geometric properties, 211
Rotation matrix, 516
Row rank, 626
Row regular, 626
Row space of a matrix, 626
Row vector, 621
Running means, 460
Running median, 461

Sales of orange juice
  case study, 403, 551
Sandwich matrix, 379, 486
SAR, 526
Scatter plot, 13
  clustered, 16
Scatter plot smoothing, 468
Score function, 655
Score test, 663
Semidefinite matrix, 633
Semiparametric model, 537
Sequential logistic model, 340
Sequential model, 337
  extreme value, 340
  logistic, 340
Similar matrices, 632
Simple linear model, 22
Singular normal distribution, 650
Sir Francis Galton, 1
Smoother matrix, 469, 563
Smoothing, 44, 413
Smoothing parameter, 432, 444, 473, 478, 564
  choice, 478, 564
Smoothing splines, 448
SMSE, 145
Spatially autoregressive process, 526
Spatially varying coefficients, 546
Spectral decomposition, 633
Spherical correlation functions, 454
Spike and slab priors, 253
Splines, 415, 418
Square matrix, 622
Standardized residuals, 126, 565
Standard normal distribution, 641
STAR. *See* Structured additive regression
Stationary Gaussian process, 453
Stepwise selection, 151
Structured additive regression, 553
  Bayesian inference, 568
  boosting, 573
  case study, 576
  categorical, 556
  estimation, 561
  MCMC, 568
Studentized residuals, 126
Sum of block matrices, 628
Supermarket scanner data, 86
Survival analysis, 57
Symmetric matrix, 623

t-distribution, 644
  multivariate, 651
Tensor product bases, 503
Tensor product P-splines, 510
Tensor product splines, 503
Testing fixed effects, 380
Testing linear hypotheses, 135, 662
  categorical regression, 346
  GLM, 308
Testing random effects, 381
Test sample, 146
Thin plate spline, 513
Threshold value mechanism, 334
TP-splines, 418
Trace of a matrix, 630
Transpose of a matrix, 621
Truncated normal distribution, 641
Truncated power functions, 418
Truncated power series, 418
t-test, 131

Uniform kernel, 464
Univariate normal distribution, 641
Universal kriging, 516

Validation sample, 146
Variable coding
  dummy, 94
  effect, 97
  indicator, 26, 94
Variable selection, 139
Variable transformation, 87
Variance components model, 362, 566
Variance function, 307
Variance inflation factor, 158
Varying coefficients models, 544
Vehicle insurance, 52

Wald test, 663
Weighted least squares estimator, 180, 181
White-estimator, 190
Within-cluster effect, 353

Zellner's g-prior, 230

Printed by Printforce, the Netherlands